Credit-Risk Modelling

David Jamieson Bolder

Credit-Risk Modelling

Theoretical Foundations, Diagnostic Tools, Practical Examples, and Numerical Recipes in Python

 Springer

David Jamieson Bolder
The World Bank
District of Columbia
Washington, DC, USA

ISBN 978-3-030-06900-1   ISBN 978-3-319-94688-7 (eBook)
https://doi.org/10.1007/978-3-319-94688-7

© Springer International Publishing AG, part of Springer Nature 2018
Softcover re-print of the Hardcover 1st edition 2018
This work is subject to copyright. All rights are reserved by the Publisher, whether the whole or part of the material is concerned, specifically the rights of translation, reprinting, reuse of illustrations, recitation, broadcasting, reproduction on microfilms or in any other physical way, and transmission or information storage and retrieval, electronic adaptation, computer software, or by similar or dissimilar methodology now known or hereafter developed.
The use of general descriptive names, registered names, trademarks, service marks, etc. in this publication does not imply, even in the absence of a specific statement, that such names are exempt from the relevant protective laws and regulations and therefore free for general use.
The publisher, the authors and the editors are safe to assume that the advice and information in this book are believed to be true and accurate at the date of publication. Neither the publisher nor the authors or the editors give a warranty, express or implied, with respect to the material contained herein or for any errors or omissions that may have been made. The publisher remains neutral with regard to jurisdictional claims in published maps and institutional affiliations.

Printed on acid-free paper

This Springer imprint is published by the registered company Springer Nature Switzerland AG
The registered company address is: Gewerbestrasse 11, 6330 Cham, Switzerland

*À Thomas, mon petit loup et
la fièreté de ma vie.*

Foreword

As there are already many good technical books and articles on credit-risk modelling, one may well question why there needs to be another one. David's new book is different and will be useful to those interested in a comprehensive development of the subject through a pedagogical approach. He starts from first principles and foundations, deftly makes connections with market-risk models, which were developed first, but shows why credit-risk modelling needs different tools and techniques. Building on these foundational principles, he develops in extraordinary detail the technical and mathematical concepts needed for credit-risk modelling. This fresh, step-by-step, build-up of methodologies is almost like someone helping the readers themselves develop the logic and concepts; the technique is unique and yields a solid understanding of fundamental issues relating to credit risk and the modelling of such risk.

David goes beyond analytical modelling to address the practical issues associated with implementation. He uses a sample portfolio to illustrate the development of default models and, most interestingly, provides examples of code using the Python programming language. He includes detailed treatment of issues relating to parameter estimation and techniques for addressing the thorny problem of limited data which, in some areas of credit risk, can be severe. This is especially true when compared to market-risk models. In addition, as one who is currently working on model validation and model risk, David draws on his rich experience to elucidate the limitations and consequences of modelling choices that the practitioner is often required to make. He also provides valuable insights into the use and interpretation of model results.

This book is deeply technical, draws on David's skill and experience in educating quantitative modellers, and will appeal to serious students with a strong mathematical background who want a thorough and comprehensive understanding of the topic.

While this book is not light reading, those who put in the effort will find it well worth it; they will also be rewarded with surprising doses of humour, which lighten the path, but also further aid the understanding.

Washington, DC, USA
2018

Lakshmi Shyam-Sunder

Preface

> *It's a dangerous business, Frodo, going out your door. You step onto the road, and if you don't keep your feet, there's no knowing where you might be swept off to.*
> *(J.R.R. Tolkien)*

The genesis for this book was a series of weekly lectures provided to a collection of World Bank Group quantitative analysts. The intention of this effort was to provide an introduction to the area of credit-risk modelling to a group of practitioners with significant experience in the market-risk area, but rather less in the credit-risk setting. About half of the lectures were allocated to the models, with the remaining focus on diagnostic tools and parameter estimation. There was a strong emphasis on the *how*, but there was also an important accent on the *why*. Practical considerations thus shared equal billing with the underlying rationale. There developed a challenging and stimulating environment characterized by active discussion and reflection. A high level of participant interest was evident. This experience suggested a certain demand for knowledge and understanding of the models used in measuring and managing credit risk.

My Motivation

As sometimes happens, therefore, this innocent series of lectures evolved into something more. In my spare time, I tinkered with the organization of the various topics, expanded my notes, worked out a number of relationships from first principles, and added to my rapidly increasing Python code library. Eventually, I realized that I had gradually gathered more than enough material for a reasonably comprehensive book on the subject. At this point in most technical endeavours, every prospective author asks: do we really need another book on this topic? Hamming (1985) makes an excellent point in this regard, borrowed from de Finetti (1974), which I will, in turn, use once again,

> To add one more [book] would certainly be a presumptuous undertaking if I thought in terms of doing something better, and a useless undertaking if I were to content myself with producing something similar to the "standard" type.

Wishing to be neither presumptuous nor useless, I hope to bring a different perspective to this topic.

There are numerous excellent academic references and a range of useful introductions to credit-risk modelling; indeed, many of them are referenced in the coming chapters. What distinguishes this work, therefore, from de Finetti's (1974) *standard* type? There are *three* main deviations from the typical treatment of the underlying material. Motivated by the tone and structure of the original lectures—as well as my own practical experience—I have tried to incorporate transparency and accessibility, concreteness, and multiplicity of perspective. Each of these elements warrants further description.

Transparency and Accessibility

In virtually every subject, there are dramatic differences between introductory and professional discussion. In technical areas, this is often extreme. Indeed, advanced books and articles are typically almost incomprehensible to the uninitiated. At the same time, however, introductory works often contain frustratingly little detail. Credit-risk modelling, unfortunately, falls into this category. The quantitative practitioner seeking to gain understanding of the extant set of credit-risk models is thus trapped between relatively introductory treatment and mathematically dense academic treatises.

This book, in response to this challenge, attempts to insert itself in between these two ends of the spectrum. Details are not taken for granted and virtually every relationship is derived from either first principles or previously established quantities. The result is a text with approximately 1,000 equations. This level of transparency may, for certain highly qualified practitioners, seem like tedium or overkill. To attain the desired degree of accessibility, however, I firmly believe in the necessity of methodical, careful, and complete derivations of all models, diagnostic tools, and estimation methodologies. In this manner, it is my sincere hope that this work will help to bridge the inevitable distance between elementary introductions and highly complex, original, academic work.

Concreteness

Cautious and detailed derivation of key relationships is an important first step, but this is not a theoretical pursuit. Quite the opposite, credit-risk modelling is an applied discipline that forms a critical element of many institutions' risk-management framework. Merely discussing the material is insufficient. It is necessary to actually *show* the reader how the computations are performed. Demonstration of key concepts is precisely what is meant by the notion of concreteness.

To permit the incorporation of this practical element, a compact, but reasonably rich portfolio example is introduced in the first chapter. It is then repeatedly used—in the context of the models, diagnostics, and parameter estimations—throughout the remainder of the book. This provides a high level of continuity while both inviting and permitting intra-chapter comparisons. The practical element in this work is thus a fundamental feature and not merely a supplementary add-on. To accomplish this, the following chapters contain almost 100 Python algorithms, 130-odd figures, and more than 80 tables. Examination of how various modelling components might be implemented, investigating potential code solutions, and reviewing the results are, in fact, the central part of this work.

All of the practical computer code in the subsequent chapters employs the Python programming language. While there are many possible alternatives, this choice was not coincidental. Python offers three important advantages. First of all, it provides, among other things, a comprehensive and flexible scientific computing platform. Credit-risk models require a broad range of numerical techniques and, in my experience, Python is well suited to perform them. Second, there is a growing population of Python developers in finance. This popularity should imply a higher degree of familiarity with the proposed code found in the coming chapters. Finally, it is freely available under the Open Source License. In addition to the benefits associated with an open development community, the resource constraints this relaxes should not be underestimated.

The Python libraries employed in all of the practical examples found in this book are freely available, also under the GNU Open Source License, at the following GitHub location:

> https://github.com/djbolder/credit-risk-modelling

My sincere hope is that these libraries might help other practitioners start and extend their own modelling activities.

Multiplicity of Perspective

In recent years, after decades working as a quantitative practitioner developing and implementing models, I moved into a model oversight and governance function. In many ways, my day-to-day life has not dramatically changed. Validating complex mathematical and statistical models requires independent replication and mathematical derivations. The lion's share of my time, therefore, still involves working closely with models. What has changed, however, is my perspective. Working with a broad range of models and approaches across my institution, a fact that was already intellectually clear to me became even more evident: there are typically many alternative ways to model a particular phenomenon.

This, perhaps obvious, insight is simultaneously the source and principal mitigant of model risk. If there is no one single way to model a particular object of interest, then one seeks the *best* model from the universe of alternatives. Best, however, is a notoriously slippery concept. Determination of a superior model is both an objective and subjective task. In some areas, the objective aspect dominates. This is particularly the case when one can—as in the case of forecasting or pricing—compare model and actual outcomes. In other cases, when actual outcomes are infrequent or only indirectly observed, model choice becomes more subjective.

What is the consequence of this observation? The answer is that model selection is a difficult and complex decision, where the most defensible choice varies depending on the application and the context. In all cases, however, multiple models are preferable to a single model. Model validation thus involves extensive use of alternative approaches—often referred to as benchmark models—and gratuitous use of sensitivity analysis. Not only do these techniques enhance model understanding, but they also help highlight the strengths and weaknesses of one's possible choices. This work adopts this position. Rarely, if at all, is a single approach presented for any task. Instead, multiple methodologies are offered for every model, diagnostic tool, or estimation technique. In some cases, the alternative is quite simple and involves some questionable assumptions. Such models, despite their shortcomings, are nonetheless useful as bounds or sanity checks on more complex alternatives. In other cases, highly technically involved approaches are suggested. They may be excessively complex and difficult to use, but still offer variety of perspective.

The punchline is that, in real-life settings, we literally never know the *right* model. Moreover, we do know that our model is likely to be flawed, often in important ways. To mitigate this depressing fact, the best approach is to examine many models with differing assumptions, fundamental structures, or points of entry. By examining and using a multiplicity of approaches, we both decrease our reliance on any one model and increase the chances of making useful insights about our underlying problem. This is nevertheless *not* a book on model risk, but it would be naive and irresponsible—in light of the advice provided by Derman (1996), Rebonato (2003), Morini (2011) and many others—to ignore this perspective. Recognition of this key fact, through my recent experience in model oversight, has resulted in an explicit incorporation of many competing perspectives in this book.

Some Important Caveats

This is *not* an academic work; indeed, there is relatively little new or original—beyond perhaps organization and presentation—in the following chapters. One should not, however, conclude that the following pages are devoid of academic concepts. On the contrary, it is chock-full of them. This is a challenging, technical area of study. The point I am trying to make is that this work is practitioner literature. Written by a practitioner for other practitioners, the principal focus is on the practicalities of implementing, using, and communicating credit-risk models. Ultimately, the following chapters were designed to support quantitative financial analysts—working in banking, portfolio management, and insurance—in measuring and managing default risk in their portfolios. This practical focus will also hopefully appeal to advanced undergraduate and graduate students studying this area of finance.

The following chapters cover a significant amount of ground in substantial detail. This has two important consequences. First, the more material and the greater the depth of discussion, the higher the probability of errors. Much time and effort was invested carefully reviewing the text to ensure clear, correct, and cogent treatment of all topics, but it is inevitable that some small errors and inconsistencies remain. I ask for the reader's patience and understanding and note that cautious use of any source is an integral part of every quantitative analyst's working life.

The second, and final, point is that, no matter how large or detailed a work, it is literally impossible to cover all aspects of a given field. This book is no exception. Models, diagnostic tools, and parameter estimation cover a lot of ground, but not everything. Two topics, in particular, receive rather meagre attention: severity of default and migration risk. Both are addressed and all of the necessary tools are provided to move much further into these areas, but these themes often receive much greater scrutiny. To a large extent, underplaying these areas was a conscious choice. Both are, in my view, generalizations of the basic framework and add complexity without enormous additional insight. This is not to say that these are unimportant areas, far from it, but rather that I have downplayed them somewhat for pedagogical reasons.

Washington, DC, USA   David Jamieson Bolder
2018

References

Bolder, D. J. (2015). *Fixed income portfolio analytics: A practical guide to implementing, monitoring and understanding fixed-income portfolios*. Heidelberg, Germany: Springer.
de Finetti, B. (1974). *Theory of probability*. New York: Wiley.
Derman, E. (1996). Model risk. Goldman Sachs Quantitative Strategies Research Notes.
Hamming, R. W. (1985). *Methods of mathematics applied to calculus, probability, and statistics*. Upper Saddle River, NJ, USA: Prentice-Hall, Inc.
Morini, M. (2011). *Understanding and managing model risk: A practical guide for quants, traders, and validators*. West Sussex, UK: Wiley.
Rebonato, R. (2003). Theory and practice of model risk management. Quantitative Research Centre (QUARC) Technical Paper.

Acknowledgements

I would, first of all, like to sincerely thank Ivan Zelenko for his valuable support. This project was made possible by his decision to bring me into his team and through his ability to create a consistent atmosphere of intellectual curiosity. I would also like to thank my colleague, Marc Carré, for offering a steady stream of fascinating modelling challenges that stimulated my interest and the ultimate choice of subject matter for this work. I also owe a heartfelt debt of gratitude to Jean-Pierre Matt, a former colleague and mentor, for introducing me to the Python programming language and gently advocating its many benefits.

I would also like to warmly acknowledge the World Bank Group participants who attended my lecture series in early 2017. Despite the early time slots, these dedicated individuals provided the key motivation for the production of this work and helped, in many ways, to lend it a pragmatic tone. I would like to particularly single out Mallik Subbarao and Michal Certik, who participated in numerous weekly discussions, replicated some of the examples, and proofread large sections of the text. These contributions were invaluable and have certainly led to a higher-quality work.

Finally, and perhaps most importantly, I would like to thank my wife and son for their patience and understanding with the long hours required to produce this work. After Bolder (2015), I promised my wife that I would never write another book. The dual lesson, it seems, is that she is a patient woman and that it would be best to refrain from making promises that I cannot keep. I should also probably, only half facetiously, thank the many ice-hockey rinks and arenas where much of this book was written. Through my son, an avid ice-hockey player, we've travelled to a myriad of places over the last few years along the East Coast of the United States for both games and practice. My laptop—along with notes and drafts of the forthcoming chapters—has been my constant companion on these travels.

All of my thanks and acknowledgement are, quite naturally, entirely free of implication. All errors, inconsistencies, shortcomings, coding weaknesses, design flaws, or faults in logic are to be placed entirely on my shoulders.

Contents

1 **Getting Started** 1
  1.1 Alternative Perspectives 3
    1.1.1 Pricing or Risk-Management? 3
    1.1.2 Minding our $\mathbb{P}$'s and $\mathbb{Q}$'s 6
    1.1.3 Instruments or Portfolios? 7
    1.1.4 The Time Dimension 9
    1.1.5 Type of Credit-Risk Model 10
    1.1.6 Clarifying Our Perspective 11
  1.2 A Useful Dichotomy 11
    1.2.1 Modelling Implications 13
    1.2.2 Rare Events 15
  1.3 Seeing the Forest 18
    1.3.1 Modelling Frameworks 19
    1.3.2 Diagnostic Tools 21
    1.3.3 Estimation Techniques 23
    1.3.4 The Punchline 24
  1.4 Prerequisites 24
  1.5 Our Sample Portfolio 26
  1.6 A Quick Pre-Screening 27
    1.6.1 A Closer Look at Our Portfolio 27
    1.6.2 The Default-Loss Distribution 29
    1.6.3 Tail Probabilities and Risk Measures 30
    1.6.4 Decomposing Risk 33
    1.6.5 Summing Up 37
  1.7 Final Thoughts 37
  References 38

Part I Modelling Frameworks

Reference 40

2 A Natural First Step 41
  2.1 Motivating a Default Model 42
    2.1.1 Two Instruments 45
    2.1.2 Multiple Instruments 46
    2.1.3 Dependence 50
  2.2 Adding Formality 50
    2.2.1 An Important Aside 53
    2.2.2 A Numerical Solution 55
    2.2.3 An Analytical Approach 62
  2.3 Convergence Properties 67
  2.4 Another Entry Point 73
  2.5 Final Thoughts 82
  References 82

3 Mixture or Actuarial Models 85
  3.1 Binomial-Mixture Models 86
    3.1.1 The Beta-Binomial Mixture Model 92
    3.1.2 The Logit- and Probit-Normal Mixture Models 101
  3.2 Poisson-Mixture Models 113
    3.2.1 The Poisson-Gamma Approach 115
    3.2.2 Other Poisson-Mixture Approaches 125
    3.2.3 Poisson-Mixture Comparison 129
  3.3 CreditRisk+ 131
    3.3.1 A One-Factor Implementation 131
    3.3.2 A Multi-Factor CreditRisk+ Example 141
  3.4 Final Thoughts 147
  References 148

4 Threshold Models 149
  4.1 The Gaussian Model 150
    4.1.1 The Latent Variable 150
    4.1.2 Introducing Dependence 152
    4.1.3 The Default Trigger 154
    4.1.4 Conditionality 155
    4.1.5 Default Correlation 158
    4.1.6 Calibration 161
    4.1.7 Gaussian Model Results 162
  4.2 The Limit-Loss Distribution 165
    4.2.1 The Limit-Loss Density 170
    4.2.2 Analytic Gaussian Results 172
  4.3 Tail Dependence 175
    4.3.1 The Tail-Dependence Coefficient 176
    4.3.2 Gaussian Copula Tail-Dependence 179

		4.3.3	t-Copula Tail-Dependence	180
	4.4	The t-Distributed Approach		182
		4.4.1	A Revised Latent-Variable Definition	182
		4.4.2	Back to Default Correlation	186
		4.4.3	The Calibration Question	188
		4.4.4	Implementing the t-Threshold Model	190
		4.4.5	Pausing for a Breather	193
	4.5	Normal-Variance Mixture Models		193
		4.5.1	Computing Default Correlation	197
		4.5.2	Higher Moments	198
		4.5.3	Two Concrete Cases	200
		4.5.4	The Variance-Gamma Model	201
		4.5.5	The Generalized Hyperbolic Case	202
		4.5.6	A Fly in the Ointment	204
		4.5.7	Concrete Normal-Variance Results	206
	4.6	The Canonical Multi-Factor Setting		211
		4.6.1	The Gaussian Approach	211
		4.6.2	The Normal-Variance-Mixture Set-Up	214
	4.7	A Practical Multi-Factor Example		218
		4.7.1	Understanding the Nested State-Variable Definition	219
		4.7.2	Selecting Model Parameters	221
		4.7.3	Multivariate Risk Measures	224
	4.8	Final Thoughts		225
	References			226
5	**The Genesis of Credit-Risk Modelling**			**229**
	5.1	Merton's Idea		230
		5.1.1	Introducing Asset Dynamics	233
		5.1.2	Distance to Default	236
		5.1.3	Incorporating Equity Information	238
	5.2	Exploring Geometric Brownian Motion		240
	5.3	Multiple Obligors		245
		5.3.1	Two Choices	246
	5.4	The Indirect Approach		247
		5.4.1	A Surprising Simplification	249
		5.4.2	Inferring Key Inputs	251
		5.4.3	Simulating the Indirect Approach	252
	5.5	The Direct Approach		255
		5.5.1	Expected Value of $A_{n,T}$	257
		5.5.2	Variance and Volatility of $A_{n,T}$	259
		5.5.3	Covariance and Correlation of $A_{n,T}$ and $A_{m,T}$	261
		5.5.4	Default Correlation Between Firms n and m	263
		5.5.5	Collecting the Results	265
		5.5.6	The Task of Calibration	265
		5.5.7	A Direct-Approach Inventory	270

	5.5.8	A Small Practical Example	270
5.6		Final Thoughts	280
References			282

Part II Diagnostic Tools
References 286

6 A Regulatory Perspective 287
- 6.1 The Basel Accords 288
  - 6.1.1 Basel IRB 290
  - 6.1.2 The Basic Structure 292
  - 6.1.3 A Number of Important Details 295
  - 6.1.4 The Full Story 300
- 6.2 IRB in Action 303
  - 6.2.1 Some Foreshadowing 306
- 6.3 The Granularity Adjustment 309
  - 6.3.1 A First Try 311
  - 6.3.2 A Complicated Add-On 312
  - 6.3.3 The Granularity Adjustment 317
  - 6.3.4 The One-Factor Gaussian Case 318
  - 6.3.5 Getting a Bit More Concrete 325
  - 6.3.6 The CreditRisk+ Case 328
  - 6.3.7 A Final Experiment 340
  - 6.3.8 Final Thoughts 346
- References 348

7 Risk Attribution 351
- 7.1 The Main Idea 352
- 7.2 A Surprising Relationship 355
  - 7.2.1 The Justification 357
  - 7.2.2 A Direct Algorithm 361
  - 7.2.3 Some Illustrative Results 364
  - 7.2.4 A Shrewd Suggestion 366
- 7.3 The Normal Approximation 368
- 7.4 Introducing the Saddlepoint Approximation 372
  - 7.4.1 The Intuition 373
  - 7.4.2 The Density Approximation 376
  - 7.4.3 The Tail Probability Approximation 378
  - 7.4.4 Expected Shortfall 381
  - 7.4.5 A Bit of Organization 382
- 7.5 Concrete Saddlepoint Details 384
  - 7.5.1 The Saddlepoint Density 388
  - 7.5.2 Tail Probabilities and Shortfall Integralls 391
  - 7.5.3 A Quick Aside 392
  - 7.5.4 Illustrative Results 393
- 7.6 Obligor-Level Risk Contributions 394

		7.6.1	The VaR Contributions	395
		7.6.2	Shortfall Contributions	400
	7.7	The Conditionally Independent Saddlepoint Approximation		406
		7.7.1	Implementation	412
		7.7.2	A Multi-Model Example	415
		7.7.3	Computational Burden	420
	7.8	An Interesting Connection		421
	7.9	Final Thoughts		426
	References			426
8	**Monte Carlo Methods**			429
	8.1	Brains or Brawn?		430
	8.2	A Silly, But Informative Problem		431
	8.3	The Monte Carlo Method		439
		8.3.1	Monte Carlo in Finance	440
		8.3.2	Dealing with Slowness	444
	8.4	Interval Estimation		445
		8.4.1	A Rough, But Workable Solution	445
		8.4.2	An Example of Convergence Analysis	447
		8.4.3	Taking Stock	450
	8.5	Variance-Reduction Techniques		450
		8.5.1	Introducing Importance Sampling	451
		8.5.2	Setting Up the Problem	453
		8.5.3	The Esscher Transform	457
		8.5.4	Finding θ	460
		8.5.5	Implementing the Twist	462
		8.5.6	Shifting the Mean	467
		8.5.7	Yet Another Twist	474
		8.5.8	Tying Up Loose Ends	476
		8.5.9	Does It Work?	479
	8.6	Final Thoughts		485
	References			486

Part III Parameter Estimation

9	**Default Probabilities**			491
	9.1	Some Preliminary Motivation		492
		9.1.1	A More Nuanced Perspective	493
	9.2	Estimation		498
		9.2.1	A Useful Mathematical Object	499
		9.2.2	Applying This Idea	506
		9.2.3	Cohort Approach	508
		9.2.4	Hazard-Rate Approach	511
		9.2.5	Getting More Practical	512
		9.2.6	Generating Markov-Chain Outcomes	513
		9.2.7	Point Estimates and Transition Statistics	518

		9.2.8	Describing Uncertainty	523
		9.2.9	Interval Estimates	538
		9.2.10	Risk-Metric Implications	541
	9.3	Risk-Neutral Default Probabilities		544
		9.3.1	Basic Cash-Flow Analysis	544
		9.3.2	Introducing Default Risk	547
		9.3.3	Incorporating Default Risk	553
		9.3.4	Inferring Hazard Rates	557
		9.3.5	A Concrete Example	561
	9.4	Back to Our $\mathbb{P}$'s and $\mathbb{Q}$'s		567
	9.5	Final Thoughts		571
	References			571
10	**Default and Asset Correlation**			**575**
	10.1	Revisiting Default Correlation		576
	10.2	Simulating a Dataset		580
		10.2.1	A Familiar Setting	581
		10.2.2	The Actual Results	587
	10.3	The Method of Moments		589
		10.3.1	The Threshold Case	593
	10.4	Likelihood Approach		595
		10.4.1	The Basic Insight	596
		10.4.2	A One-Parameter Example	598
		10.4.3	Another Example	602
		10.4.4	A More Complicated Situation	606
	10.5	Transition Likelihood Approach		615
		10.5.1	The Elliptical Copula	616
		10.5.2	The Log-Likelihood Kernel	622
		10.5.3	Inferring the State Variables	626
		10.5.4	A Final Example	628
	10.6	Final Thoughts		633
	References			634
A	**The t-Distribution**			**637**
	A.1	The Chi-Squared Distribution		638
	A.2	Toward the t-Distribution		639
	A.3	Simulating Correlated t Variates		643
	A.4	A Quick Example		647
B	**The Black-Scholes Formula**			**649**
	B.1	Changing Probability Measures		649
	B.2	Solving the Stochastic Differential Equation		653
	B.3	Evaluating the Integral		655
	B.4	The Final Result		658

C	**Markov Chains**		659
	C.1	Some Background	659
	C.2	Some Useful Results	661
	C.3	Ergodicity	663
D	**The Python Code Library**		667
	D.1	Explaining Some Choices	668
	D.2	The Library Structure	669
	D.3	An Example	671
	D.4	Sample Exercises	672
	References		673
Index			675
Author Index			681

List of Figures

Fig. 1.1	A (stylized) default-loss distribution	5
Fig. 1.2	Pricing credit risk	5
Fig. 1.3	Market vs. credit risk	13
Fig. 1.4	Brownian motion in action	15
Fig. 1.5	Brownian motion in credit risk	16
Fig. 1.6	A stylized comparison	17
Fig. 1.7	A sample portfolio	26
Fig. 1.8	A bit more perspective	28
Fig. 1.9	Default-loss distributions	30
Fig. 1.10	Tail probabilities	31
Fig. 1.11	Obligor VaR contributions	34
Fig. 1.12	VaR contribution perspectives	36
Fig. 2.1	Four-instrument outcomes	48
Fig. 2.2	Pascal's triangle	49
Fig. 2.3	Many instruments	49
Fig. 2.4	A schematic perspective	57
Fig. 2.5	The loss distribution	61
Fig. 2.6	Impact of homogeneity assumptions	67
Fig. 2.7	Identifying λ_n	75
Fig. 2.8	Indistinguishable	77
Fig. 2.9	The law of rare events	81
Fig. 3.1	The role of default dependence	98
Fig. 3.2	Simulating beta random variables	99
Fig. 3.3	The logistic and probit functions	102
Fig. 3.4	Comparing default-probability densities	105
Fig. 3.5	Comparing default probability mass functions	108
Fig. 3.6	The gamma function	116
Fig. 3.7	Comparing tail probabilities	130

Fig. 3.8	CreditRisk+ conditional default probabilities		134
Fig. 3.9	A default-correlation surface		136
Fig. 3.10	CreditRisk+ tail probabilities		140
Fig. 4.1	Conditional default probability		156
Fig. 4.2	$p_n(G)$ dynamics		157
Fig. 4.3	Comparative tail probabilities		164
Fig. 4.4	Different limit-loss densities		172
Fig. 4.5	The analytic fit		175
Fig. 4.6	Tail-dependence coefficients		181
Fig. 4.7	Calibrating outcomes		188
Fig. 4.8	Various tail probabilities		192
Fig. 4.9	The modified Bessel function of the second kind		202
Fig. 4.10	Visualizing normal-variance-mixture distributions		209
Fig. 4.11	Multiple threshold tail probabilities		210
Fig. 5.1	A model zoology		230
Fig. 5.2	Default-event schematic		233
Fig. 5.3	ODE behaviour		242
Fig. 5.4	SDE behaviour		244
Fig. 5.5	Inferred quantities		252
Fig. 5.6	Comparing tail probabilities		255
Fig. 5.7	Inferring the capital structure		276
Fig. 5.8	Equity and asset moments		276
Fig. 5.9	Direct Merton-model default-loss distribution		280
Fig. 6.1	Basel asset correlation		297
Fig. 6.2	Maturity correction		299
Fig. 6.3	Risk-capital coefficient		301
Fig. 6.4	Obligor tenors		303
Fig. 6.5	Comparative tail probabilities		305
Fig. 6.6	Ordered risk-capital contributions		308
Fig. 6.7	Basel II's pillars		310
Fig. 6.8	Gaussian granularity adjustment functions		324
Fig. 6.9	Concentrated vs. diversified tail probabilities		326
Fig. 6.10	Testing the granularity adjustment		327
Fig. 6.11	Choosing $\delta_\alpha(a)$		343
Fig. 6.12	$\mathcal{G}_\alpha(L)$ accuracy		345
Fig. 6.13	Revisiting economic-capital contributions		347
Fig. 7.1	Monte-Carlo obligor contributions		365
Fig. 7.2	Regional contributions		366
Fig. 7.3	VaR-matched comparison		368
Fig. 7.4	Accuracy of normal VaR approximation		372
Fig. 7.5	Saddlepoint positivity		380
Fig. 7.6	Cumulant generating function and derivatives		389
Fig. 7.7	Independent-default density approximation		390

Fig. 7.8	Independent-default tail-probability and shortfall-integral approximations	394
Fig. 7.9	Independent-default VaR-contribution approximation	399
Fig. 7.10	VaR contribution perspective	399
Fig. 7.11	Independent-default shortfall-contribution approximation	405
Fig. 7.12	Integrated saddlepoint density approximation	408
Fig. 7.13	Multiple-model visual risk-metric comparison	416
Fig. 7.14	Saddlepoint risk-measure contributions	418
Fig. 7.15	Proportional saddlepoint shortfall contributions	419
Fig. 7.16	Back to the beginning: The t-threshold model	419
Fig. 8.1	A silly function	432
Fig. 8.2	A simple, but effective, approach	433
Fig. 8.3	A clever technique	434
Fig. 8.4	A randomized grid	436
Fig. 8.5	The eponymous location	440
Fig. 8.6	The role of M	447
Fig. 8.7	Understanding convergence	448
Fig. 8.8	Variance-reduction techniques	451
Fig. 8.9	Twisting probabilities	463
Fig. 8.10	Conditional VaR	468
Fig. 8.11	$\mathcal{J}_1$ behaviour	471
Fig. 8.12	Twisting and shifting	473
Fig. 8.13	The shifted mean	474
Fig. 8.14	Raw vs. IS VaR comparison	481
Fig. 8.15	A tighter comparison	482
Fig. 9.1	VaR sensitivity to default probabilities	493
Fig. 9.2	Obligor differentiation	497
Fig. 9.3	Calculating forward Markov-Chain probabilities	503
Fig. 9.4	Simulating a Markov chain	515
Fig. 9.5	Transition overview	518
Fig. 9.6	Type of transitions	519
Fig. 9.7	Simulated transition counts	522
Fig. 9.8	The profile likelihood	524
Fig. 9.9	Likelihood curvature	525
Fig. 9.10	Binomial likelihood	534
Fig. 9.11	Binomial bootstrap distribution	536
Fig. 9.12	Comparing approaches	539
Fig. 9.13	Default-probability confidence bounds	540
Fig. 9.14	Bootstrap results	541
Fig. 9.15	An assumed risk-free spot curve	562
Fig. 9.16	Key ingredients and outputs	566
Fig. 9.17	Key CDS spreads	568
Fig. 9.18	Approximate risk-premium evolution	569

Fig. 10.1	Creating transition data	583
Fig. 10.2	Choosing ρ_G	588
Fig. 10.3	Aggregate default data	588
Fig. 10.4	Single-parameter likelihood function	599
Fig. 10.5	Single-parameter estimates	602
Fig. 10.6	The CreditRisk+ likelihood surface	603
Fig. 10.7	Fixed a CreditRisk+ likelihood functions	605
Fig. 10.8	Regional default data	610
Fig. 10.9	The regional likelihood surface	611
Fig. 10.10	Regional profile likelihoods	612
Fig. 10.11	Regional estimates	615
Fig. 10.12	Multiple log-likelihoods	630
Fig. 10.13	Averaging log-likelihoods	630
Fig. 10.14	Lost in translation	633
Fig. A.1	Bivariate scatterplots	648
Fig. D.1	The library structure	670
Fig. D.2	A sample library	671

List of Tables

Table 1.1	Topic overview	19
Table 1.2	Summary risk measures	31
Table 1.3	VaR-contribution breakdown	35
Table 2.1	A single instrument example	43
Table 2.2	A two-instrument example	45
Table 2.3	Four instruments	47
Table 2.4	Numerical independent default results	61
Table 2.5	Analytic vs. numerical results	66
Table 2.6	Numerical results	77
Table 2.7	Analytical independent-default results	80
Table 3.1	Analytic beta-binomial results	97
Table 3.2	Numerical beta-binomial mixture results	100
Table 3.3	Binomial calibration results	110
Table 3.4	Analytic results	111
Table 3.5	Numerical mixture results	112
Table 3.6	Analytic Poisson-Gamma results	123
Table 3.7	Numerical Poisson-Gamma mixture results	124
Table 3.8	Poisson calibration results	127
Table 3.9	Analytic Poisson-mixture results	129
Table 3.10	Numerical Poisson-mixture results	130
Table 3.11	Possible CreditRisk+ calibrations	138
Table 3.12	A one-factor CreditRisk+ model	140
Table 3.13	Possible multi-factor CreditRisk+ parameterization	145
Table 3.14	Multi-factor CreditRisk+ default-correlation matrix	146
Table 3.15	A multi-factor CreditRisk+ model	147
Table 4.1	Gaussian threshold results	164
Table 4.2	Gaussian threshold analytic-numerical comparison	174
Table 4.3	Threshold calibration comparison	190
Table 4.4	t-threshold results	192

Table 4.5	One-factor summary	194
Table 4.6	Normal-variance mixture calibration comparison	207
Table 4.7	Simulated vs. theoretical moments	208
Table 4.8	Multiple threshold results	210
Table 4.9	Canonical multi-factor summary	217
Table 4.10	Sample parameter structure	222
Table 4.11	Default correlation matrix	223
Table 4.12	Multivariate threshold results	225
Table 5.1	Indirect K_n	248
Table 5.2	Our portfolio in the Merton world	251
Table 5.3	Indirect Merton results	254
Table 5.4	Challenges of the direct approach	256
Table 5.5	Key Merton quantities	271
Table 5.6	A low-dimensional example	272
Table 5.7	Asset correlations	275
Table 5.8	Default correlations	277
Table 5.9	Direct Merton results	279
Table 5.10	A litany of default triggers	282
Table 6.1	Benchmark models	304
Table 6.2	Benchmarking results	306
Table 6.3	Risk-capital contributions	307
Table 6.4	Two differing portfolios	325
Table 6.5	The numerical reckoning	328
Table 6.6	CreditRisk+ parameter choices	343
Table 6.7	Testing the CreditRisk+ granularity adjustment	346
Table 7.1	Key saddlepoint approximations	383
Table 7.2	Computational direction	392
Table 7.3	Numerical saddlepoint results	394
Table 7.4	VaR contribution results	398
Table 7.5	Shortfall contribution results	405
Table 7.6	Multiple-model saddlepoint comparison	416
Table 7.7	Multiple-model top-ten VaR contributions	417
Table 7.8	Comparing computational times	420
Table 8.1	Integration results	438
Table 8.2	Convergence analysis	449
Table 8.3	A different perspective	465
Table 8.4	A first try	466
Table 8.5	Variance-reduction report card	483
Table 9.1	Ranking VaR default-probability sensitivities	496
Table 9.2	Counting Markov-Chain outcomes	500
Table 9.3	Numerical parameter uncertainty	540
Table 9.4	Categorizing obligors	542

List of Tables

Table 9.5	Parameter uncertainty and VaR	543
Table 9.6	Bootstrapping results	566
Table 9.7	Comparing $\mathbb{P}$ and $\mathbb{Q}$ estimates	569
Table 10.1	Simulated data at a glance	589
Table 10.2	Mixture method of moments results	592
Table 10.3	Threshold method of moments results	595
Table 10.4	Single-dataset results	600
Table 10.5	Single-parameter simulation study results	601
Table 10.6	CreditRisk+ MLE results	606
Table 10.7	Regional single-dataset results	612
Table 10.8	Regional simulation study results	614
Table 10.9	$\mathcal{T}$-threshold model results	631
Table 10.10	t-threshold model simulation study results	631
Table 10.11	t-threshold model full information estimators	632
Table D.1	Function count	670

List of Algorithms

2.1	Independent defaults by Monte Carlo	59
2.2	Numerically computing risk measures	59
2.3	Monte Carlo binomial model	60
2.4	Analytic binomial implementation	64
2.5	Approximated expected shortfall in the analytic model	65
2.6	Independent Poisson defaults by Monte Carlo	76
2.7	Analytic Poisson implementation	79
3.1	Analytic beta-binomial	94
3.2	Simulation beta-binomial	100
3.3	The logit- and probit-normal mixture density functions	106
3.4	Analytic logit- and probit-normal	107
3.5	Logit- and probit-normal model calibration system	109
3.6	Logit- and probit-normal model Monte Carlo	112
3.7	Analytic Poisson-Gamma mixture	117
3.8	Analytic Poisson-Gamma moments	120
3.9	Poisson-Gamma model calibration system	120
3.10	Poisson-Gamma numerical implementation	124
3.11	Poisson-mixture integral	126
3.12	Poisson-mixture analytic model	126
3.13	Poisson-mixture model moments	128
3.14	One-factor CreditRisk+ Monte-Carlo implementation	139
3.15	Multi-factor CreditRisk+ Monte-Carlo implementation	146
4.1	Computing $p_n(G)$	157
4.2	The Gaussian joint default probability	159
4.3	A one-factor calibration objective function	162
4.4	Generating Gaussian latent variables	163
4.5	The numerical Gaussian one-factor threshold implementation	163
4.6	An analytical approximation to the Gaussian threshold model	173
4.7	An analytical expected-shortfall computation	173

4.8	Computing the t-distribution tail coefficient		182
4.9	Calibrating the t-threshold model		189
4.10	Computing the joint default probability from the t-threshold model		190
4.11	Generating t-threshold state variables		191
4.12	The t-threshold Monte-Carlo implementation		191
4.13	The normal-variance mixture density		205
4.14	The normal-variance mixture distribution function		205
4.15	The normal-variance mixture inverse distribution function		206
4.16	Generating normal-variance mixture state variables		208
4.17	Constructing the regional default-correlation matrix		223
5.1	Determining K_n and $\delta_{n,T-t}$		249
5.2	Simulating the indirect Merton model		254
5.3	Finding $A_{n,t}$ and σ_n		275
5.4	Computing default correlations		277
5.5	Simulating the direct Merton model		278
6.1	Asset-correlation and maturity adjustment functions		300
6.2	Basel IRB risk capital functions		302
6.3	A simple contribution computation		307
6.4	The granularity adjustment		318
6.5	The density functions for the Gaussian threshold granularity adjustment		319
6.6	Conditional-expectation functions for the Gaussian threshold granularity adjustment		322
6.7	Conditional variance functions for the Gaussian threshold granularity adjustment		324
6.8	Key CreditRisk+ granularity adjustment functions		339
6.9	Computing $\delta_\alpha(a)$		342
6.10	Simulating the one-factor CreditRisk+ with stochastic loss-given-default		344
6.11	CreditRisk+ granularity adjustment		344
7.1	Monte-Carlo-based VaR and expected-shortfall decompositions		364
7.2	The $\mathcal{J}_k$ function for $j = 0, 1, 2$		384
7.3	Moment and cumulant generating functions		386
7.4	First and second derivatives of $K_L(t)$		387
7.5	Finding the saddlepoint: $\tilde{t}_\ell$		390
7.6	The saddlepoint density		390
7.7	Computing the saddle-point tail probabilities and shortfall integrals		392
7.8	Identifying a specific saddlepoint VaR estimate		393
7.9	Computing the independent-default VaR contributions		397
7.10	Computing the independent-default shortfall contributions		404
7.11	A general saddlepoint function		412
7.12	A general integral function		413

7.13	Integrated densities, tail probabilities and shortfall integrals	413
7.14	Computing integrated VaR risk contributions	414
7.15	Determining the numerator for integrated VaR contributions	414
8.1	Evaluating our silly function	432
8.2	Two numerical integration options	435
8.3	Monte-Carlo integration	437
8.4	Identifying θ	462
8.5	Computing twisted probabilities	463
8.6	Computing the Esscher-transform Radon-Nikodym derivative	463
8.7	Importance sampling in the independent default case	464
8.8	Finding the mean-shift parameter	473
8.9	The *Full Monte* threshold-model importance-sampling algorithm	480
9.1	Simulating a four-state Markov chain	517
9.2	Counting transitions	520
9.3	Cohort transition-matrix estimation	521
9.4	Generating our bootstrap distribution	537
9.5	Hazard-function implementation	563
9.6	The survival probability	563
9.7	Theoretical CDS prices	564
9.8	Pricing the protection payment	565
10.1	Computing Q and Δ	585
10.2	Simulating correlated transition data	586
10.3	Accounting for counterparties, defaults, and probabilities	589
10.4	Simultaneous system of mixture model method of moments equations	592
10.5	Computing threshold model moments	594
10.6	Simultaneous system of threshold model method of moments equations	594
10.7	Single-parameter log-likelihood functions	599
10.8	Two-parameter regional log-likelihood functions	611
10.9	t-copula, transition, log-likelihood kernel	625
10.10	Building a one-parameter, correlation matrix	625
10.11	Inferring latent-state variables	629
A.1	Simulating correlated Gaussian and t-distributed random variates	647

Chapter 1
Getting Started

> *If I cannot do great things, I can do small things in a great way.*
>
> (Martin Luther King, Jr.)

Risk-management teams, in financial institutions around the world, are classically organized by functional responsibility. This implies distinct credit-risk and market-risk departments. Increasingly, we observe additional sub-groups responsible for operational risk, but the two solitudes of market and credit risk still remain more often the rule than the exception.

Such a distinction is *not* necessarily a poor way to organize a risk-management function. Market and credit risk do have some important differences and, in many cases, require different skill sets. For the modern risk manager, however, there is a significant advantage to being familiar with these *two* key dimensions. A risk manager who looks at the world solely through his own, be it credit- or market-, risk lens, is more likely to miss something important. A complete risk manager appreciates both of these key dimensions of financial risk. This is particularly true given that the distinction between these two areas is not as clear cut as many pretend.

There is already a large number of excellent works, however, in the area of credit-risk modelling. Why, the reader is entirely justified for asking, do we need another one? This book, written by a risk-management practitioner, takes an inherently practical approach to modelling credit risk. Much of the extant literature is either extremely mathematically complex and specialized or relatively general and vague on the technical details. In this work, the focus is on the tools and techniques required to actually perform a broad range of day-to-day, credit-risk computations. These include, but are not restricted to, estimating economic-capital computations, decomposing the contribution of each obligor to overall risk, comparing the results in the context of alternative underlying mathematical models, and identifying key model parameters. There are, thankfully, many works that may be used by the credit-risk modeller seeking more insight into the world of market risk.[1] Fewer works exist

---

[1] Jorion (2006), for example, is an excellent starting point.

for the market-risk manager looking to familiarize herself with the practical details of the credit-risk world. This book seeks, in part, to rectify this situation.

These ideas can all be found in other sources, in varying degrees of complexity, but are rarely combined together in a single source and developed from first principles. The unifying concept behind this work is the provision of a road-map to the quantitative analyst who is trying to assess the credit riskiness of a given asset portfolio. To this end, we will use a common portfolio example throughout the entire book. Moreover, in addition to the mathematical development and motivation associated with different modelling techniques, we will also provide tangible illustration of how to implement these ideas. This will take the form of concrete computer-code snippets from the Python programming language. The objective is to thereby shorten the distance between textbook and practical calculation. Python also offers many tools to visually supplement our analysis. Visualization of quantitative data is of the utmost importance and we seek to incorporate this aspect—inspired heavily by Tufte (2001)—in all aspects of the modelling discussion.

Credit-risk modelling is a vast area of practitioner and academic research. The existing collection of models are both rich and complex in terms of technical detail and application. A single work, of manageable size, cannot hope to cover it all. Some focus is, therefore, necessary. We have, consequently, chosen to emphasize the risk-management dimension. While many of the models in this work can be extended to price credit-risky instruments, our focus, however, will be on the construction of empirical default-loss distributions under the physical probability measure. Once this distribution is in hand, it can be used to compute tail probabilities, risk measures such as VaR and expected shortfall, and to approximate the contribution of individual credit obligors to these measures.

Another area that perhaps makes this work somewhat unique is its incorporation of a dimension of growing importance in the world of risk management: model risk. The risk of model misspecification, selection of inappropriate parameters, or implementation errors is significant with credit-risk models. Indeed, it is typically judged to be significantly higher than in the market-risk setting. Unlike the high-frequency movements of market-risk factors, default is a rare event. This paucity of default data creates challenges for model selection, calibration, and estimation; back-testing one's model, in most cases, is literally impossible. Written by a practitioner, who has spent much of his career on the market-risk side of the business, this book has a keen appreciation for the data shortcomings in the world of high-quality credit obligors. As a consequence, it consistently seeks to explore modelling alternatives, the use of diagnostic tools, and the direct incorporation of the model-risk dimension into the credit-risk modelling framework. The aim is to thereby enhance the usefulness of the discussion for risk-management practitioners.

In this first chapter, we highlight and motivate the key questions addressed, in a risk-management setting, by credit-risk models. This exercise provides useful insight into what makes credit-risk modelling such a challenging exercise. A comprehensive comparison with the market-risk problem also helps to identify the unique elements involved on the credit-risk side. We also introduce our sample portfolio and provide an illustrative, and highly graphical, demonstration of what we seek to accomplish. As a first step, however, we need to provide a greater degree of precision regarding how these specific issues will be addressed in the latter chapters. That is, we begin by detailing the perspective adopted in this book.

1.1 Alternative Perspectives

As a starting point, it is important to have a clear idea of what precisely one means by credit risk. Credit risk stems, ultimately, from a counterparty, or obligor, defaulting on their obligation. It can be an outright default or a deterioration of an obligor's ability to pay, which ultimately makes default more probable. The former situation is termed default risk, whereas the latter is typically referred to as migration risk. Our principal focus in the following chapters is upon the default event, but we will see that this perspective is readily generalized to the migration setting. It is, in fact, precisely in this latter case that the credit perspective begins to overlap with market-risk analysis.

Default or migration events, which are the key elements of credit risk, lead to financial losses. The scope of the loss can range from a small fraction to the entirety of one's investment. This type of risk can arise with almost any asset that requires ongoing future payment. Typically, however, we think of loans, bonds, swaps, or deposits. The notion of default, however, can be broadly applied to credit-card obligations, student loans, mortgages, car loans, or complex financial derivative contracts. One can even use—by defining default as the occurrence of a rare, non-financial, risk event such as a fire, a car accident, or an earthquake—credit-risk modelling techniques in assessing insurance claims.[2] Thus, default risk is inherent in two key pillars of the finance industry: banking and insurance.

1.1.1 Pricing or Risk-Management?

An assessment of the uncertainty surrounding the payment of future asset claims is thus central to informing important decisions about the structure of financial claims. Credit-risk models act as an important tool in performing such an assessment. There are, at least, *two* broad applications of credit-risk models:

1. pricing credit risk; and
2. measuring the riskiness of credit exposures.

---

[2]Indeed, there exist financial contracts, such as credit-default swaps, that provide insurance against the default of specific obligor.

The first element seeks to estimate the price at which one would be willing—in the face of uncertainty regarding a counterpart's ability to pay—to purchase an asset.[3] It looks to determine, on average, what is the cost of credit riskiness associated with a given position. This price is usually treated as a credit charge, typically as a spread over the risk-free rate, on the loan obligation. Equivalently, this amounts to a discount on the value of the asset. Pricing naturally leads to the notion of mean, or expected, default loss.[4]

Riskiness is not so much about what happens on average, but rather what is the worst case event that might be experienced. In risky physical activities, such as sky-diving and race-car driving, worst-case events are typically quite gruesome. In a financial setting, however, we are worried about how much money one might lose. It is insufficient, however, to simply say: we could lose our entire investment. While entirely true, this is missing a key ingredient. We need to consider the interaction of a range of possible losses and the associated probability of incurring these losses. In the combination of outcomes and likelihood of negative events lies the assessment of risk. Marrying these two ideas together naturally leads us to statistics and the notion of a default-loss distribution.

The distribution of default losses is the object of central interest for credit-risk modellers. Pricing and risk-management, therefore, both share this interest. In the former case, the central aspect is of more interest, whereas the latter focuses on its tails. Figure 1.1 describes this situation. Let us denote the default loss as the random variable, L, for a given portfolio, over a given time interval. The default-loss density, $f_L(\ell)$, summarizes the relative probability across a range of possible loss outcomes, denoted ℓ. The form of this, admittedly schematic, default density function is quite telling. It places a large amount of the probability mass on small levels of loss and a correspondingly modest amount of probability on large losses. This is very typical of credit-loss distributions. In a phrase, the probability of default is typically quite low, but when it occurs, losses can be extensive.

Figure 1.1 also highlights a few aspects of the default-loss distribution, which provide additional precision on the distinction between pricing and risk-management. The average default loss, denoted as $\mathbb{E}(L)$, is the principal input for the determination of the *price* of credit risk. Conversely, the value-at-risk, or VaR, is one possible—and quite useful—measure of the tail of the distribution. It is used extensively by risk managers to assess extreme, or worst-case, outcomes. It would be an oversimplification, however, to state that valuation experts care

---

[3] In an insurance setting, one examines the uncertainty of the occurrence of a particular risky event and asks: how much would be required, as a premium, to insure that risk. In both cases, we seek to find how much compensation is required to accept the financial risk. The slight additional complexity in the non-life insurance setting is that a single contract may simultaneously handle a number of perils.

[4] When computing prices, one is also typically interested in hedging these positions. This leads to a related problem of estimating the sensitivity of a given price to a change in an underlying risk factor. Since this is also typically performed under the pricing measure, we consider this part of the pricing problem.

1.1 Alternative Perspectives

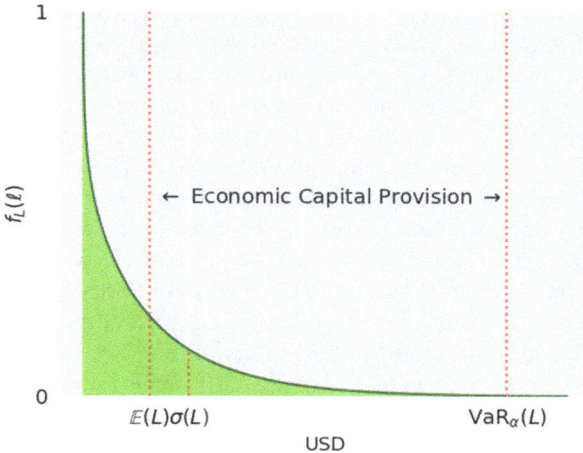

Fig. 1.1 *A (stylized) default-loss distribution*: Here is a popular schematic that reveals a number of key points about the default-loss distribution.

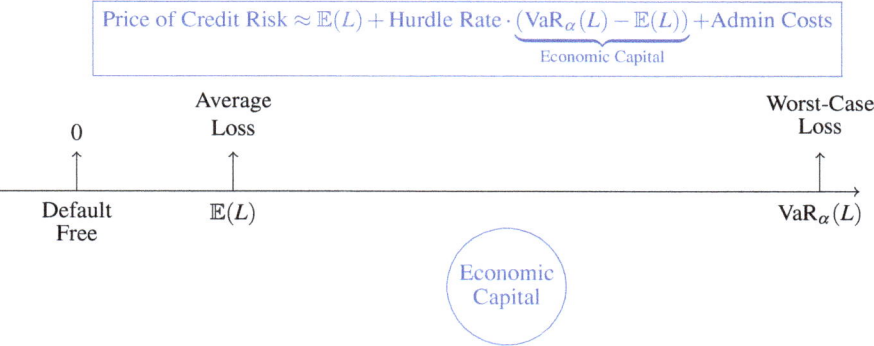

Fig. 1.2 *Pricing credit risk*: For pricing credit risk, both the expectation and the (expensive) capital requirement are typically involved.

solely about the central part of the default-loss distribution, whereas risk managers focus only on its tail. Both perspectives routinely employ multiple aspects of the statistical description of default. Figure 1.2 provides an, again highly schematic, description of how credit-risk is priced. It holds that the price of credit risk is the expected default plus some additional terms. One additional term is a reflection of administrative costs, which fall outside of the realm of modelling, but another term seeks to incorporate tail information.

Financial institutions, as we've discussed, take default risk in many ways. As a first step, it needs to be properly priced so that proper compensation for risk is provided. These institutions also set aside some additional funds—which act as a kind of reserve or rainy-day fund—to cover any unexpected losses arising from default outcomes. Figure 1.1, after all, clearly indicates that the distance between average and extreme outcomes can be quite substantial. These additional funds are often referred to as economic capital. There are different ways to estimate economic capital, but it generally amounts to the difference between a worst-case, or tail, risk measure and the expected loss. It is thus occasionally referred to as *unexpected losses*.[5] Regulators of financial institutions are very concerned about the magnitude of one's economic capital and the methodology required for its computation.

How does this idea impact the pricing of credit risk? Setting aside funds in the amount of one's economic capital is expensive. It implies that these funds cannot be used to make additional loans and generate profits for the firm. Many practical pricing models, therefore, add an explicit charge to compensate the lender for the costs of holding economic capital. A natural way to estimate this expense is to multiply the approximated economic-capital amount with the firm's cost of capital, or hurdle rate. In this way, the computation of a price for credit risk uses multiple aspects of the default-loss distribution. In a phrase, therefore, the price of credit risk associated with a security can depend on both the central and extreme aspects of the default-loss distribution.

1.1.2 Minding our $\mathbb{P}$'s and $\mathbb{Q}$'s

One of the major insights of modern mathematical finance is that the price of an instrument is well described by the discounted expected future value of its pay-outs. The classic example is the celebrated Black and Scholes (1973) option-pricing formula. This is not, however, an arbitrary expected value, but rather the expectation taken with respect to the equivalent martingale measure, $\mathbb{Q}$, induced using an appropriate numeraire asset.[6] When we select the money-market account as our numeraire asset, this is popularly referred to as the risk-neutral probability measure. This result is termed the *fundamental theorem of asset pricing*. While this can get quite mathematically involved—the interested reader is referred to Harrison and Kreps (1979) and Harrison and Pliska (1981) for the foundations—the intuition is quite appealing. Discounting future cash-flows is difficult, because the discount rate depends on their riskiness. The assumption of lack of arbitrage implies the

---

[5]To confuse matters, some authors refer to the standard deviation of the default-loss distribution, or $\sigma(L)$, as the unexpected loss. Other use this term to describe economic capital. To avoid unnecessary confusion, we will try to avoid this term and refer to loss volatility.

[6]An expectation, in its most fundamental sense, is an integral taken with respect to a probability measure.

existence of an equivalent martingale measure, $\mathbb{Q}$, which takes risk considerations out of the equation. Expected pay-outs under this measure can thus be discounted using the risk-free rate.

The probability measure thus plays a critical role in mathematical finance. In practitioner and regulatory settings, such as BIS (2001, 2004, 2005), the probability measure is rarely mentioned. It is thus relatively easy to get confused. It is nevertheless, if not explicitly mentioned, always present. A bit of clarification is, therefore, useful.

Valuation analysts spend the vast majority of their time working under the $\mathbb{Q}$ measure to price various types of financial instruments. Credit-risky securities are no exception. In credit-risk pricing work, like say Bielecki and Rutkowski (2002), the equivalent martingale measure plays a starring role. For risk managers, however, the situation is rather different. Risk managers are less interested in understanding the asset prices at a given point in time, but instead are focused on how the value of instruments move dynamically over time. While the $\mathbb{Q}$ measure is required to price each instrument at each static point in time, it provides little insight into the intertemporal dynamics of these valuations. This is captured by the true, or physical, probability measure, which is generally denoted as $\mathbb{P}$. This is often referred to as the real-world probability measure.

This topic is extremely technical and subtle; this work does not intend to add to this important and fundamental area. Duffie (1996) is, for interested readers, an excellent place to start to understand the complexities of asset pricing. We seek only to raise the important point that, in the large, pricing applications employ the equivalent martingale, or risk-neutral, measure, $\mathbb{Q}$. Risk managers, conversely, typically use the physical measure, $\mathbb{P}$. This is, of course, something of a simplification, but nevertheless represents a reasonable rule of thumb.

The key point is that the differences between these two measures are intimately related to aggregate market attitudes toward risk. Valuation works best when risk preferences can be extracted from the computation. Risk managers, by the very definition of their task, are not able to ignore risk attitudes. It is at the heart of what they do. Risk-management, to be relevant and useful, has no choice but to operate under the physical probability measure.[7] Given our risk-management focus, this book will also predominately focus on the real-world probability measure. Any deviation from this perspective will be explicitly mentioned and justified.

1.1.3 Instruments or Portfolios?

To this point, it is unclear whether we refer to a specific credit-risky security or a portfolio of such instruments. The simple answer is that the default-loss distribution found in Fig. 1.1 could equally apply to either a single instrument or a portfolio.

---

[7]Risk managers thus operate in the same world as statisticians, econometricians, and forecasters.

The choice is, however, not without importance. Typically, we imagine that the default or migration risk associated with a single instrument is determined by the creditworthiness of a sole entity. Portfolio default risk, conversely, is determined by the interdependence of the credit worthiness of multiple entities. More technically, the former involves a single marginal distribution, whereas the latter encompasses a full-blown, joint-default distribution. We can all agree that joint distributions of collections of random variables are, on the whole, more involved than any of the individual marginal distributions.

Once again, this is something of an oversimplification. Some complex instruments—such as collateralized debt obligations or asset-backed securities—do actually depend on the creditworthiness of large number of distinct obligors. Such instruments are, in fact, actually portfolios of other simpler securities. The message, however, remains the same. Understanding, and modelling, the credit risk of a portfolio is a more involved undertaking than dealing with a single security.

In this work, we will tend to focus on portfolios rather than individual securities. This viewpoint allows us to place more emphasis on the modelling of default and migration events. It places the credit-loss distribution at the forefront of our discussion. This is not to say that the characteristics of individual securities in one's portfolio do not matter. They matter a great deal. For our purposes, however, it will be sufficient to understand the size of the exposure associated with each obligor, the unconditional probability of its default or credit migration, and the magnitude of the loss in the event of a default outcome. We will, in some cases, require additional information—such as the tenor of the individual security—but information such as day-count conventions, coupon rates, payment indices, or implied volatilities will not not explicitly considered.

It is not that these elements are not employed in credit-risk analysis. They do, in fact, play an important role. There is an entire literature allocated to determining the worst-case exposure that one might have in the event of a default event. The ideas of *potential future exposure* (PFE) and *expected positive exposure* (EPE) arise in this setting. This is, in many ways, the flip side of the typical market-risk problem. Market-risk models generally seek to identify the worst-case loss—for a given portfolio over a given time interval with a particular level of confidence—that one might experience due to market-risk movements. In a credit-risk setting, the greater the market value of an instrument, the larger its exposure to default, and the worse the ultimate default-loss outcome. PFE and EPE models, therefore, are conceptually in the business of estimating the best-case VaR measures for one's portfolio. While these are an integral part of credit-risk modelling, they do not describe the heart of the undertaking. Indeed, for the most part, they fall into the category of market-risk modelling.

1.1.4 The Time Dimension

Time is always a complicating factor. Risk-management usually involves computing high-frequency estimates—daily, weekly, or monthly—of the risk associated with one's current portfolio over a pre-defined time horizon. For market-risk practitioners, the time horizon is typically quite short, usually ranging from a few days to about a month. In the credit-risk world, the approach is similar, although the time horizon is generally somewhat longer—one-year is probably the most popular, but multiple years are not uncommon.

This raises an important question: what kind of assumptions does one make about the evolution of the portfolio over a given horizon? The shortness of the market-risk perspective makes it reasonably defensible to assume a static portfolio. When the horizon is multiple years, however, this is less easy to defend. In the credit-risk setting, there are, at least, two aspects to this challenge: changes in the portfolio and movements in the creditworthiness of the individual obligors.

It is common practice, in both market- and credit-risk settings, to assume that the portfolio does not change over the horizon of analysis. In some cases, this leads to silly results. If one's horizon is two years, but some instruments possess shorter tenors, it is often clearly unrealistic to treat the portfolio as unchanged. Moreover, if the portfolio is managed relative to a benchmark, it should be expected to change frequently to reflect benchmark movements. The problem, however, is that trying to describe—in a sensible, tractable, and realistic way—these structural or tactical portfolio movements is very difficult. It basically amounts to forecasting future changes in your portfolio. In many cases, this may prove to be virtually impossible.

To the extent that one is trying to measure the amount of risk—either from a credit or market perspective—it may not even be desirable to model the dynamic evolution of one's portfolio. Ongoing risk monitoring and management often employ a snapshot approach to portfolio surveillance. Computing the credit-risk, over a given horizon, based on a static portfolio assumption allows one to compare today's portfolio to tomorrow's and to last week's. A static, one-period perspective is thus, in many ways, the risk manager's default setting. In the following chapter, for the most part, we will also adopt this viewpoint.

This is not to say that the dynamic perspective is uninteresting. Static long-term credit analysis, as suggested, often assumes that the credit worthiness of the individual obligors is also constant. This is a strong assumption and can be relaxed by permitting credit migration. This turns out to be a reasonably straightforward extension of the static, one-period approach. Incorporating losses, and gains, associated with non-default migration, however, is a bit more work. We will, in the upcoming chapters, discuss the essence of this idea and provide appropriate references for moving further in this direction. Our main focus will be, to be clear, on static, one-period default models. As will soon become evident, this is already a sufficiently complicated undertaking.

1.1.5 Type of Credit-Risk Model

When embarking upon, even a cursory, overview of the credit-risk modelling literature, one will be confronted with two competing approaches: the so-called structural and reduced-form methodologies. The difference is relatively easy to describe. Structural models are descriptive; they say something about the reason behind the default event. In this setting, default is endogenous to the model. This is consistent with the way the term *structural* is used in other modelling frameworks. It denotes the imposition of specific model behaviour as suggested by theory.

Reduced-form models take an alternative perspective. These methods treat the default event as exogenous. That is, there is no structural description of the underlying cause of default. It simply occurs with a particular probability. The driving idea is that the default event is modelled in an empirical sense, rather than informing it from a structural or theoretical perspective.

A simple example might help the reader understand the differences between these two models. Imagine that your boss is trying to construct a model of your, perhaps difficult-to-predict, morning arrival time. A reduced-form model might simply involve collecting some data and describing your office entry with a classic Poisson-arrival technique. This is a purely statistical approach. A structural approach, conversely, would seek to describe the main reason for your tardiness. The structurally minded boss might model your arrival as conditional upon the weather, the time of year, the state of public transportation, or the day of the week. Clearly, there is some room for overlap, but there is a basic difference. Reduced-form models are generally agnostic about the underlying reasons and are informed by data, whereas structural model seek to impose certain relationships often imposed by theory or logic.

There are advantages and disadvantages to each approach and, as such, it is not particularly useful to describe one method as superior to the other. In the following chapters, we will examine both methodologies. This provides a useful insight into the differences between the techniques. Comparing and contrasting alternative model assumptions is also a powerful tool in managing and mitigating model risk. As famously stated by the statistician George Box, "all models are wrong, but some of them are useful." This, slightly depressing mantra for quantitative analysts, suggests that we should welcome competing model approaches. It allows us to test the robustness of a result to their underlying, and differing, assumptions. Moreover, an awareness of the fallibility of our models, along with understanding of (hopefully) realistic alternatives, are the first defence against over-reliance on a specific modelling technique and the associated model risk such a reliance creates.

1.1.6 Clarifying Our Perspective

As we've seen in the previous discussion, there is a wide variety of possible perspectives one might reasonably adopt for credit-risk modelling. Some focus is, therefore, required. We may, for example, choose among:

Application	Pricing or risk-management;
Granularity	Instrument or portfolio level;
Time	Dynamic or static perspective; or
Model Type	Structural or reduced-form approaches.

To summarize, in a concise manner, the choices taken in this book, we will consider static structural and reduced-form credit-risk models in a portfolio setting from a predominately risk-management perspective. Although our principal application is risk management, most of the ideas, with a few changes such as choice of probability measure, are readily applied to the pricing context.

First and foremost, however, our focus in this work is on alternative techniques to describe the default-loss distribution and associated summary statistics for portfolios of financial assets. This is the essence of all credit-risk modelling approaches and requires us to think carefully about the very nature of credit risk and how it can be effectively modelled with mathematical and statistical methods. Much can be learned about the challenges and techniques associated with addressing this question by examining the market-risk perspective. This is precisely our focus in the following section.

1.2 A Useful Dichotomy

The value of asset portfolios rise and fall with the passage of time. These changes in value can occur for a variety of reasons. It is useful, however, to decompose these movements into two (very) broad categories:

1. general market movements making certain groups of assets more or less valuable; and
2. idiosyncratic movements in specific assets that lead to relative changes in their individual value.

This disaggregation is *not* new; it has been a key element in the finance literature for decades. Indeed, this is the intuition behind the celebrated capital asset-pricing model (CAPM). Much has been written on the CAPM; a good start, for the reader interested in more specifics, would include Sharpe (1963, 1964); Treynor (1961) and Lintner (1965). The CAPM is not used explicitly in credit-risk models, but it appears to motivate some of the fundamental ideas.

The distinction between general, or systematic, and specific, or idiosyncratic, risks is an immensely useful breakdown. Let us consider a few examples. When the general level of interest rates rises in an economy, it affects many assets. This is a general, systematic movement. The value of fixed-income assets fall and, depending on the context, it may lead to a positive upward shift in equity prices. If, conversely, a particular company has internal difficulties, for reasons that are unique to the firm, it typically has a more limited impact. The value of its debt and equity claims may fall, but we would not expect a concurrent impact on the value of other assets in the economy. This is a specific, or idiosyncratic, type of risk.

These different types of risk have different implications for how we model their associated portfolio-value movements. Moreover, in a very broad sense, these two perspectives capture the distinction between market and credit risk. To maintain that market risk arises from systematic factors and credit risk from idiosyncratic elements is sweeping generalization. It does not always hold. Some aspects of market risk can be quite specific to industries or regions, whereas credit risk can arise from general macroeconomic trends or events. There is, however, some truth to this statement and it offers a useful first-order approach to organizing the differences between these two key risk dimensions.

Let us start by considering the market-risk setting in a bit more detail. Market risk describes the portfolio P&L implications associated with movements in underlying market-risk factors. A specific financial instrument can be influenced to different degrees by a broad range of possible risk factors. The same risk factor, however, can impact multiple classes of financial instruments. Movements in these factors can lead to gains or losses in an asset portfolio. Ang (2013) offers a clever and helpful analogy of the link between asset classes and risk factors as strongly resembling the connection between food and nutrients. Portfolios (meals) are constructed from combinations of asset classes (foods). Diet experts do not focus on foods, but rather on their underlying nutrients. Nutrients are what really matter in our diets and, analogously, risk factors are what really drives asset returns. As a financial risk-expert, one is expected to delve into what really determines price movements in one's portfolios: risk factors.

In a market-risk setting, these underlying market risk factors typically include interest rates, exchange rates, equity and commodity prices, market liquidity, volatility, break-even inflation and credit spreads. Broadly speaking, therefore, these market risk factors relate to general, or systematic, quantities. As indicated, it is not always quite this simple. Credit spreads tend to move in unison, for example, but there is often also an idiosyncratic element. Moreover, while a security's value may depend on a general risk factor, its factor loading, or sensitivity to this factor, may be defined in a unique way.

The credit-risk domain is, in many ways, more difficult to manage. Credit risk, as has been indicated, describes the inability of a credit counterparty to meet their obligation. The ability to pay is, in a general sense, influenced by a rather different array of factors including, but not restricted to: indebtedness and capital structure, success of business model, management ability, competition, ongoing litigation, and willingness to pay. While these criteria appear to be predominately specific to the credit obligor, many of them can, however, be influenced by general economic conditions. As such, they also possess, to some extent, a general or systematic element.

1.2 A Useful Dichotomy

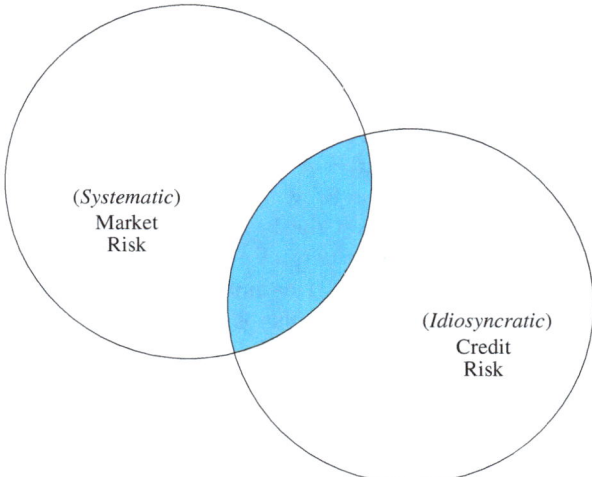

Fig. 1.3 *Market vs. credit risk*: Credit and market risk are *not* completely independent, but they are typically treated separately.

It would thus be a mistake to suggest that market risk depends on a underlying factor structure, whereas credit risk does not. The use of observable or latent risk factors, or state variables, is an integral aspect of modern finance. They arise in term-structure, risk-management, macroeconomics, and strategic analysis. They are simply too convenient and useful—in terms of dimension reduction and the ensuing mathematical parsimony—to not employ them in multiple settings. Both credit and market-risk computations depend on underlying risk factors. The point, however, is that the nature of these factors are often quite different in a credit-risk setting.

It is also an oversimplification to state that credit-risk factors are uniformly idiosyncratic in nature, but an important part of credit risk does indeed stem from specific factors. Figure 1.3 attempts, in a schematic and illustrative manner, to summarize this key message. Market risk is predominately systematic in nature, whereas much of credit risk arises from specific, idiosyncratic factors. The intersection of these two types of risk is *not* empty. There is an overlap of market and credit risk. It is sufficiently small that it is expedient to treat them separately; hence, the separation of risk-management functions into credit- and market-risk departments. It is nonetheless sufficiently large that it cannot be completely ignored.

1.2.1 Modelling Implications

Underlying driving risk factors are not the only difference between credit and market risk. In the market-risk setting, we often think of value changes to instruments and portfolios as occurring incrementally in a gradual fashion. Risk managers are, of course, worried about large jumps in market-risk factors leading to catastrophic losses. Such events certainly do occur, but not often. To this end, market-risk managers frequently employ stress-testing techniques to assess this type of risk.

Most market-risk computations, implicitly assuming continuous and gradual risk-factor dynamics, make use of a rather different modelling approach. In particular, it is common to assume that market-risk factors follow a stochastic process with continuous sample paths. Geometric Brownian motion is one of the most common choices. An examination of many of the key market-risk sources—for example, Mina and Xiao (2001); Morgan/Reuters (1996), and Mina (2005)—reveals the fundamental aspect of this assumption. This amounts to describing the infinitesimal dynamics with a stochastic differential equation of the following form,

$$\text{Change in Factor} = \text{Drift} + \text{Diffusion}, \quad (1.1)$$

$$dX_t = f(X_t)dt + g(X_t)dW_t,$$

where X_t is defined on $(\Omega, \mathcal{F}, \mathbb{P})$, $f(X_t)$ and $g(X_t)$ are measurable functions and $\{W_t, t \geq 0\}$ is a standard Wiener process. A key feature of the Wiener process is that,

$$W_t - W_s \sim \mathcal{N}(0, t-s), \quad (1.2)$$

for all $t \geq s$.[8] In other words, the increments of a Wiener process are independent, and identically normally distributed with a variance that is directly proportional to the size of the time step.

What is the rationale behind this choice? The path of a Brownian motion is noisy, bumpy, but ultimately continuous. Moreover, Brownian motion is inherently unpredictable. The same can be said, most of the time, for the movement of market risk factors. Changes in exchange rates, interest rates, and credit spreads (typically) occur gradually over time—they are also very difficult to predict. Brownian motions, with the appropriate parametrization, are also readily correlated. This is highly convenient for market-risk factors, which are generally expected to possess some degree of dependence or interaction. The implicit assumption embedded in the use of this approach is that market uncertainty is reasonably well described by a collection of correlated Brownian motions.

Is this a reasonable assumption? Figure 1.4 tries to motivate this choice. It displays two graphics: one displaying the actual daily changes in the USD-JPY exchange rate, a market-risk factor, and another simulated from a theoretical Brownian motion. Which is the real, observed risk factor? Ultimately, it does not matter and will remain a secret, since visually these two processes appear essentially the same. Does this mean that market-risk factors are equivalent to geometric Brownian motions? No, not at all. It does, however, motivate why

---

[8] See Karatzas and Shreve (1998, Chapter 2) or Billingsley (1995, Chapter 37) for much more thorough and rigorous discussion on Brownian motions, Wiener processes, and their applications.

1.2 A Useful Dichotomy

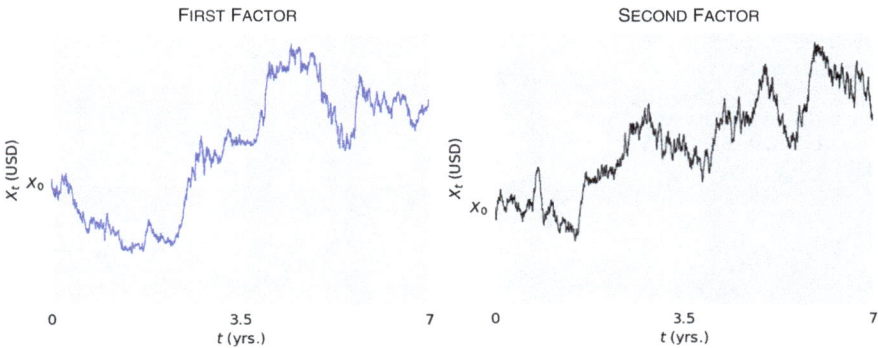

Fig. 1.4 *Brownian motion in action*: Examine the underlying risk factors. Which is a real, observed market-risk factor and which is a simulated path from a Brownian motion?

financial modellers opted, and continue to opt, for geometric Brownian motion as an approximation to reality. There is a significant amount of evidence indicating that the inherent Gaussianity and continuity of geometric Brownian motion are not perfectly suited for financial markets. This has led to numerous extensions and improvements—such as through the addition of stochastic-volatility, imposition of more complex transition densities with heavier tails, or jump-diffusion models—but the choice still has significant conceptual merit.

1.2.2 Rare Events

Default risk is also unpredictable. It rarely occurs, however, in a continuous and gradual manner. There is typically some element of surprise in a default outcome. Moreover, defaults are, for the most part, rare events. Precisely how rare and surprising a default event is considered, depends, of course, on the credit quality of the underlying obligor. The default of a highly creditworthy and financially solid counterparty would be considered surprising. As a consequence, such an event would typically be assigned a low probability. Default of a highly risky credit counterpart with poor creditworthiness, conversely, would probably not be considered such a rare and surprising event.

Although the degree of rarity and surprise depends on the credit quality of the obligor, the nature of the default event is inherently different than in the market-risk setting. Changes in a portfolio's value occur constantly, as market-risk factors move, and may have either positive or negative profit-and-loss implications. Default is a binary event. It either happens or it doesn't. Once it happens, default is also typically a permanent state. After arrival in default, exit is, in general, a slow and messy process. This is a fundamentally different type of risk.

Fig. 1.5 *Brownian motion in credit risk*: Geometric Brownian motion can, and often does, arise in the credit-risk setting. It plays, however, a rather different role.

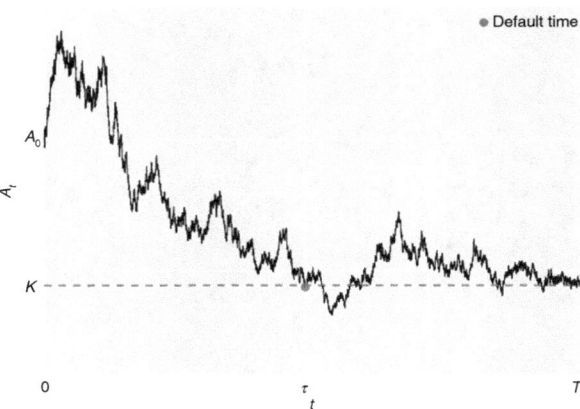

This does not necessarily imply that continuous sample-path stochastic processes cannot play a role in the credit-risk setting. The canonical credit-risk reference, and genesis of much of the current state of credit modelling, Merton (1974), is built on a foundation of geometric Brownian motion. In this approach, a firm's asset-value process, $\{A_t, t \geq 0\}$, evolves over time according to geometric Brownian motion. K is the lower bound, or threshold, denoting the value of the firm's outstanding liabilities. If, or when, $A_t \leq K$ for the first time, default is triggered. After this point, the firm is considered to be in default. Figure 1.5 graphically illustrates this situation.

This is, in technical vernacular, referred to as a first-passage time. That is, in the Merton (1974) model, the first time that the firm's asset value falls below its liabilities, default occurs. While this remarkably useful intuition supports much of credit-risk modelling, a few points are worth mentioning. First of all, it is *two* rather different things to assume that over a predefined time interval:

1. our investment's value changes with the evolution of a stochastic process; and
2. we experience a significant loss on our investment if the same stochastic process falls below a particular threshold.

The nature of the financial risk—and, by extension, the modelling tools used to estimate this risk—is different in these two cases. We should expect links between the credit- and market-risk dimensions, but there will also be important contrasts.

The second critical point is that, from a default perspective, the process may never cross the barrier over a particular time interval; or, if it does, it occurs only hundreds of years in the future. This brings us back to the binary aspect of credit risk. Default events are rare. Market-risk movements, by way of contrast, occur virtually at every instant in time. For some classes of assets—such as credit cards and consumer loans—default losses may, of course, occur with a high degree of regularity. Defaults may also be observed more frequently with car loans and mortgages, but since they are backed with assets, the magnitude of the

1.2 A Useful Dichotomy

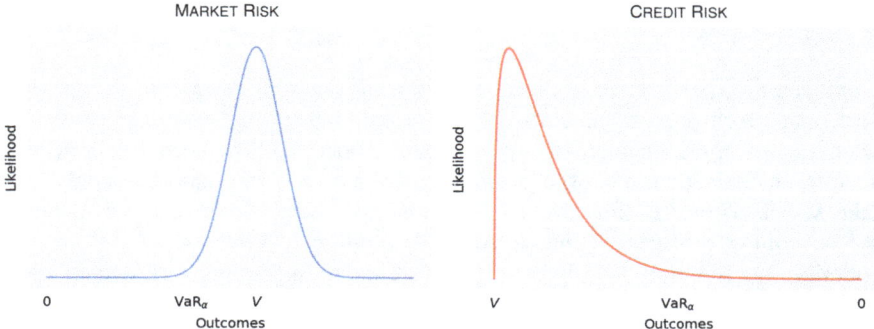

Fig. 1.6 *A stylized comparison*: Due to their fundamental differences, profit-and-loss outcomes and their associated likelihoods are combined in alternative ways in the market- and credit-risk settings.

losses are typically less important. In a more typical investment setting—corporate loans, derivative contracts, insurance policies, and high-grade investments—default typically occurs with much less frequency.[9] At a one-year horizon, for example, the probability of default for investment-grade debt approaches zero. Indeed, it is still only a few percent for non-junk bonds below the invest-grade line. Default can, therefore—from a modelling perspective, at least—be considered to be a rare, low-probability event. In these cases, it is usually considered to be a *surprise*.

It is important to appreciate this point, because it has *two* important implications. First, the structure of credit-risk models will differ in fundamental ways from the models employed in a market-risk setting. This idea is illustrated, in a stylized manner, by Fig. 1.6. Market risk is, generally speaking, the combination of numerous, symmetric and systematic risk-factor movements, whereas credit risk is associated with low-probability events of large magnitude. The result is two rather different loss distributions. Alternative tools, concepts, and methods will, therefore, be required in each of the two settings. Conceptually, however, the task is the same: combining profit-and-loss outcomes along with an assessment of their relative likelihoods to estimate the magnitude of total risk.

The second aspect has a larger impact on the analyst's life. The rarity of default events implies a shortage of useful data for the estimation and calibration of model parameters. A multitude of market risk-factor movements are observed every single trading day. Conversely, only a handful of high-quality credit defaults are observed over the course of a decade. This creates a number of challenges—many of which are addressed in the coming chapters—and represents a core reality for the credit-risk modeller.

---

[9] In an insurance setting, we can think of default as being basically equivalent to a claim.

1.3 Seeing the Forest

Our principal perspective is that of a risk manager. In this context, we will seek to perform credit Value-at-Risk, expected shortfall or, almost equivalently, economical-capital calculations for an asset portfolio. Whenever one selects a specific mathematical model to compute metrics or perform important analysis to take key decisions, it should not be done lightly. Analysis is required to assess the wisdom of the specific choice. At a minimum, it requires the analyst, and his superiors, to examine the key aspects of the modelling framework and to opine on the reasonableness of the overall approach. This is easier said than done. An examination of the credit-risk literature reveals a broad range of competing models and a high-degree of complexity.

To make matters a bit worse, in most credit-risk settings, empirical questions cannot be easily answered using statistical techniques. The principal reason, as already mentioned, is a dearth of default-event data. The limited amount of available data will not take us extremely far. Its principal use is in the estimation of model parameters and we will demonstrate how, in many cases, this can be very challenging.

Compare this to the market-risk world. Market-risk managers compute quantile measures, such as Value-at-Risk, of portfolio profit-and-loss movements on a daily basis. They may then compare these estimates to actual observed valuation shocks. If they are patient enough to collect a few years of data, there are a battery of useful statistical tests that can be usefully employed to assess the reasonableness of one's estimates. This exercise, known as back-testing, is an invaluable part of an analyst's toolkit.[10] It provides comfort and perspective in selecting a specific model; back-testing can also play an important objective role in model selection and calibration. Credit-risk analysts, due to the shortage of default data, cannot typically use back-testing techniques. Credit-risk modellers must thus learn to manage a correspondingly higher degree of subjectivity.

Critical assessment, and a general aversion to subjectivity, is nevertheless an essential part of the model-selection process. We seek to build this directly into our discussion rather than try to layer in onto our thinking at the end. We thus employ alternative techniques and metrics to assess model appropriateness. In particular, our analysis will highlight key modelling decisions, discuss alternatives, and perform analysis to motivate and challenge each choice.

The analyst's model-selection challenge is thus an important motivating factor behind the organization of this book. There is, of course, a significant amount of material to cover to help the quantitative analyst, risk manager, or graduate student to become more acquainted with this area and help them take defensible choices. To facilitate this, the following chapters are organized into three distinct themes:

---

[10] See Christoffersen (1998), Kupiec (1995), or Bolder (2015, Chapter 12) for a more detailed discussion of this area.

1.3 Seeing the Forest

Table 1.1 *Topic overview*: This table illustrates topics addressed in the following chapters and technical appendices.

Chapter	Topic	Theme
2	A Natural First Step	Modelling Frameworks
3	Mixture or Actuarial Models	
4	Threshold Models	
5	The Genesis of Credit-Risk Modelling	
6	A Regulatory Perspective	Diagnostic Tools
7	Risk Attribution	
8	Monte Carlo Methods	
9	Default Probabilities	Estimation Techniques
10	Default and Asset Correlation	
A	The t Distribution	Technical Appendices
B	The Black-Scholes Formula	
C	Markov Chains	
D	The Python Code Library	

modelling frameworks, diagnostic tools, and estimation techniques. Table 1.1 summarizes the organization of the following chapters.

1.3.1 Modelling Frameworks

Our thematic categorization is intended to inform *three* distinct questions. The first question, associated with the initial theme, is: which credit-risk model should one use? There is, naturally, no single correct answer for all applications. It will depend upon one's situation and needs. We will try to help by allocating *four* chapters to the consideration of three alternative modelling frameworks including:

1. an independent-default model;
2. a dependent-default reduced-form approach; and
3. a variety of alternatives within the structural setting.

There are numerous interlinkages between these models, which will help the analyst better understand the collection of extant models. We will not cover every possible modelling approach, but hope to provide a sufficiently broad overview so that further models will be easily accessed and classified.

There are a number of common elements, albeit handled in different ways, among these three modelling approaches. The first aspect is default dependence. Dependence between default events is a critical aspect of this discussion. Dependence of risk-factor movements is almost always an essential aspect of financial risk management, but in many ways its influence is even larger for the credit-risk modeller than in the market-risk setting. The reason is that the addition of default-dependence, induced in alternative ways, has an enormous influence on the ultimate default-loss distribution. Understanding the relative advantages and disadvantages of one's model choice, therefore, basically reduces to confronting issues of default dependence.

We also consider how the heterogeneity of exposures and default probabilities interact with the choice of model. In general, as the portfolio increases in size and homogeneity, the better behaved the model, the easier the estimation effort, and the stronger the performance of analytic approximations. This is an important consideration for the many organizations facing, for underlying structural business reasons, portfolios with concentrated and heterogeneous exposure and credit-quality profiles. Diversification thus inevitably also plays a role in the credit-risk setting.

It is also natural to ask how many factors are required, or are desirable, in a credit-risk model. The choice ranges from one to many and represents the classical trade-off between parsimony and goodness of fit. Goodness of fit is a tricky criterion for model selection in this setting, however, given the lack of data for calibration. There is some inevitable circularity, since with various parametric choices, it is typically possible to calibrate a one-factor model to yield broadly similar results to its multi-factor equivalent. Logical considerations based on modelling advantages and disadvantages thus play an important role in the ultimate model selection.

The advantages of a one-factor approach relate principally to the availability of a broad range of analytic formulae for its solution and computation of ancillary model information. On the negative side, many one-factor models may just be too simple to describe more complex credit portfolios. Multi-factor approaches may, in certain settings, provide a higher degree of realism and conceptual flexibility. The ability to categorize default-risk and dependence by, for example, industrial or regional dimensions is useful and valuable. Indeed, specific organization of one's factor structure provides a powerful lens through which one can consider, discuss, and take decisions on one's risk profile.

More factors imply, however, a higher degree of complexity. Typically, for example, model computations are performed through stochastic simulation. This can be either complex, computationally expensive or both. Moreover, the larger the number of factors, the broader the range of possible model parameters. These parameters must be either estimated from historical data, calibrated to current market data, or selected through expert judgement. In virtually all cases, this is a non-trivial undertaking. We will work, in the coming chapters, predominantly with one-factor models due to their greater ease of exposition and illustration. We will, however, also explore multi-factor implementations to provide critical perspective. Often, the results generalize quite naturally, but there are also unique challenges arising only in multi-factor situations. Where appropriate, these will be highlighted and discussed.

The marginal risk-factor distributions in these models may take alternative forms. Irrespective of the number of model factors selected and their distributions, a joint distribution must be imposed to proceed toward economic-capital calculations. The simplest choice would be to assume independence, which would permit straightforward construction of the joint default distribution. This is equivalent to an independent-default approach where each default event is treated as a Bernoulli trial implying, in the limit, a Gaussian loss distribution.[11] Much can be gleaned from this approach, but ultimately it is necessary to add default dependence. As dependence is introduced, even in the simplest case of a Gaussian joint distribution for the underlying risk factor (or factors), the loss distribution is no longer Gaussian. In fact, it is highly skewed and heavily fat-tailed.

The specific distributional choice has, not incidentally, important implications for the size and dependence of outcomes in the tail of the default-loss distribution. Tail dependence describes the probability, as we move very far out into the tail of a distribution, that a given counterparty defaults conditional on the fact another counterparty has already defaulted. This is an essential feature in a credit-risk model, particularly since one routinely uses the 99.50th or 99.97th quantile of the loss distribution to make economic-capital decisions.

Chapters 2 to 5 will thus explore a variety of credit-risk models and, in each case, consider dependence, distributional, factor, and portfolio homogeneity considerations. The objective is to provide a useful frame of reference for selecting and defending a specific modelling choice.

1.3.2 Diagnostic Tools

Once a specific credit-risk model has been selected and implemented, the work of the conscientious analyst is far from over. Models occasionally break down or, in some circumstances, they can provide conflicting or non-intuitive results. This seems, from personal experience, to typically happen at the worst possible time. The question addressed by the second theme is: how can we mitigate this risk? Sanity checks and diagnostic tools are helpful and necessary in resolving breakdowns or explaining incongruities in model outputs. Moreover, it is also essential to understand and communicate modelling results to other stakeholders in one's organization. Understanding and communication are intimately linked. Without a strong intuition of why and how one's model works, it is a tall order to communicate its outputs to one's colleagues or superiors. These two perspectives argue strongly for the inclusion of model diagnostics in one's modelling framework. We find this sufficiently important to allocate three chapters to this notion.

Credit-risk modelling is not performed in a vacuum. Most institutions performing credit-risk management operate under the control of regulatory regimes. Chapter 6 explores the regulatory guidance from the Basel Committee on Banking Supervision's (BCBS) Accords with an eye to link the main ideas to the modelling discussion found in Chaps. 2 to 5. While the ideas may seem relatively simple, they form a critical backdrop for other credit-risk modelling activities. The Basel Internal-Ratings Based approach is constructed based on an ongoing careful review

---

[11] The collection of joint events follows a binomial distribution, which in the limit by the central limit theorem, converges to a Gaussian law.

of existing models. A thorough understanding of the regulatory framework will help one better understand, and compare, existing choices to the decisions made by the BCBS. It will also help analysts answer inevitable questions from senior management on the differences between the internal and regulatory model results. These issues are the focus of Chap. 6.

Credit-risk economic-capital is, up to a constant, a quantile of the credit-loss distribution. This is a useful measurement, but others, such as the expected shortfall, are also possible. In isolation, these measures are nonetheless silent on the contribution of each obligor, or perhaps region or industry, to the total risk. Decomposing a specific risk measure into individual contributions along a predetermined dimension is referred to as risk attribution. It is reasonably well known that, in general, the risk-metric contribution from a given obligor is the product of the counterparty (or regional or industry) weight and its corresponding marginal-risk value.[12] This marginal quantity is only available in analytic form, however, under certain model assumptions. If one deviates from these assumptions, it may still be computed, but one must resort to simulation methods.

Despite these challenges, risk-metric decompositions are an invaluable tool. They are critical for trouble-shooting model results by permitting high-dimensional comparison to previous periods. They also aid in understanding how the results vary as model inputs—such as parameter values, default probabilities, and exposures—are adjusted. Finally, they are indispensable for communication of results to other stakeholders. Their regular computation and use should ultimately permit better decision making. In short, computation of individual obligor contributions to one's risk metrics may be the most powerful diagnostic at the modeller's disposition. Chapter 7 addresses this topic in detail.

It is almost inevitable that, in the course of one's credit-risk model implementation, one will make use of Monte Carlo methods. Credit-risk metrics typically involve taking a value from the very extreme of the default-loss distribution, such as the 99.9th percentile. When this rare event is estimated using stochastic simulation, it is natural and healthy to ask how many simulations are necessary to provide a reasonable and robust approximation. Such an assessment requires significant analysis and diagnostic tools. In particular, the following quantities prove instrumental in answering this question:

1. standard-error estimates, used for the construction of confidence intervals, for various quantile measures;
2. an analysis of the sensitivity of model results to different parameter choices; and
3. an examination of the stability of the marginal VaR computations.

This type of analysis provides important insights into the stability and robustness of one's estimates and will suggest both the number of simulations required and other supporting numerical techniques that may be employed. Chapter 8 addresses a variety of issues related to the Monte Carlo implementation. Throughout this

---

[12] For the reader unaware of this fact, we will demonstrate this claim in Chap. 7.

discussion, a concerted effort is provided to understand the speed and uncertainty of all estimates. It also introduces a technique for the reduction of the variance of credit-risk-based quantile estimators. This so-called variance-reduction technique is both an effective tool and diagnostic for the quantitative analyst.

1.3.3 Estimation Techniques

Models require parameters. The final theme addresses the question of how these important values are estimated and managed. We may mathematically incorporate default dependence into our model, for example, but the magnitude of this dependence is summarized by the value of a model parameter. What specific parameter to select is an empirical question. In a pricing context, these parameter values can often be inferred from other market instruments. Such an approach is less satisfactory when computing risk-management metrics.[13] The reason is that risk managers seek parameter values under the physical probability measure, $\mathbb{P}$. This requires the use of statistical techniques to infer parameter values from time-series data. In addition to parametric descriptions of the degree of default dependence, critical parameters include unconditional default probabilities for individual obligors. Default probabilities characterize the relative creditworthiness of a specific obligor. As such, they are absolutely central to one's assessment of default risk.

Both default-dependence and probability parameters can be estimated using the method of maximum likelihood. This requires significant understanding of one's model structure, which explains why this discussion occurs only in Chaps. 9 and 10. The form of the log-likelihood functions is mathematically complex and it can sometimes be difficult, given its non-linear form, to numerically maximize. Moreover, the degree of informativeness of the underlying data for this purpose is often somewhat limited. To investigate this, we consider:

1. tools to enhance our understanding of the actual meaning and implications of particular parameter choices;
2. a formal review of alternative standard-error estimation techniques to help us construct confidence intervals, which provide some insight on the robustness of our estimates; and
3. consideration of multiple techniques to reduce our reliance on any one specific method.

We also pose the following question: is there any scope for the use of expert judgement in the selection of these parameters? Or, at the very least, can expert judgement be employed to enhance or validate the estimates coming from the quantitative analysis? This question is difficult to answer, but it argues for the

---

[13]This market information can, and is, still used in a risk-management setting, but it requires the analyst to grapple with thorny risk-preference questions.

collection of the maximum amount of information. To this end, we also examine, in significant detail, the extraction of so-called risk-neutral default probabilities from credit-default-swap markets. While not immediately applicable to risk-management analysis, these estimates nonetheless contain information. While Chaps. 9 and 10 include significant technical detail, this final question is of great importance. We need to understand how far quantitative techniques can take us and when they need to be supplemented with more qualitative inputs or when quantitative estimates, from alternative sources, might profitably be combined.

1.3.4 The Punchline

To summarize, model choice and ongoing usage are always difficult. In the credit-risk setting, this is particularly true. Model risk—cogently identified by Derman (1996)—is thus a real problem and it arises in many ways. One of its most important sources arises through use of a single model without broader context. By context, we are referring model benchmarking, sensitivity analysis, and liberal use of model diagnostics. This is not a book about model risk, but it would be irresponsible and naive to ignore this dimension. We seek to mitigate it, from the outset, by consistently considering, in significant detail, a variety of modelling frameworks, tools and diagnostics, and estimation techniques. This is the motivation behind the organization of the book.

As a final point, an effort has been made to make this a reasonably self-contained work. It naturally includes numerous references to relevant works for readers interested in delving further into the details and seeing an alternative perspective. Additionally, reasonably comprehensive, if not completely rigorous, demonstrations will be provided for key results. Our goal is to motivate the use of alternative techniques and provide useful intuition behind them; we do not seek to create a mathematics textbook. Practical numerical examples and sample Python code are also provided for all of the methods introduced in the following chapters. Additionally, there are *three* technical appendices covering some key ideas that arise in our discussion. A fourth, and final, technical appendix summarizes the structure of the Python-code libraries used, and displayed, throughout the following chapters. This code can also be found on GitHub.

1.4 Prerequisites

Credit-risk modelling is surprisingly difficult and mathematically rich. It requires subtle analytic results, extensive use of numerical methods, and competence with statistical techniques. The following mathematical methods, for example, are used

1.4 Prerequisites

(some quite frequently) over the course of the coming chapters:

1. Raw and variance-reduced Monte Carlo methods;
2. Gaussian, t, χ^2, beta, Poisson, gamma, uniform, Weibull, variance-gamma, generalized-hyperbolic, binomial, logit, and probit distributions;
3. numerical non-linear optimization and root-solving;
4. the method of Lagrange multipliers and Taylor-series expansions;
5. point and interval estimation with the method of maximum-likelihood;
6. the statistical re-sampling technique termed bootstrapping;
7. Laplace and Fourier transforms in the context of characteristic and moment-generating functions;
8. some basic results from the stochastic calculus and change-of-measure techniques; and
9. liberal use of numerical quadrature methods.

Despite the richness of the mathematical tools used in this area, you do not need to be a professional mathematician to use and profit from this work. These are simply tools employed to help better understand and measure the risk embedded in our portfolios. We will strive to use, to the extent possible, concrete examples and make extensive use of graphics to support the basic intuition.

The reader does require a solid mastery of integral and differential calculus, linear algebra, and familiarity with the basic concepts in probability and statistics. Anyone with an undergraduate education in finance, economics, engineering, computer science, physics, or mathematics should thus find the material entirely accessible.

To implement these ideas, a familiarity with programming methods is also helpful. Under duress, one could implement many of these models in a spreadsheet environment. While possible, it would nonetheless be awkward. One could also use a professional programming language, such as C, C++, or Java. In this case, the effort would be more complicated. We've opted for a middle ground and have selected the, increasingly popular, Python programming language for our exposition. Python offers multiple high-performance libraries for scientific computing, while, at the same time, is also an easy-to-use, interpreted language. One need not be a professional computer programmer to use it, but the code does not suffer from lack of libraries and slowness.[14] Moreover, it is available on GitHub under the open-source license, which enhances its accessibility. We add to these libraries with the presentation of many code snippets, or algorithms, throughout the text. Our hope is that this will aid understanding and make this work more practical.

---

[14] All programming language choices, like most things in life, involve making trade-offs between relative advantages and disadvantages. See Aruoba and Fernández-Villaverde (2014) for an interesting and useful benchmarking exercise among scientific computing languages.

1.5 Our Sample Portfolio

Over the course of the coming chapters, we will consider a wide range of techniques and models. Our stated objective of practicality leads us to the use of concrete examples. We do not, however, wish to invent a new and different practical illustration for each method. Not only does this waste time, but it robs us of a frame of reference. As a result, we opted to create a single—and completely fictitious—sample portfolio. Virtually every technique in this book will be applied to this portfolio, permitting a direct comparison of the implications of different choices.

It is not *a priori* obvious what a sample portfolio should look like. Should it, for example, be large or small; heterogeneous or homogeneous; low or high risk? Ultimately, we decided to simulate a reasonably heterogeneous portfolio of moderate size with a relatively sizable amount of credit risk.

To be specific, our portfolio is comprised of 100 distinct credit counterparties. The total size of the portfolio has been normalized to \$1,000, implying an average exposure of \$10. The actual exposure sizes were, however, drawn from a Weibull distribution and range from less than \$1 to almost \$50. The empirical distribution of these exposure outcomes is summarized in the right-hand graphic of Fig. 1.7. The average unconditional default probability of the obligors is 1%. Again, although these values are centred around 1%, they are drawn from a χ^2 distribution. The left-hand graphic in Fig. 1.7 summarizes the default probabilities, which range from close to zero to roughly 7%. Both the Weibull and χ^2 distributions were selected for their positive support and skew; this permitted the construction of a relatively interesting sample portfolio. Otherwise, there was no particular reason for their choice.

This portfolio choice is not overly small, but also not enormous. It has a broad range of exposures and default probabilities, which should make it interesting to analyze. It is sufficiently small that we can focus on individual counterparties, but not too small to be trivial. Real-world portfolios can, however, be much larger with 1,000s of individual obligors. This portfolio is also readily simplified. We

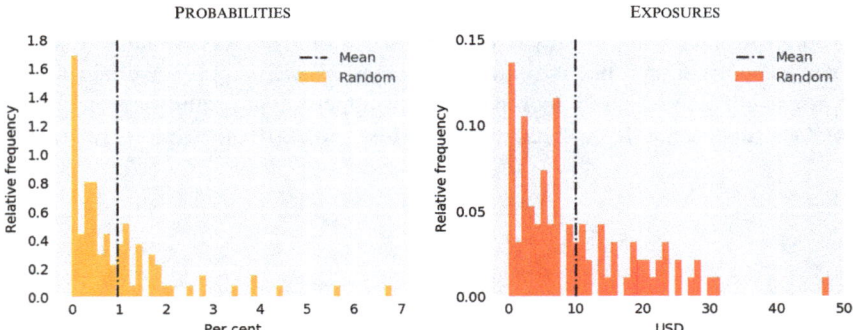

Fig. 1.7 *A sample portfolio*: The underlying graphics illustrate the individual exposures and unconditional default probabilities associated with our 100 obligor asset portfolio.

can, and will, examine the implications associated with a portfolio of 100 obligors each with the common exposure of $10 and 1% default probability. In this way, we can toggle back and forth between heterogeneity and homogeneity. Finally, we will, on different occasions, also use the existing exposures, but assign new default probabilities.

1.6 A Quick Pre-Screening

In this final part of our introductory discussion, we will use the previously defined sample portfolio to demonstrate the type of analysis examined in this book. It should not be considered an exhaustive review of the techniques considered in subsequent chapters, but rather as something of an *hors-d'oeuvre*. This implies that we will also skip over virtually all of the technical details—this means no programming and very little mathematics—and focus on the high-level intuition and motivation these ideas can add to one's credit-risk analysis. Our first step is to take a closer look at our sample portfolio.

1.6.1 A Closer Look at Our Portfolio

A famous quote, attributed to Napoleon Bonaparte, states that "war is ninety percent information." While the author is not particularly well placed to comment on the veracity of this claim in the military milieu, a similar sentiment certainly applies to the realm of quantitative analysis. One is invariably rewarded for taking the time to carefully examine the structure of one's data; or, more simply, attempting to transform data into information. This, therefore, will be our starting point.

Figure 1.7 provided some useful insight into the default probabilities and exposures associated with our 100-obligor sample portfolio. It does not, however, tell the whole story. Figure 1.7 is, to be more specific, completely silent on the interaction of these two important dimensions. How, one is justified for asking, are the exposures related to the default probabilities? Clearly, this matters. If the largest exposures are judged to have an extremely strong degree of credit quality, we may feel more confident about the riskiness of our portfolio.

Figure 1.8 displays two separate graphics to address this important point. The left-hand entry illustrates a scatterplot of each individual exposure against its unconditional default probability. The size of each individual dot, to ease interpretation, is a function of the size of the exposure. The larger the exposure, the larger the dot. With exposure on the horizontal axis, this implies that the dots get larger as one moves to the right of the origin. The majority of the large positions appears to be quite creditworthy, but there are a few exceptions. One large exposure, in particular, has a magnitude of about $30 and a default probability of roughly

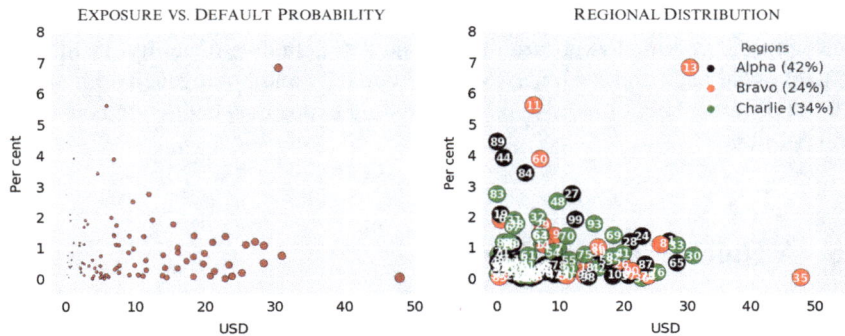

Fig. 1.8 *A bit more perspective*: These graphics examine the, extremely important, interaction between default probabilities and exposures. They also introduce a regional classification for our set of credit obligors.

7%. No other low-quality credit obligor has an exposure exceeding about $10. The largest single position, at almost $50, has a vanishingly small default probability.

The right-hand entry introduces a regional dimension. This is a common feature of credit portfolios. These regions can, in a broader sense, be extended to geographical, industrial, or obligor-size categorizations. In our case, we have randomly allocated our 100 credit counterparties into *three* fictional regions with names inspired by the first three characters of the NATO alphabet: Alpha, Bravo, and Charlie. Each is highlighted with a different colour. Again, to enhance our understanding, each dot includes a distinct number. In a real-world setting, these would be counterparty names. It allows us to identify certain specific obligors. We can make a number of observations:

- there does not, visually at least, appear to be any regional pattern in the data;
- the, ominously numbered, obligor 13 is our large high-risk position;
- counterparty 35 is the largest position, albeit it is currently assigned a relatively low risk of default;
- there are a cluster of positions—obligors 1, 8, 30, 33, and 65—with sizable exposures and default probabilities in the neighbourhood of 1%;
- another group—obligors 11, 44, 60, 84, and 89—have significant default probabilities, but modest exposures; and
- there is a jumble of, difficult to identify, small and relatively low-risk positions.

While very specific and detailed, this information will provide us with some useful perspective as we proceed with our analysis. One point is clear, we should expect a reasonable amount of credit risk in this portfolio.

1.6 A Quick Pre-Screening

1.6.2 The Default-Loss Distribution

To assess the default risk embedded in this portfolio, we ultimately need to construct a default-loss distribution. This is easier said than done. Construction of such a distribution requires adoption of a specific credit-risk model. In line with our stated objective of managing model risk, we will use *three* distinct models to construct a default-loss distribution. In particular, these include:

Independent This is the simplest possible credit-risk model that assumes each obligor default event is independent. Although simple, this model provides many insights into the underlying default-risk problem. It is introduced in Chap. 2.

Mixture This class of models relaxes the assumption of default independence by randomizing the default probabilities. It also, by construction, offers a rich array of possible distributional assumptions.[15] This approach is addressed in Chap. 3.

Threshold This popular modelling technique endogenizes the default event, permits default dependence, and allows for a number of alternative distributional choices. As of the writing of this book, it is reasonably safe to claim that this is the most popular model among credit-risk practitioners.[16] Chapters 4 and 5 discuss threshold models in detail.

While we will be examining a rather wider set of possible models, these three methodologies reasonably span the modelling frameworks provided in the following chapters.

In keeping with our high-level objectives, we will avoid any further technical description of these three approaches. Although there is an inevitable amount of overlap, numerous assumptions differ among these competing approaches. This is desirable situation. It allows us to meaningfully compare and contrast the results thereby building intuition about our problem and simultaneously acting to mitigate model risk.

Figure 1.9 jumps to the solution and provides two alternative perspectives on the one-period default-loss distribution associated with each model choice. The left-hand graphic displays the empirical distribution of default outcomes—these outcomes are each generated using Monte-Carlo methods with 10 million iterations. A few points can be observed. First, the independent-default approach appears to have a greater probability of low-magnitude losses, but a shorter tail; that is, low probability of large losses. Differences between the mixture and threshold models are difficult to identify. Both have a thinner waist and a longer tail. The mixture model also appears to have a smoother form relative to the other models. This is

---

[15] The specific implementation is the so-called beta-binomial mixture model; this approach, along with a variety of others are described in Chap. 3.

[16] The specific model choice is the Gaussian threshold method; the family of threshold models are introduced in Chap. 4.

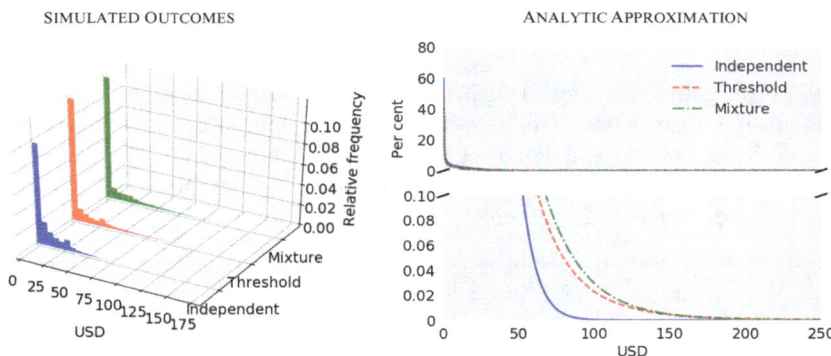

Fig. 1.9 *Default-loss distributions*: These graphics highlight—for our sample portfolio—the default-loss distribution estimated with three alternative credit-risk models.

directly related to a key modelling assumption. In the mixture setting, typically only average, not individual, default probabilities are employed. As we'll see in latter discussion, this has important implications.

The rarity of default events makes the visual examination of default-loss distributions quite difficult. Almost all of the probability mass is at low-loss levels, with very small, and difficult to see, probabilities of large losses. It is, however, precisely these low-frequency, high-magnitude losses that are of interest to the risk manager. Different visualization strategies, therefore, are required. The right-hand graphic of Fig. 1.9—computed incidentally using the analytic approximations discussed in Chap. 7—seeks to help in this regard. It provides, for the tail of the distribution, an enormous magnification of the outcomes. This allows us to see the scope of this high-magnitude losses and distinguish between the implications of the different models.

Figure 1.9 clearly indicates that worst-case outcomes are significantly lower in the independent-default setting. Extreme events appear, by comparison, to be roughly double for the mixture and threshold models. We thus see, for the first time, some concrete evidence of the importance of default dependence. This is the major difference between these two modelling approaches. It also highlights the nature of large-magnitude, low-probability events examined in the credit-risk setting.

1.6.3 Tail Probabilities and Risk Measures

Armed with the default-loss distribution, we have all the ingredients required to actually compute measures of portfolio riskiness. As suggested by Fig. 1.9, it is not generally useful to examine the full default-loss distribution. Instead, we focus our attention on its tail. Figure 1.10 provides another perspective, by examining the range of outcomes over the interval, $[0.95, 1.00]$. While this provides, in fact,

1.6 A Quick Pre-Screening

Fig. 1.10 *Tail probabilities*: In this figure, we examine the tail of the loss distribution under a high degree of magnification: these are termed tail probabilities. As in Fig. 1.9, three alternative models are examined.

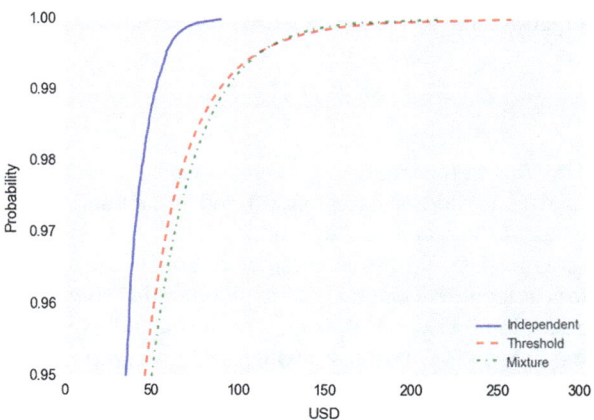

Table 1.2 *Summary risk measures*: This table summarizes a number of summary statistics—expected loss, loss volatility, VaR, and expected-shortfall ($\mathcal{E}_\alpha(L)$)—for three different model choices across a variety of degrees of confidence.

Quantile (α)	Independent		Threshold		Mixture	
	$\text{VaR}_\alpha(L)$	$\mathcal{E}_\alpha(L)$	$\text{VaR}_\alpha(L)$	$\mathcal{E}_\alpha(L)$	$\text{VaR}_\alpha(L)$	$\mathcal{E}_\alpha(L)$
95.00th	$35.2	$45.5	$46.2	$73.5	$50.2	$76.6
97.00th	$40.4	$50.8	$58.8	$87.9	$63.5	$90.2
99.00th	$52.6	$61.2	$89.0	$122.1	$92.7	$119.7
99.50th	$58.8	$66.8	$110.5	$145.8	$111.4	$138.4
99.90th	$72.5	$80.4	$167.9	$206.9	$155.6	$182.3
99.97th	$84.7	$92.3	$229.1	$270.0	$197.3	$222.9
99.99th	$90.0	$97.7	$257.5	$299.3	$216.2	$241.3
Other statistics						
Expected loss	$9.2		$9.2		$9.5	
Loss volatility	$12.8		$19.3		$19.8	
Total iterations	10,000,000					

nothing more than a repetition of the lower part of the right-hand graphic in Fig. 1.9, it is a useful segue into risk measurement. In the following chapters, the perspective from Fig. 1.10 will be used extensively to compare individual models.

Each extreme quantile of the default-loss distribution is determined by simply selecting a confidence level from the vertical axis and reading off the appropriate dollar loss on the horizontal axis. The 99th quantile associated with the independent-default model is thus approximately $50, but closer to $90 for the other two approaches.

Repeating this procedure for a handful of different quantiles leads naturally to the results summarized in Table 1.2. What is the significance of these quantile measures? This requires a bit of explanation. Morgan/Reuters (1996) originally suggested the following construct for a risk measure,

$$\text{VaR}_\alpha(L) = \inf\left(x : \mathbb{P}(L \leq x) \geq 1 - \alpha\right). \tag{1.3}$$

This function depends on *two* parameters: α a threshold for the probability on the right-hand side of equation 1.3 and the default loss, L. The first is a parameter, whereas the second is our random default-loss variable of interest. Imagine that we set α to 0.95. With this parameter choice, $\text{VaR}_\alpha(L)$, describes the largest default-loss outcome exceeding 95% of all possible losses, but is itself exceeded in 5% of all cases. In other words, it is the $(1 - \alpha)$-quantile of the portfolio's return distribution. By convention, this loss, which is by definition negative, is typically presented as a positive number and termed the value-at-risk (or VaR) of the portfolio. It is perhaps the most common risk measure employed in portfolio management—when computed for default or migration risk, it is often referred to as credit VaR.

A second, and increasingly common measure, is defined as the following conditional expectation,

$$\mathcal{E}_\alpha(L) = \mathbb{E}\left(L \,\Big|\, L \geq \text{VaR}_\alpha(L)\right), \tag{1.4}$$

$$= \frac{1}{1-\alpha} \int_{\text{VaR}_\alpha(L)}^\infty \ell f_L(\ell) d\ell,$$

where $f_L(\ell)$, as before, represents the default-loss density function. In words, therefore, equation 1.4 describes the expected default-loss given that one finds oneself at, or beyond, the $(1 - \alpha)$-quantile, or $\text{VaR}_\alpha(L)$, level. This quantity is, for this reason, often termed the conditional VaR, the tail VaR, or the expected shortfall. We will predominately use the term expected shortfall and refer to it symbolically as $\mathcal{E}_\alpha(L)$ to explicitly include the desired quantile defining the tail of the return distribution.

These expressions are fairly abstract. With a bit of additional effort, we may simplify the first one somewhat. From equation 1.3, the VaR measure, by definition, satisfies the following relation,

$$\int_{\text{VaR}_\alpha(L)}^\infty f_L(\ell) d\ell = \alpha, \tag{1.5}$$

$$1 - \mathbb{P}\left(L \geq \text{VaR}_\alpha(L)\right) = 1 - \alpha,$$

$$\mathbb{P}\left(L \leq \text{VaR}_\alpha(L)\right) = 1 - \alpha,$$

$$F_L\left(\text{VaR}_\alpha(L)\right) = 1 - \alpha,$$

1.6 A Quick Pre-Screening 33

where $F_L(\cdot)$ denotes the cumulative default-loss distribution function. The solution is merely,

$$\text{VaR}_\alpha(L) = F_L^{-1}(1-\alpha). \tag{1.6}$$

The explains quite clearly our interest in the quantiles of the default-loss distribution. They are, in fact, merely VaR measures for different levels of confidence.

If one is lucky enough to have a closed-form expression for the default-loss density, then equations 1.4 and 1.6 can often be manipulated to derive closed-form expressions for the VaR and expected-shortfall measures. We will, unfortunately, rarely find ourselves in this happy situation. Instead, as is discussed in detail in Chaps. 7 and 8, we will make use of Monte-Carlo methods and analytical approximations.

With this understanding, let us return to Table 1.2. We observe that, as we would expect, the VaR measure is an increasing function of the confidence level. Moreover, in all cases, the expected shortfall exceeds the VaR estimates. This must be the case; one measure is a cut-off, whereas the other is the average loss behind this cut-off point. The expected-shortfall measure, in this particular case, exceeds the VaR by roughly 10 to 30% beyond the 99th quantile. The distance between these two will, in general, depend on the form of the default distribution and, most particularly, the heaviness of its tail.

We also observe important differences between the models. The threshold model estimates, beyond the 99th quantile, approach three times the size of the independent-default equivalents. This can be considered a significant deviation. Despite very similar expected losses, therefore, the alternative models offer divergent descriptions of extreme outcomes. This variety of tail estimates stem directly from differing distributional and dependence assumptions.

Which model is superior? This is a difficult question to answer. Empirical techniques are, sadly, not extremely useful in informing this question. Unlike the market-risk setting, we simply do not have a sufficient amount of data to make definitive statements. This argues, in our view, for a multiplicity of models with a variety of underlying assumptions. This is a motivating factor behind our consideration, in this work, of a relatively diverse set of modelling frameworks.

1.6.4 Decomposing Risk

Computation of risk measures is a critical part of risk management, but it is by no means the last step. Understanding, interpreting, verifying the reasonableness of these measures, and communicating the results to senior decision makers are also an essential part of the task. To accomplish this effectively, the modern quantitative risk manager requires model diagnostics. These typically include benchmark models, confidence intervals, stress testing, and sensitivity analysis. While we will attempt to address all of these issues in the coming chapters, there is one additional model diagnostic that warrants immediate consideration: the decomposition of our risk-measure estimates.

In the market-risk setting, this typically means either an instrument or risk-factor level additive decomposition of a given risk measure. One might allocate a $100 market risk-metric estimate into $50 from interest rates, $30 from foreign-exchange rates, and the remaining $20 to spread movements. This particular type of breakdown is not precisely relevant in the credit-risk setting, but the basic idea still holds. We desire instead a risk-measure decomposition on the basis of individual credit obligors, regions, or industrial sectors. This type of information can lead to valuable insights into the inner workings of our models.

Computing these quantities is indeed possible, but either computationally or analytically complex. The details are thoroughly discussed in Chaps. 7 and 8. At this point, we simply skip to the results. Figure 1.11 summarizes the USD 99.9th quantile VaR contributions by individual obligor for each of our three models. Although, given the abundance of obligors, it is not extremely easy to read, we can extract a few interesting observations. The independent-default and threshold models appear, for example, to agree on the largest individual contribution. The mixture model assigns this honour to another obligor. There is also significant variation in the amount of VaR contribution, within each model, assigned to different credit counterparties. There are also interesting differences between models. The independent-default model seems, for instance, to assign the lion's share to a handful of obligors. The other two models, however, spread the VaR contribution more widely across counterparties.

Table 1.3 adds some additional colour by presenting the ten largest individual contributions for each model. Since the order differs by model, we used the rankings

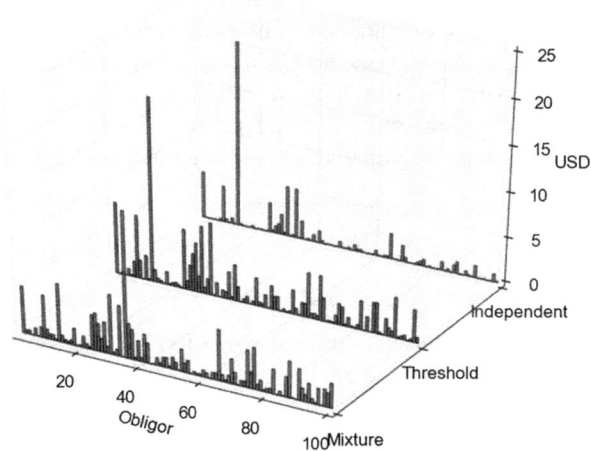

Fig. 1.11 *Obligor VaR contributions*: This figures outlines, for our three modelling choices, the contribution of each obligor to the overall 99.9th quantile VaR measure.

1.6 A Quick Pre-Screening

Table 1.3 *VaR-contribution breakdown*: This table illustrates the 99.9th quantile VaR contribution by credit counterparty—for the ten largest contributors—across each of our three different model choices.

	Obligor			Model		
Rank	#	Default probability	Exposure	Independent	Threshold	Mixture
1	13	6.82%	$30.4	$20.5	$20.2	$6.4
2	30	0.74%	$30.9	$5.6	$7.5	$6.6
3	33	1.07%	$28.4	$5.6	$8.1	$5.6
4	1	1.19%	$27.1	$5.2	$8.0	$5.2
5	8	1.09%	$25.8	$4.1	$7.1	$4.8
6	24	1.36%	$22.9	$3.3	$6.7	$3.9
7	65	0.50%	$28.3	$2.8	$5.0	$5.6
8	28	1.18%	$21.0	$2.3	$5.5	$3.4
9	35	0.02%	$47.8	$2.0	$1.5	$15.9
10	69	1.39%	$18.4	$1.8	$5.0	$2.7
Other obligors			$718.9	$19.3	$93.4	$95.6
Total			$1,000.0	$72.5	$167.9	$155.6

from the independent-default model to identify the top-ten candidates. The results are fascinating. In particular, we note that:

- the ten largest 99.9th-quantile VaR contributors account for about 70% of the independent-default-model VaR estimate, but only around 40% for the other two models;
- the ordinality for the independent-default and threshold models are similar, whereas the mixture model makes different choices;
- more specifically, the independent default and threshold models assign the largest contribution to our large high-risk obligor 13;
- the mixture model reserves this honour for counterparty 35 with a large exposure, but a relatively small default probability;
- the remaining contributions seem to stem principally from the cluster of large exposure, modest default probability positions; and
- the small-exposure, high default-probability positions do not make our top-ten list.

These observations are consistent with the basic assumption structure associated with these alternative approaches. The independent-default and threshold models, for example, make full use of both the exposure and default probability estimates to construct a default-loss distribution. The specific mixture model considered, conversely, only employs the average creditworthiness; this is because it essentially randomizes the default probability. This may, at first glance, seem like an inferior choice. After all, we certainly want to explicitly take into account the creditworthiness of each obligor. While very true, the mixture model's alternative perspective is also quite valuable. By assigning the highest VaR contribution to obligor 35, it

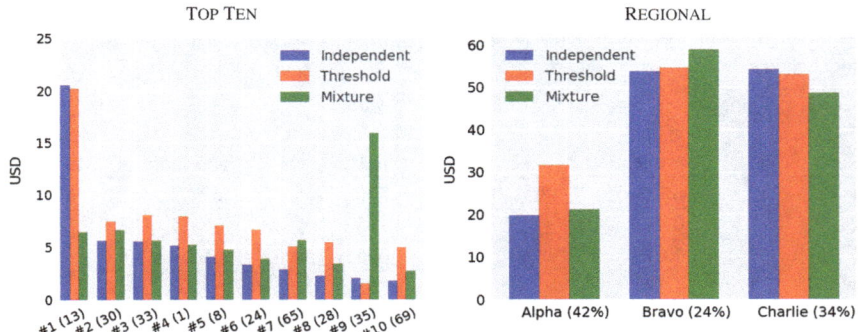

Fig. 1.12 *VaR contribution perspectives*: These two graphics demonstrate two alternative perspectives that might be applied to the obligor-level 99.9th quantile contributions: the top-ten contributors and the regional breakdown.

is focusing more on the size of its exposure and less on its low-risk status. This reminds us that this is, by virtue of its size, a risky position. It is precisely this type of incongruence of model assumption that, as a professional modeller, one seeks to help benchmark and organize one's thinking.

An additional, but related, point is the difference between high-exposure, low-risk and low-exposure, high-risk positions. Upon reflection, the former are more dangerous. Despite their rarity, when default does occur, the profit-and-loss consequences are high. This is, virtually by definition, more of a risk for one's portfolio than small positions in a poorly rated entity. We expect that, as we move into the tail, that our model will capture these high-magnitude, low probability events. The fact that it does, and is clearly apparent in our VaR-contribution analysis, is encouraging and comforting. Furthermore, careful and regular examination of these decompositions can signal if something has gone wrong. This might arise from data-input errors or coding problems. They can also be a significant aid in communicating model results and answering difficult questions from senior management.

Figure 1.12 provides the final word on our preliminary analysis by graphically examining our top-ten VaR contributions and also adding a regional VaR decomposition.[17] The right-hand graphic underscores, perhaps more clearly, the important diversity of view of the mixture model relative to the other two approaches. To reiterate, although one may not agree with the specific assumptions, it does make this model a useful candidate for benchmarking purposes.

The regional perspective, in the right-hand graphic, offers something of a surprise. The Alpha region, with the largest share of total exposure, accounts for the smallest total contribution to our 99.9th quantile VaR estimate; this result holds for

---

[17] It is, indeed, possible to also perform this decomposition for the expected-shortfall measure. While we have excluded it in this discussion, because it is quite repetitive, the necessary techniques along with many practical examples are addressed in Chap. 7.

all three models. Bravo region, the smallest in terms of total exposure, looks by all accounts to make the largest contribution to risk. This did not seem obvious, to the author at least, by inspecting the regional distribution found in Fig. 1.8 on page 28. It is possible that the clever reader might have guessed this result by observing that the infamous obligors 13 and 35 both fall into the Bravo region. Nevertheless, the regional structure is not, in our view, *a priori* evident. It is only through the lens of a risk-measure decomposition, or attribution, that is this relationship clear. This is yet another compelling reason for the ongoing use of these risk attributions as a model diagnostic.

1.6.5 Summing Up

We have, hopefully, succeeded in providing a useful and insightful demonstration of the type of credit-risk modelling addressed in this work. As we will soon see, the underlying mathematical and computational structure required to perform these calculations are actually quite involved. The reader can expect to find, in the coming pages, detailed practical direction regarding how to perform similar analysis on portfolios of their choice. There are, of course, a number of additional analyses that we might have considered in the context of our demonstration: sensitivity to parameter choices and confidence bounds are two obvious examples. Adding these dimensions would, however, require us to jump into more detail regarding the specific form of the model and the numerical methods used to estimate them. This work is left for the upcoming chapters.

1.7 Final Thoughts

Our goal, in this first chapter, was to highlight and motivate the key questions addressed by credit-risk models in a risk-management setting. This involved giving the reader a greater degree of precision regarding our specific angle on the credit-risk problem, providing some insight into what makes credit-risk modelling challenging, and a comprehensive comparison with the market-risk problem. This latter element helps to identify the unique elements involved on the credit-risk side. This was followed by a review of our thematical organization by modelling framework, diagnostic tools, and estimation techniques. Finally, we also introduced our sample portfolio and, in the form of an appetizer, provided an illustrative, and highly graphical, demonstration of what we seek to accomplish in the coming chapters.

References

Ang, A. (2013). *Asset management: A systematic approach to factor investing*. Madison Avenue, New York: Oxford University Press.

Aruoba, S. B., & Fernández-Villaverde, J. (2014). *A comparison of programming languages in economics*. University of Maryland.

Bielecki, T. R., & Rutkowski, M. (2002). *Credit risk: Modeling, valuaton and hedging* (1st edn.). Berlin: Springer-Verlag.

Billingsley, P. (1995). *Probability and measure* (3rd edn.). Third Avenue, New York, NY: Wiley.

BIS. (2001). The internal ratings-based approach. *Technical report*. Bank for International Settlements.

BIS. (2004). International convergence of capital measurement and capital standards: A revised framework. *Technical report*. Bank for International Settlements.

BIS. (2005). An explanatory note on the Basel II IRB risk weight functions. *Technical report*. Bank for International Settlements.

Black, F., & Scholes, M. S. (1973). The pricing of options and corporate liabilities. *Journal of Political Economy, 81*, 637–654.

Bolder, D. J. (2015). *Fixed income portfolio analytics: A practical guide to implementing, monitoring and understanding fixed-income portfolios*. Heidelberg, Germany: Springer.

Christoffersen, P. F. (1998). Evaluating interval forecasts. *International Economic Review, 39*(4), 841–862.

Derman, E. (1996). Model risk. Goldman Sachs Quantitative Strategies Research Notes.

Duffie, D. (1996). *Dynamic asset pricing theory* (2nd edn.). Princeton, NJ: Princeton University Press.

Harrison, J., & Kreps, D. (1979). Martingales and arbitrage in multiperiod security markets. *Journal of Economic Theory, 20*, 381–408.

Harrison, J., & Pliska, S. (1981). Martingales and stochastic integrals in the theory of continuous trading. *Stochastic Processes and Their Applications, 11*, 215–260.

Jorion, P. (2006). *Value at risk: The new benchmark for managing financial risk*. New York, NY: McGraw-Hill Ryerson Limited.

Karatzas, I., & Shreve, S. E. (1998). *Methods of mathematical finance*. New York, NY: Springer-Verlag.

Kupiec, P. H. (1995). Techniques for verifying the accuracy of risk measurement models. *The Journal of Derivatives, 3*, 73–84.

Lintner, J. (1965). The valuation of risky asset and the selection of risky investments in stock portfolios and capital budgets. *The Review of Economics and Statistics, 47*(1), 13–37.

Merton, R. (1974). On the pricing of corporate debt: The risk structure of interest rates. *Journal of Finance, 29*, 449–470.

Mina, J. (2005). Risk budgeting for pension plans. Riskmetrics Group Working Paper.

Mina, J., & Xiao, J. Y. (2001). *Return to RiskMetrics: The evolution of a standard*. New York: RiskMetrics Group Inc.

Morgan/Reuters, J. (1996). *RiskMetrics — technical document*. New York: Morgan Guaranty Trust Company.

Sharpe, W. F. (1963). A simplified model for portfolio analysis. *Management Science, 9*(2), 277–293.

Sharpe, W. F. (1964). Capital asset prices: A theory of market equilibrium under conditions of risk. *The Journal of Finance, 19*(3), 425–442.

Treynor, J. L. (1961). Market value, time, and risk. Unpublished Manuscript.

Tufte, E. R. (2001). *The visual display of quantitative information* (2nd edn.). Chesire, CT: Graphics Press LLC.

Part I
Modelling Frameworks

Models are the central element of this book, which is hardly a surprise considering its title. Given the stated emphasis on model selection and multiplicity of perspective, it is also unsurprising that a substantial number of possibilities are examined. Chapter 2 begins with the simple, but illuminating, independent-default setting. The resulting binomial and Poisson models may be unrealistic, but they form an excellent lower bound on one's credit-risk estimates and thus work well as benchmark comparators. They also act as a kind of foundation for alternative methods. With this in mind, two competing modelling families are then considered. The first, the collection of mixture models, are reduced-form in nature. Through randomization of default—which ultimately amounts in binomial or Poisson mixture distributions—one induces default independence. Practical choices examined in Chap. 3 involve introduction of logit, probit, Weibull, gamma, beta, and lognormally distributed systematic state variables to accomplish this task. The second family, often alluded to as the group of threshold models, are variations of Merton (1974)'s structural approach. That is, they seek to describe the underlying reason for a default event. These models are best examined within the context of the collection of so-called normal-variance mixture models. In Chap. 4, the Gaussian, t, variance-gamma, and generalized hyperbolic approaches are comprehensively investigated. No discussion of credit-risk models, however, would be complete without a careful examination of the original Merton (1974) method, which motivated this entire literature. Understanding when, and when not, this approach coincides with the threshold model is quite useful and forms the backbone of Chap. 5. Overall, in the following four chapters, more than a dozen modelling choices are presented to the reader. While other choices are possible, the hope is that this variety of choice will nevertheless prove useful in the management, interpretation, and comparison of one's own models.

Reference

Merton, R. (1974). On the pricing of corporate debt: The risk structure of interest rates. *Journal of Finance, 29,* 449–470.

Chapter 2
A Natural First Step

> *The purpose of computing is insight, not numbers.*
>
> (Richard Hamming)

Default is, quite simply, an inability, or unwillingness, of an obligor to meet its obligations. Imagine you are given a blank sheet of paper and accorded 15 minutes to write down all of the factors that might contribute toward a default event. The specific type of credit obligor will, of course, play a role. An individual, a household, a firm, an international organization, or a sovereign entity can be expected to behave differently. This fact, therefore, must be included in your list. What else might it include? Certainly, it will involve some notion of financial strength: earning power, revenues or the size of a country's tax base. It would also probably include the level of indebtedness, future earnings prospects, business model or political agenda, industry membership, competitiveness of one's environment, geographical location, and overall size. A good list would also mention general factors such as the state of the global and regional macroeconomies, the level of interest rates and exchange rates, and perhaps even commodity prices. Finally, one might also add some notion of stability such as management prowess for a firm, political leadership for a sovereign, or education and experience for an individual.

Your list is thus already quite lengthy. Indeed, our illustrative collection of points, while a useful start, probably significantly underestimates the number of distinct aspects considered by a credit analyst in judging the creditworthiness of a specific obligor. While a credit analyst desires all of these inputs and could not properly do her job without it, this dimensionality is an extreme challenge for the quantitative modeller. Models, by their very definition, seek to reduce dimensionality and get to the essence of a problem. Often, they go too far and need to be enriched and extended. No mathematical model, however, will ever capture the full complexity of the real world.

Modelling is essentially about dimension reduction since we cannot possibly hope to reasonably incorporate all of the aforementioned items. A good first step for the credit-risk modelling problem is to construct an extremely simple, but fairly realistic description of the default outcome. Risk management involves describing the intersection of two main quantities: events and likelihoods. Since a probability distribution captures this relationship in many different ways, this explains the extensive use of applied statistics in modern financial risk management.

In a market-risk setting, there are an infinite number of possible events; that is, the range of possible portfolio value changes is well approximated by an infinite set. Each portfolio-change event requires the assignment of an associated probability. In the credit-risk setting, conversely, the set of possible events can be dramatically reduced. Consider a particular time interval; a week, a month, or a year. *Two* things can happen: either the obligor defaults or it does not. If it does not default, we term this survival. Default is thus a binary outcome like black or white, on or off, win or lose, yin or yang, Dr. Jekyll or Mr Hyde: in this case, the two sides are default or survival.

The set of risk events, therefore, is small and finite. The market-risk world, by comparison, has a sample space with with uncountably many elements. One might be tempted, on this basis, to conclude that credit-risk modelling is consequently easier. We'll see that this is not actually true; the complications arise in other areas. What it does mean is, to begin, we have a shift in our perspective. In particular, it moves us from a continuous realm of events to a discrete set. We still use statistical distributions to marry events and their likelihoods, but the setting is different. We will be working, to start at least, with discrete probability distributions. Later, we will see that there are other ways to introduce the structural break between default and survival.

2.1 Motivating a Default Model

Most readers' introduction to probability distributions likely began in one's late teenage years with a coin-flipping example. Although admittedly somewhat dry, this well-used situation communicates a powerful idea that encompasses many phenomena including default risk. Let's try to extend the coin-toss notion to our setting by considering a single investment. Imagine that you, in an attempt to be financially responsible, forgo the purchase of tickets to see your favourite band and instead acquire a bond with a value of $c = \$100$. You are quite pleased with your investment, but subsequently realize that it is not entirely without risk. If the issuer of the bond, your counterparty, defaults, then you could lose the entire amount.

Having made this depressing realization—and perhaps deeply regretting not buying the concert tickets—your first question is likely: how might we assess this risk? As previously hinted, risk measurement requires the combination of possible outcomes and their likelihood. Further, we have established that only one of two outcomes is possible: default or survival. In short, you either lose c, or you don't. We know the outcome. What is missing, at this point, is an assessment of its relative likelihood. You do a bit of research on the Internet and find an external rating agency who claims that the default probability over the interval, $[0, T]$ is $p = 0.02$.

How might we interpret this? Well, this brings us to our original coin-flipping example. If we flip a fair coin, the probability of heads (and tails) is $p = 0.5$. In this

2.1 Motivating a Default Model

default-risk situation, conversely, the coin is deeply unfair. If we flip it 100 times, for example, we expect to have only two heads, but 98 tails. This is perhaps pushing the idea of coin tossing a bit too far. If you prefer, you can instead imagine an urn with 100 equally sized stones. 98 of the stones are yellow and two are red. Imagine that it is well mixed and you can't see inside. We can replace our coin flip with reaching into the urn and selecting a stone. Default occurs only when you extract one of the two red stones.

A Bit of Structure

This is the basic idea, but to avoid dragging around a crazy coin or a large urn full of stones, we will need to add some mathematical structure. We'll begin by denoting a default event as $\mathcal{D}$. This is the subset of our sample space where default occurs. We further introduce a default indicator variable as

$$\mathbb{I}_\mathcal{D} = \begin{cases} 1 : \text{ defaults occurs before time } T \text{ with probability } p \\ 0 : \text{ survival until time } T \text{ with probability } 1 - p \end{cases}. \quad (2.1)$$

This mathematical summary of our binary default event is quite helpful. It allows us to, in a straightforward and parsimonious manner, describe a number of important quantities. For instance, the default loss, denoted L, has the following convenient form,

$$L = c \cdot \mathbb{I}_\mathcal{D}. \quad (2.2)$$

It is simply the product of our credit exposure, c, and our default indicator. It is easy to see that the default loss, $L = c \cdot \mathbb{I}_\mathcal{D}$, can only take two possible discrete values, 0 and c. In other words, the default loss is, as we would expect, properly bounded: $L \in [0, c]$.

This simple object—our unfair coin flip or stone extracted from an urn—is a commonplace statistical object termed a Bernoulli trial. It is a common, and quite useful, way to describe binary events such as coin tossing.

Table 2.1 summarizes our notation as well as these two outcomes and likelihoods. We can use this information to extend our analysis and better understand some of the properties of the default-loss event. A natural place to start is with the expected

Table 2.1 *A single instrument example*: This table highlights the outcome and likelihood inputs required to compute the risk associated with a single-instrument situation.

Outcome	$\mathbb{I}_\mathcal{D}$	Loss		Probability	
Case 1	1	c	$100	p	0.02
Case 2	0	0	$0	$(1-p)$	0.98

loss, which is readily computed as

$$\mathbb{E}(L) = p \cdot c + \underbrace{(1-p) \cdot 0}, \tag{2.3}$$
$$= 0.02 \cdot \$100,$$
$$= \$2.$$

The volatility of our default loss is also easily computed from first principles,

$$\text{var}(L) = \underbrace{p \cdot c^2 + \underbrace{(1-p) \cdot 0^2}}_{\mathbb{E}(L^2)} - \underbrace{c^2 p^2}_{\mathbb{E}(L)^2}, \tag{2.4}$$
$$= c^2 p(1-p),$$
$$\sigma(L) = \sqrt{\text{var}(L)} = c\sqrt{p(1-p)},$$
$$= \$100\sqrt{0.02 \cdot 0.98},$$
$$= \$14.$$

On average, therefore, we lose about $2, but there is significant uncertainty around this amount. The volatility is, in this case, seven times larger than the expected outcome. A volatility exceeding the expectation by this magnitude suggests, for a positive-valued random variable, a large dispersion of outcomes. In other words, there appears to be a substantial amount of risk. To really describe risk, however, we would like to talk about worst-case outcomes. This naturally brings us to the idea of a quantile of the loss distribution. In this case, with only two possible outcomes, this doesn't make much sense. As we add additional instruments, we will explore this perspective.

We have also been completely agnostic about the reason default occurs. It simply happens or it doesn't: there are only *two* possible outcomes. All of the risk stems from some unspecified and unobservable idiosyncratic risk factor associated with a single counterparty. Our first model, therefore, can be categorized as a reduced-form model. This assumption will continue to hold for the rest of this chapter, but will ultimately be relaxed as we broaden our modelling perspective.

These ideas, of course, only really make sense if we consider repeating the experiment many times. In reality, our counterparty will only either default or survive. This statistical perspective, where we consider the relative frequency of different outcomes, is the key tool in risk measurement and analysis. It underlies our ultimate goal of constructing a loss distribution. For a single instrument, we have, in fact, succeeded. The probability mass function for a Bernoulli trial can be written, following from equation 2.1, as

$$f(k|p) = p^k (1-p)^{1-k}, \tag{2.5}$$

2.1 Motivating a Default Model

for $k = 0, 1$ in our single-period, one-obligor example. Although we have a complete statistical description of the situation, it is hard to get too excited. While the statistics are technically correct, the situation is too simple. A single exposure to a credit obligor is certainly well represented by an unfair coin flip, but rarely do financial institutions expose themselves to a single bet. Instead, they diversify their exposures by making a host of investments with a broad range of credit counterparties. Thus, although we've made a tentative first step, we need to extend our analysis to incorporate multiple obligors.

2.1.1 Two Instruments

Having captured a single instrument setting, the logical next step is to consider two instruments. Although not exactly a diversified portfolio, it is tractable and helps us get to the key ideas. So, to gain some diversification, it is again necessary to forgo some additional consumption; this time, to raise the necessary funds, you skip a nice Friday night meal at your favourite restaurant. After calling your broker, we now have *two* bonds with exposures, $c_1 = \$100$ and $c_2 = \$75$. Some additional Internet browsing leads to default-probability estimates of $p_1 = 0.02$ and $p_2 = 0.10$. The second obligor is thus significantly less creditworthy than our first choice.

How might we characterize this situation? Conceptually, it's analogous to flipping two (unfair) coins simultaneously. Or, if you prefer, drawing stones from two distinct urns. The coins, or urns, have slightly different characteristics. We thus need to update our default events: $\mathcal{D}_n$ denotes the default of the nth counterparty. We expand our default indicator to $\mathbb{I}_{\mathcal{D}_n}$. This simple setting has $n = 1, 2$, but it is a start.

A bit of reflection reveals that there are, in fact, *four* possible outcomes. The extremes include both obligors defaulting or surviving. The other two cases include one surviving and one default. These four cases are outlined, along with the base exposure and default probability information, in Table 2.2.

Table 2.2 makes a critical, and difficult to defend, assumption. The relative probabilities of each case are estimated as the product of the probabilities of the individual events. That is, the default of our two default obligors are estimated as p_1 and p_2, respectively. Table 2.2 implies that the joint default probability is merely

Table 2.2 *A two-instrument example*: This table extends our discussion from Table 2.1 to include a second credit-risky instrument. In this setting, there are *four* different default cases.

Outcome	$\mathbb{I}_{\mathcal{D}_1}$	$\mathbb{I}_{\mathcal{D}_2}$	Loss		Probability	
Case 1	1	1	$c_1 + c_2$	$175	$p_1 \cdot p_2$	0.002
Case 2	1	0	c_1	$100	$p_1 \cdot (1 - p_2)$	0.018
Case 3	0	1	c_2	$75	$(1 - p_1) \cdot p_2$	0.098
Case 4	0	0	0	$0	$(1 - p_1) \cdot (1 - p_2)$	0.882

$p_1 \cdot p_2$. Such a claim is, in general, quite false. It is only true, if we make the additional assumption that each default event is statistically independent.

While independence may be unrealistic, it certainly makes our life easier. We will explore the consequences of this choice and, in following chapters, relax it. For the moment, however, let us examine the implications of this decision. We begin with the moments of our default loss. The expected loss is,

$$\mathbb{E}(L) = \underbrace{p_1 \cdot c_1 + (1-p_1) \cdot 0}_{\text{Obligor 1}} + \underbrace{p_2 \cdot c_2 + (1-p_2) \cdot 0}_{\text{Obligor 2}}, \tag{2.6}$$

$$= 0.02 \cdot \$100 + 0.10 \cdot \$75,$$

$$= \$9.5.$$

The loss volatility, exploiting the pattern from equation 2.4, is simply

$$\sigma(L) = \underbrace{c_1 \sqrt{p_1(1-p_1)}}_{\text{Obligor 1}} + \underbrace{c_2 \sqrt{p_2(1-p_2)}}_{\text{Obligor 2}}, \tag{2.7}$$

$$= \$100 \cdot \sqrt{0.02 \cdot 0.98} + \$75 \cdot \sqrt{0.10 \cdot 0.90},$$

$$= \$36.5.$$

The average loss is thus about \$10, but the uncertainty is about four times as large at roughly \$37. This is an interesting result. We added a second risky instrument and, by doing so, increased the absolute magnitude of the risk, but reduced the relative dispersion. Much of this effect can be explained by the increased expected default loss. Part of the story, however, is the vanishingly small probability of extreme outcomes. An extreme outcome, in this simple example, is a double default event. Assumptions regarding the interdependence between the default of each counterparty, of course, play a crucial role in this regard.

2.1.2 Multiple Instruments

At the risk of testing the reader's patience, we will extend our example to consider *four* instruments. While there is essentially nothing new—save complexity— relative to the two-instrument example, it will help us to identify a pattern and correspondingly construct a more general model. To ease the initial calculations, assume that our four risky assets share the same exposure and relative default probability. That is, where $c_n = \$10$ and a relatively sizable $p_n = 0.10$ for $n = 1, \ldots, 4$. While not particularly realistic, this will make the graphic and tabular results a bit easier to identify.

2.1 Motivating a Default Model

Table 2.3 *Four instruments*: This table broadens the previous analysis to consider four distinct instruments. While this is becoming somewhat repetitive, it does help us extract a pattern from the default outcomes.

Defaults	Loss		How often?	Probability	
0	$0 \cdot c$	$0	$\left(\frac{4!}{(4-0)!0!}\right) = 1$	$p^0(1-p)^{4-0}$	0.6561
1	$1 \cdot c$	$10	$\left(\frac{4!}{(4-1)!1!}\right) = 4$	$p^1(1-p)^{4-1}$	0.2916
2	$2 \cdot c$	$20	$\left(\frac{4!}{(4-2)!2!}\right) = 6$	$p^2(1-p)^{4-2}$	0.0486
3	$3 \cdot c$	$30	$\left(\frac{4!}{(4-3)!3!}\right) = 4$	$p^3(1-p)^{4-3}$	0.0036
4	$4 \cdot c$	$40	$\left(\frac{4!}{(4-4)!4!}\right) = 1$	$p^4(1-p)^{4-4}$	0.0001

To describe our default situation, we define the number of defaults as,

$$\mathbb{D}_N = \sum_{n=1}^{N} \mathbb{I}_{\mathcal{D}_n}. \qquad (2.8)$$

This is nothing other than the sum of our default indicators. There are *five* possible outcomes: 0, 1, 2, 3, and 4 defaults. How can they occur? This is a familiar, albeit tedious, counting problem. The results are summarized in Table 2.3.

A clear pattern emerges from Table 2.3. Extreme outcomes—no default or complete joint default—can arrive in only a single way. Intermediate outcomes—different combinations of default and survival for the various obligors—can occur in multiple ways. Given N credit counterparts, the number of different combinations of ways that k defaults might occur are described by the binomial coefficient,

$$\binom{N}{k} = \frac{N!}{(N-k)!k!}. \qquad (2.9)$$

Moreover, the probability of each of these k defaults can be summarized by a simple extension of equation 2.5 or

$$p^k(1-p)^{N-k}. \qquad (2.10)$$

This leads us to the binomial distribution. It is a well-known fact—see, for example, Casella and Berger (1990)—that a collection of independent Bernoulli trials follows the binomial law. The probability of observing k defaults, in this setting, has the following form:

$$\mathbb{P}(\mathbb{D}_N = k) = \underbrace{\left(\frac{N!}{(N-k)!k!}\right)}_{\text{\# of outcomes}} \underbrace{\left(p^k(1-p)^{N-k}\right)}_{\text{Probability of outcome}}. \qquad (2.11)$$

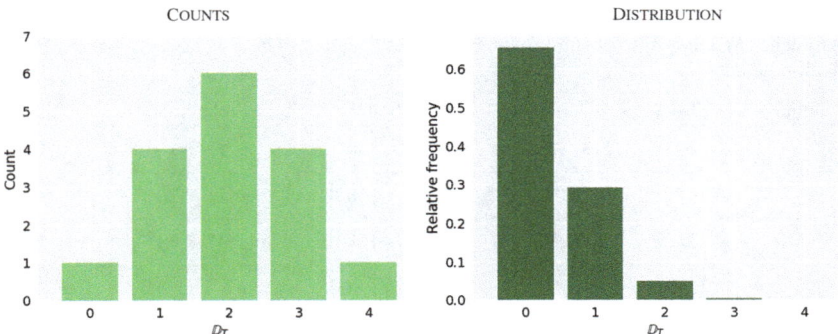

Fig. 2.1 *Four-instrument outcomes*: This figure illustrates the individual default counts along with the default-loss distribution for the four-instrument example detailed in Table 2.3.

This is, more formally, the probability mass function for the binomial distribution with parameters N and p. It is sufficient, under the assumption of independence, to know the total number of obligors and their (common) probability of default to specify the associated loss distribution. It nonetheless requires the assumption of a common default probability for all credit counterparties and independence of default. Both of these unrealistic choices will be relaxed: the former in the subsequent section and the latter in the next chapter.

It is instructive, even for this four-instrument example, to examine the implications of equation 2.11 in more detail. Figure 2.1 assists in this regard by illustrating the key values from Table 2.3. The left-hand graphic demonstrates the number of different ways that a specific number of defaults can be achieved. There are, for example, six different paths to two defaults, but only one way where all obligors survive or default. The right-hand graphic displays the default-loss distribution. The majority of the probability mass is placed on $k = 0, 1$; indeed, roughly 95% of the cases involve less than a single default. This left-handed skew is very typical of credit-loss distributions.

The dispersion of the individual default counts has a familiar symmetric form. Indeed, the familiarity becomes even stronger as we increase the value of N. Figure 2.2 graphically illustrates the sequence of binomial coefficients—that is, $\binom{50}{k}$ for $k = 0, \ldots, 50$. This (triangular) array of binomial coefficients is a very old result with many applications; indeed, the pattern in Fig. 2.2 is the 50th row of, what it is often referred to as, Pascal's triangle.[1]

The greater significance of Fig. 2.2 is a well-known fact about the binomial law: as the parameter N grows sufficiently large, it converges to the Gaussian distribution. In other words, for large portfolios, the loss distribution tends toward the normal distribution. Figure 2.3 displays this fact illustrating the default-loss

---

[1] This is in reference to French mathematician, Blaise Pascal. For an enjoyable high-level description of this interesting mathematical object, see Smith (1973).

2.1 Motivating a Default Model

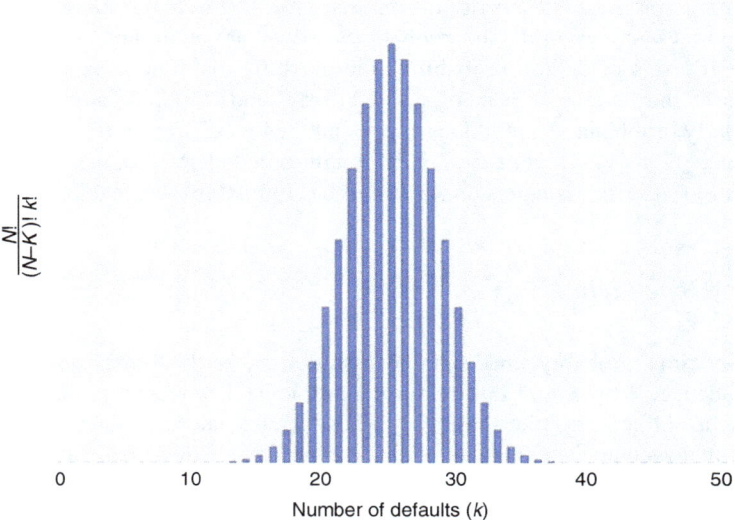

Fig. 2.2 *Pascal's triangle*: This figure illustrates the sequence of binomial-coefficient values associated with $\binom{50}{k}$ for $k = 0, \ldots, 50$; this is the 50th row of what is often referred to as Pascal's triangle. Its basic shape should appear relatively familiar.

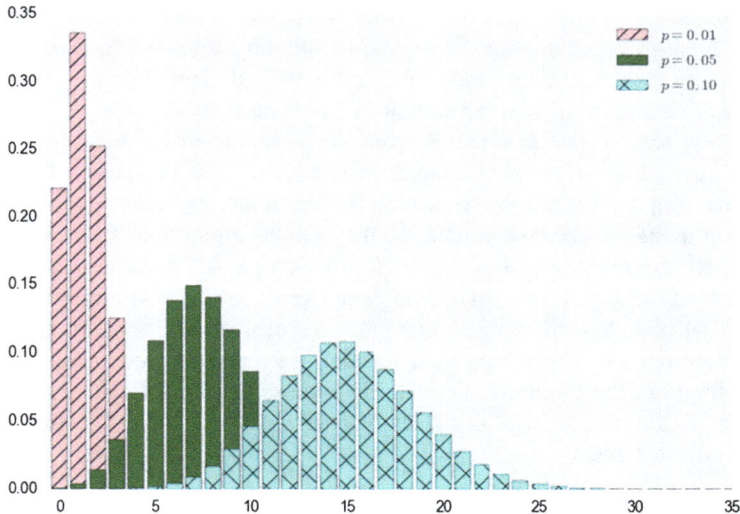

Fig. 2.3 *Many instruments*: This figure describe the distribution of default losses for a portfolio of 150 equally sized investments. Three possible common default-probability values are examined: 1, 5, and 10%.

distribution for a sizable portfolio comprised of 150 equally sized investments. Three possible common default-probability values are examined: 1, 5, and 10%. For small levels of default probability, the portfolio distribution is skewed to the left, but as the parameter p increases, it clearly tends to Gaussianity. This is not particularly problematic, but it does imply that the loss distribution—irrespective of the choice of p—will inherit the relatively thin-tailed characteristics of the normal distribution. As risk managers, this may not be completely acceptable.

2.1.3 Dependence

Another, quite probably unacceptable assumption, is the imposition of default independence. Why would this be such a problem? The reason is that credit risk and credit obligors, despite the highly idiosyncratic nature of these risks, do not operate in a vacuum. There are numerous compelling reasons to expect dependence of default events. There may exist important industry interlinkages between specific credit obligors. Consider, for example, supply-chain relationships or competition. Poor performance could, therefore, lead to a deterioration or an improvement of other possible credit counterparts. In the financial industry, there is certainly an interdependence of indebtedness—this is one, among many, justifications for treating some financial institutions as too big to fail. We should also expect shared regional or country risk exposures and, in the most general sense, common exposure to global economic conditions. If a particular region, country, or the entire planet faces difficult macroeconomic conditions, we should expect to observe correlation in degradation of creditworthiness and, ultimately, default. In brief, there is no shortage of reasons to dislike the default-independence assumption.

The best reason, perhaps, stems from the principle of conservatism. As will become very clear in the next chapter, dependence is at the heart of credit-risk modelling. Imposition of even quite modest amounts of default dependence has dramatic effects on our risk estimates; it is fair to argue that dependence is the single *most* important aspect of credit-risk modelling. All realistic and reasonable credit-risk models, therefore, place the dependence structure at the centre of their model. That said, it is still the case that many lessons can be drawn from the simple independent-default model; this model underpins many of the extant default-risk models found in the literature. One builds a house from the foundation upwards; with that in mind, we now proceed to generalize and formalize the previous motivational examples.

2.2 Adding Formality

This section seeks to properly introduce the independent-default binomial model. There are numerous useful sources that motivate the following exposition and represent useful supplemental information for the reader including Frey and McNeil

2.2 Adding Formality

(2003); Schönbucher (2000a,b), and Bluhm et al. (2003). While not a realistic model, it is worthwhile investing a reasonable amount of effort highlighting the basic properties of this model. Most particularly, we seek to clearly and concisely understand its shortcomings. This effort will pay dividends later as we seek to establish links with other common credit-risk modelling frameworks. The following work will help us see why certain, more complicated, modelling assumptions are generally selected.

The formal model will require a small amount of repetition. We begin, as before, with a collection of independent default-indicator functions,

$$\mathbb{I}_{\mathcal{D}_n} = \begin{cases} 1 : \text{ when firm } n\text{'s default occurs before time } T \\ 0 : \text{ when firm } n \text{ survives to time } T \end{cases}. \quad (2.12)$$

for $n = 1, \ldots, N$. This is the simplest, and most natural, description of a default event. $\mathbb{I}_{\mathcal{D}_n}$, as we've already seen, is a binary variable: it takes the value of unity in the event of default and zero otherwise.

This simple structure can take us quite far. Indeed, we will use variations on this convenient indicator definition for virtually all of the models considered in this book. We may, for instance, using the basic properties of indicator variables, easily determine the expected default of the nth issuer as,

$$\mathbb{E}(\mathbb{I}_{\mathcal{D}_n}) = \underbrace{1 \cdot \mathbb{P}\left(\mathbb{I}_{\mathcal{D}_n} = 1\right)}_{\text{Default}} + \underbrace{0 \cdot \mathbb{P}\left(\mathbb{I}_{\mathcal{D}_n} = 1\right)}_{\text{Survival}}, \quad (2.13)$$

$$= \underbrace{\mathbb{P}\left(\mathbb{I}_{\mathcal{D}_n} = 1\right)}_{p_n}.$$

Default occurs with probability, p_n, whereas survival occurs with probability, $1 - p_n$. By working from first principles, the variance associated with each default event can also be determined as,

$$\text{var}(\mathbb{I}_{\mathcal{D}_n}) = \mathbb{E}\left(\mathbb{I}_{\mathcal{D}_n}^2 - \left(\underbrace{\mathbb{E}(\mathbb{I}_{\mathcal{D}_n})}_{p_n}\right)^2\right), \quad (2.14)$$

$$= \mathbb{E}\left(\mathbb{I}_{\mathcal{D}_n}^2\right) - p_n^2,$$

$$= \underbrace{1^2 \cdot \mathbb{P}\left(\mathbb{I}_{\mathcal{D}_n} = 1\right) + 0^2 \cdot \mathbb{P}\left(\mathbb{I}_{\mathcal{D}_n} = 0\right)}_{\mathbb{E}(\mathbb{I}_{\mathcal{D}_n}^2)} - p_n^2,$$

$$= \underbrace{\mathbb{P}\left(\mathbb{I}_{\mathcal{D}_n} = 1\right)}_{p_n} - p_n^2,$$

$$= p_n(1 - p_n).$$

This demonstrates that each default event is indeed a Bernoulli trial with parameter, p_n.

How are the default outcomes of two counterparties related? A good starting point is the covariance between two default events. This is described for two arbitrary obligors n and m, using the definition of covariance, as

$$\operatorname{cov}(\mathbb{I}_{\mathcal{D}_n}, \mathbb{I}_{\mathcal{D}_m}) = \mathbb{E}\left(\left(\mathbb{I}_{\mathcal{D}_n} - \underbrace{\mathbb{E}\left(\mathbb{I}_{\mathcal{D}_n}\right)}_{p_n}\right)\left(\mathbb{I}_{\mathcal{D}_m} - \underbrace{\mathbb{E}\left(\mathbb{I}_{\mathcal{D}_m}\right)}_{p_m}\right)\right), \quad (2.15)$$

$$= \mathbb{E}\left(\mathbb{I}_{\mathcal{D}_n}\mathbb{I}_{\mathcal{D}_m} - \mathbb{I}_{\mathcal{D}_n}p_m - \mathbb{I}_{\mathcal{D}_m}p_n + p_n p_m\right),$$

$$= \mathbb{E}\left(\mathbb{I}_{\mathcal{D}_n}\mathbb{I}_{\mathcal{D}_m}\right) - 2 p_n p_m + p_n p_m,$$

$$= \underbrace{\overbrace{\mathbb{E}\left(\mathbb{I}_{\mathcal{D}_n}\right)}^{p_n} \overbrace{\mathbb{E}\left(\mathbb{I}_{\mathcal{D}_m}\right)}^{p_m}}_{\text{By independence}} - p_n p_m,$$

$$= 0.$$

Since each of the default indicators are independent, they are naturally uncorrelated. This is exactly as we expected, but it is nonetheless comforting to see it hold explicitly.

Our interest is the range of possible portfolio losses associated with each potential constellation of default events. If we denote, as usual, the exposure to counterparty n at default as c_n, then the total portfolio loss over the time interval, $[0, T]$, is

$$L_N = \sum_{n=1}^{N} c_n \mathbb{I}_{\mathcal{D}_n}. \quad (2.16)$$

L_N is thus nothing other than the sum of all the exposures associated with those counterparties defaulting in the interval, $[0, T]$.[2] Without some probabilistic statement attached to it, this is a reasonably vacuous statement. We have, however, added a bit of structure. We can, in fact, directly compute the expected loss,

$$\mathbb{E}(L_N) = \mathbb{E}\left(\sum_{n=1}^{N} c_n \mathbb{I}_{\mathcal{D}_n}\right), \quad (2.17)$$

---

[2] This basic notation, stolen (or rather borrowed) from Glasserman (2004a), is quite convenient. We will use it across a broad range of models.

2.2 Adding Formality

$$= \sum_{n=1}^{N} c_n \underbrace{\mathbb{E}\left(\mathbb{I}_{\mathcal{D}_n}\right)}_{p_n},$$

$$= \sum_{n=1}^{N} c_n p_n.$$

The expected portfolio loss is thus the sum of the probability weighted counterparty exposures. While interesting and intuitive, this is not a terribly helpful quantity. Credit defaults are rare events, which means that the individual default probabilities (i.e., the p_n's) are typically very small. This means that, even when the exposures are quite large, the expected loss relative to a worst-case notion of loss, is also generally quite small. Our principal interest, therefore, is in the entire distribution of portfolio losses. We are particularly interested in extreme outcomes relating to multiple defaults of large exposures. In other words, our principal focus lies in the tail of the distribution.

2.2.1 An Important Aside

Before preceding, a quick precision on the counterparty exposure is warranted. We have treated, and will continue to handle, the counterparty exposure as a simple constant, denoted c_n. In reality, the nth exposure is written more precisely as,

$$c_n = \mu_n(1 - \mathcal{R}). \tag{2.18}$$

The quantity μ_n is often referred to as the *exposure-at-default*—this is the amount you risk losing at default. It may, or may not, be treated as a constant or a random variable. Indeed, for many derivative securities, this exposure is estimated through other models attempting to determine potential future exposure (or PFE). These are often large-scale market-risk models. This effort can also be, even if deterministic, quite complex involving the aggregation and netting of multiple instruments associated with a single obligor. Ruiz (2015) provides an comprehensive discussion of these ideas.

$\mathcal{R} \in (0, 1)$ is typically termed the *recovery rate*; again, this is, in principle, an unknown random variable. The quantity, $1 - \mathcal{R}$, is generally termed the *loss-given-default* or the proportional severity of default. This is how much you *actually* stand to lose, since you generally expect to be able to recover a portion of μ_n. If we denote severity as S, then recovery and severity are linked by the simple identity, $S = 1 - \mathcal{R}$.

While the default probabilities are the only random inputs in our current model, a different structure is entirely possible. One alternative would be to make the exposure-at-default, μ_n, a function of the initial exposure, in units of currency, and an assumed severity. This severity amount can then readily be interpreted as a random variable. Imagine, for example, that we describe the severity associated with the nth country as,

$$S_n \sim B(a_n, b_n). \qquad (2.19)$$

That is, the severity of each obligor's default is described by a beta distribution with parameters, $a_n, b_n \in \mathbb{R}^+$.[3] There are many different possible distributions that might be employed, but the beta distribution is a popular choice since its support, like the default severity, is restricted to the unit interval.

If we use equation 2.19, we may, following from the development in equation 2.17, update the expected loss as follows,

$$L_N = \sum_{n=1}^{N} \mu_n \cdot S_n \cdot \mathbb{I}_{D_n}, \qquad (2.20)$$

$$\mathbb{E}(L_N) = \mathbb{E}\left(\sum_{n=1}^{N} \mu_n \cdot S_n \cdot \mathbb{I}_{D_n}\right)$$

$$= \sum_{n=1}^{N} \mu_n \mathbb{E}\left(S_n \mathbb{I}_{D_n}\right).$$

Here, we arrive at an impasse. The quantity, $\mathbb{E}\left(S_n \mathbb{I}_{D_n}\right)$, is, without additional information or assumptions, not exactly straightforward to evaluate. It requires, to be more specific, the joint distribution of default severity and probability. Given the paucity of default data, this is a rather challenging distribution to parameterize. As a consequence, it is common to assume that default severity and probability are independent. While a very strong assumption, it leads us to readily solve our expectation from equation 2.20,

$$\mathbb{E}(L_N) = \sum_{n=1}^{N} \mu_n \mathbb{E}\left(S_n \mathbb{I}_{D_n}\right), \qquad (2.21)$$

$$= \sum_{n=1}^{N} \mu_n \mathbb{E}\left(S_n\right) \mathbb{E}\left(\mathbb{I}_{D_n}\right),$$

$$= \sum_{n=1}^{N} \mu_n \left(\frac{a_n}{a_n + b_n}\right) p_n.$$

---

[3] For more information on the beta distribution, see Casella and Berger (1990) or Stuart and Ord (1987).

2.2 Adding Formality

Since we can certainly identify parameters such that,

$$s_n = \frac{a_n}{a_n + b_n}, \qquad (2.22)$$

the expected loss computations are entirely consistent with random, but independent, default severities. This is true when we restrict our attention to the first moment of the default-loss distribution. In an economic-capital setting, this would no longer hold true.

Much has been written about these quantities and their determination can be quite complex—Gregory (2010), for example, provides another excellent treatment of many of the issues involved. For the lion's share of our analysis, we take them as given. This does not mean, however, that they are unimportant or that treating them as constants is a defensible assumption.

2.2.2 A Numerical Solution

Without additional assumptions, it is difficult to derive a closed-form expression for the distribution of L_N. One easy way to see the analytical challenges is to try to compute the variance of L_N.

$$\text{var}(L_N) = \mathbb{E}\left(L_N^2 - \mathbb{E}(L_N)^2\right), \qquad (2.23)$$

$$= \mathbb{E}\left(\left(\sum_{n=1}^{N} c_n \mathbb{I}_{\mathcal{D}_n}\right)^2 - \mathbb{E}\left(\sum_{n=1}^{N} c_n p_n\right)^2\right),$$

$$= \mathbb{E}\left((c_1 \mathbb{I}_{\mathcal{D}_1} + \cdots + c_N \mathbb{I}_{\mathcal{D}_N})^2\right) - (c_1 p_1 + \cdots + c_N p_N)^2.$$

Equation 2.23, when expanded, generates an unwieldy number of separate terms. For a relatively small model portfolio comprised of only 100 distinct counterparties, this would already yield an unmanageable number of elements. This set of possible combinations of defaults is even more difficult to analytically organize. Brute force algebraic manipulations are not going to provide us with a meaningful solution.

At this point, however, the good news is that we have most of the necessary ingredients to actually put this model into action. There are *two* main approaches to handling any credit-risk model:

1. numerical methods; or
2. analytic approximation.

Neither approach dominates the other in terms of usefulness. Numerical methods are, in general, practical and allow much more generality. For this reason, they are much more popular among practitioners. Analytic approximations conversely are typically faster and, most importantly, help us gain important intuition. They do, however, often involve making certain—sometimes quite unrealistic—assumptions. In the following discussion we will, in fact, consider both approaches. We'll start with the general numerical approach to describe the loss distribution. Once this is complete, we'll turn to examine the analytic version to learn more about the model and its shortcomings. Fixing these shortcomings will be the principal task for more complex models in the upcoming chapters.

To make this situation less abstract and aid our understanding, let's return to the example introduced in Chap. 1. This portfolio incorporates 100 obligors characterized by heterogeneous exposures and default probabilities. Recall that the default probabilities and exposures were drawn from χ^2 and Weibull distributions, respectively. The average default probability is set to 1% and the total portfolio exposure amounts to USD 1,000. The exposure and default probability distributions are summarized graphically in Chap. 1.

The main numerical technique that we will use in this setting is termed Monte-Carlo simulation. It is sufficiently important that we will dedicate the entirety of Chap. 8 to its discussion. Before delving into these technical details, however, we will repeatedly use this method to approximate default-loss distributions. The basic idea is easily explained. We use our computer to randomly generate a set of Bernoulli trials and actually compute the associated default losses. If we do this only a few times, the results will be underwhelming, but if we repeat this millions of time, we can reliably trace out the form of the default-loss distribution.

2.2.2.1 Bernoulli Trials

The generation of independent Bernoulli trials is, computationally, straightforward. It is, in many ways, similar to a huge number of extremely unfair, simultaneous coin flips.[4] Each flip can be simulated by drawing a uniform random number in the unit interval, U. If $U \in [0, p]$ then it is considered a default; if $U \in (p, 1]$, then the counterparty survives. If the interval, $[0, p]$, is very small, it is easy to see that the random uniform number will not fall into it very often and, as a consequence, default will not occur frequently. The problem, however, is that to arrive at a reasonable estimate, we need to perform this unfair coin flip many, many times. This would be the height of tedium, not to mention glacially slow. Thankfully, computers are very good at this kind of repeated operation and can do it literally millions of times in short periods of time. Nevertheless, as we will see in Chap. 8, there are limits to what we can reasonably achieve, even with a powerful computer.

Although the uniform-variate approach is quite intuitive, like all things it requires a bit of practice. Figure 2.4 attempts to clarify the situation with a simple, albeit poorly scaled, schematic. To simulate the (relatively rare) default event, you simply

---

[4]Or, if you prefer, pulling stones from an urn.

2.2 Adding Formality

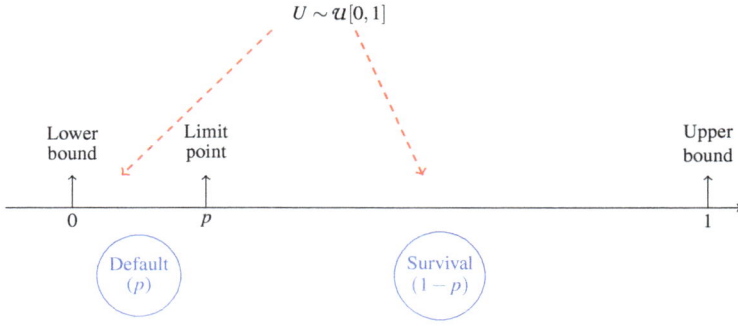

Fig. 2.4 *A schematic perspective*: This figure illustrates the exposure and default-probability composition of our sample 100-obligor portfolio.

partition the unit interval, draw a random uniform variate, $U \sim \mathcal{U}[0, 1]$, and determine where it falls. Repeat this a few millions time and you're done.

With the intuition covered, it is time for a more technical description. We particularly need to see how this operation can be used to construct our desired risk measures. To be more specific, therefore, one generates $N \times M$ standard uniform random variates on the unit interval of the form, $U_n^{(m)} \sim \mathcal{U}[0, 1]$ for $n = 1, \ldots, N$ and $m = 1, \ldots, M$. M, in this setting, represents the number of Monte-Carlo iterations; imagine, for example, that $M = 1,000,000$. Then, for each counterparty n and each choice of m, if the random variable $U_n^{(m)}$ is found in the interval $[0, p_n]$, a default has occurred. If, however, $U_n^{(m)}$ lands in the remaining part of the interval, $(p_n, 1]$, then the counterparty survives. Mathematically, it is described as

$$\mathbb{I}_{\mathcal{D}_n}^{(m)} = \begin{cases} 1 : U_n^{(m)} \in [0, p_n] \\ 0 : U_n^{(m)} \in (p_n, 1] \end{cases}, \quad (2.24)$$

for $n = 1, \ldots, N$ and $m = 1, \ldots, M$. Thus we have a simulated descriptor of the default indicator in equation 2.12.

For each iteration, m, we have an approximation of the portfolio loss,

$$L_N^{(m)} = \sum_{n=1}^{N} c_n \mathbb{I}_{\mathcal{D}_n}^{(m)}. \quad (2.25)$$

An unbiased estimator for the expected loss is,

$$\widehat{\mathbb{E}(L_N)} = \frac{1}{M} \sum_{m=1}^{M} L_N^{(m)}. \quad (2.26)$$

The loss volatility is similarly estimated by,

$$\widehat{\sigma(L_N)} = \sqrt{\frac{1}{M-1} \sum_{m=1}^{M} \left(L_N^{(m)} - \widehat{\mathbb{E}(L_N)}\right)^2}. \tag{2.27}$$

For sufficiently large M, these estimates will provide reasonably sharp estimates of the true underlying values.

Expected loss and loss volatility are relatively easy to handle, since no sorting is involved. VaR and expected shortfall require a bit more effort. To estimate the $\text{VaR}_\alpha(L_N)$, one begins by placing the collection of portfolio losses, $\{L_N^{(m)}; m = 1, \ldots, M\}$ in ascending order. This means that $L_N^{(1)}$ is the smallest observed default loss—certainly equal to zero—and $L_N^{(M)}$ is the largest simulated loss, which is presumably significantly lower than the total exposure.

If we denote this ordered vector of simulated portfolio losses as $\tilde{L}_N$, then our VaR estimator is

$$\widehat{\text{VaR}_\alpha(L_N)} = \tilde{L}_N(\lceil \alpha \cdot M \rceil), \tag{2.28}$$

where $\lceil x \rceil$, denotes the smallest integer greater or equal to x.[5]

We use the same ordered loss vector for the computation of expected shortfall. As the average loss beyond a given quantile, it is computed as

$$\widehat{\mathcal{E}_\alpha(L_N)} = \frac{1}{M - \lceil \alpha \cdot M \rceil} \sum_{m=\lceil \alpha \cdot M \rceil}^{M} L_N^{(m)}. \tag{2.29}$$

The difficult part, $\frac{1}{M-\lceil \alpha \cdot M \rceil}$, is merely the number of observations in the tail. With this value and the ordered vector, the actual computation is trivial.

Equations 2.25 to 2.29 are thus a collection of unbiased estimators that apply generally to any Monte Carlo method. Differences between methods arise due to alternative approaches toward generating the loss outcomes. That is, once you have generated the vector of default losses in $\mathbb{R}^M$, the risk measures are all computed in the same way independent of the model choice. As we'll see shortly, this has some fortunate implications for our computer implementation.

2.2.2.2 Practical Details

Our very first Python code snippet parsimoniously generates an entire loss distribution for the independent-default model using our simple uniform-variate trick.

---

[5] This is sometimes referred to as a ceiling operator.

2.2 Adding Formality 59

Algorithm 2.1 Independent defaults by Monte Carlo

```
def independentBinomialLossDistribution (N,M,p,c,alpha):
    U = np.random.uniform(0,1,[M,N])
    defaultIndicator = 1*np.less(U,p)
    lossDistribution = np.sort(np.dot(defaultIndicator,c),axis=None)
    return lossDistribution
```

Algorithm 2.2 Numerically computing risk measures

```
def computeRiskMeasures(M,lossDistribution,alpha):
    expectedLoss = np.mean(lossDistribution)
    unExpectedLoss = np.std(lossDistribution)
    expectedShortfall = np.zeros([len(alpha)])
    var = np.zeros([len(alpha)])
    for n in range(0,len(alpha)):
        myQuantile = np.ceil(alpha[n]*(M-1)).astype(int)
        eShortfall[n] = np.mean(lossDistribution[myQuantile:M-1])
        var[n] = lossDistribution[myQuantile]
    return expectedLoss, unExpectedLoss, var, eShortfall
```

Algorithm 2.1 provides a Python sub-routine which—given the number of counterparties, N, the desired number of simulations, M, the default probabilities and exposures (i.e., p and c)—proceeds to compute a number of risk measures using the previously described Monte Carlo approach. It generates the requisite uniform variates in a single step and constructs a large matrix of default indicators. The dot product of this matrix and the exposure vector leads to a vector of M default losses.

The defaultIndicator $\in \mathbb{R}^{M \times N}$ array in Algorithm 2.1 is essentially a huge matrix with mostly zeros—indicating survival of obligor n for simulation m—and a sparse collection of ones indicating default. We need to transform this enormous collection of zeros and ones into meaningful risk measures. The first step is the dot product with the exposure vector, c, to create a vector of losses called, quite appropriately, lossDistribution $\in \mathbb{R}^M$. All of the heavy lifting thus reduces to a single line of code: the actually default outcomes are determined by np.less(U,p). This is basically a logical condition representing the indicator variable in equation 2.24.

This M-dimensional array is used to compute the expected loss, loss density, the expected shortfall, and the VaR figures for the specified quantile values found in alpha. Since we'll be doing this for many models, it makes sense to do it only once. This is the job of the computeRiskMeasures sub-routine described in Algorithm 2.2. It is, in fact, simply a practical implementation of equations 2.25 to 2.29. The key job is to order the losses. It can be used, for any collection of

Algorithm 2.3 Monte Carlo binomial model

```
def independentBinomialSimulation(N,M,p,c,alpha):
    lossDistribution = independentBinomialLossDistribution(N,M,p,c,alpha)
    el,ul,var,es = util.computeRiskMeasures(M,lossDistribution,alpha)
    return el,ul,var,es
```

numerically generated loss outcomes, to estimate a range of risk measures. This generic function requires the number of iterations, M, the M-dimensional array of default-losses, lossDistribution, and the degree of confidence, alpha. To repeat, Algorithm 2.2 is agnostic regarding the source of the loss distribution—the underlying model is unimportant. As you can see, the length and complexity of the code is quite low.

Algorithms 2.1 and 2.2 are sub-routines used to estimate the independent-default binomial model. The actual simulation function is now summarized in Algorithm 2.3 and comprises only a couple of lines. Since it will be used frequently and is model independent, the computeRiskMeasures function will be placed in the util library.[6]

2.2.2.3 Some Results

The numerical results for our practical example are summarized in Table 2.4. The total number of iterations, M, was set at 10 million and the total computational time associated with Algorithm 2.3 was in the neighbourhood of 16 seconds. This implies that the computational expense is not particularly high for a small portfolio. It is also not, however, completely negligible; this is particularly striking when considering the simplicity of the model. More importantly, perhaps, we have not yet addressed the question of model convergence. Speed and convergence, in the practitioner world, are critical and these questions will be duly addressed in Chap. 8.

For a range of quantile levels, Table 2.4 indicates both the VaR and expected shortfall values. The expected-shortfall figures are, of course, larger than the VaR estimates. This is because the VaR is the loss at a given quantile, while the expected shortfall is the average of all outcomes beyond the specified quantile. We also observe that, as we might hope, the loss volatility exceeds, although not by much, the expected loss. This suggests a modest amount of default-loss dispersion.

---

[6]See Appendix D for a more detailed description of the code organization.

2.2 Adding Formality

Table 2.4 *Numerical independent default results*: This table outlines the quantiles of the loss distribution—or VaR estimates—for the Monte Carlo estimates of the independent-default models.

Quantile	$\widehat{\text{VaR}_\alpha}(L_N)$	$\widehat{\mathcal{E}_\alpha}(L_N)$
95.00th	$35.3	$45.5
97.00th	$40.4	$50.8
99.00th	$52.6	$61.2
99.50th	$58.8	$66.8
99.90th	$72.3	$80.3
99.97th	$82.1	$89.5
99.99th	$90.2	$97.8
Other statistics		
Expected loss	$9.2	
Loss volatility	$12.8	
Total iterations	10,000,000	
Computational time	16.2 s	

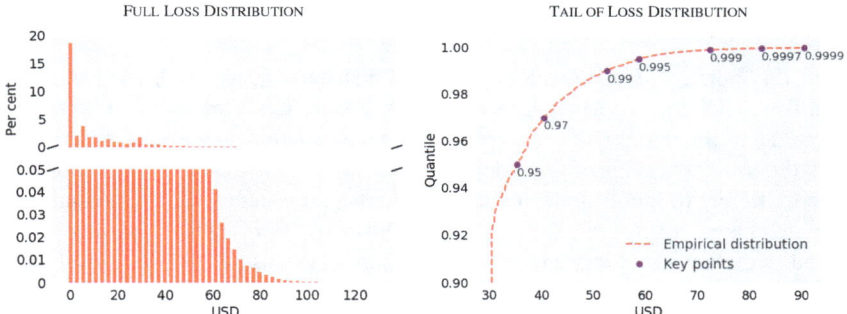

Fig. 2.5 *The loss distribution*: This figure outlines the full loss distribution for our simple 100-counterparty portfolio along with its tail from the 90th to the 99.99th quantile. The results from Table 2.4 are indicated.

Figure 2.5 provides an illuminating perspective on the results from Table 2.4. It shows the tail of the loss distribution from the 90th to 99.99th quantile. The extreme outcomes are clearly much larger, in this case by a factor of 9 to 10, than the average expected loss. It is precisely the tail of the loss distribution that is used to determine the economic-capital requirements of an organization in the face of default risk.

Indeed, when examining the entire loss distribution, as in the right-hand graphic of Fig. 2.5, it is necessary to split the y-axis to magnify the tail outcomes. Otherwise, they are simply not visible. This provides some indication of the rarity of the events under consideration. The 99.99th percentile, for instance, is dangerously close to the worst outcome even with 10,000,000 iterations. With this sizable simulation, there are only 1,000 observations at or about the 99.99th quantile.

While this is a relatively small number of observations upon which to base our estimates, it could be worse. 10 million iterations is, of course, a very substantial number. Often, analysts will seek, in the interests of managing the computational expense, to use something much smaller. The implications, however, can be important. A simulation employing 100,000 iterations, for example, would generate only 10 relevant observations to estimate the 99.99th quantile. The expected shortfall, in this setting, would be the average of only 10 numbers. This reasoning underscores the criticality of using large numbers of iterations in our numerical computations. We will, as we move forward in Chap. 8, take great pains to understand the number of iterations required for a given problem.

While generally straightforward to handle numerically in the case of heterogeneous default probabilities and exposures, the overall approach does not provide much insight into the independent-default model. By making a few simplifying assumptions, we can derive the analytic form of the default-loss distribution and learn much more about the implications of this modelling choice.

2.2.3 An Analytical Approach

Up to this point, to tease out a default-loss distribution, we have used numerical brute force. This is not a negative comment. Monte Carlo methods allowed us to compute a useful set of risk measures. They have revealed, however, relatively little about the inner workings of the independent-default credit-risk model. Instinctively, given its failure to incorporate default dependence, it is safe to conclude that we do not much care for this approach. This may, however, not be the only shortcoming. Indeed, we want to understand precisely where else this simple model might fail and why. Only then can we appreciate how more complex models act to improve it. This is the first step toward an intelligent model dialogue. To accomplish this, we need to adjust our approach and replace numerical brawn with analytic cleverness.

Analytic approaches, very often, require simplifying assumptions. This case is no exception. To create a tractable set-up, we are forced to make *two* simplifying assumptions. Specifically, we hold that,

1. $p = p_n$ for all $n = 1, \ldots, N$; and
2. $c = c_n$ for all $n = 1, \ldots, N$,

In words, therefore, we force all individual credit obligors in our portfolio to share common default probabilities and exposures. In some cases, such as large well-diversified portfolios, this might be reasonable. For the majority of settings, this is entirely unrealistic. Our objective, however, is *not* to replace the sensible and quite accurate numerical default-loss estimators from the previous section, but rather to gain some insight into the key assumptions behind the model. Thus, even though these choices are indefensible, they will still help us attain our goal.

The first implication of these assumptions is that $\mathbb{I}_{\mathcal{D}_1}, \cdots, \mathbb{I}_{\mathcal{D}_N}$ are now a collection of independent, identically distributed Bernoulli trials, each sharing the

2.2 Adding Formality

common parameter, p. The credit loss can now be written as,

$$L_N = c \underbrace{\sum_{n=1}^{N} \mathbb{I}_{\mathcal{D}_n}}_{\text{Number of defaults}}, \qquad (2.30)$$

$$= c\mathbb{D}_N,$$

where, as previously, $\mathbb{D}_N$ denotes the total number of defaults in the portfolio over the time interval, $[0, T]$. This is the first step toward a description of the loss distribution. Each credit loss, by virtue of homogeneous exposures, has the same implication—a loss of c. The entire loss distribution can thus be characterized by describing the sequence of losses, $\{c \cdot 0, c \cdot 1, \cdots, c \cdot N\}$. That is, the smallest loss is zero and the largest is the sum of all exposures, or $c \cdot N$. The loss outcomes are both discrete and easily described. The associated probability mass function is simply,

$$f_{L_N}(c \cdot k) = \mathbb{P}(L_N = c \cdot k), \qquad (2.31)$$

$$= \mathbb{P}\left(\underbrace{\not{c}\mathbb{D}_N}_{\text{Equation 2.30}} = \not{c} \cdot k\right),$$

$$= \mathbb{P}(\mathbb{D}_N = k),$$

$$= f_{\mathbb{D}_N}(k),$$

for arbitrary $k = 0, 1, \ldots, N$. To understand the loss distribution, therefore, it is entirely sufficient to consider the distribution of the total number of defaults.

The expectation of $\mathbb{D}_N$ is easily determined from equation 2.13 as,

$$\mathbb{E}(\mathbb{D}_N) = \mathbb{E}\left(\sum_{n=1}^{N} \mathbb{I}_{\mathcal{D}_n}\right), \qquad (2.32)$$

$$= \sum_{n=1}^{N} \underbrace{\mathbb{E}\left(\mathbb{I}_{\mathcal{D}_n}\right)}_{p},$$

$$= Np,$$

while the variance follows from equation 2.14,

$$\text{var}(\mathbb{D}_N) = \text{var}\left(\sum_{n=1}^{N} \mathbb{I}_{\mathcal{D}_n}\right), \qquad (2.33)$$

$$= \sum_{n=1}^{N} \underbrace{\text{var}\left(\mathbb{I}_{\mathcal{D}_n}\right)}_{p(1-p)},$$

$$= Np(1-p).$$

Indeed, as these two quantities strongly suggest, and as previously indicated, a collection of Bernoulli trials with a common parameter follow a binomial distribution with parameters N and p. The probability mass function is written as,

$$f_{\mathbb{D}_N}(k) = \mathbb{P}(\mathbb{D}_N = k), \tag{2.34}$$

$$= \binom{N}{k} p^k (1-p)^{N-k},$$

and the cumulative distribution function is now naturally written as,

$$F_{\mathbb{D}_N}(m) = \mathbb{P}(\mathbb{D}_N \leq m), \tag{2.35}$$

$$= \sum_{k=0}^{m} \binom{N}{k} p^k (1-p)^{N-k}.$$

This is entirely consistent with our previous discussion.

2.2.3.1 Putting It into Action

The use of these results to construct the distribution of the number of losses is summarized in Algorithm 2.4. It literally requires a few short lines of code and marginal amounts of computational time.

Algorithm 2.4 Analytic binomial implementation

```
def independentBinomialAnalytic(N,p,c,alpha):
    pmfBinomial = np.zeros(N+1)
    for k in range(0,N+1):
        pmfBinomial[k] =   util.getBC(N,k)*(p**k)*((1-p)**(N-k))
    cdfBinomial = np.cumsum(pmfBinomial)
    varAnalytic = c*np.interp(alpha,cdfBinomial,np.linspace(0,N,N+1))
    esAnalytic  = util.analyticExpectedShortfall(N,alpha,pmfBinomial,c)
    return pmfBinomial,cdfBinomial,varAnalytic,esAnalytic
```

2.2 Adding Formality   65

Algorithm 2.5 Approximated expected shortfall in the analytic model

```
def analyticExpectedShortfall(N,alpha,pmf,c):
    cdf = np.cumsum(pmf)
    numberDefaults = np.linspace(0,N,N+1)
    expectedShortfall = np.zeros(len(alpha))
    for n in range(0,len(alpha)):
        myAlpha = np.linspace(alpha[n],1,1000)
        loss = c*np.interp(myAlpha,cdf,numberDefaults)
        prob = np.interp(loss,numberDefaults,pmf)
        expectedShortfall[n] = np.dot(loss,prob)/np.sum(prob)
    return expectedShortfall
```

Algorithm 2.4 simply exploits equations 2.34 and 2.35 to identify analytic VaR estimates. It does require the binomial coefficient, which, due to its simplicity, is not explicitly provided. The utility function, `getBC`, is embedded in a separate library, `util`, and follows directly from equation 2.9.

The estimation of the VaR and expected shortfall—described by the variables `varAnalytic` and `esAnalytic`, respectively—merits some additional comment. Since the binomial distribution is *not* continuous, but rather has probability mass only on distinct discrete points, it is not obvious as to how to determine a specific quantile. That is, $\alpha = 0.99$, may lie in between two discrete points. To achieve the precise quantile values we permit fractional numbers of defaults. In Algorithm 2.4, we have opted to use Python's `numpy` interpolation function to linearly determine the continuously defined number of defaults and loss associated with a specific quantile. Although this is stretching things somewhat, hopefully the reader will not view it as a capital crime.

Approximation of the expected shortfall for our analytic implementation of the independent-default binomial model is also a bit challenging. It suffers from the same problems as the VaR measure, since we need to compute the quantile, but it also involves averaging. Our solution, summarized in Algorithm 2.5, again uses the interpolation function. It partitions the space beyond the quantile and then estimates the losses and their probabilities for each grid point. This permits the computation of the expected value, which is, as a final step, normalized by the size of the quantile. Again this is far from perfect; it merely seeks to reasonably capture the magnitude of the expected-shortfall measure.[7]

---

[7]Again, given its general usefulness, it is placed in the `util` library.

2.2.3.2 Comparing Key Assumptions

Equipped with these practical algorithms, we may now return to our previous example. As before, we have 100 counterparties, but now we fix the default probability to the average value of 1%. Each counterparty shares a common exposure of $10—this is equal to the average exposure in the previous example. Total portfolio exposure of $1,000 remains unchanged. This is performed for both the numerical and analytic calculations so as to permit an *apples-to-apples* comparison.

Table 2.5 illustrates a selected number of quantiles with the analytic results both in dollar terms and numbers of defaults. The final column restates the numerical values computed with heterogeneous default probabilities and exposures. In all cases, the numerical values dominate the analytic results. The moments match perfectly, however, and the differences are quite modest. The reason we observe any difference is because the simulation algorithm does not permit fractional defaults; this fact likely makes it more realistic than our analytic approximation. It nevertheless makes for an interesting contrast.

In the final two columns of Table 2.5, the *true* portfolio VaR and expected shortfall figures are repeated from Table 2.4, for ease of comparison. By *true* portfolio, we imply the results computed without assuming homogeneous exposures and default probabilities. The contrast is quite striking. The homogeneity assumptions, required for the analytic results, clearly lead to an understatement of the portfolio risk. Heterogeneity of default probabilities and exposures thus matters. It is precisely the large exposures that occur with relatively higher probability that contribute to extreme tail outcomes. In the homogeneous setting, the identity of the obligor does

Table 2.5 *Analytic vs. numerical results*: This table compares the analytic VaR quantiles—both in dollar terms and number of defaults—to the associated numerical results. To permit a fair comparison, both analytical and numerical computations assume that the default probabilities and exposures are fixed at 1% and $10 for all counterparties. The results for the true heterogeneous exposures and default probabilities are also provided.

| | | | | Numerical | | | |
| | Analytic | | | Homogeneous | | Heterogeneous | |
Quantile	$\mathbb{D}_N$	$\text{VaR}_\alpha(L_N)$	$\mathcal{E}_\alpha(L_N)$	$\widehat{\text{VaR}_\alpha(L_N)}$	$\widehat{\mathcal{E}_\alpha(L_N)}$	$\widehat{\text{VaR}_\alpha(L_N)}$	$\widehat{\mathcal{E}_\alpha(L_N)}$
95.00th	2.4	$23.8	$24.1	$30.0	$33.8	$35.3	$45.5
97.00th	2.7	$27.4	$27.7	$30.0	$36.3	$40.4	$50.8
99.00th	3.4	$34.5	$34.7	$40.0	$43.3	$52.6	$61.2
99.50th	3.8	$38.3	$38.6	$40.0	$46.6	$58.8	$66.8
99.90th	4.8	$47.6	$47.8	$50.0	$54.7	$72.3	$80.3
99.97th	5.3	$53.3	$53.5	$60.0	$62.0	$82.1	$89.5
99.99th	5.9	$58.7	$58.9	$60.0	$65.9	$90.2	$97.8
Other statistics							
Expected loss	1.0	$9.5		$9.5		$9.2	
Loss volatility	1.0	$9.7		$9.7		$12.8	
Computational time		0.01		16.15			

2.3 Convergence Properties

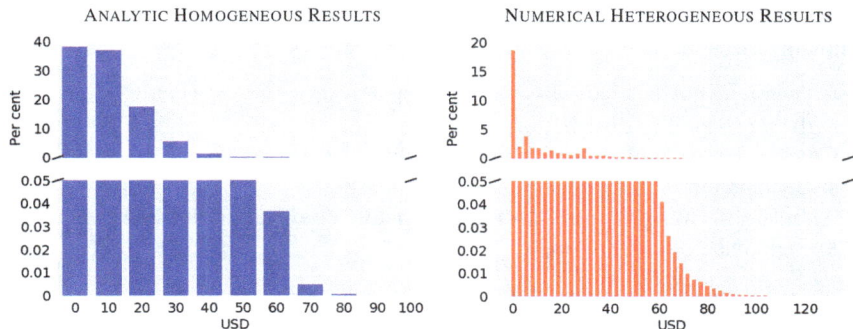

Fig. 2.6 *Impact of homogeneity assumptions*: This figure outlines the full loss distribution for the analytic solution, with its embedded homogeneity assumptions for default probabilities, against the heterogeneous numerical results found in Fig. 2.5.

not, in fact, play a role. All obligor defaults generate, quite unreasonably, the same losses with the same relative probabilities. By adding some complexity into the exposure and creditworthiness profile of our credit counterparts, a much richer range of tail outcomes are observed. This important lesson will be revisited on a number of occasions and, as it turns out, has wide-reaching implications in the regulatory setting.

Figure 2.6 compares the corresponding analytic-homogeneous and numerical-heterogeneous loss distributions. These two results converge if one imposes homogeneous default probabilities and exposures on the numerical model. It nevertheless provides a strong indication that, given the more realistic nature of the heterogeneous case, the numerical approach is a more flexible solution.

Credit-risk is, however, often a difficult problem to handle. Use of extreme quantiles implies that, even with very large number of simulations, there are only handfuls of observations at or above the desired level. Convergence and the asymptotic behaviour of a given model, therefore, are *key* issues for all credit-risk approaches. Numerical techniques can only take us so far down the road to understanding the implications of our assumptions. In the next section, therefore, we will try to build the general case using our analytical model.

2.3 Convergence Properties

Our task in this section is to gain some insight into the convergence properties of the binomial independent-default model. What do we mean by convergence? We are, in a technical sense, trying to get a handle on the asymptotic behaviour of our model. With N credit counterparties in a portfolio, convergence is concerned with what happens as N becomes very large. Let's consider *three* notions of convergence by imagining sequential random experiments with increasing N. This yields the following three ideas:

1. Convergence in distribution—the situation gets progressively better described by a given probability distribution.
2. Convergence in probability—the probability of an unusual outcome gets progressively smaller.
3. Almost-sure convergence—the result basically converges pointwise to a particular outcome.

Although a bit tedious, these notions are critical, since they inform us about the terminal behaviour of our model.[8]

Convergence in Probability

Despite its general inappropriateness, the analytic solution does help us better understand the independent default model. Consider the following random variable, $\frac{\mathbb{D}_N}{N}$, which denotes the proportion of defaults in the portfolio. It is interesting, and useful, to understand what happens to this proportion as the portfolio grows very large. To gain this understanding, we use a common result employed in proving convergence in probability: Chebyshev's inequality. It holds that,

$$\mathbb{P}\left(|X - \mathbb{E}(X)| \geq \epsilon\right) \leq \frac{\mathrm{var}(X)}{\epsilon^2}, \qquad (2.36)$$

for any random variable, X, where the mean and variance are defined, and for $\epsilon \in \mathbb{R}_+$. Applying this to $\frac{\mathbb{D}_N}{N}$, the proportion of defaults in the portfolio, we have

$$\mathbb{P}\left(\left|\frac{\mathbb{D}_N}{N} - \mathbb{E}\left(\frac{\mathbb{D}_N}{N}\right)\right| \geq \epsilon\right) \leq \frac{\mathrm{var}\left(\frac{\mathbb{D}_N}{N}\right)}{\epsilon^2}, \qquad (2.37)$$

$$\mathbb{P}\left(\left|\frac{\mathbb{D}_N}{N} - \frac{1}{N}\underbrace{\mathbb{E}\left(\mathbb{D}_N\right)}_{Np}\right| \geq \epsilon\right) \leq \frac{\overbrace{\mathrm{var}\left(\mathbb{D}_N\right)}^{Np(1-p)}}{N^2\epsilon^2},$$

$$\mathbb{P}\left(\left|\frac{\mathbb{D}_N}{N} - p\right| \geq \epsilon\right) \leq \frac{p(1-p)}{N\epsilon^2},$$

---

[8] See Billingsley (1995); Durrett (1996) or DasGupta (2008) for much more detail on these fundamental ideas from probability theory.

2.3 Convergence Properties

from which it is clear that as $N \to \infty$, that

$$\mathbb{P}\left(\left|\frac{\mathbb{D}_N}{N} - p\right| \geq \epsilon\right) \to 0, \tag{2.38}$$

as $N \to \infty$ for all $\epsilon \in \mathbb{R}_+$. Thus, by Chebyshev's inequality, we have that the proportion of defaults in the portfolio, $\frac{\mathbb{D}_N}{N}$, tends to the fixed parameter, p, as the portfolio gets very large. Thus, we have that $\frac{\mathbb{D}_N}{N}$ converges to p in probability. This is termed the weak law of large numbers.

Almost-Sure Convergence

There are a number of sources—see, for example, DasGupta (2008)—that demonstrate almost-sure convergence of the so-called binomial proportion. It is not entirely straightforward, requiring derivation of the first four moments of the binomial distribution, the Markov's inequality and the Borel-Cantelli theorem. It is nonetheless a useful exercise, which we will repeat.

Let's begin with the determination of the moments of the binomial distribution. Their computation using the moment-generating function is straightforward, if tedious. Moreover, they are well described in many sources.[9] The moment generating function, for discrete random variable X, is defined as follows,

$$M_X(t) = \sum_{k=0}^{\infty} e^{tk} \mathbb{P}(X = k), \tag{2.39}$$

$$= \mathbb{E}\left(e^{tX}\right).$$

Moreover, the nth moment of X can be determined by computing the nth derivative of $M_X(t)$ and evaluating it at $t = 0$. The raw form of the moment-generating function for the binomial distribution is, following from equations 2.34 and 2.35, written as,

$$M_{\mathbb{D}_N}(t) = \sum_{k=0}^{N} e^{tk} \mathbb{P}(\mathbb{D}_N = k), \tag{2.40}$$

$$= \sum_{k=0}^{N} \binom{N}{k} e^{tk} p^k (1-p)^{N-k}.$$

---

[9] See, for example, Casella and Berger (1990, Chapter 2).

This is not immediately simplified. The binomial theorem—see Harris and Stocker (1998, Chapter 1)—holds, however, that

$$\sum_{k=0}^{N} \binom{N}{k} a^k b^{N-k} = (a+b)^N. \tag{2.41}$$

If we set $a = pe^t$ and $b = 1 - p$, we can re-write equation 2.40 as

$$M_{\mathbb{D}_N}(t) = \sum_{k=0}^{N} \binom{N}{k} \underbrace{\left(pe^t\right)}_{a}^{k} \underbrace{\left(1-p\right)}_{b}^{N-k}, \tag{2.42}$$

$$= \left(pe^t + (1-p)\right)^N.$$

To provide an idea of how this result is employed, we will use it to compute the second moment of the binomial distribution. First we require the second derivative of $M_{\mathbb{D}_N}(t)$. It has the following form,

$$\frac{\partial M_{\mathbb{D}_N}(t)}{\partial t^2} = \frac{\partial \left(pe^t + (1-p)\right)^N}{\partial t^2}, \tag{2.43}$$

$$= \frac{\partial}{\partial t} \left(N \left(pe^t + (1-p)\right)^{N-1} pe^t\right),$$

$$= N(N-1)p^2 \left(pe^t + (1-p)\right)^{N-2} e^{2t} + Np \left(pe^t + (1-p)\right)^{N-1} e^t.$$

If we evaluate this result at $t = 0$, we have

$$\left.\frac{\partial M_{\mathbb{D}_N}(t)}{\partial t^2}\right|_{t=0} = N(N-1)p^2 \underbrace{\left(pe^0 + (1-p)\right)^{N-2}}_{=1} e^0 + Np \underbrace{\left(pe^0 + (1-p)\right)^{N-1}}_{=1} e^0,$$

$$= N(N-1)p^2 + Np. \tag{2.44}$$

The variance of the binomial distribution is now readily solved using this result as follows,

$$\mathbb{E}\left((\mathbb{D}_N - Np)^2\right) = \mathbb{E}\left(\mathbb{D}_N^2 - 2\mathbb{D}_N Np + (Np)^2\right), \tag{2.45}$$

$$= \mathbb{E}\left(\mathbb{D}_N^2\right) - 2Np \underbrace{\mathbb{E}(\mathbb{D}_N)}_{Np} + N^2 p^2,$$

$$= N(N-1)p^2 + Np - 2N^2 p^2 + N^2 p^2,$$

$$= (N^2 - N)p^2 + Np - N^2 p^2,$$

2.3 Convergence Properties

$$= \cancel{N^2 p^2} - Np^2 + Np - \cancel{N^2 p^2},$$
$$= Np(1-p).$$

After a number of painful computations, with enough patience, one can use precisely the same technique to show that,

$$\mathbb{E}\left((\mathbb{D}_N - Np)^4\right) = Np(1-p)\left(1 + 3(N-2)p(1-p)\right). \quad (2.46)$$

This allows us to perform the following computation with the same starting point as when we used Chebyshev's inequality in equation 2.36 and equation 2.37,

$$\mathbb{P}\left(\left|\frac{\mathbb{D}_N}{N} - p\right| \geq \epsilon\right) = \mathbb{P}\left(|\mathbb{D}_N - Np| \geq N\epsilon\right), \quad (2.47)$$

$$= \mathbb{P}\left((\mathbb{D}_N - Np)^4 \geq (N\epsilon)^4\right),$$

$$\leq \underbrace{\frac{\mathbb{E}\left((\mathbb{D}_N - Np)^4\right)}{N^4 \epsilon^4}}_{\text{Markov's inequality}},$$

$$= \frac{Np(1-p)(1 + 3(N-2)p(1-p))}{N^4 \epsilon^4},$$

$$= \frac{Np(1-p) + (3N^2 - 6N)p^2(1-p)^2}{N^4 \epsilon^4},$$

$$\leq \frac{Np(1-p) + 3N^2 p^2(1-p)^2}{N^4 \epsilon^4},$$

$$= \frac{Np(1-p)}{N^4 \epsilon^4} + \frac{3N^2 p^2(1-p)^2}{N^4 \epsilon^4},$$

$$= \frac{p(1-p)}{N^3 \epsilon^4} + \frac{3p^2(1-p)^2}{N^2 \epsilon^4},$$

$$\leq \infty.$$

The consequence of this manipulation is that we have demonstrated that,

$$\mathbb{P}\left(\left|\frac{\mathbb{D}_N}{N} - p\right| \geq \epsilon\right) \leq \infty, \quad (2.48)$$

which implies, using Borel-Cantelli theorem, that $\frac{\mathbb{D}_N}{N}$ converges, in an almost-sure sense, to the Bernoulli parameter, p.[10] This is referred to as the strong law of large numbers.

Cutting to the Chase

In the previous discussion, we demonstrated—by Chebyshev's inequality—that for large N, the proportion of defaults in the portfolio tends to p. In other words, $\frac{\mathbb{D}_N}{N}$ converges to p in probability. We also showed—by Markov's inequality and the Borel-Cantelli theorem—that $\frac{\mathbb{D}_N}{N}$ converges almost surely to p. This are referred to as the *weak* and *strong* laws of large numbers, respectively. The names themselves suggest the relative strength of these convergence results.

Why should we care? Almost-sure convergence amounts to a very strong assumption embedded in the independent-default model. For large portfolios, the proportion of defaults will converge to the default probability. So, if your parameter, p, is equal to 1%, then for large portfolios, you will *not* observe sizable deviations in average default loss outcomes from about that level. There is just very little in the way of randomness. In other words, it means the model is very well behaved.

We already knew that the binomial independent-default approach was *not* a particularly defensible modelling choice. Now we understand better the specific reasons. The first is the assumption of default independence; this is economically unrealistic. Second, the limiting distribution of the binomial distribution converges—if we recall Pascal's triangle—to the Gaussian law. The tails of the normal distribution are, in general, relatively thin. All else equal, as conservative risk managers, we would prefer to have more probability mass allocated to extreme events. This is particularly important in the credit risk setting where our focus is almost entirely on rare events. The final nail on the coffin of the binomial model is its overly smooth asymptotic behaviour. Real-life financial markets are just not this well-behaved. Indeed, for sufficiently large N, model behaviour is completely described by a single parameter.

Improving our approach to credit risk will thus involve us attacking each of these shortcomings and finding reasonable alternatives. This will motivate many of the models considered in subsequent chapters. Before moving to this task, we will finish this discussion by demonstrating the equivalence of the binomial model to another, quite useful, approach that is used extensively in the literature. Understanding the links between these two methodologies—and appreciating their relative equivalence—is very helpful in following the extant models in both academia and practice.

---

[10] See Durrett (1996) and Billingsley (1995) for more colour on the Borel-Cantelli theorem.

2.4 Another Entry Point

We saw that treating the default event as a coin flip, or a stone drawn from an urn, naturally leads us to a Bernoulli trial and then, ultimately, to the binomial distribution. The question is: can another distributional choice be taken? Since we seek a binary outcome, it is perhaps natural to select a random variable taking only two values. While convenient, this is not strictly speaking necessary. One need only make astute use of an indicator variable. If, for example, we define a random variable, X, then the following binary default event is readily defined as

$$\mathbb{I}_{\{X > \tilde{x}\}} = \begin{cases} 1 : \text{default} \\ 0 : \text{survival} \end{cases}. \tag{2.49}$$

Default occurs on the set, $\{X > \tilde{x}\}$, where the random variable exceeds a particular level, denoted $\tilde{x}$. In this case, X, can take a variety of values and may be either discretely or continuously defined. This clever trick allows us to define the default event in a much more general way.

If we are looking to replace the binomial distribution, a natural choice is the Poisson distribution. This distribution is typically used to model the number of times a particular event occurs over a predefined time interval. These events could be depressing, like heart attacks or industrial accidents, or more encouraging, such as the arrival of the subway or an e-mail from a friend. As a discrete distribution, it also plays a central role similar to the Gaussian distribution among continuous distributions. That is, loosely speaking, the sum of independent Poisson random variables converge to a standard normal distribution. This limiting role, in addition to its flexibility, makes it an natural candidate.

To model a collection of default events, we will need to define a large number of X's. As in the binomial setting, we will continue to assume independence of default. To accomplish this, let us define $X_n \sim$ i.i.d. $\mathcal{P}(\lambda_n)$ for $n = 1, \ldots, N$—that is, we have a set of independent, identically distributed Poisson[11] random variables, each with its own parameter, λ_n. The probability of observing X_n taking any positive integer value k is described by the following probability mass function,

$$\mathbb{P}(X_n = k) = \frac{e^{-\lambda_n} \lambda_n^k}{k!}. \tag{2.50}$$

The Poisson distribution's parameter, $\lambda_n \in \mathbb{R}$, is often referred to as the arrival rate or intensity. It basically governs the probability of event occurrence per unit of time. Note that, in principle, X_n may take any positive integer value. That is, $X_n \in \mathbb{N}_+$ for all $n = 1, \ldots, N$.

---

[11] We have opted to denote the Poisson distribution with the symbol, $\mathcal{P}$. While not a common notation, we find it useful.

To use this distribution, we need to define, following the trick from equation 2.49, a default event. Since X can take any positive integer value, plus zero, this means finding a reasonable value of $\tilde{x}$. In our case, the most natural choice for the nth default event is,

$$\mathcal{D}_n \equiv \{X_n \geq 1\}. \tag{2.51}$$

This might not seem particularly useful, at first glance, but equation 2.50 allows us to define the probability of this event in a much more specific manner. In particular, it follows logically that

$$\mathbb{P}\{X_n \geq 1\} = 1 - \mathbb{P}\{X_n \leq 0\}, \tag{2.52}$$
$$= 1 - \underbrace{\frac{e^{-\lambda_n}\lambda_n^0}{0!}}_{\text{Equation 2.50}},$$
$$= 1 - e^{-\lambda_n}.$$

This has some interesting consequences. The unconditional default probability, p_n, is equivalent to the probability of the default event. This directly implies that,

$$p_n = \mathbb{P}\{X_n \geq 1\}, \tag{2.53}$$
$$= 1 - e^{-\lambda_n},$$
$$\lambda_n = -\ln(1 - p_n).$$

Thus, we have a clear and exploitable mathematical relationship between the unconditional default probability and the arrival intensity. Incidentally, other choices are possible. One could, in equation 2.52, for example, define the default event to $\{X_n = 1\}$. For the typical values of λ_n, this is not unreasonable since $\{X_n \geq 2\}$ is vanishingly small.[12] In this setting, it also makes sense to set $p_n = \lambda_n$. Figure 2.7 compares the two methods and indicates that the difference, for small values of p_n, is indeed quite small. For most practical implementations, it is reasonably safe, therefore, to use either approach. Although completely defensible, we still prefer, when possible, to use the higher degree of precision afforded by equations 2.52 and 2.53.

---

[12]For an arrival rate of $p_n \approx \lambda_n = 0.01$, the probability that X_n is greater than or equal to two is less than 0.0001. Not zero, but not particularly material. This is sufficiently close for many applications.

2.4 Another Entry Point

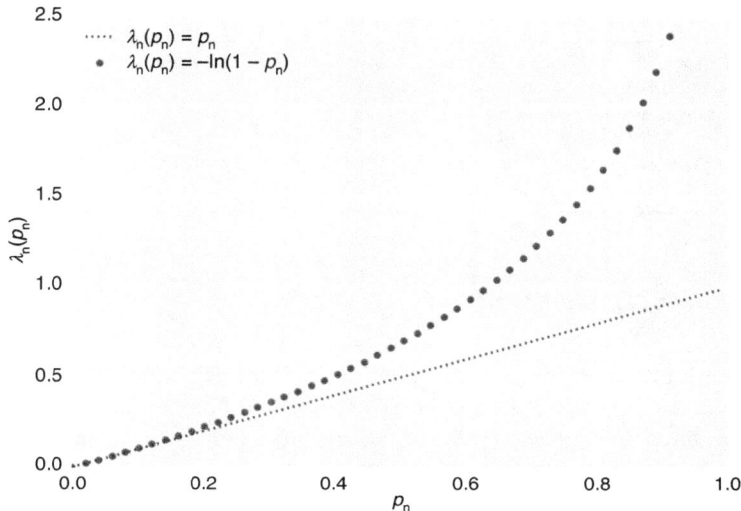

Fig. 2.7 *Identifying λ_n*: This figure outlines two different approaches toward identifying λ_n using the unconditional default probability, p_n. One is exact, whereas the other is an approximation. For small values of p_n, the differences are very small.

The link between the arrival intensity and default probability allows us to proceed. In particular, our default indicator is now simply,

$$\mathbb{I}_{\mathcal{D}_n} \equiv \mathbb{I}_{\{X_n \geq 1\}} = \begin{cases} 1 : \text{defaults occurs before time } T \text{ with probability } p_n = 1 - e^{-\lambda_n} \\ 0 : \text{survival until time } T \text{ with probability } e^{-\lambda_n} \end{cases}, \quad (2.54)$$

which is profitably compared to equation 2.49. The default loss, for a portfolio of N obligors is consequently,

$$L_N = \sum_{n=1}^{N} c_n \mathbb{I}_{\{X_n \geq 1\}}. \quad (2.55)$$

It is left as a straightforward exercise for the reader to demonstrate that the expected loss remains the sum of the exposure-weighted default probabilities (or arrival intensities).

This is enough to proceed with this model toward the numerical approximation of the default-loss distribution.

Algorithm 2.6 Independent Poisson defaults by Monte Carlo

```
def independentPoissonLossDistribution(N,M,p,c,alpha):
    lam = -np.log(1-p)
    H = np.random.poisson(lam,[M,N])
    defaultIndicator = 1*np.greater_equal(H,1)
    lossDistribution = np.sort(np.dot(defaultIndicator,c),axis=None)
    return lossDistribution
```

A Numerical Implementation

To implement this aproach, the only requirement is the generation of large numbers of independent Poisson random variables. There are different tricks to perform this task—the interested reader is referred to Fishman (1995)—but we will use the embedded algorithms from Python's numpy library.

Algorithm 2.6 summarizes the code to simulate the associated default-loss distribution. It is virtually identical, up to three small differences, with the approach followed in Algorithm 2.1. The first difference is that, although the unconditional default probabilities (i.e., p) are passed to the function, they are no longer used directly. Instead, we use the relationship from equation 2.53 to transform them into the corresponding Poisson arrival rates. The second difference is that independent Poisson, and not independent Bernoulli trials, are generated. The final distinction is the form of the default indicator: np.greater_equal(H,1). Despite these small practical differences, the two algorithms are conceptually identical.

The actual risk measures, for this model, are computed by passing the ordered lossDistribution array to the function util.computeRiskMeasures; this is the same as in Algorithm 2.3. Figure 2.8 compares the resulting loss distributions and tail probabilities—associated with our usual sample portfolio—of the numerical independent-default binomial and Poisson distributions. The results are visually indistinguishable.

Is this a coincidence? It is hard to expect them to be exactly the same. A bit more information would be helpful. Table 2.6 provides detailed comparison of the various quantiles and moments of the two simulated loss distributions. The expected loss and loss volatilities as well as the VaR and expected shortfall measures up to the 99.9th quantile reveal no differences. There is some divergence very far out into the tail—particularly with regard to the expected shortfall—but it is very small. The only important deviation is that the independent-default Poisson model requires about twice the computational time as its binomial equivalent; this is due to the incremental computational expense associated with generating Poisson relative to uniform random variables.

This is certainly too much of a similarity to be attributed to coincidence. The simple fact is that there is a fundamental link between these two models and their underlying distributional assumptions. It is critical to understand this link to

2.4 Another Entry Point

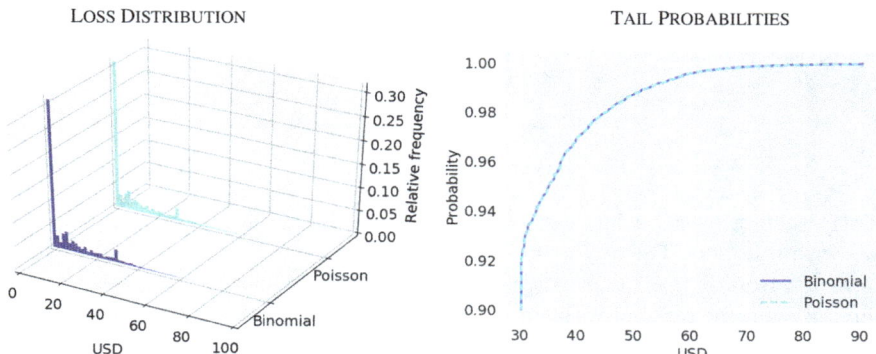

Fig. 2.8 *Indistinguishable*: This figure compares both the loss distribution and its tail probabilities for identically calibrated binomial and Poisson independent-default models. The results, while not exact, are visually indistinguishable.

Table 2.6 *Numerical results*: This table compares key risk measures for the numerical estimation of the independent-default binomial and Poisson models. Although not exact, the differences are very small.

	Binomial		Poisson	
Quantile(α)	VaR$_\alpha(L_N)$	$\mathcal{E}_\alpha(L_N)$	VaR$_\alpha(L_N)$	$\mathcal{E}_\alpha(L_N)$
95.00th	$35.3	$45.5	$35.3	$45.5
97.00th	$40.4	$50.8	$40.4	$50.8
99.00th	$52.7	$61.2	$52.5	$61.1
99.50th	$58.8	$66.8	$58.8	$66.8
99.90th	$72.2	$80.3	$72.2	$80.3
99.97th	$82.1	$89.4	$82.0	$89.6
99.99th	$90.1	$97.5	$90.2	$97.9
Other statistics				
Expected loss	9.2		9.2	
Loss volatility	12.8		12.8	
Computational time	17.8		34.1	

appreciate extant credit-risk models. As in the previous discussion, we will explore this link by examining the analytic version of the independent Poisson model.

The Analytic Model

The usual unrealistic assumptions apply for the analytic version of the independent Poisson model. That is, $c = c_n$ and $p = p_n$ for all $n = 1, \ldots, N$. The corollary of this assumption, from equation 2.53, is that $\lambda = \lambda_n$ for $n = 1, \ldots, N$. It should, after all of our previous effort, be clear that $L_N = c \mathbb{D}_N$. The issue is that $\mathbb{D}_N$ is now the sum of independent, and identically distributed, Poisson random variables. That is, we have that

$$\mathbb{D}_N = \sum_{n=1}^{N} \mathbb{I}_{\{X_n \geq 1\}}. \tag{2.56}$$

This structure turns out to be relatively inconvenient for the type of development we wish to follow. The problem is the presence of the indicator variable. Remark, however, from Fig. 2.7 that we can dispense with the indicator variable and work directly with the random variables themselves. This is possible, because for small values of λ, we have that

$$\mathbb{I}_{\{X_n \geq 1\}} \approx X_n. \tag{2.57}$$

In other words, although each X_n can take the value of any positive integer, for all practical purposes when λ is small, it behaves like a binary variable. The consequence of equation 2.57 is that we may restate the total number of defaults as,

$$\mathbb{D}_N = \sum_{n=1}^{N} \mathbb{I}_{\{X_n \geq 1\}}, \tag{2.58}$$

$$\approx \sum_{n=1}^{N} X_n.$$

This is a more tractable situation, since we know that the sum of independent Poisson random variables themselves follows a Poisson distribution.[13] Indeed, it is true that,

$$\mathbb{D}_N \approx \sum_{n=1}^{N} X_n, \tag{2.59}$$

$$\sim \mathcal{P}(N\lambda).$$

This is a very nice result. We now know the distribution of the number of defaults, $\mathbb{D}_N$, and by extension, the default-loss distribution. In fact, the probability mass function is simply

$$f_{\mathbb{D}_N}(k) = \mathbb{P}(\mathbb{D}_N = k) = \frac{e^{-N\lambda}(N\lambda)^k}{k!}, \tag{2.60}$$

---

[13] For a more formal discussion of this fact, please see Casella and Berger (1990).

2.4 Another Entry Point

and the cumulative density function is now naturally written as,

$$F_{\mathbb{D}_N}(m) = \mathbb{P}(\mathbb{D}_N \leq m) = \sum_{k=0}^{m} \frac{e^{-N\lambda}(N\lambda)^k}{k!}. \tag{2.61}$$

This can be compared with the independent-default binomial distribution described in equations 2.34 and 2.35.

The expectation and variance of the number of defaults are also quite straightforward, since the Poisson distribution has the rare feature of having an expectation and variance that coincide. As a result, we have that

$$\mathbb{E}(\mathbb{D}_N) = \text{var}(\mathbb{D}_N) = N\lambda. \tag{2.62}$$

For our sample portfolio, this permits a simple back-of-the-envelope calculation. We have $N = 100$ counterparties, an average default probability of about $p = 0.01$, and it is easily verified from equation 2.53 that $\lambda \approx p$. We thus expect, for our portfolio, to observe a single default with an equal amount of average dispersion around this estimate. This translates into about \$10.

Once again, equations 2.60 and 2.61 permit us to implement the model and compute our usual collection of risk measures. Algorithm 2.7 summarizes the practical Python implementation.

Structurally, the approach follows directly from Algorithm 2.4. Two points are worth highlighting. First, the input argument, myLam, is nothing other than $N\lambda$. Second, our self-constructed util library is used to construct the Poisson density.

The application of Algorithm 2.6 to our 100-obligor portfolio leads to the results summarized in Table 2.7. Once again, as with the numerical analysis, the results are, while not exact, extremely similar. Comparison of the fractional defaults reveals, across basically every quantile, essentially no difference between the two models. The VaR and expected-shortfall figures deviate slightly; this nonetheless appears to

Algorithm 2.7 Analytic Poisson implementation

```
def independentPoissonAnalytic (N,c ,myLam, alpha ):
    pmfPoisson = np.zeros(N+1)
    for k in range(0,N+1):
        pmfPoisson[k] = util.poissonDensity(myLam,k)
    cdfPoisson = np.cumsum(pmfPoisson)
    varAnalytic = c*np.interp(alpha ,cdfPoisson ,np.linspace(0,N,N+1))
    esAnalytic = util.analyticExpectedShortfall(N, alpha , pmfPoisson ,c)
    return pmfPoisson ,cdfPoisson ,varAnalytic ,esAnalytic
```

Table 2.7 *Analytical independent-default results*: This table contrasts the binomial and Poisson independent-default analytical model risk measures for our 100-obligor example portfolio. As expected, although not precisely the same, the differences are economically immaterial.

	Binomial			Poisson		
Quantile(α)	$\mathbb{D}_N$	$\text{VaR}_\alpha(L_N)$	$\mathcal{E}_\alpha(L_N)$	$\mathbb{D}_N$	$\text{VaR}_\alpha(L_N)$	$\mathcal{E}_\alpha(L_N)$
95.00th	2.4	$23.8	$24.1	2.4	$24.1	$24.4
97.00th	2.7	$27.4	$27.7	2.8	$27.6	$27.9
99.00th	3.4	$34.5	$34.7	3.5	$34.9	$35.2
99.50th	3.8	$38.3	$38.6	3.9	$38.6	$38.8
99.90th	4.8	$47.6	$47.8	4.8	$48.0	$48.4
99.97th	5.3	$53.3	$53.5	5.4	$54.3	$54.6
99.99th	5.9	$58.7	$58.9	5.9	$59.1	$59.4
Other statistics						
Expected loss	1.0	$9.5		1.0	$9.6	
Loss volatility	1.0	$9.7		1.0	$9.8	
Computational time	0.009			0.003		

be a result of magnification of small differences.[14] The punchline of this analysis is that we have both functional analytic and numerical equivalence between our independent-default and binomial models.

The Law of Rare Events

The simple question remains: why do these seemingly quite different models yield virtually interchangeable default-loss distributions? The reason relates to a well-known relationship between the binomial and Poisson distributions. More specifically, in certain cases, they experience a form of convergence.

Since the binomial distribution takes two parameters—p and N—and the Poisson only one (i.e., λ), we need to perform some combination for them to be comparable. It turns out that, in the previous section, this combination was already performed. The Poisson arrival-intensity parameter is thus defined as $\lambda = Np$. That is, we compare $\mathcal{B}(p, N)$ to $\mathcal{P}(Np)$.[15] The result is simple and powerful; for sufficiently small p and large N, these two distributions coincide with one another.

To see this more clearly, Fig. 2.9 provides, for a fixed $N = 100$, a comparison of the binomial and Poisson probability-mass functions for two choices of p. In the first case, summarized in the right-hand graphic, we set $p = 0.01$, which leads to an arrival rate of $N\lambda = 1$. There is very little difference between the two distributions;

---

[14] As we'll see in the next section, these methods are only truly equivalent as N tends to infinity.

[15] One could, of course, go in the other direction by considering a binomial distribution with parameters N and $p = \frac{\lambda}{N}$.

2.4 Another Entry Point

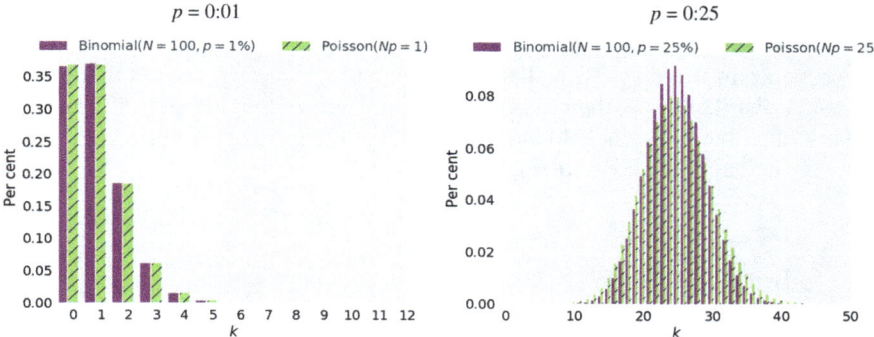

Fig. 2.9 *The law of rare events*: This figure compares, for different choices of p and N, the binomial and Poisson distributions. For sufficiently small $\lambda = pN$, the results are almost indistinguishable; this is referred to as the law of rare events.

they are not exact, but the deviations are extremely small. The right-hand graphic, however, tells a different story. When $p = 0.25$, the two distributions no longer agree. While they have similar symmetric shapes, the tails of the Poisson distribution are distinctly fuller than the binomial.

For those who are not as convinced by graphic arguments, this can also be shown mathematically. Let us begin with the binomial distribution's probability mass function for our random default-loss, L_N, insert the Poisson parameter, $p = \frac{\lambda}{N}$, and evaluate the limit as N tends to infinitely as follows,

$$f_{L_N}(k) = \mathbb{P}(L_N = k), \qquad (2.63)$$

$$= \binom{N}{k} p^k (1-p)^{N-k},$$

$$= \frac{N!}{(N-k)!k!} \left(\frac{\lambda}{N}\right)^k \left(1 - \frac{\lambda}{N}\right)^{N-k},$$

$$= \frac{N!}{(N-k)!k!} \left(\frac{\lambda^k}{N^k}\right) \left(1 - \frac{\lambda}{N}\right)^N \left(1 - \frac{\lambda}{N}\right)^{-k},$$

$$= \frac{N!}{(N-k)!N^k} \left(\frac{\lambda^k}{k!}\right) \left(1 - \frac{\lambda}{N}\right)^N \left(1 - \frac{\lambda}{N}\right)^{-k},$$

$$\lim_{N \to \infty} \mathbb{P}(L_N = k) = \lim_{N \to \infty} \underbrace{\frac{N!}{(N-k)!N^k}}_{=1} \left(\frac{\lambda^k}{k!}\right) \underbrace{\left(1 - \frac{\lambda}{N}\right)^N}_{=e^{-\lambda}} \underbrace{\left(1 - \frac{\lambda}{N}\right)^{-k}}_{=1},$$

$$= \frac{e^{-\lambda} \lambda^k}{k!}.$$

This result, widely known as the law of rare events, shows us that for large N and small p the binomial and Poisson distributions are roughly equivalent. What is the consequence of this fact? We will see, in the coming chapters, models that use the binomial distribution as their starting point and others that, in contrast, take the Poisson distribution as their foundation. This analysis shows us that we can view these choices as being approximately equivalent.

2.5 Final Thoughts

The independent-default binomial and the roughly equivalent independent Poisson models are not quite legitimate approaches; they have numerous shortcomings. These include thin tails, overly stable asymptotic behaviour, and a lack of default dependence. They are, however, *not* completely without use. In particular, they are easy to understand, use, and implement. They can also act as a lower bound and point of comparison for other, more complex models. In short, the independent-default binomial and Poisson models are very useful benchmarks. They can be interpreted as a lower bound for default risk. Bounds are useful. Indeed, if one is serious about managing model risk, then such simple, easy to understand benchmark models can play an essential role. They can act to provide context and diagnostics for more complex and realistic models.

In the coming chapters, we will examine a range of other possible approaches. We will, however, also keep these independent models in the background. Indeed, we will explicitly build upon them to solve some of their more important weaknesses. In the next chapter, we will cover actuarial (or mixture) default-risk models popularized in the finance world by Frey and McNeil (2003). These approaches extend model realism through the randomization of the default probability or arrival intensity. This change introduces a second, so-called mixture distribution, into the credit-risk model thereby inducing default dependence and permitting larger probabilities of extreme events.

References

Billingsley, P. (1995). *Probability and measure* (3rd edn.). Third Avenue, New York, NY: Wiley.
Bluhm, C., Overbeck, L., & Wagner, C. (2003). *An introduction to credit risk modelling* (1st edn.). Boca Raton: Chapman & Hall, CRC Press.
Casella, G., & Berger, R. L. (1990). *Statistical inference*. Belmont, CA: Duxbury Press.
DasGupta, A. (2008). *Asymptotic theory of statistics and probability*. New York: Springer-Verlag.
Durrett, R. (1996). *Probability: Theory and examples* (2nd edn.). Belmont, CA: Duxbury Press.
Fishman, G. S. (1995). *Monte Carlo: Concepts, algorithms, and applications*. 175 Fifth Avenue, New York, NY: Springer-Verlag. *Springer series in operations research*.
Frey, R., & McNeil, A. (2003). Dependent defaults in model of portfolio credit risk. *The Journal of Risk, 6*(1), 59–92.

References

Glasserman, P. (2004a). *Monte Carlo methods in financial engineering* (1st edn.). Berlin: Springer.

Gregory, J. (2010). *Counterparty credit risk: The new challenge for global financial markets*. Chicester, UK: John Wiley & Sons Ltd.

Harris, J. W., & Stocker, H. (1998). *Handbook of mathematics and computational science*. New York, NY: Springer-Verlag.

Ruiz, I.: 2015, *XVA Desks: A New Era for Risk Management*, Palgrave Macmillan, New York, NY.

Schönbucher, P. J. (2000a). Credit risk modelling and credit derivatives. Rheinischen Freidrich-Wilhelms-Universität Bonn—Dissertation.

Schönbucher, P. J. (2000b). Factor models for portfolio credit risk. Department of Statistics, University of Bonn.

Smith, K. J. (1973). Pascal's triangle. *The Two-Year College Mathematics Journal, 4*(1), 1–13.

Stuart, A., & Ord, J. K. (1987). *Kendall's advanced theory of statistics*. New York: Oxford University Press.

Chapter 3
Mixture or Actuarial Models

> *The major difference between a thing that might go wrong and a thing that cannot possibly go wrong is that when a thing that cannot possibly go wrong goes wrong, it usually turns out to be impossible to get at or repair.*
>
> (Douglas Adams)

In the previous chapter, we identified a number of shortcomings in the binomial and Poisson independent-default setting. The most important failure stems from its failure to capture the inherent interaction between the default outcomes of multiple counterparties. Credit counterparties—private firms, countries, or international agencies—operate in a common economic environment. Their success and, more importantly, ability to repay their obligations should thus exhibit some degree of dependence. In some cases, there are important and obvious inter-linkages—one may, for example, operate above or below another in the supply chain for a given industry or one country might act as a key trade partner for another. They would, therefore, likely be creditors or debtors for one another. Should one entity default, it is perhaps more probable that the other might default. The opposite may also be true. Should one credit counterparty default, the other may actually profit from its absence and, as a consequence, be less likely to default.

There are, therefore, many reasons to expect some degree of correlation between the default of one's credit counterparties. Not least of which is the fact that the assumption of default independence is not, from a risk perspective, particularly conservative. Assuming default independence is actually a strong assumption on the potential for diversification in one's credit portfolio. Indeed, the Capital-Asset Pricing Model (CAPM)—see, for example, Sharpe (1963, 1964)—suggests that, for sufficiently large portfolios, all idiosyncratic risk can be diversified away. In the limit, therefore, the risk of independent-default models can be made arbitrarily small by spreading out the specific risk of each obligor among many other equally small positions.

With positive default correlation, a systematic element is introduced, which, following from the logic of the CAPM, cannot be diversified away. One would thus expect more risk given dependence among individual obligor default events. Thus, from conceptual, economic, and risk-management perspectives, it is strongly advisable to add default correlation to one's credit model. The focus of this chapter, the class of so-called mixture models, essentially adjusts the broadly inappropriate independent-default model to incorporate default dependence.

Mixture models are an entire family of methodologies with different possible entry points. Consistent with the previous chapter, we consider *two* possible starting points: the binomial and the Poisson. Each are considered separately in detail with a broad range of possible implementations. In the final section, we consider a commercial model—first introduced by Wilde (1997)—and use the preceding ideas to investigate how one might reasonably proceed to implement a multi-factor mixture model.

3.1 Binomial-Mixture Models

The general idea behind the class of mixture models is to randomize the default probability. Practically, this involves making the default probability dependent on a random variable, Z. That is, the conditional default probability for each issuer is denoted as,

$$\mathbb{P}\left(\mathbb{I}_{\mathcal{D}_n} \mid Z\right) \equiv p(Z) \in (0, 1), \tag{3.1}$$

where Z is an unspecified random variable with cumulative distribution function, $F_Z(z)$ and probability density function, $f_Z(z)$. Since p is now a function of Z, it follows that $p(Z)$ is also a random variable. Let us denote its cumulative distribution and probability density functions as $F_{p(Z)}(z)$ and $f_{p(Z)}(z)$, respectively.

We begin with full generality since the specifics depend on the assumptions for the distribution of Z and the mathematical form of p. In subsequent sections, we will examine a number of possible choices, derive the necessary quantities, discuss the selection of parameters, and consider numerous practical examples.

What are the principle implications for $p(Z)$? The expectation is,

$$\begin{aligned}\mathbb{E}\left(p(Z)\right) &= \int_{-\infty}^{\infty} p(z) f_Z(z) dz, \\ &= \bar{p},\end{aligned} \tag{3.2}$$

assuming, for the moment, that Z has infinite support. We will not know the precise form of the expectation of $p(Z)$ until we specify the distribution of Z along with its parameters. For the moment, we will denote it as $\bar{p}$.

3.1 Binomial-Mixture Models

Conditional Independence

What changes now, in this adjusted setting, is that our default indicators, $\mathbb{I}_{\mathcal{D}_1}, \cdots, \mathbb{I}_{\mathcal{D}_N}$ remain identically distributed, but are no longer independent. The trick is that, given Z, the default events are independent. By construction, we have that,

$$\mathbb{E}\left(\mathbb{I}_{\mathcal{D}_n} \mid Z\right) = 1 \cdot \underbrace{\mathbb{P}\left(\mathbb{I}_{\mathcal{D}_n} = 1\right)}_{p(Z)} + 0 \cdot \underbrace{\mathbb{P}\left(\mathbb{I}_{\mathcal{D}_n} = 0\right)}_{1-p(Z)}, \quad (3.3)$$

$$= p(Z).$$

This is the conditional expectation of default and a key modelling tool; in the coming sections, it will take a variety of forms. It is useful to compare this quantity to the independent default case where $\mathbb{P}\left(\mathbb{I}_{\mathcal{D}_n} = 1\right) = p_n$. In this adjusted setting, the probability of default is no longer a constant. Instead, it depends on the stochastic value of Z. This raises the question: what is the unconditional probability of default? This is given as,

$$\mathbb{E}\left(\mathbb{I}_{\mathcal{D}_n}\right) = \underbrace{\mathbb{E}\left(\overbrace{\mathbb{E}\left(\mathbb{I}_{\mathcal{D}_n} \mid Z\right)}^{\text{Equation 3.3}}\right)}_{\text{Iterated expectations}}, \quad (3.4)$$

$$= \mathbb{E}\left(p(Z)\right),$$

$$= \bar{p}.$$

This is a sensible outcome. If one integrates out the random variable, Z, the most reasonable guess for the unconditional default probability is the average across all outcomes of Z or, as we defined in equation 3.2, $\bar{p}$.

It is also natural to enquire about the unconditional variance of the default indicator. It can be derived as,

$$\text{var}\left(\mathbb{I}_{\mathcal{D}_n}\right) = \mathbb{E}\left(\mathbb{I}_{\mathcal{D}_n}^2 - \underbrace{\mathbb{E}\left(\mathbb{I}_{\mathcal{D}_n}\right)^2}_{\bar{p}}\right), \quad (3.5)$$

$$= \underbrace{\mathbb{E}\left(\mathbb{I}_{\mathcal{D}_n}^2\right)}_{\mathbb{P}(\mathbb{I}_{\mathcal{D}_n}=1)} - \bar{p}^2,$$

$$= \underbrace{\mathbb{P}\left(\mathbb{I}_{\mathcal{D}_n} = 1\right)}_{\bar{p}} - \bar{p}^2,$$

$$= \bar{p}(1 - \bar{p}).$$

Again, this is analogous to the independent-default setting, but with the average default probability replacing the model parameter. The Bernoulli-trial structure is, unconditionally at least, still present.

What about the conditional variance? We would expect that the conditional default variance is no longer constant, but like the expectation, it is random. It is easily determined as,

$$\text{var}\left(\mathbb{I}_{\mathcal{D}_n} \mid Z\right) = \mathbb{E}\left(\mathbb{I}_{\mathcal{D}_n}^2 - \underbrace{\mathbb{E}\left(\mathbb{I}_{\mathcal{D}_n} \mid Z\right)^2}_{p(Z)^2} \mid Z\right), \tag{3.6}$$

$$= \mathbb{E}\left(\mathbb{I}_{\mathcal{D}_n}^2 \mid Z\right) - p(Z)^2,$$

$$= 1^2 \cdot \underbrace{\mathbb{P}\left(\mathbb{I}_{\mathcal{D}_n} = 1 \mid Z\right)}_{p(Z)} + 0^2 \cdot \underbrace{\mathbb{P}\left(\mathbb{I}_{\mathcal{D}_n} = 0 \mid Z\right)}_{1-p(Z)} - p(Z)^2,$$

$$= p(Z)(1 - p(Z)).$$

The conditional variance, therefore, can take a range of possible values each depending on the outcome of Z.

Default-Correlation Coefficient

Of particular interest is the interaction between two default events. Recall, from Chap. 2, that the linear correlation between default events in the independent case was zero. In the mixture-model framework, this is no longer true. Let's directly compute the correlation between the default indicators of arbitrarily selected credit counterparties n and m,

$$\text{cov}\left(\mathbb{I}_{\mathcal{D}_n}, \mathbb{I}_{\mathcal{D}_m}\right) = \mathbb{E}\left(\left(\mathbb{I}_{\mathcal{D}_n} - \underbrace{\mathbb{E}\left(\mathbb{I}_{\mathcal{D}_n}\right)}_{\bar{p}}\right)\left(\mathbb{I}_{\mathcal{D}_m} - \underbrace{\mathbb{E}\left(\mathbb{I}_{\mathcal{D}_m}\right)}_{\bar{p}}\right)\right), \tag{3.7}$$

$$= \mathbb{E}\left(\mathbb{I}_{\mathcal{D}_n}\mathbb{I}_{\mathcal{D}_m} - \bar{p}\mathbb{I}_{\mathcal{D}_n} - \bar{p}\mathbb{I}_{\mathcal{D}_m} + \bar{p}^2\right),$$

$$= \mathbb{E}\left(\mathbb{I}_{\mathcal{D}_n}\mathbb{I}_{\mathcal{D}_m}\right) - \bar{p}\underbrace{\mathbb{E}\left(\mathbb{I}_{\mathcal{D}_n}\right)}_{\bar{p}} - \bar{p}\underbrace{\mathbb{E}\left(\mathbb{I}_{\mathcal{D}_m}\right)}_{\bar{p}} + \bar{p}^2,$$

$$= \mathbb{E}\left(\mathbb{I}_{\mathcal{D}_n}\mathbb{I}_{\mathcal{D}_m}\right) - \bar{p}^2,$$

$$= \underbrace{\mathbb{E}\left(\mathbb{E}\left(\mathbb{I}_{\mathcal{D}_n}\mathbb{I}_{\mathcal{D}_m} \mid Z\right)\right)}_{\text{Iterated expectations}} - \bar{p}^2,$$

3.1 Binomial-Mixture Models

$$
\begin{aligned}
&= \mathbb{E}\left(\underbrace{\mathbb{E}\left(\mathbb{I}_{\mathcal{D}_n}\,|\, Z\right) \mathbb{E}\left(\mathbb{I}_{\mathcal{D}_m}\,|\, Z\right)}_{\text{By conditional independence}} \right) - \bar{p}^2, \\
&= \mathbb{E}\left(p(Z)^2\right) - \bar{p}^2, \\
&= \mathbb{E}\left(p(Z)^2 - \mathbb{E}\left(p(Z)\right)^2\right), \\
&= \text{var}\left(p(Z)\right).
\end{aligned}
$$

The last line follows from the definition of variance. In words, therefore, the covariance between any two credit counterparties reduces to the variance of the random default probability. Since each counterparty also depends in the same manner on Z, the covariance and variance are equivalent. The specific form of this variance will not, of course, be known until we specify the distributional choice for Z and the structure of p.

The Distribution of $\mathbb{D}_N$

We have all of the basic ingredients required to construct the loss distribution and investigate its convergence properties. As before, the loss associated with a portfolio of N counterparties is described as $L_N = c \sum_{n=1}^{N} \mathbb{I}_{\mathcal{D}_n} = c\mathbb{D}_N$, where $\mathbb{D}_N$ denotes the total number of defaults in $[0, T]$. This is predicated on the assumption that all issuers share the same exposure—that is, $c_n = c$ for $n = 1, \ldots, N$. Again, as we found in Chap. 2, we have that $f_{L_N}(c \cdot k) = f_{\mathbb{D}_N}(k)$; or, rather that to describe the loss distribution, it is sufficient to examine the distribution of the number of defaults.

Description of $\mathbb{P}(\mathbb{D}_N = k)$ is, however, not as convenient to compute since the default indicators are no longer independent. To derive sensible analytic formulae, we will need to exploit the property of conditional independence. We know, by construction, that *conditional* on Z, the total number of defaults follows a binomial distribution. More specifically,

$$\mathbb{P}\left(\mathbb{D}_N = k\,|\, Z\right) = \binom{N}{k} p(Z)^k \left(1 - p(Z)\right)^{N-k}. \tag{3.8}$$

Although perfectly true, equation 3.8 is not immediately useful. We may, however, eliminate the conditioning variable, Z, by simply integrating it out. By iterated expectations,

$$\mathbb{P}(\mathbb{D}_N = k) = \mathbb{E}\left(\underbrace{\mathbb{P}(\mathbb{D}_N = k | Z)}_{\text{Equation 3.8}}\right), \quad (3.9)$$

$$= \mathbb{E}\left(\binom{N}{k} p(Z)^k (1 - p(Z))^{N-k}\right),$$

$$= \binom{N}{k} \int_{-\infty}^{\infty} p(z)^k (1 - p(z))^{N-k} f_Z(z) dz,$$

again assuming infinite support of Z. This integral may, or may not, be analytically tractable. It will depend on the choice of the form of $p(Z)$ and the density, $f_Z(z)$. In the coming sections, we will examine choices that permit the analytic solution of equation 3.8 and others where it must be solved numerically. Since it is only a one-dimensional integral, numerical solution is, in principle, neither particularly difficult nor computationally expensive.

The lack of independence of the default variables also has implications for the mean and variance of the total number of defaults. The expected number of defaults is,

$$\mathbb{E}(\mathbb{D}_N) = \mathbb{E}\left(\sum_{n=1}^{N} \mathbb{I}_{\mathcal{D}_n}\right), \quad (3.10)$$

$$= \sum_{n=1}^{N} \underbrace{\mathbb{E}\left(\mathbb{I}_{\mathcal{D}_n}\right)}_{\bar{p}},$$

$$= N\bar{p}.$$

The difference with the independent-default setting is not dramatic. The variance, however, is another story.

$$\text{var}(\mathbb{D}_N) = \text{var}\left(\sum_{n=1}^{N} \mathbb{I}_{\mathcal{D}_n}\right), \quad (3.11)$$

$$= \sum_{n=1}^{N} \sum_{m=1}^{N} \text{cov}\left(\mathbb{I}_{\mathcal{D}_n}, \mathbb{I}_{\mathcal{D}_m}\right),$$

$$= \sum_{n=1}^{N} \underbrace{\text{var}\left(\mathbb{I}_{\mathcal{D}_n}\right)}_{\bar{p}(1-\bar{p})} + \underbrace{\sum_{n=1}^{N} \sum_{\substack{m=1 \\ n \neq m}}^{N} \text{cov}\left(\mathbb{I}_{\mathcal{D}_n}, \mathbb{I}_{\mathcal{D}_m}\right)}_{\text{var}(p(Z))},$$

$$= N\bar{p}(1-\bar{p}) + N(N-1)\text{var}(p(Z)).$$

This is a rather more complex expression.

3.1 Binomial-Mixture Models 91

Convergence Properties

What does it imply for convergence? We make use of a popular convergence tool also used in Chap. 2: Chebyshev's inequality. Applying it to $\frac{\mathbb{D}_N}{N}$, we have

$$\mathbb{P}\left(\left|\frac{\mathbb{D}_N}{N} - \mathbb{E}\left(\frac{\mathbb{D}_N}{N}\right)\right| \geq \epsilon\right) \leq \frac{\text{var}\left(\frac{\mathbb{D}_N}{N}\right)}{\epsilon^2}, \quad (3.12)$$

$$\mathbb{P}\left(\left|\frac{\mathbb{D}_N}{N} - \frac{1}{N}\underbrace{\mathbb{E}\left(\mathbb{D}_N\right)}_{N\bar{p}}\right| \geq \epsilon\right) \leq \frac{\overbrace{N\bar{p}(1-\bar{p}) + N(N-1)\text{var}(p(Z))}^{\text{Equation 3.11}}}{N^2\epsilon^2},$$

$$\mathbb{P}\left(\left|\frac{\mathbb{D}_N}{N} - \bar{p}\right| \geq \epsilon\right) \leq \frac{\bar{p}(1-\bar{p})}{N\epsilon^2} + \frac{(N-1)\text{var}(p(Z))}{N\epsilon^2},$$

from which it is clear that as $N \to \infty$, that

$$\mathbb{P}\left(\left|\frac{\mathbb{D}_N}{N} - \bar{p}\right| \geq \epsilon\right) \to \frac{\text{var}(p(Z))}{\epsilon^2}, \quad (3.13)$$

for all $\epsilon \in \mathbb{R}+$. The proportion of defaults in the portfolio no longer converges to a constant value as the number of credit counterparties becomes very large. This is an important difference with the independent-default models. It implies that the behaviour of $p(Z)$ will have an important influence on the default loss distribution.

If we condition, as before, on the state variable, Z, then all of the convergence results from the independent-default model apply. A second application of Chebyshev's inequality to $\frac{\mathbb{D}_N}{N}\big|Z$ yields

$$\mathbb{P}\left(\left|\frac{\mathbb{D}_N}{N} - \mathbb{E}\left(\frac{\mathbb{D}_N}{N}\bigg|Z\right)\right| \geq \epsilon\bigg|Z\right) \leq \frac{\text{var}\left(\frac{\mathbb{D}_N}{N}\bigg|Z\right)}{\epsilon^2}, \quad (3.14)$$

$$\mathbb{P}\left(\left|\frac{\mathbb{D}_N}{N} - \frac{1}{N}\underbrace{\mathbb{E}\left(\mathbb{D}_N|Z\right)}_{Np(Z)}\right| \geq \epsilon\bigg|Z\right) \leq \frac{Np(Z)(1-p(Z))}{N^2\epsilon^2},$$

$$\mathbb{P}\left(\left|\frac{\mathbb{D}_N}{N} - p(Z)\right| \geq \epsilon\bigg|Z\right) \leq \frac{p(Z)(1-p(Z))}{N\epsilon^2}.$$

As $N \to \infty$, therefore, the conditional proportion of defaults converges to $p(Z)$.

We can also, as in Chap. 2, use the fourth conditional moment of the binomial proportion, Markov's inequality, and the Borel-Cantelli theorem to show that given Z, the quantity $\frac{\mathbb{D}_N}{N}$ also converges almost surely to $p(Z)$. Almost-sure convergence implies convergence in distribution for large N. Using the law of iterated expectations we may conclude that,

$$\mathbb{P}\left(\frac{\mathbb{D}_N}{N} \leq x \bigg| Z\right) \to \mathbb{I}_{p(Z) \leq x}, \qquad (3.15)$$

$$\mathbb{E}\left(\mathbb{P}\left(\frac{\mathbb{D}_N}{N} \leq x \mid Z\right)\right) \to \mathbb{E}\left(\mathbb{I}_{p(Z) \leq x}\right),$$

$$\mathbb{P}\left(\frac{\mathbb{D}_N}{N} \leq x\right) \to \mathbb{P}\left(p(Z) \leq x\right),$$

$$\mathbb{P}\left(\frac{\mathbb{D}_N}{N} \leq x\right) \to F_{p(Z)}(x),$$

as N tends to infinity. In short, for large portfolios, the distribution of the proportion of defaults converges to the distribution of $p(Z)$. This is a further demonstration that the specification of the form and structure of $p(Z)$ is the key choice for the implementation of these models. A judicious choice of $p(Z)$ will permit a fat-tailed, strongly dependent loss distribution. In the next section, we examine, in detail, a few possible choices along with a number of practical examples.

3.1.1 The Beta-Binomial Mixture Model

A judicious choice of the conditioning, or mixing, variable is essential to creating a useful model. While the randomized default probability, $p(Z)$, is a random variable, it is also intended to represent a probability. As a consequence, it may only take values in the unit interval, $[0, 1]$. This restricts our possible choices. We may either use a random variable confined to $[0, 1]$ or employ a random variable, Z, with infinite support, but map it through the choice of function, p, into $[0, 1]$. In this section, following from Frey and McNeil (2003), we will consider both possible choices through *three* concrete implementations: the beta, logit-normal, and probit-normal binomial mixture models.

In the first implementation, we assume that the state variable, Z, follows a beta distribution with parameters α and β. More formally,

$$Z \sim \text{Beta}(\alpha, \beta). \qquad (3.16)$$

The density of the beta distribution is given as,

$$f_Z(z) = \frac{1}{B(\alpha, \beta)} z^{\alpha-1}(1-z)^{\beta-1}, \qquad (3.17)$$

3.1 Binomial-Mixture Models

for $z \in (0, 1)$ and $\alpha, \beta \in \mathbb{R}_+$. $B(\alpha, \beta)$ denotes the so-called beta function, which is described as,

$$B(\alpha, \beta) = \int_0^1 t^{\alpha-1}(1-t)^{\beta-1} dt, \tag{3.18}$$

which bears a strong resemblance to the form of the beta density. The beta function need not be computed directly since there exists a convenient and direct link to the gamma function.[1] In particular,

$$B(\alpha, \beta) = \frac{\Gamma(\alpha)\Gamma(\beta)}{\Gamma(\alpha + \beta)}, \tag{3.19}$$

where we recall that the gamma function, which is a continuous analogue of the discrete factorial function, is given as,

$$\Gamma(t) = \int_0^\infty x^{t-1} e^{-x} dx. \tag{3.20}$$

The beta function, given its finite support, is often used in statistics and engineering to model proportions. Interestingly, if one sets $\alpha = \beta = 1$, then the beta distribution collapses to a continuous uniform distribution. Moreover, it can easily be constructed from the generation of gamma random variates.[2] Finally, the mean and variance of a beta-distributed random variable, Z, with parameters α and β are,

$$\mathbb{E}(Z) = \frac{\alpha}{\alpha + \beta}, \tag{3.21}$$

and

$$\text{var}(Z) = \frac{\alpha\beta}{(\alpha + \beta)^2 (\alpha + \beta + 1)}. \tag{3.22}$$

The choice of p is the identity function. That is,

$$p(Z) \equiv Z, \tag{3.23}$$

for all $Z \in (0, 1)$. We now have all the information required to specify the distribution of the total number of defaults. If we return to equation 3.9, we may

---

[1] See Abramovitz and Stegun (1965, Chapter 6) for a detailed description of the gamma function.
[2] See Casella and Berger (1990, Chapter 3) and Fishman (1995, Chapter 3) for more information on the beta distribution.

simplify as follows,

$$\mathbb{P}(\mathbb{D}_N = k) = \binom{N}{k} \int_0^1 p(z)^k (1-p(z))^{N-k} f_Z(z)dz, \qquad (3.24)$$

$$= \binom{N}{k} \int_0^1 z^k (1-z)^{N-k} \underbrace{\left(\frac{1}{B(\alpha,\beta)} z^{\alpha-1}(1-z)^{\beta-1}\right)}_{\text{Equation 3.17}} dz,$$

$$= \frac{1}{B(\alpha,\beta)} \binom{N}{k} \underbrace{\int_0^1 z^{(\alpha+k)-1}(1-z)^{(\beta+N-k)-1} dz}_{B(\alpha+k,\beta+N-k)},$$

$$= \binom{N}{k} \frac{B(\alpha+k,\beta+N-k)}{B(\alpha,\beta)}.$$

This yields a closed-form expression for the probability mass and cumulative distribution functions associated with the default loss distribution. The cumulative distribution function, following from equation 3.9, is merely

$$\mathbb{P}(\mathbb{D}_N \leq m) = \sum_{k=0}^{m} \mathbb{P}(\mathbb{D}_N = k). \qquad (3.25)$$

Algorithm 3.1 highlights the Python code required to compute the probability mass and cumulative distribution functions for the beta-binomial mixture model. a and b denote the beta-distribution coefficients, α and β, respectively. The function, `computeBeta`, is merely a practical implementation of the trick described in equation 3.19. Otherwise, the structure is highly analogous to independent-default binomial implementation.

Algorithm 3.1 Analytic beta-binomial

```
def betaBinomialAnalytic (N,c,a,b,alpha):
    pmfBeta = np.zeros(N+1)
    den = util.computeBeta(a,b)
    for k in range(0,N+1):
        pmfBeta[k] = util.getBC(N,k)*util.computeBeta(a+k,b+N-k)/den
    cdfBeta = np.cumsum(pmfBeta)
    varAnalytic = c*np.interp(alpha,cdfBeta,np.linspace(0,N,N+1))
    esAnalytic = util.analyticExpectedShortfall(N,alpha,pmfBeta,c)
    return pmfBeta,cdfBeta,varAnalytic,esAnalytic
```

3.1 Binomial-Mixture Models

3.1.1.1 Beta-Parameter Calibration

The next question relates to the choice of parameters, α and β. We wish to calibrate them to specific choices of average default probability and default correlation. The default correlation between issuers n and m, as derived in equations 3.5 and 3.7, has the following parsimonious form

$$\rho_{\mathcal{D}} = \rho\left(\mathbb{I}_{\mathcal{D}_n}, \mathbb{I}_{\mathcal{D}_m}\right) = \frac{\overbrace{\mathrm{cov}\left(\mathbb{I}_{\mathcal{D}_n}, \mathbb{I}_{\mathcal{D}_m}\right)}^{\text{Equation 3.7}}}{\underbrace{\sqrt{\mathrm{var}\left(\mathbb{I}_{\mathcal{D}_n}\right)}}_{\text{Equation 3.5}} \underbrace{\sqrt{\mathrm{var}\left(\mathbb{I}_{\mathcal{D}_m}\right)}}_{\text{Equation 3.5}}}, \tag{3.26}$$

$$= \frac{\mathrm{var}(p(Z))}{\sqrt{\mathbb{E}(p(Z))(1 - \mathbb{E}(p(Z)))}\sqrt{\mathbb{E}(p(Z))(1 - \mathbb{E}(p(Z)))}},$$

$$= \frac{\mathrm{var}(Z)}{\sqrt{\mathbb{E}(Z)(1 - \mathbb{E}(Z))}\sqrt{\mathbb{E}(Z)(1 - \mathbb{E}(Z))}},$$

$$= \frac{\mathrm{var}(Z)}{\mathbb{E}(Z)(1 - \mathbb{E}(Z))},$$

$$= \frac{\overbrace{\left(\dfrac{\alpha\beta}{(\alpha+\beta)^2(\alpha+\beta+1)}\right)}^{\text{Equation 3.22}}}{\underbrace{\left(\dfrac{\alpha}{\alpha+\beta}\right)}_{\mathbb{E}(Z)}\left(1 - \underbrace{\dfrac{\alpha}{\alpha+\beta}}_{\mathbb{E}(Z)}\right)},$$

$$= \frac{\left(\dfrac{\alpha\beta}{(\alpha+\beta)^2(\alpha+\beta+1)}\right)}{\left(\dfrac{\alpha\beta}{(\alpha+\beta)^2}\right)},$$

$$= \left(\frac{\cancel{\alpha\beta}}{\cancel{(\alpha+\beta)^2}(\alpha+\beta+1)}\right)\left(\frac{\cancel{(\alpha+\beta)^2}}{\cancel{\alpha\beta}}\right),$$

$$= \frac{1}{(\alpha+\beta+1)}.$$

This is one linear equation. If we add a second linear equation,

$$\mathbb{E}(Z) = \bar{p} = \frac{\alpha}{\alpha+\beta}, \tag{3.27}$$

then we may—given pre-specified values for the default probability and correlation, $\rho_\mathcal{D}$ and $\bar{p}$—solve for the appropriate values of α and β. This is a simple algebraic exercise of solving two equations for two unknowns. Putting the two equations in standard form, we have

$$\rho_\mathcal{D}\alpha + \rho_\mathcal{D}\beta = 1 - \rho_\mathcal{D}, \tag{3.28}$$
$$(\bar{p} - 1)\alpha + \bar{p}\beta = 0.$$

Placing this simple system into matrix format and solving it, we arrive at

$$\begin{bmatrix} \rho_\mathcal{D} & \rho_\mathcal{D} \\ \bar{p} - 1 & \bar{p} \end{bmatrix} \begin{bmatrix} \alpha \\ \beta \end{bmatrix} = \begin{bmatrix} 1 - \rho_\mathcal{D} \\ 0 \end{bmatrix}, \tag{3.29}$$

$$\begin{bmatrix} \alpha \\ \beta \end{bmatrix} = \left(\begin{bmatrix} \rho_\mathcal{D} & \rho_\mathcal{D} \\ \bar{p} - 1 & \bar{p} \end{bmatrix} \right)^{-1} \begin{bmatrix} 1 - \rho_\mathcal{D} \\ 0 \end{bmatrix},$$

$$= \frac{1}{\rho_\mathcal{D}} \begin{bmatrix} \bar{p} & -\rho_\mathcal{D} \\ 1 - \bar{p} & \rho_\mathcal{D} \end{bmatrix} \begin{bmatrix} 1 - \rho_\mathcal{D} \\ 0 \end{bmatrix},$$

$$= \begin{bmatrix} \dfrac{\bar{p}(1 - \rho_\mathcal{D})}{\rho_\mathcal{D}} \\ \dfrac{(1 - \bar{p})(1 - \rho_\mathcal{D})}{\rho_\mathcal{D}} \end{bmatrix}.$$

We thus have in equation 3.29, for a given set of target properties of the loss distribution, the appropriate associated parameter choices.

3.1.1.2 Back to Our Example

If we return to our simple 100 issuer portfolio with total exposure of $1,000, we can compare the results of the beta-binomial mixture model to the independent-default approach. This analysis will provide a concrete demonstration of the role of default dependence.

To do this effectively, we want to minimize the number of moving parts. Thus, each counterparty has an equal exposure of $10. The common, and fixed, default probability from the independent-default setting remains 1% and, by construction, there is no dependence between default outcomes. The average default probability, $\bar{p}$, in the beta-binomial mixture model is 1%. Moreover, the assumed default correlation coefficient (i.e., $\rho_\mathcal{D}$) is set to 0.05. This implies, using the results from equation 3.29, that $\alpha \approx 0.2$ and $\beta \approx 18.8$.[3]

---

[3]The associated standard deviation of Z is about 0.02, which implies that it can deviate fairly importantly from the average of 1%.

3.1 Binomial-Mixture Models

Table 3.1 *Analytic beta-binomial results*: This table compares the analytic independent-default and beta-binomial model results. In both cases, the models are calibrated for a 1% average default probability. The beta-binomial model also targets a default correlation of 0.05.

Quantile	Binomial			Beta-binomial		
	$\mathbb{D}_N$	VaR_α	$\mathcal{E}_\alpha$	$\mathbb{D}_N$	VaR_α	$\mathcal{E}_\alpha$
95.00th	2.5	$24.8	$25.1	5.1	$51.4	$54.1
97.00th	2.8	$28.1	$28.4	7.1	$70.5	$72.4
99.00th	3.6	$35.6	$35.8	11.5	$114.8	$156.2
99.50th	3.9	$39.0	$39.3	14.3	$143.2	$184.6
99.90th	4.8	$48.4	$48.6	20.9	$209.3	$249.5
99.97th	5.5	$55.1	$55.4	25.8	$257.7	$296.3
99.99th	5.9	$59.4	$59.6	30.0	$300.2	$337.5
Other statistics						
Expected loss	1.0	$10.0		1.0	$10.0	
Loss volatility	1.0	$9.9		2.2	$22.2	
Default correlation	0.00			0.05		

The results, for the usual range of quantiles, are found in Table 3.1. The differences are striking. The beta-binomial model, with these parameter choices, has the same number of expected defaults as the independent-default equivalent. The standard deviation of total defaults, however, is only a single default in the independent setting, but increases to about 2.5 in the beta-binomial approach. The same trend is evident in the quantile estimates. At the 99.9th quantile, the beta-binomial VaR estimate is roughly five times larger than in the independent model. Instead of about six defaults, it estimates roughly 30 default events increasing the VaR estimate from approximately $60 to about $300. This impact is a direct consequence of the imposition of default dependence on the counterparties.

Figure 3.1 provides a visual perspective. Both loss distributions, zoomed in on the extreme outcomes, are illustrated. The independent-default distribution has more probability mass at low numbers of defaults—with significant probability of 0 to three defaults. There is, however, very little probability associated with larger numbers (i.e., 8 to 10) of defaults.

The beta-binomial mixture, conversely, has a larger probability of a single default—it is roughly 70%—with lower probability of 1 to five defaults. The probability of multiple defaults, again in the 8 to 10 category, is significantly higher. Indeed, the requisite magnification of Fig. 3.1, is 20 times larger for the beta-binomial model compared to the base independent-default approach. In summary, this parameterization of the beta distribution is fat-tailed and this property is inherited by the default-loss distribution. The loss outcomes are classically leptokurtotic with a thin waist and fat tails. The ultimate result is a more conservative, and more realistic, assessment of the risk associated with this simple portfolio.

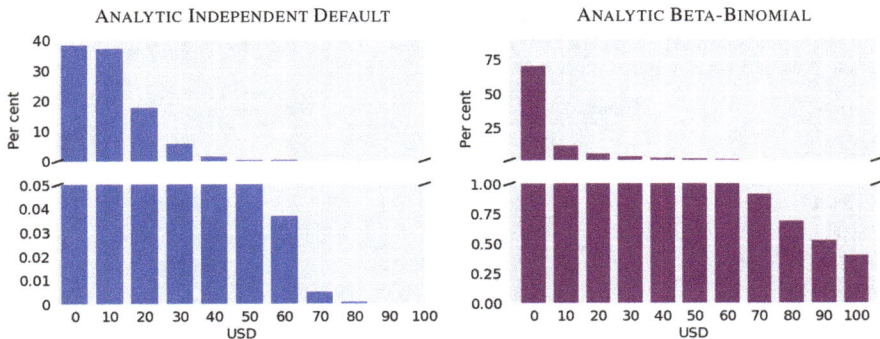

Fig. 3.1 *The role of default dependence*: This figure compares the loss distributions from the independent-default and beta-binomial mixture models. Both models are nested. The beta-binomial has, however, randomized the default probability and induced a default dependence on the order of 0.05. The differences are striking.

3.1.1.3 Non-homogeneous Exposures

One of the principal shortcomings of the analytic solution to the beta-binomial model, described in the previous sections, is that it necessitates common exposures to each of the credit counterparties. This is clearly unrealistic. If we are willing to use Monte-Carlo techniques, however, this problematic assumption is easily relaxed.

Stochastic simulation of default events in the beta-binomial mixture model requires generation of beta-distributed random variates. Not all scientific software has functionality for creating beta random variables. A simple, if perhaps not optimally fast, method involves the use of gamma-distributed variables.[4] Indeed, it turns out that if $X \sim \Gamma(\alpha, 1)$ and $Y \sim \Gamma(\beta, 1)$ are independent random variables, then

$$\frac{X}{X+Y} \sim \text{Beta}(\alpha, \beta). \tag{3.30}$$

Figure 3.2 illustrates 10,000 draws from a Beta(0.2, 18.9) distribution using the insight from equation 3.30. Visually, for these parameter choices, the distribution is highly leptokurtotic. Roughly 50% of the outcomes have a vanishingly small default probability, whereas the largest default probability outcome—across the 10,000 draws—is almost 0.30. The average default probability is very close to the 1% theoretical value and the variability of the default probabilities is on the order of 2%.

The generation of dependent beta-binomial mixture outcomes is very similar to the approach used for the numerical implementation of the independent-default

---

[4] See Fishman (1995, Chapter 3) for more detail and a number of more efficient algorithms.

3.1 Binomial-Mixture Models

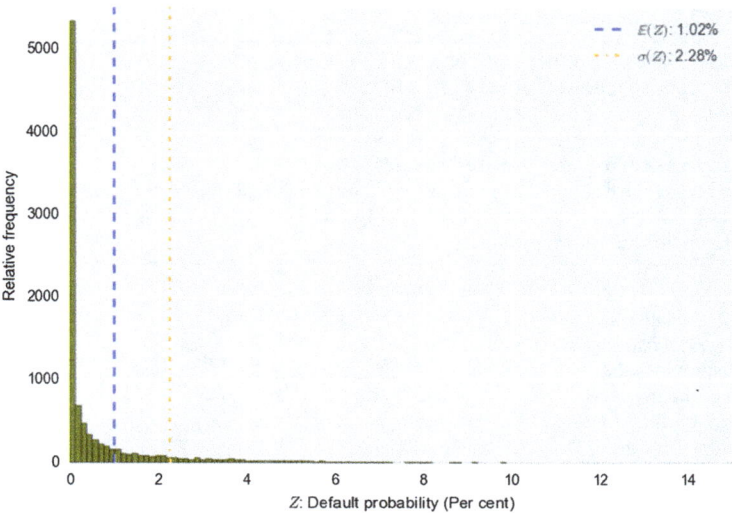

Fig. 3.2 *Simulating beta random variables*: This figure illustrates a histogram of 10,000 beta-distributed random variates with parameters $\alpha = 0.2$ and $\beta = 18.9$. The outcomes were generated from the relationship found in equation 3.30.

model. As before, one generates $N \times M$ standard uniform random variates on the unit interval of the form, $U_n^{(m)} \sim \mathcal{U}[0, 1]$ for $n = 1, \ldots, N$ and $m = 1, \ldots, M$. One additionally generates $Z^{(m)} \sim \text{Beta}(\alpha, \beta)$ for $m = 1, \ldots, M$. If the random variable $U_n^{(m)}$ is found in the interval $[0, p(Z^{(m)})]$, a default has occurred. If, conversely, $U_n^{(m)}$ falls in, $(p(Z^{(m)}), 1]$, then the counterparty survives. Mathematically, we merely update our default indicator as

$$\mathbb{I}_{\mathcal{D}_n}^{(m)} = \begin{cases} 1 : U_n^{(m)} \in [0, p(Z^{(m)})] \\ 0 : U_n^{(m)} \in (p(Z^{(m)}), 1] \end{cases}, \quad (3.31)$$

for $n = 1, \ldots, N$ and $m = 1, \ldots, M$. The remaining loss distribution, expected loss, unexpected loss, VaR, and expected shortfall computations have precisely the same form as introduced in Chaps. 1 and 2. In summary, the basic structure from the independent-default model is preserved, but a randomized default probability has been added.

Algorithm 3.2 summarizes the Python sub-routine used to simulate the beta-binomial model with non-constant counterparty exposures. It is structurally very similar to numerical independent-default binomial model code. It takes as input the number of counterparties, N, the desired number of simulations, M, and the exposures, c. Unlike the independent-default setting, it does not take default probabilities as an input argument. Instead, it takes the two beta-distribution model parameters—a and b—and randomly computes them. Each iteration involves the generation of

Algorithm 3.2 Simulation beta-binomial

```
def betaBinomialSimulation (N,M,c,a,b,alpha):
    Z = util.generateGamma(a,b,M)
    pZ = np.transpose(np.tile(Z,(N,1)))
    U = np.random.uniform(0,1,[M,N])
    lossIndicator = 1*np.less(U,pZ)
    lossDistribution = np.sort(np.dot(lossIndicator,c),axis=None)
    el,ul,var,es=util.computeRiskMeasures(M,lossDistribution,alpha)
    return el,ul,var,es
```

Table 3.2 *Numerical beta-binomial mixture results*: This table provides the quantiles of the loss distribution for the Monte Carlo estimates of the independent-default and beta-binomial mixture models. For the beta-binomial mixture, the default probabilities are generated randomly.

Quantile	Binomial		Beta-binomial	
	$\widehat{\text{VaR}}_\alpha$	$\widehat{\mathcal{E}}_\alpha$	$\widehat{\text{VaR}}_\alpha$	$\widehat{\mathcal{E}}_\alpha$
95.00th	$35.3	$45.5	$59.3	$101.4
97.00th	$40.4	$50.7	$80.1	$123.1
99.00th	$52.5	$61.1	$127.0	$170.9
99.50th	$58.8	$66.8	$157.6	$201.3
99.90th	$72.3	$80.3	$228.6	$271.1
99.97th	$82.2	$89.6	$279.5	$321.3
99.99th	$90.3	$97.4	$326.4	$366.6
Other statistics				
Expected loss	$9.2		$10.0	
Loss volatility	$12.8		$25.8	
Total iterations	10,000,000			
Computational time	16.8		21.4	

a random variable, $Z \sim \text{Beta}(\alpha, \beta)$, that is shared by each credit counterparty.[5] It then generates the requisite uniform variates and then computes the associated set of risk-measure estimates for the specified quantile values found in `alpha`.

Revisiting our example and, employing the heterogeneous exposures, we continue to assume an average default probability of 1%, and a default indicator-variable correlation coefficient of 0.05. The results, along with the comparable binomial independent-default model results from Chap. 2, are summarized in Table 3.2. In the numerical setting, it is also straightforward to compute the expected shortfall measure.

The results in Table 3.2 are broadly similar to the analytic results in Table 3.1. The only difference is the heterogeneity of counterparty exposures. The main consequence, however, is that the overall risk is heightened. The impact, interestingly, is more significant for the independent-default model as compared to the beta-binomial mixture. The expected and unexpected loss statistics are virtually identical to the analytic results—this underscores the notion that concentration of default exposures has a larger impact on the tail of the loss distribution.

---

[5]The `generateGamma` function merely exploits the relationship from equation 3.30.

A slight adjustment to the independent-default model—the randomization of the default probabilities though a mixture with the beta distribution—enhances its realism and leads to significantly higher risk estimates. It permits convenient analytic formulae, is readily calibrated to key model assumptions, and is easily implemented in a stochastic-simulation setting. Its parsimonious structure and general realism makes this a sensible benchmark model to include in a suite of comparator approaches for one's risk-metric calculations.

3.1.2 The Logit- and Probit-Normal Mixture Models

The beta-binomial mixture model created a leptokurtotic loss distribution through the choice of underlying default-probability distribution. This is not the only approach. The underlying distribution need not be fat-tailed, but leptokurtosis can be induced by the appropriate choice of the function, $p(\cdot)$. We will examine two additional, but related, choices. The first option has the following form,

$$p_1(Z) = \frac{1}{1 + e^{-(\mu_1 + \sigma_1 Z)}}, \quad (3.32)$$

where $Z \sim \mathcal{N}(0, 1)$ and $\sigma_1 > 0$. $p_1(\cdot)$, with its S-shaped form, is termed the *logistic* function. While its domain encompasses $\mathbb{R}$, the logistic function only takes values in the unit interval, $[0, 1]$. The ratio $-\frac{\mu_1}{\sigma_1}$ denotes the mid-point of the S on the horizontal axis, while σ_1 regulates the steepness of the overall S. The larger the value of σ, the tighter—or more squeezed together—the S shape.[6]

An alternative choice is,

$$p_2(Z) = \int_{-\infty}^{\mu_2 + \sigma_2 Z} \frac{1}{\sqrt{2\pi}} e^{-\frac{x^2}{2}} dx \quad (3.33)$$
$$= \Phi(\mu_2 + \sigma_2 Z),$$

where, again, $Z \sim \mathcal{N}(0, 1)$ and $\sigma_2 > 0$. The mapping, in this case, is nothing other than the cumulative distribution function for the standard normal random variable. This is termed the probit-normal mapping. A bit of reflection reveals that any cumulative distribution function is a possible candidate for $p(\cdot)$, because it, by definition, takes only values in the unit interval. Indeed, this mapping also has an S-shaped form with the parameter σ_2 playing a very similar role to the σ_1-parameter in the logistic setting.

---

[6]Mathematically, it is possible for $\sigma_1 < 0$, but it then takes an inverted S-form—for this reason, we restrict our attention to $\mathbb{R}_+$.

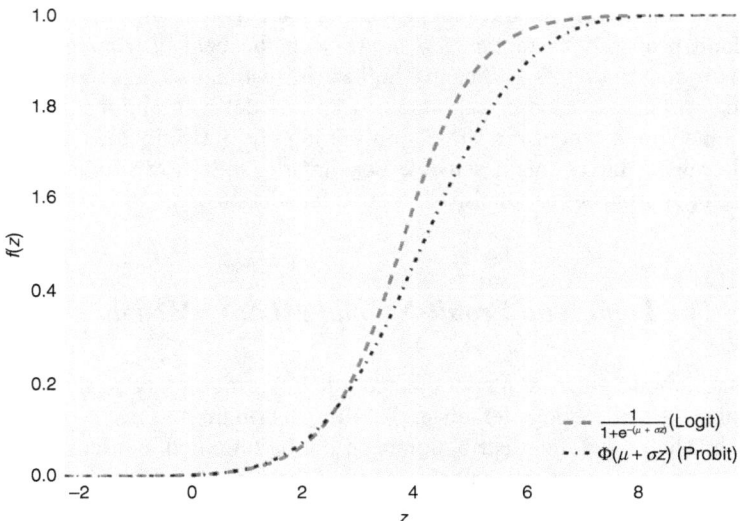

Fig. 3.3 *The logistic and probit functions*: This figure provides a graphical representation of the S-shaped, logistic and probit functions described in equations 3.32 and 3.33. Both curves are calibrated to have similar behaviour.

We will use $p(\cdot)$ to refer to a generic mapping of some latent variable, Z. To avoid any confusion, $p_1(Z)$ and $p_2(Z)$ will denote the logistic and probit mappings, respectively. Figure 3.3 provides a graphical illustration of the two p mappings—both calibrated to have similar characteristics, a notion which will be made more precise in a moment—over a reasonable range of values for the standard-normal distribution. Despite the common calibration, these alternative approaches have slightly different implications for a given value of the conditioning variable, Z.

The mappings in equations 3.32 and 3.33 are a convenient choice for the stochastic description of default probabilities. Our principal task, as with the previously examined beta-binomial mixture model, is to now identify the precise mathematical form of the probability mass and cumulative distribution functions associated with a generic choice of p. The development is virtually identical for both choices of $p(\cdot)$. We begin by recalling that Z is a standard normal variate and then work from the first-principle definition of the distribution of $p(Z)$,

$$F_{p(Z)}(z) = \mathbb{P}\left(p(Z) \leq z\right), \tag{3.34}$$
$$= \mathbb{P}\left(p^{-1}(p(Z)) \leq p^{-1}(z)\right),$$
$$= \mathbb{P}\left(Z \leq p^{-1}(z)\right),$$

… 3.1 Binomial-Mixture Models

$$= \int_{-\infty}^{p^{-1}(z)} f_Z(x)dx,$$

$$= \int_{-\infty}^{p^{-1}(z)} \frac{1}{\sqrt{2\pi}} e^{-\frac{x^2}{2}} dx,$$

$$= \Phi\left(p^{-1}(z)\right).$$

This convenient form is presumably the reason for applying this function to a standard normal variate, Z.

3.1.2.1 Deriving the Mixture Distributions

To go any further, we require the form of $p^{-1}(z)$ for each choice of mapping. Let's examine each in turn. We begin using equation 3.32 to find $p_1^{-1}(z)$ for the logistic approach. We need only solve the following expression for y,

$$z = \frac{1}{1+e^{-(\mu_1+\sigma_1 y)}}, \tag{3.35}$$

$$e^{-(\mu_1+\sigma_1 y)} = \frac{1}{z} - 1,$$

$$-(\mu_1+\sigma_1 y) = \ln\left(\frac{1-z}{z}\right),$$

$$\sigma_1 y = -\ln\left(\frac{1-z}{z}\right) - \mu_1,$$

$$p_1^{-1}(z) \equiv y = \frac{1}{\sigma_1}\left(\ln\left(\frac{z}{1-z}\right) - \mu_1\right).$$

The probit-normal inverse also has a tractable form. Once again, we exploit the form of equation 3.33 to solve the underlying equation for y,

$$z = \Phi(\mu_2 + \sigma_2 y), \tag{3.36}$$

$$\Phi^{-1}(z) = \Phi^{-1}\left(\Phi(\mu_2 + \sigma_2 y)\right),$$

$$p_2^{-1}(z) \equiv y = \frac{\Phi^{-1}(z) - \mu_2}{\sigma_2}.$$

This implies that the corresponding distribution functions of $\{p_i(Z); i = 1, 2\}$, are determined by combining the previous results as follows,

$$F_{p_1(Z)}(z) = \Phi\left(\underbrace{\frac{1}{\sigma_1}\left(\ln\left(\frac{z}{1-z}\right) - \mu_1\right)}_{\text{Equation 3.35}}\right), \qquad (3.37)$$

for the logistic function and,

$$F_{p_2(Z)}(z) = \Phi\left(\underbrace{\frac{\Phi^{-1}(z) - \mu_2}{\sigma_2}}_{\text{Equation 3.36}}\right). \qquad (3.38)$$

To determine the probability mass function associated with each particular choice of $p(Z)$, we require, as outlined in equation 3.9, the density function, $f_{p(Z)}(z)$. This is obtained by differentiating equation 3.37 with respect to z. This is basically just a calculus exercise. Let's start with the logistic function,

$$dF_{p_1(Z)}(z) = \frac{\partial}{\partial z}\left(\Phi\left(\frac{1}{\sigma_1}\left(\ln\left(\frac{z}{1-z}\right) - \mu_1\right)\right)\right), \qquad (3.39)$$

$$f_{p_1(Z)}(z) = \frac{\partial}{\partial z}\left(\int_{-\infty}^{\frac{1}{\sigma_1}\left(\ln\left(\frac{z}{1-z}\right)-\mu_1\right)} \frac{1}{\sqrt{2\pi}} e^{-\frac{x^2}{2}} dx\right),$$

$$= \phi\left(\frac{1}{\sigma_1}\left(\ln\left(\frac{z}{1-z}\right) - \mu_1\right)\right) \frac{\partial}{\partial z}\left(\frac{1}{\sigma_1}\left(\ln\left(\frac{z}{1-z}\right) - \mu_1\right)\right),$$

$$= \phi\left(\frac{1}{\sigma_1}\left(\ln\left(\frac{z}{1-z}\right) - \mu_1\right)\right)\left(\frac{1}{\sigma_1 z} + \frac{1}{\sigma_1(1-z)}\right),$$

$$= \frac{\phi\left(\frac{1}{\sigma_1}\left(\ln\left(\frac{z}{1-z}\right) - \mu_1\right)\right)}{\sigma_1 z(1-z)},$$

$$= \frac{1}{z(1-z)\sqrt{2\pi\sigma_1^2}} e^{-\frac{\left(\ln\left(\frac{z}{1-z}\right)-\mu_1\right)^2}{2\sigma_1^2}}.$$

This is the density of the so-called logit-normal, or logistic-normal, distribution.

3.1 Binomial-Mixture Models

We may now repeat this tedious exercise for the probit-normal model. One need only recall that the derivative of the inverse cumulative standard normal distribution function, $\Phi^{-1}(x)$ is $\frac{1}{\phi(\Phi^{-1}(x))}$ where ϕ denotes the density of the standard normal distribution. This leads to,

$$dF_{p_2(Z)}(z) = \frac{\partial}{\partial z}\left(\frac{\Phi^{-1}(z) - \mu_2}{\sigma_2}\right), \tag{3.40}$$

$$f_{p_1(Z)}(z) = \frac{\partial}{\partial z}\left(\int_{-\infty}^{\frac{\Phi^{-1}(z)-\mu_2}{\sigma_2}} \frac{1}{\sqrt{2\pi}} e^{-\frac{x^2}{2}} dx\right),$$

$$= \phi\left(\frac{\Phi^{-1}(z) - \mu_2}{\sigma_2}\right) \frac{\partial}{\partial z}\left(\frac{\Phi^{-1}(z) - \mu_2}{\sigma_2}\right),$$

$$= \phi\left(\frac{\Phi^{-1}(z) - \mu_2}{\sigma_2}\right)\left(\frac{1}{\sigma_2 \phi(\Phi^{-1}(z))}\right),$$

$$= \frac{1}{\sigma_2} e^{\frac{\sigma_2^2 \left(\Phi^{-1}(z)\right)^2 - \left(\Phi^{-1}(z) - \mu_2\right)^2}{2\sigma_2^2}}.$$

Figure 3.4 seeks to provide a visual comparison between the mathematical densities derived thus far in our analysis. The independent-default binomial model is

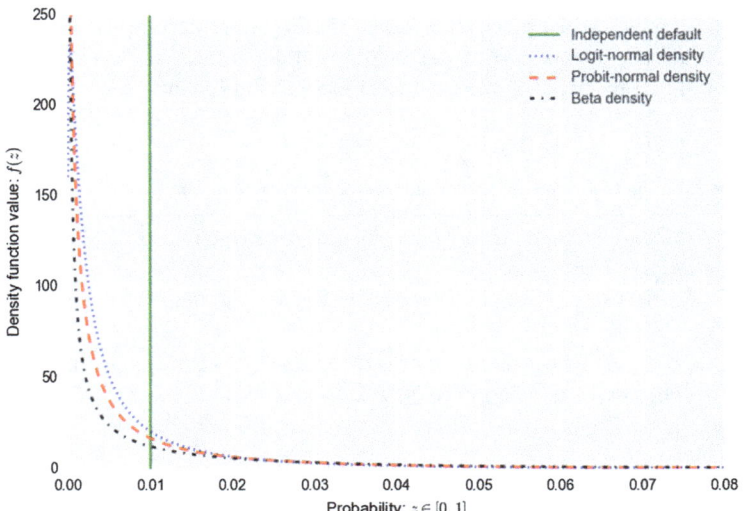

Fig. 3.4 *Comparing default-probability densities*: This figure compares the four calibrated default probability density function—including the trivial independent-default constant-parameter binomial model. Despite a common calibration, each choice has slightly different implications for default probabilities and hence one's risk-measure estimates.

not a terribly meaningful density, since it places all of the probability mass at a single point. The other models—beta-binomial, logit-normal, and probit-normal—have quite similar density functions when identically calibrated. Despite their common calibration, however, each choice has slightly different implications for default probabilities and hence one's risk-measure estimates.

3.1.2.2 Numerical Integration

Using the final results in equations 3.39 and 3.40—along with recalling the general form from equation 3.9—we can describe the probability mass function of the total number of defaults under the logit-normal and probit-normal binomial mixture models as,

$$\mathbb{P}(\mathbb{D}_N \leq k) = \binom{N}{k} \int_0^1 z^k (1-z)^{N-k} f_{p_i(Z)}(z) dz, \qquad (3.41)$$

for $i = 1, 2$. This integral cannot, as was the case for the beta-binomial model, be solved analytically. Equation 3.41 needs to be solved using numerical techniques.

Numerical integration may seem inconvenient, but in this setting, it is reasonably straightforward. One starts with a sub-routine describing the function of interest. Algorithm 3.3 is a direct implementation of the logit- and probit-normal mixture density function summarized in equation 3.41. For a given value of Z, a choice of model $i = 1, 2$, number of counterparties, target number of defaults, and the model parameters μ_i and σ_i, it will determine the value of the integrand. No numerical integration has yet taken place, but this is the key first step. Since these are very similar functions, it makes sense to combine them into a single sub-routine; the `isLogit` flag tells Algorithm 3.3 which model to implement.

Algorithm 3.3 The logit- and probit-normal mixture density functions

```
def logitProbitMixture(z,N,k,mu,sigma,isLogit):
    if isLogit==1:
        density = util.logitDensity(z,mu,sigma)
    else:
        density = util.probitDensity(z,mu,sigma)
    probTerm = util.getBC(N,k)*(z**k)*((1-z)**(N-k))
    f = probTerm*density
    return f
```

3.1 Binomial-Mixture Models

Algorithm 3.4 Analytic logit- and probit-normal

```
def logitProbitBinomialAnalytic(N,c,mu,sigma,alpha,isLogit):
    pmf = np.zeros(N+1)
    for k in range(0,N+1):
        pmf[k],err=nInt.quad(logitProbitMixture,0,1,args=(N,k,mu,sigma,isLogit))
    cdf = np.cumsum(pmf)
    varAnalytic = c*np.interp(alpha,cdf,np.linspace(0,N,N+1))
    esAnalytic = util.analyticExpectedShortfall(N,alpha,pmf,c)
    return pmf,cdf,varAnalytic,esAnalytic
```

The second step, found in Algorithm 3.4, is to use the numerical integration functionality to actually compute the probability mass and cumulative probability distribution functions. Logically, this works in basically the same manner as found in the independent-default binomial model in Chap. 2. The only difference is that the integral providing the default probability for each number of defaults is evaluated numerically using the integrate package from Python's scipy libraries.[7] Each evaluation is extremely efficient and the entire computation requires about the same minimal amount of computation time as the beta-binomial mixture approach.

It is interesting to compare the implications for the probability of any given number of defaults associated with each of our models. Figure 3.5 illustrates the default probability mass function for the independent-default and the three mixture models. The length of the tail associated with the independent-default binomial approach is visibly much shorter than in the mixture models. Although they are clearly not identical, it is not particularly easy to identify important differences between the mixture models.

3.1.2.3 Logit- and Probit-Normal Calibration

As with the beta-binomial mixture model, we require a technique for the determination of the model parameters—μ_i and σ_i for $i = 1, 2$—that are consistent with desired model behaviour. Unlike the beta-binomial setting, the calibration will not yield straightforward closed-form expressions. Instead, it is necessary to numerically determine the appropriate parameter choices.

We begin by returning to equations 3.4 and 3.7. The first equation holds that $\mathbb{E}(p(Z)) = \bar{p}$. That is, for calibration, we require that the first moment of the logistic

---

[7]This, in turn, is based on the Fortran 77 library function QUADPACK, which employs the Gauss-Kronrod quadrature formulae. For more information on this approach, in particular or numerical integration in general, see Ralston and Rabinowitz (1978, Chapter 4) or Press et al. (1992, Chapter 4).

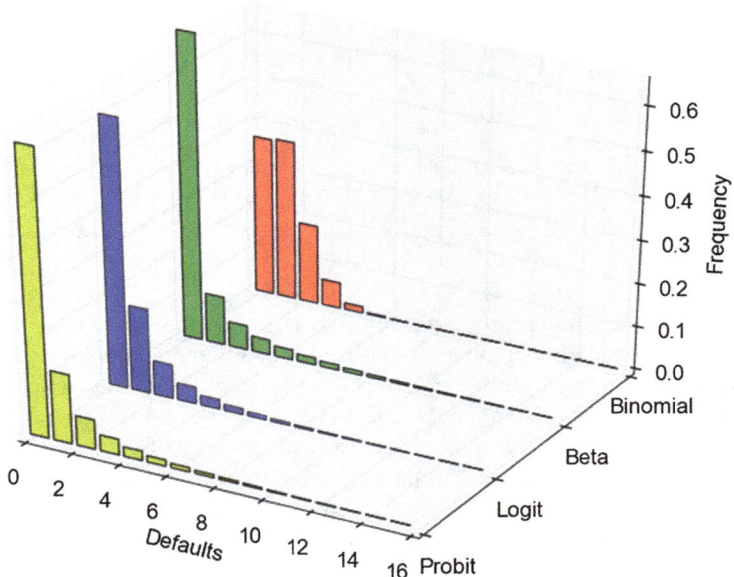

Fig. 3.5 *Comparing default probability mass functions*: This figure compares the four calibrated default probability mass functions: independent-default, beta-binomial, logit-normal, and probit-normal. All follow from the basic structure of equation 3.41.

distribution takes our desired value of $\bar{p}$. That is,

$$\bar{p} = \mathbb{E}(p(Z)), \qquad (3.42)$$
$$= \int_{-\infty}^{\infty} p(z) f_Z(z) dz,$$
$$= \mathcal{M}_1,$$

where $\mathcal{M}_1$ denotes the first moment of $p(Z)$. Although the form of this moment is slightly different for the logit and probit models, the basic structure is extremely similar.

Equation 3.7 provides the general expression for the covariance between two arbitrarily selected counterparties, n and m. We use it as a starting point for a series of simple manipulations,

$$\text{cov}\left(\mathbb{I}_{\mathcal{D}_n}, \mathbb{I}_{\mathcal{D}_m}\right) = \text{var}(p(Z)), \qquad (3.43)$$

$$\rho\left(\mathbb{I}_{\mathcal{D}_n}, \mathbb{I}_{\mathcal{D}_m}\right) = \frac{\text{var}\left(p(Z)\right)}{\sqrt{\mathbb{E}\left(p(Z)\right)\left(1 - \mathbb{E}\left(p(Z)\right)\right)}\sqrt{\mathbb{E}\left(p(Z)\right)\left(1 - \mathbb{E}\left(p(Z)\right)\right)}},$$

3.1 Binomial-Mixture Models

$$\rho_{\mathcal{D}} = \frac{\overbrace{\mathbb{E}\left(p(Z)^2\right) - (\mathbb{E}(p(Z)))^2}^{\text{By definition}}}{\mathbb{E}(p(Z))(1 - \mathbb{E}(p(Z)))},$$

$$= \frac{\mathbb{E}(p(Z)^2) - \mathcal{M}_1^2}{\mathcal{M}_1(1 - \mathcal{M}_1)},$$

$$= \frac{\overbrace{\int_{-\infty}^{\infty} p(z)^2 f_Z(z)dz}^{\mathcal{M}_2} - \mathcal{M}_1^2}{\mathcal{M}_1(1 - \mathcal{M}_1)},$$

$$= \frac{\mathcal{M}_2 - \mathcal{M}_1^2}{\mathcal{M}_1 - \mathcal{M}_1^2}.$$

This yields two simultaneous non-linear equations,

$$\bar{p} - \mathcal{M}_1 = 0, \qquad (3.44)$$
$$(1 - \rho_{\mathcal{D}})\mathcal{M}_1^2 + \rho_{\mathcal{D}}\mathcal{M}_1 - \mathcal{M}_2 = 0.$$

The moments in the non-linear system described in equation 3.44 depend on the choice of model parameters, μ_i and σ_i. Algorithm 3.5 summarizes a Python implementation of this non-linear systems of equations.[8]

Algorithm 3.5 computes the two moments by numerical integration and then computes two function values, f1 and f2. It takes as arguments the target average

Algorithm 3.5 Logit- and probit-normal model calibration system

```
def logitProbitCalibrate(x,pTarget,rhoTarget,isLogit):
    if x[1]<=0:
        return [100, 100]
    M1,err = nInt.quad(logitProbitMoment,-8,8,args=(x[0],x[1],1,isLogit))
    M2,err = nInt.quad(logitProbitMoment,-8,8,args=(x[0],x[1],2,isLogit))
    f1 = pTarget - M1
    f2 = rhoTarget*(M1 - (M1**2)) - (M2 - (M1**2))
    return [f1, f2]
```

---

[8] See Press et al. (1992, Chapter 9) for a detailed discussion of numerically solving non-linear systems of equations.

Table 3.3 *Binomial calibration results*: This table summarizes the parameter values determined by calibrating all models to an average default probability of 1% and a default correlation coefficient of 0.05.

Model	Targets				Parameters	
	$\mathcal{M}_1$ or $\bar{p}$	$\mathcal{M}_2$	$\sqrt{\text{var}(p(Z))}$	$\rho_\mathcal{D}$	α, μ_i	β, σ_i
Beta-binomial	0.01	0.0006	0.02	0.05	0.19	18.81
Logit-normal	0.01	0.0006	0.02	0.05	−5.66	1.50
Probit-normal	0.01	0.0006	0.02	0.05	−2.81	0.68
Binomial	0.01	0.0000	0.00	0.00	n/a	n/a

default probability $\bar{p}$=pTarget and the default correlation, $\rho_\mathcal{D}$=rhoTarget. To find the roots to this system, we make the following scipy.optimize function call,

$$r = \text{fsolve(logitProbitCalibrate,x0,args=(pTarget,rhoTarget,isLogit))} \quad (3.45)$$

fsolve is a Python function, which is essentially a wrapper around MINPACK's hybrd algorithms—these are a collection of numerical FORTRAN libraries made available to the public by the Mathematics and Computer Science division of the Argonne National Laboratory in the 1980s.[9] x0 are the starting values for the algorithm.

Table 3.3 highlights the results of the calibration exercise for the logit- and probit-normal models along with the beta-binomial and independent default approaches. It is precisely these parameter values that were used in generating Figs. 3.3 to 3.5. We see clearly that, although the parameter values vary dramatically, there is perfect agreement with the first two moments and a number of associated statistics of the default probability distributions.

3.1.2.4 Logit- and Probit-Normal Results

We now have all of the necessary elements required to apply the logit- and probit-normal models to our simple example. Table 3.4 compares—for the common calibration summarized in Table 3.3—the analytic independent-default to the beta-binomial, logit-normal, and probit-normal mixture model results. The results are

---

[9] Although we are basically consumers of this algorithm, a few details are nonetheless useful. hybrd determines the roots of a system of N non-linear functions in N unknowns using the Powell-hybrid, or dogleg, method—this approach tries to minimize the sum of the squared function values using Newton's method. The Jacobian is calculated numerically, employing user-provided function routines, with a forward finite-difference approximation.

3.1 Binomial-Mixture Models

Table 3.4 *Analytic results*: This table compares the analytic independent-default to the beta-binomial, logit-normal, and probit-normal mixture model results. In all cases, the models are calibrated for a 1% average default probability. The mixture approaches also target a default correlation coefficient of 0.05.

Quantile	Binomial			Beta-binomial			Logit-normal			Probit-normal		
	$\mathbb{D}_N$	VaR_α	$\mathcal{E}_\alpha$	$\mathbb{D}_N$	VaR_α	$\mathcal{E}_\alpha$	$\mathbb{D}_N$	VaR_α	$\mathcal{E}_\alpha$	$\mathbb{D}_N$	VaR_α	$\mathcal{E}_\alpha$
95.00th	2.5	$25	$25	5.1	$51	$54	4.0	$40	$47	4.5	$45	$50
97.00th	2.8	$28	$28	7.1	$71	$72	5.7	$57	$63	6.3	$63	$68
99.00th	3.6	$36	$36	11.5	$115	$156	10.6	$106	$174	11.2	$112	$168
99.50th	3.9	$39	$39	14.3	$143	$185	14.6	$146	$224	14.8	$148	$209
99.90th	4.8	$48	$49	20.9	$209	$249	27.0	$270	$363	24.5	$245	$311
99.97th	5.5	$55	$55	25.8	$258	$296	38.2	$382	$475	32.4	$324	$390
99.99th	5.9	$59	$60	30.0	$300	$337	48.8	$488	$575	39.7	$397	$462
Other statistics												
Expected loss	1.0	$10.0		1.0	$10.0		1.0	$10.0		1.0	$10.0	
Loss volatility	1.0	$9.9		2.2	$22.2		2.2	$22.2		2.2	$22.2	
Default correlation	0.00			0.05			0.05			0.05		

striking. Up to about the 99th quantile, there is relatively little difference between the mixture models. In each case, the number of defaults varies between about 10 or 11. Beyond that point, however, they begin to diverge. The most conservative estimate comes from the logit-normal model, predicting roughly 49 defaults at the 99.99th quantile. This is followed by about 40 defaults for the probit-normal approach and about 30 for the beta-binomial choice. This implies that at the 99.99th quantile, the VaR estimates range from $300 to $500. This compares to only about $60 for the independent-default binomial model.

In summary, therefore, the three mixture models generate qualitatively similar results up to about the 99th percentile. Beyond that point, the results diverge with differing degrees of risk-metric conservatism. Naturally, it is also possible to relax the assumption of equal counterparty exposures with the use of Monte Carlo methods. This requires a simple extension of Algorithm 3.2. One merely generates a collection of standard-normal random variates, applies the necessary mapping to create default probabilities, and then employs the usual trick with uniform random variates to simulate Bernoulli trials. As in the previous algorithms, the same sub-routine can be used—through our isLogit flag—to handle both logit- and probit-normal mixture models. These steps are summarized in Algorithm 3.6.

Algorithm 3.6 Logit- and probit-normal model Monte Carlo

```
def logitProbitBinomialSimulation(N,M,c,mu,sigma,alpha,isLogit):
    Z = np.random.normal(0,1,M)
    if isLogit==1:
        p = np.reciprocal(1+np.exp(-(mu+sigma*Z)))
    else:
        p = norm.cdf(mu+sigma*Z)
    pZ = np.transpose(np.tile(p,(N,1)))
    U = np.random.uniform(0,1,[M,N])
    lossIndicator = 1*np.less(U,pZ)
    lossDistribution = np.sort(np.dot(lossIndicator,c),axis=None)
    el,ul,var,es=util.computeRiskMeasures(M,lossDistribution,alpha)
    return el,ul,var,es
```

Table 3.5 *Numerical mixture results*: This table provides the quantiles of the loss distribution for the Monte Carlo estimates of the independent-default model relative to the beta-binomial, logit-normal, and probit-normal mixture approaches. All models have an average default probability of 1% and each mixture model is calibrated to a 0.05 default-indicator correlation coefficient.

Quantile	Binomial		Beta-binomial		Logit-normal		Probit-normal	
	$\widehat{\text{VaR}_\alpha}$	$\widehat{\mathcal{E}_\alpha}$	$\widehat{\text{VaR}_\alpha}$	$\widehat{\mathcal{E}_\alpha}$	$\widehat{\text{VaR}_\alpha}$	$\widehat{\mathcal{E}_\alpha}$	$\widehat{\text{VaR}_\alpha}$	$\widehat{\mathcal{E}_\alpha}$
95.00th	$35.3	$45.5	$59.3	$101.4	$49.8	$94.9	$53.7	$98.4
97.00th	$40.4	$50.7	$80.1	$123.1	$67.4	$119.9	$73.2	$122.5
99.00th	$52.5	$61.1	$127.0	$170.9	$117.8	$186.5	$123.8	$181.4
99.50th	$58.8	$66.8	$157.6	$201.3	$159.5	$237.5	$161.3	$222.7
99.90th	$72.3	$80.3	$228.6	$271.1	$284.2	$377.1	$260.2	$326.7
99.97th	$82.2	$89.6	$279.5	$321.3	$395.9	$491.0	$339.7	$405.7
99.99th	$90.3	$97.4	$326.4	$366.6	$503.8	$592.7	$414.3	$477.7
Other statistics								
Expected loss	$9.2		$10.0		$10.0		$10.0	
Loss volatility	$12.8		$25.8		$25.9		$25.8	
Total iterations	10,000,000							
Computational time	16.8		21.4		20.1		20.7	

The final numerical results, for all actuarial models considered in this section, are found in Table 3.5. As before, due to the heterogeneous exposures, the results are somewhat larger than the analytic estimates summarized in Table 3.4. Once again, in the numerical setting, the differences between mixture models are marginal up to about the 99th percentile. The expected shortfall estimates provide some additional insight into the range of tail outcomes associated with the different models. The logit-distribution—under a common calibration—clearly exhibits the most conservative worst-case default outcomes with a 99.99th quantile expected shortfall estimate just slightly less than $600 or 60% of the total portfolio. This is approaching twice the beta-binomial model estimate.

As a final note, none of the simulations was particularly computationally intensive. For a total of 10 million iterations, the computation time ranges about

20 seconds across all binomial-mixture models; a slight increase in computational expense is observed relative to the independent-default approach.

3.2 Poisson-Mixture Models

To this point, we have examined a collection of mixture models using the binomial distribution as the departure point. It is possible to construct alternative mixture models using a different jumping-off point: the Poisson distribution. This is entirely consistent with the rough equivalence of the binomial and Poisson approaches when default probabilities and, correspondingly, arrival rates are small. It is useful and important to examine these Poisson-mixture models for, at least two reasons. First of all, they offer another set of possible models based on different assumptions and incorporating alternative distributions. From a model risk perspective, due to the rarity of default events and the consequent inability to perform meaningful back-testing, the more possible modelling choices, the better. The second reason is that one of the most popular extant credit-risk models—the so-called CreditRisk+ model introduced by Wilde (1997) and extensively described in Gundlach and Lehrbass (2004)—is a multi-factor implementation of a specific choice of Poisson-mixture model.

We will begin with a one-factor setting and then, to give a flavour of what is possible, consider the multi-factor approach. As in the binomial setting, there are numerous possible implementations that one might consider. Some are more mathematically tractable than others—such as the beta-binomial approach discussed in the previous sections—which leads to greater popularity. We will begin with what is perhaps the most popular approach: the Poisson-gamma mixture model.

Let us first recall a few simple facts from the previous chapter. The use of a Poisson distributed random variable to describe the nth default event may feel, at first glance, somewhat odd. The Poisson distribution is defined on the positive integers, whereas default is a binary event. This choice works, however, when the Poisson arrival rate is sufficiently small, because it leads to the assignment of the vast majority of the probability mass to the values 0 and 1. This gives it a binary profile and, at its heart, is the intuition behind the link between the Poisson and binomial distributions; this relationship is often referred to as the law of rare events.

The independent-default Poisson model involves the introduction of N independent and identically distributed Poisson random variables, $X_n \sim \mathcal{P}(\lambda_n)$. The default indicator is re-defined as,

$$\mathbb{I}_{\mathcal{D}_n} \equiv \mathbb{I}_{\{X_n \geq 1\}} = \begin{cases} 1 : \text{defaults occurs before time } T \text{ with probability } 1 - e^{-\lambda_n} \\ 0 : \text{survival until time } T \text{ with probability } e^{-\lambda_n} \end{cases},$$
(3.46)

It follows directly from this definition that the unconditional default probability is,

$$p_n = 1 - e^{-\lambda_n}. \tag{3.47}$$

This link between the arrival intensity and default probability allows us to calibrate the model by setting the individual Poisson parameters to,

$$\lambda_n = -\ln(1 - p_n). \tag{3.48}$$

The default loss, for a portfolio of N obligors, is consequently,

$$L_N = \sum_{n=1}^{N} c_n \mathbb{I}_{\{X_n \geq 1\}}. \tag{3.49}$$

Chapter 2 uses these relationships to explore, from both analytic and numerical perspectives, the implementation of this model choice. Ultimately, we found that the independent-default binomial and Poisson models were, for practical purposes, essentially indistinguishable. The key idea of the Poisson-mixture approach is to randomize the arrival intensity and, indirectly, the default probability. Quite simply, we assume that

$$\lambda_n \equiv \lambda(S), \tag{3.50}$$

for $n = 1, \ldots, N$ where $S \in \mathbb{R}_+$ is a random variable with density $f_S(s)$. The immediate consequence is that the probability of k defaults is written as,

$$\mathbb{P}(\mathbb{D}_N = k | S) = \frac{e^{-\lambda(S)} \lambda(S)^k}{k!}. \tag{3.51}$$

This is, of course, a conditional probability since the value of S needs to be revealed before we can evaluate equation 3.51. The structure is, however, quite similar to the binomial-mixture setting. We assume that, given a particular value of S, the default events are independent. This conditional independence assumption is, in fact, a common feature across virtually all credit-risk models. Given that each individual obligor conditions on S, we again have the potential to induce default dependence.

Ultimately, we require the unconditional default probability. This can, as before, be identified by integrating over the conditioning variable. More specifically, we have that

$$\mathbb{P}(\mathbb{D}_N = k) = \int_{\mathbb{R}_+} \frac{e^{-\lambda(s)} \lambda(s)^k}{k!} f_S(s) ds. \tag{3.52}$$

In the one-dimensional case, this integral is readily evaluated numerically. In some cases, depending on the choice of distribution for S, we can actually identify analytic formulae for the total number of defaults.

3.2.1 The Poisson-Gamma Approach

It is common, in actuarial circles, to assume that the conditioning variable, S, follows a gamma distribution; that is, $S \sim \Gamma(a,b)$. Moreover, we further assume that the arrival intensity for each of our N Poisson random variables is $\lambda_n = \lambda(S) \equiv S$ for $n = 1, \ldots, N$. More simply put, the random variable S is the common arrival rate for each default event. This can, and will, be modified. For the moment, however, it allows to gain an understanding of the model structure with a minimum of clutter.

Our distributional choice for S implies that the density function from equation 3.52 is described as,

$$f_S(s) = \frac{b^a e^{-bs} s^{a-1}}{\Gamma(a)}, \tag{3.53}$$

where $\Gamma(\cdot)$ denotes the Gamma function introduced in equation 3.20.[10]

We may now proceed to use the density from equation 3.53 and the unconditional default-count probability to simplify things somewhat.

$$\mathbb{P}(\mathbb{D}_N = k) = \int_{\mathbb{R}_+} \frac{e^{-s} s^k}{k!} \underbrace{\frac{b^a e^{-bs} s^{a-1}}{\Gamma(a)}}_{\text{Equation 3.53}} ds, \tag{3.54}$$

$$= \frac{b^a}{k!\Gamma(a)} \int_{\mathbb{R}_+} e^{-s(b+1)} s^{(k+a)-1} ds,$$

$$= \frac{b^a}{k!\Gamma(a)} \underbrace{\left(\frac{(b+1)^{a+k}}{(b+1)^{a+k}}\right)}_{=1} \underbrace{\left(\frac{\Gamma(a+k)}{\Gamma(a+k)}\right)}_{=1} \int_{\mathbb{R}_+} e^{-s(b+1)} s^{(k+a)-1} ds,$$

$$= \frac{b^a \Gamma(a+k)}{k!\Gamma(a)(b+1)^{a+k}} \underbrace{\int_{\mathbb{R}_+} \overbrace{\frac{(b+1)^{a+k} e^{-s(b+1)} s^{(k+a)-1}}{\Gamma(a+k)}}^{\Gamma(a+k, b+1)} ds}_{=1},$$

$$= \frac{\Gamma(a+k)}{k!\Gamma(a)} \left(\frac{b}{b+1}\right)^a \left(\frac{1}{b+1}\right)^k,$$

$$= \frac{\Gamma(a+k)}{\Gamma(k+1)\Gamma(a)} \left(\frac{b}{b+1}\right)^a \left(1 - \frac{b}{b+1}\right)^k,$$

$$= \frac{\Gamma(a+k)}{\Gamma(k+1)\Gamma(a)} q_1^a (1-q_1)^k,$$

---

[10] The gamma density function can be written in alternative formats; we have opted for the shape-rate parametrization. Ultimately, although the moments are different, the choice is immaterial.

where we define,

$$q_1 = \frac{b}{b+1}. \tag{3.55}$$

A few tricks were involved in resolving equation 3.54. First, as is often the case in manipulations of the gamma distribution, we were able to re-arrange the integrand so that it amounted to a $\Gamma(a+k, b+1)$ density. Since a density integrated over its support must yield unity, this dramatically simplifies our calculation. One of the defining features of the gamma function is that $(n-1)! = \Gamma(n)$ for any positive integer. We may, therefore, safely replace $k!$ with $\Gamma(k+1)$. Figure 3.6 provides a visual comparison of the link between the gamma function and the factorial operator for the first few integers.

Finally, the introduction of the parameter, q_1, as described in equation 3.55 does more than just streamline our unconditional probability. It also identifies the Poisson-gamma mixture experiment as a well-known named distribution in its own right: the negative binomial distribution.

The convenient form of equation 3.55 lends itself readily to implementation. Algorithm 3.7 summarizes our Python function for the computation of the usual risk measures. It has essentially the same input arguments as the binomial-mixture models presented in the previous section, albeit with some slight changes in its structure to accommodate the new distributions.

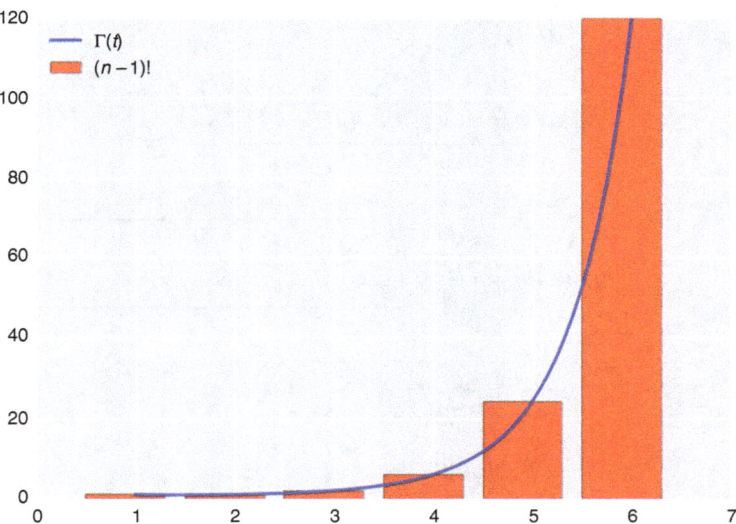

Fig. 3.6 *The gamma function*: This figure provides a very rough visual justification for the relationship between the gamma function and the factorial operator. For a fascinating description of the origins and history of this special function, see Davis (1959).

3.2 Poisson-Mixture Models

Algorithm 3.7 Analytic Poisson-Gamma mixture

```
def poissonGammaAnalytic(N,c,a,b,alpha):
    pmfPoisson = np.zeros(N+1)
    q = np.divide(b,b+1)
    den = math.gamma(a)
    for k in range(0,N+1):
        num = np.divide(math.gamma(a+k),scipy.misc.factorial(k))
        pmfPoisson[k] = np.divide(num,den)*np.power(q,a)*np.power(1-q,k)
    cdfPoisson = np.cumsum(pmfPoisson)
    varAnalytic = c*np.interp(alpha,cdfPoisson,np.linspace(0,N,N+1))
    esAnalytic = util.analyticExpectedShortfall(N,alpha,pmfPoisson,c)
    return pmfPoisson,cdfPoisson,varAnalytic,esAnalytic
```

3.2.1.1 Calibrating the Poisson-Gamma Mixture Model

As before, while the actual implementation is relatively straightforward—requiring only several lines of code as summarized in Algorithm 3.7—we need some idea of the gamma-distribution parameters, a and b. We will present two distinct approaches: one formal and the other something of a short-cut. Ultimately, the differences are not particularly large, which provides some useful insight into the underlying model.

Both ideas follow from the technique used with the binomial-mixture approaches, albeit with a twist provided by equations 3.47 and 3.48. In particular, we have indirectly assumed that

$$p(S) = 1 - e^{-S}. \qquad (3.56)$$

To perform the calibration, we need to compute the expected value and variance of $p(S)$; this will allow us to set a and b to be consistent with estimated unconditional default probabilities (i.e., the p_n's) and the desired level of default correlation. More specifically, we require the first two moments of $p(S)$, which we have been denoting $\mathcal{M}_1$ and $\mathcal{M}_2$. We will identify these quantities along the way.

It turns out that both of these quantities can be determined analytically. Although a bit tedious, we can identify the expectation as,

$$\mathcal{M}_1 = \mathbb{E}(p(S)) = \mathbb{E}\left(\underbrace{1 - e^{-S}}_{\text{Equation 3.56}}\right), \qquad (3.57)$$

$$= 1 - \mathbb{E}\left(e^{-S}\right),$$

$$= 1 - \int_{\mathbb{R}_+} e^{-s} f_S(s) ds,$$

$$= 1 - \int_{\mathbb{R}_+} e^{-s} \frac{b^a e^{-bs} s^{a-1}}{\Gamma(a)} ds,$$

$$= 1 - b^a \frac{(b+1)^a}{\underbrace{(b+1)^a}_{=1}} \int_{\mathbb{R}_+} \frac{e^{-s(b+1)} s^{a-1}}{\Gamma(a)} ds,$$

$$= 1 - \frac{b^a}{(b+1)^a} \underbrace{\int_{\mathbb{R}_+} \frac{\overbrace{(b+1)^a e^{-s(b+1)} s^{a-1}}^{\Gamma(a,b+1)}}{\Gamma(a)} ds}_{=1},$$

$$= 1 - \left(\frac{b}{b+1}\right)^a,$$

$$= 1 - q_1^a.$$

A useful fact, from equation 3.57, is that $\mathbb{E}(e^{-S}) = q_1^a$. This helps in identifying the variance, which is determined in a similar manner as,

$$\text{var}(p(S)) = \text{var}\left(\underbrace{1 - e^{-S}}_{\text{Equation 3.56}}\right), \tag{3.58}$$

$$= \mathbb{E}\left(\left(1 - e^{-S}\right)^2 - \underbrace{\mathbb{E}\left(1 - e^{-S}\right)^2}_{\text{Equation 3.57}}\right),$$

$$= \mathbb{E}\left(\left(1 - e^{-S}\right)^2\right) - \mathcal{M}_1^2,$$

$$= \mathbb{E}\left(1 - 2e^{-S} + e^{-2S}\right) - \mathcal{M}_1^2,$$

$$= 1 - 2\underbrace{\mathbb{E}\left(e^{-S}\right)}_{q_1^a} + \mathbb{E}\left(e^{-2S}\right) - \mathcal{M}_1^2,$$

$$= 1 - 2q_1^a + \int_{\mathbb{R}_+} e^{-2s} \frac{b^a e^{-bs} s^{a-1}}{\Gamma(a)} ds - \mathcal{M}_1^2,$$

3.2 Poisson-Mixture Models

$$= 1 - 2q_1^a + b^a \underbrace{\frac{(b+2)^a}{(b+2)^a}}_{=1} \int_{\mathbb{R}_+} \frac{e^{-s(b+2)}s^{a-1}}{\Gamma(a)}ds - \mathcal{M}_1^2,$$

$$= 1 - 2q_1^a + \left(\frac{b}{b+2}\right)^a \underbrace{\int_{\mathbb{R}_+} \overbrace{\frac{(b+2)^a e^{-s(b+2)}s^{a-1}}{\Gamma(a)}}^{\Gamma(a,b+2)} ds}_{=1} - \mathcal{M}_1^2,$$

$$= \underbrace{1 - 2q_1^a + q_2^a}_{\mathcal{M}_2} - \mathcal{M}_1^2,$$

where $\mathcal{M}_2 = \mathbb{E}\left((1-e^{-S})^2\right)$ and the parameter,

$$q_2 = \frac{b}{b+2}. \tag{3.59}$$

Thus we have relatively manageable expressions for the expectation and variance—as well as the first two moments—of $p(S)$.

Recalling equation 3.26, we have already established that the default correlation is a function of the expectation and variance of $p(S)$. Equations 3.57 and 3.58 allow us to write this default correlation as,

$$\rho_\mathcal{D} = \rho\left(\mathbb{I}_{\mathcal{D}_n}, \mathbb{I}_{\mathcal{D}_m}\right) = \frac{\text{var}(Z)}{\sqrt{\mathbb{E}(Z)(1-\mathbb{E}(Z))}\sqrt{\mathbb{E}(Z)(1-\mathbb{E}(Z))}}, \tag{3.60}$$

$$= \frac{\mathcal{M}_2 - \mathcal{M}_1^2}{\mathcal{M}_1 - \mathcal{M}_1^2},$$

$$= \frac{1 - 2q_1^a + q_2^a - (1-q_1^a)^2}{(1-q_1^a) - (1-q_1^a)^2},$$

$$= \frac{\cancel{1} - \cancel{2q_1^a} + q_2^a - (\cancel{1} - \cancel{2q_1^a} + q_1^{2a})}{\cancel{1} - q_1^a - (\cancel{1} - \cancel{2q_1^a} + q_1^{2a})},$$

$$= \frac{q_2^a - q_1^{2a}}{q_1^a(1-q_1^a)}.$$

We thus have a succinct definition of the default correlation between any two obligors in our portfolio.

Algorithm 3.8 Analytic Poisson-Gamma moments

```
def poissonGammaMoment(a,b,momentNumber):
    q1 = np.divide(b,b+1)
    q2 = np.divide(b,b+2)
    if momentNumber==1:
        myMoment = 1-np.power(q1,a)
    if momentNumber==2:
        myMoment = 1 - 2*np.power(q1,a) + np.power(q2,a)
    return myMoment
```

Algorithm 3.9 Poisson-Gamma model calibration system

```
def poissonGammaCalibrate(x,pTarget,rhoTarget):
    if x[1]<=0:
        return [100, 100]
    M1 = poissonGammaMoment(x[0],x[1],1)
    M2 = poissonGammaMoment(x[0],x[1],2)
    f1 = pTarget - M1
    f2 = rhoTarget*(M1 - (M1**2)) - (M2 - (M1**2))
    return [f1, f2]
```

Algorithm 3.9 follows directly from the logit and probit approaches described in Algorithm 3.5—the only difference is that we do not need to perform any numerical integration to arrive at the moments of the default probabilities: these are provided in closed form in Algorithm 3.8. The code structure and the conceptual task is identical, but computationally it takes a bit more time to solve for the calibrated parameters.

3.2.1.2 A Quick and Dirty Calibration

The preceding approach involved a significant amount of effort. Is there an easier way? With a very simple assumption, we can indeed dramatically simplify the calibration process. We merely assume that,

$$\lambda(S) = S \approx p(S) = S. \tag{3.61}$$

In this case, by the properties of the Gamma distribution that

$$\mathbb{E}(p(S)) = \mathbb{E}(S) = \frac{a}{b}, \tag{3.62}$$

3.2 Poisson-Mixture Models

and

$$\text{var}(p(S)) = \text{var}(S) = \frac{a}{b^2}. \tag{3.63}$$

The default correlation is now determined directly as,

$$\rho_D = \frac{\text{var}(S)}{\mathbb{E}(S)(1 - \mathbb{E}(S))}, \tag{3.64}$$

$$= \frac{\frac{a}{b^2}}{\frac{a}{b}\left(1 - \frac{a}{b}\right)},$$

$$= \frac{1}{b - a}.$$

This leads to the following two equations in two unknowns,

$$\bar{p} = \frac{a}{b}, \tag{3.65}$$

$$\rho_D = \frac{1}{b - a}$$

As we did with the beta-binomial mixture model, we can easily place this system into matrix format. Doing so, and solving it, we arrive at

$$\begin{bmatrix} -\rho_D & \rho_D \\ -1 & \bar{p} \end{bmatrix} \begin{bmatrix} a \\ b \end{bmatrix} = \begin{bmatrix} 1 \\ 0 \end{bmatrix}, \tag{3.66}$$

$$\begin{bmatrix} a \\ b \end{bmatrix} = \left(\begin{bmatrix} -\rho_D & \rho_D \\ -1 & \bar{p} \end{bmatrix} \right)^{-1} \begin{bmatrix} 1 \\ 0 \end{bmatrix},$$

$$= \frac{1}{\rho_D(1 - \bar{p})} \begin{bmatrix} \bar{p} & -\rho_D \\ 1 & -\rho_D \end{bmatrix} \begin{bmatrix} 1 \\ 0 \end{bmatrix},$$

$$= \begin{bmatrix} \dfrac{\bar{p}}{\rho_D(1 - \bar{p})} \\ \dfrac{1}{\rho_D(1 - \bar{p})} \end{bmatrix}.$$

For a given set of target properties of the loss distribution, we now have two different approaches—one exact and the other approximate—for the identification of the gamma distribution parameters, a and b.

If we perform the exact computation for $\bar{p} = 0.01$ and $\rho_\mathcal{D} = 0.05$, as in the binomial setting, we arrive at a value of $a = 0.19$ and $b = 18.41$. If we employ the approximation from equation 3.66, the results are very similar: $a = 0.20$ and $b = 20.20$. This indicates the close link between the arrival rate and the default probability for small numbers. Assuming equivalence of the default probability and the Poisson arrival rate does not appear to involve, in this case at least, dramatic costs in terms of accuracy.

3.2.1.3 Poisson-Gamma Results

Armed with our calibrated values, we may now use Algorithm 3.7 to apply the base Poisson-Gamma mixture model to our usual 100-obligor example. If we use the calibrated a and b parameters directly for the analytic implementation, without any adjustment, we will obtain erroneous results. The reason is simple: in the analytic setting, our variable of interest is $\mathbb{D}_N$, the total number of defaults. Since this is the sum of N conditionally independent Poisson random variables—$X_n \sim \mathcal{P}(S)$—its parameter is actually $N \cdot S$. It has, however, been calibrated to $\bar{p} = 0.01$.

The trick is to adjust the parameters, a and b, such that the mean and variance of S are scaled appropriately by N. If we start with the mean, we have

$$\mathbb{E}(S) = \frac{a}{b}, \qquad (3.67)$$

$$N\mathbb{E}(S) = N\frac{a}{b},$$

$$\mathbb{E}(N \cdot S) = \frac{a}{b/N}.$$

It may seem more natural to merely increase the a parameter by N, until we look at the variance

$$\text{var}(S) = \frac{a}{b^2}, \qquad (3.68)$$

$$N^2 \text{var}(S) = N^2 \frac{a}{b^2},$$

$$\text{var}(N \cdot S) = \frac{a}{(b/N)^2}.$$

It is, therefore, sufficient to divide the inverse-scale or rate parameter, b, by the total number of obligors. Thus, to be crystal clear about our approach for the analytic implementation of the Poisson-Gamma model, we set $a = 0.19$ and $b = 0.18$.

3.2 Poisson-Mixture Models

Table 3.6 *Analytic Poisson-Gamma results*: This table compares the analytic Poisson independent-default and Poisson-Gamma model results. In both cases, in line with the previous analysis, the latter models is calibrated for a 1% average default probability and a target default correlation of 0.05.

Quantile	Poisson			Poisson-Gamma		
	$\mathbb{D}_N$	VaR$_\alpha$	$\mathcal{E}_\alpha$	$\mathbb{D}_N$	VaR$_\alpha$	$\mathcal{E}_\alpha$
95.00th	2.4	$24.1	$24.4	5.3	$52.9	$57.7
97.00th	2.8	$27.6	$27.9	7.3	$73.4	$80.8
99.00th	3.5	$34.9	$35.2	12.1	$121.4	$170.2
99.50th	3.9	$38.6	$38.8	15.4	$154.0	$204.0
99.90th	4.8	$48.0	$48.4	23.3	$233.4	$285.5
99.97th	5.4	$54.3	$54.6	29.5	$295.2	$348.3
99.99th	5.9	$59.1	$59.4	35.3	$352.8	$406.7
Other statistics						
Expected loss	1.0	$10.0		1.0	$10.3	
Loss volatility	1.0	$10.0		2.4	$23.7	
Default correlation	0.00			0.05		

The results of the analytic approximation of the Poisson-Gamma model are outlined in Table 3.6 along with the original Poisson model outcomes for comparison. As with the binomial model, the introduction of default correlation makes a dramatic difference in the VaR and expected shortfall estimates. At the highest levels of confidence, the mixture model with positive default correlation yields results that are five or six times larger than their independent-default equivalents. This is slightly larger than the beta-binomial model, but somewhat less than the logit- and probit-mixture methods.

One of the beneficial consequences of mixing the Poisson model with another distribution is that it breaks one of the standard, and unwanted, features of the Poisson distribution: equal variance and expected value. The loss volatility—and since it takes a value of unity, the loss variance—and the expected loss are both equal to a single default or $10 in the independent-default Poisson setting. As we incorporate the Gamma distribution for the arrival intensity, we observe that the loss volatility grows to almost 2.4 times the expected loss. This is not merely a happy coincidence, but rather one of the principal reasons for the introduction of the mixing variable. Not only does the base Poisson methodology fail to capture default dependence, it is also, much like its sister approach, the binomial model, far too well behaved.

The next step is to examine the numerical implementation of the Poisson-Gamma model. The code for this approach is summarized in Algorithm 3.10. It has a straightforward structure. One generates M gamma-variates using Python's numpy

Algorithm 3.10 Poisson-Gamma numerical implementation

```
def poissonGammaSimulation (N,M,c,a,b,alpha):
    lam = np.random.gamma(a,1/b,M)
    H = np.zeros([M,N])
    for m in range(0,M):
        H[m,:] = np.random.poisson(lam[m],[N])
    lossIndicator = 1*np.greater_equal(H,1)
    lossDistribution = np.sort(np.dot(lossIndicator,c),axis=None)
    el,ul,var,es=util.computeRiskMeasures(M,lossDistribution,alpha)
    return el,ul,var,es
```

Table 3.7 Numerical Poisson-Gamma mixture results: This table provides the quantiles of the loss distribution for the Monte Carlo estimates of the independent-default Poisson and Poisson-gamma mixture models.

Quantile	Poisson		Poisson-Gamma	
	$\widehat{VaR_\alpha}$	$\widehat{\mathcal{E}_\alpha}$	$\widehat{VaR_\alpha}$	$\widehat{\mathcal{E}_\alpha}$
95.00th	$35.3	$45.5	$59.5	$101.6
97.00th	$40.4	$50.8	$80.3	$123.5
99.00th	$52.6	$61.2	$127.6	$171.5
99.50th	$58.8	$66.9	$158.3	$201.8
99.90th	$72.5	$80.6	$228.7	$270.8
99.97th	$82.4	$89.8	$280.6	$319.8
99.99th	$90.4	$97.8	$323.6	$361.9
Other statistics				
Expected loss	$9.2		$10.0	
Loss volatility	$12.8		$25.8	
Total iterations	10,000,000			
Computational time	33.2		48.6	

library.[11] It then uses these gamma variates to simulate the requisite Poisson random variables. From this point, it is quite familiar. We create the loss indicator, build the loss distribution, and construct our risk measures. The a and b parameters can be used directly and require no adjustment as described in equations 3.67 and 3.68. The numerical estimation works on the obligor level and, as a consequence, it is entirely appropriate to use the unadjusted calibrated values.

The associated numerical results for our usual example are summarized in Table 3.7. The values are broadly consistent with the binomial-mixture models and the analytic Poisson-gamma figures. Interestingly, unlike the binomial-mixture setting, the Poisson-gamma numerical results are slightly lower than the analytic estimates. The differences are not dramatic and are presumably related to the underlying nature of the model. As a final note, the simulation algorithm is relatively slow for both Poisson implementations. The Poisson-Gamma mixture

---

[11] The Python implementation uses the shape-scale representation of the gamma distribution. This means that we need to send the reciprocal of the b parameter to their random-number engine.

3.2 Poisson-Mixture Models

model requires almost 50 seconds to generate 10 million iterations. For large portfolios, this could easily become quite computationally expensive.

3.2.2 Other Poisson-Mixture Approaches

As with the binomial setting, the gamma distribution is not the only possible choice to mix with the Poisson approach. In this section, we will briefly explore *two* other alternative approaches. Broadly speaking, while not equivalent, these can be considered to be analogous to the logit- and probit-mixtures within the base binomial model.

Since the arrival rate or intensity parameter in the Poisson model must take a positive value, one important condition for any other candidate mixture distribution is a support limited to the positive real numbers. This somewhat restricts our choice, but does not pose any serious problem. Many distributions are defined only for positive values. We also require, to perform a sensible parametrization, a two-parameter distribution. This further constrains our choice.

After a bit of reflection and searching through Stuart and Ord (1987) and Casella and Berger (1990), we identified and selected the log-normal and Weibull distributions. Both have two parameters and are defined only for positive real values. This allows us to re-define equation 3.52 as,

$$\mathbb{P}(\mathbb{D}_N = k) = \int_{\mathbb{R}_+} \frac{e^{-s} s^k}{k!} f_{S_i}(s \mid a_i, b_i) \, ds, \qquad (3.69)$$

for $i = 1, 2$. In the first case, we assume that $\lambda(S) \equiv S$ where $S \sim \ln-\mathcal{N}(a_1, b_1^2)$. The log-normal density is defined as,

$$f_{S_1}(s \mid a_1, b_1) = \frac{1}{s\sqrt{2\pi b_1^2}} e^{\frac{(\ln s - a_1)^2}{2b_1^2}}. \qquad (3.70)$$

In the second case, $S \sim \mathcal{W}(a_2, b_2)$, where the Weibull density has the following form,

$$f_{S_2}(s \mid a_2, b_2) = \frac{a_2}{b_2} \left(\frac{s}{b_2}\right)^{a_2 - 1} e^{-\left(\frac{s}{b_2}\right)^{a_2}}. \qquad (3.71)$$

It may be possible, with some clever tricks, to analytically evaluate equation 3.69. Since, as they say, discretion is the better part of valour, we opted to use numerical integration to resolve it.

Algorithm 3.11 Poisson-mixture integral

```
def poissonMixtureIntegral(s,k,a,b,N,whichModel):
    pDensity = util.poissonDensity(N*s,k)
    if whichModel==0:
        mixDensity = util.logNormalDensity(s,a,b)
    elif whichModel==1:
        mixDensity = util.weibullDensity(s,a,b)
    f = pDensity*mixDensity
    return f
```

Algorithm 3.12 Poisson-mixture analytic model

```
def poissonMixtureAnalytic(N,myC,a,b,alpha,whichModel):
    pmf = np.zeros(N+1)
    for k in range(0,N+1):
        pmf[k],err = nInt.quad(poissonMixtureIntegral,0,k+1,\
            args=(k,a,b,N,whichModel))
    cdf = np.cumsum(pmf)
    varAnalytic = myC*np.interp(alpha,cdf,np.linspace(0,N,N+1))
    esAnalytic = util.analyticExpectedShortfall(N,alpha,pmf,myC)
    return pmf,cdf,varAnalytic,esAnalytic
```

Practically, as with the logit- and probit-mixture models, this can be accomplished in a single sub-routine. Algorithm 3.11 describes the integral found in equation 3.69. Both models share the Poisson form, denoted as pDensity, and the appropriate mixture density is determined by the whichModel flag. The final return value, f, is the product of the Poisson and mixture densities. The Poisson density, since this will be used from the analytic implementation, has the value of the mixing variable, s, grossed up by the number of obligors. This is the adjustment associated with equations 3.67 and 3.68.

The final step is to use Python's quad routine again to construct the probability-mass function for the total number of defaults. The now familiar code is summarized in Algorithm 3.12.

This is certainly beginning to become somewhat repetitive, but with each addition we are adding to our toolkit of possible model implementations. Specific model choices have become popular as commercial software solutions. Sales and marketing staff might make one believe that these are unique approaches. With all due respect to the model developers, the vast majority of commercial methodologies are special cases of more general models. By examining a large number of different possible implementations, we gain both breadth and depth of insight into the universe of credit-risk models. This insight is invaluable in implementing or selecting specific models for use in one's analysis.

3.2.2.1 A Calibration Comparison

Table 3.8 collects the calibrated parameters, for each of our Poisson-based models, associated with an average default probability of 0.01 and default correlation of 0.05. These parameter choices, following from the calibration algorithm, yield exactly the desired portfolio characteristics. It is useful and interesting to compare these results to the binomial-mixture values found in Table 3.3 on page 110.

The actual calibration of log-normal and Weibull Poisson-mixture models has precisely the same form as the Poisson-gamma and logit-probit mixtures in algorithms 3.4 and 3.9. The only additional element of interest is the computation of the necessary moments. The interesting point is that we seek the first two moments of $p(S) = 1 - e^{-S}$. This leaves us with two choices. We may compute the desired moment as,

$$\mathcal{M}_k(p(S)) = \underbrace{\int_{\mathbb{R}_+} p(s)^k f_S(s) ds}_{\text{Option \#1}} = \underbrace{\int_{\mathbb{R}_+} s^k f_{p(S)}(s) ds}_{\text{Option \#2}}. \tag{3.72}$$

In other words, you may compute the moment of a function of a random variable in two ways. In both cases, it is an integral. In the first case, the integrand is the product of the function and the underlying random variable's density: $p(S)$ and $f_S(s)$. In the latter case, one identifies the density of the transformed random variable: $f_{p(S)}(s)$. This latter approach necessitates use of the so-called change-of-variables formula to determine the density of $p(S)$. Since we used the (easier) first approach with the logit-probit binomial mixture models, we will use the second for the Poisson models. This will provide the reader with an overview of both techniques.

A bit of work, however, is first required. We have that S has density $f_S(s)$. We define the function,

$$Y = p(S) = 1 - e^{-S}. \tag{3.73}$$

Table 3.8 *Poisson calibration results*: This table summarizes the parameter values determined by calibrating all models to an average default probability of 1% and a default correlation coefficient of 0.05.

	Targets				Parameters	
Model	$\mathcal{M}_1$ or $\bar{p}$	$\mathcal{M}_2$	$\sqrt{\text{var}(p(S))}$	$\rho_{\mathcal{D}}$	a	b
Poisson-Gamma	0.01	0.0006	0.02	0.05	0.19	18.413
Poisson-lognormal	0.01	0.0006	0.02	0.05	−5.60	1.432
Poisson-Weibull	0.01	0.0006	0.02	0.05	0.48	0.005
Poisson	0.01	0.0000	0.00	0.00	n/a	n/a

To find the density of Y, it is necessary for it to have an inverse. While this is not always true, in our example, this is easily found as,

$$\mu(Y) = -\ln(1-Y). \tag{3.74}$$

By the change-of-variables formulae—which is well described in Billingsley (1995, Chapter 16) with all the necessary terms and conditions—the density of Y can be written as,

$$f_Y(y) = f_S(\mu(Y)) \cdot |\mu'(Y)|, \tag{3.75}$$

where $|\mu'(Y)|$ is typically referred to, in higher dimensions, as the determinant of the Jacobian matrix. If we then specialize this relationship to our specific case, it leads to

$$f_{p(S)}(y) = f_S(-\ln(1-y)) \cdot \left| -\frac{1}{1-y} \right|, \tag{3.76}$$

$$= \frac{f_S(-\ln(1-y))}{1-y}.$$

Thus we have a concrete definition of the density of the transformed random variable, $p(S)$, to evaluate its moments.

Although a bit messy, it is nonetheless readily implemented. Algorithm 3.13 is a general sub-routine that easily handles both the lognormal and Weibull cases. It essentially uses equation 3.76 to construct $f_{p(S)}(s)$, which is creatively termed psDensity. The actual calibration then follows from an fsolve function call of the form described in equation 3.45 on page 110. It does, we should mention, appear to be somewhat more sensitive to the starting values than the corresponding binomial logit- and probit-mixture model calibrations.

Algorithm 3.13 Poisson-mixture model moments

```
def poissonTransformedMixtureMoment(s,a,b,momentNumber,whichModel):
    vy = -np.log(1-s)
    jacobian = np.divide(1,1-s)
    if whichModel==0:
        psDensity = util.logNormalDensity(vy,a,b)*jacobian
    elif whichModel==1:
        psDensity = util.weibullDensity(vy,a,b)*jacobian
    myMoment = (s**momentNumber)*psDensity
    return myMoment
```

3.2 Poisson-Mixture Models

Table 3.9 *Analytic Poisson-mixture results*: This table compares the analytic Poisson independent-default outcomes to the Poisson gamma, log-normal, and Weibull model results. In all cases, the models are calibrated for a 1% average default probability. The mixture approaches also target a default correlation coefficient of 0.05.

Quantile	Poisson			Poisson-Gamma			Poisson log-normal			Poisson Weibull		
	$\mathbb{D}_N$	VaR_α	$\mathcal{E}_\alpha$	$\mathbb{D}_N$	VaR_α	$\mathcal{E}_\alpha$	$\mathbb{D}_N$	VaR_α	$\mathcal{E}_\alpha$	$\mathbb{D}_N$	VaR_α	$\mathcal{E}_\alpha$
95.00th	2.5	$25	$25	5.3	$53	$58	3.8	$38	$52	4.7	$47	$63
97.00th	2.8	$28	$28	7.3	$73	$81	5.3	$53	$82	6.6	$66	$111
99.00th	3.6	$36	$36	12.1	$121	$170	9.6	$96	$165	11.8	$118	$184
99.50th	3.9	$39	$39	15.4	$154	$204	13.3	$133	$218	15.8	$158	$233
99.90th	4.8	$48	$49	23.3	$233	$285	26.1	$261	$390	27.4	$274	$367
99.97th	5.5	$55	$55	29.5	$295	$348	40.9	$409	$561	38.2	$382	$486
99.99th	5.9	$59	$60	35.3	$353	$407	59.7	$597	$720	49.6	$496	$604
Other statistics												
Expected loss	1.0	$10.0		1.0	$10.0		1.0	$10.0		1.0	$10.0	
Loss volatility	1.0	$10.0		2.2	$22.2		2.2	$21.7		2.2	$22.2	
Default correlation	0.00			0.05			0.05			0.05		

3.2.3 Poisson-Mixture Comparison

We now have *four* distinct Poisson-based models: the independent-default, Poisson-gamma, Poisson-lognormal, and Poisson-Weibull versions. It is natural, and useful, to examine their implications for our simple example. Table 3.9, in a dizzying array of numbers, provides the fractional number of defaults, the VaR, and expected-shortfall figures for each flavour of analytically evaluated Poisson models.

In a manner very similar to the binomial-mixture models, the Poisson-mixture estimates are highly similar out to the 99.9th quantile. After that point, which is admittedly quite far into the tail of the loss distribution, they start to diverge. The lognormal approach is the most conservative, followed by the Weibull and Poisson-Gamma approaches, respectively. The magnitude of the results do not appear to differ importantly from the binomial case; the one exception is the Poisson-lognormal model, which appears to yield the highest estimate across all eight models.

Table 3.10 provides, for completeness, the numerical VaR and expected-shortfall estimates for each of the Poisson-based models.[12] Again, the sheer volume of numbers is somewhat overwhelming. Figure 3.7 on page 130 provides a graphical summary of the tail probabilities and, in an attempt to permit comparison of the

---

[12] We have dispensed with showing the `poissonMixtureSimulation` algorithm, because it is almost identical to Algorithm 3.10 and this chapter is already sufficiently long and repetitive. The interested reader can easily locate it, however, in the `mixtureModels` library. See Appendix D for more details.

Table 3.10 *Numerical Poisson-mixture results*: This table provides the quantiles of the loss distribution for the Monte Carlo estimates of the independent—default Poisson model relative to the Poisson-gamma, log-normal, and Weibull mixture approaches. All models have an average default probability of 1% and each mixture model is calibrated to a 0.05 default-indicator correlation coefficient.

Quantile	Poisson		Poisson-Gamma		Poisson-lognormal		Poisson-Weibull	
	$\widehat{\text{VaR}}_\alpha$	$\widehat{\mathcal{E}}_\alpha$	$\widehat{\text{VaR}}_\alpha$	$\widehat{\mathcal{E}}_\alpha$	$\widehat{\text{VaR}}_\alpha$	$\widehat{\mathcal{E}}_\alpha$	$\widehat{\text{VaR}}_\alpha$	$\widehat{\mathcal{E}}_\alpha$
95.00th	$35.3	$45.5	$59.5	$101.6	$49.1	$93.4	$54.5	$98.9
97.00th	$40.4	$50.8	$80.3	$123.5	$66.0	$118.1	$74.1	$122.7
99.00th	$52.6	$61.2	$127.6	$171.5	$114.2	$185.6	$124.1	$180.5
99.50th	$58.8	$66.9	$158.3	$201.8	$155.3	$239.6	$160.6	$221.0
99.90th	$72.5	$80.6	$228.7	$270.8	$286.9	$399.2	$257.9	$324.4
99.97th	$82.4	$89.8	$280.6	$319.8	$418.4	$541.3	$338.3	$404.9
99.99th	$90.4	$97.8	$323.6	$361.9	$557.1	$675.9	$412.4	$477.4
Other statistics								
Expected loss	$9.2		$10.0		$10.0		$10.0	
Loss volatility	$12.8		$25.8		$25.9		$25.9	
Total iterations	10,000,000							
Computational time	33.2		48.6		47.6		48.0	

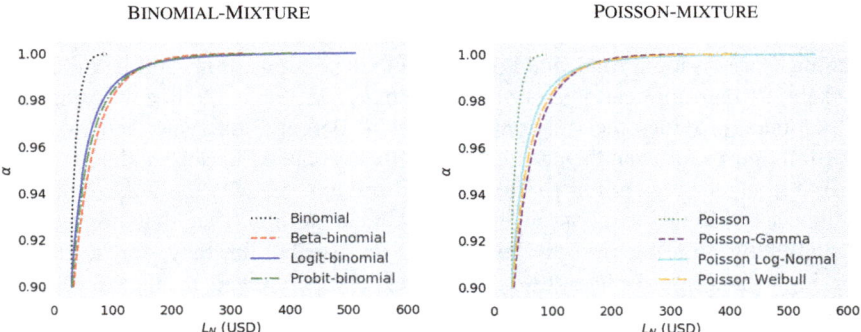

Fig. 3.7 *Comparing tail probabilities*: The underlying figures illustrate, for *eight* distinct binomial- and Poisson-mixture models, the structure of tail probabilities for our base portfolio example. As we incorporate default dependence and employ a common calibration, the general form is similar across all models. As we move further into the tail, however, significant differences arise.

binomial and Poisson approaches, includes all eight of the models considered thus far in the text.

The scale on both graphics in Fig. 3.7 are identical to ease comparison. Given that all models have been calibrated to the same desired default probability and correlation characteristics, and have been applied to the same dataset, the only differences stem from the structure of the underlying models. This allows us to roughly categorize the various versions. We have, for example, already identified the close relationship between the Poisson and binomial models. Visually, however, the beta-binomial and Poisson-gamma implementations also appear to have a similar form; both seem to be the least extreme of the mixture models. Indeed, we can also pair up the remaining models. The most extreme models are the logit-binomial and Poisson-lognormal approaches: both generate 99.99th VaR estimates in excess of $500. This leads, by process of elimination, to the two intermediate choices: the probit-binomial and Poisson-Weibull methodologies.

This is not to suggest that these convenient model pairings imply model equivalence—there are certainly numerous subtle differences between each implementation associated with their own unique distributional assumptions. Moreover, with an alternative portfolio, the results might vary significantly. Nevertheless, it is an interesting exercise and provides some insight into the behaviour of these various model choices.

Although we have considered an impressive array of actuarial models to this point, relatively few of these models are used by practitioners in actual applications. In the next section, we will build on these ideas and demonstrate a popular model quite widely used in commercial applications. The reader will quickly see, however, that it is nothing other than an, admittedly clever, twist on the preceding ideas.

3.3 CreditRisk+

There are, at least, two important shortcomings of all of binomial- and Poisson-mixture models addressed in the previous section. In particular,

1. all obligors are forced to share a common default probability and thus information on individual creditworthiness is lost; and
2. it is not immediately obvious how additional systematic factors might be introduced into the framework.

The perhaps best known, and most widely used commercial mixture-model addresses both of these issues. Roughly 20 year ago, as of the time of this book's publication, Wilde (1997) released the details of Credit-Suisse First Boston's (CSFB) CreditRisk+ model. In the remainder of this chapter, we will consider it in detail, apply it to our standard portfolio example, discuss how it might be calibrated, and show how it relates to the general class of mixture models.

3.3.1 A One-Factor Implementation

The full-blown CreditRisk+ approach is a multi-factor implementation. It is easier to understand, however, in its one-factor version. This will form our starting point and will address the first shortcoming of the mixture models examined thus far.

At its heart, the CreditRisk+ model is basically a Poisson-gamma mixture model. Or, equivalently as we saw in the preceding sections, a negative-binomial model. In the standard Poisson-gamma setting, the conditional probability of default for the n obligor is simply,

$$\mathbb{P}\left(\mathbb{I}_{\mathcal{D}_n}\big|\, S\right) \equiv \underbrace{\mathbb{P}\left(\mathbb{I}_{\{X_n \geq 1\}}\big|\, S\right)}_{p(S)}, \qquad (3.77)$$
$$= 1 - e^{-S},$$

where $S \sim \Gamma(a, b)$. In the CreditRisk+ setting, the key innovation is a different, and quite useful, specification of the conditional default probability. Specifically, Wilde (1997) opted to give it the following form,

$$p_n(S) \equiv \mathbb{P}\left(\mathbb{I}_{\mathcal{D}_n} = 1\big|\, S\right) \qquad (3.78)$$
$$= p_n\left(\omega_0 + \omega_1 S\right),$$

where p_n is the usual unconditional default probability associated with the nth credit counterparty, $\omega_0, \omega_1 \in \mathbb{R}$, and, as before, $S \sim \Gamma(a, b)$. We denote it as $p_n(S)$ to clearly indicate that it depends on the individual obligor, which is not the case for the more generic Poisson and binomial mixture models.

The default event remains $\mathbb{I}_{\mathcal{D}_n} = \{X_n \geq 1\}$, where $X_n \sim \mathcal{P}(p_n(S))$. The rather complex term in equation 3.78 is the Poisson arrival rate or intensity. Unlike the Poisson-gamma setting, where we derive a relationship between $\lambda(S)$ and $p(S)$—it is directly imposed. To be clear, the conditional default probability and the Poisson arrival intensity are identically equal in this setting.

At first blush, this neither seems to be an improvement nor a simplification. Two additional assumptions are required to gain some traction. First, we force the parameters, ω_0 and ω_1, to sum to unity or rather, $\omega_0 + \omega_1 = 1$. Second, we select the gamma-distributional parameters, a and b, such that $\mathbb{E}(S) = 1$. Furnished with these assumptions, we can proceed to evaluate the expected value of $p_n(S)$ as,

$$\mathbb{E}\left(\underbrace{\mathbb{P}\left(\mathbb{I}_{\mathcal{D}_n} = 1\big|\, S\right)}_{p_n(S)}\right) = \mathbb{E}\left(p_n\left(\omega_0 + \omega_1 S\right)\right), \qquad (3.79)$$

$$\mathbb{E}\left(\mathbb{E}\left(\mathcal{D}_n = 1\big|\, S\right)\right) = p_n\left(\omega_0 + \omega_1 \underbrace{\mathbb{E}(S)}_{=1}\right),$$

$$\mathbb{E}\left(\mathcal{D}_n = 1\right) = p_n \underbrace{\left(\omega_0 + \omega_1\right)}_{=1},$$

$$\mathbb{P}\left(\mathbb{I}_{\mathcal{D}_n}\right) = p_n.$$

3.3 CreditRisk+

These judicious choices for the model parameters, and the functional form of the conditional default probability, imply that the individual obligor's unconditional default probabilities are recovered. In other words, instead of a common, portfolio-wide unconditional default probability of $\bar{p}$, this approach allows each obligor to, in principle, have its own characterization of its creditworthiness.

In accomplishing this task, it also maintains the presence of a systematic factor to permit default correlation. If we revisit equation 3.78, we observe an interesting decomposition of the conditional default probability,

$$p_n(S) = p_n \left(\underbrace{\omega_0}_{\substack{\text{Idio-}\\\text{syncratic}}} + \underbrace{\omega_1 S}_{\substack{\text{Syste-}\\\text{matic}}} \right). \qquad (3.80)$$

This returns us to the powerful general ideas of the CAPM: idiosyncratic and systematic risk factors. Much depends on the outcome of S. It is easy to see that if $S = 0$, then this form collapses to an independent-default model. If $S > 1$, then the conditional default probability exceeds its unconditional equivalent. More simply, if we think of S as a common global economic factor, a large outcome scales up the default probability for all obligors. It accomplishes this in a manner commensurate with their original creditworthiness. S, in essence, acts as a kind of proxy for global economic conditions.

The condition on the factor loadings is easy to respect, but some additional structure is required to force $\mathbb{E}(S) = 1$. Since for $S \sim \Gamma(a, b)$, the expectation is $\frac{a}{b}$, then it follows that we must force $a = b$. This essentially amounts to the elimination of a parameter. The generic variance of a gamma-distributed random variable is $\frac{a}{b^2}$, which directly implies that the $\text{var}(S) = \frac{1}{a}$. To respect the $\mathbb{E}(S) = 1$ condition, therefore, we draw the gamma-distributed variates from $S \sim \Gamma(a, a)$.

The value of the parameters also plays an important role. ω_0 and ω_1 can be viewed as the loadings on idiosyncratic and systematic risk. If $\omega_0 = 1$, we again return to the independent-default setting. In the opposite case of $\omega_1 = 1$, all risk becomes systematic—albeit always with regard to the credit quality of each individual counterpart.

Figure 3.8 illustrates 100 random simulations of this one-factor CreditRisk+ conditional default probability. We set $\omega_1 = 0.2$, $a = 0.01$, and $p_n = 0.01$. The consequence is that $\omega_0 = 0.8$ and thus 80% of the weight is placed on the idiosyncratic factor, which implies that $p_n(S)$ is often around 80 basis points. Any S realization in the unit interval can adjust the systematic contribution by only zero to 20 basis points. Occasionally, however, a large outcome of S occurs and pulls it up above this level. In some cases, for this parameterization, the conditional default probability reaches the 0.2 level.

While a clever set-up, we still need to be able to calibrate this model to yield any desired amount of default correlation. This requires computing a number of

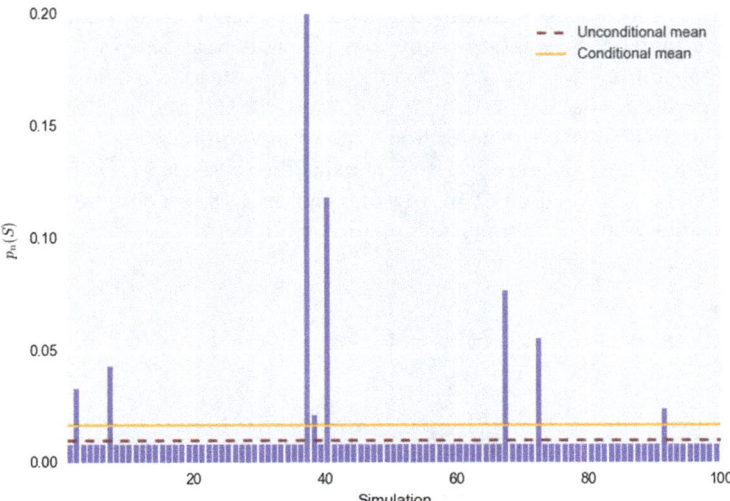

Fig. 3.8 *CreditRisk+ conditional default probabilities*: This graphic summarizes 100 random simulations of the CreditRisk+ conditional default probability with $\omega_1 = 0.2$, $a = 0.01$, and $p_n = 0.01$. 80% of the weight is on the idiosyncratic factor, which implies that $p_n(S)$ is often around 80 basis points, but occasionally a large outcome of S pulls it up above this level.

relationships. The conditional variance, for example, becomes,

$$\operatorname{var}(p_n(S)) = \operatorname{var}(p_n(\omega_0 + \omega_1 S)), \tag{3.81}$$
$$= \operatorname{var}(p_n \omega_1 S),$$
$$= p_n^2 \omega_1^2 \underbrace{\operatorname{var}(S)}_{1/a},$$
$$= \frac{p_n^2 \omega_1^2}{a},$$
$$\sigma(p_n(S)) = p_n \sqrt{\frac{\omega_1^2}{a}}.$$

The default correlation needs to be determined from first principles as,

$$\rho_{\mathcal{D}} = \rho\left(\mathbb{I}_{\mathcal{D}_n}, \mathbb{I}_{\mathcal{D}_m}\right) = \frac{\mathbb{E}\left(\left(\mathbb{I}_{\{X_n \geq 1\}} - \mathbb{E}\left(\mathbb{I}_{\{X_n \geq 1\}}\right)\right)\left(\mathbb{I}_{\{X_m \geq 1\}} - \mathbb{E}\left(\mathbb{I}_{\{X_m \geq 1\}}\right)\right)\right)}{\sqrt{\operatorname{var}\left(\mathbb{I}_{\{X_n \geq 1\}}\right)}\sqrt{\operatorname{var}\left(\mathbb{I}_{\{X_m \geq 1\}}\right)}},$$
$$= \frac{\mathbb{E}\left(\left(\mathbb{I}_{\{X_n \geq 1\}} - p_n\right)\left(\mathbb{I}_{\{X_m \geq 1\}} - p_m\right)\right)}{\sqrt{p_n(1-p_n)}\sqrt{p_m(1-p_m)}},$$
$$= \frac{\mathbb{E}\left(\mathbb{I}_{\{X_n \geq 1\}} \mathbb{I}_{\{X_m \geq 1\}}\right) - p_n p_m}{\sqrt{p_n(1-p_n)}\sqrt{p_m(1-p_m)}}. \tag{3.82}$$

3.3 CreditRisk+

To take this further, we need to evaluate the integral, $\mathbb{E}\left(\mathbb{I}_{\{X_n\geq 1\}}\mathbb{I}_{\{X_m\geq 1\}}\right)$. This is a bit messy, but with the use of iterated expectations and our definition of conditional default probability, we can reduce it to something meaningful. In particular, we have that

$$\mathbb{E}\left(\mathbb{I}_{\{X_n\geq 1\}}\mathbb{I}_{\{X_m\geq 1\}}\right) = \mathbb{E}\bigg(\underbrace{\mathbb{E}\left(\mathbb{I}_{\{X_n\geq 1\}}\mathbb{I}_{\{X_m\geq 1\}}\big|\, S\right)}_{\text{By iterated expectations}}\bigg), \quad (3.83)$$

$$= \mathbb{E}\bigg(\underbrace{\mathbb{E}\left(\mathbb{I}_{\{X_n\geq 1\}}\big|\, S\right) \mathbb{E}\left(\mathbb{I}_{\{X_m\geq 1\}}\big|\, S\right)}_{\text{By conditional independence}}\bigg),$$

$$= \mathbb{E}\left(p_n(S)p_m(S)\right),$$

$$= \mathbb{E}\left(p_n(\omega_0+\omega_1 S)p_m(\omega_0+\omega_1)S\right),$$

$$= p_n p_m \mathbb{E}\left((\omega_0+\omega_1 S)^2\right),$$

$$= p_n p_m \mathbb{E}\left(\omega_0^2 + 2\omega_0\omega_1 S + \omega_1^2 S^2\right),$$

$$= p_n p_m \left(\omega_0^2 + 2\omega_0\omega_1 + \omega_1^2 \mathbb{E}\left(S^2\right)\right).$$

To find $\mathbb{E}(S^2)$, we need to start with the variance of S and work backwards,

$$\underbrace{\text{var}(S)}_{1/a} = \mathbb{E}\bigg(S^2 - \underbrace{\mathbb{E}(S)^2}_{=1}\bigg), \quad (3.84)$$

$$\frac{1}{a} = \mathbb{E}(S^2) - 1,$$

$$\mathbb{E}(S^2) = 1 + \frac{1}{a}.$$

Substituting this into equation 3.83 and simplifying, we arrive at

$$\mathbb{E}\left(\mathbb{I}_{\{X_n\geq 1\}}\mathbb{I}_{\{X_m\geq 1\}}\right) = p_n p_m \bigg(\omega_0^2 + 2\omega_0\omega_1 + \omega_1^2 \underbrace{\left(1+\frac{1}{a}\right)}_{\mathbb{E}(S^2)}\bigg), \quad (3.85)$$

$$= p_n p_m \bigg(\underbrace{\omega_0^2 + 2\omega_0\omega_1 + \omega_1^2}_{(\omega_0+\omega_1)^2=1} + \frac{\omega_1^2}{a}\bigg),$$

$$= p_n p_m \left(1 + \frac{\omega_1^2}{a}\right).$$

Returning to equation 3.82, we can finally describe the default correlation in the one-factor CreditRisk+ model as,

$$\rho(\mathcal{D}_n, \mathcal{D}_m) = \rho(\mathbb{I}_{\mathcal{D}_n}, \mathbb{I}_{\mathcal{D}_m}) = \frac{\overbrace{p_n p_m \left(1 + \frac{\omega_1^2}{a}\right)}^{\mathbb{E}(\mathbb{I}_{\{X_n \geq 1\}} \mathbb{I}_{\{X_m \geq 1\}})} - p_n p_m}{\sqrt{p_n(1-p_n)}\sqrt{p_m(1-p_m)}}, \qquad (3.86)$$

$$= \left(\frac{\omega_1^2}{a}\right)\left(\frac{p_n p_m}{\sqrt{p_n(1-p_n)}\sqrt{p_m(1-p_m)}}\right).$$

Default correlation is thus determined by both the systematic factor's loading parameter, ω_1, and its single parameter, a. Warning signs are probably going off in many readers minds as we see the a parameter in the denominator. As we make a arbitrarily small, unfortunately, we can increase default correlation to whatever levels we desire. Since the correlation coefficient should not exceed unity, this creates some practical problems. Given that ω_1^2 takes its maximum value at $\omega_1 = 1$, a conservative lower bound for a is $\frac{p_n p_m}{\sqrt{p_n(1-p_n)}\sqrt{p_m(1-p_m)}}$. A better constraint can be formulated, depending on one's choice of ω_1^2, but we need to be aware of this issue.

Figure 3.9 outlines the relationship between the default correlation described in equation 3.86 and its parameters. To construct this graphic, we have, for simplicity,

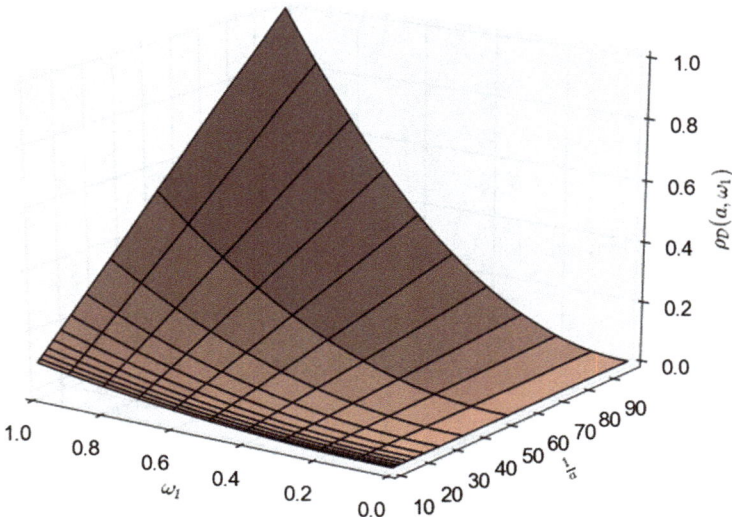

Fig. 3.9 *A default-correlation surface*: The underlying figure outlines the relationship between the default correlation described in equation 3.86 and its parameters. The highest levels of default correlation are achieved when ω_1 and $\frac{1}{a}$ are simultaneously large. In all cases, we have assumed that $p_n = p_m = 0.01$. Understanding this fact helps us sensibly calibrate this model.

3.3 CreditRisk+

assumed that $p_n = p_m = 0.01$. The highest levels of default correlation are achieved when ω_1 and $\frac{1}{a}$ are simultaneously large.[13] The punchline is that there is no unique way to target a given amount of default correlation. If the analyst wishes this model to generate an average default-correlation coefficient of 0.05, for example, she will, in principle, need to wrestle with both ω_1 and a.

We nevertheless cannot, as before, find two equations in two unknowns and solve for a and ω_1. The unconditional default expectation in equation 3.79 is free from either parameter. The unconditional default variance and the default correlation—found in equations 3.81 and 3.86, respectively—contain both variables, but they always enter as a quotient of one another. This means that we have basically two equations in a single unknown. Overall, we have three possible moving parts: the average default volatility, $\sigma_{\bar{p}}$, the average default correlation, $\rho_\mathcal{D}$, and the term $\frac{\omega_1^2}{a}$. Calibration thus involves fixing some subset of these values and then determining the others. Breaking down the quotient of our two parameters, however, is somewhat arbitrary.

Let's see how this might work. Imagine that you wish to have the average default correlation equal to 0.05. This means that, for equation 3.86, we replace p_n and p_m with $\bar{p}$. This gives us the following base calibration equation,

$$0.05 = \rho_\mathcal{D} = \frac{\omega_1^2}{a} \frac{p_n p_m}{\sqrt{p_n(1-p_n)p_m(1-p_m)}}, \qquad (3.87)$$

$$= \frac{\omega_1^2}{a} \frac{\bar{p}^2}{\bar{p}(1-\bar{p})},$$

$$= x \frac{\bar{p}}{(1-\bar{p})},$$

where $x = \frac{\omega_1^2}{a}$. That is, we have combined the two parameters into a single value. The desired default correlation is $\rho_\mathcal{D} = 0.05$ and setting, consistent with our running example, $\bar{p} = 0.01$. The direct consequence is that,

$$x = \frac{\rho_\mathcal{D} \cdot (1-\bar{p})}{\bar{p}} = \frac{0.05 \cdot (1-0.01)}{0.01} = 4.95. \qquad (3.88)$$

We only know the necessary value of our parameters up to their quotient. If you fix one, then you find the other. Suppose that we assume, or estimate from empirical data, that a, the unconditional default volatility, is 0.01. In that case, we may immediately conclude that,

$$\omega_1 = \sqrt{x \cdot a} = \sqrt{4.96 \cdot 0.01} = 0.22. \qquad (3.89)$$

---

[13] Or, in an equivalent manner, as a becomes small.

Table 3.11 *Possible CreditRisk+ calibrations*: This table shows the interaction between different parameter choices and the average default correlation and volatility values. The values were computed by fixing a and varying the value of ω_1—one could, of course, work in the opposite direction.

Parameter choices				
Selected			Inferred	
ω_1	a	$x = \frac{\omega_1^2}{a}$	$\rho_\mathcal{D}$	$\sigma_{\bar{p}}$
0.00	0.01	0.00	0.00	0.00
0.05	0.01	0.25	0.00	0.00
0.15	0.01	2.25	0.02	0.01
0.20	0.01	4.00	0.04	0.02
0.25	0.01	6.25	0.06	0.02
0.30	0.01	9.00	0.09	0.03
0.40	0.01	16.00	0.15	0.04

Naturally, it could go in the other direction. If you assume that $\omega_1 = 0.3$, then it follows that $a = 0.018$. In both cases, the average default volatility is $\sqrt{x}\bar{p} = \sqrt{4.5} \cdot 0.01 = 0.022$ or 2.2%.

In their excellent reference on CreditRisk+, Gundlach and Lehrbass (2004) suggest that the key input to model calibration is one's estimate of default-probability volatility. This is entirely consistent with the previously described approach, albeit with default-probability volatility replacing the default correlation; see, for example, the similarities with equation 3.81. We have focused on the default correlation in this discussion to ensure consistency with the previous model calibrations and, quite simply, because it also feels more natural.[14] The default-probability volatility will be considered again, however, in Chap. 10. Ultimately, this is more a question of personal preference since manipulation of our previous definitions lead to the following identity linking both quantities,

$$\rho_\mathcal{D} = \sqrt{\frac{\omega_1^2}{a} \frac{\bar{p}}{(1-\bar{p})}} \equiv \frac{\sigma(\bar{p})}{1-\bar{p}}. \tag{3.90}$$

Table 3.11 describes the relationship between different parameter choices and the average default correlation and default-probability volatility values. The parameter settings were computed by fixing a and varying the value of ω_1. One then infers the other average default correlation and volatility quantities.[15] The default correlation thus ranges from 0 to 0.15 as we increase the systematic-factor loading from 0 to 0.40. We have thus established a slightly fragile, but clear, relationship between the systematic factor loading and the overall level of default dependence.

We may now use these results to compute the usual risk measures for our 100-obligor example. At this point, both Wilde (1997) and Gundlach and Lehrbass (2004) invest a significant amount of time and effort to demonstrate how one might

---

[14] It also helps, as we see in the next chapter, to create a closer conceptual link to the family of threshold models.

[15] As we've stressed so far, one could, of course, work in the opposite direction.

analytically solve the CreditRisk+ model. The basic idea is to use the so-called probability generating function—which is, loosely speaking, the discrete analogue of the Laplace transform or moment-generating function used in the continuous setting—to find an expression for the default-loss distribution. In the end, however, one has to use a recursion relationship—known as the Panjer recursion, see Panjer (1981)—to numerically solve it. Gundlach and Lehrbass (2004), among others, suggest that this approach is not numerically stable and offer a number of alternatives approaches. Moreover, it is a fairly involved mathematical technique.

For our purposes, this model is easily and readily implemented using Monte-Carlo methods. Moreover, the probability generating function approach does not provide any important insight into the underlying problem. It can, at the expense of significant complexity, reduce the computational expense somewhat. In our view, this is not a price worth paying.[16]

Algorithm 3.14 summarizes the Python sub-routine used to generate one-factor CreditRisk+ estimates. With the exception of the additional parameter, w, which is ω_1, it is essentially a Poisson-gamma model. The numpy gamma-distributed random variable generator uses the alternative definition of the gamma distribution; as a consequence, we draw the state variable $S \sim \Gamma\left(a, \frac{1}{a}\right)$. This difference stems merely from a definitional choice and has no influence on the final results.

Applying Algorithm 3.14 to our example, for a few of the different calibrations found in Table 3.11, is summarized in Table 3.12. If we set $\omega_1 = 0$, this implies default independence and we recover the Poisson model; the CreditRisk+ model has, therefore, the pleasant property of nesting the independent-default approach. As we scale up ω_1, and thereby the level of default correlation, we observe a gradual increase in the overall magnitude of our risk estimates. Interestingly, as we increase the default correlation, we see a decrease in the level of expected loss, but an increase in loss volatility. This is entirely consistent with a shift of weight into the tails of the default-loss distribution.

Algorithm 3.14 One-factor CreditRisk+ Monte-Carlo implementation

```
def crPlusOneFactor(N,M,w,p,c,v,alpha):
    S = np.random.gamma(v, 1/v, [M])
    wS = np.transpose(np.tile(1-w + w*S,[N,1]))
    pS = np.tile(p,[M,1])*wS
    H = np.random.poisson(pS,[M,N])
    lossIndicator = 1*np.greater_equal(H,1)
    lossDistribution = np.sort(np.dot(lossIndicator,c),axis=None)
    el,ul,var,es=util.computeRiskMeasures(M,lossDistribution ,alpha)
    return el,ul,var,es
```

---

[16]The reader is naturally free to disagree; for those that do, Gundlach and Lehrbass (2004) is the place to start.

Table 3.12 *A one-factor CreditRisk+ model*: This table summarizes the results associated with four implementations of the one-factor CreditRisk+ model to our simple 100-obligor portfolio: each has a different weight, ω, on the systematic risk factor, S. We observe that as ω increases, the loss volatility and the tail outcomes increase in magnitude.

Quantile	$\rho_\mathcal{D} = 0.00$		$\rho_\mathcal{D} = 0.02$		$\rho_\mathcal{D} = 0.06$		$\rho_\mathcal{D} = 0.15$	
	$\widehat{\text{VaR}_\alpha}$	$\widehat{\mathcal{E}_\alpha}$	$\widehat{\text{VaR}_\alpha}$	$\widehat{\mathcal{E}_\alpha}$	$\widehat{\text{VaR}_\alpha}$	$\widehat{\mathcal{E}_\alpha}$	$\widehat{\text{VaR}_\alpha}$	$\widehat{\mathcal{E}_\alpha}$
95.00th	$35.0	$45.2	$34.9	$55.6	$33.3	$63.8	$30.9	$74.8
97.00th	$40.1	$50.5	$41.6	$67.6	$41.2	$81.7	$39.6	$101.6
99.00th	$52.2	$60.9	$60.3	$104.7	$69.4	$143.6	$92.7	$195.8
99.50th	$58.6	$66.5	$83.8	$140.3	$118.4	$198.0	$165.7	$267.3
99.90th	$71.9	$80.0	$177.5	$232.6	$250.3	$319.5	$335.2	$414.4
99.97th	$81.7	$89.4	$245.9	$295.5	$337.5	$397.0	$438.0	$498.3
99.99th	$90.2	$97.8	$303.3	$346.8	$406.6	$457.4	$509.4	$558.0
Other statistics								
Expected loss	$9.1		$8.9		$8.6		$8.1	
Loss volatility	$12.7		$16.1		$19.4		$24.1	
Total iterations	10,000,000							
Computational time	40.8		40.6		40.5		40.8	

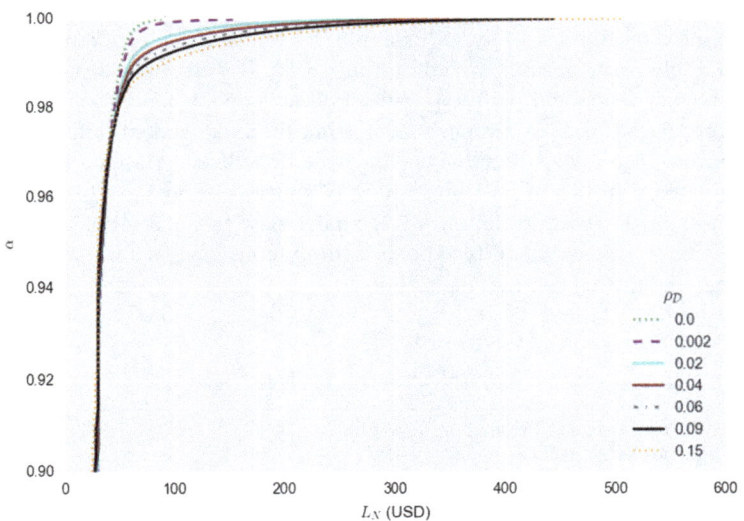

Fig. 3.10 *CreditRisk+ tail probabilities*: This figure highlights the one-factor CreditRisk+ tail probabilities for different possible choices of the correlation parameter, ω.

Figure 3.10 provides the tail probabilities, over the interval [0.9500, 0.9999], for each of the seven calibrations in Table 3.11. We see a collection of CreditRisk+ models all indexed to a particular systematic risk-factor loading. It carries all of the advantages of the other binomial and Poisson mixture models—fat tails, more complex asymptotic behaviour, and default dependence—but also has the important additional advantage of incorporating the obligor-specific default probabilities.

This combination of numerous advantageous characteristics explains the popularity of the CreditRisk+ model among practitioners and its 20-year longevity as a widely sought-out default-risk model. In practice, however, it is rarely used in its one-factor form. Instead, it is generalized to include an arbitrary number of systematic risk factors. We consider this setting in the following, and final, section of this chapter.

3.3.2 A Multi-Factor CreditRisk+ Example

To say that the multi-factor version of the CreditRisk+ is an extension of the one-factor model in the previous section is, while strictly true, a bit flippant. The basic structure is familiar, but the overall level of complexity grows quite substantially. The plan for this final section is to work through the details in an effort to understand its subtleties and motivate its use and calibration. Once we have the basic structures in place, we will then use the regional breakdown from our concrete portfolio example to see how it works in practice.

The first step is to add additional systematic factors. We introduce therefore, K distinct gamma-distributed random variables,

$$\mathbf{S} = \begin{bmatrix} S_1 & \cdots & S_K \end{bmatrix}, \tag{3.91}$$

where $S_k \sim \Gamma(a_k, a_k)$ for $k = 1, \ldots, K$. A critical assumption is that each S_k is independent and, given $\mathbf{S}$, each default event is also independent. The independence of these individual systematic state variables dramatically simplifies what is already a complex undertaking. This choice is, however, viewed by some as an important weakness. Gundlach and Lehrbass (2004) suggest a few, rather more complicated, alternatives.

Equation 3.91 allows us to define the conditional default probability—or equivalently the Poisson arrival rate—for the nth obligor as,

$$p_n(\mathbf{S}) = p_n \left(\omega_{n,0} + \sum_{k=1}^{K} \omega_{n,k} S_k \right), \tag{3.92}$$

$$= p_n \left(\sum_{k=0}^{K} \omega_{n,k} S_k \right),$$

where $S_0 \equiv 1$. This specification permits each obligor to maintain its own idiosyncratic risk, described by $\omega_{n,0}$, and also load uniquely onto any of the other remaining K systematic factors. The result is a theoretical limit of $(K+1) \cdot N$ model parameters. In practical settings, the actual number of parameters is significantly smaller, but the general development is not the appropriate place to impose such restrictions.

As before, we have selected the gamma parameters such that $\mathbb{E}(S_k) = 1$. In the one-factor case, it was sufficient to set $\omega_0 + \omega_1 = 1$ to ensure that the expected Poisson arrival intensity for counterparty n was equal to p_n. In this case, not surprisingly, it is a bit more involved. In particular, if we evaluate the expectation, we have

$$\mathbb{E}(p_n(\mathbf{S})) = \mathbb{E}\left(p_n\left(\sum_{k=0}^{K} \omega_{n,k} S_k\right)\right), \tag{3.93}$$

$$= p_n\left(\sum_{k=0}^{K} \omega_{n,k} \underbrace{\mathbb{E}(S_k)}_{=1}\right),$$

$$= p_n\left(\sum_{k=0}^{K} \omega_{n,k}\right),$$

$$= p_n.$$

This final step is only true if we impose the following N equality constraints,

$$\sum_{k=0}^{K} \omega_{n,k} = 1, \tag{3.94}$$

for $n = 1, \ldots, N$. The factor loadings, for each individual obligor, to the idiosyncratic and K systematic factors must sum to unity. Imposition of this constraint permits us to have the expected arrival intensity coincide with the unconditional default probability.

The unconditional variance (and volatility) of $p_n(\mathbf{S})$ also has a familiar, if rather more elaborate, form,

$$\text{var}(p_n(\mathbf{S})) = \text{var}\left(p_n\left(\sum_{k=0}^{K} \omega_{n,k} S_k\right)\right), \tag{3.95}$$

$$= p_n^2 \sum_{k=0}^{K} \omega_{n,k}^2 \text{var}(S_k),$$

$$\sigma(p_n(\mathbf{S})) = p_n \sqrt{\sum_{k=0}^{K} \frac{\omega_{n,k}^2}{a_k}}.$$

3.3 CreditRisk+

This is the multivariate result associated with equation 3.81. Arrival-intensity volatility now depends, however, on the standard deviation of all systematic factors and their relative loadings. This increases the flexibility and potential realism of the model.

The final requirement is an expression for the default correlation. This provides insight into the model and some ideas for how to actually calibrate it. We learned from equation 3.82 that,

$$\rho\left(\mathbb{I}_{\mathcal{D}_n}, \mathbb{I}_{\mathcal{D}_m}\right) = \frac{\mathbb{E}\left(\mathbb{I}_{\{X_n \geq 1\}} \mathbb{I}_{\{X_m \geq 1\}}\right) - p_n p_m}{\sqrt{p_n(1-p_n)} \sqrt{p_m(1-p_m)}}. \tag{3.96}$$

This result applies in both the univariate and multivariate settings. The key term, of course, is $\mathbb{E}\left(\mathbb{I}_{\{X_n \geq 1\}} \mathbb{I}_{\{X_m \geq 1\}}\right)$. Simplifying it to a meaningful and useful expression, however, requires a bit of effort.

$$\mathbb{E}\left(\mathbb{I}_{\{X_n \geq 1\}} \mathbb{I}_{\{X_m \geq 1\}}\right) = \mathbb{E}\left(\mathbb{E}\left(\mathbb{I}_{\{X_n \geq 1\}}\right)\left(\mathbb{I}_{\{X_m \geq 1\}}\right)\right), \tag{3.97}$$

$$= \mathbb{E}\left(p_n(\mathbf{S}) \, p_m(\mathbf{S})\right),$$

$$= \mathbb{E}\left(p_n\left(\sum_{k=0}^{K} \omega_{n,k} S_k\right) p_m\left(\sum_{j=0}^{K} \omega_{m,j} S_j\right)\right),$$

$$= p_n p_m \mathbb{E}\left(\sum_{k=0}^{K} \sum_{j=0}^{K} \omega_{n,k} \omega_{m,j} S_k S_j\right),$$

$$= p_n p_m \left(\underbrace{\sum_{k=0}^{K} \sum_{j=0}^{K} \omega_{n,k} \omega_{m,j}}_{k \neq j} + \sum_{k=0}^{K} \omega_{n,k} \omega_{m,k} \underbrace{\mathbb{E}\left(S_k^2\right)}_{1+\frac{1}{a_k}}\right),$$

$$= p_n p_m \left(\underbrace{1 - \sum_{k=0}^{K} \omega_{n,k} \omega_{m,k}}_{\text{Equation 3.98}} + \sum_{k=0}^{K} \omega_{n,k} \omega_{m,k} \left(1 + \frac{1}{a_k}\right)\right),$$

$$= p_n p_m \left(1 + \sum_{k=1}^{K} \frac{\omega_{n,k} \omega_{m,k}}{a_k}\right).$$

We adjust the index from $k = 0$ to $k = 1$ in the final step, because the idiosyncratic factor, $S_0 = 1$, as a constant does not contribute to the covariance; quite simply,

$\frac{1}{a_0} = 0$. This is a dramatic simplification and allows us to find a reasonably tractable definition of default correlation.[17] Plugging this result into equation 3.97, we have a general default correlation result as,

$$\rho\left(\mathbb{I}_{\mathcal{D}_n}, \mathbb{I}_{\mathcal{D}_m}\right) = \frac{\overbrace{p_n p_m \left(1 + \sum_{k=1}^{K} \frac{\omega_{n,k}\omega_{m,k}}{a_k}\right) - p_n p_m}^{\text{Equation 3.97}}}{\sqrt{p_n(1-p_n)}\sqrt{p_m(1-p_m)}}, \quad (3.99)$$

$$= \left(\sum_{k=1}^{K} \frac{\omega_{n,k}\omega_{m,k}}{a_k}\right)\left(\frac{p_n p_m}{\sqrt{p_n(1-p_n)}\sqrt{p_m(1-p_m)}}\right),$$

which can be directly compared to equation 3.86. Again, some caution is required with the specification of each a_k. If a very small value is provided, the default correlation can become arbitrarily high. The biggest difference with the one-factor setting is that default correlation depends, in principal, on the full set of systematic factors and their relative loadings. This makes logical sense and, again, is the source of significant flexibility.

Up to this point, we have treated the individual factors as abstract. To implement this model, however, they likely need to be given some identity. They typically relate to commonalities among the different credit obligors in one's portfolio. These might include industrial, regional, obligor creditworthiness, or national classifications. They could also include other distinctions such as size of institution (or a country's GDP) or whether the obligor is a growth or a value firm. The credit analyst is free to make this selection based on her needs, beliefs, and data availability to inform the parameter values.

Recall from Chap. 1 that each obligor in our portfolio example has been assigned to one of three regions: Alpha, Charlie, and Bravo. Ignoring the nonsensical names,

---

[17] An important step in the previous development stems from our constraint in equation 3.94, which implies that,

$$\underbrace{\sum_{\substack{k=0 \\ k \neq j}}^{K} \sum_{j=0}^{K} \omega_{n,k}\omega_{m,j}}_{} + \sum_{k=0}^{K} \omega_{n,k}\omega_{m,k} = \sum_{k=0}^{K}\sum_{j=0}^{K} \omega_{n,k}\omega_{m,j}, \quad (3.98)$$

$$= \sum_{k=0}^{K} \omega_{n,k} \underbrace{\sum_{j=0}^{K} \omega_{m,j}}_{=1},$$

$$= 1.$$

3.3 CreditRisk+

Table 3.13 *Possible multi-factor CreditRisk+ parameterization*: This table summarizes the factor-loading choices for each regional factor. These have not been estimated, but rather selected in an arbitrary manner to demonstrate the application of our multi-factor CreditRisk+ model to our practical portfolio example. The inverse of the gamma-distribution parameter, $\frac{1}{a_k}$, is also provided for each regional factor.

Region	Idiosyncratic factor	Systematic parameters		
		S_1	S_2	S_3
Alpha	0.75	0.20	0.03	0.02
Bravo	0.65	0.08	0.10	0.12
Charlie	0.65	0.02	0.03	0.30
$1/a_k$	0.00	100.00	100.00	100.00

it is not unreasonable to expect varying degrees of integration within and between individual regions. To see how this works, we will create three systematic factors—one for each region—and apply the associated multi-factor CreditRisk+ to our usual example. Gamma-distributed factors, S_1 to S_3, can thus be interpreted as systematic factors pertaining to Alpha, Charlie, and Bravo regions, respectively.

Determining the magnitude of desired default correlation within and between each region is an empirical question—that will be addressed in Chap. 10. Absent empirical data to inform these choices, one could also rely on expert judgement to identify the factor loadings. While subjective, this can still be very effective; particularly, in the context of sensitivity analysis and stress testing. Table 3.13 attempts, however, to demonstrate both the CreditRisk+ model and the potential role of factor loadings. We have thus created, relying on our best internal experts, three different regional profiles:

Alpha   This region has a reasonable amount of intraregional default correlation, but relatively modest interaction with the other regions. It also has the highest amount of idiosyncratic obligor risk.

Beta   Obligors in this region have a modest amount of default correlation among themselves, but their default events are quite highly correlated with the other two regions.

Charlie   This region is also relatively independent of the other geographical areas, but has a sizable degree of default correlation within the region.

Although these factor choices were arbitrarily selected, each choice is intended to demonstrate the flexibility of the multi-factor setting. Table 3.14 takes these values and equation 3.99 to determine the *average* level of default correlation within our three-factor system. Let's start with the diagonal. Alpha, Bravo, and Charlie have intraregional default correlation of 0.03, 0.04, and 0.09, respectively. Clearly, the interlinkages are quite strong in Charlie region. This is consistent with their profiles. The interaction between Bravo and Charlie regions is strong, whereas it is relatively weak between the Bravo-Alpha and Alpha-Charlie pairs. The average overall level of default correlation, both within and between regions, is roughly 0.035.

Table 3.14 *Multi-factor CreditRisk+ default-correlation matrix*: The underlying table uses equation 3.99 and the parameters from Table 3.13 to summarize the average default correlation between the various regions. The average default correlation, across all regions, is roughly 0.035.

	Alpha	Bravo	Charlie
Alpha	0.03	0.02	0.01
Bravo	0.02	0.04	0.04
Charlie	0.01	0.04	0.09

Algorithm 3.15 Multi-factor CreditRisk+ Monte-Carlo implementation

```
def crPlusMultifactor(N,M,wMat,p,c,aVec,alpha,rId):
    K = len(aVec)
    S = np.zeros([M,K])
    for k in range(0,K):
        S[:,k] = np.random.gamma(aVec[k], 1/aVec[k], [M])
    W = wMat[rId,:]
    wS = np.tile(W[:,0],[M,1]) + np.dot(S,np.transpose(W[:,1:]))
    pS = np.tile(p,[M,1])*wS
    H = np.random.poisson(pS,[M,N])
    lossIndicator = 1*np.greater_equal(H,1)
    lossDistribution = np.sort(np.dot(lossIndicator,c),axis=None)
    el,ul,var,es=util.computeRiskMeasures(M,lossDistribution,alpha)
    return el,ul,var,es
```

Whatever the factor-loading choices, however, a default-correlation matrix, as found in Table 3.14, is an essential tool. Given the complexity of the interactions between the factor loadings and the systematic factors, it very difficult to predict the final results. It is, in fact, an example of a simple, but quite powerful, model diagnostic. Good practice would be to furnish the implied default correlations along with one's risk-measure estimates to provide context and ease interpretation.

The actual estimation of risk using the CreditRisk+ model proceeds in a conceptually similar manner to the one-factor approach. Algorithm 3.15 illustrates a Python implementation; it is usefully compared to Algorithm 3.14. Some additional inputs are required: the gamma-distribution parameters, `aVec`, the factor loadings, `wMat`, and the regional identifiers, `rId`.

The new aspect of Algorithm 3.15, relative to the one-factor code, is the generation of K systematic factors and application of the necessary factor loadings. Once this is accomplished and we have the $M \times N$ matrix of Poisson arrival rates, the remainder of the algorithm is, at this point in the proceedings, both familiar and quite possibly tiresome.

Applying Algorithm 3.15 to our portfolio example with the parametrization from Table 3.13 leads to the results summarized in Table 3.15. It compares three different approaches: the independent-default Poisson model and the one- and multi-factor CreditRisk+ implementations. The one-factor CreditRisk+ model is calibrated to $\rho_D = 0.02$ as illustrated in Table 3.12.

Table 3.15 *A multi-factor CreditRisk+ model*: This table compares, for our running portfolio example, the independent-default Poisson model to the one- and multi-factor CreditRisk+ implementations. The parameters and default correlation associated with the multi-factor model are summarized in Tables 3.13 and 3.14. The one-factor CreditRisk+ model is calibrated to $\rho_D = 0.02$ as illustrated in Table 3.12.

	Poisson		CreditRisk+			
			One-factor		Multi-factor	
Quantile	$\widehat{\text{VaR}}_\alpha$	$\widehat{\mathcal{E}}_\alpha$	$\widehat{\text{VaR}}_\alpha$	$\widehat{\mathcal{E}}_\alpha$	$\widehat{\text{VaR}}_\alpha$	$\widehat{\mathcal{E}}_\alpha$
95.00th	$35.0	$45.2	$34.9	$55.6	$35.2	$62.4
97.00th	$40.1	$50.5	$41.6	$67.6	$44.4	$78.0
99.00th	$52.2	$60.9	$60.3	$104.7	$76.2	$122.3
99.50th	$58.6	$66.5	$83.8	$140.3	$107.3	$155.1
99.90th	$71.9	$80.0	$177.5	$232.6	$185.9	$230.0
99.97th	$81.7	$89.4	$245.9	$295.5	$241.2	$280.3
99.99th	$90.2	$97.8	$303.3	$346.8	$286.9	$319.7
Other statistics						
Expected loss	$9.1		$8.9		$8.5	
Loss volatility	$12.7		$16.1		$17.4	
Total iterations	10,000,000					
Computational time	40.8		40.6		41.4	

The results are quite interesting. Despite an average default correlation of 0.035 for the multi-factor model—compared to 0.02 for the one-factor setting—the more extreme VaR and expected-shortfall estimates are smaller. This is not completely surprising given the broad range of default-correlation values on both the inter- and intraregional levels. It is simply not correct to compare average default correlation levels, because multi-factor models are more complex and involved. A multiplicity of factors provides the modeller with much more flexibility permitting description of the interlinkages between obligors in one's portfolio. Such flexibility is important— and in some cases essential—but, it also comes at the cost of increased complexity. Multi-factor models, even more so than their one-factor equivalents, absolutely require the use of model diagnostics for effective interpretation. Default-correlation matrices are just the start of this process; we will discuss this topic in much greater detail in latter chapters.

3.4 Final Thoughts

In this chapter, we have examined a broad-range of binomial and Poisson motivated mixture models. In the classic setting, the default probability, or arrival intensity, conditions on a common random variable to create more realistic asymptotic behaviour, fatter tails, and default dependence. In all cases, conditional independence is a key feature. These models can, for judicious choices of the conditioning variable's distribution, lead to analytic expressions for the default-loss distribution. In general, however, numerical integration can be employed should the relationships become less tractable. In all cases, however, Monte-Carlo methods are available and readily applicable to generate the loss distribution and associated risk measures.

It may seem that an excessive number of model choices have been considered. As we indicated in Chap. 1, however, credit-risk model selection is, due to the rarity of default, a very challenging task. In this context, more rather than fewer models are preferable. Indeed, in the next chapter, we consider a second family of credit-risk methodologies: the threshold models. Again, to provide the analyst with choice and perspective, we will attempt to consider a variety of possible implementations. Moreover, we have only touched the surface on the larger topic of model diagnostics.

As a final note, this is probably not the standard development that one would find for mixture models. Other sources may only examine a subset of the approaches considered and gloss over many of the details; leaving them as an exercise for the reader. Our view is that foundational work is critical and effort at this point pays dividends in the form of enhanced model understanding and communication.

References

Abramovitz, M., & Stegun, I. A. (1965). *Handbook of mathematical functions*. New York: Dover Publications.
Billingsley, P. (1995). *Probability and measure* (3rd edn.). Third Avenue, New York, NY: Wiley.
Casella, G., & Berger, R. L. (1990). *Statistical inference*. Belmont, CA: Duxbury Press.
Davis, P. J. (1959). Leonhard euler's integral: A historical profile of the gamma function. *The American Mathematical Monthly, 66*(10), 849–869.
Fishman, G. S. (1995). *Monte Carlo: Concepts, algorithms, and applications*. 175 Fifth Avenue, New York, NY: Springer-Verlag. *Springer series in operations research*.
Frey, R., & McNeil, A. (2003). Dependent defaults in model of portfolio credit risk. *The Journal of Risk, 6*(1), 59–92.
Gundlach, M., & Lehrbass, F. (2004). *CreditRisk+ in the banking industry* (1st edn.). Berlin: Springer-Verlag.
Panjer, H. H. (1981). Recursive evaluation of a family of compound distributions. *Astin Bulletin*, (12), 22–26.
Press, W. H., Teukolsky, S. A., Vetterling, W. T., & Flannery, B. P. (1992). *Numerical recipes in C: The art of scientific computing* (2nd edn.). Trumpington Street, Cambridge: Cambridge University Press.
Ralston, A., & Rabinowitz, P. (1978). *A first course in numerical analysis* (2nd edn.). Mineola, NY: Dover Publications.
Sharpe, W. F. (1963). A simplified model for portfolio analysis. *Management Science, 9*(2), 277–293.
Sharpe, W. F. (1964). Capital asset prices: A theory of market equilibrium under conditions of risk. *The Journal of Finance, 19*(3), 425–442.
Stuart, A., & Ord, J. K. (1987). *Kendall's advanced theory of statistics*. New York: Oxford University Press.
Wilde, T. (1997). *CreditRisk+: A credit risk management framework*. Credit Suisse First Boston.

Chapter 4
Threshold Models

> *First learn to be a craftsman, it won't keep you from being a genius.*
>
> (Eugène Delacroix)

In the previous chapters, we examined two independent-default and six distinct binomial and Poisson mixture models. After investigation of these eight possible credit-default approaches, one might conclude that we have amassed a sufficient array of models. In reality, however, we have only considered two separate modelling approaches; the individual models are, in effect, instances or special cases. As an analyst, a multiplicity of possible model implementations is valuable, but we fundamentally seek a broad range of modelling approaches. It is, quite simply, highly desirable to have a collection of alternative methodologies to attack the same problem. Indeed, arriving at a solution from two different directions is a powerful method to corroborate one's results. The surfeit of default data and its associated empirical challenges makes this particularly important in the credit-risk setting.

In this chapter, we add to our reservoir of modelling approaches by considering the family of threshold models. We begin, quite naturally, with the original Gaussian setting, which may be familiar to many readers. It represents the underlying idea behind the popular commercial approach offered by CreditMetrics—see, for example, Gupton et al. (2007). We then proceed to consider a range of extensions offered by the family of so-called normal-variance mixture models. While mathematically more involved, the advantages include heavy default-loss distribution tails and the imposition of tail dependence. We then, in the final section, consider the canonical multi-factor threshold model and examine a practical multivariate implementation. Along the way, we will address issues related to model calibration and make extensive use of our portfolio example.

4.1 The Gaussian Model

The threshold approach, first suggested by Vasicek—see, for example, Vasicek (1987, 1991, 2002)—is a clever combination of a pragmatic, latent-variable approach and the classic Merton (1974) model.[1] In the practitioner world, it has become the *de facto* credit-risk model and is responsible for most of the heavy-lifting in economic capital and credit VaR computations.[2] Its popularity stems in part from its ease of implementation, but also the existence of multi-factor versions and possible variations on the joint distribution used to induce default dependence among the credit counterparties.

4.1.1 The Latent Variable

The starting point for this model varies by author and application. In all cases, however, it involves, in one way or another, a state-variable definition. In some cases, one defines the state variable explicitly as the counterparty's asset value or the normalized log-return of its assets. In others, it is simply treated as a latent variable. Let's simply denote it as y_n and give it the following form,

$$y_n = \underbrace{\text{Systematic Element}}_{a \cdot G} + \underbrace{\text{Idiosyncratic Element}}_{b \cdot \epsilon_n}, \quad (4.1)$$
$$= aG + b\epsilon_n,$$

where G denotes a systematic global factor and ϵ_n is an idiosyncratic factor. Again, we find ourselves returning to Sharpe (1963, 1964)'s CAPM structure. The exposure to, or the loading onto, the global factor is described by the coefficient a, while the exposure to the idiosyncratic factor is summarized by the b coefficient. Moreover, both G and ϵ_n are independent and identically distributed standard normal variates. That is, $G, \epsilon_n \sim \mathcal{N}(0, 1)$ for all $n \in \{1, \ldots, N\}$.

The threshold approach is essentially a factor model. The use of factors to reduce dimensionality, complexity and permit a parsimonious description of one's problem has a long and rich history in finance. Sometimes the factors, or state variables, are explicit and observable. In other settings, they are latent, statistically derived, and unobservable. Factor representations are present in the CAPM setting and play an important role, to name just a few concrete settings, in asset pricing, market risk, and interest-rate modelling. It should, therefore,

---

[1] Merton (1974) is sufficiently important and fundamental that we assign the entirety of Chap. 5 to its consideration.

[2] It also acts, as we'll see in Chap. 6, as the foundation for the Basel bank-capital requirements. See Fok et al. (2014) and BIS (2006a).

4.1 The Gaussian Model

be no surprise to see their appearance in approaches to describe credit risk. It nevertheless takes some time, patience, and tedious statistical investigation to arrive at the point where the role of the underlying state variables becomes completely clear.

Since the normal distribution is closed under addition, it follows that y_n is also Gaussian; it need not, however, be standard normal. Let's, therefore, compute its mean and variance. The expectation of Y_n is,

$$\mathbb{E}(y_n) = \mathbb{E}(aG + b\epsilon_n), \tag{4.2}$$
$$= a\underbrace{\mathbb{E}(G)}_{=0} + b\underbrace{\mathbb{E}(\epsilon_n)}_{=0},$$
$$= 0.$$

It has the correct location. Now, if we compute the variance, we have by independence that

$$\mathrm{var}(y_n) = \mathrm{var}(aG + b\epsilon_n), \tag{4.3}$$
$$= a^2 \underbrace{\mathrm{var}(G)}_{=1} + b^2 \underbrace{\mathrm{var}(\epsilon_n)}_{=1},$$
$$= a^2 + b^2.$$

It would be much more convenient, as we will see shortly, if y_n was also standard normally distributed. To accomplish this, we need to place some restrictions on a and b. Indeed, we must identify coefficients a and b such that:

$$a^2 + b^2 = 1, \tag{4.4}$$
$$b = \sqrt{1 - a^2}.$$

b is not a free parameter, so we have

$$Y_n = a \cdot G + \underbrace{\sqrt{1 - a^2}}_{b}\,\epsilon_n. \tag{4.5}$$

This satisfies the constraint and ensures that y_n has unit variance. Moreover, to avoid complex numbers, we require that

$$1 - a^2 \geq 0, \tag{4.6}$$
$$|a| \leq 1.$$

A number of functions of a will do the job, but aesthetically we often choose

$$y_n = (\sqrt{a}) \cdot G + \sqrt{1 - (\sqrt{a})^2} \epsilon_n, \qquad (4.7)$$
$$= \sqrt{a} \cdot G + \sqrt{1 - a} \epsilon_n.$$

In the latter case, we naturally require $a \in [0, 1]$.

For the purposes of our analysis, we will use this latter parametrization of the state variable as,

$$y_n = \underbrace{\sqrt{\rho} \, G}_{a} + \underbrace{\sqrt{1 - \rho} \, \epsilon_n}_{b}, \qquad (4.8)$$

for $n = 1, \ldots, N$. We replace the symbol a with the Greek character, ρ. ρ is classically used by statisticians to describe the correlation coefficient; the reasons for this choice, foreshadowing aside, will soon become more obvious.

4.1.2 Introducing Dependence

It is straightforward exercise—following from equations 4.2 and 4.3—to show that $\mathbb{E}(y_n) = 0$ and $\text{var}(y_n) = 1$. Each obligor in our portfolio is assigned its own state variable, y_n for $n = 1, \ldots, N$. The idea is to use this collection of random variables to say something about portfolio default events. To incorporate default dependence, however, we need to establish a clear relationship between each of these variables. It turns out, quite happily, that one already exists. To see this, let's now compute the covariance between the state variable of two arbitrary counterparties, y_n and y_m,

$$\text{cov}(y_n, y_m) = \mathbb{E}\bigg((y_n - \underbrace{\mathbb{E}(y_n)}_{=0})(y_m - \underbrace{\mathbb{E}(y_m)}_{=0})\bigg), \qquad (4.9)$$
$$= \mathbb{E}(y_n y_m),$$
$$= \mathbb{E}\bigg(\Big(\sqrt{\rho} G + \sqrt{1-\rho} \epsilon_n\Big)\Big(\sqrt{\rho} G + \sqrt{1-\rho} \epsilon_m\Big)\bigg),$$
$$= \mathbb{E}(\rho G^2),$$
$$= \rho \text{var}(G),$$
$$= \rho.$$

The intermediate terms—for example, $\mathbb{E}(G\epsilon_n) = \mathbb{E}(G)\mathbb{E}(\epsilon_n)$—vanish, because the expectation of each individual random variable is zero. Moreover, since both y_n and y_m have unit variance, it follows that the covariance and correlation of y_n and y_m are identical. Thus, $\rho(y_n, y_m) = \rho$.

4.1 The Gaussian Model

The choice of notation for the coefficients associated with the global and idiosyncratic factors was, therefore, not without foundation. The two defaults are no longer, as in Chap. 2, independent. Conditioning on the same global variable through the parameter, ρ, creates default dependence. A value of $\rho = 0$ would imply complete independence, whereas a value of $\rho = 1$ would imply perfect default correlation.

Let's look at the collection of latent default variables as a system. We have established that

$$\underbrace{\begin{bmatrix} y_1 \\ y_2 \\ \vdots \\ y_N \end{bmatrix}}_{Y} \sim \mathcal{N}\left(\underbrace{\begin{bmatrix} 0 \\ 0 \\ \vdots \\ 0 \end{bmatrix}}_{\mathbf{0}}, \underbrace{\begin{bmatrix} 1 & \rho & \cdots & \rho \\ \rho & 1 & \cdots & \rho \\ \vdots & \vdots & \ddots & \vdots \\ \rho & \rho & \cdots & 1 \end{bmatrix}}_{\Omega} \right). \quad (4.10)$$

This looks fairly trivial, since there is common covariance between all state variables. It is still useful for, at least, *two* reasons. First, it provides a perspective on the overall system. We have constructed a system of multivariate normally distributed latent variables. Both the joint and marginal distributions are thus Gaussian. Second, it helps us better understand the generalization. In latter sections, we will adjust this joint distribution: this is the jumping-off point.

Looking at equation 4.10, we identify an overly simplistic pattern in the correlation structure. All obligors depend on the global factor, G, in the same way. As such, they all share the same pairwise correlation between latent variables. This can, of course, be relaxed. We may permit each obligor have its own correlation parameter. In this case, we have

$$y_n = \sqrt{\rho_n} G + \sqrt{1 - \rho_n} \epsilon_n. \quad (4.11)$$

In this case, each credit counterparty is affected by the global state variable in a different way. If we work through the implications, we have

$$\begin{bmatrix} y_1 \\ y_2 \\ \vdots \\ y_N \end{bmatrix} \sim \mathcal{N}\left(\begin{bmatrix} 0 \\ 0 \\ \vdots \\ 0 \end{bmatrix}, \begin{bmatrix} 1 & \sqrt{\rho_1 \rho_2} & \cdots & \sqrt{\rho_1 \rho_N} \\ \sqrt{\rho_2 \rho_1} & 1 & \cdots & \sqrt{\rho_2 \rho_N} \\ \vdots & \vdots & \ddots & \vdots \\ \sqrt{\rho_N \rho_1} & \sqrt{\rho_N \rho_2} & \cdots & 1 \end{bmatrix} \right). \quad (4.12)$$

This is rarely done since it leads to a dizzying $N + 1$ parameters making estimation very challenging. For now, we will remain in the simplest setting since we now see that full generality is always possible.

4.1.3 The Default Trigger

We have already performed a significant number of computations and introduced extensive notation, but the basic form of the model remains unclear. This will now be rectified. The next step is to return to the default event. An obligor is deemed to find itself in a default state if,

$$\mathcal{D}_n \triangleq \{y_n \leq K_n\}, \tag{4.13}$$

for an abitrary firm, n. In other words, if the latent variable, y_n, falls below a threshold, then default occurs. Unlike the models considered so far, this approach attempts to describe the nature of the default event. The threshold family is thus a structural model.

The indicator variable associated with the default event is now written as,

$$\mathbb{I}_{\mathcal{D}_n} \equiv \mathbb{I}_{\{y_n \leq K_n\}} = \begin{cases} 1 : \text{ when firm } n \text{ defaults before time } T \\ 0 : \text{ otherwise} \end{cases}. \tag{4.14}$$

This might appear somewhat awkward, since neither the value of K_n nor the probability of default have yet been introduced. Following from previous chapters, the (unconditional) T-period probability of default, denoted p_n, is one of our model primitives. This value is separately estimated from empirical data.[3] Following from first principles, therefore, we have that

$$\begin{aligned} p_n &= \mathbb{E}(\mathbb{I}_{\mathcal{D}_n}), \\ &= \mathbb{P}(\mathcal{D}_n), \\ &= \mathbb{P}(y_n \leq K_n), \\ &= F_{y_n}(y_n \leq K_n), \\ &= \frac{1}{\sqrt{2\pi}} \int_{-\infty}^{K_n} e^{-\frac{u^2}{2}} du, \\ &= \Phi(K_n), \end{aligned} \tag{4.15}$$

where $\Phi(\cdot)$ denotes the Gaussian cumulative distribution function. The result in equation 4.15 follows simply due to the fact that y_n is a standard normal variable. It thus also follows directly from equation 4.15 that

$$\begin{aligned} p_n &= \Phi(K_n), \\ K_n &= \Phi^{-1}(p_n). \end{aligned} \tag{4.16}$$

---

[3] Techniques to estimate these important probabilities are addressed in Chap. 9.

4.1 The Gaussian Model

The threshold, K_n, is thus determined by the unconditional default probability or, rather, our assessment of the relative creditworthiness of an obligor. This is a compelling result. Credit counterparties who are judged to have precarious financial health are assigned a more aggressive barrier than their strong, creditworthy peers. It also permits to redefine the default indicator more practically as,

$$\mathbb{I}_{\mathcal{D}_n} \equiv \mathbb{I}_{\{y_n \leq K_n\}} \equiv \mathbb{I}_{\{y_n \leq \Phi^{-1}(p_n)\}}. \tag{4.17}$$

This is a useful, and concrete, working definition of obligor default.

4.1.4 Conditionality

The systematic global variable, G, drives the dependence between each of the credit counterparties. Practically, it plays the same role as the Z and S conditioning variables in the binomial- and Poisson-mixture models examined in Chap. 3. The actual role of G is not entirely clear. How, for example, does an outcome of G impact the probability of default for a given obligor? One way to understand this is to consider the conditional expectation of the default of the nth counterparty given the value of the global factor, G. This is readily computed as,

$$\begin{aligned} p_n(G) &= \mathbb{E}\left(\mathbb{I}_{\mathcal{D}_n} \,\middle|\, G\right), \tag{4.18} \\ &= \mathbb{P}\left(\mathcal{D}_n \,\middle|\, G\right), \\ &= \mathbb{P}\left(y_n \leq \Phi^{-1}(p_n) \,\middle|\, G\right), \\ &= \mathbb{P}\left(\underbrace{\sqrt{\rho}G + \sqrt{1-\rho}\epsilon_n}_{\text{Equation 4.8}} \leq \Phi^{-1}(p_n) \,\middle|\, G\right), \\ &= \mathbb{P}\left(\epsilon_n \leq \frac{\Phi^{-1}(p_n) - \sqrt{\rho}G}{\sqrt{1-\rho}} \,\middle|\, G\right), \\ &= \Phi\left(\frac{\Phi^{-1}(p_n) - \sqrt{\rho}G}{\sqrt{1-\rho}}\right). \end{aligned}$$

This is a monotonically increasing function of $-G$, which implies that large negative realizations of G induce high conditional probabilities of default across all credit counterparties. Given the value of G, however, each of the default events associated with each counterparty is independent. This effect is visually demonstrated in Fig. 4.1.

This is a helpful quantity. $p_n(G)$ is also usefully compared to the conditional default probabilities, $p(Z)$, $p(S)$, and $p_n(S)$ used in the binomial-mixture, Poisson-

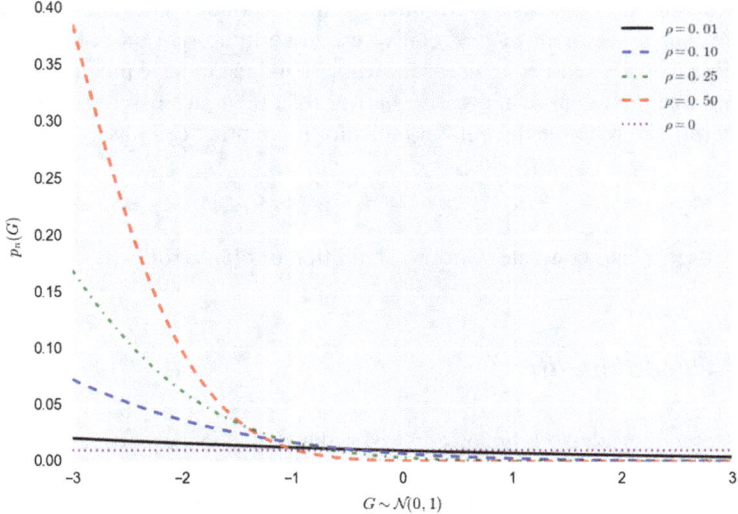

Fig. 4.1 *Conditional default probability*: This figure uses the conditional default probability, as described in equation 4.18, to demonstrate the impact of the value of G. As G takes on large negative values, it leads to monotonically greater conditional default probabilities. The effect increases for increasing values of ρ.

mixture, and CreditRisk+ settings. It is interesting to compute the expected value of $p_n(G)$. If we do so, we arrive at,

$$\mathbb{E}\Big(p_n(G)\Big) = \mathbb{E}\Big(\mathbb{E}\left(\mathbb{I}_{\mathcal{D}_n}\,\big|\, G\right)\Big), \tag{4.19}$$

$$= \mathbb{E}\Big(\mathbb{E}\left(\mathbb{I}_{\{y_n \leq \Phi^{-1}(p_n)\}}\,\Big|\, G\right)\Big),$$

$$= \mathbb{E}\Big(\mathbb{I}_{\{y_n \leq \Phi^{-1}(p_n)\}}\Big),$$

$$= \mathbb{P}\Big(y_n \leq \Phi^{-1}(p_n)\Big),$$

$$= \Phi\Big(\Phi^{-1}(p_n)\Big),$$

$$= p_n.$$

We find, therefore, that as in the CreditRisk+ model, the individual creditworthiness of each obligor is recovered when we compute the average conditional default probability. Adjusting the parameters of y_n to force each latent variable to be standard normally distributed is thus, conceptually at least, the same trick as

4.1 The Gaussian Model

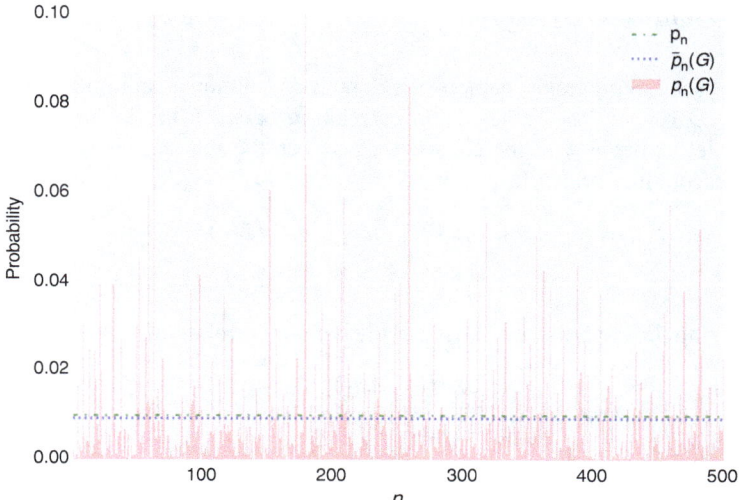

Fig. 4.2 $p_n(G)$ *dynamics*: This figure assumes a value of $p_n = 0.01$ and $\rho = 0.2$. It then simulates 500 random draws of $G \sim \mathcal{N}(0, 1)$ and computes the associated conditional default probabilities. We see, quite clearly, that the default probability has been randomized.

forcing the gamma-distributed mixing variables in the CreditRisk+ model to have an expectation of unity.

To make this notion even more concrete, Fig. 4.2 demonstrates the role of the conditional default probability. It assumes a value of $p_n = 0.01$ and $\rho = 0.2$ and simulates 500 random draws of $G \sim \mathcal{N}(0, 1)$. It then proceeds to compute the associated conditional default probabilities; both the unconditional and average conditional default probabilities are displayed. All of this is accomplished using Algorithm 4.1, which is a practical implementation of equation 4.18. We see, quite clearly, that the default probability has been randomized. Depending on the outcome of G, it can reach levels exceeding 0.10 or fall close to zero. The average outcome, denoted as $\bar{p}_n(G)$, even over 500 trials, remains quite close to the unconditional input of 0.01.

Algorithm 4.1 Computing $p_n(G)$

```
def computeP(p,rho,g):
    num = norm.ppf(p)-np.multiply(np.sqrt(rho),g)
    pG = norm.cdf(np.divide(num,np.sqrt(1-rho)))
    return pG
```

4.1.5 Default Correlation

As we've seen in previous chapters, the default correlation is an important quantity. It is, moreover, not the same as the correlation between the latent variables. To see this, let us begin with the covariance between the nth and mth default events. Working from first principles, we have

$$\operatorname{cov}\left(\mathbb{I}_{\mathcal{D}_n}, \mathbb{I}_{\mathcal{D}_m}\right) = \mathbb{E}\left(\mathbb{I}_{\mathcal{D}_n} \mathbb{I}_{\mathcal{D}_m}\right) - \mathbb{E}\left(\mathbb{I}_{\mathcal{D}_n}\right) \mathbb{E}\left(\mathbb{I}_{\mathcal{D}_m}\right), \quad (4.20)$$
$$= \mathbb{P}\left(\mathcal{D}_n \cap \mathcal{D}_m\right) - \mathbb{P}\left(\mathcal{D}_n\right) \mathbb{P}\left(\mathcal{D}_m\right).$$

Normalizing the covariance to arrive at the default correlation,

$$\rho\left(\mathbb{I}_{\mathcal{D}_n}, \mathbb{I}_{\mathcal{D}_m}\right) = \frac{\operatorname{cov}\left(\mathbb{I}_{\mathcal{D}_n}, \mathbb{I}_{\mathcal{D}_m}\right)}{\sqrt{\operatorname{var}(\mathbb{I}_{\mathcal{D}_n})} \sqrt{\operatorname{var}(\mathbb{I}_{\mathcal{D}_m})}}, \quad (4.21)$$
$$= \frac{\mathbb{P}\left(\mathcal{D}_n \cap \mathcal{D}_m\right) - \mathbb{P}\left(\mathcal{D}_n\right) \mathbb{P}\left(\mathcal{D}_m\right)}{\sqrt{\mathbb{P}(\mathcal{D}_n)(1 - \mathbb{P}(\mathcal{D}_n))} \sqrt{\mathbb{P}(\mathcal{D}_m)(1 - \mathbb{P}(\mathcal{D}_m))}}.$$

Recalling that the variance of an indicator variable coincides with a Bernoulli trial permits a return to our simpler notation. We thus have

$$\rho\left(\mathbb{I}_{\mathcal{D}_n}, \mathbb{I}_{\mathcal{D}_m}\right) = \frac{\mathbb{P}\left(\mathcal{D}_n \cap \mathcal{D}_m\right) - p_n p_m}{\sqrt{p_n p_m (1 - p_n)(1 - p_m)}}. \quad (4.22)$$

Comparing this result with equation 4.9, we verify that, while certainly related, the default correlation is indeed *not* the same as the correlation between the latent variables. Default correlation depends not only on the unconditional default probabilities, but also on the joint probability of default between counterparties n and m. By definition, the model assumes standard normal marginal distributions for the global and idiosyncratic factors, G and ϵ_n. This is extremely convenient and permits a number of analytic simplifications.

The natural consequence of our development thus far is that the joint distribution of y_n and y_m is also Gaussian.[4] In this case, we have a bivariate normal distribution for the joint probability, $\mathbb{P}\left(\mathcal{D}_n \cap \mathcal{D}_m\right)$. Specifically, it has the following form,

$$\mathbb{P}\left(\mathcal{D}_n \cap \mathcal{D}_m\right) = \mathbb{P}\left(y_n \leq \Phi^{-1}(p_n), y_m \leq \Phi^{-1}(p_m)\right), \quad (4.23)$$
$$= \frac{1}{2\pi\sqrt{1-\rho^2}} \int_{-\infty}^{\Phi^{-1}(p_n)} \int_{-\infty}^{\Phi^{-1}(p_m)} e^{-\frac{(u^2 - 2\rho u v + v^2)}{2(1-\rho^2)}} du\, dv,$$
$$= \Phi\left(\Phi^{-1}(p_n), \Phi^{-1}(p_m); \rho\right).$$

---
[4]This will, we assure the reader, be relaxed in latter sections.

4.1 The Gaussian Model

Algorithm 4.2 The Gaussian joint default probability

```
def jointDefaultProbability(x1,x2,rho):
    support = [[-7, norm.ppf(x1)],[-7, norm.ppf(x2)]]
    f,err = nInt.nquad(bivariateGDensity,support,args=(rho,))
    return f
def bivariateGDensity(x1,x2,rho):
    S = np.array([[1,rho],[rho,1]])
    t1 = 2*math.pi*np.sqrt(anp.det(S))
    t2 = np.dot(np.dot(np.array([x1,x2]),anp.inv(S)),np.array([x1,x2]))
    return np.divide(1,t1)*np.exp(-t2/2)
```

This expression easily permits us to directly compute the default correlation between counterparties n and m using equation 4.22. It can be evaluated using the nquad numerical integration function as described in Algorithm 4.2. The joint default probability is found by merely calling jointDefaultProbability, which in turn calls bivariateGDensity, for the inverse cumulative distribution function values of p_n and p_m.

It is also possible to determine this probability in another way by using the conditional default probability from equation 4.18, exploiting conditional independence, and integrating out the global variable, G. While a bit less direct, it is always handy to have multiple approaches to solve the same problem. This conditional-probability approach, which we will develop later, can be useful when working with some of the latter threshold models, since it avoids the need to work directly with the bivariate state-variable distributions.

It is convenient, and quite informative, to put our joint distribution in equation 4.23 into matrix form. Define the correlation matrix as,

$$\Omega_{nm} = \begin{bmatrix} 1 & \rho \\ \rho & 1 \end{bmatrix}. \tag{4.24}$$

The determinant of Ω_{nm} is given as,

$$\det(\Omega_{nm}) = |\Omega_{nm}| = 1 - \rho^2, \tag{4.25}$$

and the inverse is defined as,

$$\Omega_{nm}^{-1} = \frac{1}{|\Omega_{nm}|} \begin{bmatrix} 1 & -\rho \\ -\rho & 1 \end{bmatrix}, \tag{4.26}$$

$$= \begin{bmatrix} \frac{1}{1-\rho^2} & \frac{-\rho}{1-\rho^2} \\ \frac{-\rho}{1-\rho^2} & \frac{1}{1-\rho^2} \end{bmatrix}.$$

If we define the vector $x = \begin{bmatrix} u & v \end{bmatrix}^T$, then

$$x^T \Omega_{nm}^{-1} x = \begin{bmatrix} u & v \end{bmatrix} \begin{bmatrix} \frac{1}{1-\rho^2} & \frac{-\rho}{1-\rho^2} \\ \frac{-\rho}{1-\rho^2} & \frac{1}{1-\rho^2} \end{bmatrix} \begin{bmatrix} u \\ v \end{bmatrix}, \quad (4.27)$$

$$= \frac{u^2 - 2\rho u v + v^2}{1 - \rho^2}.$$

This implies that the expression in equation 4.23 may be succinctly rewritten as,

$$\Phi\left(\Phi^{-1}(p_n), \Phi^{-1}(p_m); \rho\right) = \frac{1}{2\pi\sqrt{1-\rho^2}} \int_{-\infty}^{\Phi^{-1}(p_n)} \int_{-\infty}^{\Phi^{-1}(p_m)} e^{-\frac{\left(u^2-2\rho uv+v^2\right)}{2(1-\rho^2)}} du dv,$$

$$= \frac{1}{2\pi\sqrt{|\Omega_{nm}|}} \int_{-\infty}^{\Phi^{-1}(p_n)} \int_{-\infty}^{\Phi^{-1}(p_m)} e^{-\frac{x^T \Omega_{nm}^{-1} x}{2}} dx. \quad (4.28)$$

There is a reason for this gratuitous use of mathematics. The result should look familiar—it is referred to as the Gaussian copula.[5] Statisticians often use the notion of copula functions to describe dependence between random variables. This notion—originally introduced by Sklar (1959)—is a powerful, and perhaps the definitive, way to describe dependence. The main role of a copula function is to map a collection of marginals into a joint distribution.

Equation 4.28 thus characterizes the specific structure of dependence between each pair default risks in this model. As we will see in subsequent sections, it is *not* the only—or even most defensible—choice. The Gaussian distribution—be it univariate or multivariate—and, by extension, its copula, is often considered somewhat too well behaved in the tails.

Introduction of the copula function, at this point in the proceedings, might seem like overkill. We will see relatively soon, however, that it can prove useful when we seek to understand the implications of our distributional choices on the ultimate properties of the default-loss distribution. Credit risk is predominately about the behaviour of one's portfolio far out in the tails of the loss distribution.

We are also naturally interested in the joint dependence across all of our credit counterparties. Equation 4.28 can be generalized to N counterparties, where Ω is the general correlation matrix between the N counterparties, then

$$\Phi\left(\Phi^{-1}(p_1), \cdots, \Phi^{-1}(p_N); \Omega\right) = \frac{1}{(2\pi)^{\frac{N}{2}}\sqrt{|\Omega|}} \int_{-\infty}^{\Phi^{-1}(p_1)} \cdots \int_{-\infty}^{\Phi^{-1}(p_N)} e^{-\frac{x^T \Omega^{-1} x}{2}} dx.$$

(4.29)

---

[5]The upper limits of integration in the classic description are typically standard normal inverse cumulative distribution functions of standard normal variates. That is, $u_i \sim \mathcal{U}(0, 1)$. Since the domain of each $\Phi^{-1}(p_i)$ is also [0, 1], the form in equation 4.28 is mathematically equivalent to the Gaussian copula.

4.1 The Gaussian Model

This is a function, defined on the unit cube $[0, 1]^N$, that transforms a set of marginal distributions into a single joint distribution.[6] It is commonly employed to characterize default dependence, but it is *not* the only possible choice in this setting. Indeed, as we'll see shortly, it is a potentially questionable choice for a credit-risk setting.

4.1.6 Calibration

The lever to adjust the default correlation is the parameter, ρ. As in the binomial- and Poisson-mixture settings discussed in Chap. 3, it is useful to be able to calibrate the level of ρ to attain a given degree of default correlation. In principle, as described in equations 4.22 and 4.23, the Gaussian threshold model default correlation has the following form,

$$\rho\left(\mathbb{I}_{\mathcal{D}_n}, \mathbb{I}_{\mathcal{D}_m}\right) = \frac{\overbrace{\Phi\left(\Phi^{-1}(p_n), \Phi^{-1}(p_m); \rho\right)}^{\mathbb{P}(\mathcal{D}_n \cap \mathcal{D}_m)} - p_n p_m}{\sqrt{p_n p_m (1 - p_n)(1 - p_m)}}. \qquad (4.30)$$

For each pair of obligors n and m, a different correlation value is possible depending on one's estimates of p_n and p_m. Not only do we not have sufficient parameters to fit a complicated pairwise default-correlation structure, we lack the empirical evidence to force a particular correlation between each pair of credit counterparties.

At this point, the analyst has some choice. She could organize the obligors by some external factors—such as region or industry—and permit different values of ρ for each sub-group. In this way, she could target varying degrees of default correlation both within and without these sub-groupings. This natural idea is the rationale behind, as we'll see later in the discussion, multi-factor threshold implementations.

A simpler approach—and the one we will take to maintain comparability with the previous mixture-model analysis—will be to target a single average level of default correlation across the entire portfolio. This involves replacing each p_n and p_m with the average default probability, $\bar{p}$, in equation 4.30 as,

$$\rho\left(\mathbb{I}_{\mathcal{D}_n}, \mathbb{I}_{\mathcal{D}_m}\right) = \frac{\Phi\left(\Phi^{-1}(\bar{p}), \Phi^{-1}(\bar{p}); \rho\right) - \bar{p}^2}{\bar{p}(1 - \bar{p})}, \qquad (4.31)$$

for all $n, m = 1, \ldots, N$. This provides a single target to be matched with a single parameter.

---

[6]In this simple setting, the correlation matrix, $\Omega \in \mathbb{R}^{N \times N}$, has ones along the main diagonal and the parameter ρ populates all off-diagonal elements; as in equation 4.10.

Algorithm 4.3 A one-factor calibration objective function

```
def calibrateGaussian(x,myP,targetRho):
    jointDefaultProb = jointDefaultProbability(myP,myP,x)
    defaultCorrelation = np.divide(jointDefaultProb -myP**2,myP*(1-myP))
    return np.abs(defaultCorrelation -targetRho)
```

This leads to the fairly obvious calibration Algorithm 4.3. The objective function, `calibrateGaussian`, takes in the average default probability, `myP`, along with the target default correlation level, `targetRho`. We will use the level of $\rho_\mathcal{D} = 0.05$ employed in Chap. 3, which will make our computations more directly comparable with the mixture-model approach. The objective function is thus the absolute difference between Equation 4.31 and one's desired value of $\rho_\mathcal{D}$. To find the optimal level of ρ, we use `scipy`'s minimization functionality to solve the non-linear optimization problem. The function call is merely,

$$r = \text{scipy.optimize.minimize(calibrateGaussian,r0,args=(0.01,0.05))} \quad (4.32)$$

The optimal choice—using $\bar{p} = 0.01$ and $\rho_\mathcal{D} = 0.05$—of ρ turns out to be 0.315. Imposition of this single parameter for the one-factor Gaussian threshold model should make it broadly comparable to the one-factor mixture models considered in Chap. 3.

4.1.7 Gaussian Model Results

Let us begin our application of the Gaussian threshold model to our usual example with a numerical Monte-Carlo implementation. This stochastic-simulation approach has the strong advantage that it can be applied to all threshold models—irrespective of the number of factors and assumptions on the distribution of the y_n's. Different expressions may be used, but the sequence and general logic of the simulation will not vary.

There are *two* main steps in the practical implementation: the generation of the latent state variables and the estimation of the associated default-loss distribution. While these can easily be performed in a single step, we have broken them into separate tasks. The key reason is that it is, occasionally, useful to generate the latent state variables without actually building the entire default-loss distribution. In this case, it is helpful to have a distinct sub-routine for their generation. Algorithm 4.4 illustrates the three lines of code required to simulate M by N realizations of the system of latent state variables. The first line creates the global systematic variable, the second builds the idiosyncratic outcomes, and the final step combines them together following equation 4.8.

4.1 The Gaussian Model

Algorithm 4.4 Generating Gaussian latent variables

```
def getGaussianY (N,M,p,rho):
    G = np.transpose(np.tile(np.random.normal(0,1,M),(N,1)))
    e = np.random.normal(0,1,[M,N])
    Y = math.sqrt(rho)*G + math.sqrt(1-rho)*e
    return Y
```

Algorithm 4.5 The numerical Gaussian one-factor threshold implementation

```
def oneFactorGaussianModel (N,M,p,c,rho,alpha):
    Y = getGaussianY (N,M,p,rho)
    K = norm.ppf(p)*np.ones((M,1))
    lossIndicator = 1*np.less(Y,K)
    lossDistribution = np.sort(np.dot(lossIndicator,c),axis=None)
    el,ul,var,es=util.computeRiskMeasures(M,lossDistribution,alpha)
    return el,ul,var,es
```

The actual estimation of the risk factors, summarized in Algorithm 4.5 should look extremely familiar after reading Chap. 3. Given the latent state variables, Y, called from Algorithm 4.4, it then constructs the thresholds K using equation 4.16. The default-loss indicator is then summarized by the simple condition, np.less(Y,K). The same computeRiskMeasures sub-routine is employed to approximate the desired risk metrics.

Let us now, as a final step, turn to use our calibrated parameter choice and Algorithm 4.5 to apply the Gaussian threshold model to our 100-obligor example. The results, along with the independent-default and beta-binomial mixture models, are summarized in Table 4.1. Naturally, both the beta-mixture and Gaussian threshold models dramatically exceed, due to the presence of default correlation, the binomial model. Interestingly, while the beta-binomial model generates slightly higher VaR estimates across all quantiles, the Gaussian approach has higher expected-shortfall figures at the larger quantiles.

The key point from Table 4.1 is that, despite the Gaussian distribution being famously thin-tailed, it produces results that are broadly comparable to the fat-tail mixture models addressed in Chap. 3. Ultimately, therefore, even though the latent state variables follow a Gaussian distribution, the default-loss distribution is, quite definitively, *not* normally distributed. Figure 4.3 demonstrates this effect in graphical format. The beta-binomial mixture model dominates its Gaussian-threshold competitor for virtually all quantiles, but they seem to converge as we move further into the tail of the default-loss distribution. The specific form of this distribution will be derived in the next section when we seek to find an analytic approximation to the basic Gaussian-threshold methodology.

Table 4.1 *Gaussian threshold results*: This table compares the Gaussian threshold VaR and expected shortfall estimates to the binomial independent-default and beta-binomial results. The beta-binomial model is calibrated for a 1% average default probability, while both the mixture and threshold method target a default correlation of 0.05.

Quantile	Binomial		Beta-mixture		Gaussian	
	$\widehat{\text{VaR}_\alpha}$	$\widehat{\mathcal{E}_\alpha}$	$\widehat{\text{VaR}_\alpha}$	$\widehat{\mathcal{E}_\alpha}$	$\widehat{\text{VaR}_\alpha}$	$\widehat{\mathcal{E}_\alpha}$
95.00th	$35.3	$45.5	$57.5	$99.2	$48.4	$82.7
97.00th	$40.4	$50.8	$78.0	$120.8	$63.5	$100.9
99.00th	$52.6	$61.2	$124.9	$168.5	$101.9	$145.6
99.50th	$58.8	$66.9	$155.2	$198.9	$130.1	$177.1
99.90th	$72.5	$80.5	$227.0	$269.5	$206.7	$258.2
99.97th	$84.9	$92.7	$293.4	$334.4	$286.9	$340.1
99.99th	$90.4	$98.2	$323.3	$363.9	$323.8	$377.7
Other statistics						
Expected loss	$9.2		$9.5		$9.2	
Loss volatility	$12.8		$25.2		$21.7	
Total iterations	10,000,000					
Computational time	13.9		18.5		49.0	

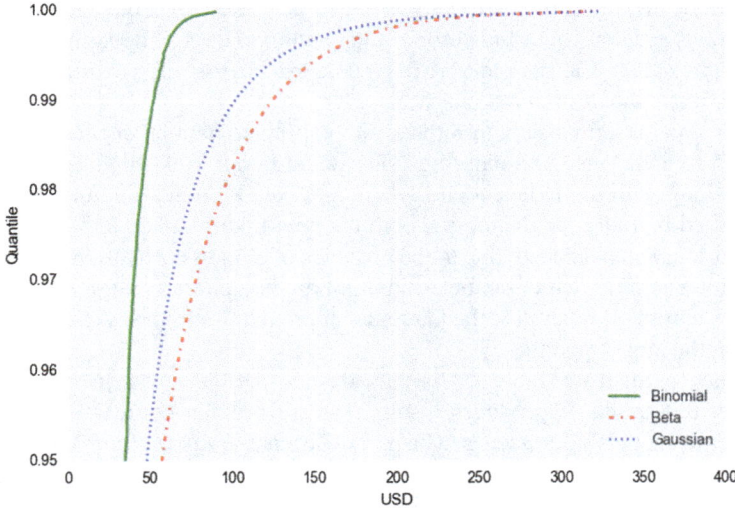

Fig. 4.3 *Comparative tail probabilities*: This figure provides a graphic description of the results found in Table 4.1. Observe how the beta-binomial and Gaussian threshold models converge in the tail.

The final point relates to the speed of the underlying simulation. The threshold approach is more than $2\frac{1}{2}$ times slower than its mixture-model equivalent. The reason is not due to the need to generate more random variables, both require basically the same amount. It is, however, significantly less costly to generate $M \times N$ uniform random variates, as required by the mixture models, relative to the same number of standard-normal outcomes. This is the source of the additional computational expense.

4.2 The Limit-Loss Distribution

Vasicek (1991) makes some additional assumptions and proceeds to derive an analytic expression for the portfolio default loss. Since it is both interesting and edifying, it deserves additional scrutiny. It essentially involves replacing the number of total defaults with the proportion of defaults in a portfolio with N counterparties. The derivation also, quite importantly, assumes that no single exposure dominates the others—that is, the exposure profile is approximately homogeneous. In this section we will examine its form and derivation, which, when complete, can be directly employed to approximate any desired quantile. Specifically, we derive the limiting loss distribution as the number of counterparties in the portfolio, N, become extremely large. A rather good reference for this development is Schönbucher (2000b).

The first assumption is that $p_n = p$ for all $n = 1, \ldots, N$. In other words, as usual, we hold that all of the issuers are homogeneous. We also assume that each of the portfolios has the same exposure, recovery and loss given default.[7] This is often referred to as an infinitely grained, or granular, portfolio. Quite clearly, this type of analysis lends itself to large, diversified portfolios of credit counterparties. For small, concentrated portfolios, these assumptions will be strongly violated.[8]

Let's begin and see how far we can go without any additional assumptions. The one-factor Gaussian Vasicek (1987, 1991, 2002) model describes the default event for the n counterparty, $\mathcal{D}_n$, conditional on the global systemic factor. Indeed, it is necessary to condition on the global factor to say something about the default event. For example, recall that the *unconditional* expectation of the default event is given as,

$$\mathbb{E}(\mathbb{I}_{\mathcal{D}_n}) = \Phi\left(\frac{\Phi^{-1}(p) - \sqrt{\rho}G}{\sqrt{1-\rho}}\right). \tag{4.33}$$

---

[7]This can be slightly relaxed as long as none of the counterparties dominates the portfolio.
[8]Gordy (2002) offers concentration adjustment for such, real-world, non-infinitely grained portfolios. This interesting and pertinent issue will be addressed in Chap. 6.

While interesting, it is difficult, without knowledge of G, to proceed much further with equation 4.33. What is possible, and indeed quite common, is to consider the *conditional* expectation of $\mathcal{D}_n$, or $p_n(G)$ as derived in equation 4.18. Since there is a single, common default probability, we will refer to this as $p(G)$. This quantity, analogous to the mixture-model setting, will prove a key building block for the construction of the analytic loss distribution.

The loss distribution is summarized by the following familiar variable,

$$\mathbb{D}_N = \sum_{n=1}^{N} \mathbb{I}_{\{y_n \leq \Phi^{-1}(p)\}}, \tag{4.34}$$

$$= \sum_{n=1}^{N} \mathbb{I}_{\mathcal{D}_n},$$

which is nothing other than the number of defaults in the portfolio. We are quite naturally interested in $\mathbb{P}(\mathbb{D}_N = k)$. We make use of the law of iterated expectations, from which the following statement holds,

$$\mathbb{E}(\mathbb{I}_{\mathbb{D}_N = k}) = \mathbb{E}(\mathbb{E}(\mathbb{I}_{\mathbb{D}_N = k}|G)), \tag{4.35}$$

$$\mathbb{P}(\mathbb{D}_N = k) = \mathbb{E}(\mathbb{P}(\mathbb{D}_N = k|G)),$$

$$= \int_{-\infty}^{\infty} \mathbb{P}(\mathbb{D}_N = k|g)\phi(g)dg,$$

where, as usual, $\phi(\cdot)$ denotes the standard normal density. Equation 4.35 implies that if we want to understand the probability of a given number of defaults, we need to consider the probability of the number of defaults integrated, or averaged if you prefer, over the range of possible values of the global risk factor, G.

The number of defaults, conditional on a given value of G, is actually independent. $\mathbb{D}_N$ is, as we learned in previous chapters, the sum of a number of independent Bernoulli trials, which implies $\mathbb{D}_N$ has a binomial distribution. Moreover each conditionally independent trial has a conditional probability of $p(G)$. Together this implies that,

$$\mathbb{P}(\mathbb{D}_N = k|G) = \frac{N!}{k!(N-k)!}\mathbb{P}(\mathcal{D}_k|G)^k (1-\mathbb{P}(\mathcal{D}_k|G))^{N-k}, \tag{4.36}$$

$$= \frac{N!}{k!(N-k)!}\left(\underbrace{\mathbb{P}(y_k \leq \Phi^{-1}(p)|G)}_{\text{Equation 4.18}}\right)^k \left(1-\underbrace{\mathbb{P}(y_k \leq \Phi^{-1}(p)|G)}_{\text{Equation 4.18}}\right)^{N-k},$$

$$= \binom{N}{k}p(G)^k(1-p(G))^{N-k}.$$

4.2 The Limit-Loss Distribution

Plugging equation 4.36 into equation 4.35, we have

$$\mathbb{P}(\mathbb{D}_N = k) = \underbrace{\int_{-\infty}^{\infty} \binom{N}{k} p(g)^k (1 - p(g))^{N-k} \phi(g) dg}_{\text{Equation 4.36}}. \quad (4.37)$$

Thus, we have the probability mass function for the total number of defaults in the Gaussian threshold model. Naturally, the more common object for the loss distribution is,

$$\mathbb{P}(\mathbb{D}_N \leq m) = \sum_{k=0}^{m} \mathbb{P}(\mathbb{D}_N = k), \quad (4.38)$$

$$= \sum_{k=0}^{m} \binom{N}{k} \int_{-\infty}^{\infty} p(g)^k (1 - p(g))^{N-k} \phi(g) dg.$$

This is not an analytic expression, but it can be readily solved through numerical integration. This would bring us to a similar situation as we found ourselves in the mixture-model setting. It turns out, however, that if one is willing to make a further assumption, which is quite consistent with the previous homogeneity assumptions, then one can take an additional step toward analytic tractability. The premise is that we move to proportional defaults.

This part of the derivation requires deriving a limiting loss distribution as the number of counterparties (i.e., N) tends to infinity. This requires *two* insights. The first key insight is that $p(G)$, as described in equation 4.18, is a strictly decreasing function of G. Decreases in G lead to an increased probability of default, whereas increases in the common risk factor generate lower default probabilities. The consequence of this fact is that,

$$p(-G) \geq x, \quad (4.39)$$
$$p^{-1}(p(-G)) \geq p^{-1}(x),$$
$$-G \geq p^{-1}(x),$$
$$G \leq -p^{-1}(x),$$

for an arbitrary choice of x. Note that $p^{-1}(x)$ is defined by inverting equation 4.18. Practically, this means solving the right-hand side of the following expression for y,

$$x = \Phi\left(\frac{\Phi^{-1}(p) - \sqrt{\rho} y}{\sqrt{1 - \rho}}\right), \quad (4.40)$$

$$\Phi^{-1}(x) = \frac{\Phi^{-1}(p) - \sqrt{\rho} y}{\sqrt{1 - \rho}},$$

$$\frac{\Phi^{-1}(p) - \sqrt{1-\rho}\Phi^{-1}(x)}{\sqrt{\rho}} = y,$$

$$p^{-1}(x) = \frac{\Phi^{-1}(p) - \sqrt{1-\rho}\Phi^{-1}(x)}{\sqrt{\rho}}.$$

Pulling it all together, we have that

$$G \leq -p^{-1}(x), \qquad (4.41)$$

$$G \leq \frac{\sqrt{1-\rho}\Phi^{-1}(x) - \Phi^{-1}(p)}{\sqrt{\rho}},$$

for an arbitrary choice of x.

The second step begins by redefining the variable used to summarize the loss distribution. This, basically implicit assumption, amounts to the elimination of the number of counterparties and movement to a proportion perspective. Let's define,

$$\check{\mathbb{D}}_N = \frac{1}{N}\sum_{n=1}^{N}\mathbb{I}_{\{y_n \leq \Phi^{-1}(p)\}}, \qquad (4.42)$$

$$= \frac{1}{N}\sum_{N=1}^{N}\mathbb{I}_{\mathcal{D}_n},$$

$$= \frac{\mathbb{D}_N}{N},$$

which is the *proportion* of defaults in the portfolio. This transformation moves us from a discrete number of defaults to a continuum of proportional losses over the unit interval. While clever and very helpful, we have lost the notion of individual obligor exposures. Although perhaps not so dramatic at this point, it will have important implications later in the regulatory setting discussed in Chap. 6.

Observe that we may now compute the conditional expectation of $\check{\mathbb{D}}_N$ as follows,

$$\mathbb{E}(\check{\mathbb{D}}_N | G) = \mathbb{E}\left(\frac{1}{N}\sum_{n=1}^{N}\mathbb{I}_{\mathcal{D}_n}\bigg| G\right), \qquad (4.43)$$

$$= \frac{1}{N}\sum_{n=1}^{N}\mathbb{E}\left(\mathbb{I}_{\mathcal{D}_n}\big| G\right),$$

$$= \frac{1}{N}\sum_{n=1}^{N}\mathbb{P}\left(\mathcal{D}_n | G\right),$$

4.2 The Limit-Loss Distribution

$$= \frac{1}{\cancel{N}}\left(\underbrace{\cancel{N}\Phi\left(\frac{\Phi^{-1}(p)-\sqrt{\rho}G}{\sqrt{1-\rho}}\right)}_{\text{Equation 4.18}}\right),$$

$$= p(G).$$

The conditional variance is also easily determined,

$$\text{var}(\check{\mathbb{D}}_N | G) = \text{var}\left(\frac{1}{N}\sum_{N=1}^{N}\mathbb{I}_{\mathcal{D}_n}\bigg| G\right), \qquad (4.44)$$

$$= \frac{1}{N^2}\sum_{n=1}^{N}\text{var}\left(\mathbb{I}_{\mathcal{D}_n}\big| G\right),$$

$$= \frac{1}{N}\sum_{n=1}^{N}\mathbb{P}(\mathcal{D}_n|G)(1-\mathbb{P}(\mathcal{D}_n)|G),$$

$$= \frac{1}{N\cancel{?}}\left(\cancel{N}p(G)(1-p(G))\right),$$

$$= \frac{p(G)(1-p(G))}{N}.$$

The moments of $\check{\mathbb{D}}_N$ have a familiar binomial form that is reminiscent of the mixture-model case.

Although the moments are interesting, it would be even better to have the full conditional distribution. Indeed, it was initially derived by a clever observation in Vasicek (1991), which noted that for a very large portfolio, following from the weak law of large numbers, conditional on the common risk factor, the fraction of counterparties actually defaulting tends toward the individual default probability. Mathematically, this condition is written as follows,

$$\mathbb{P}\left(\lim_{n\to\infty}\check{\mathbb{D}}_N = p(G)\bigg| G\right) = 1. \qquad (4.45)$$

This result is provided without proof, but informally looking at the form of the conditional variance in equation 4.44, it is not difficult to see that it tends to zero as $N \to \infty$. Since, we are interested in the loss distribution of $\check{\mathbb{D}}_N$, it is enormously helpful to know that equation 4.45 implies that,

$$F(x) = \mathbb{P}(\check{\mathbb{D}}_N \leq x) = \mathbb{P}(p(G) \leq x), \qquad (4.46)$$

$$= \mathbb{P}(G \leq -p^{-1}(x)),$$

$$= \mathbb{P}\left(G \le \frac{\sqrt{1-\rho}\Phi^{-1}(x) - \Phi^{-1}(p)}{\sqrt{\rho}}\right),$$

$$= \Phi\left(\frac{\sqrt{1-\rho}\Phi^{-1}(x) - \Phi^{-1}(p)}{\sqrt{\rho}}\right).$$

This is the distribution of the proportion of default loss on the portfolio, which gives us essentially everything we require. We must only be cautious of the fact that this readily computed approximation works for sufficiently large N. The appropriate size of N for the responsible use of analytic formulae is an empirical question. We will address this question shortly.

4.2.1 The Limit-Loss Density

If we differentiate equation 4.46 with respect to x and simplify, we arrive at the default-loss density, which can be directly employed to approximate any desired quantile by selecting the appropriate value of x. It is computed as,

$$\frac{\partial F(x)}{\partial x} = \frac{\partial}{\partial x}\left(\Phi\left(\frac{\sqrt{1-\rho}\Phi^{-1}(x) - \Phi^{-1}(p)}{\sqrt{\rho}}\right)\right). \tag{4.47}$$

This is a relatively simple, albeit slightly tedious, calculus exercise. Nevertheless, it is a useful quantity to possess, so it warrants evaluation. Let's start with a few useful facts,

$$\frac{\partial \Phi(x)}{\partial x} = \frac{\partial}{\partial x}\left(\int_{\infty}^{x} \frac{1}{\sqrt{2\pi}} e^{-\frac{u^2}{2}} du\right), \tag{4.48}$$

$$= \frac{1}{\sqrt{2\pi}} e^{-\frac{x^2}{2}}.$$

We can rewrite equation 4.46 in a more meaningful form,

$$F(x) = h(b(x)), \tag{4.49}$$

where

$$h(x) = \int_{\infty}^{x} \frac{1}{\sqrt{2\pi}} e^{-\frac{u^2}{2}} du, \tag{4.50}$$

and

$$b(x) = \frac{\sqrt{1-\rho}\Phi^{-1}(x) - \Phi^{-1}(p)}{\sqrt{\rho}}. \tag{4.51}$$

4.2 The Limit-Loss Distribution

Thus, by the chain rule, we have that,

$$F'(x) = h'(b(x))b'(x). \tag{4.52}$$

Thus we need to evaluate $h'(x)$ and $b'(x)$. $h'(x)$ is, basically by definition, summarized in equation 4.48,

$$h'(x) = \frac{1}{\sqrt{2\pi}} e^{-\frac{x^2}{2}}. \tag{4.53}$$

while $b'(x)$ is

$$b'(x) = \frac{\sqrt{1-\rho}}{\sqrt{\rho}} (\Phi^{-1})'(x), \tag{4.54}$$

$$= \frac{\sqrt{1-\rho}}{\sqrt{\rho}} \frac{1}{\Phi'(\Phi^{-1}(x))},$$

$$= \frac{\sqrt{1-\rho}}{\sqrt{\rho}} \sqrt{2\pi} e^{\frac{(\Phi^{-1}(x))^2}{2}},$$

which is essentially another application of the chain rule. Combining equations 4.53 and 4.54, we have

$$F'(x) = h'(b(x))b'(x), \tag{4.55}$$

$$= \underbrace{\frac{1}{\sqrt{2\pi}} e^{-\frac{b(x)^2}{2}}}_{\text{Equation 4.53}} \underbrace{\frac{\sqrt{1-\rho}}{\sqrt{\rho}} \sqrt{2\pi} e^{\frac{\Phi^{-1}(x)^2}{2}}}_{\text{Equation 4.54}},$$

$$= \sqrt{\frac{1-\rho}{\rho}} e^{\frac{(\Phi^{-1}(x))^2}{2} - \frac{b(x)^2}{2}},$$

$$f_{\breve{\mathbb{D}}_N}(x; p, \rho) = \sqrt{\frac{1-\rho}{\rho}} \exp\left(\frac{(\Phi^{-1}(x))^2}{2} - \frac{\left(\sqrt{1-\rho}\Phi^{-1}(x) - \Phi^{-1}(p)\right)^2}{2\rho}\right),$$

for $x \in (0, 1)$.[9] Thus we have the expression as found in Vasicek (1991).

---

[9]We've restricted the support of this density, because inspection of equation 4.55 reveals that the density is undefined for $\rho = 0$ and takes the value of zero when $\rho = 1$.

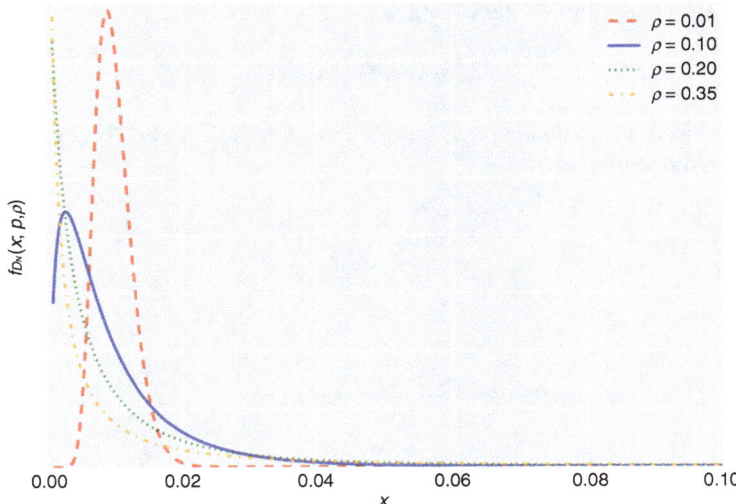

Fig. 4.4 *Different limit-loss densities*: This figure graphs the limit-loss density derived in equation 4.55 for different settings of the parameter, ρ. In all cases, $p = 0.01$.

The result, despite being based on a Gaussian model, is, depending on the choice of ρ, far from Gaussian. Figure 4.4 examines the limit-loss density derived in equation 4.55 for different settings of the parameter, ρ. In all cases, to make the curves more readily comparable, we have held $p = 0.01$. When $\rho \approx 0$, the density has a symmetric, Gaussian-like form. As we increase the level of ρ, the density takes on a skewed, asymmetric shape with a relatively fat tail.

This approach is also referred to—as coined by Gordy (2002)—as the *asymptotic single risk-factor* (ASRF) model. It will play a critical role in the discussion of regulatory models in Chap. 6. For now, let's continue to simply think of it as an analytical approximation of the one-factor Gaussian threshold model. Later, we'll learn its role in the regulatory world.

4.2.2 Analytic Gaussian Results

The next logical step is to use the limit-loss distribution and density from equations 4.46 and 4.55 to produce risk estimates for our running example. In principle, this is conceptually the same as the analytic formulae used in Chaps. 2 and 3 for the total number of defaults, $\mathbb{D}_N$. The only difference, in this case, is that we have slightly transformed the loss variable to the continuous proportional amount of default, $\mathbb{D}_N$.

The actual implementation is relatively straightforward. Algorithm 4.6 provides the details; it builds the cumulative distribution function from equation 4.46 on the

4.2 The Limit-Loss Distribution

Algorithm 4.6 An analytical approximation to the Gaussian threshold model

```
def asrfModel(myP,rho,c,alpha):
    myX = np.linspace(0.0001,0.9999,10000)
    num = np.sqrt(1-rho)*norm.ppf(myX)-norm.ppf(myP)
    cdf = norm.cdf(num/np.sqrt(rho))
    pdf = util.asrfDensity(myX,myP,rho)
    varAnalytic = np.sum(c)*np.interp(alpha,cdf,myX)
    esAnalytic = asrfExpectedShortfall(alpha,myX,cdf,pdf,c,rho,myP)
    return pdf,cdf,varAnalytic,esAnalytic
```

Algorithm 4.7 An analytical expected-shortfall computation

```
def asrfExpectedShortfall(alpha,myX,cdf,pdf,c,rho,myP):
    expectedShortfall = np.zeros(len(alpha))
    for n in range(0,len(alpha)):
        myAlpha = np.linspace(alpha[n],1,1000)
        loss = np.sum(c)*np.interp(myAlpha,cdf,myX)
        prob = np.interp(loss,myX,pdf)
        expectedShortfall[n] = np.dot(loss,prob)/np.sum(prob)
    return expectedShortfall
```

support defined by myX. The VaR estimates are then simply interpolated values of the cumulative distribution function based on the desired quantiles in alpha.

The expected shortfall requires a modestly different implementation to the generic function used in the independent-default and mixture-model cases. This approach is summarized in Algorithm 4.7. The reason for the change is that, because $\mathbb{D}_N$ is a continuous variable, we can no longer simply compute the probability (mass) density function by summing over the cumulative distribution function. Instead, we use the analytic expression in equation 4.55. This is accomplished using the asrfDensity function, which we've placed in our util library.

The remainder of the effort involves partitioning the space beyond the quantile and estimating the losses and their probabilities for each grid point. This permits the computation of the expected value, which is normalized by the size of the quantile to produce our final estimate. If one is patient and careful, it may even be possible to compute analytic formulae for the VaR and expected shortfall. In this work, however, we've opted for a more direct approach.

Table 4.2 compares the Monte-Carlo-based numerical implementation to this analytical approximation. The analytic approach provides results that are surprisingly close to the base model. The mean and volatility of the loss are extremely close. Both VaR and expected shortfall exhibit relatively few differences.

Table 4.2 *Gaussian threshold analytic-numerical comparison*: This table provides the analytic approximation of the one-factor Gaussian model from Vasicek (1991) for our usual example. It also replicates, for comparison, the Monte-Carlo implementation from Table 4.1.

Quantile	Numeric		Analytic	
	$\widehat{\text{VaR}}_\alpha$	$\widehat{\mathcal{E}}_\alpha$	$\widehat{\text{VaR}}_\alpha$	$\widehat{\mathcal{E}}_\alpha$
95.00th	$48.4	$82.7	$43.7	$84.4
97.00th	$63.5	$100.9	$60.3	$106.1
99.00th	$101.9	$145.6	$106.2	$160.6
99.50th	$130.1	$177.1	$140.6	$199.5
99.90th	$206.7	$258.2	$234.0	$298.7
99.97th	$286.9	$340.1	$311.2	$376.8
99.99th	$323.8	$377.7	$383.5	$447.9
Other statistics				
Expected loss	$9.2		$9.5	
Loss Volatility	$21.7		$21.7	
Total iterations	10,000,000		0	
Computational time	49.0		0.0	

The analytic implementation does, however, for most quantiles, provide slightly higher VaR and expected shortfall estimates. Generally, we would expect that the numerical approach—which fully captures the multiplicity and concentration of default probabilities and exposures—to be slightly higher. The asymptotic method uses a common default probability and assumes small and equal contributions of each obligor to total exposure. The first assumption would appear to push the risk measures upwards, whereas the second has a downward impact. It would appear that, all else equal, the impact of an average default probability dominates for this portfolio. Nevertheless, we conclude that the analytic approximation, based on numerous assumptions, does an entirely reasonable job for our 100-obligor portfolio. This is confirmed by the remarkably close visual relationship displayed in Fig. 4.5.[10]

Up to this point, we have established a fully-fledged framework for the Gaussian-threshold model. It includes incorporation of individual default probabilities and exposures, comfortably handles default correlation, permits both numerical and analytical implementations, and allows for relatively straightforward calibration. Moreover, despite its Gaussian foundations, the limiting distribution is both asymmetric and leptokurtotic.

The reader may reasonably wonder why other threshold models need be considered. The reason stems from two main shortcomings. The first is the inability to control the probability mass in the tails of the default distribution. The second relates to the nature of default-event correlation as we move arbitrarily far into the tail of the loss distribution. The Gaussian model has some deficiencies in both these areas—moreover, both are addressed by extending it. First, however, it is important to understand the scope of the problem. This is the focus on the next section.

---

[10]This fact presumably explains, at least in part, its popularity in regulatory circles.

4.3 Tail Dependence

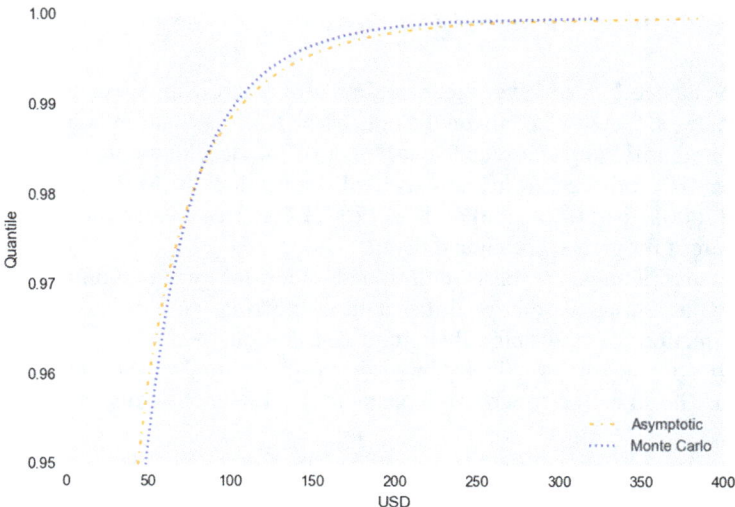

Fig. 4.5 *The analytic fit*: This figure compares the tail probabilities from the limit-loss analytic approximation and the full Monte-Carlo computation of the Gaussian threshold model. It confirms visually our conclusions from Table 4.2.

4.3 Tail Dependence

The Gaussian-threshold model has an important weakness in the area of tail dependence. This notion, quite familiar to statisticians and probabilists, is less well known and employed in finance. Tail dependence describes, from a technical perspective, the limiting conditional probability that a given random variable will exceed a quantile given that another random variable has already exceeded it. *Limiting* in this context implies that the quantile is taken to an extreme value, such as 0 or 1. In simpler language, as the name suggests, it describes the dependence between two, or more, random variables in the extremes (i.e., tails) of their joint distribution.

In a credit-risk setting, the notion of tail dependence is clearly important and, as such, it is completely natural to wish to understand this quantity in some detail. The joint state-variable distribution describing the default process, as we will see, has important implications for the nature of tail dependence. In this section, we will review the basic ideas in tail dependence and consider a few practical examples. This will pave the way for a number of concrete extensions to the basic Gaussian-threshold framework.

4.3.1 The Tail-Dependence Coefficient

Tail dependence is a relatively technical area of study. Since we, as quantitative credit analysts, focus so keenly on the tails of the loss distribution, it is important to gain a familiarity with these concepts. Most of the development in the following discussion is motivated from the excellent descriptions in McNeil et al. (2015), Nelsen (2006), Joe (1997), and de Kort (2007). Readers are encouraged to turn to these sources for greater rigour and detail.

Let's start with some basic definitions. Consider two random variables, X and Y, with continuous marginal distribution functions, $F_Y(y)$ and $F_X(x)$, and a copula function, C, describing their joint distribution. While there are alternative approaches to measuring tail dependence, the following appears to be the most common. The so-called upper-tail-dependence coefficient has this basic form,

$$\Lambda_{\text{Upper}} = \lim_{q \to 1} \mathbb{P}\left(Y > q\text{-quantile of } Y \mid X > q\text{-quantile of } X\right). \tag{4.56}$$

In words, therefore, it is the limit of the conditional probability of Y exceeding its qth quantile given X exceeds its qth quantile as q tends toward unity. That is, what is the interdependence of X and Y as we move arbitrarily far into the tail of their joint distribution?

Equation 4.56 is a bit loose, but it can be written more precisely as,

$$\begin{aligned}\Lambda_{\text{Upper}} &= \lim_{q \uparrow 1} \mathbb{P}\left(Y > F_Y^{-1}(q) \mid X > F_X^{-1}(q)\right), \tag{4.57}\\ &= \lim_{q \uparrow 1} \underbrace{\frac{\mathbb{P}\left(Y > F_Y^{-1}(q), X > F_X^{-1}(q)\right)}{\mathbb{P}(X > F_X^{-1}(q))}}_{\text{By Bayes' Rule}}, \\ &= \lim_{q \uparrow 1} \frac{1 - F_Y\left(F_Y^{-1}(q)\right) - F_X\left(F_X^{-1}(q)\right) + F\left(Y > F_Y^{-1}(q), X > F_X^{-1}(q)\right)}{1 - F_X\left(F_X^{-1}(q)\right)}, \\ &= \lim_{q \uparrow 1} \frac{1 - 2q + F\left(Y > F_Y^{-1}(q), X > F_X^{-1}(q)\right)}{1 - q}, \\ &= \lim_{q \uparrow 1} \frac{1 - 2q + C(q, q)}{1 - q},\end{aligned}$$

where $C(\cdot, \cdot)$ denotes the copula function for Y and X. This measure describes the probability of two random variates taking values in the extreme upper end of the distribution. In statistical parlance, if $\Lambda_{\text{Upper}} \neq 0$, then the copula function, C, is said to possess upper-tail dependence.

4.3 Tail Dependence

It is also possible to look at the other extreme of the distribution. The lower-tail-dependence coefficient has the following form,

$$\Lambda_{\text{Lower}} = \lim_{q \downarrow 0} \mathbb{P}\left(Y \leq F_Y^{-1}(q) \middle| X \leq F_X^{-1}(q)\right), \quad (4.58)$$

$$= \lim_{q \downarrow 0} \underbrace{\frac{\mathbb{P}\left(Y \leq F_Y^{-1}(q), X \leq F_X^{-1}(q)\right)}{\mathbb{P}(X \leq F_X^{-1}(q))}}_{\text{By Bayes' Rule}},$$

$$= \lim_{q \downarrow 0} \frac{F\left(Y \leq F_Y^{-1}(q), X \leq F_X^{-1}(q)\right)}{F_X\left(F_X^{-1}(q)\right)},$$

$$= \lim_{q \downarrow 0} \frac{C(q, q)}{q}.$$

Analogous to equation 4.57, if $\Lambda_{\text{Lower}} \neq 0$, then the copula function C—or equivalently, the joint distribution of X and Y—is said to possess lower-tail dependence.

For symmetric distributions—which is our ultimate focus in this area—there is no difference between upper and lower-tail dependence. In fact, we have that,

$$\Lambda = \Lambda_{\text{Upper}} = \Lambda_{\text{Lower}} = \lim_{q \downarrow 0} \frac{C(q, q)}{q}, \quad (4.59)$$

and thus we are free to use the more convenient representation in equation 4.58. In this case, we consider a symmetric copula function, C, to have tail dependence if $\Lambda \neq 0$.

Evaluating the limit in equation 4.59 appears, at first glance, somewhat daunting. Since it is a ratio of two functions, it seems reasonable to apply something like L'Hôpital's rule. This yields the following simplification of equation 4.59,

$$\Lambda = \lim_{q \downarrow 0} \frac{C(q, q)}{q}, \quad (4.60)$$

$$= \lim_{q \downarrow 0} \frac{dC(q, q)}{dq},$$

$$= \lim_{q \downarrow 0} dC(u_1, u_2),$$

$$= \lim_{q \downarrow 0} \left(\frac{\partial C(u_1, u_2)}{\partial u_1} + \frac{\partial C(u_1, u_2)}{\partial u_2}\right).$$

There exists an unexpected link between the partial derivatives of a copula function and the underlying random variables' conditional distributions. In particular, it can be shown that,

$$\frac{\partial C(u_1, u_2)}{\partial u_2} = \mathbb{P}(U_1 \leq u_1 | U_2 = u_2). \tag{4.61}$$

A detailed (and lengthy) proof of the result in equation 4.61 can be found in Schmitz (2003, Section 2.2).[11] It does, however, permit us to dramatically simplify the tail-dependence coefficient,

$$\begin{aligned} \Lambda &= \lim_{q \downarrow 0} \frac{C(q, q)}{q}, \\ &= \lim_{q \downarrow 0} \left(\frac{\partial C(u_1, u_2)}{\partial u_1} + \frac{\partial C(u_1, u_2)}{\partial u_2} \right), \\ &= \lim_{q \downarrow 0} \left(\mathbb{P}(U_1 \leq u_1 | U_2 = u_2) + \mathbb{P}(U_2 \leq u_2 | U_1 = u_1) \right), \\ &= 2 \lim_{q \downarrow 0} \mathbb{P}(U_1 \leq q | U_2 = q), \end{aligned} \tag{4.63}$$

where the last step follows from exchangeability of the copula function—a condition which is a form of strong symmetry and is met by both Gaussian and t-copulas.[12]

---

[11] An intuitive, rough-and-ready justification from Schmitz (2003, Section 2.2) involves,

$$\begin{aligned} \frac{\partial C(F_X(y), F_Y(x))}{\partial F_Y(y)} &= \lim_{h \to 0} \frac{C(F_X(x), F_Y(y) + \epsilon(h)) - C(F_X(x), F_Y(y))}{\epsilon(h)}, \\ &= \lim_{h \to 0} \frac{C\left(F_X(x), \cancel{F_Y(y)} + \overbrace{F_Y(y+h) - \cancel{F_Y(y)}}^{\epsilon(h)}\right) - C(F_X(x), F_Y(y))}{\underbrace{F_Y(y+h) - F_Y(y)}_{\epsilon(h)}}, \\ &= \lim_{h \to 0} \frac{F_{XY}(x, y+h) - F_{XY}(x, y)}{F_Y(y+h) - F_Y(y)}, \\ &= \lim_{h \to 0} \mathbb{P}\left(X \leq x \,\middle|\, Y \in (y, y+h)\right), \\ &= \mathbb{P}(X \leq x | Y = y). \end{aligned} \tag{4.62}$$

This is neither precise, nor probably even exactly true, but it is a useful bit of motivation for the preceding result.

[12] See McNeil et al. (2015, Chapter 7) for more details.

4.3.2 Gaussian Copula Tail-Dependence

Equation 4.63 provides a workable definition for the tail-dependence coefficient. We may now proceed to evaluate its value for two specific distributional choices. Consider first the bivariate Gaussian distribution. Let Y and X be standard normal variates with (Y, X) following a bivariate Gaussian distribution with correlation coefficient, ρ. In this case, the tail-dependence coefficient may be written as,

$$\Lambda = 2 \lim_{x \downarrow -\infty} \mathbb{P}(Y \leq x | X = x), \tag{4.64}$$

where the limit changes due to the infinite support of the Gaussian distribution. To evaluate this expression, we need to know the moments of the bivariate conditional distribution. We know that,

$$Y|X = x \sim \mathcal{N}\left(\mu_Y + \frac{\sigma_Y}{\sigma_X}\rho(x - \mu_X), (1 - \rho^2)\sigma_Y^2\right), \tag{4.65}$$

and, for the *standard* conditional bivariate normal distribution, we have $Y|X = x \sim \mathcal{N}(\rho x, 1 - \rho^2)$. Returning to equation 4.65, we can finally evaluate the desired limit,

$$\begin{aligned}
\Lambda_{\mathcal{N}} &= 2 \lim_{x \downarrow -\infty} \mathbb{P}(Y \leq x | X = x), \\
&= 2 \lim_{x \downarrow -\infty} F_{Y|X=x}(x), \\
&= 2 \lim_{x \downarrow -\infty} \Phi\left(\frac{x - \rho x}{\sqrt{1 - \rho^2}}\right), \\
&= 2 \lim_{x \downarrow -\infty} \Phi\left(\frac{x(1 - \rho)}{\sqrt{(1 - \rho)(1 + \rho)}}\right), \\
&= 2 \lim_{x \downarrow -\infty} \Phi\left(\frac{x\sqrt{1 - \rho}}{\sqrt{1 + \rho}}\right), \\
&= 0,
\end{aligned} \tag{4.66}$$

if $\rho < 1$. The consequence is that the Gaussian distribution does *not* exhibit tail dependence, since the tail-dependence coefficient has a zero value. This is an important result. It implies that use of the Gaussian copula leads to, at least from an asymptotic perspective, independence between random variables as one tends to the extremes of the distribution. Moreover, this is independent of the value of the correlation coefficient.[13] Since credit-risk computations routinely examine outcomes at the 99.97th quantile of the loss distribution or further, this is not merely

---

[13] With the exception of perfect correlation—that is, the uninteresting case when $\rho = 1$.

a theoretical exercise. This is, therefore, an inherent shortcoming of the Gaussian copula as it is employed in the threshold-model setting.

4.3.3 t-Copula Tail-Dependence

Let us now consider the alternative bivariate t-distribution. As before, we now let Y and X follow standard t marginal distributions with ν degrees of freedom and (Y, X) follow a bivariate t-distribution with correlation coefficient, ρ, and ν degrees of freedom. The tail-dependence coefficient has the same form as found in equation 4.64, albeit with a different conditional distribution. We know—although tedious to show, see Roth (2013, Appendix A.6) for the gory details—that the conditional t distribution is,

$$Y|X = x \sim F_{\mathcal{T}_{\nu+1}}\left(\mu_Y + \frac{\rho}{\sigma_X}(x - \mu_X), \frac{\nu + \frac{(x-\mu_X)(x-\mu_Y)}{\sigma_X}}{\nu + 1}\left(\sigma_Y - \frac{\rho^2}{\sigma_X^2}\right)\right), \tag{4.67}$$

and, for the *standard* conditional bivariate t-distribution, we have $Y|X = x \sim F_{\mathcal{T}_{\nu+1}}\left(\rho x, \frac{\nu+x^2}{\nu+1}(1 - \rho^2)\right)$. Returning to equation 4.66, we can finally evaluate the desired limit,

$$\Lambda_{\mathcal{T}_{\nu}} = 2 \lim_{x \downarrow -\infty} \mathbb{P}(Y \le x | X = x), \tag{4.68}$$

$$= 2 \lim_{x \downarrow -\infty} F_{Y|X=x}(x),$$

$$= 2 \lim_{x \downarrow -\infty} F_{\mathcal{T}_{\nu+1}}\left(\frac{x - \rho x}{\sqrt{\frac{\nu+x^2}{\nu+1}(1-\rho^2)}}\right),$$

$$= 2 \lim_{x \downarrow -\infty} F_{\mathcal{T}_{\nu+1}}\left(\sqrt{\frac{\nu+1}{\nu+x^2}} \frac{x(1-\rho)}{\sqrt{(1-\rho)(1+\rho)}}\right),$$

$$= 2 \lim_{x \downarrow -\infty} F_{\mathcal{T}_{\nu+1}}\left(\sqrt{\frac{\nu+1}{\nu+x^2}} \frac{x\sqrt{1-\rho}}{\sqrt{1+\rho}}\right),$$

$$= 2 \lim_{x \downarrow -\infty} F_{\mathcal{T}_{\nu+1}}\left(\frac{x}{\sqrt{\nu+x^2}}\sqrt{\frac{(\nu+1)(1-\rho)}{1+\rho}}\right),$$

$$= 2 F_{\mathcal{T}_{\nu+1}}\left(-\sqrt{\frac{(\nu+1)(1-\rho)}{1+\rho}}\right).$$

4.3 Tail Dependence

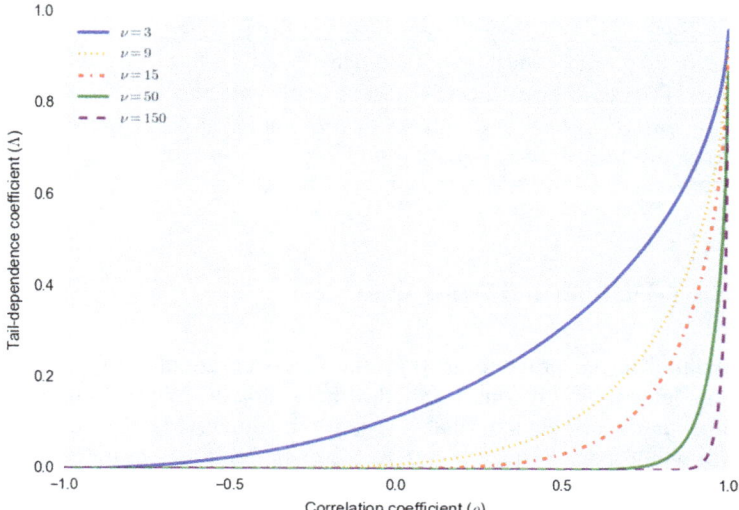

Fig. 4.6 *Tail-dependence coefficients*: This figure illustrates, for alternative choices of degrees-of-freedom parameters, the interaction between the correlation coefficient and the tail-dependence coefficient for the t-distribution summarized in equation 4.68. Observe that tail-dependence is an increasing function of the correlation coefficient, but a decreasing function of the degrees-of-freedom parameter.

This implies that for all values of $\rho \in (-1, 1)$ and $\nu > 0$, the tail-dependence coefficient is non-zero for the t-distribution. This suggests that the multivariate t distribution may represent a more judicious choice for credit-risk measurement applications.

Figure 4.6 highlights, for alternative choices of degrees-of-freedom parameters, the interaction between the correlation coefficient and the tail-dependence coefficient for the t-distribution summarized in equation 4.68. It demonstrates clearly that tail-dependence for the multivariate t-distribution is an increasing function of the correlation coefficient, but a decreasing function of the degrees-of-freedom parameter. This makes logical sense, because as the degrees-of-freedom parameter gets large, the multivariate t-distribution converges to the multivariate Gaussian distribution, which, as we've just demonstrated, does *not* exhibit tail dependence.

To perform this relatively simple computation, we implement equation 4.68 in Algorithm 4.8. When, in the subsequent sections, we implement the t-threshold model, this function will prove useful for its calibration. It takes two inputs arguments, rho and nu, which naturally correspond to the distributional parameters. Python's scipy statistics library, stats, does the hard work of computing the univariate t cumulative distribution function.

Algorithm 4.8 Computing the *t*-distribution tail coefficient

```
from scipy.stats import t as myT
def tTailDependenceCoefficient(rho,nu):
    a = -np.sqrt(np.divide((nu+1)*(1-rho),1+rho))
    return 2*myT.cdf(a,nu+1)
```

4.4 The *t*-Distributed Approach

As highlighted in the previous section, the Gaussian copula is employed in the generic implementation of one-factor threshold models. While Gaussian models have come under criticism of late—see, for example, MacKenzie and Spears (2012)—it would be a mistake to assign these shortcomings to the relatively thin tails associated with the Gaussian distribution. The loss distribution, as we've already seen, in the one-factor Gaussian threshold model is, for positive values of the parameter ρ, both highly skewed and leptokurtotic.

A rather more compelling reason for moving away from the Gaussian distribution relates to the notion of tail dependence. We have demonstrated that the tail-dependence coefficient tends to zero for the Gaussian distribution. For the *t*-distribution, not only does tail dependence not disappear as we move arbitrarily far into the tails of the joint distribution, it is also, to a certain extent, governed by the choice of the asset-correlation parameter, ρ.

Since economic-capital computations routinely estimate VaR at confidence levels of 99.97% and higher, the notion of tail dependence is not a merely theoretic question in probability theory. It has strong practical implications for the behaviour of our risk measurements. Moreover, this analysis offers a strong argument for the employment of a *t*-copula for the joint distribution of the latent threshold variables. This choice, while defensible and desirable, is not without costs. It leads to a rather more complex model implementation.

4.4.1 A Revised Latent-Variable Definition

We will, therefore, work through the details of the *t*-distributed extension of the Gaussian threshold model.[14] Once this is complete, we will generalize the development and illustrate two additional possible extensions; at the same time, we will present a framework that embeds all threshold models. Practically, the vast majority the *t*-distributed implementation agrees with the Gaussian version described in a preceding section. A number of small details differ. It begins with

---

[14] See Kuhn (2005) and Kostadinov (2005) for more detail on the *t*-distributed case.

4.4 The t-Distributed Approach

the latent variable, y_n, which has a slightly different form

$$y_n = \sqrt{\frac{v}{W}} \left(\sqrt{\rho} G + \sqrt{1-\rho}\, \epsilon_n \right), \tag{4.69}$$

for $n = 1, \ldots, N$ where G, ϵ_n are still the same independent, identically distributed standard normal variables. The new character in this production, however, is W, an independent $\chi^2(v)$ random variable, where $v \in \mathbb{R}$ denotes the degrees-of-freedom parameter. This implies that $y_n \sim \mathcal{T}_v\left(0, \frac{v}{v-2}\right)$ for all $n = 1, \ldots, N$. In words, the coefficient $\sqrt{\frac{v}{W}}$ transforms each y_n into a univariate standard t-distributed random variable.[15] This implies that $\mathbb{E}(y_n) = 0$ and $\text{var}(y_n) = \frac{v}{v-2}$ for all $n = 1, \ldots, N$. These final two points merit some demonstration. The expected value of equation 4.69 is,

$$\mathbb{E}(y_n) = \mathbb{E}\left(\sqrt{\frac{v}{W}} \left(\sqrt{\rho} G + \sqrt{1-\rho}\, \epsilon_n \right) \right), \tag{4.70}$$

$$= \sqrt{\rho} \cdot \overbrace{\mathbb{E}\left(\sqrt{\frac{v}{W}}\right) \underbrace{\mathbb{E}(G)}_{=0}}^{\text{By independence}} + \sqrt{1-\rho} \cdot \overbrace{\mathbb{E}\left(\sqrt{\frac{v}{W}}\right) \underbrace{\mathbb{E}(\epsilon_n)}_{=0}}^{\text{By independence}},$$

$$= 0.$$

The next step is the variance of y_n. This requires, while quite similar, a bit more effort,

$$\text{var}(y_n) = \mathbb{E}\left(y_n^2 - \underbrace{\mathbb{E}(y_n)^2}_{=0} \right), \tag{4.71}$$

$$= \mathbb{E}\left(\left(\sqrt{\frac{v}{W}} \left(\sqrt{\rho} G + \sqrt{1-\rho}\, \epsilon_n \right) \right)^2 \right),$$

$$= \mathbb{E}\left(\frac{v}{W} \cdot \rho \cdot G^2 \right) + \mathbb{E}\left(\frac{v}{W} \cdot (1-\rho) \cdot \epsilon_n^2 \right),$$

$$= \rho \cdot v \mathbb{E}\left(\frac{1}{W} \right) \underbrace{\mathbb{E}(G^2)}_{=1} + (1-\rho) \cdot v \cdot \mathbb{E}\left(\frac{1}{W} \right) \underbrace{\mathbb{E}(\epsilon_n^2)}_{=1},$$

$$= v\mathbb{E}\left(\frac{1}{W} \right).$$

---

[15] See Appendix A for a detailed description of the construction of the t-distribution.

This leads us to something of an impasse, since the expectation of the reciprocal of a chi-squared distribution with ν degrees of freedom is not an obvious fact. It can, however, be determined directly by solving the following integration problem,

$$\mathbb{E}\left(\frac{1}{W}\right) = \int_{\mathbb{R}_+} \frac{1}{w} f_W(w) dw, \qquad (4.72)$$

$$= \int_{\mathbb{R}_+} \frac{1}{w} \underbrace{\frac{1}{2^{\frac{\nu}{2}} \Gamma\left(\frac{\nu}{2}\right)} w^{\frac{\nu}{2}-1} e^{-\frac{w}{2}}}_{W \sim \chi^2(\nu)} dw,$$

$$= \frac{1}{2^{\frac{\nu}{2}} \Gamma\left(\frac{\nu}{2}\right)} \int_{\mathbb{R}_+} w^{\frac{\nu-2}{2}-1} e^{-\frac{w}{2}} dw,$$

$$= \frac{2^{\frac{\nu-2}{2}} \Gamma\left(\frac{\nu-2}{2}\right)}{2^{\frac{\nu}{2}} \Gamma\left(\frac{\nu}{2}\right)} \underbrace{\int_{\mathbb{R}_+} \overbrace{\frac{1}{2^{\frac{\nu-2}{2}} \Gamma\left(\frac{\nu-2}{2}\right)} w^{\frac{\nu-2}{2}-1} e^{-\frac{w}{2}}}^{\chi^2(\nu-2)} dw}_{=1},$$

$$= \frac{1}{2} \frac{\Gamma\left(\frac{\nu-2}{2}\right)}{\Gamma\left(\frac{\nu}{2}\right)},$$

$$= \frac{1}{2} \frac{\left(\frac{\nu-2}{2}-1\right)!}{\left(\frac{\nu}{2}-1\right)!},$$

$$= \frac{1}{2} \frac{1}{\left(\frac{\nu}{2}-1\right)},$$

$$= \frac{1}{\nu - 2},$$

which directly implies, from equation 4.71 that, indeed as we claimed, $\text{var}(y_n) = \frac{\nu}{\nu-2}$. The last line of equation 4.72 follows from the fact that $\nu \in \mathbb{N}_+$ and $\Gamma(n) = (n-1)!$.

The marginal systematic global and idiosyncratic variable distributions of the t-distributed model remain Gaussian—it is the joint and marginal distributions of the latent default-state variables (i.e., $\{y_n; n = 1, \ldots, N\}$) that follow a t-distribution. Consequently, the covariance—and correlation—between the latent state variables y_n and y_m are also slightly transformed.

4.4 The t-Distributed Approach

We repeat the analysis in equation 4.9 on page 152 using the new definition in equation 4.69,

$$\text{cov}(y_n, y_m) = \mathbb{E}(y_n y_m), \tag{4.73}$$

$$= \mathbb{E}\left(\underbrace{\sqrt{\frac{\nu}{W}}\left(\sqrt{\rho}G + \sqrt{1-\rho}\epsilon_n\right)}_{y_n}\underbrace{\sqrt{\frac{\nu}{W}}\left(\sqrt{\rho}G + \sqrt{1-\rho}\epsilon_m\right)}_{y_m}\right),$$

$$= \mathbb{E}\left(\left(\frac{\nu}{W}\right)\left(\sqrt{\rho}G + \sqrt{1-\rho}\epsilon_n\right)\left(\sqrt{\rho}G + \sqrt{1-\rho}\epsilon_m\right)\right),$$

$$= \mathbb{E}(\rho\left(\frac{\nu}{W}\right)G^2),$$

$$= \rho\nu\mathbb{E}\left(\frac{1}{W}\right)\underbrace{\mathbb{E}(G^2)}_{=1},$$

$$= \rho\left(\frac{\nu}{\nu-2}\right).$$

The final few steps are also justified if we recall that when $W \sim \chi^2(\nu)$, then $\frac{1}{W} \sim \text{Inv-}\chi^2(\nu)$—consequently, we have that $\mathbb{E}\left(\frac{1}{W}\right) = \frac{1}{\nu-2}$.[16] In English, the reciprocal of a chi-squared random variable follows the so-called inverse-Gamma distribution.

To transform this into correlation, we need to make the following adjustment,

$$\rho(y_n, y_m) = \frac{\text{cov}(y_n, y_m)}{\sqrt{\text{var}(y_n)}\sqrt{\text{var}(y_m)}}, \tag{4.74}$$

$$= \frac{\rho\left(\frac{\nu}{\nu-2}\right)}{\sqrt{\frac{\nu}{\nu-2}}\sqrt{\frac{\nu}{\nu-2}}},$$

$$= \frac{\rho\left(\cancel{\frac{\nu}{\nu-2}}\right)}{\cancel{\frac{\nu}{\nu-2}}},$$

$$= \rho,$$

which is the same result as in the Gaussian model. This is good news, because it implies that we can continue to interpret the ρ parameter in the same manner.

---

[16] This logic could have saved us the effort, although we would have missed the edification, in equation 4.72. Naturally, this is only really sensible for values of $\nu > 2$.

The default indicator, $\mathcal{D}_n$, has the same conceptual definition as in Gaussian case, but the default threshold is slightly different. Specifically,

$$p_n = \mathbb{E}(\mathbb{I}_{\mathcal{D}_n}), \qquad (4.75)$$
$$= \mathbb{P}(\mathcal{D}_n),$$
$$= \mathbb{P}(Y_n \leq K_n),$$
$$= F_{\mathcal{T}_\nu}(K_n),$$

implying directly that $K_n = F_{\mathcal{T}_\nu}^{-1}(p_n)$. We use $F_{\mathcal{T}_\nu}$ and $F_{\mathcal{T}_\nu}^{-1}$ to denote the cumulative and *inverse* cumulative distribution functions of the standard t-distribution with ν degrees of freedom, respectively. This small difference seems quite sensible given that we've changed the underlying marginal and joint distributions.

4.4.2 Back to Default Correlation

To calibrate the model, we need to gain some insight into the default-correlation coefficient. It has, of course, the same basic form as derived previously,

$$\rho\left(\mathbb{I}_{\mathcal{D}_n}, \mathbb{I}_{\mathcal{D}_m}\right) = \frac{\mathbb{P}\left(\mathcal{D}_n \cap \mathcal{D}_m\right) - p_n p_m}{\sqrt{p_n p_m (1 - p_n)(1 - p_m)}}. \qquad (4.76)$$

Nevertheless, despite the similar form, the models again diverge. The joint distribution of y_n and y_m is now assumed to follow a bivariate t-distribution with ν degrees of freedom. It has the following form,

$$\mathbb{P}\left(\mathcal{D}_n \cap \mathcal{D}_m\right) = \mathbb{P}(y_n \leq F_{\mathcal{T}_\nu}^{-1}(p_n), y_m \leq F_{\mathcal{T}_\nu}^{-1}(p_m)), \qquad (4.77)$$
$$= \frac{\Gamma\left(\frac{\nu+d}{2}\right)}{\Gamma\left(\frac{\nu}{2}\right)(\nu\pi)^{\frac{d}{2}}|\Omega_{nm}|^{\frac{1}{2}}} \int_{-\infty}^{F_{\mathcal{T}_\nu}^{-1}(p_n)} \int_{-\infty}^{F_{\mathcal{T}_\nu}^{-1}(p_m)} \left(1 + \frac{x^T \Omega_{nm}^{-1} x}{\nu}\right)^{-\left(\frac{\nu+d}{2}\right)} dx,$$

for Ω_{nm} as defined in equation 4.24, $x \in \mathbb{R}^2$ with, of course, $d = 2$. This permits computation of the default correlation between counterparties n and m using equation 4.76. This is the classic, direct form. There is another approach, which we will examine since it can prove helpful. Both approaches are mathematically equivalent, although depending on the context, one may prove more convenient than the other to actually implement.

The alternative description of the joint default probability requires the conditional default probability in the t-threshold setting. It is analogous to the Gaussian setting

4.4 The t-Distributed Approach

in equation 4.18 on page 155, but it requires a second conditioning variable. That is,

$$\begin{aligned}p_n(G, W) &= \mathbb{E}\left(\mathbb{I}_{\mathcal{D}_n} \big| G, W\right), &(4.78)\\ &= \mathbb{P}\left(\mathcal{D}_n | G, W\right),\\ &= \mathbb{P}\left(y_n \leq F_{\mathcal{T}_n}^{-1}(p_n) \bigg| G, W\right),\\ &= \mathbb{P}\Bigg(\underbrace{\sqrt{\frac{v}{W}}\left(\sqrt{\rho}G + \sqrt{1-\rho}\epsilon_n\right)}_{\text{Equation 4.69}} \leq F_{\mathcal{T}_n}^{-1}(p_n) \bigg| G, W\Bigg),\\ &= \mathbb{P}\left(\epsilon_n \leq \frac{\sqrt{\frac{W}{v}}F_{\mathcal{T}_n}^{-1}(p_n) - \sqrt{\rho}G}{\sqrt{1-\rho}} \bigg| G, W\right),\\ &= \Phi\left(\frac{\sqrt{\frac{W}{v}}F_{\mathcal{T}_n}^{-1}(p_n) - \sqrt{\rho}G}{\sqrt{1-\rho}}\right).\end{aligned}$$

To determine the conditional probability of y_n, both the global systematic factor, G, and the mixing random variate, W, must be revealed.

This might not seem particularly useful for the determination of the joint default probability, but let's work with the basic definition and see where it takes us,

$$\begin{aligned}\mathbb{P}\left(\mathcal{D}_n \cap \mathcal{D}_m\right) &= \mathbb{E}\left(\mathbb{I}_{\mathcal{D}_n} \cap \mathbb{I}_{\mathcal{D}_m}\right), &(4.79)\\ &= \underbrace{\mathbb{E}\left(\mathbb{E}\left(\mathbb{I}_{\mathcal{D}_n} \cap \mathbb{I}_{\mathcal{D}_m}\right) \bigg| G, W\right)}_{\text{By iterated expectations}},\\ &= \mathbb{E}\bigg(\underbrace{\mathbb{E}\left(\mathbb{I}_{\mathcal{D}_n} \big| G, W\right) \cdot \mathbb{E}\left(\mathbb{I}_{\mathcal{D}_m} \big| G, W\right)}_{\text{By conditional independence}}\bigg),\\ &= \mathbb{E}\left(p_n(G, W) \cdot p_m(G, W)\right),\\ &= \int_{-\infty}^{\infty}\int_{0}^{\infty} p_n(g, w) \cdot p_m(g, w) \cdot f_W(w) \cdot \phi(g) dw dg.\end{aligned}$$

In other words, we have exploited the conditional independence of the latent state variables to find an alternative, but entirely equivalent, representation of the joint default probability. In both cases, a two-dimension integral must be resolved and the two parameters ρ and ν play an important role. With the question of default correlation resolved, we now turn our attention to the calibration of the t-threshold model.

4.4.3 The Calibration Question

Let's put aside the ν parameter for a moment and consider, for a given level of ν, the relationship between the average default correlation and ρ. Figure 4.7 illustrates the results of such an experiment. It finds, for a given ν, the value of ρ, that is consistent with an average default correlation coefficient of 0.05. We already know that this value is approximately 0.31 for the Gaussian model. When $\nu = 10$, the appropriate ρ t-threshold parameter is about 0.16. As we increase ν, it moves gradually closer to the Gaussian value. This makes complete sense, of course, since as $\nu \to \infty$, the t-distribution converges to the normal.

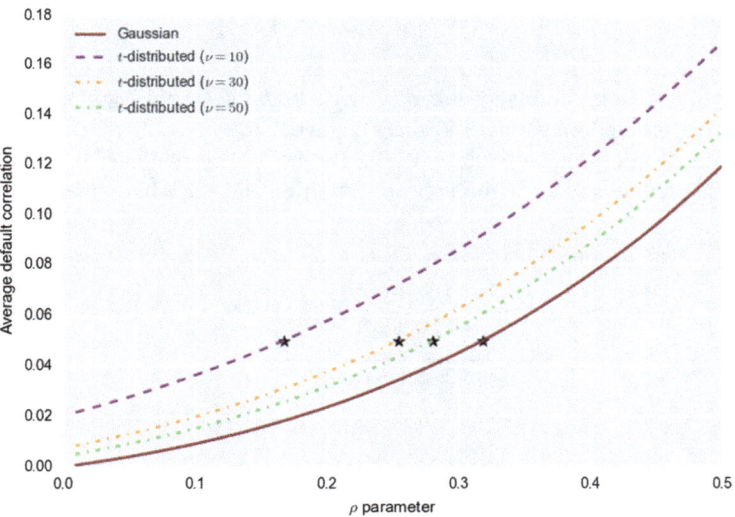

Fig. 4.7 *Calibrating outcomes*: This figure illustrates—for the Gaussian and selection of t-distributed models each with a differing degrees-of-freedom parameter, ν—the relationship between the average default correlation and the single model parameter, ρ. As $\nu \to \infty$, the t-distributed calibration converges to the Gaussian. The black stars denote the optimal ρ value associated with our target of $\rho_\mathcal{D} = 0.05$.

4.4 The *t*-Distributed Approach

Algorithm 4.9 Calibrating the *t*-threshold model

```
def tCalibrate(x,myP,rhoTarget,tdTarget):
    if (x[0]<=0) | (x[1]<=0):
        return [100, 100]
    jointDefaultProb = jointDefaultProbabilityT(myP,myP,x[0],x[1])
    rhoValue = np.divide(jointDefaultProb-myP**2,myP*(1-myP))
    tdValue = tTailDependenceCoefficient(x[0],x[1])
    f1 = rhoValue - rhoTarget
    f2 = tdValue - tdTarget
    return [f1, f2]
```

The choice of ν, however, provides one additional dimension—that is, the relative shape of the latent-variable distributions—not available in the Gaussian setting. To calibrate the *t*-distributed models thus requires two targets. We already seek an average default correlation of 0.05. A natural second candidate is the tail-dependence coefficient from equation 4.68. Given a target amount of tail dependence, which we will denote as $\Lambda_\mathcal{D}$, we can construct the following two non-linear equations in two unknowns,

$$\rho_\mathcal{D} - \frac{\int_{-\infty}^{\infty}\int_0^{\infty} p_n(g,w) \cdot p_m(g,w) \cdot f_W(w) \cdot \phi(g) dw dg - \bar{p}^2}{\bar{p}(1-\bar{p})} = 0, \quad (4.80)$$

$$\Lambda_\mathcal{D} - 2F_{\mathcal{T}_{\nu+1}}\left(-\sqrt{\frac{(\nu+1)(1-\rho)}{1+\rho}}\right) = 0.$$

Both of these expressions depend on our choice of ν and ρ. Given attainable target values for the default-correlation and tail-dependence coefficients—$\rho_\mathcal{D}$ and $\Lambda_\mathcal{D}$— we can find the two *t*-threshold parameters satisfying equation 4.80.

Algorithm 4.9 outlines the code implementation of equation 4.80. This should look quite similar to the technique used to determine the two parameters in the binomial- and Poisson-mixture models from Chap. 3. While relatively straightforward, the heavy-lifting is embedded in the `jointDefaultProbabilityT` sub-routine. Algorithm 4.10 summarizes the details behind the two-dimensional numerical integration from equation 4.79. We make use of the `nquad` function from `scipy.integrate`. Some caution is required on the bounds of integration for the chi-squared variable, w, to ensure that it does not wander into negative territory. It also requires, by virtue of the additional dimension, rather more computational time.

All that remains is to select concrete values for our two targets and run the calibration algorithm. To maintain maximum comparability, we will continue to set the average default correlation, or $\rho_\mathcal{D}$, to 0.05. The level of the tail-dependence coefficient is less obvious. After a bit of trial and error, we opted to set $\Lambda_\mathcal{D} = 0.02$. This is an admittedly modest choice and can readily be adjusted upwards. Even if it

Algorithm 4.10 Computing the joint default probability from the t-threshold model

```
def jointDefaultProbabilityT(p,q,myRho,nu):
    lowerBound = np.maximum(nu-40,2)
    support = [[-10,10],[lowerBound,nu+40]]
    pr,err=scipy.integrate.nquad(jointIntegrandT,support,args=(p,q,myRho,nu))
    return pr
def jointIntegrandT(g,w,p,q,myRho,nu):
    p1 = computeP_t(p,myRho,g,w,nu)
    p2 = computeP_t(q,myRho,g,w,nu)
    density1 = util.gaussianDensity(g,0,1)
    density2 = util.chi2Density(w,nu)
    f = p1*p2*density1*density2
    return f
```

Table 4.3 *Threshold calibration comparison*: This table compares the results of the Gaussian and t-distributed threshold model calibrations. The key difference is that the Gaussian model has a single parameter, whereas the t-distributed model target both the default-correlation and tail-dependence coefficients.

Quantile	Parameters		Default correlation	Tail dependence	Calibration time
	ρ	ν			
Gaussian	0.319	∞	0.05	0.00	3.3
t-distributed	0.173	9.6	0.05	0.02	36.4

is set to zero, it will not numerically reduce to a Gaussian threshold model.[17] The tail of the t-distributed model remains fatter than its Gaussian equivalent.

The Gaussian and t-threshold parameter calibrations, for our current targets, are provided in Table 4.3. The t-threshold ρ parameter is roughly 0.17, whereas $\nu \approx 10$. Both targets are met, although roughly 10 times more computational effort was required to solve the system of t-threshold non-linear calibrating equations. In general, as one increases the tail-dependence coefficient, the value of ρ increases, while the level of ν will tend to fall. The punchline is that, in the extended model, more subtle behaviour and nuanced calibration is possible.

4.4.4 *Implementing the t-Threshold Model*

The Python code used to implement the t-threshold approach, happily, turns out to be both compact and conceptually quite similar to the Gaussian setting. Algorithm 4.11 describes the generation of the collection of $M \times N$ t-threshold

---

[17]Indeed, when $\Lambda_\mathcal{D} \approx 0.00$, we have $\rho = 0.29$ and $\nu = 71$. This is still, at the extreme tails, quite far from the Gaussian distribution.

4.4 The t-Distributed Approach

Algorithm 4.11 Generating t-threshold state variables

```
def getTY(N,M,p,rho,nu):
    G = np.transpose(np.tile(np.random.normal(0,1,M),(N,1)))
    e = np.random.normal(0,1,[M,N])
    W = np.transpose(np.sqrt(nu/np.tile(np.random.chisquare(nu,M),(N,1))))
    Y = np.multiply(W,math.sqrt(rho)*G + math.sqrt(1-rho)*e)
    return Y
```

Algorithm 4.12 The t-threshold Monte-Carlo implementation

```
def oneFactorTModel(N,M,p,c,rho,nu,alpha):
    Y = getTY(N,M,p,rho,nu)
    K = myT.ppf(p,nu)*np.ones((M,1))
    lossIndicator = 1*np.less(Y,K)
    lossDistribution = np.sort(np.dot(lossIndicator,c),axis=None)
    el,ul,var,es=util.computeRiskMeasures(M,lossDistribution,alpha)
    return el,ul,var,es
```

state variables. Indeed, it only requires a single additional line to generate the mixing variable, W, using numpy's chisquare function.[18]

This t-threshold risk-measure estimation approach—below in Algorithm 4.12— is also virtually identical to the Gaussian version outlined in Algorithm 4.5. The only visible difference is the use of the t inverse cumulative distribution function, ppf. Otherwise, it has the same structure as all the other Monte-Carlo implementations considered to this point in the discussion.

After all this effort, we may now compute the t-threshold risk metrics for our running 100-obligor example. The results, with a comparison to the beta-mixture and Gaussian threshold approaches, are found in Table 4.4. The slightly higher tail dependence leads the t-distributed to dominate the Gaussian model at virtually all observed quantiles—moreover, the difference increases as we move further into the tail. The deviations are even more marked for the expected-shortfall measure. The t-distributed approach also generates higher tail estimates than the beta-binomial mixture model. It has a lower loss volatility and, consequently, appears to push a greater proportion of the probability mass into the extremes of the default-loss distribution.

Figure 4.8 provides some additional colour on the numerical values in Table 4.4. It summarizes the tail probabilities for the *two* main techniques considered thus far in this chapter: the Gaussian and t-threshold models. The beta-binomial and independent default models are included for comparison. The significantly longer tail of the t-distributed approach is immediately evident, as is the alternative form

---

[18] Additional details, for the interested reader, on the generation of multivariate t-distributed random variables are found in Appendix A.

Table 4.4 *t-threshold results*: This table compares the *t*-threshold VaR and expected-shortfall estimates to the beta-binomial mixture and Gaussian-threshold results. The beta-binomial model is calibrated for a 1% average default probability, while both the mixture and threshold methods target a default correlation of 0.05. The *t*-threshold is also fitted to a 0.02 tail-dependence coefficient.

Quantile	Beta-mixture		Gaussian		*t*-distributed	
	$\widehat{\text{VaR}}_\alpha$	$\widehat{\mathcal{E}}_\alpha$	$\widehat{\text{VaR}}_\alpha$	$\widehat{\mathcal{E}}_\alpha$	$\widehat{\text{VaR}}_\alpha$	$\widehat{\mathcal{E}}_\alpha$
95.00th	$57.5	$99.2	$48.4	$82.7	$48.4	$86.4
97.00th	$78.0	$120.8	$63.5	$100.9	$64.6	$107.0
99.00th	$124.9	$168.5	$101.9	$145.6	$107.7	$158.0
99.50th	$155.2	$198.9	$130.1	$177.1	$140.0	$194.4
99.90th	$227.0	$269.5	$206.7	$258.2	$228.9	$289.0
99.97th	$293.4	$334.4	$286.9	$340.1	$322.8	$385.2
99.99th	$323.3	$363.9	$323.8	$377.7	$366.2	$429.5
Other statistics						
Expected loss	$9.5		$9.2		$9.2	
Loss volatility	$25.2		$21.7		$22.9	
Total iterations	10,000,000					
Computational time	18.5		49.0		55.9	

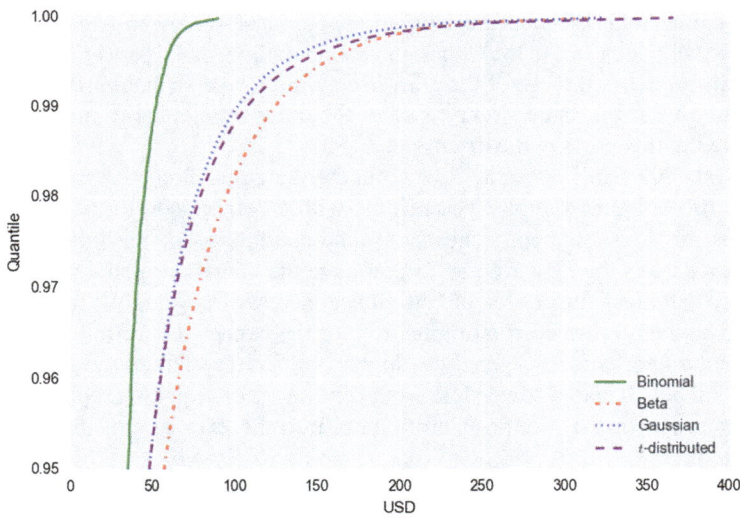

Fig. 4.8 *Various tail probabilities*: This figure summarizes the tail probabilities for the three models considered thus far in this chapter: the Gaussian threshold, its analytical approximation, and the *t*-threshold models. The beta-binomial and independent default models are included for comparison.

of the beta-binomial model. While broadly similar, the mixture and threshold techniques do ultimately generate rather different default-loss distributions.

4.4.5 Pausing for a Breather

What have we learned so far? As dependence is introduced, even in the case of a Gaussian joint distribution for the risk factor, the loss distribution is no longer Gaussian. In fact, it is highly skewed and heavily fat-tailed. It is not true, therefore, that the specification of a Gaussian copula leads to a Gaussian loss distribution. This is not a compelling argument for the use of an alternative copula function.

A better argument for the use of an alternative copula relates to the notion of *tail dependence*. A mathematical measure of this notion can be constructed. One finds, however, that for a Gaussian copula, this measure is zero. That is, as one moves far enough into the tail, the Gaussian copula implies independence of default. This is not a desirable feature in a credit-risk model, particularly since one routinely uses the 99.97th quantile of the loss distribution to make economic-capital decisions. Not all copulas imply tail independence. The t-distribution is an example of a member of the elliptical family that exhibits tail dependence. Others are possible.

A quantitative analyst can learn much from the careful comparison of various modelling approaches. This is the reasoning behind our detailed discussion of the various alternatives and focus on their technical differences. Table 4.5 aids in this process by comparing and contrasting, in the one-factor setting, a number of key objects from the threshold setting. Model understanding is the first key step toward managing the inherent risk associated with modelling activities. To this end, we will take the threshold-modelling framework a step further to examine a generalization of the construction of the t-distributed model. This will lead to a family of models with a wide range of rather exotic resulting distributions.

4.5 Normal-Variance Mixture Models

Other mixing variables are also possible. If we define the latent variable, y_n, as

$$y_n = \sqrt{V}\left(\sqrt{\rho}G + \sqrt{1-\rho}\epsilon_n\right), \tag{4.81}$$

for $n = 1, \ldots, N$. G and ϵ_n remain unchanged as independent, identically distributed standard Gaussian variates. V is an independent, but as yet undefined, random variable. This generic mixing variable, V, transforms the marginal and joint distributions into a general class of so-called normal-variance mixture distribution. The density of the resulting distribution, first introduced, as a special case, into the finance literature by Madan and Seneta (1990) to model equity returns, is

Table 4.5 *One-factor summary*: This table summarizes the main differences, and similarities, between the one-factor Gaussian and t-distributed threshold models. The principal differences arise in the structure of the joint distribution of the latent variables and the associated default correlations.

Variable	Definition	
	Gaussian	t-distributed
G, ϵ_n	Identical, independent $\mathcal{N}(0,1)$ random variates	
ν	Not applicable	Degrees-of-freedom parameter
W	Not applicable	$\chi^2(\nu)$ random variable
y_n	$\sqrt{\rho}G + \sqrt{1-\rho}\epsilon_n \sim \mathcal{N}(0,1)$	$\sqrt{\dfrac{\nu}{W}}\left(\sqrt{\rho}G + \sqrt{1-\rho}\epsilon_n\right) \sim \mathcal{T}_\nu(0,1)$
$\text{var}(y_n)$	1	$\dfrac{\nu}{\nu-2}$
$\text{cov}(y_n, y_m)$	ρ	$\left(\dfrac{\nu}{\nu-2}\right)\rho$
$\rho(y_n, y_m)$	$\rho \in (-1,1)$	
$\mathcal{D}_n$	$\{y_n \leq K_n\}$	
p_n	$\mathbb{P}(\mathcal{D}_n) = \Phi(K_n)$	$\mathbb{P}(\mathcal{D}_n) = F_{\mathcal{T}_\nu}(K_n)$
K_n	$\Phi^{-1}(p_n)$	$F_{\mathcal{T}_\nu}^{-1}(p_n)$
$p_n(G, W)$	$\Phi\left(\dfrac{\Phi^{-1}(p_n) - \sqrt{\rho}G}{\sqrt{1-\rho}}\right)$	$\Phi\left(\dfrac{\sqrt{\dfrac{W}{\nu}}F_{\mathcal{T}_\nu}^{-1}(p_n) - \sqrt{\rho}G}{\sqrt{1-\rho}}\right)$
$\rho(\mathbb{I}_{\mathcal{D}_n}, \mathbb{I}_{\mathcal{D}_m})$	$\dfrac{\mathbb{P}(\mathcal{D}_n \cap \mathcal{D}_m) - p_n p_m}{\sqrt{p_n p_m (1-p_n)(1-p_m)}}$	
$\mathbb{P}(\mathcal{D}_n \cap \mathcal{D}_m)$	$\dfrac{1}{2\pi\sqrt{\|\Omega_{nm}\|}} \int_{-\infty}^{\Phi^{-1}(p_n)} \int_{-\infty}^{\Phi^{-1}(p_m)} e^{-\frac{x^T \Omega_{nm}^{-1} x}{2}} dx$	$\dfrac{\Gamma\left(\frac{\nu+2}{2}\right)}{\Gamma\left(\frac{\nu}{2}\right)\nu\pi\sqrt{\|\Omega_{nm}\|}} \int_{-\infty}^{F_{\mathcal{T}_\nu}^{-1}(p_n)} \int_{-\infty}^{F_{\mathcal{T}_\nu}^{-1}(p_m)} \left(1 + \dfrac{x^T \Omega_{nm}^{-1} x}{2}\right)^{-\left(\frac{\nu+2}{2}\right)} dx$
Ω_{nm}	$\begin{bmatrix} 1 & \rho \\ \rho & 1 \end{bmatrix}$	

4.5 Normal-Variance Mixture Models

determined by,

$$f(x) = \int_{\mathbb{R}_+} \frac{1}{\sigma\sqrt{2\pi v}} e^{-\frac{(x-\mu)^2}{2\sigma^2 v}} g(v) dv, \qquad (4.82)$$

where $g(v)$ is the mixing density.[19] The idea is conceptually quite simple, even if it gives rise to rather complex results. The variance of the normal distribution is replaced with another random variable, V. Depending on the choice of this mixing variable, various distributions with different properties can be constructed. Indeed, this approach can, in fact, be considered a type of stochastic-volatility model.

We will consider two practical flavours of this model—indeed, without realizing it, we have already considered two special cases. For the moment, however, we will proceed in full generality. While the solution to equation 4.82 is, in principle, quite messy, we can rather easily derive the necessary basic model quantities examined in the Gaussian and t-distributed settings. The expectation of y_n remains

$$\mathbb{E}(y_n) = \mathbb{E}\left(\sqrt{V}\left(\sqrt{\rho}G + \sqrt{1-\rho}\epsilon_n\right)\right), \qquad (4.83)$$

$$= \sqrt{\rho} \cdot \overbrace{\mathbb{E}\left(\sqrt{V}\right) \underbrace{\mathbb{E}(G)}_{=0}}^{\text{By independence}} + \sqrt{1-\rho} \cdot \overbrace{\mathbb{E}\left(\sqrt{V}\right) \underbrace{\mathbb{E}(\epsilon_n)}_{=0}}^{\text{By independence}},$$

$$= 0.$$

The variance of y_n becomes

$$\mathrm{var}(y_n) = \mathbb{E}\left(y_n^2 - \underbrace{\mathbb{E}(y_n)^2}_{=0}\right), \qquad (4.84)$$

$$= \mathbb{E}\left(\left(\sqrt{V}\left(\sqrt{\rho}G + \sqrt{1-\rho}\epsilon_n\right)\right)^2\right),$$

$$= \mathbb{E}\left(V \cdot \rho \cdot G^2\right) + \mathbb{E}\left(V \cdot (1-\rho) \cdot \epsilon_n^2\right),$$

$$= \rho \cdot \mathbb{E}(V) \underbrace{\mathbb{E}\left(G^2\right)}_{=1} + (1-\rho) \cdot \mathbb{E}(V) \underbrace{\mathbb{E}\left(\epsilon_n^2\right)}_{=1},$$

$$= \mathbb{E}(V).$$

This is a fascinating, but perfectly logical result. The latent state variable has a mean of zero, but a variance equal to the expected value of the mixing variable, V. The mixing variable thus determines the variance of the resulting default-risk factor. In

---

[19] A more general framework was provided a few years later by Eberlein and Keller (1995).

principle, therefore, the greater the expected value of V, the larger the dispersion around the latent state variable, y_n.

The covariance between two arbitrary obligors n and m reduces to,

$$\text{cov}(y_n, y_m) = \mathbb{E}(y_n y_m), \qquad (4.85)$$

$$= \mathbb{E}\left(\underbrace{\sqrt{V}\left(\sqrt{\rho}G + \sqrt{1-\rho}\epsilon_n\right)}_{y_n} \underbrace{\sqrt{V}\left(\sqrt{\rho}G + \sqrt{1-\rho}\epsilon_m\right)}_{y_m}\right),$$

$$= \mathbb{E}\left(V\left(\sqrt{\rho}G + \sqrt{1-\rho}\epsilon_n\right)\left(\sqrt{\rho}G + \sqrt{1-\rho}\epsilon_m\right)\right),$$

$$= \mathbb{E}(\rho V G^2),$$

$$= \rho \mathbb{E}(V) \underbrace{\mathbb{E}(G^2)}_{=1},$$

$$= \rho \mathbb{E}(V).$$

The transformation of this, into state-variable correlation, merely requires the usual adjustment

$$\rho(y_n, y_m) = \frac{\text{cov}(y_n, y_m)}{\sqrt{\text{var}(y_n)}\sqrt{\text{var}(y_m)}}, \qquad (4.86)$$

$$= \frac{\rho \mathbb{E}(V)}{\sqrt{\mathbb{E}(V)}\sqrt{\mathbb{E}(V)}},$$

$$= \frac{\rho \cancel{\mathbb{E}(V)}}{\cancel{\mathbb{E}(V)}},$$

$$= \rho,$$

which, quite happily, is the same result as seen in the Gaussian and t-threshold cases.

The default indicator, $\mathcal{D}_n$, also takes a slightly different, albeit very familiar, form. Specifically,

$$p_n = \mathbb{E}(\mathbb{I}_{\mathcal{D}_n}), \qquad (4.87)$$

$$= \mathbb{P}(\mathcal{D}_n),$$

$$= \mathbb{P}(Y_n \leq K_n),$$

$$= F_{\mathcal{NV}}(K_n),$$

implying directly that $K_n = F_{\mathcal{NV}}^{-1}(p_n)$ using, of course, $F_{\mathcal{NV}}$ and $F_{\mathcal{NV}}^{-1}$ to denote the cumulative and *inverse* cumulative distribution functions of the resulting normal-variance mixture random variable.

4.5 Normal-Variance Mixture Models

There are many possible choices of V and naturally not all of them generate sensible results. Our interest will lie with those selections of V, which give rise to sensible marginal and joint distributions. Before we get to the specific choices of V, however, it is important to address the question of calibration.

4.5.1 Computing Default Correlation

It is essential, for the purposes of calibration and general model diagnostics, to determine the form of default correlation in the normal-mixture setting. Since the resulting distributions can be rather unwieldy and difficult to handle, it is preferable to work with the conditional default probability as outlined in equation 4.79 on page 187. The resulting conditional default probability is,

$$
\begin{aligned}
p_n(G, V) &= \mathbb{E}\left(\mathbb{I}_{\mathcal{D}_n} \mid G, V\right), \quad &(4.88)\\
&= \mathbb{P}\left(\mathcal{D}_n \mid G, V\right),\\
&= \mathbb{P}\left(y_n \leq F_{\mathcal{N}\mathcal{V}}^{-1}(p_n) \,\Big|\, G, V\right),\\
&= \mathbb{P}\left(\underbrace{\sqrt{V}\left(\sqrt{\rho}G + \sqrt{1-\rho}\epsilon_n\right)}_{\text{Equation 4.81}} \leq F_{\mathcal{N}\mathcal{V}}^{-1}(p_n) \,\Big|\, G, V\right),\\
&= \mathbb{P}\left(\epsilon_n \leq \frac{\sqrt{\frac{1}{V}} F_{\mathcal{N}\mathcal{V}}^{-1}(p_n) - \sqrt{\rho}G}{\sqrt{1-\rho}} \,\Big|\, G, V\right),\\
&= \Phi\left(\frac{\sqrt{\frac{1}{V}} F_{\mathcal{N}\mathcal{V}}^{-1}(p_n) - \sqrt{\rho}G}{\sqrt{1-\rho}}\right).
\end{aligned}
$$

As in the previous discussion, both underlying variables G and V must be known to determine the conditional default probability. Following from our previous development, the actual default-correlation coefficient is now simply,

$$
\begin{aligned}
\mathbb{P}\left(\mathcal{D}_n \cap \mathcal{D}_m\right) &= \underbrace{\mathbb{E}\left(\mathbb{E}\left(\mathbb{I}_{\mathcal{D}_n} \cap \mathbb{I}_{\mathcal{D}_m}\right) \,\Big|\, G, V\right)}_{\text{By iterated expectations}}, \quad &(4.89)\\
&= \mathbb{E}\left(\underbrace{\mathbb{E}\left(\mathbb{I}_{\mathcal{D}_n} \mid G, V\right) \cdot \mathbb{E}\left(\mathbb{I}_{\mathcal{D}_m} \mid G, V\right)}_{\text{By conditional independence}}\right),
\end{aligned}
$$

$$= \mathbb{E}\left(p_n(G, V) \cdot p_m(G, V)\right),$$

$$= \int_{-\infty}^{\infty} \int_{0}^{\infty} p_n(g, v) \cdot p_m(g, v) \cdot f_V(v) \cdot \phi(g) dv dg.$$

Although the form is familiar, this still represents a two-dimension integral that will need to be solved numerically. The alternative approach involving the use of the multivariate normal-variance mixture distribution—through its copula—is possible, but it is quite heavy work. Nevertheless, the approach in equation 4.89 is not completely straightforward. The inverse cumulative distribution function, $F_{\mathcal{N}\mathcal{V}}^{-1}(\cdot)$ from equation 4.88, while (perhaps) conceptually obvious, will prove to be fairly demanding from a computational perspective.

4.5.2 Higher Moments

It turns out that tail dependence in the normal-variance mixture setting is a complicated affair. von Hammerstein (2016) and McNeil et al. (2015) indicate that the resulting general distribution from these mixture models has either extreme or Gaussian tail dependence.[20] This means that, unlike the t-distributed case, use of the tail dependence coefficient is sadly not a reliable criterion for the calibration of these models. It turns out—see McNeil et al. (2015, Examples 7.39, 16.9)—that the mixing variable determines tail dependence of the normal-variance mixture. That is, if V possesses a power tail, then the subsequent normal-variance mixture will be positive tail dependent. A power, polynomial, or Pareto tail is a technical definition of a heavy-tailed distribution; practically speaking, a distribution is termed heavy-tailed if its density function decays more slowly than the exponential distribution. This is intimately related to the class of subexponential distributions.[21]

Another possible calibration target, therefore, would be on the higher moments of the distribution of our latent state variable. Mean and variance are known. We can thus consider the skewness, $\mathcal{S}(\cdot)$, and kurtosis, $\mathcal{K}(\cdot)$. These are, although tedious, fairly readily computed as,

$$\mathcal{S}(y_n) = \mathbb{E}\left(\left(\frac{y_n - \overbrace{\mathbb{E}(y_n)}^{=0}}{\sqrt{\text{var}(y_n)}}\right)^3\right), \quad (4.90)$$

$$= \frac{1}{\mathbb{E}(V)^{\frac{3}{2}}} \mathbb{E}\left(y_n^3\right),$$

---

[20]Tail dependence is, in general, a fairly complex area. See Schmidt (2003) for a useful description of the concept in very general setting of elliptical copulae.

[21]See, for example, Embrechts et al. (1999, Chapter 1) for vastly more detail on the notion of heavy tails. See also , Kotz et al. (2001) or Barndorff-Nielsen et al. (1982) for more information on variance-mixture distributions and tail dependence.

4.5 Normal-Variance Mixture Models

$$\begin{aligned}
&= \frac{1}{\mathbb{E}(V)^{\frac{3}{2}}} \mathbb{E}\left(\left(\sqrt{V}\left(\sqrt{\rho}G + \sqrt{1-\rho}\epsilon_n\right)\right)^3\right), \\
&= \frac{1}{\mathbb{E}(V)^{\frac{3}{2}}} \mathbb{E}\left(V^{\frac{3}{2}}\rho^{\frac{3}{2}}G^3 + V^{\frac{3}{2}}(1-\rho)^{\frac{3}{2}}\epsilon_n^3\right), \\
&= \frac{1}{\mathbb{E}(V)^{\frac{3}{2}}} \left(\rho^{\frac{3}{2}}\mathbb{E}\left(V^{\frac{3}{2}}\right)\underbrace{\mathbb{E}(G^3)}_{=0} + (1-\rho)^{\frac{3}{2}}\mathbb{E}\left(V^{\frac{3}{2}}\right)\underbrace{\mathbb{E}(\epsilon_n^3)}_{=0}\right), \\
&= 0.
\end{aligned}$$

Given that the skewness of the standard normal variates vanishes, this implies that our latent state-variable distributions are symmetric.[22] The skewness will thus offer no clues for the calibration of our model.

The kurtosis, however, is another story. Another fairly painful computation brings us to,

$$\mathcal{K}(y_n) = \mathbb{E}\left(\left(\frac{y_n - \overbrace{\mathbb{E}(y_n)}^{=0}}{\sqrt{\text{var}(y_n)}}\right)^4\right), \quad (4.91)$$

$$= \frac{1}{\mathbb{E}(V)^2} \mathbb{E}\left(y_n^4\right),$$

$$= \frac{1}{\mathbb{E}(V)^2} \mathbb{E}\left(\left(\sqrt{V}\left(\sqrt{\rho}G + \sqrt{1-\rho}\epsilon_n\right)\right)^4\right),$$

$$= \frac{1}{\mathbb{E}(V)^2} \mathbb{E}\left(V^2\rho^2 G^4 + 6\rho(1-\rho)V^2 G^2\epsilon_n^2 + V^2(1-\rho)^2\epsilon_n^4\right),$$

$$= \frac{\mathbb{E}(V^2)}{\mathbb{E}(V)^2}\left(\rho^2 \underbrace{\mathbb{E}(G^4)}_{=3} + 6\rho(1-\rho)\underbrace{\mathbb{E}(G^2)}_{=1}\underbrace{\mathbb{E}(\epsilon_n)^2}_{=1} + (1-\rho)^2 \underbrace{\mathbb{E}(\epsilon_n^4)}_{=3}\right),$$

$$= \frac{\mathbb{E}(V^2)}{\mathbb{E}(V)^2}\underbrace{\left(3\rho^2 + 6\rho(1-\rho) + 3(1-\rho)^2\right)}_{=3},$$

$$= 3 \cdot \frac{\mathbb{E}(V^2)}{\mathbb{E}(V)^2}.$$

---

[22] The cubic in equation 4.90 gives rise to many more terms than shown. Any term including G or ϵ_n not raised to a power, however, is equal to zero and can be ignored.

This surprisingly elegant expression will be a function of the parameters of the mixing variable, V. To summarize, the first four moments of our default state-variable function are 0, $\mathbb{E}(V)$, 0, and $3 \cdot \frac{\mathbb{E}(V^2)}{\mathbb{E}(V)^2}$.

4.5.3 Two Concrete Cases

To this point, we have proceed in the general setting. To proceed further, however, we need to make a more concrete choice of V. We will investigate two choices:

$V \sim \Gamma(a, b)$ This choice creates the so-called variance-gamma threshold model. In other words, the marginal and joint latent state variables follow a variance-gamma distribution. It was introduced, as previously indicated, by Madan and Seneta (1990).

$V \sim \text{GIG}(c, \sigma, \theta, \nu)$ In this second case, V follows the generalized inverse-Gaussian distribution. We will set the parameter $\nu = 1$, which leads to marginal and joint generalized hyperbolic distributions.[23]

We wish to underscore that many other choices are possible. In principle, any positive-valued random variable can be used.[24] Since neither the gamma nor the generalized inverse-Gaussian laws possess a power tail, sadly these models do not exhibit positive tail dependence.

It is not incredibly meaningful to explicitly write out the variance-gamma and generalized hyperbolic distributions. They are detailed, complex, and, most importantly, we will not be working with them directly. Instead, we will use equation 4.81 to construct the state-variable outcomes and use the moments of V to determine the requisite moments for calibration of both models. Indeed, to avoid use of these unwieldy joint distributions is, in fact, the driving reason behind the use of the conditional default probabilities to define default correlation in equation 4.89.

Neither of these models are particularly popular, nor do they appear often in the academic literature. The principal reason is presumably their lack of analytic tractability. We will need, for example, to work with some specialized functions and perform a large number of numerical computations, which have the impact of slowing the overall determination of risk measures. Although they are not commonly addressed in textbooks, we have opted to include a detailed implementation to demonstrate the range of possible threshold choices. Chapter 3 provided the same perspective in the mixture-model setting. We hope that a similar service in

---

[23] See Hu and Kercheval (2007) for more details.

[24] Another possible choice is to set V to an exponentially-distributed random variable, which gives rise to a Laplace distribution. Kotz et al. (2001, Chapter 2) provides more details.

4.5.4 The Variance-Gamma Model

In previous chapters, we have become quite acquainted with the gamma distribution. In particular, if $V \sim \Gamma(a,b)$ then $\mathbb{E}(V) = \frac{a}{b}$, $\text{var}(V) = \frac{a}{b^2}$ and $\mathbb{E}(V^2) = \frac{a(a+1)}{b^2}$. A bit of reflection, however, reveals that we have three parameters—a, b, and ρ—and three targets: the variance and kurtosis of y_n as well as the default correlation. We can, with a bit of thought and imagination, find reasonable target values for default correlation and kurtosis. The variance of y_n is rather different. We may, by using the trick from Wilde (1997), force V to have unit variance. This is accomplished, in the shape-rate formulation of the gamma distribution, by simply setting $a = b$. Thus, $\mathbb{E}(V) = 1$, $\text{var}(V) = \frac{1}{a}$ and $\mathbb{E}(V^2) = \frac{a+1}{a}$. Most importantly, we have two targets—default correlation and kurtosis—and two parameters, a and ρ, to calibrate our model.

The only remaining required quantity is the kurtosis of our default state variable. Using equation 4.91, it is readily determined as,

$$\mathcal{K}(y_n) = 3 \cdot \frac{\mathbb{E}(V^2)}{\mathbb{E}(V)^2}, \qquad (4.92)$$

$$= 3 \cdot \left(\frac{a+1}{a}\right).$$

The system of calibrating equations is thus the following two non-linear equations in two unknowns,

$$3 \cdot \left(\frac{a+1}{a}\right) = \mathcal{K}_\mathcal{D} \qquad (4.93)$$

$$\frac{\overbrace{\int_{-\infty}^{\infty}\int_{0}^{\infty} p_n(g,v) \cdot p_m(g,v) \cdot f_V(v) \cdot \phi(g) dv dg - \bar{p}^2}^{\text{Equation 4.89}}}{\bar{p}(1-\bar{p})} = \rho_\mathcal{D},$$

where $\mathcal{K}_\mathcal{D}$ and $\rho_\mathcal{D}$ are the target state-variable kurtosis and default correlation values. $\bar{p}$ denotes, as usual, the average unconditional default probability. One could, of course, by setting $a = \frac{3}{\mathcal{K}_\mathcal{D}-3}$, reduce this to a single non-linear equation in one unknown. Since, in full generality, the kurtosis could depend on more variables, we've decided upon this system of calibrating equations. It also allows us, as we'll see in the next section, to use the same computer code for both of our concrete cases.

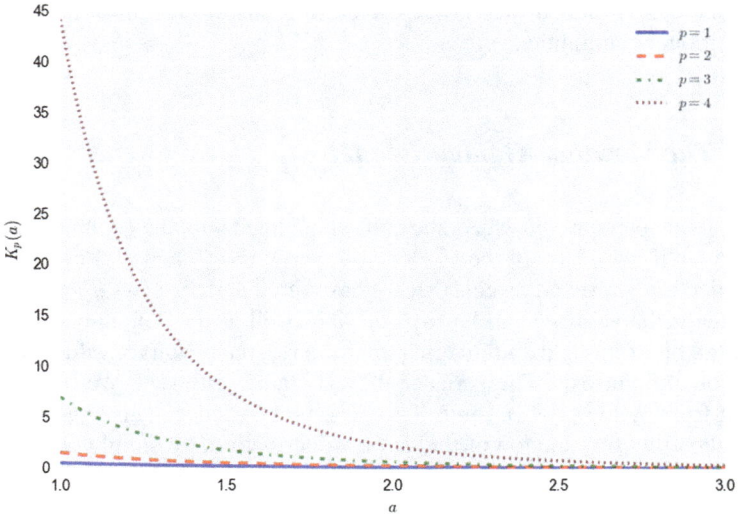

Fig. 4.9 *The modified Bessel function of the second kind*: This figure illustrates, for different choices of parameters a and p, the form of the modified Bessel function of the second kind.

4.5.5 The Generalized Hyperbolic Case

The second concrete example is a bit more involved. The mixing variable obeys the generalized inverse-Gaussian law. While not terribly well-known in finance circles, it is well described in Johnson et al. (1994, Chapter 15). There are two- and three-parameter versions, but we will use the two-parameter case. The latter case involves setting the first two parameters—often designated as a and b—equal to one another thus eliminating b.[25] It has the following density function,

$$f(v) = \frac{1}{2K_p(a)} v^{p-1} e^{\frac{a}{2}\left(v + \frac{1}{v}\right)}, \qquad (4.94)$$

where $v \in \mathbb{R}_+$, $a > 0$, $p \in \mathbb{R}$, and $K_p(\cdot)$ is a modified Bessel function of the second kind.[26] The role of $K_p(a)$ is basically as a constant to ensure that the density in equation 4.94 integrates to unity. The actual formal definition of the modified Bessel function is complicated and not particularly easy to use. Fortunately, most mathematical softwares have a convenient and efficient implementation. Python is no exception. We employ scipy's special.kn(p,a) function. Figure 4.9

---

[25] More specifically, if X follows a three-parameter generalized inverse Gaussian distribution, it is written $X \sim \text{GIG}(a, b, p)$. In the two-parameter version, $X \sim \text{GIG}(a, p) \equiv \text{GIG}(a, a, p)$.

[26] These specialized functions arise naturally as the solution to various differential equations. The interested reader is referred to Abramovitz and Stegun (1965, Chapter 9) for more details.

4.5 Normal-Variance Mixture Models

outlines, for different choices of parameters a and p, the form of the modified Bessel function of the second kind. For our purposes, however, we can think of it as a glorified constant.

To ensure that the joint and marginal distributions of our normal-variance mixture models follow a generalized hyperbolic distribution, it is essential to set the parameter p to unity. This implies that we have reduced equation 4.94 to a single-parameter distribution. This is helpful for model calibration. The expectation, and first moment, of the generalized inverse-Gaussian distribution—with $p = 1$—is given as,

$$\mathbb{E}(V) = \frac{K_2(a)}{K_1(a)}. \tag{4.95}$$

In other words, it is the ratio of two modified Bessel functions of the second kind. The variance has a similar, albeit somewhat more involved, form

$$\text{var}(V) = \frac{K_3(a)K_1(a) - K_2(a)^2}{K_1(a)^2}. \tag{4.96}$$

To determine the kurtosis of our state variable, y_n, we also require $\mathbb{E}\left(V^2\right)$.[27] This quantity follows from the definition of variance and equations 4.95 and 4.96 as,

$$\mathbb{E}\left(V^2\right) = \text{var}(V) + \mathbb{E}(V)^2, \tag{4.97}$$

$$= \underbrace{\frac{K_3(a)K_1(a) - K_2(a)^2}{K_1(a)^2}}_{\text{Equation 4.96}} + \underbrace{\left(\frac{K_2(a)}{K_1(a)}\right)^2}_{\text{Equation 4.95}},$$

$$= \frac{K_3(a)K_1(a) - \cancel{K_2(a)^2}}{K_1(a)^2} + \frac{\cancel{K_2(a)^2}}{K_1(a)^2},$$

$$= \frac{K_3(a)\cancel{K_1(a)}}{K_1(a)^{\cancel{2}}},$$

$$= \frac{K_3(a)}{K_1(a)}.$$

---

[27] We could, of course, require that $\text{var}(V) = 1$ as in the variance-gamma case. This would, however, require an additional parameter, which we would prefer to avoid.

This implies that the kurtosis is now easily determined as,

$$\mathcal{K}(y_n) = 3 \cdot \frac{\mathbb{E}(V^2)}{\mathbb{E}(V)^2}, \quad (4.98)$$

$$= 3 \cdot \left(\frac{\left(\frac{K_3(a)}{K_1(a)} \right)}{\left(\frac{K_2(a)^2}{K_1(a)^2} \right)} \right),$$

$$= 3 \cdot \left(\frac{K_3(a) K_1(a)}{K_2(a)^2} \right).$$

Once again, we have a straightforward expression, which leads to a similar set of calibrating equations to those found in equation 4.93. This two-equation format meets our needs quite well.

4.5.6 A Fly in the Ointment

A close examination of the conditional default probabilities embedded in the integral of equation 4.89 reveal a slightly disturbing term: $F_{\mathcal{NV}}^{-1}$. This is the inverse cumulative distribution function of the appropriate normal-variance mixture distribution. In the Gaussian and t-distributed cases, this is straightforward. In fact, it is not mathematically particularly easy, but the heavy-lifting has been performed by others and the requisite functionality is embedded conveniently in virtually every scientific software. For this application, unfortunately, we need to construct an algorithm to compute $F_{\mathcal{NV}}^{-1}$.

Happily, this is conceptually straightforward. We have the density of the normal-variance mixture in equation 4.82. Let's restate it with $\mu = 0$ and $\sigma = 1$ as follows,

$$f_{\mathcal{NM}}(x) = \int_{\mathbb{R}_+} \frac{1}{\sqrt{2\pi v}} e^{-\frac{x^2}{2v}} g(v) dv, \quad (4.99)$$

where $g(v)$ takes the gamma, or generalized inverse-Gaussian distribution, as the situation requires. Equation 4.99 has to be solved with numerical integration. Algorithm 4.13 describes two sub-routines for this purpose. The first, nvmDensity, recreates equation 4.99 and incorporates some logic to handle both models. The second function, nvmPdf, performs the numerical integration.

4.5 Normal-Variance Mixture Models

Algorithm 4.13 The normal-variance mixture density

```
def nvmDensity(v,x,myA,whichModel):
    t1 = np.divide(1,np.sqrt(2*math.pi*v))
    t2 = np.exp(-np.divide(x**2,2*v))
    if whichModel==0:
        return t1*t2*util.gammaDensity(v,myA,myA)
    elif whichModel==1:
        return t1*t2*util.gigDensity(v,myA)
def nvmPdf(x,myA,whichModel):
    f,err = nInt.quad(nvmDensity,0,50,args=(x,myA,whichModel))
    return f
```

Algorithm 4.14 The normal-variance mixture distribution function

```
def nvmCdf(x,myA,whichModel):
    F,err = nInt.quad(nvmPdf,-8,x,args=(myA,whichModel))
    return F
```

By definition, the cumulative distribution function of the normal-variance mixture model follows from first principles as,

$$F_{\mathcal{NM}}(x) = \int_{-\infty}^{x} f_{\mathcal{NM}}(y)dy, \qquad (4.100)$$

$$= \int_{-\infty}^{x} \underbrace{\int_{\mathbb{R}_{+}} \frac{1}{\sqrt{2\pi v}} e^{-\frac{y^2}{2v}} g(v) dv}_{\text{Equation 4.99}} dy.$$

Unlike most popular cases, therefore, neither the density nor the cumulative distribution function is known in closed form. Determination of the distribution function thus requires resolving a double integral. While not terribly difficult, it is somewhat computationally intensive.

The Python solution is summarized in Algorithm 4.14. Creatively termed nvmCdf, it requires only a single line of code. In this instance, however, the quad function is numerically integrating a function, nvmPdf, which is itself a numerical integral.

Armed with the probability density and cumulative distribution functions, it is possible to determine the inverse cumulative distribution function. This is also often referred to as the quantile function. We have a random variable, X, with probability distribution function, $F_{\mathcal{NM}}(\cdot)$. Given a probability, p, we seek $F_{\mathcal{NM}}^{-1}(p)$. That is, we want the value of the random variable, $X = x$, associated with the choice of

Algorithm 4.15 The normal-variance mixture inverse distribution function

```
def nvmTarget(x,myVal,myA,whichModel):
    F,err = nInt.quad(nvmPdf,-8,x,args=(myA,whichModel))
    return F-myVal
def nvmPpf(myVal,myA,whichModel):
    r = scipy.optimize.fsolve(nvmTarget,0,args=(myVal,myA,whichModel))
    return r[0]
```

probability, p. In other words, the pth quantile. Pulling this together, we want

$$x = F_{\mathcal{NM}}^{-1}(p). \tag{4.101}$$

This doesn't seem to help very much, since $F_{\mathcal{NM}}^{-1}(\cdot)$ is unknown to us. But we do know $F_{\mathcal{NM}}(\cdot)$, or at least, we have a way to determine it numerically. If we rearrange equation 4.101, therefore, we arrive at

$$F_{\mathcal{NM}}(x) = F_{\mathcal{NM}}\left(F_{\mathcal{NM}}^{-1}(p)\right), \tag{4.102}$$

$$F_{\mathcal{NM}}(x) - p = 0.$$

If we numerically identify the x that solves equation 4.102, then we have identified our quantile. Finding the inverse of a cumulative distribution function thus basically reduces to a root-finding exercise.

Algorithm 4.15 shows us how this is performed. Again, it requires two subroutines: `nvmTarget` and `nvmPpf`. The former builds the function from equation 4.102 whose root we seek. The latter uses `scipy`'s `fsolve` routine to find the desired value. `nvmPpf` is thus the internally constructed function that we will employ, whenever necessary, to determine $F_{\mathcal{NV}}^{-1}(\cdot)$.

This is, to be quite honest, an inelegant, brute-force solution to our problem. Professional computations of the inverse of well-known distributions, unavailable in closed form—such as the Gaussian or t-distribution—make use of clever series expansions to quickly generate the desired results. With apologies to the reader for the gracelessness of our approach, we nonetheless confirm that, if fairly slow and awkward, it works quite well.

4.5.7 Concrete Normal-Variance Results

The final step is to apply these two practical variance-normal mixture models to our usual portfolio example. Using the basic form of equation 4.93 to represent our two non-linear equations in two unknowns, we again employ the Python `fsolve` function to determine the appropriate model parameters. As before, quite naturally,

4.5 Normal-Variance Mixture Models

Table 4.6 *Normal-variance mixture calibration comparison*: This table compares the results of the variance-gamma and generalized hyperbolic model calibrations. They have a common average default correlation target, but elect, given their alternative forms, for different level of kurtosis.

Quantile	Parameters		Default correlation	Kurtosis measure	Calibration time
	ρ	a			
Variance-gamma	0.04	1.00	0.05	6.00	30.5
Generalized hyperbolic	0.14	2.60	0.05	4.00	38.3

we target a 0.05 default correlation coefficient to make the risk estimates comparable with previous analysis. The kurtosis targets, however, vary between the two models. The variance-gamma model is calibrated to a kurtosis value of 6, whereas the generalized hyperbolic method is fitted to a value of 4. Targeting the same kurtosis value, which seems quite sensible, and was our first instinct, leads to wildly differing risk outcomes. This suggests that the underlying distributional structures of these two possible choices are quite dissimilar.

The resulting ρ and mixing-density parameter estimates are found in Table 4.6. While the parameter values are dramatically different—the variance-gamma ρ is less than one third of the generalized hyperbolic value—they both generate average default correlation of 0.05 and their desired levels of kurtosis. Incidentally, due to all of the numerical integration and root-finding, both approaches require more than 30 seconds for their calibration.

The actual computation of our risk measures requires the use of the basic threshold structure, but also that we simulate both gamma and generalized inverse-Gaussian random variates. The former is straightforward—indeed, we've already performed this task in Chap. 3—but the latter is rather more difficult. It is not easy to find random-number generators for the generalized-inverse Gaussian distribution. In fact, none are natively available in Python.

This is a potential headache. There are, however, *two* possible solutions. We could construct our own implementation by following work by Devroye (2012) or Leydold and Hörmann (2013). This is tedious, difficult, and quite outside of the field of expertise for the typical financial analyst. The easier alternative, which we have cheerfully adopted, is to make use of an R package named `GIGrvg`, which was developed for precisely this purpose.[28] Leydold and Hörmann (2017) details how it is applied and, very conveniently, it is easy to use within the Python environment through the use of the `rpy2` package.[29]

---

[28]There are, incidentally, many useful R packages that might be employed. The variance-gamma and generalized-hyperbolic distributions are, for example, handled in Scott (2015a,b).

[29]This package was developed by Laurent Gautier. It does not seem possible to find any formal paper written on this package. There are, however, numerous useful blogs and online documentation available on the Internet for the interested reader.

Algorithm 4.16 Generating normal-variance mixture state variables

```
from rpy2.robjects.packages import importr
gig = importr('GIGrvg')
def getNVMY(N,M,rho,myA,whichModel):
    G = np.transpose(np.tile(np.random.normal(0,1,M),(N,1)))
    e = np.random.normal(0,1,[M,N])
    if whichModel==0:
        V = np.transpose(np.sqrt(np.tile(np.random.gamma(myA,1/myA,M),(N,1))))
    elif whichModel==1:
        V = np.transpose(np.sqrt(np.tile(gig.rgig(M,1,myA,myA),(N,1))))
    Y = np.multiply(V,math.sqrt(rho)*G + math.sqrt(1-rho)*e)
    return Y
```

Table 4.7 *Simulated vs. theoretical moments*: There is a significant distance, in the normal-variance-mixture setting, between the actual construction and theory of the latent state variables. The underlying table compares, and verifies the agreement, of the latent state-variable theoretical and simulated moments.

Moment	Variance-gamma		Gen.-hyperbolic	
	Simulated	Theory	Simulated	Theory
Mean	0.000	0.000	−0.000	0.000
Variance	1.000	1.000	1.616	1.619
Volatility	1.000	1.000	1.271	1.272
Skew	−0.000	0.000	−0.000	0.000
Kurtosis	6.008	6.000	3.999	4.000

Algorithm 4.16 illustrates the Python-code used to generate our variance-gamma and generalized hyperbolic state variables. It uses a flag, whichModel, to determine the desired model. The gamma random variates are produced in the typical way, whereas the generalized inverse-Gamma outcomes are created using the rgig function from the GIGrvg package. The remainder the Algorithm 4.16 follows the, by now, quite familiar threshold pattern.

The incredulous reader might legitimately enquire, given the circuitous path for the construction of the state-variable outcomes, if it actually performs as advertised. Indeed, there are many things that could go wrong. It is thus sensible and responsible to confirm that the latent default state-variable expression from equation 4.81 does indeed, when applied, actually create variance-gamma or generalized hyperbolic outcomes with the required characteristics. To investigate this question, we generated 10 million outcomes of each model and computed the resulting moments. Table 4.7 summarizes the results and compares them to the theoretical targeted moments. The simulated and theoretical results are virtually indistinguishable. While not a definitive proof, it is nonetheless reassuring.

4.5 Normal-Variance Mixture Models

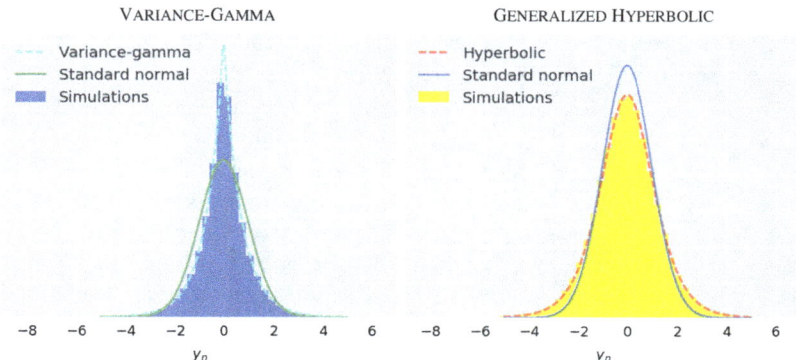

Fig. 4.10 *Visualizing normal-variance-mixture distributions*: This figure summarizes the theoretical and simulated variance-gamma and generalized-hyperbolic density functions. A standard normal density is also overlaid for comparison purposes.

An additional visual verification is also performed. Figure 4.10 provides histograms of the simulation results overlaid with the theoretical distribution using the parameters in Table 4.6 and the density function described in Algorithm 4.13. The results, once again, strongly agree. Our implementation thus passes the eyeball test. Figure 4.10 also provides powerful insight into the general characteristics of the marginal state-variable distributions. The variance-gamma distribution has a very thin waist with probability mass extending far into the tails. It has a form that differs dramatically from the overlaid Gaussian distribution. The generalized hyperbolic distribution, conversely, is also leptokurtotic, but it has a form that is significantly closer in shape to the Gaussian distribution. Selecting a kurtosis parameter of six, to match the gamma-variance approach, would, however, lead to enormous amounts of excess kurtosis.

Table 4.8 and Fig. 4.11 bring our analysis of one-factor threshold models to a close by collecting the results for the four approaches considered in this chapter: the Gaussian, t-distributed, gamma-variance, and generalized-hyperbolic implementations. A number of fascinating differences can be identified. First of all, the normal-variance mixture models generate essentially identical risk estimates at the lower quantiles. The 95th and 99th quantile VaR figures are in the neighbourhood of $50 and $100 for all models, respectively. As we move past the 99.9th percentile, the results begin to diverge. The most conservative estimates stem from the t-distributed model, followed, in order, by the Gaussian, generalized hyperbolic, and variance-gamma approaches.

Each of these threshold models has the same degree of default correlation, but yet acts quite differently as we move out into the tail. These observations suggest that the tail-dependence properties of the default-state variable distribution play an important role in the extreme quantile of the default-loss distribution. Much like

Table 4.8 *Multiple threshold results*: This table compares the four distinct one-factor threshold models: the Gaussian, t-distributed, variance-gamma, and generalized hyperbolic approaches. Each is calibrated to have an average default correlation of 0.05.

Quantile	Gaussian $\widehat{\text{VaR}}_\alpha$	$\widehat{\mathcal{E}}_\alpha$	t-distributed $\widehat{\text{VaR}}_\alpha$	$\widehat{\mathcal{E}}_\alpha$	Variance-gamma $\widehat{\text{VaR}}_\alpha$	$\widehat{\mathcal{E}}_\alpha$	Gen.-hyperbolic $\widehat{\text{VaR}}_\alpha$	$\widehat{\mathcal{E}}_\alpha$
95.00th	$48.4	$82.7	$48.4	$86.4	$51.9	$82.1	$48.8	$81.4
97.00th	$63.5	$100.9	$64.6	$107.0	$67.0	$97.7	$63.7	$98.5
99.00th	$101.9	$145.6	$107.7	$158.0	$100.7	$131.5	$100.3	$139.2
99.50th	$130.1	$177.1	$140.0	$194.4	$122.2	$152.8	$126.1	$166.7
99.90th	$206.7	$258.2	$228.9	$289.0	$172.5	$202.4	$192.4	$235.7
99.97th	$286.9	$340.1	$322.8	$385.2	$218.8	$247.5	$261.3	$304.9
99.99th	$323.8	$377.7	$366.2	$429.5	$239.7	$267.9	$293.3	$336.8
Other statistics								
Expected loss	$9.2		$9.2		$9.2		$9.2	
Loss volatility	$21.7		$22.9		$21.2		$21.3	
Total iterations	10,000,000							
Computational time	49.0		55.9		120.8		155.8	

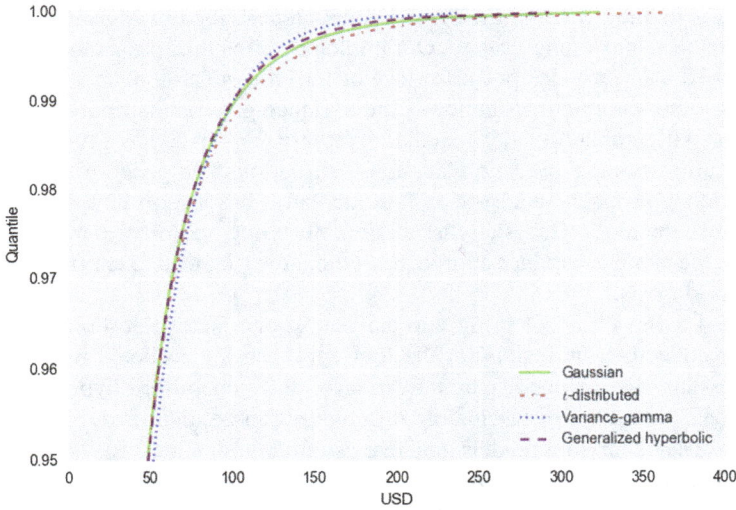

Fig. 4.11 *Multiple threshold tail probabilities*: This figure summarizes the tail probabilities for the four one-factor threshold models considered in this chapter: the Gaussian, t, variance-gamma, and generalized-hyperbolic approaches. Visually, the models generate quite similar results.

the Poisson log-normal and Weibull implementations in Chap. 3, it is fair to say that the variance-gamma and generalized hyperbolic approaches have not found widespread application in the practitioner community. Popularity is not, however, the determining criterion for the usefulness of a particular model. Model risk is always a danger for quantitative analysts. Even if we do not use these approaches, pushing mixture and threshold models to their limits aids in understanding them and, ultimately, acts to mitigate model risk. It also offers legitimate alternatives that, depending on one's specific modelling challenges, may prove both welcome and practical.

4.6 The Canonical Multi-Factor Setting

Once one decides to move to a multi-factor setting, there is an infinity of possible variations. It is thus important to understand the general case. We will thus begin with the canonical model and then proceed to examine a specific multivariate implementation to fix ideas. The specific choice of model varies depending on one's desired joint distribution for the default state variables. We will, to keep the clutter to a minimum, consider both the Gaussian and, more general, normal-variance-mixture implementations.

4.6.1 The Gaussian Approach

In its canonical form, the K-dimensional Gaussian Vasicek (1987, 1991, 2002) threshold model has a default state variable for the nth credit counterparty with the following form,

$$y_n = \underbrace{a_{n,K+1}\epsilon_n}_{\text{Idiosyncratic}} + \underbrace{\sum_{k=1}^{K} a_{n,k} Z_k}_{\text{Systematic}}. \tag{4.103}$$

As in the one-factor setting, there is a separation between idiosyncratic and systematic latent variables.[30] The key difference with the multi-factor model, however, is the greater number of systematic variables and naturally the overall number of parameters. The factor weights, or loadings, are summarized by the collection of parameters, $\{a_{n,k}; k = 1, \ldots, K+1; n = 1, \ldots, N\}$. As before, there are N idiosyncratic latent variables, but only $K \ll N$ systematic variables.

In the multi-factor environment, we observe the common risk-factor approach employed by many financial models. These systematic elements are, in principle, shared by all credit counterparties, but the magnitude of their influence is governed by the factor-loadings. The description and structure of these variables is, of course,

---

[30]The very convenient notation used in this section has been borrowed, naturally without implication, from Glasserman (2006).

the modeller's choice; typically they include global, regional, industry, or country-specific factors. The only limit, however, is one's imagination, and one's ability to convince others of its relevance. The modeller may also use the full set of parameters or restrict their values in some way. This adds to model flexibility and realism, but it naturally increases the complexity.

As in the one-factor Gaussian setting, all of the latent variables are standard normal variates. This implies, which is easily verified, that $\mathbb{E}(y_n) = 0$. The variance is a bit more complex. Let's evaluate the variance of y_n and set what is required for it to take a value of unity,

$$1 = \text{var}(y_n), \tag{4.104}$$

$$= \text{var}\left(a_{n,K+1}\epsilon_n + \sum_{k=1}^{K} a_{n,k} Z_k\right),$$

$$= \text{var}\left(a_{n,K+1}\epsilon_n\right) + \sum_{k=1}^{K} \text{var}\left(a_{n,k} Z_k\right),$$

$$= a_{n,K+1}^2 \underbrace{\text{var}\left(\epsilon_n\right)}_{=1} + \sum_{k=1}^{K} a_{n,k}^2 \underbrace{\text{var}\left(Z_k\right)}_{=1},$$

$$a_{n,K+1} = \sqrt{1 - \sum_{k=1}^{K} a_{n,k}^2}.$$

This suggests that, if we wish to maintain each y_n as a standard normal variate, the values $\{a_{n,K+1}; n = 1, \ldots, N\}$ are *not* free parameters. This acts a natural, and fairly welcome, constraint on the size of our parameter space.

Moreover, to ensure that equation 4.104 does not lead to complex values, we require that

$$1 - \sum_{k=1}^{K} a_{n,k}^2 \geq 0, \tag{4.105}$$

$$\sum_{k=1}^{K} a_{n,k}^2 \leq 1,$$

$$a_n^T a_n \leq 1,$$

for $n = 1, \ldots, N$ where,

$$a_n = \begin{bmatrix} a_{n,1} \\ \vdots \\ a_{n,K} \end{bmatrix}. \tag{4.106}$$

4.6 The Canonical Multi-Factor Setting

A bit more caution with parameter specification is thus clearly required in the multi-factor setting.

What does this multi-factor version of model imply for correlation between the default state variables? This can be determined directly by resolving,

$$\mathrm{cov}(y_n, y_m) = \mathbb{E}(y_n y_m), \tag{4.107}$$

$$= \mathbb{E}\left(\left(a_{n,K+1}\epsilon_n + \sum_{k=1}^{K} a_{n,k} Z_k\right)\left(a_{m,K+1}\epsilon_m + \sum_{k=1}^{K} a_{m,k} Z_k\right)\right),$$

$$= \mathbb{E}\left(\sum_{k=1}^{K} a_{n,k} a_{m,k} Z_k^2\right),$$

$$= \sum_{k=1}^{K} a_{n,k} a_{m,k} \mathbb{E}\left(Z_k^2\right),$$

$$= \sum_{k=1}^{K} a_{n,k} a_{m,k} \underbrace{\mathrm{var}\left(Z_k^2\right)}_{=1},$$

$$= a_n^T a_m.$$

Since y_n and y_m have unit variance, $\mathrm{cov}(y_n, y_m) = \rho(y_n, y_m) = a_n^T a_m$.[31] As before, the model parameters induce correlation between the default-state variables. It does so, however, in a significantly more complex manner. Instead of a ρ or $\sqrt{\rho_n \rho_m}$, it depends on each factor loading of both the nth and mth obligors.

The default event remains $\mathcal{D}_n = \{y_n \leq \Phi^{-1}(p_n)\}$ and we maintain the use of the default indicator, $\mathbb{I}_{\mathcal{D}_n}$. Since y_n remains $\mathcal{N}(0,1)$, the default threshold is still $\Phi^{-1}(p_n)$, where p_n denotes the exogeneously determined unconditional default probability. If we condition on the vector of systematic latent variables, $Z = \begin{bmatrix} Z_1 & \cdots & Z_K \end{bmatrix}^T$, then the conditional default probability is,

$$p_n(Z) = \mathbb{E}\left(\mathbb{I}_{\mathcal{D}_n} \mid Z\right), \tag{4.108}$$

$$= \mathbb{P}\left(\mathcal{D}_n \mid Z\right),$$

$$= \mathbb{P}\left(y_n \leq \Phi^{-1}(p_n) \mid Z\right),$$

---

[31] To resolve equation 4.107, it helps to recall that for independent random variables, $x_1, \ldots, x_N$, each with $\mathbb{E}(x_i) = 0$, their product $\mathbb{E}(x_n x_m) = \mathbb{E}(x_n) \cdot \mathbb{E}(x_m) \neq 0$ only when $n = m$.

$$= \mathbb{P}\left(\underbrace{a_{n,K+1}\epsilon_n + \sum_{k=1}^{K} a_{n,k} Z_k}_{\text{Equation 4.103}} \leq \Phi^{-1}(p_n) \Bigg| Z\right),$$

$$= \mathbb{P}\left(\epsilon_n \leq \frac{\Phi^{-1}(p_n) - \sum_{k=1}^{K} a_{n,k} Z_k}{a_{n,K+1}} \Bigg| Z\right),$$

$$= \Phi\left(\frac{\Phi^{-1}(p_n) - a_n^T Z}{a_{n,K+1}}\right).$$

As in the one-factor setting, the vector of systematic latent variables, Z, drives the dependence between the threshold state variables, $y_1, \ldots, y_n$. Again, the strength of the dependence is governed by the systematic variable parameters, $\{a_{n,k}; n = 1, \ldots, N; k = 1, \ldots, K\}$. For a given value of Z, however, the default events are independent.

The remainder of the model structure is virtually identical to the one-factor set-up. The default correlations and the joint density have precisely the same general form. If one opts to use the bivariate normal distribution to compute the joint default probability, the only difference is that the correlation matrix, Ω, has a more complex form. In general, the full covariance (and correlation) matrix is

$$\Omega = \begin{bmatrix} 1 & a_1^T a_2 & \cdots & a_1^T a_{N-1} & a_1^T a_N \\ a_2^T a_1 & 1 & \cdots & a_2^T a_{N-1} & a_2^T a_N \\ \vdots & \vdots & \cdots & \ddots & \vdots \\ a_N^T a_1 & a_N^T a_2 & \cdots & a_N^T a_{N-1} & 1 \end{bmatrix}, \quad (4.109)$$

where the diagonals are equal to one by the restriction in equation 4.104. In the model implementation, this does not play such an important role. It is of critical importance, however, when one uses credit-quality transition data to estimate the model parameters. This key question is addressed in Chap. 10.

4.6.2 The Normal-Variance-Mixture Set-Up

Although it might seem tedious, it is nonetheless useful to examine the normal-variance-mixture approach in a general setting.[32] In the canonical multivariate

---

[32] It can, of course, get significantly more involved. See Daul et al. (2005) for additional extensions in this area with a particular focus on the t-distribution.

4.6 The Canonical Multi-Factor Setting

normal-variance mixture model, the default state variable is a modification of equation 4.103,

$$y_n = \sqrt{h(V)} \left(a_{n,K+1}\epsilon_n + \sum_{k=1}^{K} a_{n,k} Z_k \right), \qquad (4.110)$$

where, as in the one-factor setting, V is a positive-valued independent random variable. This construction nests a family of possible models. We have considered *three* choices, which we summarize as,

$$y_n \stackrel{\triangle}{=} \begin{cases} t\text{-distributed}: & h(V) = \dfrac{\nu}{V} \text{ where } V \sim \chi^2(\nu) \\ \text{variance-gamma}: & h(V) \equiv V \sim \Gamma(a, a) \\ \text{generalized-hyperbolic}: & h(V) \equiv V \sim \text{GIG}(a, 1) \end{cases} \qquad (4.111)$$

It is easy to see that if $h(V) \equiv 1$, we recover the multivariate Gaussian threshold model.

A little bit of algebra leads us to conclude that,

$$\text{cov}(y_n, y_m) = \mathbb{E}\Big(h(V)\Big) a_n^T a_m, \qquad (4.112)$$

This looks familiar to the Gaussian setting. Since, for the normal-variance-mixture models, $\text{var}(y_n) = \mathbb{E}(h(V))$, it directly follows that

$$\rho(y_n, y_m) = \frac{\text{cov}(y_n, y_m)}{\sqrt{\text{var}(y_n)}\sqrt{\text{var}(y_m)}}, \qquad (4.113)$$

$$= \frac{\mathbb{E}\big(h(V)\big) a_n^T a_m}{\sqrt{\mathbb{E}\big(h(V)\big)}\sqrt{\mathbb{E}\big(h(V)\big)}},$$

$$= a_n^T a_m,$$

which directly implies that the correlation between the state variables of counterparties n and m is $a_n^T a_m$—this is identical to the Gaussian model. This is an important point. The correlation structure, and hence correlation matrices, are identical in the Gaussian and normal-variance-mixture versions of the canonical model.

We saw in the Gaussian setting that certain constraints on the parameters values needed to be imposed. We must now also determine the nature of these constraints in the normal-variance backdrop. The development is analogous to the Gaussian setting, but we now wish our parameter choices to be consistent not with unit variance, but with the expectation of the transformation of the mixing variable,

$h(V)$. This leads to,

$$\mathbb{E}\Big(h(V)\Big) = \text{var}(y_n), \tag{4.114}$$

$$= \text{var}\left(\sqrt{h(V)}\left(a_{n,K+1}\epsilon_n + \sum_{k=1}^{K} a_{n,k} Z_k\right)\right),$$

$$= \mathbb{E}\left(\left(\sqrt{h(V)}\left(a_{n,K+1}\epsilon_n + \sum_{k=1}^{K} a_{n,k} Z_k\right)\right)^2 - \underbrace{\left(\mathbb{E}\left(\sqrt{h(V)}\left(a_{n,K+1}\epsilon_n + \sum_{k=1}^{K} a_{n,k} Z_k\right)\right)\right)^2}_{=0}\right),$$

$$= \mathbb{E}\Big(h(V)\Big)\mathbb{E}\left(\left(a_{n,K+1}\epsilon_n + \sum_{k=1}^{K} a_{n,k} Z_k\right)^2\right),$$

$$\cancel{\mathbb{E}\big(h(V)\big)} = \cancel{\mathbb{E}\big(h(V)\big)}\left(a_{n,K+1}^2 + \underbrace{\sum_{k=1}^{T} a_{n,k}^2}_{a_n^T a_n}\right),$$

$$a_{n,K+1} = \sqrt{1 - a_n^T a_n},$$

with $a_n^T a_n \leq 1$ for all $n = 1, \ldots, N$. This verifies that the parametric constraints are identical in the Gaussian and normal-variance-mixture versions of the canonical multi-factor model.

As in the one-factor setting, the default threshold is $K_n = F_{\mathcal{N}\mathcal{V}}^{-1}(p_n)$. This has implications for the conditional default probability,

$$p_n(Z, V) = \mathbb{P}(y_n \leq K_n | Z, V), \tag{4.115}$$

$$= \mathbb{P}\left(\underbrace{\sqrt{h(V)}\left(a_{n,K+1}\epsilon_n + \sum_{k=1}^{K} a_{n,k} Z_k\right)}_{\text{Equation 4.110}} \leq F_{\mathcal{N}\mathcal{V}}^{-1}(p_n) \middle| Z, V\right),$$

$$= \Phi\left(\frac{\sqrt{\frac{1}{h(V)}} F_{\mathcal{N}\mathcal{V}}^{-1}(p_n) - a_n^T Z}{a_{n,K+1}}\right).$$

Again, the structure of the multivariate normal-variance-mixture model has much the same form as its Gaussian equivalent. Table 4.9 provides a detailed comparison of the identities associated with these two modelling choices.

4.6 The Canonical Multi-Factor Setting

Table 4.9 *Canonical multi-factor summary*: This table summarizes the main differences, and similarities, between the canonical multi-factor Gaussian and the generic normal-variance-mixture threshold models. The principal differences arise in the structure of the joint distribution of the latent variables and the associated default correlations.

Variable	Definition	
	Gaussian	Normal-variance mixture
$Z = \begin{bmatrix} Z_1 & \cdots & Z_K \end{bmatrix}^T, \epsilon_n$	Identical, independent $\mathcal{N}(0,1)$ random variates	
V	Not applicable	Positive-valued random variable
y_n	$a_{n,K+1}\epsilon_n + \sum_{k=1}^{K} a_{n,k} Z_k$	$\sqrt{h(V)} \left(a_{n,K+1}\epsilon_n + \sum_{k=1}^{K} a_{n,k} Z_k \right)$
a_n	$\begin{bmatrix} a_{n,1} & \cdots & a_{n,K} \end{bmatrix}^T$	
$\mathrm{var}(y_n)$	1	$\mathbb{E}\left(h(V)\right)$
$\mathrm{cov}(y_n, y_m)$	$a_n^T a_m$	$\mathbb{E}\left(h(V)\right) a_n^T a_m$
$\rho(y_n, y_m)$	$a_n^T a_m \in (0,1)$	
$\mathcal{D}_n$	$\{y_n \leq K_n\}$	
p_n	$\mathbb{E}(\mathbb{I}_{\mathcal{D}_n}) \equiv \mathbb{P}(\mathcal{D}_n) = \Phi(K_n)$	$\mathbb{E}(\mathbb{I}_{\mathcal{D}_n}) \equiv \mathbb{P}(\mathcal{D}_n) = F_{\mathcal{N}(y)}(K_n)$
K_n	$\Phi^{-1}(p_n)$	$F_{\mathcal{N}(y)}^{-1}(p_n)$
$p_n(Z,V)$	$\Phi\left(\dfrac{\Phi^{-1}(p_n) - a_n^T Z}{a_{n,K+1}} \right)$	$\Phi\left(\sqrt{\dfrac{1}{h(V)}} \dfrac{F_{\mathcal{N}(y)}^{-1}(p_n) - a_n^T Z}{a_{n,K+1}} \right)$
$\rho(\mathbb{I}_{\mathcal{D}_n}, \mathbb{I}_{\mathcal{D}_m})$	$\dfrac{\mathbb{P}(\mathcal{D}_n \cap \mathcal{D}_m) - p_n p_m}{\sqrt{p_n p_m (1-p_n)(1-p_m)}}$	
$\mathbb{P}(\mathcal{D}_n \cap \mathcal{D}_m)$	$\dfrac{1}{2\pi\sqrt{|\Omega_{nm}|}} \int_{-\infty}^{\Phi^{-1}(p_n)} \int_{-\infty}^{\Phi^{-1}(p_m)} e^{-\frac{x^T \Omega_{nm}^{-1} x}{2}} dx$	$\int_{-\infty}^{\infty} \int_{0}^{\infty} p_n(z,v) \cdot p_m(z,v) \cdot f_V(v) \cdot \phi(g) dv dz$
Ω_{nm}	$\begin{bmatrix} 1 & a_n^T a_m \\ a_m^T a_n & 1 \end{bmatrix}$	

4.7 A Practical Multi-Factor Example

The canonical multi-factor model, derived in the previous section, is useful for its generality, but is not particularly handy for practical implementation. Given N obligors in one's portfolio, it possesses $\frac{N^2-N}{2}$ asset-correlation pairs.[33] For $N = 100$ obligors and 10 factors, therefore, this matrix is summarized by about 1,000 parameters. Such a rich parametrization would require enormous amounts of data. In the credit-risk setting, where in most cases such data is quite simply unavailable, a model of this nature would be impossible to estimate and nightmarish to manage.

Parsimony is thus critical, but simplification restricts the model dynamics. There is, as is often the case, an inherent trade-off between flexibility and practicality. Two concrete steps can be taken. First, it is advisable, unless there is a strong reason, to avoid a large number of systematic variables. Consider the limiting case of a regional factor structure. If each obligor was treated as a separate region, this would somewhat defeat the purpose of the model.[34] The second aspect involves restricting the nature of the factor loadings. Each region or industry, for example, need not have a unique loading onto the state variable.

In this final section, taking this advice to heart, we consider a specific multi-factor threshold choice that seeks, quite reasonably, to reduce the dimensionality of the parameter space by dramatically restricting the model relative to the canonical approach. An infinity of defensible choices are possible; this is only one of them. The ultimate selection will depend on your specific application, your underlying data-set, and what you are trying to accomplish. For better or for worse, therefore, our proposed default state variable has the following form,

$$y_n(k) = \sqrt{h(V)}\left(\sqrt{a}G + \sqrt{1-a}\left(\sqrt{b_k}R_k + \sqrt{1-b_k}\epsilon_n\right)\right), \qquad (4.116)$$

$$= \sqrt{h(V)} \left(\underbrace{\sqrt{a}G + \sqrt{(1-a)b_k}R_k}_{\text{Systematic element}} + \underbrace{\sqrt{(1-a)(1-b_k)}\epsilon_n}_{\text{Idiosyncratic element}} \right),$$

for $n = 1, \ldots, N$, and $k = 1, \ldots, K$. We let y_n denote the nth credit counterparty, but add the region as an argument to the state variable. Thus, $y_n(k)$ denotes the nth counterparty, exposed to the kth region.[35] This is a bit clumsy, but it nevertheless clearly denotes the key parameter relationships and restrictions.

---

[33] This excludes an obligors's correlation with itself.

[34] The idea is to induce heightened correlation among regions and were each credit counterpart to form its own region, the effect would be indistinguishable from the idiosyncratic factor.

[35] One could naturally add an additional index to permit varying loadings to the global variable.

4.7 A Practical Multi-Factor Example

This structure permits us to describe all approaches in the same framework. Equation 4.116 implies a single systematic global factor, G, a collection of systematic regional factors, $\{R_k; k = 1, \ldots, K\}$, and N idiosyncratic factors, $\{\epsilon_n; n = 1, \ldots, N\}$. To restrict the parameter space and ease implementation and interpretation of model results, $K \ll N$. There are, in principle, K regional b_k parameters and a single global-loading a parameter—the precise number of parameters is, of course, a model implementation choice. As before, the global, regional, and idiosyncratic factors are independent, identically distributed standard normal variates.

4.7.1 Understanding the Nested State-Variable Definition

The complex nested form might not seem particularly intuitive, but it is commonly used in practice. We'll soon see why. First, however, we need to verify that it doesn't distort any of the important modelling features. The expectation remains,

$$\mathbb{E}\left(y_n(k)\right) = \mathbb{E}\left(h(V)\left(\sqrt{a}G + \sqrt{(1-a)b_k}R_k + \sqrt{(1-a)(1-b_k)}\epsilon_n\right)\right). \quad (4.117)$$

$$= \mathbb{E}\left(h(V)\right)\left(\sqrt{a}\underbrace{\mathbb{E}(G)}_{=0} + \sqrt{(1-a)b_k}\underbrace{\mathbb{E}(R_k)}_{=0} + \sqrt{(1-a)(1-b_k)}\underbrace{\mathbb{E}(\epsilon_n)}_{=0}\right),$$

$$= 0.$$

The variance is determined as,

$$\operatorname{var}\left(y_n(k)\right) = \mathbb{E}\left(\left(\sqrt{h(V)}\left(\sqrt{a}G + \sqrt{(1-a)b_k}R_k + \sqrt{(1-a)(1-b_k)}\epsilon_n\right)\right)^2\right), \quad (4.118)$$

$$= \mathbb{E}\left(h(V)\right)\left(a\underbrace{\operatorname{var}(G)}_{=1} + (1-a)b_k\underbrace{\operatorname{var}(R_k)}_{=1} + (1-a)(1-b_k)\underbrace{\operatorname{var}(\epsilon_n)}_{=1}\right),$$

$$= \mathbb{E}\left(h(V)\right)\left(a + (1-a)b_k + (1-a)(1-b_k)\right),$$

$$= \mathbb{E}\left(h(V)\right),$$

which is precisely what we find in the canonical multivariate setting. Thus, we can see that although the nested structure in equation 4.116 appears unconventional, it maintains the requisite distributional structure of the latent model variable, $y_n(k)$.

The reason for its form can be seen in another way. Consider the general multi-factor constraint derived in equation 4.104. It restricts the choice of factor loading for the idiosyncratic factor. Our model has three factors for each choice of credit counterparty. If we map the parameter definitions from equation 4.116 into equation 4.104, we can see if they are consistent,

$$\underbrace{\sqrt{(1-a)(1-b_k)}}_{a_{n,K+1}} = \sqrt{1 - \left(\underbrace{\left(\sqrt{a}\right)^2}_{a_{n,1}^2} + \underbrace{\left(\sqrt{(1-a)b_k}\right)^2}_{a_{n,2}^2}\right)}, \qquad (4.119)$$

$$= \sqrt{(1-a) - (1-a)b_k},$$

$$= \sqrt{(1-a)(1-b_k)}.$$

This demonstrates that nested form was, in fact, selected specifically to ensure, by construction, that the general constraint from equation 4.104 is always satisfied. Although it seems odd at first glance, it sneakily adds the parameter constraint into the definition.

The next step is to define the correlation structure between two credit counterparties as proxied by two arbitrary default state variables, $y_n(k)$ and $y_m(j)$. We should expect it to be somewhat more involved. The general form will be familiar from the one-factor and canonical multi-factor models, but some caution is required with the additional parameter restrictions. The calculation is as follows,

$$\rho\left(y_n(k), y_m(j)\right) = \frac{\text{cov}(y_n(k), y_m(j))}{\sqrt{\mathbb{E}\left(h(V)\right)}\sqrt{\mathbb{E}\left(h(V)\right)}}, \qquad (4.120)$$

$$= \left(\frac{1}{\mathbb{E}\left(h(V)\right)}\right)\mathbb{E}\left(y_n(k)y_m(j)\right),$$

$$= \left(\frac{\cancel{\mathbb{E}\left(h(V)\right)}}{\cancel{\mathbb{E}\left(h(V)\right)}}\right)\mathbb{E}\left(aG^2 + \mathbb{I}_{k=j}(1-a)\sqrt{b_k b_j}R_k R_j\right),$$

$$= a + \mathbb{I}_{k=j}(1-a)\sqrt{b_k b_j}.$$

The indicator variable arises because if $k \neq j$, then $\mathbb{E}(R_k R_j) = 0$ and we are left with a—where a plays the role of the ρ coefficient in the one-factor setting. More simply, if the credit counterparties stem from different regions, then the correlation is merely the global factor loading, a, and is independent of the b_k parameters. If the two countries come from the same region, however, then we have a more complex expression of correlation between their respective state variables incorporating both the a and b_k parameters in a non-linear way.

4.7 A Practical Multi-Factor Example

We may, once again, seek assistance from the general model to verify the veracity of equation 4.120. The correlation coefficient between default state variables, y_n and y_m, in the canonical model was defined in equation 4.107 as $a_n^T a_m$. In our simplified setting, the vector a_n is, including the regional arguments, described by the following structure

$$a_n = \left[\sqrt{a} \ \sqrt{(1-a)b_{n,k}}\right]^T. \tag{4.121}$$

The dot product of a_n and a_m is thus,

$$a_n^T a_m = \left[\sqrt{a} \ \sqrt{(1-a)b_{n,k}}\right] \begin{bmatrix} \sqrt{a} \\ \sqrt{(1-a)b_{m,j}} \end{bmatrix}, \tag{4.122}$$

$$= a + (1-a)\sqrt{b_{n,k}b_{m,j}}.$$

This is almost exactly the same form as the result in equation 4.120. The only difference relates to the implicit model constraints on our specialized approach. The general framework would allow each of the countries to be impacted by the regional parameters in its own individual way. While very general, this is not a particularly practical approach. Nevertheless, the general model does provide assurance that our default-state variable correlation term is correctly specified.

The final object of interest is the conditional default probability. It has the following form,

$$p_n(G, R_k, V) = \Phi\left(\frac{\sqrt{\frac{1}{h(V)}}F_{\mathcal{N}_M}^{-1}(p_n) - \sqrt{a}G - \sqrt{(1-a)b_k}R_k}{\sqrt{(1-a)(1-b_k)}}\right), \tag{4.123}$$

for all $n = 1, \ldots, N$. This critical object now has a more complex structure. For a given obligor, it depends on the global, regional, idiosyncratic, and, depending on the model choice, the mixing variable, V. This provides significant flexibility, but naturally complicates the analysis. If, for example, we opt to use the conditional default probabilities to estimate the default correlation—as in equation 4.89—we need to numerically evaluate a three-dimensional integral. This is reaching the limits of our computational capacity and, as more factors are added, the situation will only worsen.

4.7.2 Selecting Model Parameters

If we want to actually implement this model we need to select a specific form of $h(V)$ and calibrate it. Ultimately, we have decided to demonstrate the multivariate

t-distributed threshold model. It has the benefit of positive tail dependence and is also relatively easy to handle. Practically, of course, this implies that $h(V) = \frac{v}{V}$ with $V \sim \chi^2(v)$.

Calibration, as we've seen upon numerous occasions in the preceding discussion, is mostly a matter of determining the appropriate degree of default correlation. Recall that we have *three* distinct regions in our sample portfolio: Alpha, Bravo, and Charlie. This implies three regional and one global systematic state variables. Analogous to the mixture-model discussion in Chap. 3, default correlation is more involved in the multivariate setting. We have default correlation both within and between regions.

To make things interesting—and encourage comparison across chapters and models—we have opted to select the parameters to target the same basic level of interregional default correlation as found in the multi-factor CreditRisk+ example in the previous chapter. Table 4.10 shows us the multivariate t-threshold parameter choices and the asset-correlation implications of this decision.

From the Alpha to the Charlie region, we observe increasing levels of intraregional asset correlation. All obligors, irrespective of region of origin, share relatively weak asset correlation. The implication is that, outside of one's geographical area, the interregional dependence of default events is quite low. This may, or may not, be what is desired, but the analyst is free to make these choices. With a change of the parameters, one might easily reverse this state of affairs and generate a simultaneously weak intraregional and strong interregional dependence structure.

How are the parameters in Table 4.10 determined? The first step is to be able to determine, for an arbitrary pair of obligors, the default correlation associated with a given parameterization. We've done this many times, but as hinted previously, the use of the conditional default probability trick, from equation 4.89 on page 197, is not a great idea. Equation 4.123 clearly showed us that, as we add state variables, the dimensionality of our numerical integration problem will increase. Two-dimensional numerical integration is already fairly slow, three or more dimensions are simply painful. The good news, however, is that, armed with the joint density of the state variables, the problem of computing pairwise default correlation always reduces to a two-dimensional integration of the bivariate density. Table 4.5 illustrates the formulae for the Gaussian and t-distributed cases while

Table 4.10 *Sample parameter structure*: This table highlights the asset-correlation parameter structure between the three regions of our 100-obligor portfolio example. These values were selected to match, as closely as possible, the default-correlation values found in the multivariate CreditRisk+ example in Chap. 3. The degrees-of-freedom parameter, v, is set to 30.

	Parameter		Notation		Coefficient		
Region	Global	Region	Global	Region	Global	Region	Total
Alpha	0.03	0.04	a	b_1	0.03	0.04	0.07
Bravo	0.03	0.07	a	b_2	0.03	0.07	0.10
Charlie	0.03	0.25	a	b_3	0.03	0.24	0.27

4.7 A Practical Multi-Factor Example

Algorithm 4.17 Constructing the regional default-correlation matrix

```
def buildDefaultCorrelationMatrix(a,b,pMean,regionId,nu):
    J = len(regionId)
    R = buildAssetCorrelationMatrix(a,b,regionId)
    D = np.zeros([J,J])
    for n in range(0,J):
        p_n = pMean[n]
        for m in range(0,J):
            p_m = pMean[m]
            p_nm = bivariateTCdf(norm.ppf(p_n),norm.ppf(p_m),R[n,m],nu)
            D[n,m] = (p_nm - p_n*p_m)/math.sqrt(p_n*(1-p_n)*p_m*(1-p_m))
    return D
```

Table 4.9 shows the multivariate results. Adding more state variables does *not* change this set-up, it merely makes the asset-correlation coefficient more involved.

The Python code to compute the regional default correlation matrix is found in Algorithm 4.17. For each region, we identify the average default probability, found in pMean, and then compute the 2 × 2 bivariate asset-correlation matrix, R. This can, given the parameter structure, take a variety of values. The default correlation has the usual form—see, for example, equation 4.77—where the joint default probability is provided by bivariateTCdf. This is a wrapper for the requisite two-dimensional numerical integration.

The matrix, D, from Algorithm 4.17, and for the parameters in Table 4.10, are summarized in Table 4.11. The average default correlation across all obligors is about 0.03. Intra-regionally, in the Bravo and Charlie sectors, it can, however, be significantly higher. In the Charlie sector, for example, it is three times the overall mean value. This level of flexibility is simply impossible in the one-factor setting.

The actual parameters were selected, as previously indicated, to match the regional default correlation found in the multivariate CreditRisk+ model. A perfect fit is not possible since the CreditRisk+ setting had seven parameters, while our multivariate t-distributed model has only four. Comparison of the results from Table 4.11 to the equivalent CreditRisk+ values in Table 3.14 in Chap. 3 reveals that although the diagonal values match, there is some mismatch in the interregional

Table 4.11 *Default correlation matrix*: This table uses average default correlations in each region, the bivariate t-distribution, and the parameters in Table 4.10 to present the interregional default-correlation values for our three-region example. They are a close, but not perfect, match to those found in the multivariate CreditRisk+ model in Chap. 3.

	Alpha	Bravo	Charlie
Alpha	0.03	0.02	0.02
Bravo	0.02	0.04	0.02
Charlie	0.02	0.02	0.09

values. Mathematically, this process involved selecting the parameters a, b_1, b_2, and b_3 to minimize the distance between the two default-correlation matrices.[36]

Many other possible calibration choices are, of course, possible. The multiplicity of parameters, however, makes this a challenging process. Whatever one's choice, it is essential to examine the corresponding asset- and default-correlation values. With a sizable portfolio it becomes virtually impossible to do this at the obligor level. At the regional, industrial, or rating-category level—depending on the state-variable choices of the analyst—this type of analysis is entirely feasible. Moreover, it is entirely legitimate to use expert judgement to select individual parameters. Sensitivity analysis of different choices is also useful and welcome. The bottom line is that multi-factor models offer fantastic flexibility, but are a bit hazardous. The danger arises from the difficulty in understanding precisely what is happening. This risk is well mitigated with extensive use of these model diagnostics to examine model outcomes from multiple angles.

4.7.3 Multivariate Risk Measures

Using the previous parameter choices, it is straightforward to apply the model to our running example. Table 4.12 illustrates the results along with, for comparison purposes, the one-factor Gaussian and t-distributed approaches. It is not possible, of course, to calibrate one-factor models to match all the elements of a multivariate methodology. We did our best; the one-factor Gaussian model was calibrated to the average default correlation coefficient of 0.03.[37] The t-distributed model, conversely, employs the average asset-correlation coefficient and degrees-of-freedom parameter.[38]

The differences between these three choices, ostensibly set up to match one another, are surprisingly important. The one-factor model treats all obligors in a common manner; thus the significant differences in ρ value lead to predictable divergence in the one-factor values. Asset correlation of 0.25 across all obligors is a very different story from low interregional, but high intraregional ρ parameters. The multivariate t-distributed model's results underscore the hazards of comparing average asset and default correlations across univariate and multivariate environments.

The overall risk figures in Table 4.12 are fairly modest, which appear to be principally due to the weak interregional dependence. Our review of our portfolio's exposure and default probabilities in Chap. 1 suggested that much of the risk was concentrated in the Bravo region. Breaking our portfolio into three regional pockets,

---

[36] The measure used to describe the distance between the two matrices is the Frobenius norm; see Golub and Loan (2012, Chapter 2) for more details.

[37] The ρ parameter is approximately 0.25.

[38] In this case, the ρ parameter is about 0.07 and $\nu = 30$.

4.8 Final Thoughts 225

Table 4.12 *Multivariate threshold results*: This table compares the multi-factor t-distributed model results—with the parameters and default correlations summarized in Tables 4.10 and 4.11—to similarly calibrated one-factor Gaussian and t-distributed threshold results.

| | One-factor Gaussian | | t-distributed | | | |
| | | | One-factor | | Multi-factor | |
Quantile	$\widehat{\text{VaR}}_\alpha$	$\widehat{\mathcal{E}_\alpha}$	$\widehat{\text{VaR}}_\alpha$	$\widehat{\mathcal{E}_\alpha}$	$\widehat{\text{VaR}}_\alpha$	$\widehat{\mathcal{E}_\alpha}$
95.00th	$46.1	$73.0	$41.2	$59.3	$41.3	$60.4
97.00th	$58.7	$87.2	$50.3	$68.5	$50.8	$70.3
99.00th	$88.5	$120.6	$69.7	$89.0	$71.3	$92.9
99.50th	$109.5	$143.6	$82.6	$102.6	$85.5	$108.2
99.90th	$164.0	$201.9	$114.4	$136.1	$122.0	$146.5
99.97th	$209.8	$248.6	$140.0	$162.5	$151.7	$176.4
99.99th	$251.4	$292.6	$163.7	$187.6	$179.6	$202.5
Other statistics						
Expected loss	$9.2		$9.2		$9.2	
Loss volatility	$19.2		$9.2		$16.3	
Total iterations	10,000,000					
Computational time	49.2		55.6		73.0	

each with strong interdependence, but weak overall portfolio interaction, has had a surprising effect.[39] Arriving at this conclusion, however, is the point of the exercise. This framework provides an ideal testing environment to examine the implications of different levels of dependence both within and between these regions. One could also, with relatively little additional effort, consider other approaches to categorize and parameterize the individual obligors.

4.8 Final Thoughts

In this extraordinarily lengthy and detailed chapter, we have thoroughly examined the class of threshold models. Starting with the one-factor setting, we worked through the original Gaussian model. In Chap. 5, this model will arise again and we will see how it is, in fact, the original credit-risk model. In an attempt to obtain tail-dependence, we used a statistical trick to obtain the t-distributed method, which has found favour with both practitioners and academics in recent years. The Gaussian and t-distributed models represent two key tools in the quantitative analyst's toolkit.

Having established these two common approaches, we turned to see how the clever manoeuvre employed to build the t-distributed model can actually be generalized. This led to the class of normal-variance mixture models. Two

---

[39]The reader is also encouraged to compare the results with the multivariate CreditRisk+ implementation in Table 3.15 on page 147.

examples—the variance-gamma and generalized hyperbolic models—were investigated and implemented. Although neither of these models are particularly popular, it helped us to develop a broader appreciation of the threshold methodology. Insight and potential choice are always beneficial. Model risk is a constant danger for quantitative analysts. Even if we do not use these approaches, pushing mixture and threshold models to their limits aids in understanding them and, ultimately, acts to mitigate model risk.

The final sections of this chapter were allocated to the multivariate threshold setting. Each of the one-factor models can be readily generalized to include a broad range of systematic state variables. This allows for a richer dependence structure by categorizing one's obligors into regional, industrial, national, or even credit-quality groups. The fully general canonical model is a natural starting place. Although it is far too general for practical application, it provides a solid basis for investigating and gaining understanding into the challenges of moving from the univariate to the multivariate world. We close our discussion with an examination of a practical multi-factor example, calibrating it to the regional default-correlation matrix from the CreditRisk+ model in Chap. 3, and applying it to our portfolio example.

References

Abramovitz, M., & Stegun, I. A. (1965). *Handbook of mathematical functions*. New York: Dover Publications.

Barndorff-Nielsen, O., Kent, J., & Sørensen, M. (1982). Normal variance-mean mixtures and z distributions. *International Statistical Review, 50*(2), 145–159.

BIS. (2006a). International convergence of capital measurement and capital standards: A revised framework comprehensive version. *Technical report*. Bank for International Settlements.

Daul, S., de Giorgi, E., Lindskog, F., & McNeil, A. (2005). The grouped t-copula with an application to credit risk. ETH Zürich.

de Kort, J. (2007). *Modeling tail dependence using copulas—literature review*. University of Amsterdam.

Devroye, L. (2012). Random variate generation for the generalized inverse Gaussian distribution. *Statistical Computing, 24*(1), 239–246.

Eberlein, E., & Keller, U. (1995). Hyperbolic distributions in finance. *Bernoulli, 3*(1), 281–299.

Embrechts, P., Klüppelberg, C., & Mikosch, T. (1999). *Modelling extremal events for insurance and finance* (1st edn.). New York, NY: Springer-Verlag. *Stochastic modelling and applied probability*.

Fok, P.-W., Yan, X., & Yao, G. (2014). *Analyis of credit portfolio risk using hierarchical multi-factor models*. University of Delaware.

Glasserman, P. (2006). Measuring marginal risk contributions in credit portfolios. Risk Measurement Research Program of the FDIC Center for Financial Research.

Golub, G. H., & Loan, C. F. V. (2012). *Matrix computations*. Baltimore, Maryland: The John Hopkins University Press.

Gordy, M. B. (2002). A risk-factor model foundation for ratings-based bank capital rules. Board of Governors of the Federal Reserve System.

Gupton, G. M., Finger, C. C., & Bhatia, M. (2007). *CreditMetrics — technical document*. New York: Morgan Guaranty Trust Company.

References

Hu, W., & Kercheval, A. (2007). Risk management with generalized hyperbolic distributions. *Proceedings of the fourth IASTED international conference on financial engineering and applications* (pp. 19–24).

Joe, H. (1997). *Multivariate models and dependence concepts*. Boca Raton, FL: Chapman & Hall/CRC. Monographs on statistics and applied probability (vol. 73).

Johnson, N. L., Kotz, S., & Balakrishnan, N. (1994). *Continuous univariate distributions: volume 1* (2nd edn.). New York, NY: John Wiley & Sons.

Kostadinov, K. (2005). Tail approximation for credit-risk portfolios with heavy-tailed risk factors. Zentrum Mathematik, Technische Universität München.

Kotz, S., Kozubowski, T., & Podgorski, K. (2001). *The Laplace distribution and generalizations: A revisit with applications to communications, economics, engineering, and finance*. Basel, Switzerland: Birkhäuser.

Kuhn, G. (2005). Tails of credit default portfolios. Zentrum Mathematik, Technische Universität München.

Leydold, J., & Hörmann, W. (2013). Generating generalized inverse gaussian random variates. Institute for Statistics and Mathematics, Report 123, Wirstshafts Universität Wien.

Leydold, J., & Hörmann, W. (2017). Package 'GIGrvg. R Package Documentation.

MacKenzie, D., & Spears, T. (2012). *"The Formula That Killed Wall Street?" The Gaussian copula and the material cultures of modelling*. University of Edinburgh.

Madan, D. B., & Seneta, E. (1990). The variance gamma (V.G.) model for share market returns. *The Journal of Business, 63*(4), 511–524.

McNeil, A. J., Frey, R., & Embrechts, P. (2015). *Quantitative risk management: Concepts, tools and techniques*. Princeton, NJ: Princeton University Press.

Merton, R. (1974). On the pricing of corporate debt: The risk structure of interest rates. *Journal of Finance, 29*, 449–470.

Nelsen, R. B. (2006). *An introduction to copulas*. New York, NY: Springer.

Roth, M. (2013). On the multivariate t distribution. Division of Automatic Control, Department of Electric Engineering, Linköpings Universitet, Sweden.

Schmidt, R. (2003). Credit risk modelling and estimation via elliptical copulae. In G. Bohl, G. Nakhaeizadeh, S. Rachev, T. Ridder, & K. Vollmer (Eds.), *Credit risk-measurement, evaluation and management* (pp. 267–289). Physica-Verlag.

Schmitz, V. (2003). Copulas and stochastic processes. Rheinisch-Westfällisch Technischen Hochschule Aaachen.

Schönbucher, P. J. (2000b). Factor models for portfolio credit risk. Department of Statistics, University of Bonn.

Scott, D. (2015a). Package 'GeneralizedHyperbolic. R Package Documentation.

Scott, D. (2015b). Package 'VarianceGamma. R Package Documentation.

Sharpe, W. F. (1963). A simplified model for portfolio analysis. *Management Science, 9*(2), 277–293.

Sharpe, W. F. (1964). Capital asset prices: A theory of market equilibrium under conditions of risk. *The Journal of Finance, 19*(3), 425–442.

Sklar, A. (1959). Fonctions de répartition à n dimensions et leurs marges. *Publications de l'Institut de Statistique de L'Université de Paris, 8*, 229–231.

Vasicek, O. A. (1987). *Probability of loss on loan distribution*. KMV Corporation.

Vasicek, O. A. (1991). *Limiting loan loss probability distribution*. KMV Corporation.

Vasicek, O. A. (2002). The distribution of loan portfolio value. *Risk*, (12), 160–162.

von Hammerstein, E. A. (2016). Tail behaviour and tail dependence of generalized hyperbolic distributions. In J. Kallsen, & A. Papapantoleon (Eds.), *Advanced modelling in mathematical finance* (pp. 3–40). Springer-Verlag,

Wilde, T. (1997). *CreditRisk+: A credit risk management framework*. Credit Suisse First Boston.

Chapter 5
The Genesis of Credit-Risk Modelling

> *There has been no systematic development of a theory for pricing bonds when there is a significant probability of default.*
> (Robert C. Merton)

In the preceding chapters, we have examined the independent-default model and considered two competing modelling approaches—the mixture and threshold families—incorporating default dependence and a variety of alternative default-loss distributional assumptions. This has led us to 14 distinct, albeit often conceptually related, possible credit-model implementations. Figure 5.1 provides a high level classification, or zoology, of these various approaches.

This would seem, by almost any standard, to represent a sufficient collection of credit-risk models. While we tend to agree, we are not quite finished with our focus on the foundations of credit risk models. Our objective in this chapter is not to add additional models to our list; although, our list will slightly lengthen. Rather our goal is to deepen our understanding of the current models and gain a better appreciation for the history of this branch of finance. It is often useful to go back to the roots of any endeavour and this is no exception.

To accomplish this, we return to the path-breaking paper, Merton (1974), which is widely agreed to be the first *true* credit-risk modelling work. In many discussions of credit risk, Merton (1974) is the first—and occasionally also the last—model considered. This theoretical work was first advanced more than forty years ago. One might legitimately enquire why this was not our starting point. Chronologically, this makes complete sense, but pedagogically, in our view, this is less efficient. As we'll see in the coming sections, implementation of Merton (1974) carries, superficially at least, a much higher mathematical and empirical burden relative to the approaches outlined in Fig. 5.1. It is our view that, armed with fundamental knowledge of common credit-risk models, it is much easier to understand and appreciate Merton (1974). Indeed, his work has coloured virtually the entire realm of credit-risk modelling and, as will soon become clear, has strong links to the family of threshold models.

In the coming pages, we will start with the basic intuition, notation, and mathematical structure. This will lead us to a motivating discussion regarding the notion

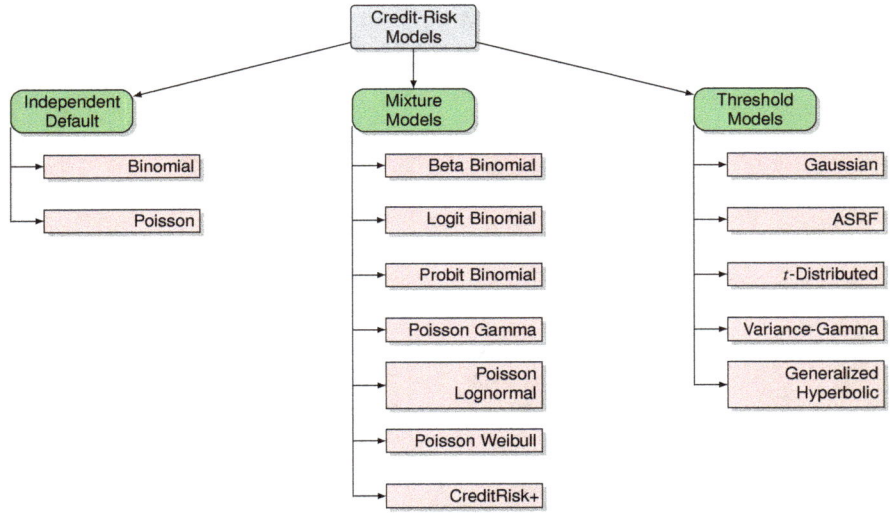

Fig. 5.1 *A model zoology*: The underlying figure categorizes and classifies the 14 distinct credit-risk models, falling into three modelling families, considered so far in our discussion.

of geometric Brownian motion. We then investigate two possible implementations of Merton (1974)'s model, which we term the indirect and direct approaches. The indirect approach will turn out to be quite familiar, whereas the direct method requires a significant amount of heavy lifting.

5.1 Merton's Idea

In the previous chapters, an obligor's default was either exogenously specified or endogenized through an unobservable state variable. Merton (1974), facing the challenge of pricing corporate debt with repayment uncertainty, took a different route. The result was a more direct, and theoretically satisfying, description of the default event. Merton (1974), at its foundation, demonstrates conceptually that a firm's equity stake is essentially equivalent to a call option on its assets with the strike price equal to its debt obligations.

This work has a broad range of applications.[1] Although, it was constructed with capital-structure decisions in mind, it has also become a keystone of credit-risk measurement. Its focus, however, is within the corporate realm. As such, when we will often refer to an obligor or counterparty as the firm. Let's remark, at the

---

[1] See Sundaresan (2013) for a useful overview of the range of applications stemming from Merton's capital-structure model.

5.1 Merton's Idea

very outset, that some of these ideas are less obviously generalized to government obligors. Nevertheless, much of the same intuition does indeed apply and the attendant challenges are primarily empirical.

As with many of the previously discussed models, a reasonable amount of set-up, introduction of notation, and mathematical development is required. In many ways, the effort is even higher in this setting. There are, however, significant benefits to this investment. The Merton (1974) approach begins with the value of the firm at time $t \in [0, T]$. By general consensus, a firm's value can be written straightforwardly as,

$$A_t = E_t + L_t, \tag{5.1}$$

where A_t, E_t, and L_t denote the assets, equity, and liabilities of the firm at time t, respectively. Equation 5.1 is consistent with the universally applied double-entry book-keeping system and echoes the treatment seen in virtually all financial statements across the world. To this end, therefore, equation 5.1 is not an assumption, but rather a notational statement.

We can use equation 5.1 to draw a few conclusions. In the event that $A_t > L_t$, the firm remains solvent. Should $A_t \leq L_t$, then the firm is deemed to find itself in default. In this latter case, the firm does *not* have sufficient means to meet its obligations. This is consistent with how credit-rating agencies assess an obligor's relative creditworthiness[2] How does the equity enter into this discussion? Merton (1974) remarked that the equity holders have a residual claim on the firm's value, once the other obligations have been met. This led to the, now ubiquitous, pay-off function for equity holders,

$$E_T = (A_T - L_T)^+, \tag{5.2}$$

where T is the terminal date of the debt.[3] While common and widely understood today, this clever construction has had a dramatic influence on corporate finance and the modelling of credit risk.

Equation 5.2 is, of course, the pay-off of a call option, where L_t denotes the strike price. Equity-holders thus hold a call-option style claim on the firm's assets. Often, to directly draw the parallel to the option-pricing setting, this is written as

$$E_T = (A_T - K)^+, \tag{5.3}$$

---

[2] As a useful example, the Standard & Poor's rating agency in S&P (2011, Paragraph 20) states further that "the most important step in analyzing the creditworthiness of a corporate or governmental obligor is gauging the resources available to it for fulfilling its obligations."

[3] Naturally, the firm's debt does not typically mature all on a single date, but this is a useful approximation of reality.

where K denotes the notional value of the debt, L_T. K is the typical symbol used to represent the strike price of the option. While L_T is, in principle, a random variable, it should be considered as reasonably slow moving and easy to predict. As a consequence, for our purposes, it is useful to treat it as a constant. Formally, therefore, equation 5.3 is a contingent claim (i.e., option) on the value of the firm's assets. We naturally seek an assessment of the equity value, E_t, at the current time, t. We may, with a bit more structure and a few assumptions, determine its theoretical price of E_t, where $t < T$, with insights from the Black and Scholes (1973) approach. One of the key contributions of Merton (1974) is a theoretical framework for the estimation of equity prices.

While interesting, equation 5.3 may not seem immediately applicable to the credit-risk setting. To relate this idea back to our previous development, consider the following

$$\{\tau \leq T\} = \{A_T \leq K\}. \tag{5.4}$$

There are two equivalent default events. The left-hand-side of equation 5.4 may, to some readers, appear somewhat less familiar. The random variable, τ, is referred to as a first-passage or hitting time.[4] Technically, the first-passage time of A_t below L_t, is the default event of the firm. Following Jeanblanc (2002), we define this first passage time as,

$$\tau = \inf\left(t > 0 : t \in [0, T] \cap \{A_t \leq L_t\}\right). \tag{5.5}$$

In other words, it is simply the first time that the assets fall below the firm's liabilities. For any single time period, $[0, T]$, and single path of A_t over this interval, default is a binary event. It either occurs or it does not. If we think of A_t as a random process, however, then the probability of the default event has more meaning. The role of the time horizon, T, is also quite important. We generally think of this T as being the time period associated with our risk computation; for example, one or two years.[5]

The right-hand-side of equation 5.4 should look quite familiar. If the firm's asset value at time T falls below the default barrier, K, then the firm is in default. This is a more direct and practical statement of the default event. Neither exogenous nor unobservable, equation 5.4 links the default event to a concrete statement about the firm's asset values. This is the incredible conceptual contribution of Merton (1974). To draw a link to the previous chapters, let us define the default event as

---

[4]The first-passage time is a special case of a more general probabilistic concept termed a stopping time.

[5]If it is extremely long, say 100 years, then the probability of default grows dramatically and the exercise becomes less meaningful.

5.1 Merton's Idea

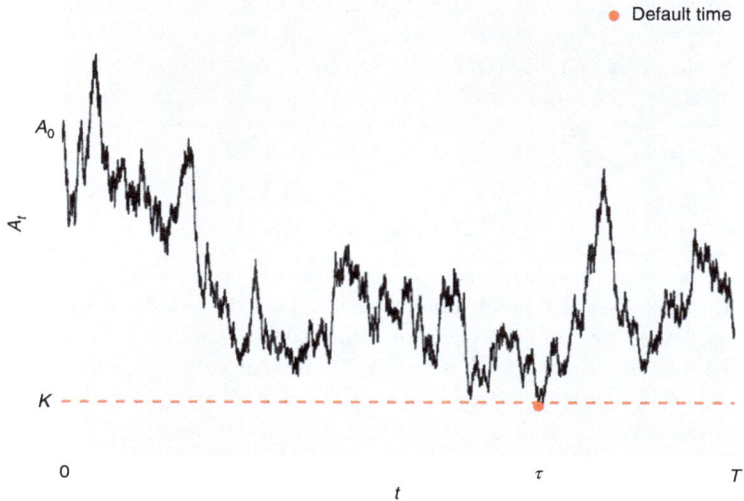

Fig. 5.2 *Default-event schematic*: The underlying figure provides an idealized description of the intuition behind the Merton (1974) default event. If the value of the firm's assets, A_t, falls below the value of its liabilities, described by K, then default is triggered.

$\mathcal{D} = \{\tau \leq T\}$, which permits us to put the default probability into the familiar form

$$\mathbb{E}(\mathbb{I}_\mathcal{D}) = \mathbb{E}\left(\mathbb{I}_{\{\tau \leq T\}}\right), \tag{5.6}$$

$$\mathbb{P}(\mathcal{D}) = \underbrace{\mathbb{P}(\tau \leq T)}_{\text{Equation 5.4}},$$

$$= \mathbb{P}(A_T \leq K).$$

Although principally notation, equation 5.5 creates a useful common ground with our previous default-event definitions.

Imagine that the asset-value process evolves stochastically over time. K is the firm's lower bound, or threshold. If, or when, $A_t \leq K$ for the first time, default is triggered. Figure 5.2 summarizes an idealized single sample path for A_t and a possible default event. The key point is that the default event is random, but once it occurs, it is final. Default cannot be undone.

5.1.1 Introducing Asset Dynamics

To make Fig. 5.2 more concrete and proceed further with this model, it is necessary to write down some sensible dynamics for the asset-value process. As previously suggested, the value of the liabilities are typically treated as a deterministic function or constant; we will continue this convention. The bad news is that Merton (1974), in the classic financial tradition, opted for a rather complex definition of the asset-value process, which requires some mathematical structure. First, it is necessary to define a probability space, $(\Omega, \mathcal{F}, \mathbb{P})$. We further assume that the asset-value has the following dynamics,

$$\frac{dA_t}{A_t} = \mu dt + \sigma d\hat{W}_t, \tag{5.7}$$

where $A_0 > 0$, $\mu \in \mathbb{R}$ denotes the instantaneous or infinitesimal mean return on the assets, under the physical probability measure $\mathbb{P}$, and $\sigma \in \mathbb{R}$ represents the instantaneous asset volatility. Equation 5.7 is a stochastic differential equation—broadly known as geometric Brownian motion—and $\{\hat{W}(t), \mathcal{F}_t\}$ is a standard scalar Wiener process on $(\Omega, \mathcal{F}, \mathbb{P})$.[6]

While it is essential to have a description of the dynamics of A_t, it is also useful to find a solution to equation 5.7. This would allow us, in closed-form, to find a representation of the default event described in equation 5.4. Solution of equation 5.7 requires the use of some results from the stochastic calculus.[7] Let us thus introduce, without proof, the following result:

Theorem 5.1 (Ito) *Let X_t be a continuous semi-martingale taking values in an open subset $U \subset \mathbb{R}$. Then, for any twice continuously differentiable function $f : U \to \mathbb{R}$, $f(X_t)$ is a semi-martingale and,*

$$f(X_T) - f(X_t) = \int_t^T \frac{\partial f}{\partial X_u} dX_u + \frac{1}{2} \int_t^T \frac{\partial^2 f}{\partial X_u^2} d\langle X_u \rangle, \tag{5.8}$$

where $d\langle X_t \rangle$ denotes the quadratic variation of the process X_t.

If we select the function $f(A_t) = \ln(A_t)$ and apply Theorem 5.1, we may solve equation 5.7, as

$$\ln(A_T) - \ln(A_t) = \int_t^T \frac{\partial \ln(A_u)}{\partial A_u} dA_u + \frac{1}{2} \int_t^T \frac{\partial^2 \ln(A_u)}{\partial A_u^2} d\langle A \rangle u, \tag{5.9}$$

$$= \int_t^T \frac{1}{A_u} \left(\mu A_u du + \sigma A_u d\hat{W}_u \right) - \frac{1}{2} \int_t^T \frac{1}{A_u^2} d\langle \mu A_u du + \sigma A_u d\hat{W}_u \rangle,$$

$$= \mu \int_t^T du + \sigma \int_t^T d\hat{W}_u - \frac{1}{2} \int_t^T \frac{1}{A_u^2} \left(\sigma^2 A_u^2 du \right),$$

---

[6] See Karatzas and Shreve (1998), Oksendal (1995), and Heunis (2011) for much more information and rigour on Brownian motion and stochastic differential equations.

[7] Readers not familiar with this area can look to the references for more background. Neftci (1996) is also a highly-recommended, gentle and intuitive introduction to this area of finance.

5.1 Merton's Idea

$$= \mu(T-t) + \sigma\left(\hat{W}_T - \hat{W}_t\right) - \frac{\sigma^2}{2}(T-t),$$

$$A_T = A_t e^{\left(\mu - \frac{\sigma^2}{2}\right)(T-t) + \sigma\left(\hat{W}_T - \hat{W}_t\right)},$$

where $\hat{W}_T - \hat{W}_t \sim \mathcal{N}(0, T-t)$ is a Gaussian-distributed independent increment of the Wiener process. Quite simply, the terminal asset value, A_T, is a kind of multiple of the starting value, A_t, and the outcome of a Gaussian random variable. Uncertainty about A_T is also an increasing function of the time elapsed over the interval, $[t, T]$.

This result is quite helpful, because it permits us to write down a closed-form expression for the unconditional probability of a firm's default at the terminal date, T. In particular, we are now interested in understanding

$$\mathbb{P}\left(\tau \leq T \mid \mathcal{F}_t\right), \tag{5.10}$$

This is *not* equivalent, it must be stressed, to conditioning on the terminal value of the assets. More specifically, it is not like conditioning on the global systematic factor, G, in the threshold setting. There is only one source of uncertainty in the Merton (1974) setting. If we are given the outcome of the Wiener increment, $\hat{W}_T - \hat{W}_t$, it is no longer a stochastic model. Working with stochastic processes, however is a bit different than working with separate random variables. A stochastic process is a, typically time-indexed, sequence of random variables. We thus need to be a bit more precise and use the filtration, $\mathcal{F}_t$, as our conditioning set. $\mathcal{F}_t$ provides a collection of all the relevant *past and current* information about our asset process and the underlying Wiener processes up until time t.[8] For this reason, in economics, the filtration is sometimes referred to as an information set. A_T is not measurable with respect to $\mathcal{F}_t$ and is, as such, unknown and random.

Using the solution to the geometric Brownian motion describing our firm's asset dynamics, we can evaluate equation 5.10 as

$$\mathbb{P}\left(\tau \leq T \mid \mathcal{F}_t\right) = \mathbb{P}\left(A_T \leq K \mid \mathcal{F}_t\right), \tag{5.11}$$

$$= \mathbb{P}\left(\overbrace{A_t e^{\left(\mu - \frac{\sigma^2}{2}\right)(T-t) + \sigma\left(\hat{W}_T - \hat{W}_t\right)}}^{\text{Equation 5.9}} \leq K \,\bigg|\, \mathcal{F}_t\right),$$

---

[8] It is, in fact, a nested sequence of σ-algebras. This is an important concept in measure theory; see Royden (1988) for more rigour and background.

$$= \mathbb{P}\left(\hat{W}_T - \hat{W}_t \le \frac{\ln\left(\frac{K}{A_t}\right) - \left(\mu - \frac{\sigma^2}{2}\right)(T-t)}{\sigma} \bigg| \mathcal{F}_t\right),$$

$$= \mathbb{P}\left(\frac{\hat{W}_T - \hat{W}_t}{\sqrt{T-t}} \le \frac{\ln\left(\frac{K}{A_t}\right) - \left(\mu - \frac{\sigma^2}{2}\right)(T-t)}{\sigma\sqrt{T-t}} \bigg| \mathcal{F}_t\right).$$

This is a bit of a mess. The final term is particularly unwieldy, so let's tidy it up somewhat with the following definition,

$$\delta_t = \frac{\ln\left(\frac{K}{A_t}\right) - \left(\mu - \frac{\sigma^2}{2}\right)(T-t)}{\sigma\sqrt{T-t}}. \tag{5.12}$$

A_t is measurable with respect to $\mathcal{F}_t$ and thus δ_t can be treated as a constant for a given choice of T. The only remaining uncertainty in the model relates to the Wiener increment. The quantity, δ_t, is a function of the liability-threshold, the initial asset value, the expected asset return, and the asset volatility. It also includes, as we would expect, the length of the time interval in question.

5.1.2 Distance to Default

We can simplify equation 5.11 beyond merely re-defining complicated terms. We also know, from the properties of the Wiener increment, that $\frac{\hat{W}_T - \hat{W}_t}{\sqrt{T-t}} \sim \mathcal{N}(0, 1)$. That is, the normalized Wiener increment is standard normally distributed. It thus naturally follows that

$$\mathbb{P}(\tau \le T \mid \mathcal{F}_t) = \mathbb{P}(A_T \le K \mid \mathcal{F}_t), \tag{5.13}$$

$$= \mathbb{P}\left(\frac{\hat{W}_T - \hat{W}_t}{\sqrt{T-t}} \le \delta_t \bigg| \mathcal{F}_t\right),$$

$$= \frac{1}{\sqrt{2\pi}} \int_{-\infty}^{\delta_t} e^{-\frac{u^2}{2}} du,$$

$$\mathbb{P}(\mathcal{D} \mid \mathcal{F}_t) = \Phi(\delta_t),$$

where $\Phi(\cdot)$ denotes the cumulative distribution function of the standard normal distribution. It has certainly not escaped the reader that this form is strongly analogous to the conditional default probability derived for the threshold models in Chap. 4.

Before proceeding, it is useful to raise a few points and consolidate some of the preceding ideas. First of all, let's lighten our notation burden. While it is formally

5.1 Merton's Idea

correct to condition the probabilities and expectations with respect to the filtration, $\mathcal{F}_t$, it is notationally somewhat heavy-handed. Moreover, it is potentially somewhat confusing since it is not a conditional expectation in the sense defined in previous chapters. As such, we will denote $\mathbb{E}(\cdot \mid \mathcal{F}_t)$ as $\mathbb{E}_t(\cdot)$—this notation is often employed in economics. We'll use the same trick for unconditional probabilities; $\mathbb{P}(\cdot \mid \mathcal{F}_t)$ as $\mathbb{P}_t(\cdot)$.

The second point is that δ_t is an important, and somewhat famous, quantity who warrants some additional consideration. Is often referred to as the *distance-to-default*. To heuristically understand this quantity, first observe that $\ln\left(\frac{K}{A_t}\right)$ is basically a return. It is roughly the percentage loss, from the current asset value, necessary to fall below the boundary level, K. If, for example, $K = 80$ and $A_t = 100$, then $\ln\left(\frac{K}{A_t}\right) \approx -20\%$.[9] Over the time interval, $[t, T]$, however, the firm's assets will generate some return—positive or negative—which is approximated by the second term, $\left(\mu - \frac{\sigma^2}{2}\right)(T - t)$. Naturally, if the expected return is quite low and the asset volatility is quite high, then this term can actually be negative. For asset volatility of 20%, the expected asset return would have to exceed 2% for this correction to be positive. The second term is a correction for the asset return—it may need to fall a bit further for positive returns, and a bit less with negative returns, to hit the default barrier. The entire numerator is defined in units of return. The denominator, $\sigma\sqrt{T-t}$, normalizes this return distance into units of volatility providing a standardized measure of distance to the default barrier.

The distance to default, δ_t, is thus a standardized random normal variate that specifies the cut-off below which default occurs. In other words, as it is so succinctly expressed in Crosbie and Bohn (2002),

> the distance-to-default is simply the number of standard deviations that the firm is away from default.

There are other rule-of-thumb based measures used to approximate this quantity, but, in this analysis, we will focus on the formal definition.[10] To actually default, the realization of A_T must be sufficiently large to cross this distance and bring the value of the firm below its liability-threshold. Since different firms have different expected asset returns, varying levels of asset-value uncertainty, and disparate assets values, it is extremely useful to perform this standardization. It allows one to compare, at a glance, the financial health of a broad range of obligors.

---

[9] In reality, it is closer to -22%, but the rough approximation still holds.
[10] One suggested quantity has the following form

$$\delta_t \approx \frac{1 - \frac{K}{A_t}}{\sigma\sqrt{T-t}}, \qquad (5.14)$$

which is computed using a number of approximations. In particular, one assumes that $\ln\left(\frac{K}{A_t}\right) \approx \frac{A_t - K}{A_t}$, $\mu(T-t) \approx \frac{A_T - A_t}{A_t}$, $\sigma^2(T-t) \approx 0$, and $\frac{A_T}{A_t} \approx 0$.

In our new notation, we may thus write equation 5.13 more succinctly as,

$$\mathbb{P}_t(\tau \leq T) = \mathbb{P}_t(A_T \leq K), \qquad (5.15)$$
$$= \Phi(\delta_t).$$

Equation 5.15 describes a convenient relationship between the conditional probability of default, the time period in question, the size of the firm's debt burden and the properties of the firm's assets. This is the logic that lies at the heart of the structural Merton (1974) approach. Working directly with default probabilities is challenging. Assets, one might be tempted to conclude, are easier to work with and, with this theoretical link, might be used as a sensible proxy. Unfortunately, as we'll soon see, a number of challenges still remain.

Computing credit risk in this setting, at least using Monte Carlo methods, should now be quite clear. We take the starting asset values, A_0, and using μ and σ—the asset return and volatility—we simulate a large number of paths for A_t over the interval $[0, T]$ using equation 5.7 or, more directly, 5.9. For each sample path, if $A_T \leq K$, then we have default. If not, the obligor survives. If we repeat this 10 million times, we build a distribution of default losses; or, more simply, zeros and ones. Multiplying these binary outcomes by the obligor's exposure leads to a default-loss distribution.

We have, interestingly, not made any use of the obligor's unconditional default probability, p. This idea has been replaced by the notion of distance-to-default. In other words, the unconditional default probability is a function of the current value of a firm's asset, its liability barrier, the expected return of these assets, and our assessment of the uncertainty regarding the future of the asset return. Implied in these quantities, and wrapped up by the distance-to-default, is the implied default probability. Quite simply, we have that

$$\mathbb{E}_t(\mathbb{I}_\mathcal{D}) = \mathbb{P}_t(\mathcal{D}), \qquad (5.16)$$
$$p = \Phi(\delta_t).$$

Given the form of δ_t, this is a fairly complicated relationship. In short, the default probability is no longer given, but must, in principle, be derived from primitive information regarding the firm.

5.1.3 Incorporating Equity Information

The major challenges of the Merton (1974) framework all relate to one simple fact: we do not observe firm asset values. At regular intervals, financial statements are produced that attempt to provide a description of the firm's assets, liabilities, and equities. These are, however, accounting figures and, in addition to their lack of timeliness, are generally not directly useful for our purposes. Publicly traded firms,

5.1 Merton's Idea

however, have equities that trade daily and provide significant insight into market views on the firm's values. These are not exactly what we seek, but they depend on assets. Equity values, therefore, provide an indirect perspective on our object of choice: the firm's assets.

We already have a relationship between a firm's equity and assets in equation 5.3; we need only evaluate it. Following from Black and Scholes (1973), this is not such an easy undertaking. In particular, it cannot be accomplished under the physical probability measure, $\mathbb{P}$. We must, therefore, change the underlying measure on our probability space such that the discounted process, A_t, is a martingale. Appendix B derives the process for A_t where $\{W_t, \mathcal{F}_t\}$ is a standard scalar Wiener process on the probability space, $(\Omega, \mathcal{F}, \mathbb{Q})$. To repeat, $\mathbb{Q}$ is the *equivalent martingale measure* generated by using the money-market account as the numeraire asset; this is often referred to as the risk-neutral probability measure.[11]

As a result of the hard work found in Appendix B, the $\mathbb{P}$-dynamics of A_t, as described in equation 5.7, have the following form under $\mathbb{Q}$,

$$dA_t = rA_t dt + \sigma A_t dW_t, \quad (5.17)$$

where r denotes the risk-free interest rate. This corresponds to the idea that, in the risk-neutral setting, all assets return the risk-free rate.

We can now value the equity claim. Modern finance theory tells us that the value of a derivative security, our contingent claim, is equal to its discounted conditional expectation under $\mathbb{Q}$. The change of measure is thus a critical step in valuing a derivative contract written upon our asset process, A_t. In our case, this is the firm's equity as introduced in equation 5.2. Consider the following expression,

$$E(t, A_t) = e^{-r(T-t)} \mathbb{E}_t^{\mathbb{Q}} \left((A_T - K)^+ \right), \quad (5.18)$$

where t indicates the current point in time and T represents the maturity date of the firm's debt payment. The attentive reader will have certainly noticed, by this point, that this is the well-known valuation identity for a plain-vanilla European call option.

Mathematically, solving equation 5.18 is far from trivial, even if the result is generally well known. Appendix B, once again, performs all of the heavy-lifting to resolve it, with the result summarized in the following celebrated formula,

$$E(t, A_t) = A_t \Phi(d_1) - e^{-r(T-t)} K \Phi(d_2), \quad (5.19)$$

---

[11] Please see Harrison and Kreps (1979) and Harrison and Pliska (1981) for the foundational work underlying these ideas.

where,

$$d_1 = \frac{\ln\left(\frac{A_t}{K}\right) + \left(r + \frac{\sigma^2}{2}\right)(T-t)}{\sigma\sqrt{T-t}}, \qquad (5.20)$$

$$d_2 = d_1 - \sigma\sqrt{T-t}.$$

From the Merton model, therefore, we have a clear default event and a relatively straightforward expression to represent the value of its equity claim as a function of its underlying assets. A close connection exists, therefore, between the Merton (1974) and Black and Scholes (1973) work.[12]

This establishes a link, but the work is not yet done. Indeed, although the basic idea is quite clear and a Monte-Carlo solution is reasonably obvious, many questions still remain. Key information—such as asset values, returns, and volatilities as well as the liability threshold—need to be identified. Equally importantly, the overall model, quite clearly, needs to incorporate multiple obligors and introduce some notion of default dependence. Given one's specific approach for the determination of this liability threshold, however, some of these concerns actually disappear. We will defer many of these questions, therefore, to our discussion of the indirect and direct approaches. Before addressing these important issues, however, we will take a brief aside to investigate, and hopefully demystify, the concept of geometric Brownian motion.

5.2 Exploring Geometric Brownian Motion

Although the Merton (1974) model looks, and feels, very much like the threshold setting, it was conceived and developed in a continuous-time mathematical finance framework. This may not be immediately familiar to all readers. Indeed, many of the mathematical tools required are subtle and complex. Nevertheless, with a bit of digging and patience, the implications of geometric Brownian motion are relatively benign for our problem. Indeed, most of the key ideas reduce to concepts that we have already seen and employed in previous chapters.

To get started, let us return to our base stochastic differential equation (SDE). Merton (1974) assumed that the asset prices followed the following generic dynamics,

$$dX_t = \underbrace{\mu X_t dt}_{\text{Drift}} + \underbrace{\sigma X_t dW_t}_{\text{Diffusion}}, \qquad (5.21)$$

---

[12]Duffie and Singleton (2003, Section 3.2) actually refer to this approach as the Black-Scholes-Merton default model.

5.2 Exploring Geometric Brownian Motion

where $\mu, \sigma \in \mathbb{R}$ refer to scalar coefficients governing the drift and diffusion terms, respectively. $\{W_t, \mathcal{F}_t\}$ is a standard scalar Wiener process on $(\Omega, \mathcal{F}, \mathbb{P})$. This, as we've already indicated, is referred to as *geometric Brownian motion*. It has many scientific applications beyond financial engineering.[13]

The first term in equation 5.21 is referred to as the drift. It is entirely deterministic and provides the basic direction for the dynamics of X_t. It is, in fact, a rather simple object. Imagine we ignore, for the moment, the worrisome dW_t term and focus solely on the drift element. Cutting out the diffusion term, therefore, we have the following ordinary differential equation (ODE):

$$dX_t = \mu X_t dt. \tag{5.22}$$

This is about as simple a differential equation as one is likely to meet. Its solution merely requires a bit of manipulation. By the fundamental theorem of calculus, we can re-write it as

$$f(X_T) - f(X_t) = \int_t^T \frac{\partial f(X_u)}{\partial X_u} dX_u. \tag{5.23}$$

Selecting $f(X_t) = \ln(X_t)$ directly yields the solution,

$$\ln(X_T) - \ln(X_t) = \int_t^T \frac{\partial \ln(X_u)}{\partial X_u} dX_u, \tag{5.24}$$

$$\ln\left(\frac{X_T}{X_t}\right) = \int_t^T \frac{1}{\cancel{X_u}} (\mu \cancel{X_u} du),$$

$$= \mu(T - t),$$

$$\frac{X_T}{X_t} = e^{\mu(T-t)},$$

$$X_T = X_t e^{\mu(T-t)}.$$

Without the diffusion, therefore, this is a simple object. It is, in fact, just an exponential growth trend.[14] Equation 5.24 implies that $\ln(X_T) = \ln(X_t) + \mu(T-t)$

---

[13] Karatzas and Shreve (1998) is the standard reference in this area, whether one is a financial analyst, mathematical physicist or electrical engineer.

[14] The solution is easily verified by differentiation,

$$\frac{dX_T}{dT} = \mu \underbrace{X_t e^{\mu(T-t)}}_{X_T}, \tag{5.25}$$

$$dX_T = \mu X_T dT.$$

This holds, of course, for any value of T and coincides with equation 5.22.

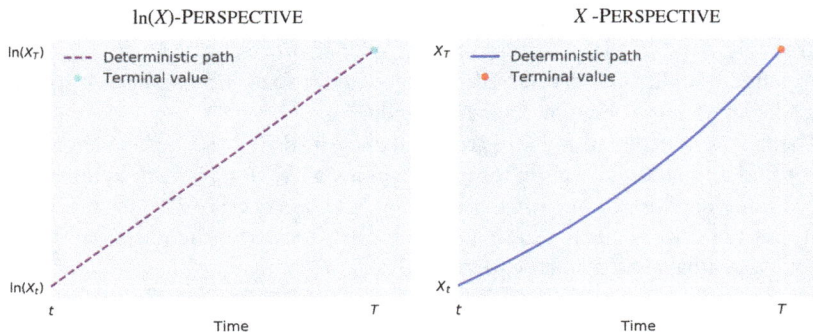

Fig. 5.3 *ODE behaviour*: These two figures graphically describe the behaviour—both with and without natural logarithms—of the embedded ordinary differential equation found in geometric Brownian motion.

is a linear function of time. Or, equivalently if one prefers to eliminate the natural logarithm, X_t is assumed to grow exponentially over time. In finance, μ is generally considered to be a return. The exponential element merely incorporates the notion of (continuous) compound interest in one's return.

Figure 5.3 provides, from both the logarithmic and exponential perspectives, the solution to the ordinary differential equation, describing the drift element in the geometric Brownian motion. The base assumption is simply that the value of the assets are continuously compounded with a return of μ over time. This is appealing, but unrealistic, since it assumes smooth and undisturbed asset growth. This would make for a great investment, but doesn't coincide with real life. In real markets, financial returns are noisy as shocks, both positive and negative, hit the firm and influence the value of its assets. This is the reason for the second, more complex, diffusion term.

Understanding the diffusion term forces us to wrangle with the dW_t quantity from equation 5.21. It has a random element, which justifies calling it a stochastic differential equation. The fundamental theorem of calculus no longer applies. Instead, we need to make use of Theorem 5.1 from the Itô calculus. This leads to an additional term in the solution of X_t. In particular,

$$f(X_T) - f(X_t) = \underbrace{\int_t^T \frac{\partial f(X_u)}{\partial X_u} dX_u}_{\text{Usual term}} + \underbrace{\frac{1}{2} \int_t^T \frac{\partial^2 f(X_u)}{\partial X_u^2} d\langle X \rangle_u}_{\text{Extra term}}. \quad (5.26)$$

where $d\langle X \rangle_u$ is termed the quadratic-variation process. It involves the usual term, used to solve the the ordinary differential equation, from equation 5.22. It also includes, however, an additional second-order term that one must either take on faith or invest a significant amount of time to fully comprehend. We opt for the former approach. Karatzas and Shreve (1991), Oksendal (1995), Durrett (1996) and

5.2 Exploring Geometric Brownian Motion

Heunis (2011) are excellent references for the reader seeking additional background and insight into this result.

As we did in the previous section, although skipping over the details, selecting $f(X_t) = \ln(X_t)$ yields the solution,

$$X_T = X_t e^{\left(\mu - \frac{\sigma^2}{2}\right)(T-t) + \sigma(W_T - W_t)}, \quad (5.27)$$

where $W_T - W_t$ is a Gaussian-distributed Wiener increment. The punchline is that geometric Brownian motion is described by a stochastic differential equation. Like ordinary differential equations, many stochastic differential equations have solutions. The additional complexity notwithstanding, the solution is somewhat different. If you solve an ordinary differential equation, for example, you have found the deterministic function, or family of functions, that satisfies it. If you solve a stochastic differential equation, conversely, you have identified the stochastic process, or family of stochastic processes, that fulfils it.

Given a stochastic process, we can explore its distributional properties. This is how we will gain insight into geometric Brownian motion. Re-arranging equation 5.27, we arrive at

$$\underbrace{\ln\left(\frac{X_T}{X_t}\right)}_{\text{Log return}} = \left(\mu - \frac{\sigma^2}{2}\right)(T - t) + \sigma\xi, \quad (5.28)$$

where we have replaced the ugly Wiener increment with its distributional equivalent, $\xi \sim \mathcal{N}(0, T - t)$. In other words, equation 5.28 tells us that the log difference of X over the interval, $[t, T]$, is normally distributed. This conclusion implies, by definition, that X is log-normally distributed. In the Merton (1974) setting—and in virtually all financial applications involving geometric Brownian motion—this is equivalent to assuming that asset returns are Gaussian, whereas the asset values themselves are log-normally distributed. The log-normal assumption is useful for assets, because typically assets cannot take negative values; they are generally bounded by zero. Furthermore, geometric Brownian motion indicates that uncertainty about asset values and returns increases the longer the time horizon under investigation. Ultimately, mathematical details aside, these are all fairly reasonable conceptual assumptions.

Let's be quite literal about these results. We now, after some pain, know much more about how X_t behaves. In particular, we have that

$$\ln\left(\frac{X_T}{X_t}\right) \sim \mathcal{N}\left(\left(\mu - \frac{\sigma^2}{2}\right)(T - t), \sigma^2(T - t)\right). \quad (5.29)$$

Asset returns are thus assumed to be Gaussian. Note, however, that the drift term now includes an additional $\frac{\sigma^2}{2}(T - t)$ term relative to the deterministic setting. Both

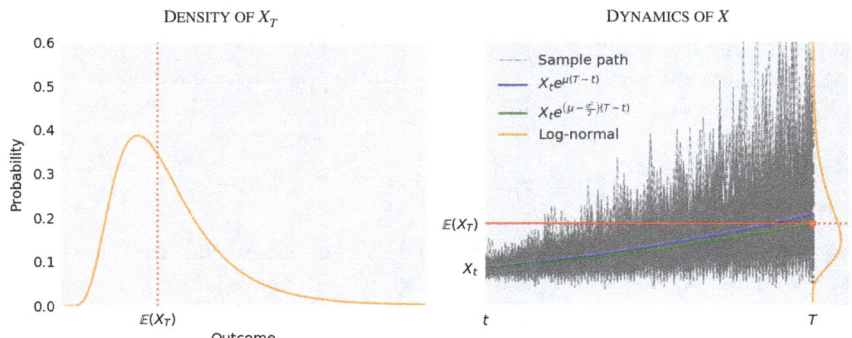

Fig. 5.4 *SDE behaviour*: Fig. 5.2 demonstrates the behaviour of the ordinary-differential element of geometric Brownian motion. These two figures describe the density and dynamics when the diffusion component is included; this dramatically changes the story.

the expected asset return and volatility contribute to the expected terminal asset outcome. The direct corollary is that

$$X_T \sim \ln-\mathcal{N}\left(\ln(X_t) + \left(\mu - \frac{\sigma^2}{2}\right)(T-t), \sigma^2(T-t)\right), \quad (5.30)$$

where,

$$f_{X_t}(x; \mu, \sigma) = \frac{1}{\sqrt{2\pi}\sigma x} e^{\frac{-(\ln(x)-\mu)^2}{2\sigma^2}}, \quad (5.31)$$

is the density of the log-normal distribution. Assets themselves obey a log-normal law. Figure 5.4 demonstrates this behaviour in a graphical manner. Compared with Fig. 5.3, there is much more going on. The left-hand graphic outlines the log-normal (positive-valued) asset distribution. The expectation of X_T is overlaid to highlight the lower bound and the right skew. The right-hand graphic examines a large number of sample paths for X over the interval, $[0, T]$. The deterministic trend—both the deterministic and stochastic-adjusted drifts—is provided for context and to invite comparison with Fig. 5.3. At the far right, we can see the log-normal return distribution. Quite naturally, the expected value of X_T coincides with its deterministic path.

At its heart, therefore, the use of geometric Brownian motion implies a deterministic asset-return trend buffeted by random noise, associated Gaussian asset returns, and positive-valued log-normal asset values. More mathematical care is required, but conceptually, we are working with a defensible object. One can, of course, legitimately argue about the defensibility of assuming normally distributed financial-asset return and many papers in the financial literature have done so. Let's treat it, however, as a reasonable first-order approximation. The Gaussianity will also turn out, as in the Gaussian threshold setting, not to play such an important role.

5.3 Multiple Obligors

After a short detour to explore the basic idea behind geometric Brownian motion, we can return to the problem at hand. It is now necessary to generalize the preceding results to consider N distinct obligors. Merton (1974) postulated that each asset-value process follows a separate stochastic differential equation on $(\Omega, \mathcal{F}, \mathbb{P})$ of the form,

$$dA_{n,t} = \mu_n A_{n,t} dt + \sigma_n A_{n,t} d\hat{W}_{n,t}, \tag{5.32}$$

for $n = 1, \ldots, N$ over the interval, $[t, T]$. The solution is the collection of stochastic processes,

$$A_{n,T} = A_{n,t} e^{\left(\mu_n - \frac{\sigma_n^2}{2}\right)(T-t) + \sigma_n \left(\hat{W}_{n,T} - \hat{W}_{n,t}\right)}, \tag{5.33}$$

for $n = 1, \ldots, N$, implying that each asset value is log-normally distributed, but the asset returns are Gaussian. We define the nth default event as,

$$\mathbb{I}_{\mathcal{D}_n} = \mathbb{I}_{\{\tau_n \leq T\}}, \tag{5.34}$$
$$= \mathbb{I}_{\{A_{n,T} \leq K_n\}},$$

for $n = 1, \ldots, N$ where K_n represents the liability threshold for the nth firm and τ_n is the nth first-passage time. Armed with knowledge of $A_{n,T}$, its distribution, and the behaviour of $\hat{W}_t$, we can move to the conditional default probability.

It might not feel that way, but we also have useful information about the default event. In particular, we have already derived the conditional default probability as a function of the distance-to-default. This is generalized to the N-firm case as,

$$\mathbb{P}_t (\tau_n = T) = \mathbb{P}_t \left(A_{n,T} \leq K_n\right), \tag{5.35}$$

$$= \mathbb{P}_t \left(\underbrace{A_{n,t} e^{\left(\mu_n - \frac{\sigma_n^2}{2}\right)(T-t) + \sigma_n \left(\hat{W}_{n,T} - \hat{W}_{n,t}\right)}}_{\text{Equation 5.33}} \leq K_n \right),$$

$$= \mathbb{P}_t \left(\underbrace{\frac{\hat{W}_{n,T} - \hat{W}_{n,t}}{\sqrt{T-t}}}_{\mathcal{N}(0,1)} \leq \underbrace{\frac{\ln\left(\frac{K_n}{A_{n,t}}\right) - \left(\mu_n - \frac{\sigma_n^2}{2}\right)(T-t)}{\sigma_n \sqrt{T-t}}}_{\text{Distance to default: } \delta_{n,T-t}} \right),$$

$$\mathbb{P}_t (\mathcal{D}_n) = \underbrace{\Phi\left(\delta_{n,T-t} \left(\mu_n, \sigma_n, A_{n,t}, K_n, T-t\right)\right)}_{\Phi(\delta_{n,T-t})}.$$

Conceptually, this is not so different from the Gaussian threshold model. Our default trigger is fairly complex and, using equation 5.35, we can extend our default-event definition from equation 5.34 to include another twist as,

$$\mathbb{I}_{\mathcal{D}_n} = \mathbb{I}_{\{A_{n,T} \leq K_n\}}, \qquad (5.36)$$
$$= \mathbb{I}_{\{\xi_n \leq \delta_{n,T-t}\}}.$$

In other words, it is equivalent to write the default event as the terminal asset-value of the nth obligor falling below its liability trigger or a standard normal variate, $\xi_n \sim \mathcal{N}(0, 1)$, taking a value below the distance to default. The former occurs in asset space, whereas the latter works with normalized values. In both cases, the uncertainty surrounding the default outcome still arises from the Wiener increment.

How does default dependence enter into this model? Since the only source of uncertainty relates to the Wiener increments, their interaction must drive default dependence. Quite simply, the diffusion elements of our asset processes are correlated. In fact, we have that

$$\left(\frac{1}{\sqrt{T-t}}\right) \begin{bmatrix} \hat{W}_{1,T} - \hat{W}_{1,t} \\ \hat{W}_{2,T} - \hat{W}_{2,t} \\ \vdots \\ \hat{W}_{N,T} - \hat{W}_{N,t} \end{bmatrix} \equiv \underbrace{\begin{bmatrix} \xi_1 \\ \xi_2 \\ \vdots \\ \xi_N \end{bmatrix}}_{\xi} \sim \mathcal{N}\left(\begin{bmatrix} 0 \\ 0 \\ \vdots \\ 0 \end{bmatrix}, \underbrace{\begin{bmatrix} 1 & \rho_{12} & \cdots & \rho_{1N} \\ \rho_{21} & 1 & \cdots & \rho_{2N} \\ \vdots & \vdots & \ddots & \vdots \\ \rho_{N1} & \rho_{N2} & \cdots & 1 \end{bmatrix}}_{\Omega_N}\right) \qquad (5.37)$$

The vector of correlated and normalized Wiener increments, ξ, drives dependence. We can think of the assets of various firms having their own expected returns, volatilities, and capital structures. The overall system of assets, however, are correlated and this creates default dependence. One can imagine a large number of systematic and idiosyncratic factors influencing these asset values. All of these interactions are summarized in Ω_N. As a consequence, it potentially requires large amounts of information for its determination. A critical part of the model implementation will involve disentangling all of the complexity embedded in this object and parametrizing Ω_N.

5.3.1 Two Choices

Whether one normalizes the default event through use of the distance-to-default or works directly with the asset values, information is required. To be more specific, asset values, expectations, volatilities, and correlations are needed. None of these values are, given the unobservability of a firm's assets, particularly easy to identify. Perhaps more important, however, is the liability threshold, K_n. In the

5.4 The Indirect Approach

threshold setting considered in Chap. 4, the threshold was indirectly identified by the unconditional probability. Given the complexity of the distance-to-default, this approach does not immediately appear to be available to us.

In the remainder of this chapter, we will examine two practical approaches for the determination of K_n—a direct and indirect approach.[15] This choice has important implications for the actual model implementation. Briefly, the two methods are summarized as:

Direct Method   One attempts to infer the value of K_n from balance-sheet information and uses market information to estimate the distance-to-default or the separate asset-based variables comprising it.[16]

Indirect Method   One could employ the unconditional default probability, computed separately using the techniques described in Chap. 9, to infer the value of K_n.

The direct approach is, quite likely, closer in spirit to what Merton (1974) had in mind. It looks carefully at the capital structure of one's obligors and extracts the necessary information to inform the credit risk in one's portfolio. This is, however, a complex, non-trivial undertaking. The indirect approach, after a bit of manipulation, is rather more straightforward. This will, therefore, be our starting point. Indeed, as we'll see in the next section, it establishes a clear and useful link with the Gaussian threshold model considered in Chap. 4.

5.4 The Indirect Approach

The distance-to-default and unconditional default probabilities, as we've previously indicated, hold essentially the same information. They merely arrive at it in different ways using alternative information sources. Default probabilities, as we'll see in Chap. 9, capture this information from rating transitions. The distance-to-default is inferred from market data. As a consequence, it would seem interesting to assume their equality. We could assume that all information was known, save the liability threshold, and infer its value. Thus, we are basically proposing an algebraic exercise to find K_n.

To be more specific, imagine that we knew the value of the assets, their volatility, their return, and the default probability. Using this information, we could work out the default barrier. To do this, we set

$$p_n = \mathbb{P}_t \left(A_{n,T} \leq K_n \right), \quad (5.38)$$
$$\equiv \Phi(\delta_{n,T-t}).$$

---

[15] There may be other possibilities, but these seem to logically incorporate the two main perspectives.

[16] This can be done directly or one might employ an external firm such as KMV or Kamakura corporation to do this on your behalf.

This is the jumping-off point. The rest is simply a bit of algebraic manipulation,

$$p_n = \mathbb{P}_t\left(A_{n,T} \leq K\right), \tag{5.39}$$

$$= \Phi\underbrace{\left(\frac{\ln\left(\frac{K_n}{A_{n,t}}\right) - \left(\mu_n - \frac{\sigma_n^2}{2}\right)(T-t)}{\sigma_n\sqrt{T-t}}\right)}_{\text{Equation 5.35}},$$

$$\Phi^{-1}(p_n) = \Phi^{-1}\left(\Phi\left(\frac{\ln\left(\frac{K_n}{A_{n,t}}\right) - \left(\mu_n - \frac{\sigma_n^2}{2}\right)(T-t)}{\sigma_n\sqrt{T-t}}\right)\right),$$

$$= \frac{\ln\left(\frac{K_n}{A_{n,t}}\right) - \left(\mu_n - \frac{\sigma_n^2}{2}\right)(T-t)}{\sigma_n\sqrt{T-t}},$$

$$\ln\left(\frac{K_n}{A_{n,t}}\right) = \left(\mu_n - \frac{\sigma_n^2}{2}\right)(T-t) + \Phi^{-1}(p_n)\sigma_n\sqrt{T-t},$$

$$K_n = A_{n,t}e^{\left(\mu_n - \frac{\sigma_n^2}{2}\right)(T-t) + \Phi^{-1}(p_n)\sigma_n\sqrt{T-t}}.$$

This makes logical sense. It is basically the distance from $A_{n,t}$ taking into account the probability of default, the volatility of the assets, and the amount of time under consideration. It resembles the solution to geometric Brownian motion without the random element. One can, if it helps, think of this quantity as the magnitude of the shock to asset values, implicit from p_n, necessary to induce default.

To make this a bit more concrete, Table 5.1 illustrates equation 5.39 through a numerical example. We consider a simple, but illustrative three-month time interval—that is, $T - t = 0.25$—with a starting asset value of \$100, an expected

Table 5.1 Indirect K_n: If you are willing to assume asset values, returns, and volatilities—$A_{n,t}$, μ_n and σ_n—then one can infer the value of K_n. The underlying table summarizes different values of K_n associated with different input values.

Assumed			Known	Derived	
$A_{n,t}$	μ_n	σ_n	$p_n \equiv \mathbb{P}_t\left(\mathbb{I}_{\mathcal{D}_n}\right)$	K_n	$\delta_{n,T-t}$
\$100	1.00%	15.00%	10 bps.	\$79.29	−3.09
\$100	1.00%	15.00%	50 bps.	\$82.41	−2.58
\$100	1.00%	15.00%	100 bps.	\$83.96	−2.33
\$100	1.00%	15.00%	200 bps.	\$85.70	−2.05
\$100	1.00%	15.00%	500 bps.	\$88.37	−1.64
\$100	1.00%	30.00%	10 bps.	\$62.36	−3.09
\$100	1.00%	30.00%	50 bps.	\$67.36	−2.58
\$100	1.00%	30.00%	100 bps.	\$69.93	−2.33
\$100	1.00%	30.00%	200 bps.	\$72.85	−2.05
\$100	1.00%	30.00%	500 bps.	\$77.45	−1.64

5.4 The Indirect Approach

Algorithm 5.1 Determining K_n and $\delta_{n,T-t}$

```
def getK(mu,sigma,dt,A,myP):
    t1 = (mu-0.5*(np.power(sigma,2)))*dt
    t2 = np.multiply(np.multiply(norm.ppf(myP),sigma),np.sqrt(dt))
    return np.multiply(A,np.exp(t1+t2))
def getDelta(mu,sigma,dt,A,K):
    t1 = np.log(K/A)
    t2 = (mu-0.5*(sigma**2))*dt
    return np.divide(t1-t2,sigma*np.sqrt(dt))
```

asset return of 1% and a variety of default probabilities and asset volatility values. We use the assumed values and known (or estimated) default probabilities to infer the liability threshold. Table 5.1 also includes the distance-to-default associated with the assumed, estimated, and derived quantities.

We have considered *two* distinct asset volatility values: 15 and 30%. As one would expect, the distance of the barrier below the starting value increases as the asset volatility increases. The idea is that if the asset value is very volatile, then large swings are more frequent. This is taken into account when setting the barrier.[17] For a given level of volatility, increases in the default probability move the barrier closer to the starting value. This is sensible, since it suggests that the greater the probability of default, the closer the firm is to default and the less tolerance the firm has for reduction in the value of its assets. It is reassuring to observe that, irrespective of the choice of σ_n, the default probabilities and distance-to-default have the same relationship to one another. This was our starting assumption from equation 5.38.

Algorithm 5.1 summarizes the two simple Python functions used, given the appropriate arguments, for the computation of the K_n and $\delta_{n,T-t}$ values. The functions getK and getDelta follow directly from equations 5.39 and 5.35. Relatively little effort is required for the implementation of these functions.

These functions, particularly getDelta, can, and will, be used in both the indirect and direct settings.

5.4.1 A Surprising Simplification

While the inference of K_n in equation 5.39 is interesting, it does not seem to immediately improve our situation. An interesting observation can, however, be made. Let us return to our expression for the default probability in equation 5.35. In the threshold setting, we placed the default trigger— defined as $\Phi^{-1}(p_n)$—directly back into the definition of the default event. This was sensible since it eliminates the

---

[17] The firm also presumably takes its asset volatility into account when taking its capital structure choices.

250   5 The Genesis of Credit-Risk Modelling

unknown K_n variable with a known quantity. Why not do the same thing here? We will substitute our new inferred definition of K_n in equation 5.39 into the default condition embedded in the default probability definition. This yields,

$$\mathbb{P}_t\left(\mathbb{I}_{\mathcal{D}_n}\right) = \mathbb{P}_t\left(\underbrace{\frac{\hat{W}_{n,T} - \hat{W}_{n,t}}{\sqrt{T-t}}}_{\xi_n} \leq \frac{\ln\left(\frac{K_n}{A_{n,t}}\right) - \left(\mu_n - \frac{\sigma_n^2}{2}\right)(T-t)}{\sigma_n \sqrt{T-t}}\right), \tag{5.40}$$

$$= \mathbb{P}_t\left(\xi_n \leq \frac{\ln\left(\overbrace{\frac{A_{n,t}e^{\left(\mu_n - \frac{\sigma_n^2}{2}\right)(T-t) + \Phi^{-1}(p_n)\sigma_n\sqrt{T-t}}}{A_{n,t}}}^{\text{Equation 5.39}}\right) - \left(\mu_n - \frac{\sigma_n^2}{2}\right)(T-t)}{\sigma_n\sqrt{T-t}}\right),$$

$$= \mathbb{P}_t\left(\xi_n \leq \frac{\ln\left(\frac{\cancel{A_{n,t}}e^{\cancel{\left(\mu_n - \frac{\sigma_n^2}{2}\right)(T-t)} + \Phi^{-1}(p_n)\sigma_n\sqrt{T-t}}}{\cancel{A_{n,t}}}\right) - \cancel{\left(\mu_n - \frac{\sigma_n^2}{2}\right)(T-t)}}{\sigma_n\sqrt{T-t}}\right),$$

$$= \mathbb{P}_t\left(\xi_n \leq \frac{\Phi^{-1}(p_n)\cancel{\sigma_n\sqrt{T-t}}}{\cancel{\sigma_n\sqrt{T-t}}}\right),$$

$$= \mathbb{P}_t\left(\xi_n \leq \Phi^{-1}(p_n)\right).$$

An enormous number of terms—including all reference to the asset value, return, and volatility—simply cancel out. What we are left with looks essentially equivalent to the Gaussian threshold model.

All of the complexity relating to the underlying stochastic process describing the asset dynamics has disappeared. The catalyst was the inference of the liability threshold through the unconditional default probability. This is thus the equivalence condition for the Merton (1974) and the Gaussian threshold models.

5.4.2 Inferring Key Inputs

Let's try to use these ideas of indirect inference of the liability threshold to our, now quite familiar, 100-counterparty example. As we've seen from equation 5.40, we do not need the asset values, returns, and volatilities. It is nevertheless enlightening to have some insight into the implied liability threshold and distance-to-default values. Since we have only unconditional default probabilities and exposures, if we wish to do some additional comparison, we will need to make some additional assumptions.

To keep things simple—and since we have no reasonable basis to do anything else—we assume fixed asset value, return and volatility settings for all obligors. More specifically, we assume that the asset return associated with all counterparties in our portfolio is 1%, the asset-return volatility is 10%, and the shared starting asset value is $100. Table 5.2 summarizes these results, along with the average outcomes across all obligors in the portfolio. Given an average unconditional default probability of about 1% and the other settings, the mean liability threshold is roughly $78. Normalized as the distance-to-default, this is equal to about 2.6 standard deviations.

Using our assumptions and the code from Algorithm 5.1, we were able to determine, for each obligor, the implied liability threshold and distance-to-default. Since we've basically pulled asset information out of thin air, the results are somewhat naive. They certainly do not represent any market insight into one's problem. That important caveat aside, Fig. 5.5 does provide some insight into the relationship between these inferred quantities and the underlying default probabilities.

The left-hand graphic of Fig. 5.5 explores the relationship between unconditional default probabilities and the distance-to-default. Recall from Chap. 1 that our default probability values range from close to zero to almost 7%; the average is approximately 1%. The associated distance-to-default value span the range from 1.5 to 4+ standard deviations. As one would expect, the lower the default probability, the greater the distance to default. The right-hand graphic conversely jointly displays default probabilities and the liability threshold. For very small default probabilities, the K_n value approaches $65. As p increases to over 0.06, the liability threshold increases to more than $85. In both cases, the relationships are non-linear. Given

Table 5.2 *Our portfolio in the Merton world*: The underlying table summarizes the average values associated with our ongoing 100-obligor portfolio example in the Merton setting.

Quantity	Notation	Mean value
Number of counterparties	N	100
Time interval	$T - t$	1 year
Counterparty exposure	c_n	$10.0
Asset value	$A_{n,0}$	$100.0
Asset return	μ_n	1.0%
Asset volatility	σ_n	10.0%
Default probability	p_n	1.0%
Default threshold	K_n	$77.7
Distance-to-default	$\delta_{n,T-t}$	−2.6

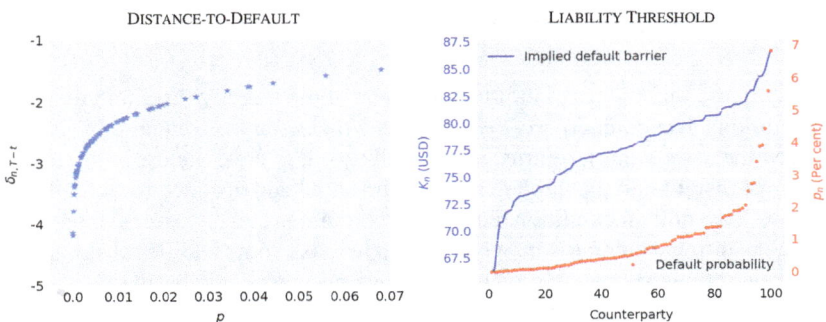

Fig. 5.5 *Inferred quantities*: Given the (assumed) asset-return information in Table 5.2, we can use p_n to infer K_n and the distance-to-default. The underlying graphics examine the relationship, on an obligor level, between these two inferred quantities and the default probabilities.

that all of the key asset inputs are fixed, these are essentially two different ways to express the unconditional default probability.

Having collected all of the basic inputs, we wish to compare the implementation of the indirect Merton (1974) model to the one-factor Gaussian threshold approach. The only missing element is an assumption regarding the correlation between the ξ_n terms in equation 5.40. This basically means populating the off-diagonal elements of Ω_N from equation 5.37. To permit ready comparison to the Gaussian-threshold setting, we simply set

$$\rho_{nm} = \rho(\xi_n, \xi_m), \qquad (5.41)$$
$$= 0.2,$$

for $n, m = 1, \ldots, N$. This convenient assumption allows us to directly compare to a set of $\rho = 0.20$ for the one-factor Gaussian threshold model. Again, it bears repeating, no market information is embedded in this choice. Use of the indirect approach has eliminated much of the flexibility, along with the complexity, of the Merton (1974) framework.

5.4.3 Simulating the Indirect Approach

Our indirect approach has reduced the computation of risk measures into an exercise in simulating a large collection of correlated multivariate Gaussian random variates. There are a number of statistical software packages available to perform this task, but it is often preferable to do the work directly. We will make use of a well-known practical statistical trick.

5.4 The Indirect Approach

Our problem is to generate $\xi \sim \mathcal{N}(0, \Omega_N)$. In virtually all scientific softwares, it is straightforward to generate $Z \sim \mathcal{N}(\mathbf{0}, I_N)$, where I_N represents the N-dimensional identity matrix.[18] Z is thus an N-dimensional standard normal random vector. We want an N-dimensional *correlated* normal random vector. Basically, we want to replace the covariance matrix, I_N, with Ω_N. This can be accomplished with a surprisingly simple computation. The statistical trick involves performing a spectral decomposition of the covariance matrix as

$$\Omega_N = UDU^T, \tag{5.42}$$

where U is a matrix of eigenvectors and D is a diagonal matrix of eigenvalues—see Golub and Loan (2012) for more information. It is also possible to use the Cholesky decomposition, but this requires the covariance matrix to be positive definite.[19] By definition Ω_N is positive definite, but occasionally, when N becomes very large, this condition can be numerically violated.[20]

We now use the outputs of the spectral decomposition to define $\xi = U\sqrt{D}Z$. This is a new N-dimensional random vector. Let's see if it fits the bill. The variance of ξ is,

$$\begin{aligned}
\text{var}(\xi) &= \text{var}\left(U\sqrt{D}Z\right), \qquad (5.43) \\
&= U\sqrt{D}\,\underbrace{\text{var}(Z)}_{I_N}\,\sqrt{D}U^T, \\
&= \underbrace{UDU^T}_{\text{Equation 5.42}}, \\
&= \Omega_N,
\end{aligned}$$

which is precisely as desired. This provides use with a simple recipe to construct $\xi \sim \mathcal{N}(\mathbf{0}, \Omega_N)$.

We now have all the necessary elements for implementation. Algorithm 5.2 summarizes the Python code. The first step is to generate a large number of standard-normal random variates: these are stored in Z. We then perform the spectral (or eigenvalue) decomposition of Ω_N and construct the correlated system of multi-variate normal random outcomes; we denote it as xi. The remainder of the code should look very familiar. We construct the liability threshold, K, check the default

---

[18] If not, one can always use the Box-Muller method. See, for example, Fishman (1995, Chapter 3).
[19] More generally, one may also use the so-called singular value decomposition; this useful alternative, for large and potentially corrupted systems, is outlined in Appendix A.
[20] For more information on the Cholesky decomposition, see Golub and Loan (2012) or Press et al. (1992, Chapter 11).

Algorithm 5.2 Simulating the indirect Merton model

```
def mertonIndirectSimulation(N,M,p,Omega,c,alpha):
    Z = np.random.normal(0,1,[M,N])
    w,v = anp.eigh(Omega)
    H = np.dot(v,np.sqrt(np.diag(w)))
    xi = np.dot(Z,np.transpose(H))
    K = norm.ppf(p)*np.ones((M,1))
    lossIndicator = 1*np.less(xi,K)
    lossDistribution = np.sort(np.dot(lossIndicator,c),axis=None)
    el,ul,var,es=util.computeRiskMeasures(M,lossDistribution,alpha)
    return el,ul,var,es
```

Table 5.3 *Indirect Merton results*: The underlying table describes the usual summary risk statistics for the indirect implementation of the Merton (1974) model. It also includes the one-factor Gaussian threshold model for comparison; there is very little difference.

Quantile	Gaussian		Merton	
	$\widehat{\text{VaR}}_\alpha$	$\widehat{\mathcal{E}_\alpha}$	$\widehat{\text{VaR}}_\alpha$	$\widehat{\mathcal{E}_\alpha}$
95.00th	$44.2	$67.0	$44.2	$67.1
97.00th	$55.4	$78.8	$55.4	$78.9
99.00th	$80.1	$106.2	$80.2	$106.4
99.50th	$97.3	$124.8	$97.6	$125.1
99.90th	$141.8	$172.5	$142.2	$172.7
99.97th	$189.5	$223.1	$190.2	$223.2
99.99th	$211.7	$246.7	$212.7	$246.8
Other statistics				
Expected loss	$9.2		$9.2	
Loss volatility	$17.7		$17.8	
Total iterations	10,000,000			
Computational time	52.5		94.3	

condition, `np.less(xi,K)`, and then call the usual `computeRiskMeasures` to generate our desired risk measures.

The results, for our sample portfolio, are summarized in Table 5.3. For comparison, we have included the comparable one-factor Gaussian threshold model with $\rho = 0.20$.

There is essentially no difference between the indirect Merton (1974) and Gaussian-threshold results. Expected loss and loss volatility are basically identical. At the 99.99th percentile, we observe some small differences of a few units of currency. This should be considered immaterial simulation noise.

Figure 5.6 takes the results in Table 5.3 a step further by plotting the tail probabilities over the interval ranging from [0.9500, 0.9999]. It requires a careful look to identify two distinct lines. Visually, therefore, the two models are indistinguishable. These are, practically speaking, the same model.

5.5 The Direct Approach

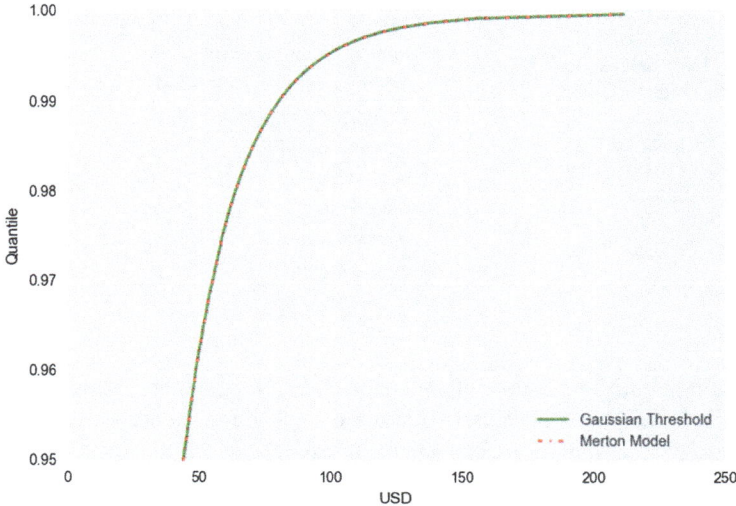

Fig. 5.6 *Comparing tail probabilities*: If one uses default probabilities to infer the default barrier, as done in Fig. 5.5, the Merton (1974) approach reduces to a one-factor threshold model. The figure below indicates that, visually, there is no difference in the tail probabilities.

We conclude, rather uncontroversially, that the Merton (1974) was, in fact, the driving motivation behind the threshold modelling framework described in Chap. 4. If, instead of using market information to inform the distance-to-default, one infers the liability threshold from the default probability, much of the mathematical complexity of Merton (1974) dissipates. What remains is a Gaussian threshold model.

One may not wish to make this simplification and, in the spirit of Merton (1974), seek to employ market and firm capital-structure information to determine one's credit risk. In this case, additional effort and thought is required. Performing this task will consume the remainder of this chapter.

5.5 The Direct Approach

The default event is, in the direct case, written as $\{A_{n,T} \leq K_n\}$ for $n = 1, \ldots, N$. $A_{n,T}$ is a function of the initial asset value, the return characteristics of the assets, and the random outcome of a Wiener process. Given this value, it must be compared to the value of the firm's liabilities. If we want to use the direct approach to the Merton (1974) model, therefore, we will need to grapple with market data. In practice, asset values, returns, and volatility are *not* observable. The level of the firm's debt, however, is observable—it can be extracted from the firm's balance sheet. In its simplest form, however, Merton (1974) makes some rather unrealistic

Table 5.4 *Challenges of the direct approach*: The underlying table details the key variables required for direct implementation of the Merton (1974) approach. It also highlights the challenges associated with their determination.

Quantity	Notation	Status
Number of counterparties	N	Known
Time interval	$T - t$	Known
Counterparty exposure	c_n	Known
Default probability	p_n	Known
Asset value	$A_{n,t}$	Unobservable
Asset return	μ_n	Unobservable
Asset volatility	σ_n	Unobservable
Asset correlation	ρ_{nm}	Unobservable
Default threshold	K_n	Unobservable

assumptions about the nature of the debt claim. It assumes a single debt issue with a fixed maturity. The capital structure of real-life firms are significantly more complex. These assumptions can all be relaxed, albeit at the cost of additional complexity.[21]

There is a significant and inevitable data collection and calibration effort. Table 5.4 provides a useful summary of the quantities used in our analysis. It also indicates their status. It reveals that *five* key elements are *not* observable. If we are going to implement the direct approach, we require a strategy for identifying, estimating, or inferring these values.

There is another, purely mathematical, challenge that we face if we are to implement the direct approach. Even if we had all the necessary data, we need to generate a system of correlated asset values. If we define,

$$A_T = \begin{bmatrix} A_{1,T} \\ \vdots \\ A_{N,T} \end{bmatrix}, \qquad (5.44)$$

then we need to be in a position to generate random draws from $A_T \sim \mathcal{N}(\bar{A}_T, \Omega_{A_T})$ where

$$\bar{A}_T = \begin{bmatrix} \mathbb{E}_t(A_{1,T}) \\ \vdots \\ \mathbb{E}_t(A_{N,T}) \end{bmatrix}, \qquad (5.45)$$

---

[21] Geske (1977), for example, is the first in a series of papers that seeks to incorporate a greater degree of realism in the Merton (1974) setting.

5.5 The Direct Approach

and

$$\Omega_{A_T} = \begin{bmatrix} \text{var}_t\left(A_{1,T}\right) & \text{cov}_t\left(A_{1,T}, A_{2,T}\right) & \cdots & \text{cov}_t\left(A_{1,T}, A_{N,T}\right) \\ \text{cov}_t\left(A_{2,T}, A_{1,T}\right) & \text{var}_t\left(A_{2,T}\right) & \cdots & \text{cov}_t\left(A_{2,T}, A_{N,t}\right) \\ \vdots & \vdots & \ddots & \vdots \\ \text{cov}_t\left(A_{N,T}, A_{1,T}\right) & \text{cov}_t\left(A_{2,T}, A_{N,T}\right) & \cdots & \text{var}_t\left(A_{N,T}\right) \end{bmatrix}. \quad (5.46)$$

This may appear relatively straightforward, but each $A_{n,T}$ follows a geometric Brownian motion and some effort is required to deduce the various expectations, variances, and covariance required to build equations 5.44 to 5.46. This will be our first order of business. Once we have derived these expressions, then we will turn our attention to the important task of identifying the necessary, and unobservable, quantities from Table 5.4.

This provides a fairly clear work program. We need to perform the effort necessary to obtain the following objects:

1. Determine the expected value for each $A_{n,t}$.
2. Compute the variance (and volatility) of each $A_{n,t}$.
3. Derive an expression for the covariance (and correlation) for each pair, $(A_{n,t}, A_{m,t})$.
4. Calculate the default correlation implicit in this model.

We will, more or less, use brute force to derive the desired objects from the first principles of the Merton (1974) model. This is, admittedly, somewhat painful, but it is rarely shown in the literature and, hopefully, will have concrete value for practitioners seeking deeper understanding of this approach.

5.5.1 Expected Value of $A_{n,T}$

The first order of business is to determine the expected value of A_n at time T. This is an important quantity since, unlike the threshold setting, it is not zero. That implies it does not vanish in our variance and covariance computations. It follows from the definition of expectation and the structure of the model,

$$\mathbb{E}_t^{\mathbb{P}}\left(A_{n,T}\right) = \mathbb{E}_t^{\mathbb{P}}\left(A_{n,t} e^{\left(\mu_n - \frac{\sigma_n^2}{2}\right)(T-t) + \sigma_n\left(\hat{W}_{n,T} - \hat{W}_{n,t}\right)}\right), \quad (5.47)$$

$$= A_{n,t} e^{\left(\mu_n - \frac{\sigma_n^2}{2}\right)(T-t)} \mathbb{E}_t^{\mathbb{P}}\left(e^{\sigma_n\left(\hat{W}_{n,T} - \hat{W}_{n,t}\right)}\right).$$

Note that we have explicitly included the physical probability measure, $\mathbb{P}$, in the expectation operator. This is not terribly common, but it underscores our risk-management perspective. When computing risk, we are not operating in the risk-neutral setting, but instead in the real world. As a consequence, we need to work with the asset moments under $\mathbb{P}$.

The second term on the right-hand side of equation 5.47 is a bit troublesome. Let us set $X = \sigma_n \left(\hat{W}_{n,T} - \hat{W}_{n,t} \right)$ and note that $X \sim \mathcal{N}(0, \sigma_n^2(T - t))$. As a consequence,

$$\mathbb{E}_t^{\mathbb{P}} \left(e^X \right) = \int_{-\infty}^{\infty} e^u f_X(u) du, \qquad (5.48)$$

where $f_X(u)$ denotes the density function of X. Consequently, we have

$$\int_{-\infty}^{\infty} e^u f_X(u) du = \int_{-\infty}^{\infty} e^u \frac{1}{\sqrt{2\pi \sigma_n^2(T-t)}} e^{\frac{-u^2}{2\sigma_n^2(T-t)}} du, \qquad (5.49)$$

$$= \int_{-\infty}^{\infty} \frac{1}{\sqrt{2\pi \sigma_n^2(T-t)}} e^{u - \frac{u^2}{2\sigma_n^2(T-t)}} du,$$

$$= \int_{-\infty}^{\infty} \frac{1}{\sqrt{2\pi \sigma_n^2(T-t)}} e^{-\frac{1}{2\sigma_n^2(T-t)} \left(u^2 - 2\sigma_n^2(T-t)u \right)} du,$$

$$= \int_{-\infty}^{\infty} \frac{1}{\sqrt{2\pi \sigma_n^2(T-t)}} e^{-\frac{1}{2\sigma_n^2(T-t)} \left(u^2 - 2\sigma_n^2(T-t)u + \underbrace{\sigma_n^4(T-t)^2 - \sigma_n^4(T-t)^2}_{=0} \right)} du,$$

$$= \int_{-\infty}^{\infty} \frac{1}{\sqrt{2\pi \sigma_n^2(T-t)}} e^{\frac{\sigma_n^4(T-t)^2}{2\sigma_n^2(T-t)}} e^{-\frac{1}{2\sigma_n^2(T-t)} \left((u - \sigma_n^2(T-t))^2 \right)} du,$$

$$= e^{\frac{\sigma_n^2(T-t)}{2}} \underbrace{\int_{-\infty}^{\infty} \frac{1}{\sqrt{2\pi \sigma_n^2(T-t)}} e^{-\frac{(u - \sigma_n^2(T-t))^2}{2\sigma_n^2(T-t)}} du}_{=1},$$

$$\mathbb{E}_t^{\mathbb{P}} \left(e^{\sigma_n (\hat{W}_{n,T} - \hat{W}_{n,t})} \right) = e^{\frac{\sigma_n^2(T-t)}{2}}.$$

This trick, referred to as completing the square in the exponent, allows one to rewrite a new density function, which, by definition, evaluates to unity. Plugging this result back into equation 5.47, we have

$$\mathbb{E}_t^{\mathbb{P}} (A_{n,T}) = A_{n,t} e^{\left(\mu_n - \frac{\sigma_n^2}{2} \right)(T-t)} \underbrace{e^{\frac{\sigma_n^2(T-t)}{2}}}_{\substack{\text{Equation} \\ 5.49}}, \qquad (5.50)$$

$$= A_{n,t} e^{\mu_n (T-t)}.$$

5.5 The Direct Approach

The asset value, in expectation, is assumed to grow exponentially with respect to time at a rate of μ_n. While not a terribly surprising result, this will prove useful in determining the other required quantities.

5.5.2 Variance and Volatility of $A_{n,T}$

The next step is to determine the variance of the asset value. Again, starting from first principles, the variance of $A_{n,T}$ can be written as,

$$\text{var}_t\left(A_{n,T}\right) = \mathbb{E}_t^{\mathbb{P}}\left(A_{n,T}^2 - \mathbb{E}_t^{\mathbb{P}}(A_{n,T})^2\right), \tag{5.51}$$

$$= \mathbb{E}_t^{\mathbb{P}}\left(A_{n,T}^2 - \underbrace{\left(A_{n,t}e^{\mu_n(T-t)}\right)^2}_{\text{Equation 5.50}}\right),$$

$$= \mathbb{E}_t^{\mathbb{P}}\left(A_{n,T}^2\right) - A_{n,t}^2 e^{2\mu_n(T-t)}.$$

The tricky element is the first term, $\mathbb{E}_t^{\mathbb{P}}\left(A_{n,T}^2\right)$. It requires some additional attention. A bit of manipulation reveals,

$$\mathbb{E}_t^{\mathbb{P}}\left(A_{n,T}^2\right) = \mathbb{E}_t^{\mathbb{P}}\left(\underbrace{\left(A_{n,t}e^{\left(\mu_n - \frac{\sigma_n^2}{2}\right)(T-t) + \sigma_n\left(\hat{W}_{n,T} - \hat{W}_{n,t}\right)}\right)^2}_{\text{Equation 5.33}}\right), \tag{5.52}$$

$$= \mathbb{E}_t^{\mathbb{P}}\left(A_{n,t}^2 e^{2\left(\mu_n - \frac{\sigma_n^2}{2}\right)(T-t) + 2\sigma_n\left(\hat{W}_{n,T} - \hat{W}_{n,t}\right)}\right),$$

$$= A_{n,t}^2 e^{2\left(\mu_n - \frac{\sigma_n^2}{2}\right)(T-t)} \mathbb{E}_t^{\mathbb{P}}\left(e^{2\sigma_n\left(\hat{W}_{n,T} - \hat{W}_{n,t}\right)}\right).$$

The second term, $\mathbb{E}_t^{\mathbb{P}}\left(e^{2\sigma_n\left(\hat{W}_{n,T} - \hat{W}_{n,t}\right)}\right)$, should look quite familiar; we solved an integral of this form in finding the expectation of $A_{n,T}$. If we write out the formal definition, we can employ the complete-the-square technique, as in equation 5.49, a second time. If we set $X = 2\sigma_n\left(\hat{W}_{n,T} - \hat{W}_{n,t}\right)$, where $X \sim \mathcal{N}(0, 4\sigma_n^2(T-t))$, then

$$\int_{-\infty}^{\infty} e^u f_X(u) du = \int_{-\infty}^{\infty} e^u \frac{1}{\sqrt{2\pi 4\sigma_n^2(T-t)}} e^{\frac{-u^2}{2\cdot 4\sigma_n^2(T-t)}} du, \quad (5.53)$$

$$= \int_{-\infty}^{\infty} \frac{1}{\sqrt{2\pi 4\sigma_n^2(T-t)}} e^{-\frac{1}{2\cdot 4\sigma_n^2(T-t)}\left(u^2 - 8\sigma_n^2(T-t)u + \underbrace{16\sigma_n^4(T-t)^2 - 16\sigma_n^4(T-t)^2}_{=0}\right)} du,$$

$$= \int_{-\infty}^{\infty} \frac{1}{\sqrt{2\pi 4\sigma_n^2(T-t)}} e^{\frac{16\sigma_n^4(T-t)^2}{8\sigma_n^2(T-t)}} e^{-\frac{1}{2\cdot 4\sigma_n^2(T-t)}\left((u-4\sigma_n^2(T-t))^2\right)} du,$$

$$= e^{2\sigma_n^2(T-t)} \underbrace{\int_{-\infty}^{\infty} \frac{1}{\sqrt{2\pi 4\sigma_n^2(T-t)}} e^{-\frac{(u-4\sigma_n^2(T-t))^2}{2\cdot 4\sigma_n^2(T-t)}} du}_{=1},$$

$$\mathbb{E}_t^{\mathbb{P}}\left(e^{2\sigma_n(\hat{W}_{n,T}-\hat{W}_{n,t})}\right) = e^{2\sigma_n^2(T-t)}.$$

Returning to equation 5.52, then we have

$$\mathbb{E}_t^{\mathbb{P}}\left(A_{n,T}^2\right) = A_{n,t}^2 e^{2\left(\mu_n - \frac{\sigma_n^2}{2}\right)(T-t)} \underbrace{\mathbb{E}_t^{\mathbb{P}}\left(e^{2\sigma_n(\hat{W}_{n,T}-\hat{W}_{n,t})}\right)}_{\text{Equation 5.53}}, \quad (5.54)$$

$$= A_{n,t}^2 e^{2\left(\mu_n - \frac{\sigma_n^2}{2}\right)(T-t)} e^{2\sigma_n^2(T-t)},$$

$$= A_{n,t}^2 e^{2\mu_n(T-t)+\sigma_n^2(T-t)}.$$

Having resolved the complicated term, we can return to the original expression in equation 5.54 and compute the final variance of the nth asset-value as of time, T. It has the following form,

$$\text{var}_t(A_{n,T}) = \underbrace{\mathbb{E}_t^{\mathbb{P}}\left(A_{n,T}^2\right)}_{\text{Equation 5.54}} - A_{n,t}^2 e^{2\mu_n(T-t)}, \quad (5.55)$$

$$= A_{n,t}^2 e^{2\mu_n(T-t)+\sigma_n^2(T-t)} - A_{n,t}^2 e^{2\mu_n(T-t)},$$

$$= A_{n,t}^2 e^{2\mu_n(T-t)}\left(e^{\sigma_n^2(T-t)} - 1\right).$$

The volatility term is thus merely the square-root of equation 5.55, which is described as

$$\sigma_t(A_{n,T}) = \sqrt{\text{var}_t(A_{n,T})}, \quad (5.56)$$

5.5 The Direct Approach

$$= \sqrt{A_{n,t}^2 e^{2\mu_n(T-t)} \left(e^{\sigma_n^2(T-t)} - 1\right)},$$

$$= A_{n,t} e^{\mu_n(T-t)} \sqrt{e^{\sigma_n^2(T-t)} - 1}.$$

While involving a painful amount of mathematics, we are making progress. We have established that $A_{n,T} \sim \mathcal{N}\left(A_{n,t} e^{\mu_n(T-t)}, A_{n,t}^2 e^{2\mu_n(T-t)} \left(e^{\sigma_n^2(T-t)} - 1\right)\right)$. The marginal distribution of the terminal asset value is known; now we need to turn our attention to the joint-asset behaviour.

5.5.3 Covariance and Correlation of $A_{n,T}$ and $A_{m,T}$

The next natural quantity that we require is the covariance between any two arbitrary asset values, call them obligors n and m, at time T. It has the following fundamental form,

$$\operatorname{cov}_t(A_{n,T}, A_{m,T}) = \mathbb{E}_t^{\mathbb{P}}\left(\left(A_{n,T} - \mathbb{E}_t^{\mathbb{P}}(A_{n,T})\right)\left(A_{m,T} - \mathbb{E}_t^{\mathbb{P}}(A_{m,T})\right)\right), \tag{5.57}$$

$$= \mathbb{E}_t^{\mathbb{P}}\left(A_{n,T} A_{m,T}\right) - \mathbb{E}_t^{\mathbb{P}}\left(A_{n,T}\right) \mathbb{E}_t^{\mathbb{P}}\left(A_{m,T}\right),$$

$$= \mathbb{E}_t^{\mathbb{P}}\left(A_{n,T} A_{m,T}\right) - A_{n,t} e^{\mu_n(T-t)} A_{m,t} e^{\mu_m(T-t)},$$

$$= \mathbb{E}_t^{\mathbb{P}}\left(A_{n,t} e^{\left(\mu_n - \frac{\sigma_n^2}{2}\right)(T-t) + X_n} A_{m,t} e^{\left(\mu_m - \frac{\sigma_m^2}{2}\right)(T-t) + X_m}\right) - A_{n,t} A_{m,t} e^{(\mu_n + \mu_m)(T-t)},$$

$$= A_{n,t} A_{m,t} e^{(\mu_n + \mu_m)(T-t)} e^{-\left(\frac{\sigma_n^2}{2} + \frac{\sigma_m^2}{2}\right)(T-t)} \mathbb{E}_t^{\mathbb{P}}(e^{X_n + X_m}) - A_{n,t} A_{m,t} e^{(\mu_n + \mu_m)(T-t)},$$

$$= A_{n,t} A_{m,t} e^{(\mu_n + \mu_m)(T-t)} \left(e^{-\left(\frac{\sigma_n^2}{2} + \frac{\sigma_m^2}{2}\right)(T-t)} \mathbb{E}_t^{\mathbb{P}}(e^{X_n + X_m}) - 1\right),$$

where $X_i = \sigma_i \left(\hat{W}_{i,T} - \hat{W}_{i,t}\right)$ for $i = n, m$. The difficult term is $\mathbb{E}_t^{\mathbb{P}}(e^{X_n + X_m})$ and warrants *yet* another application of our complete-the-square manoeuvre, albeit with a slight twist.

To accomplish this, we observe that $X_i \sim \mathcal{N}\left(0, \sigma_i^2 \left(\hat{W}_{i,T} - \hat{W}_{i,t}\right)\right)$ for $i = n, m$. This is true, because the Gaussian distribution is closed under addition.[22] The variance of $X_n + X_m$ has the following familiar form,

$$\operatorname{var}_t(X_n + X_m) = \operatorname{var}_t(X_n) + \operatorname{var}_t(X_m) + 2\operatorname{cov}_t(X_n, X_m), \tag{5.58}$$

---

[22] In other words, the sum of two Gaussian random variables is also Gaussian. This convenient property does not hold for all random variables.

$$\begin{aligned}
&= \operatorname{var}_t(X_n) + \operatorname{var}_t(X_m) + 2\rho_{nm}\sigma_t(X_n)\sigma_t(X_m),\\
&= \sigma_n^2(T-t) + \sigma_m^2(T-t) + 2\rho_{nm}\sigma_n\sigma_m(T-t),\\
&= \left(\sigma_n^2 + \sigma_m^2 + 2\rho_{nm}\sigma_n\sigma_m\right)(T-t),\\
&= \xi_{nm}.
\end{aligned}$$

Returning to the more involved expectation term $\mathbb{E}_t^{\mathbb{P}}(e^{X_n+X_m})$ from equation 5.57, we have

$$\begin{aligned}
\mathbb{E}_t^{\mathbb{P}}(e^{X_n+X_m}) &= \int_{-\infty}^{\infty} e^u f_{X_n+X_m}(u)\,du, \qquad (5.59)\\
&= \int_{-\infty}^{\infty} e^u \frac{1}{\sqrt{2\pi\xi_{nm}}} e^{-\frac{u^2}{2\xi_{nm}}}\,du,\\
&= \int_{-\infty}^{\infty} \frac{1}{\sqrt{2\pi\xi_{nm}}} e^{-\frac{1}{2\xi_{nm}}\left(u^2 - 2\xi_{nm}u + \underbrace{\xi_{nm}^2 - \xi_{nm}^2}_{=0}\right)}\,du,\\
&= \int_{-\infty}^{\infty} \frac{1}{\sqrt{2\pi\xi_{nm}}} e^{\frac{\xi_{nm}^2}{2\xi_{nm}}} e^{-\frac{1}{2\xi_{nm}}\left(u^2 - 2\xi_m u + \xi_{nm}^2\right)}\,du,\\
&= e^{\frac{\xi_{nm}}{2}} \underbrace{\int_{-\infty}^{\infty} \frac{1}{\sqrt{2\pi\xi_{nm}}} e^{-\frac{(u-\xi_m)^2}{2\xi_{nm}}}\,du}_{=1},\\
\mathbb{E}_t^{\mathbb{P}}(e^{X_n+X_m}) &= e^{\frac{\xi_{nm}}{2}},\\
&= e^{\frac{\left(\sigma_n^2 + \sigma_m^2 + 2\rho_{nm}\sigma_n\sigma_m\right)(T-t)}{2}}.
\end{aligned}$$

We may now introduce this expression into equation 5.57 and simplify to arrive at our final result. This yields

$$\begin{aligned}
\operatorname{cov}_t(A_{n,T}, A_{m,T}) &= A_{n,t} A_{m,t} e^{(\mu_n+\mu_m)(T-t)} \left(e^{-\left(\frac{\sigma_n^2}{2} + \frac{\sigma_m^2}{2}\right)(T-t)} \underbrace{\mathbb{E}_t^{\mathbb{P}}(e^{X_n+X_m})}_{\text{Equation 5.59}} - 1\right), \quad (5.60)\\
&= A_{n,t} A_{m,t} e^{(\mu_n+\mu_m)(T-t)} \left(e^{-\left(\frac{\sigma_n^2}{2} + \frac{\sigma_m^2}{2}\right)(T-t)} e^{\frac{\left(\sigma_n^2 + \sigma_m^2 + 2\rho_{nm}\sigma_n\sigma_m\right)(T-t)}{2}} - 1\right),\\
&= A_{n,t} A_{m,t} e^{(\mu_n+\mu_m)(T-t)} \left(e^{\rho_{nm}\sigma_n\sigma_m(T-t)} - 1\right).
\end{aligned}$$

5.5 The Direct Approach

We may transform this into a correlation by normalizing the covariance by the volatility of $A_{n,T}$ and $A_{m,T}$ as follows,

$$\operatorname{cov}_t(A_{n,T}, A_{m,T}) = A_{n,t} A_{m,t} e^{(\mu_n + \mu_m)(T-t)} \left(e^{\rho_{nm} \sigma_n \sigma_m (T-t)} - 1 \right), \tag{5.61}$$

$$\rho_t(A_{n,T}, A_{m,T}) = \frac{A_{n,t} A_{m,t} e^{(\mu_n + \mu_m)(T-t)} \left(e^{\rho_{nm} \sigma_n \sigma_m (T-t)} - 1 \right)}{\sigma_t(A_{n,T}) \sigma_t(A_{m,T})},$$

$$= \frac{A_{n,t} A_{m,t} e^{(\mu_n + \mu_m)(T-t)} \left(e^{\rho_{nm} \sigma_n \sigma_m (T-t)} - 1 \right)}{\cancel{A_{n,t} e^{\mu_n(T-t)}} \sqrt{e^{\sigma_n^2 (T-t)} - 1} \cancel{A_{m,t} e^{\mu_m(T-t)}} \sqrt{e^{\sigma_m^2 (T-t)} - 1}},$$

$$= \frac{e^{\rho_{nm} \sigma_n \sigma_m (T-t)} - 1}{\sqrt{\left(e^{\sigma_n^2 (T-t)} - 1 \right) \left(e^{\sigma_m^2 (T-t)} - 1 \right)}}.$$

This leads to a fairly compact expression that is, as one would expect, independent of the levels of $A_{n,T}$ and $A_{m,T}$. It depends, in fact, only on the correlation and volatility structure of the diffusions.

5.5.4 Default Correlation Between Firms n and m

As we established in previous chapters, there is an important difference between the correlation between two firm's assets—or related latent variables—and the correlation between their various default outcomes. We should also expect, given a specific correlation between the assets of two firms, that this will have implications for their default correlations. Nevertheless, the default correlations can, and will, differ. The last quantity required, in this binge of mathematics, is an (familiar-looking) analytic expression for the default correlation.

Understanding default correlation essentially amounts to determining the structure of covariance between the default events of two different credit counterparties. In particular, this amounts to

$$\operatorname{cov}_t \left(\mathbb{I}_{\mathcal{D}_n}, \mathbb{I}_{\mathcal{D}_m} \right) = \mathbb{E}_t \left(\left(\mathbb{I}_{\mathcal{D}_n} - \mathbb{E}_t \left(\mathbb{I}_{\mathcal{D}_n} \right) \right) \left(\mathbb{I}_{\mathcal{D}_m} - \mathbb{E}_t \left(\mathbb{I}_{\mathcal{D}_m} \right) \right) \right), \tag{5.62}$$

$$= \mathbb{E}_t \left(\left(\mathbb{I}_{\mathcal{D}_n} - \mathbb{P}_t \left(\mathcal{D}_n \right) \right) \left(\mathbb{I}_{\mathcal{D}_m} - \mathbb{P}_t \left(\mathcal{D}_m \right) \right) \right),$$

$$= \mathbb{E}_t \left(\mathbb{I}_{\mathcal{D}_n} \mathbb{I}_{\mathcal{D}_n} \right) - \mathbb{P}_t \left(\mathcal{D}_n \right) \mathbb{P}_t \left(\mathcal{D}_m \right),$$

$$= \mathbb{P}_t \left(\mathcal{D}_n \cap \mathcal{D}_n \right) - \mathbb{P}_t \left(\mathcal{D}_n \right) \mathbb{P}_t \left(\mathcal{D}_m \right).$$

This is the typical form investigated in a variety of forms in previous chapters. Since we are really interested in the default correlation, we need only normalize equation 5.62 as,

$$\rho_t\left(\mathbb{I}_{\mathcal{D}_n}, \mathbb{I}_{\mathcal{D}_m}\right) = \frac{\rho_t\left(\mathbb{I}_{\mathcal{D}_n}, \mathbb{I}_{\mathcal{D}_m}\right)}{\sqrt{\text{var}(\mathbb{I}_{\mathcal{D}_n})}\sqrt{\text{var}(\mathbb{I}_{\mathcal{D}_m})}}, \qquad (5.63)$$

$$= \frac{\mathbb{P}_t\left(\mathcal{D}_n \cap \mathcal{D}_n\right) - \mathbb{P}_t\left(\mathcal{D}_n\right)\mathbb{P}_t\left(\mathcal{D}_m\right)}{\sqrt{\mathbb{P}_t(\mathcal{D}_n)(1 - \mathbb{P}_t(\mathcal{D}_n))}\sqrt{\mathbb{P}_t(\mathcal{D}_m)(1 - \mathbb{P}_t(\mathcal{D}_m))}}.$$

If we recall the definition of the default event from equation 5.35, we can write this more succinctly as,

$$\rho_t\left(\mathbb{I}_{\mathcal{D}_n}, \mathbb{I}_{\mathcal{D}_m}\right) = \frac{\mathbb{P}_t\left(\mathcal{D}_n \cap \mathcal{D}_n\right) - \Phi(\delta_{n,t})\Phi(\delta_{m,t})}{\sqrt{\Phi(\delta_{n,t})\Phi(\delta_{m,t})(1 - \Phi(\delta_{n,t}))(1 - \Phi(\delta_{m,t}))}}, \qquad (5.64)$$

where $\delta_{n,t}$ is the distance-to-default.

The difficult term in this expression is, as usual, the joint default probability, $\mathbb{P}_t\left(\mathcal{D}_n \cap \mathcal{D}_n\right)$. Everything else is easily evaluated—given the required underlying data—with a number of inverse standard normal distribution function evaluations. The good news, however, is that we can manipulate this joint probability and arrive at a sensible, workable result.

Since we know, from the properties of the Wiener increment, that $\frac{\hat{W}_T - \hat{W}_t}{\sqrt{T-t}} \sim \mathcal{N}(0, 1)$, then it naturally follows that

$$\mathbb{P}_t\left(\mathcal{D}_n \cap \mathcal{D}_n\right) = \mathbb{P}_t\left(\{A_{n,T} \leq K_n\} \cap \{A_{m,T} \leq K_m\}\right), \qquad (5.65)$$

$$= \mathbb{P}\left(\left\{\frac{\hat{W}_{n,T} - \hat{W}_{n,t}}{\sqrt{T-t}} \leq \delta_{n,t}\right\} \cap \left\{\frac{\hat{W}_{m,T} - \hat{W}_{m,t}}{\sqrt{T-t}} \leq \delta_{m,t}\right\}\right),$$

$$= \frac{1}{2\pi\sqrt{1-\rho_{nm}^2}} \int_{-\infty}^{\delta_{n,t}} \int_{-\infty}^{\delta_{m,t}} e^{-\frac{(u^2 - 2\rho_{nm}uv + v^2)}{2(1-\rho_{nm})}} du dv,$$

$$= \Phi\left(\delta_{n,t}, \delta_{m,t}\right).$$

where $\Phi(\cdot, \cdot)$ denotes the cumulative distribution function of the *bivariate* standard normal distribution. This permits us to restate equation 5.64, which describes the implied default correlation in the Merton (1974) model, as the following analytic expression,

$$\rho_t\left(\mathbb{I}_{\mathcal{D}_n}, \mathbb{I}_{\mathcal{D}_m}\right) = \frac{\Phi\left(\delta_{n,t}, \delta_{m,t}\right) - \Phi(\delta_{n,t})\Phi(\delta_{m,t})}{\sqrt{\Phi(\delta_{n,t})\Phi(\delta_{m,t})(1 - \Phi(\delta_{n,t}))(1 - \Phi(\delta_{m,t}))}}. \qquad (5.66)$$

What is abundantly clear, and must be understood by the risk manager, is that asset correlations in this model are not equivalent to default correlations. Asset

5.5 The Direct Approach

correlations naturally have an impact on default correlations, but they neither take the same values nor do they have the same mathematical structure.

5.5.5 Collecting the Results

We can now return to equations 5.44 to 5.46 and fill in the missing blanks. The vector of expected asset values is simply,

$$\bar{A}_T = \begin{bmatrix} A_{1,t}e^{\mu_1(T-t)} \\ \vdots \\ A_{N,t}e^{\mu_N(T-t)} \end{bmatrix}, \tag{5.67}$$

The covariance matrix now has the underlying intimidating form,

$$\Omega_{A_T} = \begin{bmatrix} A_{1,t}^2 e^{2\mu_1(T-t)}\left(e^{\sigma_1^2(T-t)}-1\right) & \cdots & A_{1,t}A_{N,t}e^{(\mu_1+\mu_N)(T-t)}\left(e^{\rho_{1N}\sigma_1\sigma_N(T-t)}-1\right) \\ A_{2,t}A_{1,t}e^{(\mu_2+\mu_1)(T-t)}\left(e^{\rho_{21}\sigma_2\sigma_1(T-t)}-1\right) & \cdots & A_{2,t}A_{N,t}e^{(\mu_2+\mu_N)(T-t)}\left(e^{\rho_{2N}\sigma_2\sigma_N(T-t)}-1\right) \\ \vdots & \ddots & \vdots \\ A_{N,t}A_{1,t}e^{(\mu_N+\mu_1)(T-t)}\left(e^{\rho_{N1}\sigma_N\sigma_1(T-t)}-1\right) & \cdots & A_{N,t}^2 e^{2\mu_N(T-t)}\left(e^{\sigma_N^2(T-t)}-1\right) \end{bmatrix}. \tag{5.68}$$

Now when we claim that $A_T \sim \mathcal{N}(\bar{A}_T, \Omega_{A_T})$, we know precisely what we mean. What is missing at this point, and represents the last step in the journey, is to find a defensible approach to parameterize this asset-value distribution and use it to compute our collection of risk measures.

5.5.6 The Task of Calibration

Neither asset volatility, asset return, nor the value of the assets is observable. This creates an impasse for the use of the Merton (1974) approach. Fortunately, the equity of publicly traded firms is an observable quantity. One can determine the current value of equity and estimate its return and volatility. Equally importantly, we know conceptually, following from equations 5.1 and 5.2, that there is a relationship between assets, liabilities, and equities. The key task, therefore, is to identify how we can infer our desired values from equity data. It takes a bit of work, but it is entirely possible.

We have one expression, which we have already derived, that expresses the value of the firm's equity in terms of the asset values. Our starting point is thus equation 5.18, which is a representation of the equity price in terms of the underlying asset value for an arbitrary credit counterparty n.[23] We will re-write it here in its full form as,

$$E_{n,t} = A_{n,t}\Phi(d_1) - e^{-r(T-t)}K_n\Phi\left(d_1 - \sigma_n\sqrt{T-t}\right). \tag{5.69}$$

Despite its complexity, equation 5.69 explicitly includes both σ_n and $A_{n,t}$, but it is a single equation with two unknowns. Without some additional information, we cannot solve it.

There is a useful result that helps resolve this quandary. It is frequently quoted in the literature, but rarely explicitly derived or motivated. We will work through the details.[24] While not terribly complex, it does warrant a proper derivation and a bit of explanation.

The basic idea is that equation 5.69 indirectly offers an opportunity to generate an additional element describing the relationship between asset and equity volatility and the asset and equity levels. It can also, as we'll soon see, help us to create a, perhaps messy, link between expected equity and asset returns. The intuition arises from the observation that $E_{n,t}$ is a function of $A_{n,t}$. Since $A_{n,t}$ is a stochastic process, this permits us to use Theorem 5.1 to create a new stochastic process.

To accomplish this, we restate equation 5.69 as,

$$E_{n,t} = \underbrace{A_{n,t}\Phi(d_1) - e^{-r(T-t)}K_n\Phi\left(d_1 - \sigma_n\sqrt{T-t}\right)}_{f(A_{n,t})}. \tag{5.70}$$

The only point here is that we have overtly written out the firm's equity as a function of its asset value. If we plan to use Theorem 5.1 to find a more convenient representation of this function, then we will require its first two partial derivatives with respect to $A_{n,t}$. This turns out to be somewhat involved. The reason is that both d_1 and d_2 are functions of $A_{n,t}$, which makes the expressions quite cumbersome. Let's first compute a few ancillary expressions, which we will need shortly. We have that,

$$d_1(A_{n,t}) \equiv d_1 = \frac{\ln\left(\frac{A_{n,t}}{K_n}\right) + \left(r + \frac{\sigma_n^2}{2}\right)(T-t)}{\sigma_n\sqrt{T-t}}, \tag{5.71}$$

$$\frac{\partial d_1(A_{n,t})}{\partial A_{n,t}} \equiv d_1' = \frac{1}{A_{n,t}\sigma_n\sqrt{T-t}}.$$

Given their similar form, it also conveniently follows that $d_1' = d_2'$. The next step is to determine the derivative of the cumulative distribution function. We have, from

---

[23] The derivation of this expression in found in Appendix B.

[24] Tracing out the history of this idea is also a bit challenging. Hull et al. (2005) make reference to an original paper, Jones et al. (1984), which appears to sketch out the basic idea for the first time.

5.5 The Direct Approach

first principles that,

$$\Phi(d_1) = \int_{-\infty}^{d_1} \frac{1}{\sqrt{2\pi}} e^{-\frac{u^2}{2}} du, \qquad (5.72)$$

$$\Phi'(d_1) = \frac{1}{\sqrt{2\pi}} e^{-\frac{d_1^2}{2}},$$

$$= \phi(d_1).$$

The equivalent expression for $d_2 = d_1 - \sigma_n\sqrt{T-t}$ is a bit more elaborate. It has the following form,

$$\Phi'(d_1 - \sigma_n\sqrt{T-t}) = \frac{1}{\sqrt{2\pi}} e^{-\frac{(d_1 - \sigma_n\sqrt{T-t})^2}{2}}, \qquad (5.73)$$

$$= \frac{1}{\sqrt{2\pi}} e^{-\frac{(d_1^2 - 2d_1\sigma_n\sqrt{T-t} + \sigma_n^2(T-t))}{2}},$$

$$= \frac{1}{\sqrt{2\pi}} e^{-\frac{d_1^2}{2}} e^{d_1 \sigma_n \sqrt{T-t}} e^{-\frac{\sigma_n^2(T-t)}{2}},$$

$$= \frac{1}{\sqrt{2\pi}} e^{-\frac{d_1^2}{2}} e^{\left(\frac{\ln\left(\frac{A_{n,t}}{K_n}\right) + \left(r + \frac{\sigma_n^2}{2}\right)(T-t)}{\sigma_n\sqrt{T-t}}\right)\sigma_n\sqrt{T-t}} e^{-\frac{\sigma_n^2(T-t)}{2}},$$

$$= \frac{1}{\sqrt{2\pi}} e^{-\frac{d_1^2}{2}} \frac{A_{n,t}}{K_n} e^{\left(r+\frac{\sigma_n^2}{2}\right)(T-t)} e^{-\frac{\sigma_n^2(T-t)}{2}},$$

$$= \frac{A_{n,t}}{\sqrt{2\pi} K_n} e^{-\frac{d_1^2}{2}} e^{r(T-t)}.$$

We now have all the necessary pieces to construct the requisite partial derivatives for the application of Theorem 5.1. The first partial derivative is written as,

$$\frac{\partial f(A_{n,t})}{\partial A_{n,t}} = \Phi(d_1) + A_{n,t}\Phi'(d_1)d_1' - e^{-r(T-t)}K_n\Phi'(d_2)d_2', \qquad (5.74)$$

$$= \Phi(d_1) + A_{n,t}\left(\frac{1}{\sqrt{2\pi}} e^{-\frac{d_1^2}{2}}\right)\left(\frac{1}{A_{n,t}\sigma_n\sqrt{T-t}}\right)$$

$$-e^{-r(T-t)} K_n \Phi'(d_2)\left(\frac{1}{A_{n,t}\sigma_n\sqrt{T-t}}\right),$$

$$= \Phi(d_1) + \frac{1}{\sigma_n\sqrt{2\pi(T-t)}} e^{-\frac{d_1^2}{2}}$$

$$-e^{-r(T-t)}K_n\left(\frac{A_{n,t}}{\sqrt{2\pi}K_n}e^{-\frac{d_1^2}{2}}e^{r(T-t)}\right)\left(\frac{1}{A_{n,t}\sigma_n\sqrt{T-t}}\right),$$

$$=\Phi(d_1)+\frac{1}{\sigma_n\sqrt{2\pi(T-t)}}e^{-\frac{d_1^2}{2}}-\frac{1}{\sigma_n\sqrt{2\pi(T-t)}}e^{-\frac{d_1^2}{2}},$$

$$=\Phi(d_1).$$

This is also known as the option's delta.[25] The second partial derivative is,

$$\frac{\partial f^2(A_{n,t})}{\partial A_{n,t}^2}=\Phi'(d_1)d_1', \tag{5.75}$$

$$=\left(\frac{1}{\sqrt{2\pi}}e^{-\frac{d_1^2}{2}}\right)\left(\frac{1}{A_{n,t}\sigma_n\sqrt{T-t}}\right),$$

$$=\frac{\phi(d_1)}{\sigma_n A_{n,t}\sqrt{T-t}},$$

where, as usual, $\phi(\cdot)$ denotes the standard normal density function. Equation 5.75 is commonly referred to as the option's gamma.[26]

Combining equations 5.74 and 5.75 with Theorem 5.1 on page 234, we have that

$$E_{n,T}-E_{n,t}=\int_t^T\frac{\partial f(A_{n,s})}{\partial A_{n,s}}dA_{n,s}+\frac{1}{2}\int_t^T\frac{\partial f^2(A_{n,s})}{\partial A_{n,s}^2}d\langle A_n\rangle_s, \tag{5.76}$$

$$=\int_t^T\Phi(d_1)dA_{n,s}+\frac{1}{2}\int_t^T\left(\frac{\phi(d_1)}{\sigma_n A_{n,s}\sqrt{T-t}}\right)d\langle A_n\rangle_s,$$

$$=\int_t^T\Phi(d_1)dA_{n,s}+\frac{1}{2}\int_t^T\left(\frac{\phi(d_1)}{\sigma_n A_{n,s}\sqrt{T-t}}\right)\left(\sigma_n^2 A_{n,s}^2 ds\right),$$

$$=\int_t^T\Phi(d_1)\underbrace{\left(rA_{n,s}ds+\sigma_n A_{n,s}dW_{n,s}\right)}_{\mathbb{Q}\text{-dynamics of }A_{n,s}}+\int_t^T\frac{\sigma_n A_{n,s}\phi(d_1)}{2\sqrt{T-t}}ds,$$

$$E_{n,T}-E_{n,t}=\int_t^T\left(r\Phi(d_1)+\frac{\sigma_n\phi(d_1)}{2\sqrt{T-t}}\right)A_{n,s}ds+\int_t^T\Phi(d_1)\sigma_n A_{n,s}dW_{n,s},$$

$$dE_{n,t}=\left(r\Phi(d_1)+\frac{\sigma_n\phi(d_1)}{2\sqrt{T-t}}\right)A_{n,t}dt+\Phi(d_1)\sigma_n A_{n,t}dW_{n,t}.$$

---

[25] Although, generally when one talks about an option delta, it is the first partial derivative of the option price with respect to the underlying stock value, not the firm's assets.

[26] Again, the option gamma is usually the second-order sensitivity to movement in the underlying equity price.

5.5 The Direct Approach

The consequence is a stochastic differential equation for the firm's equity, written as a function of the underlying assets. We have not, as yet, specified the stochastic dynamics of $E_{n,t}$. A reasonable and non-controversial choice under $\mathbb{Q}$ would be,

$$\frac{dE_{n,t}}{E_{n,t}} = r\,dt + \sigma_{E_n} dW_{n,t}. \qquad (5.77)$$

That is, the firm's equity is assumed to follow a geometric Brownian motion with a drift, under the equivalent martingale measure, of r and the same Wiener process as the asset process. That is, the same uncertainty impacts both the firm's assets and equities. While other choices are certainly possible, this feels like a defensible assumption.

Plugging equation 5.77 into our original expression in equation 5.76, we have

$$dE_{n,t} = \left(r\Phi(d_1) + \frac{\sigma_n \phi(d_1)}{2\sqrt{T-t}}\right) A_{n,t} dt + \Phi(d_1)\sigma_n A_{n,t} dW_{n,t}, \qquad (5.78)$$

$$\underbrace{rE_{n,t}dt + \sigma_{E_n}E_{n,t}dW_{n,t}}_{\text{Equation 5.76}} = \left(r\Phi(d_1) + \frac{\sigma_n \phi(d_1)}{2\sqrt{T-t}}\right) A_{n,t} dt + \Phi(d_1)\sigma_n A_{n,t} dW_{n,t},$$

$$\underbrace{rE_{n,t}dt}_{\substack{\text{Equity}\\\text{drift}}} + \underbrace{\sigma_{E_n}E_{n,t}dW_{n,t}}_{\text{Equity diffusion}} = \underbrace{\left(r\Phi(d_1) + \frac{\sigma_n \phi(d_1)}{2\sqrt{T-t}}\right) A_{n,t} dt}_{\text{Asset drift}} + \underbrace{\Phi(d_1)\sigma_n A_{n,t} dW_{n,t}}_{\text{Asset diffusion}}.$$

This is an intuitive and elegant result. It creates a link between the drift and diffusion terms of infinitesimal dynamics of A and E under the equivalent martingale measure, $\mathbb{Q}$. Specifically, it is the diffusion term that interests us. Setting them equal to one another, we find that

$$\sigma_{E_n} E_{n,t} dW_{n,t} = \Phi(d_1) \sigma_n A_{n,t} dW_{n,t}. \qquad (5.79)$$

Since both the equity and asset processes share the same diffusion process—here we see the importance of our previous assumption—it follows that the coefficients must be equal. In particular, we conclude that

$$\sigma_{E_n} E_{n,t} = \Phi(d_1) \sigma_n A_{n,t}, \qquad (5.80)$$

This is the famous result that is frequently quoted, but rarely (if ever) explicitly derived.[27] The implication, therefore, is that equity volatility is equal to asset volatility modified by the sensitivity of equity price movements to equity price changes.[28]

---

[27] The reason is obvious: it involves some fairly disagreeable computations.
[28] From an econometric perspective, you also could interpret $\frac{\partial E_n}{\partial A}$ as a regression coefficient. In practice, it is not particularly easy to estimate since the A_t outcomes are not observable.

We now have two equations—equation 5.69 and equation 5.80—in two unknowns: $A_{n,t}$ and σ_n. This is, in principle, sufficient to solve for our two desired quantities. We do need some information: we require an estimate of the risk-free rate, r; we demand a market assessment of the firm's equity value, $E_{n,t}$; and we need to know, or approximate, K_n. Everything else follows from these values and our unknowns. Calibration of the Merton (1974) model asset value and volatility figures essentially reduces to solving two equations simultaneously. The first equation is

$$G_1(A_{n,t}, \sigma_n) \equiv \underbrace{A_{n,t} \Phi(d_1) - e^{-r(T-t)} K_n \Phi\left(d_1 - \sigma_n \sqrt{T-t}\right) - E_{n,t}}_{\text{Equation 5.69}} = 0, \quad (5.81)$$

while the second equation is,

$$G_2(A_{n,t}, \sigma_n) \equiv \underbrace{A_{n,t} \Phi(d_1) \frac{\sigma_n}{\sigma_{E_n}} - E_{n,t}}_{\text{Equation 5.80}} = 0. \quad (5.82)$$

A common and effective approach would be to solve the following optimization problem numerically,

$$\min_{A_{n,t}, \sigma_n} \sum_{i=1}^{2} \left(G_i(A_{n,t}, \sigma_n)\right)^2. \quad (5.83)$$

That is the theory; we'll return to the implementation of this calibration algorithm when we turn to a practical example.

5.5.7 A Direct-Approach Inventory

Like most 19th century Russian novels with a preponderance of characters with complicated names, the Merton (1974) model has a large number of moving parts. Table 5.5 seeks to assist us in keeping track by providing a useful summary of the key quantities that one may compute within the Merton framework. The fourth and final rows, are particularly important, since they are employed to identify the unobservable σ_n and $A_{n,t}$ values using our calibration system in equation 5.83.

5.5.8 A Small Practical Example

Our standard 100-obligor portfolio is, given its lack of detailed equity-market information, not an especially good choice for application of the direct approach.

5.5 The Direct Approach

Table 5.5 *Key Merton quantities*: This table summarizes a number of important quantities necessary for the implementation and interpretation of the Merton model.

Quantity	Remark	Definition	Location
$A_{n,T}$	Terminal asset value	$A_{n,t} e^{\left(\mu_n - \frac{\sigma_n^2}{2}\right)(T-t) + \sigma_n \left(\tilde{w}_{n,T} - \tilde{w}_{n,t}\right)}$	Equation 5.9
$\mathcal{D}_n$	Default event	$\{A_{n,T} \leq K_n\}$	Equation 5.34
$E_{n,t}$	Equity value	$A_{n,t} \Phi(d_1) - e^{-r(T-t)} K \Phi(d_2)$	Equation 5.19
d_1	Black-Scholes coefficient	$\dfrac{\ln\left(\frac{A_{n,t}}{K_n}\right) + \left(r + \frac{\sigma_n^2}{2}\right)(T-t)}{\sigma_n \sqrt{T-t}}$	Equation 5.20
d_2	Black-Scholes coefficient	$d_1 - \sigma_n \sqrt{T-t}$	Equation 5.20
$\delta_{n,t}$	Distance-to-default	$\dfrac{\ln\left(\frac{K_n}{A_{n,t}}\right) - \left(\mu_n - \frac{\sigma_n^2}{2}\right)(T-t)}{\sigma_n \sqrt{T-t}}$	Equation 5.12
$\mathbb{P}_t(\mathcal{D}_n)$	Actual default probability	$\Phi(\delta_{n,t})$	Equation 5.13
$\mathbb{Q}_t(\mathcal{D}_n)$	Risk-neutral default probability	$\Phi\left(\dfrac{\ln\left(\frac{K_n}{A_{n,t}}\right) - \left(r - \frac{\sigma_n^2}{2}\right)(T-t)}{\sigma_n \sqrt{T-t}}\right)$	Equation 5.13
K_n	Liability threshold	$A_{n,t} e^{\left(\mu_n - \frac{\sigma_n^2}{2}\right)(T-t) + \Phi^{-1}(p_n)\sigma_n \sqrt{T-t}}$	Equation 5.39
$\mathbb{E}_t^{\mathbb{P}}(A_{n,T})$	Expected terminal asset value	$A_{n,t} e^{\mu_n (T-t)}$	Equation 5.50
$\text{var}_t(A_{n,T})$	Terminal-asset variance	$A_{n,t}^2 e^{2\mu_n(T-t)} \left(e^{\sigma_n^2 (T-t)} - 1\right)$	Equation 5.55
$\sigma_t(A_{n,T})$	Terminal-asset volatility	$A_{n,t} e^{\mu_n(T-t)} \sqrt{e^{\sigma_n^2(T-t)} - 1}$	Equation 5.56
$\text{cov}_t(A_{n,T}, A_{m,T})$	Terminal-asset covariance	$A_{n,t} A_{m,t} e^{(\mu_n + \mu_m)(T-t)} \left(e^{\rho_{nm} \sigma_n \sigma_m (T-t)} - 1\right)$	Equation 5.60
$\rho_t(A_{n,T}, A_{m,T})$	Asset correlation	$\dfrac{e^{\rho_{nm}\sigma_n\sigma_m(T-t)} - 1}{\sqrt{\left(e^{\sigma_n^2(T-t)} - 1\right)\left(e^{\sigma_m^2(T-t)} - 1\right)}}$	Equation 5.61
$\text{cov}_t(\mathbb{I}_{\mathcal{D}_n}, \mathbb{I}_{\mathcal{D}_m})$	Default covariance	$\mathbb{P}_t(\mathcal{D}_n \cap \mathcal{D}_m) - \mathbb{P}_t(\mathcal{D}_n) \mathbb{P}_t(\mathcal{D}_m)$	Equation 5.62
$\rho_t(\mathbb{I}_{\mathcal{D}_n}, \mathbb{I}_{\mathcal{D}_m})$	Default correlation	$\dfrac{\Phi(\delta_{n,t}, \delta_{m,t}) - \Phi(\delta_{n,t})\Phi(\delta_{m,t})}{\sqrt{\Phi(\delta_{n,t})\Phi(\delta_{m,t})(1 - \Phi(\delta_{n,t}))(1 - \Phi(\delta_{m,t}))}}$	Equation 5.66
σ_n	Instantaneous asset volatility	$\dfrac{1}{\Phi(d_1)} \left(\dfrac{E_{n,t}}{A_{n,t}}\right) \sigma_{E_n}$	Equation 5.80
$A_{n,t}$	Starting asset value	$\dfrac{1}{\Phi(d_1)} \left(\dfrac{\sigma_{E_n}}{\sigma_n}\right) E_{n,t}$	Equation 5.80

Table 5.6 *A low-dimensional example*: Direct implementation, as suggested by Table 5.4, requires a significant data burden. This table summarizes a simplified six-obligor example intended to help us understand the key details.

Counterparty	1	2	3	4	5	6
What we observe						
Rating	BBB	AAA	AA	BBB	AA	AAA
Exposure: c_n	10	10	10	10	10	10
Debt-to-equity ratio: Δ_n	123.8%	74.8%	89.2%	122.9%	89.2%	73.8%
Sharpe ratio: S_n	0.05	0.06	0.07	0.06	0.06	0.10
Equity value: $E_{n,t}$	56	93	75	75	62	105
Equity volatility: μ_{E_n}	56.4%	42.4%	45.3%	57.9%	46.5%	43.0%
Equity return: σ_{E_n}	4.1%	3.8%	4.4%	4.7%	3.6%	5.1%
What we infer						
Default threshold: K_n	70	69	67	92	55	78
Distance to default: $\delta_{n,T-t}$	−2.3	−3.4	−3.1	−2.2	−3.0	−3.4
Actual default probability: $\mathbb{P}_t(\mathcal{D}_n)$	119.1	2.8	9.6	137.8	13.7	2.9
Risk-neutral default probability: $\mathbb{Q}_t(\mathcal{D}_n)$	137.1	3.6	12.4	161.9	16.3	4.1
Asset value: $A_{n,t}$	126	162	142	166	117	182
Asset return: μ_n	2.4%	2.6%	2.8%	2.7%	2.4%	3.4%
Asset volatility: σ_n	25.5%	24.4%	24.1%	26.4%	24.7%	24.9%

An example, however, is always useful to make some of the theoretical ideas more concrete. Instead of using our typical example, therefore, we will construct a small, easy-to-visualize portfolio for the investigation of the direct Merton (1974) approach. While not terribly realistic, it will help us examine the key points raised in the previous sections.

Table 5.6 thus describes a simple, carefully randomly generated, portfolio of *six* distinct counterparties. The top half of the table outlines key variables from the Merton model that we assume to have observed; this is information that one could expect to be able to extract from market data or financial statements. The second, and lower, half of Table 5.6 summarizes the quantities that must be inferred, using the previously described techniques, to actually implement the direct Merton (1974) model. Broadly speaking, the top half of the table contains equity data, while the lower part relates to the firm's assets.

The actual data in Table 5.6 were selected to be realistic, but are not actual data. Moreover, not being a corporate finance specialist, these values could deviate dramatically from what one normally observes in markets. The objective, however, is not to be precise, but instead to work through the practical details of this model's implementation. Moreover, some additional assumptions—and in some cases a bit of artistic license—were also taken with some aspects of the parameterization. Indeed, the logic for the creation of the portfolio requires a bit of explanation. Some miscellaneous, but important, steps include:

5.5 The Direct Approach

- The risk-free rate, r, is assumed to be 1% and the time period, $T-t$, is arbitrarily set to one year.
- The value of the portfolio is assumed to be $60 with equal weights of approximately $10 with each obligor. The portfolio thus holds a substantial portion of these firms' debt claims.
- A credit rating—ranging from AAA to BBB—was randomly selected. This gives a basic idea of the credit quality of the issuer.
- Conditional on the credit rating a debt-to-equity ratio and equity volatility were generated. The lower the credit rating, as one might expect, the higher the debt-to-equity ratio and equity volatility. The default barrier, K_n, can be derived easily from the debt-to-equity ratio. Denote the debt-to-asset ratio as $\Delta_n = \frac{D_{n,t}}{E_{n,t}}$. This is not terribly precise, given the varying tenors of the debt and the vagaries of approximating Δ_n. Nevertheless, it is a reasonable first-order approximation. In this case, the liability threshold is approximated as,

$$K_n \approx \underbrace{\left(\frac{D_{n,t}}{E_{n,t}}\right)}_{\Delta_n} E_{n,t} = D_{n,t}. \tag{5.84}$$

In other words, the lower threshold is our best approximation of the firm's overall debt level. In reality, these values must be estimated from market data and examination of a firm's balance sheet.

- An equity value and a Sharpe ratio are randomly generated. Working backwards from the Sharpe ratio, we can extract the expected equity return. From the definition of the Sharpe ratio, denoted S_n

$$S_n = \frac{\mu_E - r}{\sigma_E}, \tag{5.85}$$

$$\mu_E = r + S_n \sigma_E.$$

This is essentially the logic behind the equity risk premium. This is to ensure that the assumed equity returns are consistent with the equity volatility.

- The remaining quantities—default probabilities and the first two moments of the terminal asset values—are derived from the formulae summarized in Table 5.5.

This final point merits a bit of additional attention. We have included both the actual default probabilities and the so-called risk-neutral probabilities. The only difference is in the first case we use μ_n to compute the distance-to-default, whereas in the second we use the risk-free rate, r. In general, given risk-aversion by market participants, we expect that $\mu_n > r$. This direct implication is that actual default probabilities will be less than their risk-neutral counterparts. Logically, this is quite defensible. μ_n is a force pulling the value of the assets upwards; the greater its value, the lower the chance of breaching the liability threshold and hence the default event.

In the risk-neutral setting, $r < \mu_n$ does not pull up the asset movements with the same degree of strength.

Why does this matter? In general, as risk managers, our interest lies with the actual default probabilities. In the real world, however, it is relatively easy to identify the risk-free rate, but rather more difficult to estimate the expected return of a firm's assets. We have, for the purposes of this example, had to make some assumptions to determine the asset return, μ_n. The defensibility of these assumptions is subject to question. We are on somewhat less firm theoretical footing; matching the drift coefficients from equation 5.78 provides no real assistance. Instead, we use the same idea employed to tease out the equity return: the Sharpe ratio. Our estimate of the asset return is thus,

$$\mu_n = r + S_n \sigma_n. \tag{5.86}$$

In making this claim, we are following the same logic used to derive the relationship between equity and asset volatility. It is not obvious, however, that the same relationship holds. This explains why many market participants prefer to work with the risk-neutral probability: there is less uncertainty about the estimation of r. Nonetheless, ease of computation aside, one needs to understand clearly that risk-neutral probabilities are predicated on the assumption that agents are risk-neutral. This is perfectly acceptable for asset pricing, but poses serious problems for risk managers.

Table 5.6 provides the asset values and volatilities for each of our obligors. These need to be extracted from our set of calibrating equations. We need to find the roots of two non-linear equations—see Press et al. (1992, Chapter 9)—and this can be done in a variety of ways. We elected to solve the non-linear optimization problem in equation 5.82 using Powell's direct-search algorithm in Python. This merely requires the following function call,

`xstar = minimize(minimizeG,x,args=(r,sE[n],dt,E[n],K[n]),method='Powell')` (5.87)

where `sE` is the equity volatility, `E` is the equity value, `K` is the liability threshold, `r` is the risk-free rate, and `dt` is the time interval. The objective function, `minimizeG` is found in Algorithm 5.3; it yields $A_{n,t}$ and σ_n. The underlying functions, `getE` and `getOptionDelta`, price the option and determine the required value of d_1.

Furnished with both the asset value and the liability threshold, it makes sense to verify if they are internally consistent. Every business student can assure us that the identity, $A_t = L_t + E_t$, must hold. That is, the sum of the liabilities and equities must be equal to the assets. Table 5.7 seeks to answer this legitimate question by illustrating, as a percentage of the total assets, the obligor-level liability and equity values. This clearly respects, as one would expect, the balance-sheet constraint. In

5.5 The Direct Approach

Algorithm 5.3 Finding $A_{n,t}$ and σ_n

```
from scipy.optimize import minimize
def minimizeG(x,r,sE,dt,E,K):
    A = x[0]
    sigmaA = x[1]
    if A<=0:
        return np.inf
    else:
        G1 = getE(r,sigmaA,dt,A,K)-E
        G2 = A*getOptionDelta(r,sigmaA,dt,A,K)*(sigmaA/sE)-E
        return G1**2 + G2**2
```

Table 5.7 *Asset correlations*: This table summarizes the, randomly generated, asset correlations between the six issuers in our simple sample portfolio.

ζ	1	2	3	4	5	6
1	1.00	0.20	0.20	0.19	0.19	0.19
2	0.20	1.00	0.20	0.19	0.20	0.20
3	0.20	0.20	1.00	0.19	0.20	0.20
4	0.19	0.19	0.19	1.00	0.19	0.19
5	0.19	0.20	0.20	0.19	1.00	0.20
6	0.19	0.20	0.20	0.19	0.20	1.00

addition, as the credit quality of an issuer decreases, the proportion of liabilities appears to be larger (Fig. 5.7).

It is also interesting to compare the asset and equity expected returns and volatilities. Figure 5.8 compares, on an obligor basis, these equity and asset moments. The first observation is that, across all obligors and moments, the equity volatilities are larger than the associated asset figures. Equation 5.80 provides a hint as to why this is the case. The asset volatilities are equal to the equity values modified by the $\Phi(d_1)$, or option delta, term. Mathematically, we know that $\Phi(d_1) \in (0, 1)$, which implies that the asset values can be, at best, equal to the equity quantities. For any value of $\Phi(d_1) < 1$, which is the typical empirical case, the asset volatilities will be inferior to their equity equivalents. In the case of expected asset returns, which exhibit the same behaviour, the culprit are the combination of lower asset volatilities and the form of equation 5.86.

As we've repeatedly observed in the preceding chapters, default dependence is a key driver of credit risk. The Merton (1974) framework faces the same realities as previous models. We have derived complicated mathematical expressions for the covariance and correlation between the stochastic asset values $A_{n,T}$ and $A_{m,T}$. While helpful, ultimately, we need to estimate the correlation coefficient, ρ_{nm}. This describes the actual linear dependence between the underlying firm asset processes. Generally speaking, as we've already seen, asset time series are neither available nor are we capable of constructing a clever trick to extract asset correlations from their equity equivalents. Instead, in practice, it is a common practice to simply use equity correlations as a proxy for asset correlations. In the absence of a reasonable

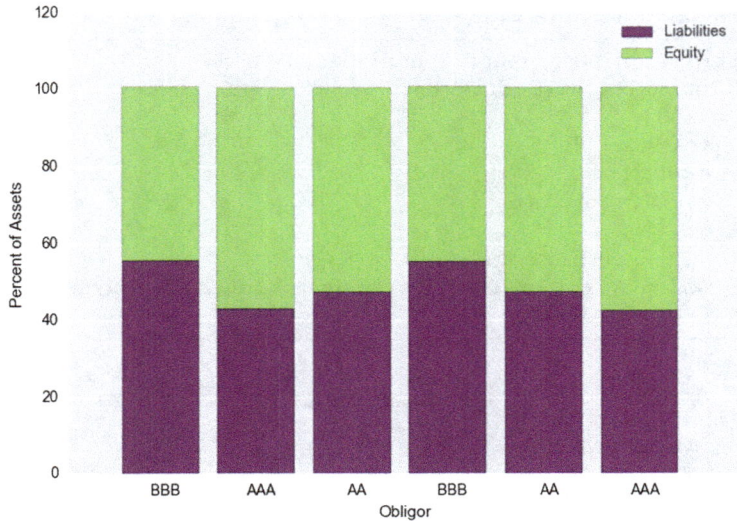

Fig. 5.7 *Inferring the capital structure*: These underlying figure provides, as a percentage of the total assets, the obligor-level liability and equity values. This clearly respects the constraint that $A_t = L_t + E_t$.

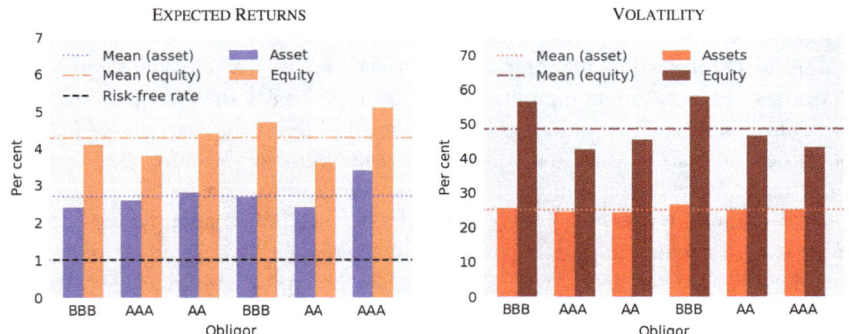

Fig. 5.8 *Equity and asset moments*: These two figures graphically compare, for each of the six obligors in Table 5.6, the asset and equity expected returns and volatility.

alternative, this seems a defensible practice.[29] de Servigny and Renault (2002) discuss this issue in detail. They claim that they "find no evidence of a downward bias in equity correlations with asset-implied correlations."

---

[29]It is also consistent with our previous assumption of assuming that the equity and asset return share a common diffusion term.

5.5 The Direct Approach

Table 5.8 *Default correlations*: Using the asset correlations in Table 5.7, the model parameters in Table 5.6 and the formulae in Table 5.5, we derive the implied default correlations. The values are summarized below.

ζ	1	2	3	4	5	6
1	1.00	0.01	0.01	0.03	0.01	0.01
2	0.01	1.00	0.00	0.01	0.00	0.00
3	0.01	0.00	1.00	0.01	0.01	0.00
4	0.03	0.01	0.01	1.00	0.01	0.01
5	0.01	0.00	0.01	0.01	1.00	0.00
6	0.01	0.00	0.00	0.01	0.00	1.00

For our practical example, we are in the business of inventing equity correlations. To simplify matters, we assume that all equities share a common correlation coefficient of 0.20. While probably a bit silly, it does allow us to easily compare our results to the one-factor Gaussian threshold model. Table 5.7 uses equation 5.61 from page 263 to transform a common $\rho_{nm} = 0.20$ into an asset-correlation matrix.

The differences are minimal. The equity and asset correlations are thus basically the same. As we've also learned, asset and default correlations are *not* equivalent. Table 5.8 takes our model investigation a step further and computes the default correlations associated with our assumed parametrization.

The default correlation is a complicated function. It depends on the default probabilities as summarized by the cumulative distribution function applied to the distance-to-default. It also include the joint probability of default—again, this involves the Gaussian distribution and the distance-to-default. Following from the discussion in equations 5.65 and 5.66, Algorithm 5.4 outlines the Python code used for its computation. In principle, it requires a two-dimensional numerical integration to evaluate the bivariate normal distribution. We have, in an effort to demonstrate alternative ways to do similar things, employed the mvn toolbox—the acronym stands, unsurprisingly, for multivariate normal—within the scipy.stats package. The function mvn.mvnun takes the ranges of integration, mean vector, and covariance matrix to evaluate the multivariate cumulative distribution function.

Algorithm 5.4 Computing default correlations

```
from scipy.stats import mvn
def getDefaultCorAB(A,B,muA,muB,sigmaA,sigmaB,rhoAB,dt,KA,KB):
    dA = getDelta(muA,sigmaA,dt,A,KA)
    dB = getDelta(muB,sigmaB,dt,B,KB)
    pA = norm.cdf(dA)
    pB = norm.cdf(dB)
    pAB,err = mvn.mvnun(np.array([-100, -100]),np.array([dA, dB]),
                    np.array([0, 0]),np.array([[1,rhoAB],[rhoAB,1]]))
    return np.divide(pAB-pA*pB,np.sqrt(pA*pB*(1-pA)*(1-pB)))
```

Algorithm 5.5 Simulating the direct Merton model

```
def mertonDirectSimulation(N,M,K,hatA,OmegaA,c,alpha):
    Z = np.random.normal(0,1,[M,N])
    w,v = anp.eigh(OmegaA)
    H = np.dot(v,np.sqrt(np.diag(w)))
    A = np.tile(hatA,(M,1)) + np.dot(Z,np.transpose(H))
    lossIndicator = 1*np.less(A,K)
    lossDistribution = np.sort(np.dot(lossIndicator,c),axis=None)
    el,ul,var,es=util.computeRiskMeasures(M,lossDistribution,alpha)
    return el,ul,var,es
```

Table 5.8 summarizes the results of applying Algorithm 5.4 to our six-obligor example. As usual, the actual level of default correlation is significantly lower than the asset values in Table 5.7. In some cases, it reaches the levels of 0.01 and 0.02, but in most situations the default correlation is quite small. The principal reason is the relatively small default probabilities—the default correlations are highest when a low-rated obligor is involved.

The final step involves using all of the information found in Table 5.6 and the asset correlations in Table 5.7 to simulate the direct model and compute our usual set of risk measures. Algorithm 5.5 summarizes the Python code; it looks virtually identical to the indirect code in Algorithm 5.2. There is, however, *one* main difference. In the indirect setting, we worked with correlated standard normal variates. In the direct setting, we are interested in simulating the multivariate normal system of terminal asset values; these are not joint standard normal.

Instead, as we highlighted in equations 5.44 to 5.46, the vector of terminal asset values is $A_T \sim \mathcal{N}(\bar{A}_T, \Omega_{A_T})$. hatA denotes the vector of expected terminal asset values, while OmegaA is the terminal-asset covariance matrix. These are passed, along with the usual information, to the simulation algorithm. To generate the correlated multivariate random variates—relying on the eigh function—we again use the eigenvalue decomposition. It is, however, necessary to add the mean to our simulated values to ensure that they have the appropriate location. Otherwise, the basic structure of the simulation methodology is conceptually the same as the indirect case and very similar to most of the preceding models examined in this book.

Table 5.9 highlights the results from the application of Algorithm 5.5 to our simple, six-obligor example. They are compared to a one-factor Gaussian model with a ρ parameter of 0.45; it also employed the same exposures and the physical-world default probabilities from Table 5.6. In the direct case, the one-factor Gaussian and direct Merton (1974) models do not generate identical risk outcomes. The reason for the difference stems from the greater complexity of the direct approach. It can no longer be considered a one-factor model. Indeed, we have treated each obligor as its own separate factor. The systematic and idiosyncratic elements of each counterparty are collected together in its equity value, volatility, debt-to-equity

5.5 The Direct Approach

Table 5.9 *Direct Merton results*: The underlying table describes the usual summary risk statistics for the direct implementation of the Merton (1974) model. It also includes the one-factor Gaussian threshold model for comparison; there are non-trivial differences. To get risk measures in the same neighbourhood, it was necessary to set the ρ parameter in the one-factor Gaussian threshold model to 0.45.

Quantile	Gaussian		Merton	
	$\widehat{\text{VaR}}_\alpha$	$\widehat{\mathcal{E}}_\alpha$	$\widehat{\text{VaR}}_\alpha$	$\widehat{\mathcal{E}}_\alpha$
95.00th	$0.0	$5.7	$10.0	$13.0
97.00th	$0.0	$9.5	$10.0	$15.0
99.00th	$10.0	$12.6	$20.0	$21.6
99.50th	$10.0	$15.2	$20.0	$23.3
99.90th	$20.0	$23.0	$30.0	$31.7
99.97th	$26.0	$30.3	$36.0	$37.7
99.99th	$30.0	$33.9	$40.0	$41.0
Other statistics				
Expected loss	$0.3		$1.3	
Loss volatility	$1.8		$1.3	
Total iterations	10,000,000			
Computational time	4.1		4.0	

ratio, and Sharpe ratio. The one-factor Gaussian model, by contrast, has a single parameter and a vector of default probabilities.

We discovered in Chaps. 3 and 4—by examining the multivariate CreditRisk+ and t-threshold models—that it is difficult to compare univariate and multivariate implementations. This is a similar case. Indeed, we note that the expected loss and loss volatility figures differ. The expected loss of the one-factor Gaussian model is significantly lower than our direct Merton (1974) implementation. The opposite is true for the standard deviation. This indicates that the moments of the two distributions are not the same. We should not, in fact, expect them to coincide. The Gaussian threshold model operates in a standard-normal setting with a single correlation variable. Even though we've forced the direct Merton model to share a common correlation, the multivariate distribution is not standard normal. Each obligor has its own asset variance and expectation, which will lead to a richer and more complex multivariate distribution.

When examining the VaR and expected-shortfall estimates, the largest difference appears to stem from a higher assessment in the Merton (1974) model of a single default. At the 95th quantile, the Merton approach already predicts a single default. The one-factor Gaussian model does not have a default until the 99th quantile. This suggests that one, or more, of the obligors possesses a sufficient level of asset volatility to generate a higher probability of a single default.

To complete our analysis, we provide a graphical depiction of the direct Merton (1974) model's default-loss distribution in Fig. 5.9. Virtually all of the probability mass—upwards of 80%—points toward no defaults at all in the portfolio. A single default has a non-trivial probability, but the remainder, as we saw in Table 5.8, have vanishingly small probability. This is consistent with our previous conclusions.

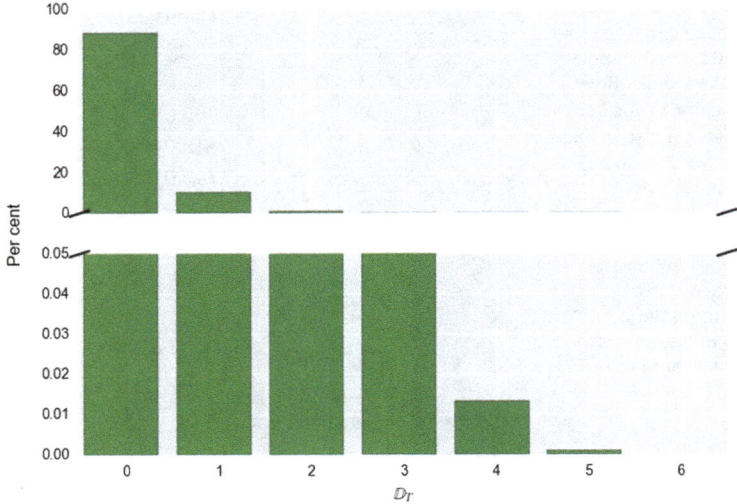

Fig. 5.9 *Direct Merton-model default-loss distribution*: For our six-counterparty example with default correlation, this graphic details the distribution of the number of losses. Almost all of the probability mass occurs at 0-1 defaults; we had to zoom in the axis to under 5 basis points to permit visibility of the distribution's tail.

Overall, the direct implementation of the Merton (1974) is more work, but it has the potential to offer us something quite similar to the multivariate mixture and threshold methodologies. Conceptually, the model is not doing anything terribly different relative to these approaches, but it offers a novel and clever approach to parametrization and incorporation of additional factors.

5.6 Final Thoughts

The Merton (1974) model provides an excellent conceptual and theoretical foundation for the family of threshold models. Practically, in its indirect form, it is basically equivalent to the one-factor Gaussian threshold model. In its direct form, it offers important links to the market and the capital structure of one's obligors—using both equity and balance-sheet information. Manipulating and implementing the Merton (1974) approach is not exactly straightforward. Implementation requires:

1. working with complex stochastic processes;
2. a complicated numerical implementation; and
3. a large amount of data.

5.6 Final Thoughts

If you feel that this approach involves an egregious amount of mathematics for a model that reduces to something close to a multivariate Gaussian threshold model, you are not wrong. In recent years, the wholesale use of continuous-time stochastic processes in practitioner finance has somewhat attenuated. A useful analogy is the literature and implementation of dynamic term-structure models. In the 1990's the only serviceable stochastic yield-curve models involved continuous-time mathematics and often fell into the affine class.[30] These highly complex models were the only game in town. Early in the 2000's, however, Diebold and Li (2003) introduced an important simplification of the overall structure. The so-called extended Nelson-Siegel model has, in recent years, become something of a standard in practitioner settings.[31] Not only was it lighter in terms of mathematical and implementation burden, it was also subsequently shown to be essentially equivalent to the affine setting.[32] This trend has occurred in other areas of finance and will likely continue. It should most assuredly, however, *not* imply that the lessons and techniques of the more complex models developed in the 1980's and 1990's should be forgotten or ignored. They contain an enormous amount of wisdom and quantitative analysts should be encouraged to examine them closely, compare them to the current frameworks, and seek to extend and improve them further. This is the fundamental nature of scholarship. The reward for such investigation is manifest: one obtains a deeper and more nuanced understanding of the underlying problem and its potential solutions.

An additional defence of the Merton (1974) model is that, in addition to setting the standard for structural credit-risk modelling methodologies, it has provided a much greater service to the finance community. It informs capital-structure decisions, provides a sound basis for pricing debt claims subject to credit risk and is used to estimate the term structure of credit spreads. It has also found application in numerous other areas.[33]

This chapter also closes out our explicit discussion of credit-risk models. The remaining chapters will turn to examine model-diagnostic tools and estimation methods to assist the quantitative analyst in understanding and using these models. We certainly do not have a shortage of possible approaches. Indeed, the Merton (1974) adds a few additional—if not completely new, but rather illuminating—entries to our model zoology in Table 5.1. Table 5.10 performs an alternative classification; it organizes our various model families by their default triggers and

---

[30] See, for more information on these models, Bolder (2001, 2006) and many of the excellent references included in it.

[31] See Bolder (2006) for some practical discussion and relative performance of these two modelling frameworks.

[32] See Diebold and Rudebusch (2013, Chapter 3) for more details.

[33] See, for example, Sundaresan (2013) and Hull et al. (2005) for a description of alternative applications of the Merton (1974) model.

Table 5.10 *A litany of default triggers*: The underlying table chronicles the default triggers associated with the main credit-risk models examined in this book.

Model	Default indicator	Factors
Binomial	$\mathbb{I}_{\mathcal{D}_n} = \begin{cases} 1 : & p_n \\ 0 : & 1-p_n \end{cases}$	None
Mixture	$\mathbb{I}_{\mathcal{D}_n} = \begin{cases} 1 : & p(S) \\ 0 : & 1-p(S) \end{cases}$	S
Threshold	$\mathbb{I}_{\mathcal{D}_n} = \begin{cases} 1 : & Y_n \leq \Phi^{-1}(p_n) \\ 0 : & Y_n > \Phi^{-1}(p_n) \end{cases}$	G and/or V
Merton model		
Indirect	$\mathbb{I}_{\mathcal{D}_n} = \begin{cases} 1 : & \xi_1 = \frac{\hat{W}_{n,T}-\hat{W}_{n,t}}{\sqrt{T-t}} \leq \Phi\left(\delta_{n,T-t}\right) \\ 0 : & \xi_N = \frac{\hat{W}_{n,T}-\hat{W}_{n,t}}{\sqrt{T-t}} > \Phi\left(\delta_{n,T-t}\right) \end{cases}$	$\frac{1}{T-t}\begin{bmatrix} \hat{W}_{1,T}-\hat{W}_{1,t} \\ \vdots \\ \hat{W}_{N,T}-\hat{W}_{N,t} \end{bmatrix}$
Direct	$\mathbb{I}_{\mathcal{D}_n} = \begin{cases} 1 : & A_{n,T} \leq K_n \\ 0 : & A_{n,T} > K_n \end{cases}$	$\begin{bmatrix} A_{1,T} \\ \vdots \\ A_{N,T} \end{bmatrix}$

the form of their underlying factors. It underscores the similarities and differences inherent in the various frameworks examined in the preceding chapters. It also acts as a useful reference as we move into the area of modelling tools and diagnostics.

References

Black, F., & Scholes, M. S. (1973). The pricing of options and corporate liabilities. *Journal of Political Economy, 81*, 637–654.
Bolder, D. J. (2001). Affine term-structure models: Theory and implementation. Bank of Canada: Working Paper 2001–15.
Bolder, D. J. (2006). Modelling term-structure dynamics for portfolio analysis: A practitioner's perspective. Bank of Canada: Working Paper 2006–48.
Crosbie, P. J., & Bohn, J. R. (2002). Modeling default risk. KMV.
de Servigny, A., & Renault, O. (2002). Default correlation: Empirical evidence. Standard and Poors, Risk Solutions.
Diebold, F. X., & Li, C. (2003). Forecasting the term structure of government bond yields. University of Pennsylvania Working Paper.
Diebold, F. X., & Rudebusch, G. D. (2013). *Yield curve modeling and forecasting: The dynamic Nelson-Siegel approach*. Princeton, NJ: Princeton University Press.
Duffie, D., & Singleton, K. (2003). *Credit risk* (1st edn.). Princeton: Princeton University Press.
Durrett, R. (1996). *Probability: Theory and examples* (2nd edn.). Belmont, CA: Duxbury Press.
Fishman, G. S. (1995). *Monte Carlo: Concepts, algorithms, and applications*. 175 Fifth Avenue, New York, NY: Springer-Verlag. *Springer series in operations research*.
Geske, R. (1977). The valuation of compound options. *Journal of Financial Economics, 7*, 63–81.
Golub, G. H., & Loan, C. F. V. (2012). *Matrix computations*. Baltimore, Maryland: The John Hopkins University Press.

References

Harrison, J., & Kreps, D. (1979). Martingales and arbitrage in multiperiod security markets. *Journal of Economic Theory, 20*, 381–408.

Harrison, J., & Pliska, S. (1981). Martingales and stochastic integrals in the theory of continuous trading. *Stochastic Processes and Their Applications, 11*, 215–260.

Heunis, A. J. (2011). *Notes on stochastic calculus*. University of Waterloo.

Hull, J. C., Nelken, I., & White, A. D. (2005). Merton's model, credit risk, and volatility skews. *Journal of Credit Risk, 1*, 3–27.

Jeanblanc, M. (2002). *Credit risk*. Université d'Evry.

Jones, E. P., Mason, S. P., & Rosenfeld, E. (1984). Contingent claims analysis of corporate capital structures: An empirical investigation. *The Journal of Finance, 39*(3), 611–625.

Karatzas, I., & Shreve, S. E. (1991). *Brownian motion and stochastic calculus* (2nd edn.). Berlin: Springer-Verlag.

Karatzas, I., & Shreve, S. E. (1998). *Methods of mathematical finance*. New York, NY: Springer-Verlag.

Merton, R. (1974). On the pricing of corporate debt: The risk structure of interest rates. *Journal of Finance, 29*, 449–470.

Neftci, S. N. (1996). *An introduction to the mathematics of financial derivatives*. San Diego, CA: Academic Press.

Oksendal, B. K. (1995). *Stochastic differential equations* (4th edn.). Berlin: Springer-Verlag.

Press, W. H., Teukolsky, S. A., Vetterling, W. T., & Flannery, B. P. (1992). *Numerical recipes in C: The art of scientific computing* (2nd edn.). Trumpington Street, Cambridge: Cambridge University Press.

Royden, H. L. (1988). *Real analysis*. Englewood Cliffs, NJ: Prentice-Hall Inc.

S&P. (2011). General criteria: Principles of credit ratings. *Technical report*. Standard & Poor's Global Inc.

Sundaresan, S. (2013). A review of merton's model of the firm's capital structure with its wide applications. *Annual Review of Financial Economics*, 1–21.

Part II
Diagnostic Tools

Box (1976) wisely suggests to the quantitative analyst that:

> Since all models are wrong the scientist must be alert to what is importantly wrong. It is inappropriate to be concerned about mice when there are tigers abroad.

To understand what is importantly right or wrong, however, one requires diagnostic tools. Multiple model implementations, based on alternative assumptions, are an excellent start. Actively seeking other sources of comparison is also highly recommended. Chapter 6 highlights a natural area of comparison by introducing the quantitative analyst to some of the methods used by regulators to assess the credit riskiness of regulated entities. The Basel Committee on Banking Supervision's internal-rating based (IRB) approach—and Gordy (2003)'s granularity adjustment extension—are helpful lenses through which one can examine this important area. Regulators have availed themselves of a variety of modelling tools to—within the context of their own constraints—meet their needs. Closely following this active and expanding area is thus an important diagnostic for any internal-modelling effort. Another key tool is the notion of risk attribution. In market-risk circles, this amounts to allocating one's risk to various market risk factors. In the credit-risk setting, however, this involves decomposing one's risk estimates into obligor or sectoral contributions. This important topic is addressed in Chap. 7. While technically demanding, and often computationally intensive, risk attribution yields enormous insight into one's measures, aids in their interpretation, and basically helps the quantitative analyst distinguish between *tigers and mice*. The final contribution in this section, found in Chap. 8, relates to the Monte-Carlo method. Stochastic simulation is a powerful technique and, quite frankly, the quantitative credit analyst would be lost without it. Like all such tools, it demands care and caution. Notions of convergence, confidence intervals, and variance reduction are thus concretely presented to enhance understanding and help avoid common pitfalls.

References

Box, G. E. P. (1976). Science and statistics. *Journal of American Statistical Association, 71*(356), 791–799.

Gordy, M. B. (2003). A risk-factor model foundation for ratings-based bank capital rules. *Journal of Financial Intermediation, 12*, 199–232.

Chapter 6
A Regulatory Perspective

> *If you have ten thousand regulations, you destroy all respect for the law.*
>
> (Winston Churchill)

Having, in the preceding chapters, covered a broad range of models, we turn our attention to the regulatory world. Let's first go back briefly to first principles and summarize some key points. The default loss, for a portfolio of N exposures, is written as

$$L_N = \sum_{n=1}^{N} c_n \mathbb{I}_{D_n}, \qquad (6.1)$$

where c_n is the exposure of the nth obligor and $\mathbb{I}_{D_n}$ is a binary indicator variable describing the nth default event. What, beyond all the technical details, have we learned so far? Although default is deceptively simple to describe, as we see in equation 6.1, its statistical distribution can be quite complex and subtle. To complicate matters, for the same portfolio, as we've seen and demonstrated in many ways, different models lead to different loss distributions for L_N. While completely understandable and reasonable from a conceptual standpoint, this fact presents a number of challenges. In particular, combined with the paucity of default data and the virtual impossibility of performing back-testing in the credit-risk setting, it makes model selection a difficult task. For this reason, therefore, we have already argued frequently for the use of a suite of models.

There are, of course, shared patterns across different default-risk model families. In other words, there are common modelling elements shared by all models. More specifically, all models require:

- a description of the marginal distribution of each default loss;
- a choice regarding the joint distribution of all default losses; and
- a decision with respect to the degree and nature of default dependence.

All models, which we have examined so far, have involved decisions along these dimensions. Now, we bring the regulator into the picture. Regulators, who suggest and, in some cases, impose common *model* standards for the determination of regulatory capital are no exception. They grapple with these key dimensions and must take some hard decisions. Furthermore, use of a range of models to inform economic-capital decisions is not a legitimate choice in the regulatory setting. Regulators thus face a number of constraints not shared by other modellers. To understand these constraints, however, it is constructive to start with what regulators are trying to accomplish.

Decisions made by regulators have an important direct impact on quantitative analysts. Despite the fact that internal models inevitably differ—as they should— from regulatory choices, the current framework employed by key regulators creates a common language for risk calculations. Regulatory models, despite ineluctable shortcomings and simplifications, represent an important benchmark for internal models. Differences are expected, but explanations and justifications of these deviations are required and helpful for the enhancement of understanding on both sides. This chapter thus seeks to discuss the current regulatory environment in detail. We do not, however, take a competing-model perspective. Instead, we look to provide a familiarity with this important activity; regulatory frameworks are, in fact, a tool for quantitative analysts in the same way that financial statements are a support for business managers. They provide important background and comparison for one's decisions. Moreover, as will become quickly apparent, virtually all of the key regulatory choices are motivated by the extant modelling literature. Fluency with both internal and regulator models is thus an important advantage for the quantitative credit analyst.

6.1 The Basel Accords

The most influential and widely used financial regulatory framework, as of the writing of this book, is undoubtedly the so-called Basel accords. While the name is familiar to many, a bit of history and context might still be beneficial. The Basel Committee on Banking Supervision (BCBS)—housed by the Bank of International Settlements (BIS)—was established in 1975. Its host organization, the BIS, was established in 1930 to oversee German war-reparation payments following WWI.[1] Its focus later shifted to fostering cooperation among member central banks— see, for example, Yago (2013). The BCBS was, therefore, established rather late in the BIS' history, but its focus is entirely consistent with this revised mandate. In fact, BCBS' stated purpose is "to encourage convergence toward common approaches and standards." It has, however, no formal authority and issues

---

[1] In the interests of full disclosure, the author was employed, although not involved with the BCBS, by the BIS from 2008 to 2016.

6.1 The Basel Accords   289

no *binding* regulation. Instead, it provides supervisory standards, guidelines and describes *best practice* for banking supervision. The idea is that these elements, if deemed satisfactory, will be adopted into national regulation. To achieve this, they have to be generally acceptable, reasonable, and implementable.

What precisely are regulators trying to achieve? The short answer is that regulators seek fairness and a level playing field. The specifics, however, are a bit more slippery. There is an enormous amount of documentation available from the BCBS, other regulators, and third parties. It is not always easy to read and represents something of a moving target; as a result, referenced documents and Internet searches do not always represent the latest information. Not being an expert on regulation, returning to first principles to clarify their role is thus a safe and helpful path. The main stated objectives of BCBS, whose advice represents our key perspective in this chapter, include:

- enhancing economic stability;
- safeguarding solvency of the banking sector; and
- supporting adequate risk-based capital.

Although this is probably not the standard way to describe BCBS objectives, they do represent important and supportable ambitions. Moreover, these objectives are clearly interlinked. Banks play an extremely important role in global and local economies. As a consequence, clear rules and a level playing field are decidedly in the public's interest. The assurance of common and reasonable rules for regulatory capital is a critical aspect of ensuring fairness and stability.

Although the BCBS has a lofty and onerous mandate, it does not operate in a vacuum. It also faces concrete constraints. We've already indicated that it has no official legal status. It also imposes numerous constraints on its own regulations and standards. Two important examples include:

- comparability across institutions; and
- data constraints.

Use of internal systems, absent a detailed prescriptive approach, naturally reduces comparability. To this point, we have already seen 15 distinct models. If we take 15 large banks and each is permitted to choose their favourite approach, harmonizing and comparing their individual risk assessments will prove arduous; such variety would also make the necessary periodic regulatory validation process significantly more difficult. This is not a statement about the efficacy of the individual internal models, nor the banks' capacity to implement them. It is rather that internal models are intended to inform internal discussion and decisions.

Regulatory models are, by construction, built for comparison. This tension between internal and regulatory models is analogous to the difference between management and financial accounting. Management accounting, although following a broad set of common principles, can, and should, be highly specialized to a given company. Financial accounting standards are, by contrast, common and any individual-firm deviations require careful scrutiny and justification. The regulators play, in this sense, a role quite similar to a firm's external auditors. Internal model developers and quantitative analysts are similar to the management accountants—their role relates more to internal decision making than global comparability. Both frames of reference—whether in the realm of accounting or modelling—are essential to the firm's well being, but they play different, albeit complementary, roles.

It is also the case that BCBS does not wish to place an overly large computational and system's burden on the financial community. Not only are complex and varied models often problematic for certain firms, they make regulatory oversight very difficult. This argues for a simplified, formulaic approach to determining appropriate levels of regulatory capital. BCBS' principal job is to find a balance between fairness, comparability, and complexity while still attaining its societally important objectives. This is not an easy road to walk and it argues for patience and understanding with the regulatory process. It may be at times slow and somewhat cumbersome, but this is understandable, since it is not a straightforward task.

The consequence of this background is that the Basel guidance for the determination of risk-capital requirements associated with default risk—which is our focal point in this chapter—is a complex and evolving area. Basel III recommendations are, as of the publication of this book, under discussion with a target implementation date of 31 March 2019. Basel II recommendations remain in force and this will be our emphasis. One of the keystones of Basel II is the recommendation that financial institutions adopt the so-called *internal-ratings based* (IRB) approach, which allow the internal specification of key model parameters such as unconditional default probabilities, exposures-at-default, and loss-given-default.

The actual capital allocation computations, as we'll see shortly, are a highly stylized collection of formulae based on the previously discussed Vasicek (1987, 1991, 2002) model. That is, they are closely linked to the one-factor Gaussian threshold model. In fact, given certain simplifying assumptions, it is closer to the asymptotic single risk-factor (ASRF) model that we derived in Chap. 4 as a limit-loss case for the one-factor Gaussian threshold model. This leads to some understatement of risk, which is accommodated, in a formal adjustment, by taking (mathematically complicated) recourse to the one-factor CreditRisk+ model. The punchline is that extensive knowledge of the field of default-risk modelling is a decided advantage in navigating the occasionally difficult terrain of Basel's framework for economic capital stemming from default-credit risk.

6.1.1 Basel IRB

There are many areas of regulatory guidance. Officially, the Basel accords cover market, credit, and operational risk. We thus need to add a bit of an accent to our discussion. As indicated, Basel II/III recommend that financial institutions adopt the IRB approach for regulatory capital. This will be our key perspective and will permit us to address a broad range of interesting and relevant issues for the quantitative credit analyst.

6.1 The Basel Accords

Given the complicated confluence of objectives and constraints facing the BCBS, it should be no surprise that the IRB approach makes a number of simplifying assumptions. There is, however, one key assumption that dominates all of the other choices. Specifically, the IRB approach makes the assumption of *portfolio invariance*. The implication of this term may not be immediately obvious, so we will define it explicitly. Portfolio invariance implies that the risk of the portfolio:

- depends only on the characteristics of the individual exposure; and
- is independent of the overall portfolio structure.

Simply put, this means that the risk of the portfolio depends only on the characteristics of the individual loans and is independent of the overall portfolio. This is the key tenet—see BIS (2001, 2004, 2005) for a (much) more detailed discussion—of the IRB approach.

Its importance warrants a bit of additional colour. Imagine that you have two portfolios with 50, and 5,000 credit counterparties, respectively. Practically, portfolio invariance means that adding a new position to each portfolio has the same additive impact on the final risk figures. The composition of the existing portfolio does not matter. This assumption is highly unrealistic. We might accept this, provided the new position is relatively small, as a first-order approximation for the large portfolio. For the smaller portfolio, however, it is much harder to countenance. In both cases, the basic idea of modern portfolio theory is that risk depends on the composition of one's portfolio. Portfolio invariance assumes this away.

The risk implications of adding a new position depend, often importantly, on the characteristics of the portfolio it is joining. While unrealistic and difficult to defend from a theoretical perspective, the assumption of portfolio invariance does have one critical advantage: it dramatically simplifies one's computations. Each position's risk contribution can be computed independently without reference to the overall portfolio and the final risk figure is simply the sum of these individual contributions.

While unrealistic, in a general sense, the Basel guidance, however, must consider political and expediency issues. As a result, the BCBS opted for an assumption of portfolio invariance to avoid the complexities associated with considering a more general portfolio context. As hinted at previously, however, this is not such a crazy assumption when considering a large portfolio. The natural conclusion is that the Basel IRB implementation was calibrated for well-diversified banks. The corollary, of course, is that it may not be well suited for institutions with concentrated exposures. Indeed, Gordy (2002) shows that VaR contributions are portfolio invariant only in the context of an asymptotic one-factor credit risk factor (ASRF) model. After reviewing the base Basel IRB framework, we will examine Gordy's result and consider the important associated notion of concentration risk in significant detail.

6.1.2 The Basic Structure

In the credit-default risk setting, it is assumed that the pricing of these instruments accounts for expected default losses. This is consistent with the basic principle of asset pricing. Risk managers and regulators, of course, worry about large unfavourable movements that are far removed from the expected outcome. The main idea of economic capital is to create a link between these unexpected losses and the amount of capital held by the institution. Quite naturally, regulators seek to establish a link between the magnitude of unexpected losses—and thus economic capital—and the overall riskiness of the assets in the organization's portfolio.

The approach is formulaic and involves, due to the assumption of portfolio invariance, the separate and independent treatment of each distinct obligor. Thus, for each credit counterparty, one computes its contribution to overall risk capital. The sum of these risk capital contributions is the overall unexpected loss and corresponds to the institution's economic-capital requirement. The risk-capital allocation for a given level of confidence is closely related—although with some slight differences—to the risk measure, $\text{VaR}_\alpha(L)$, that we have examined closely in the previous chapters. This makes it sensible to compare the Basel IRB estimates with—appropriately adjusted and calibrated—VaR results from alternative models.

We will denote the risk-capital contribution of the nth credit obligor as $\text{RC}_n(\alpha)$, where, as usual, α denotes the level of confidence. The total risk capital—or the regulatory economic capital—is thus denoted as RC_α and is defined as

$$\text{RC}_\alpha = \sum_{n=1}^{N} \text{RC}_n(\alpha), \tag{6.2}$$

where N represents the total number of credit obligors in the portfolio. Quite simply, therefore, total risk capital is the simple sum of the risk-capital contribution of each credit counterpart.

According to the current Basel IRB guidance, the risk-capital contribution of the nth obligor has the following form,

$$\text{RC}_n(\alpha) = \mu_n \mathcal{K}_\alpha(n), \tag{6.3}$$

where μ_n denotes the obligor exposure and $\mathcal{K}_\alpha(n)$ the risk-capital contribution. The former exposure term is further broken down into two components,

$$\mu_n = \underbrace{\text{Exposure-at-Default}}_{c_n} \cdot \underbrace{\text{Loss-given-default}}_{\gamma_n}, \tag{6.4}$$

$$= c_n \gamma_n.$$

6.1 The Basel Accords

c_n is the typical notation for the exposure at default.[2] Basel IRB also makes careful consideration of the loss-given-default, γ_n; in principle, each obligor has a different value. Recall that $\gamma_n = 1 - \mathcal{R}_u$, where $\mathcal{R}_u$ is the expected recovery rate for the nth credit counterparty.

The Basel guidance includes significant discussion of the exposure-at-default and loss-given-default values selected for each obligor. Exposure-at-default is intended to be the notional value of a given instrument, whereas the loss-given-default estimates should be quite conservative in nature—more specifically, they recommend a bottom-of-the-cycle perspective. While very important in practice, we will treat them essentially as data inputs, because we want to focus on the big-picture modelling elements. BIS (2001, 2004, 2005) provide, for the interested reader, more detailed discussion of these choices.

The real modelling action stems from the coefficient $\mathcal{K}_\alpha(n)$, which seeks to represent, for a given level of confidence, the unexpected component of default credit risk. Imagine that we take the one-factor Gaussian threshold model and seek to approximate the unexpected loss. This could be written, for a given level of confidence, α, as

$$\text{Unexpected Loss} = \text{Worst-Case Loss} - \text{Expected Loss}, \qquad (6.5)$$
$$= \text{VaR}_\alpha(L) - \mathbb{E}(L).$$

The unexpected loss is thus essentially the worst-case loss, for a given level of confidence, less the expected default loss. We subtract out the expected default loss, because it is, as previously mentioned, assumed to have been considered in the pricing of the underlying obligation.

Working with equation 6.5, we recall that the α default loss, all else equal, will occur when we have a large negative outcome of the systematic global variable, G. We can thus approximate the unexpected loss as,

$$\text{Unexpected Loss} \approx \mathbb{E}\left(L \,\Big|\, G = q_{1-\alpha}(G)\right) - \mathbb{E}(L), \qquad (6.6)$$

$$\approx \mathbb{E}\left(\sum_{n=1}^N c_n \mathbb{I}_{\mathcal{D}_n} \,\Big|\, G = q_{1-\alpha}(G)\right) - \mathbb{E}\left(\sum_{n=1}^N c_n \mathbb{I}_{\mathcal{D}_n}\right),$$

$$\approx \sum_{n=1}^N c_n \underbrace{\mathbb{E}\left(\mathbb{I}_{\mathcal{D}_n} \,\big|\, G = q_{1-\alpha}(G)\right)}_{p_n(q_{1-\alpha}(G))} - \sum_{n=1}^N c_n \underbrace{\mathbb{E}\left(\mathbb{I}_{\mathcal{D}_n}\right)}_{p_n},$$

---

[2] For most of our previous and subsequent analysis, we assume that $\gamma_n \equiv 1$ and thus it is equivalent to work exclusively with c_n.

$$\approx \sum_{n=1}^{N} c_n p_n(q_{1-\alpha}(G)) - \sum_{n=1}^{N} c_n p_n,$$

$$\approx \sum_{n=1}^{N} c_n \bigg(p_n(q_{1-\alpha}(G)) - p_n \bigg),$$

where $p_n(g) = \mathbb{E}(\mathbb{I}_{\mathcal{D}_n}|G) = \mathbb{P}(\mathcal{D}_n|G)$ is the conditional probability of default. The preceding development has made a heroic assumption. It assumed that the only source of default risk is systematic so that,

$$\text{VaR}_\alpha(L) \approx \mathbb{E}\bigg(L \bigg| G = q_{1-\alpha}(G) \bigg). \tag{6.7}$$

This is, in fact, the implication of assuming portfolio invariance. In general, of course, this statement is not true due to the existence of idiosyncratic risk and it is also further complicated by multiple sources of risk. In this case, we assume a single source of risk, so it is not an issue. More generally, however, it is. Equation 6.6 does, nevertheless, provide us with some insight into how the Basel-IRB risk-capital contribution was constructed.

We have defined $\mathcal{K}_\alpha(n)$ as the nth risk-capital contribution. Practically, it is a function of *two* main arguments: the unconditional default probability, p_n, and the tenor, or term to maturity, of the underlying credit obligation denoted M_n. It should be mentioned that, in the formal documentation, it is not always easy to read. These sources are regulatory documents and not designed to describe mathematical formulae. These presentational challenges should not detract from the final result; a significant amount of thought was clearly invested in the final structure. With that comment, the risk coefficient is described as

$$\mathcal{K}_\alpha(\text{Tenor, Default probability}) = \gamma_n \cdot \begin{pmatrix} \text{Conditional} & & \text{Unconditional} \\ \text{Default} & - & \text{Default} \\ \text{Probability} & & \text{Probability} \end{pmatrix} \cdot \begin{matrix} \text{Maturity} \\ \text{Adjustment} \end{matrix},$$

$$\mathcal{K}_\alpha(M_n, p_n) = \gamma_n \cdot \bigg(p_n \bigg(\Phi^{-1}(\alpha) \bigg) - p_n \bigg) \tau(M_n, p_n), \tag{6.8}$$

$$= \gamma_n \cdot \Bigg(\underbrace{\Phi \bigg(\frac{\Phi^{-1}(p_n) + \sqrt{\rho(p_n)}\Phi^{-1}(\alpha)}{\sqrt{1-\rho(p_n)}} \bigg)}_{p_n(\Phi^{-1}(\alpha))} - p_n \Bigg) \tau(M_n, p_n),$$

where γ_n is the loss-given-default assumption, $\tau(M, p_n)$ is a maturity adjustment or correction, which depends on the tenor of the credit exposure, M_n, and its unconditional default probability. $\Phi(\cdot)$ and $\Phi^{-1}(\cdot)$ denote the cumulative and inverse-cumulative Gaussian distribution functions, respectively. For the remainder

6.1 The Basel Accords

of our development, as we've done for the preceding discussion, we set $\gamma_n \equiv 1$; this does not lead to any loss of generality.

It is relatively easy to identify the similarities between equations 6.6 and 6.8. Each risk coefficient is the distance between the conditional one-factor Gaussian threshold and unconditional default probabilities multiplied by a maturity adjustment. At the heart of the risk coefficient, therefore, one finds the one-factor Gaussian threshold model; albeit a version ignoring all idea of idiosyncratic risk. While motivated by the Gaussian threshold model, we should be careful to observe that it is not a direct implementation of this approach. The Basel IRB and Gaussian threshold models are *not* directly equivalent.

A brief technical precision is also useful, because the form of the conditional default probability, $p_n(G)$, does not have precisely the same form as what was derived in Chap. 4. It is typically defined as,

$$p_n(G) = \Phi\left(\frac{\Phi^{-1}(p_n) - \sqrt{\rho}G}{\sqrt{1-\rho}}\right). \tag{6.9}$$

The numerator in equation 6.9 does not match with the equivalent term in equation 6.8. BCBS's idea, it appears, was to simplify the usage, so when g is evaluated at its worst-case α-level outcome, they have replaced $\Phi^{-1}(1-\alpha)$ with $\Phi^{-1}(\alpha)$. This changes the sign in the second term of the numerator, makes the formula easier to follow, and avoids the need to discuss the negative relationship between the systematic variable and default outcomes implicit in the one-factor Gaussian threshold model.

To return to the main point, the key element determining the degree of confidence is $p_n(G)$ evaluated at $g = \Phi^{-1}(\alpha)$. The distance

$$\left|p_n(\Phi^{-1}(\alpha)) - p_n\right|, \tag{6.10}$$

approximates unexpected losses for a highly improbably outcome of the global variable, G. Equation 6.10 is thus the principal element of the risk coefficient and, indeed, the entire Basel IRB approach. When this distance is multiplied by the counterparty's exposure, μ_n or c_n, it transforms this expected loss into units of currency. In a nutshell, that is the Basel IRB approach. What remains are essentially (important) details.

6.1.3 A Number of Important Details

The risk-capital coefficient, $\mathcal{K}_\alpha(n)$, depends, as we can see from equation 6.10 on a few additional terms: the correlation function, $\rho(p_n)$—which is typically treated as a parameter in the threshold model setting—and the maturity adjustment, $\tau(M_n, p_n)$.

While we've referred to these as details, they can exhibit an important influence on the results and require careful consideration. Let's address each in turn.

The Basel guidance refers to $\rho(p_n)$ as *asset correlation*. As we saw in Chap. 4, this parameter reduces, in the threshold model setting, to the contemporaneous linear correlation between the state-variables of each pair of credit counterparties.[3] One should imagine that, as motivated by Merton (1974) and addressed in detail in Chap. 5, these state variables proxy a firm's assets. As discussed in Chaps. 4 and 5, this parameter forms, along with one's distributional assumptions, an important link with the notion of default correlation. Indeed, a significant effort was made to calibrate this parameter to a desired or observed level of default correlation.

In a regulatory setting, it is very difficult, and generally not advisable, to allow free selection of one's input parameters. Creating incentives to understate one's risk notwithstanding, allowing each institution leeway in their parameters choices can strongly undermine comparability of results. Consequently, regulators generally provide very clear and concrete guidance on key model parameters. The Basel IRB approach is no exception to this rule.

The Basel IRB approach recommends a specific, and often distinct, level of the state-variable correlation function, $\rho(p_n)$, for each credit counterparty with the following form,

$$\rho(p_n) = \Lambda \cdot \left(\rho^- q(p_n) + \rho^+ \left(1 - q(p_n)\right) \right), \tag{6.11}$$

where

$$q(p_n) = \frac{1 - e^{hp_n}}{1 - e^h}, \tag{6.12}$$

and

$$\Lambda = \begin{cases} 1.25 & \text{If firm } n \text{ is a large and regulated financial institution} \\ 1 & \text{otherwise} \end{cases}. \tag{6.13}$$

The actual choice of ρ_n is thus a convex combination of the limiting values ρ^- and ρ^+ in terms of $q(p_n)$. The value $q(p_n)$ is confined, by construction, to the unit interval, which ensures the asset-correlation coefficient lies in the interval, $\rho(p_n) \in [\rho^-, \rho^+]$. The Λ parameter is a new addition from Basel III and will not be in effect until, as of the writing of this document, for more than another year. As such, we provide it for completeness, but do not include it in our computations.

While equations 6.11 to 6.13 are visually fairly difficult to digest, practically they are relatively straightforward. The state-variable correlation function lies in the

---

[3]Or, at least, it makes an important contribution to this relationship. Recall, from Chap. 4, that the state-variable correlation is equivalent $\rho_{nm} = \sqrt{\rho_n \rho_m}$ in the one-factor threshold model for two arbitrary obligors n and m.

6.1 The Basel Accords

interval, $\rho(p_n) \in [\rho^-, \rho^+]$. In the current guidance, $\rho^- = 0.12$ and $\rho^+ = 0.24$. Where precisely it lies along this interval depends on the default probability of the credit counterparty.

The BCBS justifies the structure of equation 6.11 through the claim that asset correlation is a decreasing function of the default probability. The logic is that if a credit counterparty has a large default probability, then much of its inherent default risk is idiosyncratic and, as such, it is less highly correlated with other credit counterparties. Conversely, very low risk counterparties predominately face systematic risk and, as such, exhibit higher levels of asset correlation. Normally, this state-variable correlation is determined pairwise for each two counterparties—while sensible, this is presumably not considered since it violates the portfolio-invariance condition. Empirically, since the range is restricted to [0.12, 0.24], the differences are not dramatic.

Figure 6.1 highlights the evolution of the functions, $q(p_n)$ and $\rho(p_n)$, for a range of plausible unconditional default probabilities. Practically, the parameter h in equation 6.12 is set to -50, which suggests that the weight on ρ^- grows very quickly as a function of p_n. For default probabilities exceeding about 10%, the correlation function is essentially equal to the lower bound. Below a default probability of 3 or 4%, for all practical purposes, the asset-correlation is conversely close to its maximum value.

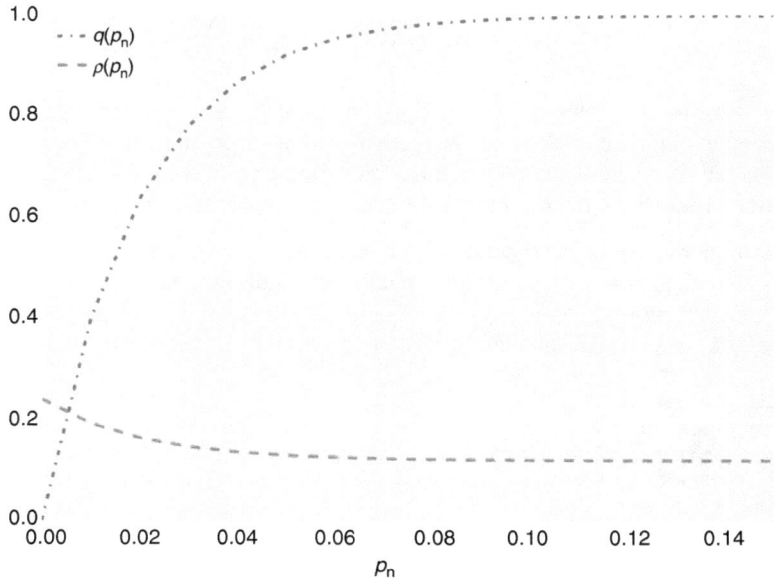

Fig. 6.1 *Basel asset correlation*: This figure outlines the weighting function, q_n, from equation 6.12 and the associated asset correlation estimate—from equation 6.11—for different levels of unconditional default probability. For values of p_n greater than about 0.12, the ρ_n parameter takes the value of ρ^- or 0.12.

The parameterization in Fig. 6.1 reflects directly the idea that high-risk credit counterparties have a greater degree of idiosyncratic risk and, as such, a lower degree of systematic correlation with other counterparties.[4] The former point is captured by making asset-correlation an *inversely* proportional function of unconditional default probability. The latter element is captured, in a limited way, by the introduction of Λ. The corollary of the first fact is that high credit-quality obligors are more highly exposed to systematic risk.[5] None of this is perfect, but it has a logical appeal and provides concrete guidance for regulated entities.

The maturity adjustment, τ, arises from the notion that long-term credit exposures are riskier than their shorter-term equivalents. As such, the Basel guidance seeks to increase the risk-capital allocation for longer-term maturities. Moreover, the maturity element is also a function of the default probability. That is, the risk impact of longer maturities is a decreasing function of the firm's unconditional default probability. The actual maturity adjustment is written as,

$$\tau(M_n, p_n) = \frac{1 + \left(M_n - \frac{5}{2}\right) b(p_n)}{1 - \frac{3}{2} b(p_n)}. \tag{6.14}$$

The form of equation 6.14, while somewhat unique, is quite clearly linear in $b(p_n)$. It is also evidently a positive function of M_n. The embedded function, $b(p_n)$, requires specification. It has been given as very specific form,

$$b(p_n) = \left(0.11852 - 0.05478 \ln(p_n)\right)^2. \tag{6.15}$$

This is occasionally referred to as a slope function. This is not immediately easy to understand and the coefficients are provided to a surprising degree of accuracy. As with the other elements of the Basel IRB framework, it is a result of careful reflection. There are *four* important elements of the maturity adjustment,

- it is an increasing linear function of the instrument's tenor, M_n;
- it is, *ceteris paribus*, larger for high-quality credit obligations;
- it is, *ceteris paribus*, smaller for low-quality credit counterparts; and
- when $M_n = 1$, it takes a neutral value of unity.[6]

---

[4]Practically, however, it should be stressed that the correlation function in the IRB setting only influences the conditional default probability used in the risk-capital coefficient computations. There is no notion of an underlying correlation matrix for all issuers given the principle of portfolio invariance.

[5]The actual values of ρ^-, ρ^+ and h were, presumably, determined through a combination of empirical work, calibration and compromise.

[6]This is easily verified,

$$\gamma(1, p_n) = \frac{1 + \left(1 - \frac{5}{2}\right) b(p_n)}{1 - \frac{3}{2} b(p_n)} = \frac{1 - \frac{3}{2}\overline{b(p_n)}}{1 - \frac{3}{2}\overline{b(p_n)}} = 1. \tag{6.16}$$

6.1 The Basel Accords

What is the logic behind this adjustment? It stems from stylized empirical evidence regarding the term structure of credit spreads, which inform the probability of default. Over long horizons, the probability of default can, and does, change. This phenomenon is termed credit migration and, in principle, it can be either positive or negative. That is, firms can experience both increases and decreases in their relative credit quality—the longer the time horizon, the greater the likelihood of migration. Empirically, the term structure of high-quality borrowers' credit spreads are typically upward sloping. This means that the credit spread is larger at longer maturities relative to shorter tenors. The reason is that for high-quality credits, migration can either stay the same or move down. Since we cannot, by definition, see any appreciable credit improvement over time, this is reflected in the credit spread.

The situation is different for low-quality credits. The term structure of poorly rated credit spreads is typically downward sloping. The risk, therefore, is high at inception. Conditional on survival, however, there is a stronger probability of positive credit migration (i.e., upgrades) and, as such, a reduction of the credit spread over time. Since the maturity correction is equal to unity for all default probabilities, the behaviour is altered for short-term tenors. The correction is corresponding larger for low-quality and smaller for high-quality credits. Again, this is consistent with both logic and empirical observation. The structure of the maturity adjustment, albeit in a stylized fashion, attempts to incorporate these stylized facts observed in actual markets into the risk-capital computation.

Figure 6.2 provides some graphical insight into the maturity correction. The left-hand graphic describes the slope term, $b(p_n)$, from equation 6.15 for different possible levels of unconditional default probability. It is a decreasing function of the unconditional default probability. The right-hand graphic describes the maturity adjustment summarized in equation 6.14 for various fixed choices of p_n. What is clear is that the maturity adjustment is significantly higher for high-quality

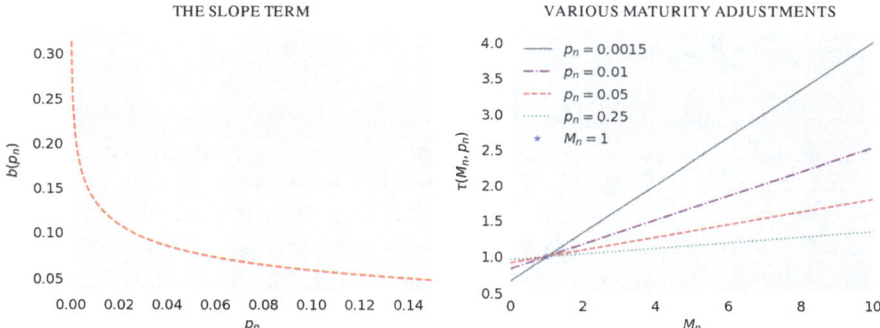

Fig. 6.2 *Maturity correction*: The first graphic in this figure describes the slope term, $b(p_n)$, from equation 6.15 for different possible levels of unconditional default probability. The second graphic describes the maturity adjustment—see equation 6.14—for various choices of p_n. There are rather different consequences for high- and low-quality credit counterparties depending on their tenor.

Algorithm 6.1 Asset-correlation and maturity adjustment functions

```
def getQ(p_n,myH=-50):
    return np.divide(1-np.exp(myH*p_n),1-np.exp(myH))
def getRho(p_n,rhoMin=0.12,rhoMax=0.24):
    myQ = getQ(p_n)
    return rhoMin*myQ + rhoMax*(1-myQ)
def getMaturitySlope(p_n,p0=0.11582,p1=-0.05478):
    return np.power(p0+p1*np.log(p_n),2)
def getMaturityAdjustment(tenor,p_n,p0=2.5):
    myB = getMaturitySlope(p_n)
    return np.divide(1+(tenor-p0)*myB,1-(p0-1)*myB)
```

long-term obligations than their low-quality counterparts. In short, as previously suggested, there are very different consequences for high- and low-quality credit counterparties depending on their tenor.

In contrast to many of the approaches discussed in previous chapters, implementation of the asset-correlation and maturity adjustment elements is not terribly complex. This is, of course, a conscious choice by the BCBS and is inherent in the design of the Basel IRB approach. Algorithm 6.1 summarizes *four* straightforward Python sub-routines for computation of the $\rho(p_n)$ and $\tau(M_n, p_n)$ functions: these include getQ, getRho, getMaturitySlope, and getMaturityAdjustment.

None of these functions require more than a few lines. Moreover, all of the BCBS-supplied numerical coefficients are provided as default arguments. They can, if so desired, be overridden in the function call. This is an attempt to minimize the poor form associated with hard-coding values into one's implementation.

6.1.4 The Full Story

We now have all of the ingredients required to revisit the risk-capital coefficient summarized in equation 6.8. Figure 6.3 provides graphical illustrations of the risk-capital coefficient (i.e., $\mathcal{K}_\alpha(M_n, p_n)$) as a function of its two arguments: tenor (i.e., M_n) and unconditional default probability (i.e., p_n). The left-hand graphic outlines the range of risk-capital coefficient outcomes across a range of plausible default probabilities for fixed tenors. We can interpret $\mathcal{K}_\alpha(n)$ as a percentage of the loss-given-default adjusted exposure-at-default. That is, if the coefficient takes a value of 0.10, the exposure-at-default is $100 and the recovery rate is 40%, then the risk-capital associated with this credit counterpart is 10% of $60 or $6. This is performed independently for all instruments in the portfolio and the sum of all the individual capital amounts forms the final IRB risk-capital requirement.

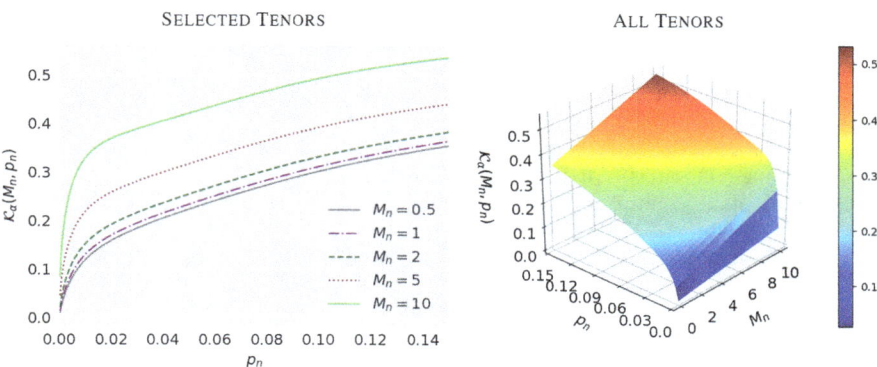

Fig. 6.3 *Risk-capital coefficient*: This figure illustrates the risk-capital coefficient as a function of its two arguments: tenor and unconditional default probability. The first graphic outlines the range of coefficient values across a range of plausible default probabilities for fixed tenors—the coefficient increases with both tenor and default likelihood. The second graphic summarizes this relationship in three-dimensional form. In all cases, the loss-given-default parameter is assumed to be equal to unity.

The right-hand graphic of Fig. 6.3 describes the interaction of the risk-capital coefficient for a broad range of tenors and unconditional default probabilities; in all cases, the loss-given-default parameter is assumed to be equal to unity. As we would expect, Fig. 6.3 demonstrates clearly that the risk-capital coefficient increases with both tenor and default likelihood. For small default probabilities and short tenors, the risk-capital coefficient typically only amounts to a few percentage points of the actual exposures. For long-tenor obligations, particularly those with low credit quality, the coefficient can exceed 0.5. Indeed, even for high-quality obligors, with tenors reaching 10 years, the risk-capital coefficient can exceed 0.30. The greatest obligor risk arises, quite reasonably, with long tenor and large default probability. All of these values, quite naturally, are scaled down linearly as one decreases the loss-given-default parameter.

Algorithm 6.2 provides the implementation of the risk-coefficient computation: it makes use of a number of functions from Algorithm 6.1. It does not require an enormous amount of effort. The function, `computeP`, is the conditional default probability, which is imported from the `thresholdModels` library used in Chap. 4. This explains, as highlighted in equation 6.9, the use of $\Phi^{-1}(1-\alpha)$ or `norm.ppf(1-alpha)`. The actual risk-capital coefficient is, in fact, very compact. It can be written as `np.multiply(pG-p_n,ma)`. This is, quite certainly, an explicit goal of the BCBS.

Algorithm 6.2 Basel IRB risk capital functions

```
import thresholdModels as th
def getBaselK(p_n,tenor,alpha):
    g = norm.ppf(1-alpha)
    rhoBasel = getRho(p_n)
    ma = getMaturityAdjustment(tenor,p_n)
    pG = th.computeP(p_n,rhoBasel,g)
    return np.multiply(pG-p_n,ma)
def getBaselRiskCapital(p_n,tenor,c,myAlpha):
    myCounter = len(myAlpha)
    riskCapital = np.zeros(myCounter)
    for n in range(0,myCounter):
        riskCapitalCoefficient=getBaselK(p_n,tenor,myAlpha[n])
        riskCapital[n] = np.dot(c,riskCapitalCoefficient)
    return riskCapital
```

The Basel IRB approach, as we've seen in this section, is a broad, formulaic, *one-size-fits-all* methodology for the computation of default-risk capital requirements. Its underlying assumption of portfolio invariance, while easing implementation for a broad range of global financial institutions and creating a common language for the discussion of risk exposures, is nevertheless not particularly well suited for serious risk management and assessment. This is entirely by construction—it is a tool for determination of economic capital. Given its importance, it nonetheless forms a critical part of the backdrop for discussion of default risk.

There is, in regulatory circles, a keen awareness of this fact, as is suggested in BIS (2005) by the following quote:

> Banks are encouraged to use whatever credit risk models best fit their internal risk assessment and risk management needs.

The IRB approach is required by regulators and supervisors, but is not used exclusively by sophisticated financial institutions for internal management of credit-default risk. It instead acts as a complement and regulatory tool. As a comparator, these computations have significant value. As a base method for the evaluation of default risk, however, they should be supplemented with more sophisticated and meaningful portfolio-variant approaches.

In the next section, we will turn to apply the Basel IRB approach to our running 100-obligor example. This will put its risk estimates into perspective and help us gain some instructive experience. Our analysis will highlight the role of idiosyncratic risk and its absence in the base methodology. This fact is well understood by the BCBS and naturally leads us to a number of (unfortunately heavy) computational adjustments, which form the final section of this chapter.

6.2 IRB in Action

As usual, we turn to consider the implications of applying the Basel IRB portfolio to the 100-obligor example introduced in Chap. 1. To perform this computation, however, we require additional information: the tenors of the underlying obligations. Without them, we are not capable of computing the necessary maturity adjustments. We've drawn the maturities uniformly from the interval, [0.2, 1.8]. This implies the average is about one year, which if you recall from equation 6.16, is essentially neutral. The results are outlined in Fig. 6.4 with an examination of the (simulated) relationship between tenors, exposures, and unconditional default probabilities.

We selected a relatively neutral choice of average tenor to permit comparison to a selection of the models considered in the preceding chapters. In these settings, the notion of the tenor of the underlying obligation is *not* considered. As has been indicated on numerous occasions, comparing a specific modelling choice to other models is perhaps our most effective tool in mitigating credit-risk model risk. The lesson also applies to gaining a better understanding of the IRB approach. To accomplish this, we include *five* alternative models. These choices are summarized in Table 6.1 with an abbreviation and some pertinent model-based information.

All comparative methodologies are one-factor models to ensure a reasonable degree of comparability. To further place the various models on a similar footing, attempts have also been made to give them a common parametrization. As we see in equation 6.11, each obligor has a different asset-correlation coefficient depending on its unconditional default probability. The average value, however, is approximately 0.19. This gives rise to a default probability of roughly 0.02 in the Gaussian threshold model. The other models have thus had their models calibrated to this level

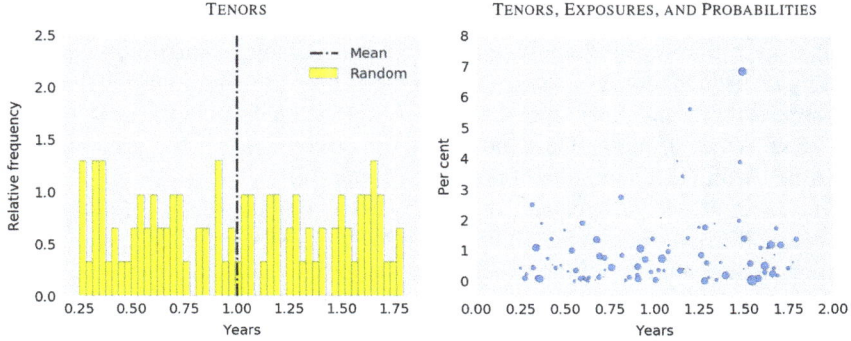

Fig. 6.4 *Obligor tenors*: The computation of the Basel IRB risk-capital allocation involves a maturity adjustment, which is a function of the underlying tenor of the counterparty exposure. The underlying left-hand figure illustrates the randomly generated tenors associated with our usual 100-obligor example portfolio. The right-hand graphic provides a comparison with the unconditional default probability and exposures—the size of the dot in the scatterplot indicates the size of the exposure.

Table 6.1 *Benchmark models*: This table summarizes the *five* benchmark models used to gain some insight into the nature of the Basel IRB risk-capital computation.

Model	Abbreviation	Location	Parameters	$\mathbb{P}(\mathcal{D})$
Independent-default binomial	B	Chap. 2	$\rho = 0$	0.00
Basel internal-rating based	IRB	Chap. 6	$\frac{1}{N}\sum_{n=1}^{N} \rho_n = 0.19$	0.02
Asymptotic single-risk factor	ASRF	Chap. 4	$\rho = 0.19$	0.02
One-factor Gaussian threshold	$1 - \mathcal{N}$	Chap. 4	$\rho = 0.19$	0.02
One-factor t-threshold	$1 - \mathcal{T}$	Chap. 4	$\rho = 0.03$ and $\nu = 9.40$	0.02
Beta-binomial mixture	$\beta - B$	Chap. 3	$\alpha = 0.41$ and $\beta = 43.03$	0.02

of default probability. The result, therefore, is a relatively high degree of parametric correspondence among the six models summarized in Table 6.1.

There is, of course, one significant difference between these models and the other choice. The Basel IRB approach is the only methodology that is founded on the notion of portfolio invariance. By virtue of its independence assumption, the binomial model does not incorporate any systematic element, but its contributions are nonetheless not additive. In this sense, the binomial approach is essentially the opposite of the Basel IRB approach: one is entirely systematic, whereas the other is completely idiosyncratic. The threshold and mixture models use the characteristics of the portfolio to tease out both the systematic and idiosyncratic elements. The only model that, conceptually, shares some notion of portfolio invariance is the ASRF approach.

If we recall, from Chap. 4, the limit-loss distribution of the one-factor Gaussian threshold model—which has also been termed the ASRF model—is derived by allowing the number of individual exposures to tend to infinity. This implies that each exposure is, in fact, infinitesimally small. Indeed, the notion of exposure disappears and one works solely with the average default probability. While not a terribly realistic model, its selection for our benchmarking exercise is not random. It so happens that—and this will be explored in significant detail in the next section— the ASRF is the *only* default-risk model that can be considered to be mathematically consistent with the assumption of portfolio invariance.

Figure 6.5 provides, for each of the six models highlighted in Table 6.1, the model tail probabilities over the range [0.9500, 0.9999]. Since the Basel IRB approach, by construction, subtracts out the expected loss, we have done the same for the risk estimates provided in this section. This has not been the general practice in our discussions to this point. Furthermore, given the small level of expected loss in our sample portfolio, it is admittedly not a large factor. Nevertheless, this was performed to permit the highest possible degree of comparability.

A number of interesting observations can be drawn from Fig. 6.5. First of all, there is a strong degree of similarity between the Basel IRB and ASRF models. Visually, it is difficult to identify a difference, which underscores our discussion of equivalence with respect to portfolio invariance. A second point is that, up

6.2 IRB in Action

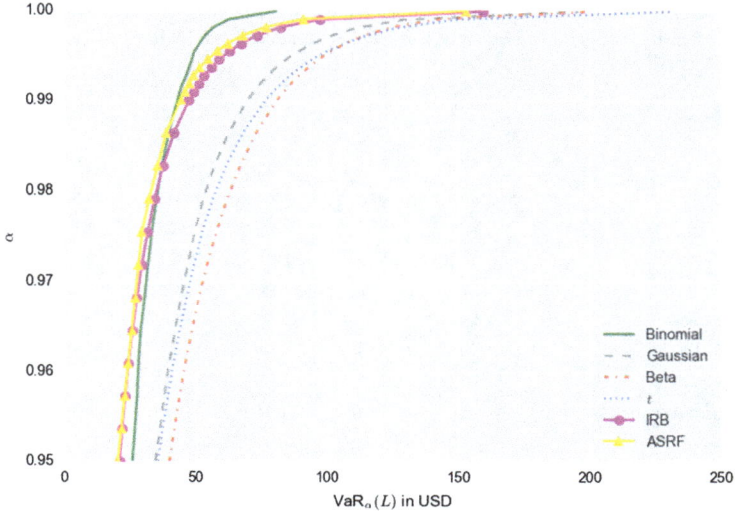

Fig. 6.5 *Comparative tail probabilities*: To put the Basel IRB estimates into context, this figure provides a comparative analysis of the tail probabilities associated with a broad range of relative model. In all cases, save the IRB and ASRF models, the expected loss has been subtracted to generate economic-capital estimates.

until about the 99.5th percentile, all models, including the independent-default approaches, exhibit a higher level of risk than the Basel IRB and ASRF models. This distance stems principally from the idiosyncratic risk explicit in the other modelling frameworks. The final point is that, at the highest level of confidence, the Basel IRB and ASRF models make a large jump. They do not, however, close the gap with the other models. That the proportional difference is significantly reduced underscores the criticality of the systematic factor for extreme levels of default loss.

Table 6.2 provides the numerical results to support Fig. 6.5. For the most part, there is nothing new in Table 6.2. The final row, however, is nonetheless quite pertinent. We have considered a relatively small and manageable 100-obligor portfolio. The simulations employed 10 million iterations and required between 15 and 60 seconds; the total time depends, of course, on the complexity of the model. The Basel IRB model, at less than one second, in stark contrast, requires virtually no computational effort. For large institutions with enormous and complex portfolios, this is a relevant and important consideration. It also has important implications for regulatory oversight and verification of Basel IRB estimates.

This is not to argue that computationally intensive models are inherently problematic. Hardware is relatively inexpensive in the current environment and there are many useful parallelization techniques to reduce computational expense.[7] In a

---

[7]See, for example, Zaccone (2015) for more detail on these techniques.

Table 6.2 *Benchmarking results*: The underlying table outlines the key results from the Basel IRB model and the *five* comparator benchmark models summarized in Table 6.1. The Basel IRB and ASRF models appear to present the largest similarities.

Quantile	IRB	B	ASRF	$1-\mathcal{N}$	$B-\beta$	$1-\mathcal{T}$
95.00th	$21	$26	$21	$35	$40	$36
97.00th	$29	$31	$28	$46	$53	$48
99.00th	$48	$43	$45	$70	$81	$77
99.90th	$98	$63	$92	$129	$142	$149
99.97th	$146	$77	$140	$181	$187	$213
99.99th	$160	$81	$153	$196	$200	$231
Expected loss	$9	$9	$5	$9	$10	$9
Loss volatility	$9	$13	$9	$18	$20	$19
Iterations	10,000,000					
Computational time (seconds)	0.0	14.0	0.2	50.8	18.8	55.9

regulatory setting, however, it is important to carefully manage the burden imposed on the regulated entities. Thus, although the Basel IRB approach lacks somewhat in modelling realism, it certainly offers benefits in terms of computational simplicity.

6.2.1 Some Foreshadowing

Computational austerity and portfolio invariance also offer benefits for the interpretation of the results associated with one's Basel IRB output. It permits one to easily determine the contribution of each individual obligor to the overall risk estimate. A bit of reflection reveals the usefulness of such a computation. In the previous chapters, we have used a number of techniques to extract a default-loss distribution. Our risk measures describe different aspects of this distribution. Since we have adopted a portfolio perspective, each risk estimate is an aggregate statement. An ability to understand how each credit counterpart contributes to these overall figures is valuable. It would permit easier interpretation, communication, and troubleshooting of one's computations.

In general, however, such contributions are not easy to compute. In the Basel IRB setting, due to the portfolio invariance assumption and the attendant additivity of obligor elements, it is trivial to determine the contribution of each position to the overall IRB economic-capital. Algorithm 6.3 summarizes the code; it is almost laughably easy. One merely loops over each exposure and calls the `getBaselRiskCapital` function from Algorithm 6.2. All of the work was, in fact, performed when building the model.

6.2 IRB in Action

Algorithm 6.3 A simple contribution computation

```
myContributions = np.zeros([len(alpha),N])
for n in range(0,N):
    myContributions[:,n]=getBaselRiskCapital(p[n],tenor[n],c[n],alpha)
```

Table 6.3 *Risk-capital contributions*: This table highlights the ten largest Basel IRB risk-capital contributions. It also provides useful obligor-level information for the interpretation and verification of the results.

Rank	#	p_n	c_n	Tenor	Contribution	
					USD	Percent
1	13	6.8%	$30	1.5	$11	7%
2	1	1.2%	$27	1.7	$7	4%
3	33	1.1%	$28	0.9	$6	4%
4	30	0.7%	$31	1.0	$6	4%
5	65	0.5%	$28	1.6	$5	3%
6	28	1.2%	$21	1.7	$5	3%
7	24	1.4%	$23	0.7	$5	3%
8	8	1.1%	$26	0.3	$5	3%
9	69	1.4%	$18	1.5	$5	3%
10	93	1.8%	$15	1.3	$4	2%
Other			$751		$101	63%
Total			$1000		$160	100%

Table 6.3 applies Algorithm 6.3 and displays the ten largest contributors at the, arbitrarily selected, 99.99th quantile. It also includes, to aid interpretation, the unconditional default probability, the size of the exposure, and its tenor. The total 99.99th quantile Basel-IRB economic-capital estimate is roughly $160. The ten largest contributions, however, contribute $59, or about 40% of the total. This is interesting and useful colour about the composition of one's portfolio. The largest contributor—accounting for $10 of economic capital—has a large default probability, a relatively long tenor, and a large exposure. All of these factors play an important role, in the Basel IRB setting, in summarizing the systematic risk associated with a given obligor.

It is thus no real surprise that the top-ten contributors all exhibit high scores on, at least, one of these three dimensions. If this were not the case, it would be a strong signal that something is amiss with one's implementation code; computations of the type found in Table 6.3 are excellent model-diagnostic tools. They are also immensely helpful in constructing internal limits on exposures or simply managing one's lending book.

Figure 6.6 restates the analysis from Table 6.3 to provide a visual perspective. It takes all of the 100 individual contributions, orders them in decreasing order, and plots them. The largest contribution—in the amount of $11 from Table 6.3—is clearly visible. The dominance of the ten or 15 largest obligors is also quite obvious. Indeed, the smallest 30 or 40 contributors play almost no role in the final risk estimate. The interpretational benefits of this computation is, for institutions and regulators alike, of significant value.

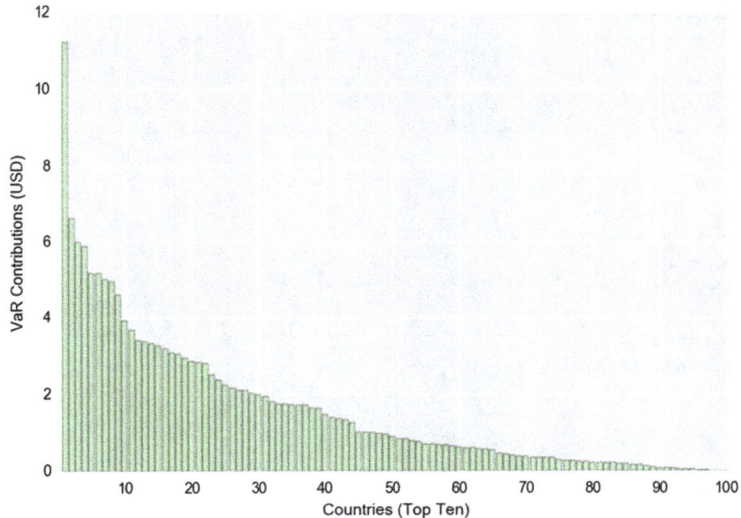

Fig. 6.6 *Ordered risk-capital contributions*: This figure provides the ordered contributions, of each of our 100 obligors, to the total Basel IRB risk-capital contribution. This is straightforward, given the assumption of portfolio invariance, but it is, in general, a difficult problem.

Obligor contributions to one's risk are a powerful model diagnostic. They are commonly used in the market-risk setting to determine the importance of various risk factors—such as interest rates, exchange rates, or credit spreads—to one's risk measures.[8] Under the typical assumptions employed in market-risk computations, this is not a terribly complex task. In the case of default-loss models, however, this is actually a fairly difficult practical problem. It requires use of either Monte Carlo methods or analytical approximations. The former is noisy and computationally expensive, while the latter is limited by dimension and demands rather complex mathematics. Obligor contributions are nevertheless sufficiently important that we assign the totality of the subsequent chapter to their discussion, review, and implementation. This important area is the principal theme of the next chapter.

Having introduced the Basel IRB approach, implemented it, and performed a detailed benchmarking exercise, we are now in an excellent position to address one of its principal shortcomings: the assumption of portfolio invariance. We have danced around the equivalence of portfolio invariance, the ASRF model, and the dominance of systematic risk in this setting. In the following section we will make these ideas much more precise and, more importantly, consider potential solutions to relax this assumption. This turns out to be a particularly thorny undertaking, given that the same general regulatory constraints still apply.

---

[8] See, for example, Bolder (2015, Chapter 11) for a more detailed discussion of this technique in the market-risk setting.

6.3 The Granularity Adjustment

The Basel-IRB approach is based upon a very important, but unrealistic, assumption: portfolio invariance.[9] Simply put, this means that the risk characteristics of the portfolio depend only on individual exposures and are independent of the overall portfolio structure. Modern portfolio theory works rather differently. If you add a new position to your portfolio—whether considering a market- or credit-risk perspective—its risk impact should, and indeed will, depend upon what you already hold.

Gordy (2003), in a very useful contribution to the literature, identifies what models, and what conditions, are consistent with the portfolio-invariance property. The list, it turns out, is not long. To get a handle on the basic concepts, he decomposes the credit-risk contribution in any model into two broad (measurable) categories. These include, for a portfolio of N distinct obligors, the following:

Unconditional Loss   This is *total* default loss, L_N.
Conditional Loss   This is systematic loss, denoted $L_N|G$.

He then poses the question: under what conditions is credit risk portfolio invariant? In essence, Gordy (2003) demonstrates that in the limit—under a number of technical conditions—when talking about portfolio invariance, only the systematic element matters. Mathematically, therefore, we can think of portfolio invariance as something like:

$$\lim_{N\to\infty} L_N \approx \mathbb{E}(L_N|G). \tag{6.17}$$

This is not terribly precisely stated, but rather more an intuitive expression.[10] The idea is that, for large, well-diversified portfolios, total default loss tends toward systematic loss. The idiosyncratic element, as in Sharpe (1963, 1964)'s CAPM structure, is diversified away. Unfortunately, in practice, not all portfolios satisfy the large, well-diversified condition.

The key finding of Gordy (2003) is that portfolio invariance is possible *only* under an asymptotic single-factor risk (ASRF) model. No other model possesses this characteristic. As discussed in Chap. 4, the ASRF is a type of limiting result, as $N \to \infty$, of the Gaussian threshold model forwarded by Vasicek (1987, 1991,

---

[9] Moreover, any work involving this assumption—such as the Standard & Poor's credit-rating agency's risk-adjusted capital ratio employed in assessing firm's credit ratings—also inherits this potential shortcoming. There is, therefore, general interest in its implications.

[10] A rigorous and axiomatic treatment is found in Gordy (2003).

2002). Gordy (2003) identifies the technical conditions under which unconditional and conditional VaR converge—this is roughly equivalent to applying the VaR operator to both sides of equation 6.17. For this to hold, *two* key assumptions are required:

1. one's portfolio must be infinitely grained; and
2. there is only a *single* systematic risk factor.

Infinitely grained, in this context, means that no single exposure may dominate total portfolio exposure. It is an asymptotic result, which does not hold for any real portfolio, but may represent a reasonable approximation for some large financial institutions. While neither assumption is particularly realistic, they nevertheless remain the price of portfolio invariance. The punchline is that both the ASRF and Basel-IRB frameworks studiously ignore the idea of idiosyncratic risk. The only source of risk in these settings stems from systematic factors impacting, to varying degrees, all credit obligors.

While large diversified portfolios might be reasonably close to this ideal, what can and should we do with real-life, concentrated portfolios? The Basel framework fully recognized this issue and have made concrete steps to incorporate concentration into their guidelines for the approximation of economic capital. Basel II is, in fact, founded on *three* so-called pillars. Figure 6.7 illustrates these three pillars; the key point is that they are not considered to operate independently, but instead include numerous points of overlap.

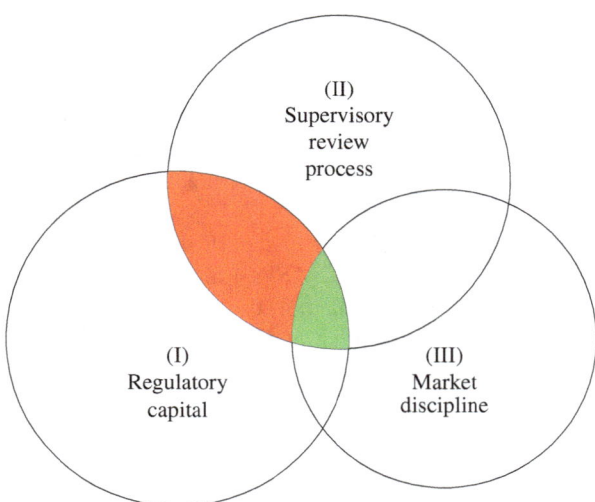

Fig. 6.7 *Basel II's pillars*: This schematic describes the three pillars in the Basel II framework. The key takeaway is that they are not considered to operate independently, but instead include numerous points of overlap.

6.3 The Granularity Adjustment

The majority of the discussion in the Basel-IRB setting has fallen under the first pillar. The simplified formulae employed in the Basel-IRB process are, as we've established, calibrated to large, well-diversified banks. Loosely speaking, therefore, the benchmark portfolio for such credit-risk capital computations is an *infinitely grained* portfolio. Portfolios with higher degrees of exposure concentration, however, will generally be riskier. This higher risk stems from non-diversified idiosyncratic risk. The entire aspect of *concentration* risk is instead considered under the *second* pillar. In particular, financial institutions are expected to address concentration risk separately under the *second* pillar.

Gordy (2003), fully aware of the inherent challenges, suggests an *add-on* to account for concentration risk. Although this sounds like a great idea, there is, of course, a bit more to the story. BIS (2006b) surveys alternative techniques for incorporating concentration risk. Concentration risk actually has *two* flavours:

Name Concentration	The residual idiosyncratic risk arising from deviation from the infinitely grained ideal.
Sector Concentration	Stems from existence of multiple systematic (sectoral) factors—arises from assumption of single underlying risk factor.

These two sources of concentration risk are *not* independent. While it is hard to state with a high-degree of certainty, it appears that many institutions address these risks separately. We will focus, in this chapter, only on the former flavour: name-concentration risk. Sector concentration adjustments typically operate in a conceptually similar *add-on* manner—albeit from a slightly different perspective—so it makes sense to focus one's examination.[11] Currently, Gordy (2003)'s idea of an add-on has gained widespread acceptance. Creating an additive adjustment for concentration risk to a diversified portfolio, however, is *not* a completely natural process. It turns out to be, mathematically at least, a surprisingly complex undertaking. In the following sections, we will describe the details behind the computation of such add-ons, making liberal use of results from Gordy (2003), Martin and Wilde (2002), Emmer and Tasche (2005), and Torell (2013) among others.

6.3.1 A First Try

Let's first start with the basic intuition behind the add-on computation before jumping into all of the technical details. Following from equation 6.17, we define the total default loss as L—we drop the N subscript to avoid notational clutter—and the systematic loss as $\mathbb{E}(L|G)$. In this case, G represents the global or systematic risk factor (or factors) impacting all credit obligors. Using a bit of algebraic sleight

---

[11] For more on sector-concentration risk, the reader is referred to Lütkebohmert (2008, Chapter 11).

of hand, we can construct a single identity, including all of the key characters, as follows,

$$L = L + \underbrace{\mathbb{E}(L|G) - \mathbb{E}(L|G)}_{=0}, \qquad (6.18)$$

$$\underbrace{L}_{\substack{\text{Total} \\ \text{Loss}}} = \underbrace{\mathbb{E}(L|G)}_{\substack{\text{Systematic} \\ \text{Loss}}} + \underbrace{L - \mathbb{E}(L|G)}_{\substack{\text{Idiosyncratic} \\ \text{Loss}}}.$$

If the total loss is L and the systematic loss is $\mathbb{E}(L|G)$, then it logically follows that $L - \mathbb{E}(L|G)$ represents the idiosyncratic risk. That is, the specific or idiosyncratic element of risk is merely the distance between total and systematic risk. This is basically a conservation-of-mass argument. As the idiosyncratic risk is diversified away, this distance $|L - \mathbb{E}(L|G)|$ becomes progressively smaller and, ultimately, tends to zero.

This only happens if, as the portfolio grows, it becomes increasingly granular. Imagine a situation with a single enormous exposure dominating 90% of a given portfolio's total size. If one dramatically increases the portfolio—maybe one increases it by a factor of 10—but this single exposure retains something close to its 90% proportion, it is certainly *not* true that this addition of new obligors will eliminate the idiosyncratic risk associated with this single dominant exposure. Far from it, the portfolio remains distinctly vulnerable to any specific shock to this single obligor's creditworthiness. This is a clearly extreme and somewhat artificial case, but it underscores the basic logic. Only when each obligor is small relative to the others, does this limiting argument create full diversification of the idiosyncratic dimension.

For small and concentrated portfolios, idiosyncratic risk is probably quite large and requires one's consideration. Measuring concentration risk, therefore, amounts to using mathematical techniques—either computational or analytic—to approximate the size of $|L - \mathbb{E}(L|G)|$; of course, it is somewhat more involved, because we need to do this in terms of value-at-risk (VaR). That is the ultimate point of the concentration add-on; the rest is just (unfortunately quite complicated) detail.

6.3.2 A Complicated Add-On

We are now ready to jump into the actual computation. We begin in a relatively general setting, but assume that we have a generic *one-factor* stochastic credit-risk model. L, as usual, represents the default loss and G denotes the common systematic factor. We denote the VaR of our portfolio's default loss, for a given quantile α, as $q_\alpha(L)$. This is a quantile at some point in the tail of the default-loss distribution and, by construction, includes both idiosyncratic and systematic elements.

6.3 The Granularity Adjustment

We further define $q_\alpha^{\text{ASRF}}(L)$ as the diversified VaR estimate associated with the ASRF model. The *systematic* loss, to lighten somewhat the notation burden, is referred to in short form as $X = \mathbb{E}(L|G)$. Gordy (2003), in one of its many contributions, showed us that

$$q_\alpha^{\text{ASRF}}(L) = q_\alpha\left(\mathbb{E}(L|G)\right), \tag{6.19}$$
$$= q_\alpha(X).$$

This is the systematic VaR. In the limit, we know that, conditional (i.e., systematic) and unconditional (i.e., total) VaR converge. Our job, however, is to describe this difference,

$$|\text{ Real-world VaR} - \text{Infinitely grained VaR }| = |q_\alpha(L) - q_\alpha(X)|. \tag{6.20}$$

While it is clear what we need to do, it is not immediately obvious how this might be achieved. The first step involves a clever trick that helps us isolate our distance of interest. We start with $q_\alpha(L)$, which is our total α-level credit VaR, including both systematic and idiosyncratic elements. The idea is similar to the expansion from equation 6.18,

$$q_\alpha(L) = q_\alpha\left(L + \underbrace{X - X}_{=0}\right), \tag{6.21}$$
$$= q_\alpha\left(X + (L - X)\right),$$
$$= q_\alpha\left(\underbrace{X + \epsilon(L - X)}_{f(\epsilon)}\right)\Bigg|_{\epsilon=1},$$
$$= q_\alpha(f(\epsilon))|_{\epsilon=1},$$

where $L = f(1)$ conveniently contains both the idiosyncratic and systematic elements. This might not look like much—indeed, it is simply notation. It does, however, introduce both the total and systematic VaR figures into a single expression. As a first step, this proves quite beneficial.

The second step is to tease out the idiosyncratic and systematic elements. To do this, we will perform a second-order Taylor series expansion of $q_\alpha(L)$ around $\epsilon = 0$. This yields,

$$q_\alpha\left(f(\epsilon)\right)\bigg|_{\epsilon=1} \approx q_\alpha\left(f(\epsilon)\right)\bigg|_{\epsilon=0} + \underbrace{\frac{\partial q_\alpha\left(f(\epsilon)\right)}{\partial \epsilon}\bigg|_{\epsilon=0}}_{=0}(1-0)^1 + \frac{1}{2}\frac{\partial^2 q_\alpha\left(f(\epsilon)\right)}{\partial \epsilon^2}\bigg|_{\epsilon=0}(1-0)^2, \tag{6.22}$$

$$q_\alpha\left(f(1)\right) \approx q_\alpha\left(f(0)\right) + \frac{1}{2}\frac{\partial^2 q_\alpha\left(f(\epsilon)\right)}{\partial \epsilon^2}\bigg|_{\epsilon=0},$$

$$q_\alpha\left(L + 1 \cdot (L - X)\right) \approx q_\alpha\left(X + 0 \cdot (L - X)\right) + \frac{1}{2} \left.\frac{\partial^2 q_\alpha (X + \epsilon (L - X))}{\partial \epsilon^2}\right|_{\epsilon=0},$$

$$\underbrace{q_\alpha(L) - q_\alpha(X)}_{\text{Equation 6.20}} \approx \frac{1}{2} \left.\frac{\partial^2 q_\alpha (X + \epsilon (L - X))}{\partial \epsilon^2}\right|_{\epsilon=0}.$$

This is important progress. The Taylor expansion has allowed us to move from the, relatively unhelpful, VaR of the sum of two elements of risk—idiosyncratic and systematic—toward the (approximate) difference of the VaRs of these quantities. In other words, we have, using a mathematical trick, happily found a representation for equation 6.20. The result is that this difference is approximately one half of the second derivative of the VaR of $f(\epsilon)$ evaluated at $\epsilon = 0$. This an elegant result.

Elegance aside, further progress depends importantly on our ability to evaluate the second derivative of this VaR expression. Fortunately, this is a known result provided by Gourieroux et al. (2000). The careful reader may also question why the first derivative vanishes. It is somewhat messy and depends on the structure of the first derivative of VaR—also found in Gourieroux et al. (2000). The formal development is,

$$\left.\frac{\partial q_\alpha (X + \epsilon(L - X))}{\partial \epsilon}\right|_{\epsilon=0} = \mathbb{E}\left(L - X \bigg| X = q_\alpha(X)\right), \quad (6.23)$$

$$= \mathbb{E}\left(L - \mathbb{E}(L|G) \bigg| \mathbb{E}(L|G) = q_\alpha(\mathbb{E}(L|G))\right),$$

$$= \mathbb{E}\left(L - \mathbb{E}(L|G) \bigg| \underbrace{G = q_{1-\alpha}(G)}_{\substack{X \text{ is monotonic} \\ \text{in } G}}\right),$$

$$= \mathbb{E}\left(L \bigg| G = q_{1-\alpha}(G)\right) - \mathbb{E}\left(\mathbb{E}(L|G) \bigg| G = q_{1-\alpha}(G)\right),$$

$$= \mathbb{E}\left(L \bigg| G = q_{1-\alpha}(G)\right) - \underbrace{\mathbb{E}\left(L \bigg| G = q_{1-\alpha}(G)\right)}_{\text{By iterated expectations}},$$

$$= 0.$$

The first critical line stems from Gourieroux et al. (2000).[12] It turns out that there is a close link between conditional expectation and the derivatives of VaR. We provide this result without derivation, but the interested reader is referred to the derivation provided in Chap. 7. The second point, where the condition variable $X = q_\alpha(X)$ is replaced by $G = q_{1-\alpha}(G)$, stems from the relationship between default loss and the

---

[12] Rau-Bredow (2004) also addresses this tricky issue in substantial detail.

6.3 The Granularity Adjustment

systematic risk variable. L and $\mathbb{E}(L|G)$ are both strictly monotonically decreasing in G.[13] For this reason, the two conditioning sets are equivalent.

Having established that the first derivative of VaR vanishes in this setting, we now face the challenge of evaluating the second derivative. This requires a bit of additional heavy labour. Starting from the object of interest, we may perform the following manipulations, using the definition of the second derivative of VaR, from Gouriéroux et al. (2000) as our jumping-off point,

$$\frac{\partial^2 q_\alpha (f(\epsilon))}{\partial \epsilon^2}\bigg|_{\epsilon=0} = \frac{\partial^2 q_\alpha (X + \epsilon (L - X))}{\partial \epsilon^2}\bigg|_{\epsilon=0}, \qquad (6.24)$$

$$= -\text{var}(L - X|X = x) \frac{\partial \ln f_X(x)}{\partial x}\bigg|_{x=q_\alpha(X)} - \frac{\partial \text{var}(L - X|X = x)}{\partial x}\bigg|_{x=q_\alpha(X)},$$

$$= \left(-\text{var}(L - X|X = x) \left(\frac{1}{f_X(x)}\right) \frac{\partial f_X(x)}{\partial x} - \frac{\partial \text{var}(L - X|X = x)}{\partial x}\right)\bigg|_{x=q_\alpha(X)},$$

$$= -\frac{1}{f_X(x)} \left(\text{var}(L - X|X = x) \frac{\partial f_X(x)}{\partial x} + \frac{\partial \text{var}(L - X|X = x)}{\partial x} f_X(x)\right)\bigg|_{x=q_\alpha(X)},$$

$$= -\frac{1}{f_X(x)} \frac{\partial}{\partial x} \underbrace{\left(\text{var}(L - X|X = x) f_X(x)\right)}_{\text{By product rule}}\bigg|_{x=q_\alpha(X)},$$

$$= -\frac{1}{f_X(x)} \frac{\partial}{\partial x} \left(\text{var}(L|X = x) f_X(x)\right)\bigg|_{x=q_\alpha(X)},$$
$$\underbrace{}_{\text{var}(X|X=x)=0}$$

where the manipulation in the final step arises, because $\text{var}(X|X = x)$ is equal to zero—given the value of the random variable, X, there is no remaining uncertainty. This is a useful simplification, but we'd like to take it a step further. The conditioning, and differentiating, variable is $x = q_\alpha(X)$. We would like, as we did in equation 6.23, to perform a transformation of variables so that we replace X with the global systematic variable, G. We have established already, through the negative monotonicity of X in G, that the following two conditioning statements are equivalent,

$$\{x = q_\alpha(X)\} \equiv \{g = q_{1-\alpha}(G)\}. \qquad (6.25)$$

A critical consequence of this fact is that the α-quantile of the systematic VaR—that is, $q_\alpha(X)$—coincides with the expected loss when the systematic variable falls to its

---

[13] This is specialized to the Gaussian threshold setting, but in any one-factor setting this behaviour is prevalent. The systematic factor may be, of course, monotonically increasing in G. The key is that, in all one-factor models, default loss is a monotonic function of the systematic factor; the direction of this relationship does not create any loss of generality.

$1-\alpha$ quantile. Specifically, it is true that

$$q_\alpha(X) = \mathbb{E}\left(L|G = q_{1-\alpha}(G)\right). \tag{6.26}$$

If we plan to change variables, then we will move from the density of X to G. This will require the change-of-variables formula.[14] The density of X can now be written as,

$$f_G(q_{1-\alpha}(G)) = f_X(q_\alpha(X)) \left| \frac{\partial q_\alpha(X)}{\partial q_{1-\alpha}(G)} \right|, \tag{6.27}$$

$$= f_X(q_\alpha(X)) \left| \frac{\partial \overbrace{\mathbb{E}(L|G = q_{1-\alpha}(G))}^{\text{Equation 6.26}}}{\partial q_{1-\alpha}(G)} \right|,$$

$$f_X(q_\alpha(X)) = f_G(q_{1-\alpha}(G)) \left(\left| \frac{\partial \mathbb{E}(L|G = q_{1-\alpha}(G))}{\partial q_{1-\alpha}(G)} \right| \right)^{-1}.$$

While somewhat ugly, it allows us to write the second derivative in the following form,

$$\left. \frac{\partial^2 q_\alpha(f(\epsilon))}{\partial \epsilon^2} \right|_{\epsilon=0} = -\frac{1}{f_G(g)} \frac{\partial}{\partial g} \left(\frac{f_G(g)\mathrm{var}(L|G=g)}{\frac{\partial \mathbb{E}(L|G=g)}{\partial g}} \right) \Bigg|_{g=q_{1-\alpha}(G)}, \tag{6.28}$$

This, recalling equation 6.22, leads us to the underlying approximation of the distance between total and systematic portfolio risk,

$$q_\alpha(L) - q_\alpha\left(\mathbb{E}(L|G)\right) \approx \frac{1}{2} \left. \frac{\partial^2 q_\alpha\left(\mathbb{E}(L|G) + \epsilon\left(L - \mathbb{E}(L|G)\right)\right)}{\partial \epsilon^2} \right|_{\epsilon=0}, \tag{6.29}$$

$$\mathcal{G}_\alpha(L) = -\frac{1}{2 \cdot f_G(g)} \frac{\partial}{\partial g} \left(\frac{f_G(g)\mathrm{var}(L|G=g)}{\frac{\partial \mathbb{E}(L|G=g)}{\partial g}} \right) \Bigg|_{g=q_{1-\alpha}(G)}.$$

This is commonly referred to as the granularity adjustment, which explains why we use the symbol, $\mathcal{G}$ to describe it. We have eliminated any reference, through our change of variables, to the systematic loss variable, $\mathbb{E}(L|G)$. Although it is not directly applicable in this form, equation 6.29 is now written completely in terms of the systematic global risk factor, G. Thus, with a specific choice of model, a practical add-on can be constructed to approximate idiosyncratic risk.

---

[14] See Billingsley (1995, Chapter 2) for more colour on the change-of-variables technique. Torell (2013) does this in a slightly different, but essentially, equivalent manner.

6.3.3 The Granularity Adjustment

To get to a more workable form of the granularity adjustment, we need to first evaluate the necessary derivatives. To ease the notation, let us define the following functions

$$f(g) = f_G(g), \qquad (6.30)$$
$$\mu(g) = \mathbb{E}(L|G = g),$$
$$v(g) = \text{var}(L|G = g).$$

These are the density of the systematic risk factor and the conditional expectation and variance of the default loss given a specific value of the systematic risk factor. This allows us to re-state equation 6.29 as,

$$\mathcal{G}_\alpha(L) = -\frac{1}{2 \cdot f(g)} \frac{\partial}{\partial g} \left(\frac{f(g)v(g)}{\mu'(g)} \right) \bigg|_{g=q_{1-\alpha}(G)}. \qquad (6.31)$$

Recalling the quotient and product rules, it is a simple calculus exercise to expand equation 6.31 into the following, more expedient form,

$$\mathcal{G}_\alpha(L) = -\frac{1}{2 \cdot f(g)} \left(\frac{\left(f(g)v(g) \right)' \mu'(g) - f(g)v(g)\mu''(g)}{(\mu'(g))^2} \right) \bigg|_{g=q_{1-\alpha}(G)}, \qquad (6.32)$$

$$= -\frac{1}{2 \cdot f(g)} \left(\frac{\left(f'(g)v(g) + f(g)v'(g) \right)\mu'(g) - f(g)v(g)\mu''(g)}{(\mu'(g))^2} \right) \bigg|_{g=q_{1-\alpha}(G)},$$

$$= -\frac{1}{2 \cdot f(g)} \left(\frac{f'(g)v(g)\mu'(g) + f(g)v'(g)\mu'(g) - f(g)v(g)\mu''(g)}{(\mu'(g))^2} \right) \bigg|_{g=q_{1-\alpha}(G)},$$

$$= -\frac{1}{2} \left(\frac{f'(g)v(g)}{f(g)\mu'(g)} + \frac{v'(g)}{\mu'(g)} - \frac{v(g)\mu''(g)}{(\mu'(g))^2} \right) \bigg|_{g=q_{1-\alpha}(G)},$$

$$= -\frac{1}{2} \left(\frac{1}{\mu'(g)} \left(\frac{f'(g)v(g)}{f(g)} + v'(g) \right) - \frac{v(g)\mu''(g)}{(\mu'(g))^2} \right) \bigg|_{g=q_{1-\alpha}(G)}.$$

While not particularly easy on the eyes, this expression is the starting point for the add-on to the ordinary Basel-IRB regulatory-capital amount. It is also, quite interestingly, the jumping-off point for the single-name concentration adjustment in the Standard & Poor's credit-rating agency's risk-adjusted capital ratio used in

Algorithm 6.4 The granularity adjustment

```
def granularityAdjustment(myAlpha,myP,myC,myRho):
    # Get the necessary functions and their derivatives
    f   = fG(myAlpha)
    df  = dfG(myAlpha)
    dg  = dmu(myAlpha,myP,myC,myRho)
    dg2 = d2mu(myAlpha,myP,myC,myRho)
    h   = nu(myAlpha,myP,myC,myRho)
    dh  = dnu(myAlpha,myP,myC,myRho)
    # Build and return granularity adjustment formula
    t1 = np.reciprocal(dg)
    t2 = np.divide(h*df,f)+dh
    t3 = np.divide(h*dg2,np.power(dg,2))
    return -0.5*(t1*t2-t3)
```

assessing firm's credit ratings.[15] One cannot help but be impressed and admire the cleverness—not to mention the hard work and certain trial-and-error effort—required to arrive at the expression in equation 6.32.

The actual Python implementation code for this computation is summarized in Algorithm 6.4. It requires a bit of organization, and a number of ancillary functions, but it is not too complex. The actual implementation, of course, requires a choice of underlying model. It is not possible to discuss a granularity adjustment in isolation. In the coming sections, we will examine two specific model choices: the one-factor Gaussian threshold and the one-factor CreditRisk+ model. Algorithm 6.4 will, therefore, become much more tangible as we introduce the various helper subroutines in the following pages.

6.3.4 *The One-Factor Gaussian Case*

To provide some intuition and gain experience into the form of the granularity adjustment, we will start with a relatively straightforward example. It will, which is likely no surprise to the reader at this point, still turn out to be fairly involved. The task, however, is clear. For a given model, we need to identify $f(g)$, $\mu(g)$ and $\nu(g)$ and evaluate the requisite derivatives. With these ingredients, equation 6.32 provides an easy-to-follow prescription for the estimation of the granularity adjustment. A word of warning, however, is in order. While none of the following computations is particularly difficult, this is nonetheless a tedious process.

---

[15] See S&P (2010, 2011, 2012) for more detail on this widely discussed ratio.

6.3 The Granularity Adjustment

Algorithm 6.5 The density functions for the Gaussian threshold granularity adjustment

```
def fG(myAlpha):
    return norm.pdf(norm.ppf(1-myAlpha))
def dfG(myAlpha):
    z = norm.ppf(1-myAlpha)
    return -z*fG(myAlpha)
```

In this section, we examine the one-factor Gaussian threshold model. In this setting, the systematic risk factor has a familiar form from Chap. 4: it is standard normal. That is, $G \sim \mathcal{N}(0, 1)$. The density function is thus merely,

$$f(g) = \frac{1}{\sqrt{2\pi}} e^{-\frac{g^2}{2}}, \tag{6.33}$$

while the first derivative of this density is,

$$f'(g) = -\frac{\cancel{2}g}{\cancel{2}} \frac{1}{2\pi} e^{-\frac{g^2}{2}}, \tag{6.34}$$
$$= -gf(g).$$

These two elements are thus readily computed. Equations 6.33 and 6.34 are summarized in Algorithm 6.5 in the functions fG and dfG, respectively. The argument, g, is actually the $1 - \alpha$ quantile of the Gaussian systematic risk factor; α is denoted as myAlpha in the Python implementation.

Before moving to the conditional expectation and variance of the default loss, we should recall a few facts from Chap. 4. The default loss, L, of a portfolio with N distinct obligors has the following form,

$$L = \sum_{n=1}^{N} c_n \mathbb{I}_{\mathcal{D}_n}, \tag{6.35}$$

where c_n denotes the nth exposure, $\mathbb{I}_{\mathcal{D}_n}$ is the nth default indicator, and the loss-given-default is assumed equal to unity. The default event is $\mathcal{D}_n$ with the following structure,

$$\mathcal{D}_n = \left\{ Y_n \leq \Phi^{-1}(p_n) \right\}, \tag{6.36}$$

where $\Phi^{-1}(\cdot)$ is the inverse standard normal cumulative distribution (or quantile) function and p_n is the unconditional default probability of the nth credit counterpart.

Y_n is a default state variable expressed as,

$$Y_n = \sqrt{\rho}G + \sqrt{1-\rho}\epsilon_n, \tag{6.37}$$

where $G, \epsilon_n \sim \mathcal{N}(0,1)$ are the systematic and idiosyncratic factors, respectively. $\rho \in [0,1]$ is a parameter governing the degree of state-variable—and thus indirectly default—correlation between the individual obligors.

With these preliminaries established, we can proceed to evaluate the conditional expectation of the default loss given a specific value of the systematic state variable. It can be written as,

$$\begin{aligned}\mu(g) &= \mathbb{E}\left(L\,\middle|\, G=g\right), \tag{6.38}\\ &= \mathbb{E}\left(\sum_{n=1}^{N} c_n \mathbb{I}_{\mathcal{D}_n}\,\middle|\, G=g\right),\\ &= \mathbb{E}\left(\sum_{n=1}^{N} c_n \mathbb{I}_{\{Y_n \leq \Phi^{-1}(p_n)\}}\,\middle|\, G=g\right),\\ &= \mathbb{E}\left(\sum_{n=1}^{N} c_n \mathbb{I}_{\{\sqrt{\rho}g+\sqrt{1-\rho}\epsilon_n \leq \Phi^{-1}(p_n)\}}\,\middle|\, G=g\right),\\ &= \sum_{n=1}^{N} c_n \mathbb{E}\left(\mathbb{I}_{\{\sqrt{\rho}G+\sqrt{1-\rho}\epsilon_n \leq \Phi^{-1}(p_n)\}}\,\middle|\, G=g\right),\\ &= \sum_{n=1}^{N} c_n \mathbb{P}\left(\sqrt{\rho}G+\sqrt{1-\rho}\epsilon_n \leq \Phi^{-1}(p_n)\,\middle|\, G=g\right),\\ &= \sum_{n=1}^{N} c_n \mathbb{P}\left(\epsilon_n \leq \frac{\Phi^{-1}(p_n)-\sqrt{\rho}G}{\sqrt{1-\rho}}\,\middle|\, G=g\right),\\ &= \sum_{n=1}^{N} c_n \Phi\left(\frac{\Phi^{-1}(p_n)-\sqrt{\rho}g}{\sqrt{1-\rho}}\right),\\ &= \sum_{n=1}^{N} c_n p_n(g),\end{aligned}$$

where, as usual, $\Phi(\cdot)$ denotes the standard normal cumulative distribution function. The conditional default loss is, as one might expect, the sum of the exposure-weighted *conditional* default probabilities: $p_n(g)$ for $n = 1, \ldots, N$.

6.3 The Granularity Adjustment

We need both the first and second derivatives of equation 6.38 to evaluate the granularity adjustment. The first derivative is,

$$\mu'(g) = \frac{\partial}{\partial g}\left(\sum_{n=1}^{N} c_n \Phi\left(\frac{\Phi^{-1}(p_n) - \sqrt{\rho}g}{\sqrt{1-\rho}}\right)\right), \tag{6.39}$$

$$= -\sum_{n=1}^{N} \frac{c_n\sqrt{\rho}}{\sqrt{1-\rho}} \phi\left(\frac{\Phi^{-1}(p_n) - \sqrt{\rho}g}{\sqrt{1-\rho}}\right),$$

$$= -\sum_{n=1}^{N} c_n\sqrt{\frac{\rho}{1-\rho}} \phi\left(\Phi^{-1}(p_n(g))\right),$$

where $\phi(\cdot)$ represents the standard normal density function. The last line requires a bit of explanation. We have, as originally defined in Chap. 4, described the conditional default probability more succinctly as $p_n(g)$. If we want, as is the case in equation 6.39, the argument evaluated in the Gaussian cumulative distribution function in $p_n(g)$, we merely apply the Gaussian quantile function. That is,

$$\Phi^{-1}(p_n(g)) = = \Phi^{-1}\left(\Phi\left(\frac{\Phi^{-1}(p_n) - \sqrt{\rho}g}{\sqrt{1-\rho}}\right)\right), \tag{6.40}$$

$$= \frac{\Phi^{-1}(p_n) - \sqrt{\rho}g}{\sqrt{1-\rho}}.$$

This allows us to shorten some of our expressions and recycle existing Python code. Conceptually, it makes no difference, but practically there is some efficiency to be gained.

The second derivative is a bit messier. It can, however, be written as

$$\mu''(g) = \frac{\partial}{\partial g}\left(-\sum_{n=1}^{N} \frac{c_n\sqrt{\rho}}{\sqrt{1-\rho}} \phi\left(\frac{\Phi^{-1}(p_n) - \sqrt{\rho}g}{\sqrt{1-\rho}}\right)\right), \tag{6.41}$$

$$= \frac{\partial}{\partial g}\left(-\sum_{n=1}^{N} \frac{c_n\sqrt{\rho}}{\sqrt{1-\rho}} \frac{1}{\sqrt{2\pi}} \exp\left(\frac{-\left(\frac{\Phi^{-1}(p_n)-\sqrt{\rho}g}{\sqrt{1-\rho}}\right)^2}{2}\right)\right),$$

$$= -\sum_{n=1}^{N} \frac{c_n\sqrt{\rho}}{\sqrt{1-\rho}} \left(-\frac{\sqrt{\rho}}{\sqrt{1-\rho}}\right)\left(-\frac{\Phi^{-1}(p_n) - \sqrt{\rho}g}{\sqrt{1-\rho}}\right) \frac{1}{\sqrt{2\pi}} \exp\left(\frac{-\left(\frac{\Phi^{-1}(p_n)-\sqrt{\rho}g}{\sqrt{1-\rho}}\right)^2}{2}\right),$$

Algorithm 6.6 Conditional-expectation functions for the Gaussian threshold granularity adjustment

```
import thresholdModels as th
def mu(myAlpha,myP,myC,myRho):
    pn = th.computeP(myP,myRho,norm.ppf(1−myAlpha))
    return np.dot(myC,pn)
def dmu(myAlpha,myP,myC,myRho):
    constant = np.sqrt(np.divide(myRho,1−myRho))
    ratio = norm.ppf(th.computeP(myP,myRho,norm.ppf(1−myAlpha)))
    return −constant*np.dot(myC,norm.pdf(ratio))
def d2mu(myAlpha,myP,myC,myRho):
    constant = np.divide(myRho,1−myRho)
    ratio = norm.ppf(th.computeP(myP,myRho,norm.ppf(1−myAlpha)))
    return −constant*np.dot(ratio*myC,norm.pdf(ratio))
```

$$= -\sum_{n=1}^{N} c_n \frac{\rho}{1-\rho} \frac{\Phi^{-1}(p_n) - \sqrt{\rho}g}{\sqrt{1-\rho}} \phi\left(\frac{\Phi^{-1}(p_n) - \sqrt{\rho}g}{\sqrt{1-\rho}}\right),$$

$$= -\sum_{n=1}^{N} c_n \frac{\rho}{1-\rho} \Phi^{-1}(p_n(g)) \phi\left(\Phi^{-1}(p_n(g))\right).$$

Despite their messiness, equations 6.38, 6.39 and 6.41 are readily implemented in Algorithm 6.6 in the functions mu, dmu, and d2mu, respectively. Equation 6.40 shows up, in a bit of recycled code, as we apply the norm.ppf or Gaussian quantile function to the conditional default probability routine, computeP, from the thresholdModels library introduced in Chap. 4. The practical benefit of this *recycling* approach is, should the analyst wish to change the code implementation, it need only be altered in a single location.

The final term is the conditional variance of the default loss. Using similar ideas to equation 6.38, and exploiting the conditional independence of the default events, we have

$$\nu(g) = \text{var}(L|\,G=g), \tag{6.42}$$

$$= \text{var}\left(\sum_{n=1}^{N} c_n \mathbb{I}_{\mathcal{D}_n}\,\bigg|\,G=g\right),$$

$$= \sum_{n=1}^{N} c_n^2 \text{var}\left(\mathbb{I}_{\mathcal{D}_n}\,\big|\,G=g\right),$$

$$= \sum_{n=1}^{N} c_n^2 \left(\mathbb{E}\left(\mathbb{I}_{\mathcal{D}_n}^2\,\big|\,G=g\right) - \mathbb{E}\left(\mathbb{I}_{\mathcal{D}_n}\,\big|\,G=g\right)^2\right),$$

6.3 The Granularity Adjustment

$$= \sum_{n=1}^{N} c_n^2 \bigg(\mathbb{P}(\mathcal{D}_n | G = g) - \mathbb{P}(\mathcal{D}_n | G = g)^2 \bigg),$$

$$= \sum_{n=1}^{N} c_n^2 \, \mathbb{P}(\mathcal{D}_n | G = g) \bigg(1 - \mathbb{P}(\mathcal{D}_n | G = g) \bigg),$$

$$= \sum_{n=1}^{N} c_n^2 \Phi \left(\frac{\Phi^{-1}(p_n) - \sqrt{\rho} g}{\sqrt{1-\rho}} \right) \left(1 - \Phi \left(\frac{\Phi^{-1}(p_n) - \sqrt{\rho} g}{\sqrt{1-\rho}} \right) \right),$$

$$= \sum_{n=1}^{N} c_n^2 p_n(g) \bigg(1 - p_n(g) \bigg).$$

The final step is to determine the first derivative of the conditional variance function. It can be derived as,

$$v'(g) = \frac{\partial}{\partial g} \left(\sum_{n=1}^{N} c_n^2 \Phi \left(\frac{\Phi^{-1}(p_n) - \sqrt{\rho} g}{\sqrt{1-\rho}} \right) \left(1 - \Phi \left(\frac{\Phi^{-1}(p_n) - \sqrt{\rho} g}{\sqrt{1-\rho}} \right) \right) \right), \quad (6.43)$$

$$= \frac{\partial}{\partial g} \left(\sum_{n=1}^{N} c_n^2 \left(\Phi \left(\frac{\Phi^{-1}(p_n) - \sqrt{\rho} g}{\sqrt{1-\rho}} \right) - \Phi \left(\frac{\Phi^{-1}(p_n) - \sqrt{\rho} g}{\sqrt{1-\rho}} \right)^2 \right) \right),$$

$$= \sum_{n=1}^{N} c_n^2 \left(\sqrt{\frac{\rho}{1-\rho}} \phi \left(\frac{\Phi^{-1}(p_n) - \sqrt{\rho} g}{\sqrt{1-\rho}} \right) \right.$$
$$\left. -2 \sqrt{\frac{\rho}{1-\rho}} \phi \left(\frac{\Phi^{-1}(p_n) - \sqrt{\rho} g}{\sqrt{1-\rho}} \right) \Phi \left(\frac{\Phi^{-1}(p_n) - \sqrt{\rho} g}{\sqrt{1-\rho}} \right) \right),$$

$$= \sum_{n=1}^{N} c_n^2 \sqrt{\frac{\rho}{1-\rho}} \phi \left(\frac{\Phi^{-1}(p_n) - \sqrt{\rho} g}{\sqrt{1-\rho}} \right) \left(1 - 2\Phi \left(\frac{\Phi^{-1}(p_n) - \sqrt{\rho} g}{\sqrt{1-\rho}} \right) \right),$$

$$= \sum_{n=1}^{N} c_n^2 \sqrt{\frac{\rho}{1-\rho}} \phi \bigg(\Phi^{-1}(p_n(g)) \bigg) \bigg(1 - 2 p_n(g) \bigg).$$

Again, there is nothing terribly difficult in the derivation of the previous quantities and their derivatives, but it is a painstaking and error-prone process. Nevertheless, with the appropriate caution and patience, we have identified the necessary ingredients for the granularity adjustment of the one-factor Gaussian threshold model. Algorithm 6.7 summarizes the compact, and by now quite familiar, Python-code implementation. We have thus provided all of the ancillary functions for the generic implementation found in Algorithm 6.4.

Algorithm 6.7 Conditional variance functions for the Gaussian threshold granularity adjustment

```
import thresholdModels as th
def nu(myAlpha,myP,myC,myRho):
    pn = th.computeP(myP,myRho,norm.ppf(1-myAlpha))
    return np.dot(np.power(myC,2),pn*(1-pn))
def dnu(myAlpha,myP,myC,myRho):
    pn = th.computeP(myP,myRho,norm.ppf(1-myAlpha))
    ratio = norm.ppf(pn)
    constant = np.sqrt(np.divide(myRho,1-myRho))
    return -constant*np.dot(norm.pdf(ratio)*np.power(myC,2),1-2*pn)
```

Figure 6.8 concludes this section by examining, for basic inputs found in our 100-obligor example, the form of the density, conditional-expectation, and conditional-variance functions derived in the preceding pages over the range $\alpha \in (0.95, 1.00)$. There are striking differences in scales and forms of these functions and their derivatives. All values are evaluated at the appropriate quantile value for the underlying systematic variables: $g_\alpha = q_{1-\alpha}(g)$.

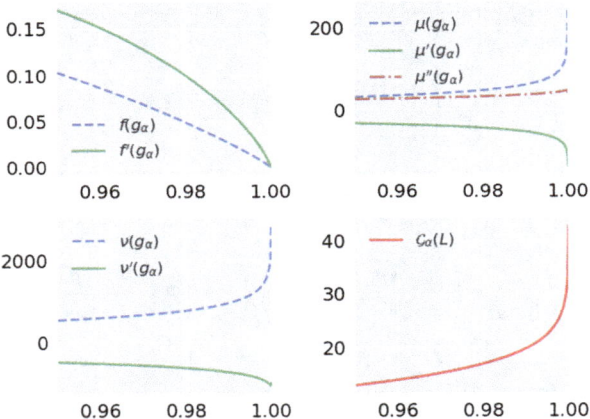

Fig. 6.8 *Gaussian granularity adjustment functions*: The underlying figure highlights, over the range $\alpha \in (0.95, 1.00)$, the form of the various functions derived in the preceding pages. There are significant differences in scales and forms of these functions and their derivatives. All values are evaluated at the appropriate quantile value for the underlying systematic variables: $g_\alpha = q_{1-\alpha}(g)$.

6.3.5 Getting a Bit More Concrete

Although essential for our implementation, the derivation of the key quantities in the previous section provides little evidence of the relative accuracy and usefulness of the granularity adjustment. To address this aspect of the analysis, we will consider a practical example. We begin with our usual 100-obligor portfolio with randomly generated default probabilities and exposures. To provide a point of comparison, we also consider another, much more *diversified*, choice. This second portfolio is larger, by a factor of 100, than the base portfolio. Both portfolios, to ensure comparability, have the same total exposure and average default probabilities. The difference is that one portfolio is concentrated, while the other approaches the infinitely grained ideal.

Table 6.4 provides a number of summary statistics for these two portfolios. The concentrated portfolio has 100 credit counterparts with an average exposure of $10, whereas the diversified portfolio has 10,000 obligors each with only $0.10.

The 99.99th VaR, estimated using the one-factor Gaussian threshold model, is significantly higher for the concentrated model. If we use, however, the ASRF, which assumes both are infinitely granular, the 99.99th VaR estimate should, in principle, be virtually identical for the two models. The objective of the granularity adjustment is to estimate the difference between the ASRF—or portfolio invariant—estimate and the full-blown model approximation incorporating the portfolio's level of concentration. Figure 6.9 illustrates the tail probabilities of default-loss distributions for the concentrated and diversified portfolios; both estimates are constructed using the one-factor Gaussian threshold model with an asset-correlation

Table 6.4 *Two differing portfolios*: This table provides a number of summary statistics for two portfolios: one relatively concentrated and the other diversified. The portfolio value and average default probability have been calibrated to be identical.

Measure	Portfolio	
	Concentrated	Diversified
Number of obligors	100	10,000
Portfolio value	$1,000	
Average exposure	$10.0	$0.1
Average default probability	1.0%	
Expected default loss	$9.5	$9.6
Default loss volatility	$14.7	$19.6
ASRF 99.99th quantile VaR	$217	$217
One-factor Gaussian 99.99th quantile VaR	$254	$209
Suitable for ASRF	No	Yes
Requires granularity adjustment	Yes	No

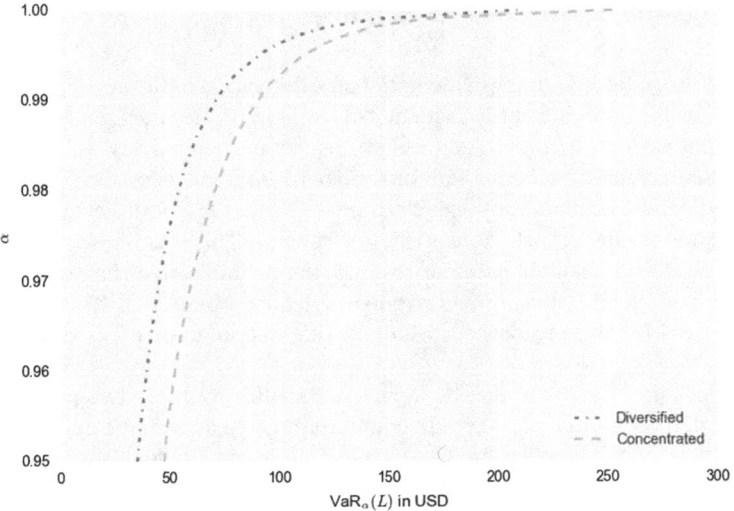

Fig. 6.9 *Concentrated vs. diversified tail probabilities*: This figure highlights the tail probabilities over the range [0.95, 0.9999] for the diversified and concentrated portfolios summarized in Table 6.4. All estimates are computed using the one-factor Gaussian threshold model with a common ρ parameter equal to 0.19.

coefficient (i.e., ρ) equal to about 0.19.[16] Having carefully controlled all other possible variables, the distance between the two curves can be attributed fully to concentration risk. It is precisely this distance that the granularity adjustment seeks to describe.

We are now in a position to determine if, in the context of this example, the granularity adjustment actually makes sense and does what it purports to do. Our experiment involves performing the following *two* different tasks:

1. we estimate the diversified portfolio with the ASRF model to see how close it is to the asymptotic result across a broad range of quantiles; and
2. we compute the granularity adjustment and add it to the ASRF estimate of the concentrated portfolio.

With regard to our first task, a small distance, for the diversified portfolio, between ASRF and one-factor Gaussian threshold model simulation results, would suggest that our 10,000 obligor example is close to infinitely grained. Adding the granularity adjustment to the ASRF (or portfolio-invariant) estimate of the concentrated portfolio, conversely, should get us close to the true one-factor threshold model

---

[16]This value is determined using equation 6.11 with an average unconditional default probability of 0.01.

6.3 The Granularity Adjustment

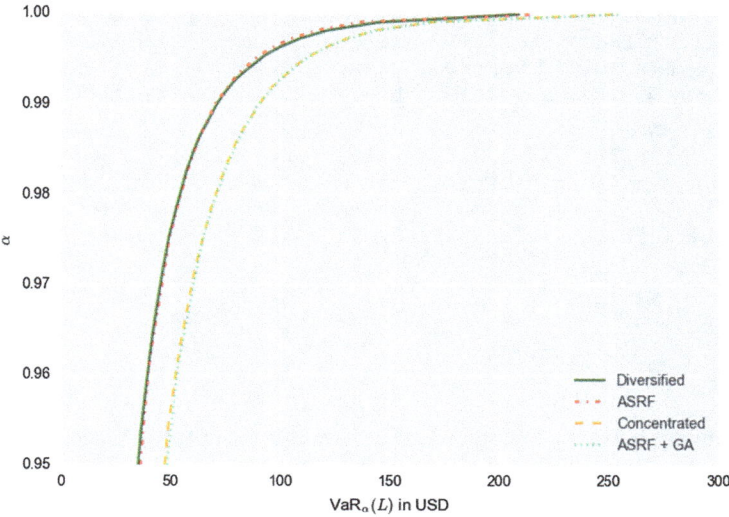

Fig. 6.10 *Testing the granularity adjustment*: This figure provides the tail probabilities for four different cases: the diversified ASRF, the concentrated ASRF, the one-factor Gaussian estimate of the diversified model, and the sum of the ASRF concentrated estimate plus the granularity adjustment. Both of our experiments are successful.

results. This is, however, an approximation; consequently, we should *not* expect it to be perfect. Our hope is that the error term is reasonably small.

Figure 6.10 summarizes our two experiments. There is, first of all, no visible difference between the ASRF and full-blown one-factor Gaussian threshold model estimates of the diversified portfolio. This strongly suggests that the diversified portfolio is reasonably close to infinitely grained and, as such, is well approximated by the ASRF model. The concentrated portfolio, conversely, given its small number of obligors and non-regular exposure pattern, is *not* very well summarized by the ASRF model. If we take the ASRF estimate of the concentrated portfolio and add the granularity adjustment, described by equation 6.32, it brings us much closer to the true model estimate of the concentrated portfolio. Although the concentrated portfolio is not *close* to infinitely grained, the granularity adjustment brings us reasonably close to the results associated by assuming the true default-loss distribution is described by a one-factor Gaussian threshold model.

Table 6.5 provides some figures to supplement the tail probabilities provided in Fig. 6.10. While the results are not perfect, the approximation is, in this case, actually quite good. There is apparently, although not particularly evident in this example, something of a tendency to overstate the size of the concentration adjustment at the higher levels of confidence. To this extent, the granularity adjustment is relatively conservative. This effect is, according to Gordy and Lütkebohmert (2007, 2013), particularly strong for small portfolios such as our 100-obligor example.

Table 6.5 *The numerical reckoning*: This table focuses on the granularity adjustment approximation in Fig. 6.9. It illustrates the ASRF, one-factor Gaussian threshold, and granularity adjustment computations for the concentrated model. The error term, which is very small, is the difference between the model VaR estimate and the ASRF value plus the granularity adjustment.

Quantile	Actual			Correction		
	ASRF	$1 - \mathcal{N}$	Δ	$\mathcal{G}_\alpha(L)$	ASRF + $\mathcal{G}_\alpha(L)$	Error
95.00th	$37	$48	$11	$12	$49	$1
97.00th	$47	$60	$14	$15	$61	$1
99.00th	$71	$91	$20	$20	$91	$-1
99.90th	$138	$169	$31	$30	$167	$-2
99.97th	$199	$235	$36	$37	$236	$1
99.99th	$217	$254	$38	$39	$256	$1

One might, however, legitimately ask: why is the one-factor Gaussian threshold model not used directly and the granularity adjustment avoided? In normal circumstances, it is advisable to select a full-blown default-risk model and perform one's computations. In the Basel-IRB setting, however, a decision was taken to assume portfolio-invariance for the implementation of their base framework. This offers a number of advantages in terms of computation speed, simplicity of calculation, and interpretation and comparison of results. It does, nevertheless, require strong assumptions about the granularity of the portfolio. When these assumptions are violated, the cost for their incorporation is the granularity adjustment.

6.3.6 The CreditRisk+ Case

Having examined a classic model in the threshold setting, it is useful to examine the implications of a common mixture model. Our specific choice—following from Gordy and Lütkebohmert (2007, 2013)—is based on the one-factor CreditRisk+ model introduced in Chap. 3. This approach—introduced by Wilde (1997) and covered in great detail in Gundlach and Lehrbass (2004)—is essentially a special case of a Poisson-Gamma model. The defaults in the portfolio of N obligors is governed by N independent and identically distributed Poisson random variables, $X_n \sim \mathcal{P}(\lambda_n(S))$. The default indicator is re-defined as,

$$\mathbb{I}_{\mathcal{D}_n} \equiv \mathbb{I}_{\{X_n \geq 1\}} = \begin{cases} 1: \text{ default occurs before time } T \text{ with probability } p_n(S) \\ 0: \text{ survival until time } T \text{ with probability } 1 - p_n(S) \end{cases}, \quad (6.44)$$

where $S \sim \Gamma(a, b)$ is an independent gamma-distributed random variable. S, in this model, plays the role of the systematic state variable; it is the analogue of the variable, G, in the one-factor threshold setting.

6.3 The Granularity Adjustment

Given the relatively small probability of default, it is further assumed that,

$$p_n(S) \equiv \mathbb{E}\left(\mathbb{I}_{\mathcal{D}_n} \mid S\right), \qquad (6.45)$$
$$= \mathbb{P}(\mathcal{D}_n \mid S),$$
$$\approx \lambda_n(S).$$

In other words, the random Poisson arrival intensity of the default outcome and the conditional probability of default are assumed to coincide.[17]

Wilde (1997) gave $p_n(S)$ an ingenious functional form,

$$p_n(S) = p_n\left(\omega_{n,0} + \omega_{n,1}S\right), \qquad (6.46)$$

where $\omega_{n,0} + \omega_{n,1} = 1$, implying that we can restate equation 6.46 more succinctly as,

$$p_n(S) = p_n\left(1 - \omega_n + \omega_n S\right). \qquad (6.47)$$

We can interpret the single parameter ω_n as the factor loading on the systematic factor, while $1 - \omega_n$ denotes the importance of the idiosyncratic factor.[18] It is easy to show that, if we select the parameters of the gamma-distributed S such that $\mathbb{E}(S) = 1$, then $\mathbb{E}(p_n(S)) = p_n$. That is, in expectation, the conditional default probability reduces to the unconditional default probability. If one assumes the shape-scale parametrization of the gamma distribution, it is necessary to assume that $S \sim \Gamma\left(a, \frac{1}{a}\right)$.[19]

The best known discussion of the granularity adjustment for the one-factor CreditRisk+ model arises in Gordy and Lütkebohmert (2007, 2013). Indeed, this work forms the foundation of many Pillar-II implementations of single-name concentration risk. It is also the documented choice of S&P's single-name concentration adjustment in their risk-adjusted-capital framework implementation.[20] Given this formulation, we will work through the general CreditRisk+ granularity adjustment and touch on the simplifications and calibrations suggested by Gordy and Lütkebohmert (2007, 2013).

---

[17] Practically, for small values of $\lambda_n(S)$, the random variable X_n basically only takes two values: 0 and 1. The $\mathbb{P}(X_n > 1)$ is vanishingly small.

[18] If, for example, $\omega_n = 0$, then it reduces to an independent-default Poisson model.

[19] In the shape-rate parameterization, which we used in Chap. 3, the density has a slightly different form. In this case, we assume that $S \sim \Gamma(a, a)$. Ultimately, there is no difference in the final results; we used the shape-scale parametrization to maximize comparability with the Gordy and Lütkebohmert (2007, 2013) development.

[20] See, for example, S&P (2010, Paragraphs 148–151).

An additional twist in the Gordy and Lütkebohmert (2007, 2013) is the incorporation of random recovery. That is, the loss-given-default parameter for the nth obligor, which we will denote as γ_n, is assumed to be stochastic. Its precise distribution is not given, but it is assumed to be independent of default and any other recovery events. Moreover, its first two moments are described as,

$$\mathbb{E}(\gamma_n) = \bar{\gamma}_n, \tag{6.48}$$

$$\text{var}(\gamma_n) = \nu_{\gamma_n}.$$

The principal consequence of this modelling choice is that the default loss function is a bit more complicated as,

$$L = \sum_{n=1}^{N} \gamma_n c_n \mathbb{I}_{\mathcal{D}_n}. \tag{6.49}$$

With the model details established, all that remains is the mundane task of extracting the conditional expectation, conditional variance, and state-variable density terms and their derivatives. We will begin with the conditional expectation of the default loss given a specific value of the systematic state variable, S. Working from first principles, we have

$$\mu(s) = \mathbb{E}(L | S = s), \tag{6.50}$$

$$= \mathbb{E}\left(\sum_{n=1}^{N} \gamma_n c_n \mathbb{I}_{\mathcal{D}_n} \bigg| S = s\right),$$

$$= \sum_{n=1}^{N} c_n \, \mathbb{E}\left(\gamma_n \mathbb{I}_{\mathcal{D}_n} \big| S = s\right),$$

$$= \sum_{n=1}^{N} c_n \, \mathbb{E}(\gamma_n | S = s) \, \mathbb{E}\left(\mathbb{I}_{\mathcal{D}_n} \big| S = s\right),$$

$$= \sum_{n=1}^{N} c_n \mathbb{E}(\gamma_n) \, \mathbb{P}(\mathcal{D}_n | S = s),$$

$$= \sum_{n=1}^{N} c_n \bar{\gamma}_n p_n(s),$$

where most of the preceding manipulations are definitional. We should stress that, since each γ_n is independent of S, $\mathbb{E}(\gamma_n|S) = \mathbb{E}(\gamma_n)$; that is, there is no information provided by the conditioning set.

6.3 The Granularity Adjustment

If we recall the definition of $p_n(S)$ from equation 6.45, we can readily determine the first derivative of $\mu(s)$ as,

$$\mu'(s) = \frac{\partial}{\partial s} \left(\sum_{n=1}^{N} c_n \bar{\gamma}_n p_n(s) \right), \tag{6.51}$$

$$= \frac{\partial}{\partial s} \left(\sum_{n=1}^{N} c_n \bar{\gamma}_n p_n \left(1 - \omega_n + \omega_n s \right) \right),$$

$$= \sum_{n=1}^{N} c_n \bar{\gamma}_n p_n \omega_n.$$

It is easy to see, given $\mu(s)$ is a linear function of s, that the second derivative, $\mu''(s)$, vanishes. This simple fact has important implications since the second term in equation 6.32 on page 317 becomes zero, significantly simplifying our task.

Determining the conditional variance is rather painful. Two facts will, however, prove quite useful in its determination. First, the default event, in equation 6.44, is Poisson-distributed. An interesting feature of the Poisson distribution is that its expectation and variance are equal: that is, in this case, $\mathbb{E}(\mathbb{I}_{\mathcal{D}_n}) = \text{var}(\mathbb{I}_{\mathcal{D}_n})$. The second fact is that, from first principles, if X is a random variable with finite first two moments, then $\text{var}(X) = \mathbb{E}(X^2) - \mathbb{E}(X)^2$. This implies directly that $\mathbb{E}(X^2) = \text{var}(X) + \mathbb{E}(X)^2$. Both of these facts, which we will employ directly, apply unconditionally and conditionally.

The conditional variance can be written as,

$$v(s) = \text{var}(L|S = s), \tag{6.52}$$

$$= \text{var}\left(\sum_{n=1}^{N} \gamma_n c_n \mathbb{I}_{\mathcal{D}_n} \middle| S = s \right),$$

$$= \sum_{n=1}^{N} c_n^2 \, \text{var}\left(\gamma_n \mathbb{I}_{\mathcal{D}_n} \middle| S = s \right),$$

$$= \sum_{n=1}^{N} c_n^2 \left(\mathbb{E}\left(\gamma_n^2 \mathbb{I}_{\mathcal{D}_n}^2 \middle| S = s \right) - \mathbb{E}\left(\gamma_n \mathbb{I}_{\mathcal{D}_n} \middle| S = s \right)^2 \right),$$

$$= \sum_{n=1}^{N} c_n^2 \left(\mathbb{E}\left(\gamma_n^2 \middle| S = s \right) \mathbb{E}\left(\mathbb{I}_{\mathcal{D}_n}^2 \middle| S = s \right) - \mathbb{E}(\gamma_n | S = s)^2 \, \mathbb{E}\left(\mathbb{I}_{\mathcal{D}_n} \middle| S = s \right)^2 \right),$$

$$= \sum_{n=1}^{N} c_n^2 \left(\mathbb{E}\left(\gamma_n^2 \right) \mathbb{E}\left(\mathbb{I}_{\mathcal{D}_n}^2 \middle| S = s \right) - \mathbb{E}(\gamma_n)^2 \, \mathbb{E}\left(\mathbb{I}_{\mathcal{D}_n} \middle| S = s \right)^2 \right),$$

$$= \sum_{n=1}^{N} c_n^2 \bigg(\underbrace{\Big(\text{var}(\gamma_n) + \mathbb{E}(\gamma_n)^2\Big)}_{v_{\gamma_n} + \bar{\gamma}_n^2} \Big(\text{var}\left(\mathbb{I}_{\mathcal{D}_n} \middle| S = s\right) + \mathbb{E}\left(\mathbb{I}_{\mathcal{D}_n} \middle| S = s\right)^2 \Big) - \bar{\gamma}_n^2 p_n(s)^2 \bigg),$$

$$= \sum_{n=1}^{N} c_n^2 \bigg(\Big(v_{\gamma_n} + \bar{\gamma}_n^2\Big) \Big(\mathbb{E}\left(\mathbb{I}_{\mathcal{D}_n} \middle| S = s\right) + \mathbb{E}\left(\mathbb{I}_{\mathcal{D}_n} \middle| S = s\right)^2 \Big) - \bar{\gamma}_n^2 p_n(s)^2 \bigg),$$

$$= \sum_{n=1}^{N} c_n^2 \bigg(\Big(v_{\gamma_n} + \bar{\gamma}_n^2\Big) \Big(p_n(s) + p_n(s)^2 \Big) - \bar{\gamma}_n^2 p_n(s)^2 \bigg),$$

$$= \sum_{n=1}^{N} c_n^2 \bigg(v_{\gamma_n} p_n(s) + v_{\gamma_n} p_n(s)^2 + \bar{\gamma}_n^2 p_n(s) + \cancel{\bar{\gamma}_n^2 p_n(s)^2} - \cancel{\bar{\gamma}_n^2 p_n(s)^2} \bigg),$$

$$= \sum_{n=1}^{N} c_n^2 p_n(s) \Big(\bar{\gamma}_n^2 + v_{\gamma_n} (1 + p_n(s)) \Big).$$

We can now directly determine the required first derivative of the conditional variance. With the fairly succinct form of equation 6.52, it is readily written as

$$v'(s) = \frac{\partial}{\partial s} \left(\sum_{n=1}^{N} c_n^2 p_n(s) \Big(\bar{\gamma}_n^2 + v_{\gamma_n} (1 + p_n(s)) \Big) \right), \tag{6.53}$$

$$= \frac{\partial}{\partial s} \left(\sum_{n=1}^{N} c_n^2 \Big(v_{\gamma_n} p_n(s) + v_{\gamma_n} p_n(s)^2 + \bar{\gamma}_n^2 p_n(s) \Big) \right),$$

$$= \sum_{n=1}^{N} c_n^2 \Big(v_{\gamma_n} p_n \omega_m + v_{\gamma_n} 2 p_n(s) p_n \omega_m + \bar{\gamma}_n^2 p_n \omega_m \Big),$$

$$= \sum_{n=1}^{N} c_n^2 p_n \omega_m \Big(\bar{\gamma}_n^2 + v_{\gamma_n} (1 + 2 p_n(s)) \Big).$$

The final remaining piece of the puzzle is the density function of the systematic state variable and its first derivative. Recall that, to meet the needs of the model, we require that $S \sim \Gamma\left(a, \frac{1}{a}\right)$. The shape-scale version of the density of the gamma distribution for these parameter choices is written as,

$$f_S(s) = \frac{s^{a-1} e^{-sa}}{\Gamma(a) \left(\frac{1}{a}\right)^a}, \tag{6.54}$$

6.3 The Granularity Adjustment

where $\Gamma(a)$ denotes the gamma function.[21] The first derivative with respect to s has a convenient form,

$$f'_S(s) = \frac{\partial}{\partial s}\left(\frac{s^{a-1}e^{-sa}}{\Gamma(a)\left(\frac{1}{a}\right)^a}\right), \qquad (6.55)$$

$$= \frac{1}{\Gamma(a)\left(\frac{1}{a}\right)^a}\left((a-1)s^{a-2}e^{-sa} - as^{a-1}e^{-sa}\right),$$

$$= \underbrace{\frac{s^{a-1}e^{-sa}}{\Gamma(a)\left(\frac{1}{a}\right)^a}}_{f_S(s)}\left((a-1)s^{-1} - a\right),$$

$$= f_S(s)\left(\frac{a-1}{s} - a\right).$$

While the convenience might not be immediately clear, we will require the ratio of the first derivative of the gamma density to its raw density for the calculation of the granularity adjustment. This ratio, following from equations 6.54 and 6.55, is thus

$$\frac{f'_S(s)}{f_S(s)} = \frac{\cancel{f_S(s)}\left(\frac{a-1}{s} - a\right)}{\cancel{f_S(s)}}, \qquad (6.56)$$

$$\mathcal{A} = \frac{a-1}{s} - a.$$

This quantity depends only on the model and, more particularly, on the parametrization of the underlying gamma distribution.

We finally have all of the necessary elements to describe, at least in its raw form, the granularity adjustment for the one-factor CreditRisk+ model. Recalling equation 6.32 on page 317 we have,

$$\mathcal{G}_\alpha(L) = -\frac{1}{2}\left(\frac{1}{\mu'(s)}\left(\frac{f'(s)v(s)}{f(s)} + v'(s)\right) - \underbrace{\frac{v(s)\mu''(s)}{(\mu'(s))^2}}_{=0}\right)\Bigg|_{s=q_\alpha(S)}, \quad (6.57)$$

$$= -\frac{1}{2}\left(\frac{\mathcal{A}v(q_\alpha) + v'(q_\alpha)}{\mu'(q_\alpha)}\right),$$

$$= \frac{-\mathcal{A}v(q_\alpha) - v'(q_\alpha)}{2\mu'(q_\alpha)},$$

---

[21] See Casella and Berger (1990, Chapter 3) or Johnson et al. (1997, Chapter 17) for more information on the gamma distribution.

where $q_\alpha \equiv q_\alpha(S)$ and we use α instead of $1 - \alpha$, because the probability of default is monotonically *increasing* as a function of the state-variable in the CreditRisk+ setting.

If we select the parameter a and determine the desired degree of confidence (i.e., α), then we can input equations 6.50–6.53 and 6.56 into the revised granularity-adjustment expression in equation 6.57. This would, usually, be the end of the story. Gordy and Lütkebohmert (2007, 2013), in their desire to use this approach for regulatory purposes, quite reasonably seek to streamline the presentation and simplify the final result. This requires some additional manipulation.

To accomplish this simplification, they introduce three new terms. The first is,

$$\mathcal{R}_n = \bar{\gamma}_n p_n. \tag{6.58}$$

The second is not merely a simplification, but also an expression in its own right. It is the amount of regulatory capital for the nth obligor required as a proportion of its total exposure. More specifically, it is given as,

$$\mathcal{K}_n = \frac{\overbrace{\text{Worst-Case Loss from Obligor } n - \text{Expected Loss from Obligor } n}^{\text{Unexpected Loss from Obligor } n}}{n\text{th Exposure}}, \tag{6.59}$$

$$= \frac{\mathbb{E}\left(\gamma_n \mathcal{E}_n \mathbb{I}_{\mathcal{D}_n} \mid S = q_\alpha\right) - \mathbb{E}\left(\gamma_n \mathcal{E}_n \mathbb{I}_{\mathcal{D}_n}\right)}{\mathcal{E}_n},$$

$$= \mathbb{E}\left(\gamma_n \mathbb{I}_{\mathcal{D}_n} \mid S = q_\alpha\right) - \mathbb{E}\left(\gamma_n \mathbb{I}_{\mathcal{D}_n}\right),$$

$$= \mathbb{E}\left(\gamma_n\right) \mathbb{E}\left(\mathbb{I}_{\mathcal{D}_n} \mid S = q_\alpha\right) - \mathbb{E}\left(\gamma_n\right) \mathbb{E}\left(\mathbb{I}_{\mathcal{D}_n}\right),$$

$$= \bar{\gamma}_n p_n (1 - \omega_n + \omega_n q_\alpha) - \bar{\gamma}_n p_n,$$

$$= \cancel{\bar{\gamma}_n p_n} - \bar{\gamma}_n p_n \omega_n + \bar{\gamma}_n p_n \omega_n q_\alpha - \cancel{\bar{\gamma}_n p_n},$$

$$= \bar{\gamma}_n p_n \omega_n (q_\alpha - 1).$$

The sum of $\mathcal{R}_n$ and $\mathcal{K}_n$ also has a relatively compact form,

$$\mathcal{R}_n + \mathcal{K}_n = \underbrace{\bar{\gamma}_n p_n}_{\substack{\text{Equation} \\ 6.58}} + \underbrace{\bar{\gamma}_n p_n \omega_n (q_\alpha - 1)}_{\text{Equation 6.59}}, \tag{6.60}$$

$$= \bar{\gamma}_n p_n + \bar{\gamma}_n p_n \omega_n q_\alpha - \bar{\gamma}_n p_n \omega_n,$$

$$= \bar{\gamma}_n \underbrace{p_n \left(1 - \omega_n + \omega_n q_\alpha\right)}_{p_n(q_\alpha)},$$

$$= \bar{\gamma}_n p_n(q_\alpha).$$

6.3 The Granularity Adjustment

The final definition involves the mean and variance of the loss-given-default parameter,

$$C_n = \frac{\bar{\gamma}_n^2 + v_{\gamma_n}}{\bar{\gamma}_n}. \tag{6.61}$$

Armed with these definitions, we can follow the Gordy and Lütkebohmert (2007, 2013) approach to find a more parsimonious representation of the CreditRisk+ granularity adjustment. Starting with equation 6.57, this will take a number of steps. First, we begin with the following adjustment,

$$\mathcal{G}_\alpha(L) = \underbrace{\left(\frac{q_\alpha - 1}{q_\alpha - 1}\right)}_{=1} \frac{-\mathcal{A}v(q_\alpha) - v'(q_\alpha)}{2\mu'(q_\alpha)}, \tag{6.62}$$

$$= \frac{\delta_\alpha(a)v(q_\alpha) - (q_\alpha - 1)v'(q_\alpha)}{2(q_\alpha - 1)\mu'(q_\alpha)},$$

where the new, model-based constant, $\delta_\alpha(a)$ is written as,

$$\delta_\alpha(a) = -(q_\alpha - 1)\mathcal{A}, \tag{6.63}$$

$$= -(q_\alpha - 1)\left(\frac{a-1}{q_\alpha} - a\right),$$

$$= (q_\alpha - 1)\left(a - \frac{a-1}{q_\alpha}\right),$$

$$= (q_\alpha - 1)\left(a + \frac{1-a}{q_\alpha}\right).$$

We will return to this constant when we are ready to actually implement the model—Gordy and Lütkebohmert (2007, 2013) allocate a significant amount of time to its determination.

There are three distinct terms in equation 6.62. We will use the preceding definitions to simplify each term. Let's start with the denominator. From equation 6.51, we have that,

$$(q_\alpha - 1)\mu'(q_\alpha) = (q_\alpha - 1) \sum_{n=1}^{N} c_n \bar{\gamma}_n p_n \omega_n, \tag{6.64}$$

$$= \sum_{n=1}^{N} c_n \underbrace{\bar{\gamma}_n p_n \omega_n (q_\alpha - 1)}_{\mathcal{K}_a},$$

$$= \sum_{n=1}^{N} c_n \mathcal{K}_n,$$

$$= \mathcal{K}^*,$$

which is the total regulatory capital under the CreditRisk+ model.

The first term in the numerator of equation 6.62 is reduced, using equation 6.52, to

$$\delta_\alpha(a) v(q_\alpha) = \delta_\alpha(a) \sum_{n=1}^{N} c_n^2 p_n(q_\alpha) \left(\bar{\gamma}_n^2 + v_{\gamma_n}(1 + p_n(q_\alpha)) \right), \tag{6.65}$$

$$= \delta_\alpha(a) \underbrace{\left(\frac{\bar{\gamma}_n}{\bar{\gamma}_n} \right)}_{=1} \sum_{n=1}^{N} c_n^2 p_n(q_\alpha) \left(\bar{\gamma}_n^2 + v_{\gamma_n}(1 + p_n(q_\alpha)) \right),$$

$$= \delta_\alpha(a) \sum_{n=1}^{N} c_n^2 \underbrace{\bar{\gamma}_n p_n(q_\alpha)}_{\mathcal{R}_n + \mathcal{K}_n} \left(\underbrace{\frac{\bar{\gamma}_n^2 + v_{\gamma_n}}{\bar{\gamma}_n}}_{C_n} + \frac{v_{\gamma_n}}{\bar{\gamma}_n} \underbrace{\left(\frac{\bar{\gamma}_n}{\bar{\gamma}_n} \right)}_{=1} p_n(q_\alpha) \right),$$

$$= \delta_\alpha(a) \sum_{n=1}^{N} c_n^2 (\mathcal{R}_n + \mathcal{K}_n) \left(C_n + \frac{v_{\gamma_n}}{\bar{\gamma}_n^2} \underbrace{\bar{\gamma}_n p_n(q_\alpha)}_{\mathcal{R}_n + \mathcal{K}_n} \right),$$

$$= \sum_{n=1}^{N} c_n^2 \left(\delta_\alpha(a) C_n (\mathcal{R}_n + \mathcal{K}_n) + \delta_\alpha(a) (\mathcal{R}_n + \mathcal{K}_n)^2 \frac{v_{\gamma_n}}{\bar{\gamma}_n^2} \right).$$

Although this does not seem like a dramatic improvement, it does permit us to write everything in terms of the various definitions.

The second term in the numerator of equation 6.62, and the last term required to re-write the granularity adjustment, depends on our derivation in equation 6.53,

$$(q_\alpha - 1) v'(q_\alpha) = (q_\alpha - 1) \sum_{n=1}^{N} c_n^2 p_n \omega_m \left(\bar{\gamma}_n^2 + v_{\gamma_n}(1 + 2 p_n(q_\alpha)) \right), \tag{6.66}$$

$$= \underbrace{\left(\frac{\bar{\gamma}_n}{\bar{\gamma}_n} \right)}_{=1} \sum_{n=1}^{N} c_n^2 p_n \omega_m (q_\alpha - 1) \left(\bar{\gamma}_n^2 + v_{\gamma_n}(1 + 2 p_n(q_\alpha)) \right),$$

$$= \sum_{n=1}^{N} c_n^2 \underbrace{\bar{\gamma}_n p_n \omega_m (q_\alpha - 1)}_{\mathcal{K}_n} \left(\underbrace{\frac{\bar{\gamma}_n^2 + v_{\gamma_n}}{\bar{\gamma}_n}}_{C_n} + 2 \frac{v_{\gamma_n}}{\bar{\gamma}_n} \underbrace{\left(\frac{\bar{\gamma}_n}{\bar{\gamma}_n} \right)}_{=1} p_n(q_\alpha) \right),$$

6.3 The Granularity Adjustment

$$= \sum_{n=1}^{N} c_n^2 \mathcal{K}_u \left(C_n + 2 \frac{v_{\gamma_n}}{\bar{\gamma}_n^2} \underbrace{\bar{\gamma}_n p_n(q_\alpha)}_{\mathcal{R}_u + \mathcal{K}_u} \right),$$

$$= \sum_{n=1}^{N} c_n^2 \mathcal{K}_u \left(C_n + 2 (\mathcal{R}_u + \mathcal{K}_u) \frac{v_{\gamma_n}}{\bar{\gamma}_n^2} \right).$$

Collecting equations 6.64 to 6.66 and plugging them into equation 6.62, we have the following lengthy expression for the granularity adjustment,

$$\mathcal{G}_\alpha(L) = \frac{\delta_\alpha(a)v(q_\alpha) - (q_\alpha - 1)v'(q_\alpha)}{2(q_\alpha - 1)\mu'(q_\alpha)}, \quad (6.67)$$

$$= \frac{\overbrace{\sum_{n=1}^{N} c_n^2 \left(\delta_\alpha(a) C_n (\mathcal{R}_u + \mathcal{K}_u) + \delta_\alpha(a)(\mathcal{R}_u + \mathcal{K}_u)^2 \frac{v_{\gamma_n}}{\bar{\gamma}_n^2} \right)}^{\text{Equation 6.65}} - \overbrace{\sum_{n=1}^{N} c_n^2 \mathcal{K}_u \left(C_n + 2(\mathcal{R}_u + \mathcal{K}_u) \frac{v_{\gamma_n}}{\bar{\gamma}_n^2} \right)}^{\text{Equation 6.66}}}{2 \underbrace{\mathcal{K}^*}_{\text{Equation 6.64}}},$$

$$= \frac{1}{2\mathcal{K}^*} \sum_{n=1}^{N} c_n^2 \left(\left(\delta_\alpha(a) C_n (\mathcal{R}_u + \mathcal{K}_u) + \delta_\alpha(a)(\mathcal{R}_u + \mathcal{K}_u)^2 \frac{v_{\gamma_n}}{\bar{\gamma}_n^2} \right) - \mathcal{K}_u \left(C_n + 2(\mathcal{R}_u + \mathcal{K}_u) \frac{v_{\gamma_n}}{\bar{\gamma}_n^2} \right) \right).$$

At first blush, this does not seem like a dramatic improvement, in terms of parsimony at least, relative to the raw form found in equation 6.62. It is, however, consistent with the form summarized in equation 6 of Gordy and Lütkebohmert (2007, Page 7). More importantly, equation 6.67 is the official format quoted in most regulatory discussions of the granularity adjustment.

6.3.6.1 Some Loose Ends

Two additional precisions are required to proceed to actual implementation. First, Gordy and Lütkebohmert (2007) suggest a specific form for v_{γ_n} to "avoid the burden of a new data requirement." Their recommended choice is,

$$v_{\gamma_n} = \xi \bar{\gamma}_n (1 - \bar{\gamma}_n), \quad (6.68)$$

where $\xi \approx 0.25$. This has some precedent in previous regulatory discussions. An alternative would be to select a distribution for γ_n, select the desired mean and variance, and calibrate the parameters appropriately. The beta distribution is a popular and sensible choice. For example, if $\gamma \sim B(a_1, b_1)$, then you can construct the following two (non-linear) equations in two unknowns,

$$\mathbb{E}(\gamma) \equiv \bar{\gamma} = \frac{a_1}{a_1 + b_1}, \tag{6.69}$$

$$\mathrm{var}(\gamma) \equiv \xi \bar{\gamma}_n (1 - \bar{\gamma}_n) = \frac{a_1 b_1}{(a_1 + b_1)^2 (a_1 + b_1 + 1)}.$$

Given choices of $\bar{\gamma}$ and ξ, which in turn through equation 6.68 yields ν_γ—say, for example, the usual choices of 0.45 and 0.25—one can readily solve for the necessary values of a_1 and b_1. Specifically, if you do the algebra to solve the system in equation 6.69 for a_1 and b_1, then

$$a_1 = \frac{\bar{\gamma}(1-\xi)}{\xi}, \tag{6.70}$$

$$= \frac{0.45 \cdot (1 - 0.25)}{0.25},$$

$$= 1.35,$$

and

$$b_1 = \frac{(1-\bar{\gamma})(1-\xi)}{\xi}, \tag{6.71}$$

$$= \frac{(1 - 0.45) \cdot (1 - 0.25)}{0.25},$$

$$= 1.65.$$

Thus, in the end, drawing the necessary loss-given-default random variates from $\gamma_n \sim B(1.35, 1.65)$ for $n = 1, \ldots, N$ would be appropriate when simulating the true one-factor CreditRisk+ model.

In their quest for parsimony, Gordy and Lütkebohmert (2007) also offer a further simplification. They argue that, given the relative sizes of $\mathcal{R}_n$ and $\mathcal{K}_n$, products of these quantities are of second-order importance and can be dropped. Practically, this involves setting $(\mathcal{R}_n + \mathcal{K}_n)^2$ and $\mathcal{K}_n(\mathcal{R}_n + \mathcal{K}_n)$ equal to zero. The impact on equation 6.67 is quite substantial,

$$\mathcal{G}_\alpha(L) = \frac{1}{2\mathcal{K}^*} \sum_{n=1}^{N} c_n^2 \left(\left(\delta_\alpha(a) C_n (\mathcal{R}_n + \mathcal{K}_n) + \delta_\alpha(a) \overline{(\mathcal{R}_n + \mathcal{K}_n)^2} \frac{\nu_{\gamma_n}}{\bar{\gamma}_n^2} \right) \right.$$

$$\left. - \mathcal{K}_n \left(C_n + 2 \overline{(\mathcal{R}_n + \mathcal{K}_n)} \frac{\nu_{\gamma_n}}{\bar{\gamma}_n^2} \right) \right), \tag{6.72}$$

6.3 The Granularity Adjustment

$$= \frac{1}{2\mathcal{K}^*} \sum_{n=1}^{N} c_n^2 \Big((\delta_\alpha(a) C_n (\mathcal{R}_n + \mathcal{K}_n)) - \mathcal{K}_n C_n \Big),$$

$$= \frac{1}{2\mathcal{K}^*} \sum_{n=1}^{N} c_n^2 C_n \Big(\delta_\alpha(a) (\mathcal{R}_n + \mathcal{K}_n) - \mathcal{K}_n \Big).$$

It is occasionally simplified even further to,

$$\mathcal{G}_\alpha(L) = \frac{1}{2\mathcal{K}^*} \sum_{n=1}^{N} c_n^2 C_n Q_n, \qquad (6.73)$$

where

$$Q_n = \delta_\alpha(a)(\mathcal{R}_n + \mathcal{K}_n) - \mathcal{K}_n. \qquad (6.74)$$

Ultimately, after much effort and tedium, we have achieved a reasonable degree of simplification. Interestingly, in this approximation, the variance of the loss-given-default parameter falls out of the analysis. How good this granularity adjustment performs in an absolute sense, and relative to the more precise form in equation 6.67, is an empirical question. We will perform some experiments in the next section.

Once again, the actual implementation of these expressions is not terribly heavy. Algorithm 6.8 summarizes—in sub-routines `getDelta`, `getC`, `getK`, and `getRK`—the key objects from equations 6.67 and 6.72. Naturally these depend on knowledge of the key parameters—a and ω_n, for example—but these choices are illuminated through regulatory guidance. We'll touch on possible values when we return to our example.

Derivation of this result is, it must be admitted, something of a mathematical marathon. By assuming portfolio-invariance, a quite indefensible assumption, the Basel-IRB approach hoped to avoid some complexity and reduce the regulatory burden. The intricacies of credit-risk modelling, however, return in full force when

Algorithm 6.8 Key CreditRisk+ granularity adjustment functions

```
def getC(gBar,xi):
    gVar = xi*gBar*(1-gBar)
    return np.divide(gBar**2+gVar,gBar)
def getK(gBar,myA,myW,myP,myAlpha):
    q = gamma.ppf(myAlpha,myA,0,1/myA)
    return gBar*myP*myW*(q-1)
def getRK(gBar,myA,myW,myP,myAlpha):
    q = gamma.ppf(myAlpha,myA,0,1/myA)
    return gBar*myP*(1-myW+myW*q)
```

trying, in the context of Pillar II, to adjust for concentration risk. One could argue that it remains straightforward; one need, after all, only apply the (relatively innocuous) recipe summarized in equation 6.72—so generously derived by Gordy and Lütkebohmert (2007) and colleagues—to arrive at the result. A responsible quantitative analyst, however, is obliged to work through the details to understand what assumptions and simplifications are embedded in these regulatory formulae. This is, in fact, the principal source of value in this exercise.

6.3.7 A Final Experiment

To test the efficacy of the CreditRisk+ granularity adjustment, we will proceed in a similar fashion to the one-factor Gaussian threshold model. We will compute the infinitely granular ASRF model outcomes, simulate the full-blown one-factor CreditRisk+ model to estimate the true concentration risk, and then see how well the difference between these two values are approximated by the granularity adjustment derived in the previous section. There are, unsurprisingly, a few new challenges associated with this experiment. First, the ASRF and CreditRisk+ model are—unlike the Gaussian one-factor threshold setting—not internally consistent. We thus need to find an approach to create some, at least approximate, level of consistency between their key model parameters. The second challenge is that the loss-given-default parameter is stochastic in this analysis; we need to adjust the one-factor CreditRisk+ model to accommodate this behaviour. Finally, the CreditRisk+ model has an additional parameter, a, that needs to be identified. We will rely, as we've already heavily done thus far, on the work of Gordy and Lütkebohmert (2007, 2013) to calibrate this choice.

Let us begin with the first challenge: creating some degree of model consistency between the ASRF and one-factor CreditRisk+ models. The Basel-IRB model is, for a given parameter ρ, roughly equivalent to the ASRF model. This permits us to use the ASRF model as our starting point. The appendix of Gordy and Lütkebohmert (2007), in an effort to highlight their choice of ξ, presents an interesting method to identify the ω_n parameter. It does this by first determining the contribution to economic capital associated with the nth obligor in both the CreditRisk+ and the asymptotic one-factor Gaussian threshold (i.e., ASRF) approaches. In the CreditRisk+, we define this quantity as,

$$\mathcal{K}_n^{(1)} = \underbrace{\bar{\gamma}_n c_n p_n \left(q_\alpha(S)\right)}_{\text{Unexpected loss}} - \underbrace{\bar{\gamma}_n c_n p_n}_{\text{Expected loss}}, \qquad (6.75)$$

$$= \bar{\gamma}_n c_n \left(p_n \left(1 - \omega_n + \omega_n q_\alpha(S)\right) - p_n \right),$$

6.3 The Granularity Adjustment

$$= \bar{\gamma}_n c_n \left(\cancel{p_n} - p_n \omega_n + p_n \omega_n q_\alpha(S) - \cancel{p_n} \right),$$

$$= \bar{\gamma}_n c_n \omega_n p_n \left(q_\alpha(S) - 1 \right).$$

This is a function of obligor data, but also the parameter ω_n and the degree of confidence, α. The associated ASRF quantity is,

$$\mathcal{K}_n^{(2)} = \underbrace{\bar{\gamma}_n c_n p_n \left(q_{1-\alpha}(G) \right)}_{\text{Unexpected loss}} - \underbrace{\bar{\gamma}_n c_n p_n}_{\text{Expected loss}}, \tag{6.76}$$

$$= \bar{\gamma}_n c_n \left(\Phi \left(\frac{\Phi^{-1}(p_n) - \sqrt{\rho_n} \Phi^{-1}(1-\alpha)}{\sqrt{1-\rho_n}} \right) - p_n \right).$$

The ASRF expression depends on similar obligor data inputs, but also on the ρ_n parameter and the degree of confidence.[22]

In the Basel framework, guidance is provided for the choice of ρ_n. Indeed, it is a predefined function of the credit counterparties unconditional default probability; that is, $\rho_n = f(p_n)$. Given this choice, we seek the appropriate, and equivalent choice of ω_n in the CreditRisk+ model. Both are factor loadings on the systematic state variable. The (approximate) solution is to equate equations 6.75 and 6.76 and to solve for ω_n. The result is,

$$\mathcal{K}_n^{(1)} = \mathcal{K}_n^{(2)}, \tag{6.77}$$

$$\underbrace{\bar{\gamma}_n c_n \omega_n p_n \left(q_\alpha(S) - 1 \right)}_{\text{Equation 6.75}} = \underbrace{\bar{\gamma}_n c_n \left(\Phi \left(\frac{\Phi^{-1}(p_n) - \sqrt{\rho_n} \Phi^{-1}(1-\alpha)}{\sqrt{1-\rho_n}} \right) - p_n \right)}_{\text{Equation 6.76}},$$

$$\omega_n = \frac{\left(\Phi \left(\frac{\Phi^{-1}(p_n) - \sqrt{\rho_n} \Phi^{-1}(1-\alpha)}{\sqrt{1-\rho_n}} \right) - p_n \right)}{p_n \left(q_\alpha(S) - 1 \right)}.$$

While useful, this expression has a few drawbacks. Perhaps most importantly, it depends on the level of confidence, α. A different level of ω_n is derived for each level of α. In addition, the CreditRisk+ parameter has an additional parameter, a, summarizing the variance of the underlying gamma distribution used to describe

---

[22] Recall that the monotonicity of conditional default probabilities runs in opposite directions for the CreditRisk+ and threshold methodologies. Thus, $q_{1-\alpha}(G)$ is equivalent to $q_\alpha(S)$ where G and S denote the standard-normal and gamma distributed systematic state variables, respectively.

Algorithm 6.9 Computing $\delta_\alpha(a)$

```
from scipy.stats import gamma
def getDelta(myA,myAlpha):
    q = gamma.ppf(myAlpha,myA,0,1/myA)
    return (q-1)*(myA+np.divide(1-myA,q))
```

the Poisson arrival intensity.[23] The α quantile of S, which shows up as $q_\alpha(S)$ in equation 6.77, explicitly depends on α.

Determining a is not an obvious task. An important, and very welcome, contribution of Gordy and Lütkebohmert (2007, 2013) was to provide some concrete suggestions with regard to its value. They construct an equation where a is set equal to the default-probability variance—this is accomplished using, among other things, the trick from equation 6.77. Since a arises on both sides of the equation, they numerically find the value of a that satisfies it. In Gordy and Lütkebohmert (2007) they recommend a value of 0.25, which was revised in Gordy and Lütkebohmert (2013) to 0.125. This choice, in addition to informing the choice of ω_n, also has an important influence on the $\delta_\alpha(s)$ described in equation 6.63. Algorithm 6.9 summarizes the Python code required to compute this constant. Essentially, it merely requires using the gamma.ppf function to find the appropriate quantile of S.

Figure 6.11 highlights a variety of value for $\delta_\alpha(a)$ for α levels in the range [0.95, 0.9999] and for three possible choices of the a parameter: 0.125, 0.25, and 0.375. If one sets $a = 0.25$ and $\alpha = 0.999$, we arrive at an approximate value of 4.83 for $\delta_\alpha(a)$. This is commonly used in regulatory applications, although other values, in this range, are presumably also considered acceptable.

Table 6.6 illustrates the various parameters used in the experiment on our 100-obligor portfolio. To the extent possible, we have employed settings consistent with those recommended in Gordy and Lütkebohmert (2007, 2013). Moreover, we have imposed obligor homogeneity across all dimensions save exposures. Thus, as in the one-factor Gaussian model case, the only difference stems from concentration. Practically, this means that $p_n = p$, $\omega_n = \omega$ and $\rho_n = \rho$ for $n = 1, \ldots, N$.

Since the portfolio has homogeneous default probabilities of 0.01, the Basel-IRB asset-correlation coefficient is about 0.19. In principle, given our choice of $a = 0.25$, we can use equation 6.77 to identify ω. As mentioned, however, ω depends on the choice of confidence level. Since we consider a broad range of α parameters, we write $\omega(\alpha)$ to denote the explicit dependence on the confidence level. This gives rise to a range of values from roughly 0.70 to 0.90. Each CreditRisk+ simulation uses a different choice of ω according to the desired level of α.[24]

After computing the ASRF model, using the parameters in Table 6.6, the next order of business is to determine the concentrated CreditRisk+ estimates.

---

[23] a, in principle, also describes the expectation of S, but this has been set to unity.

[24] This is a bit time consuming, but the idea is to create fair conditions for the experiment.

6.3 The Granularity Adjustment

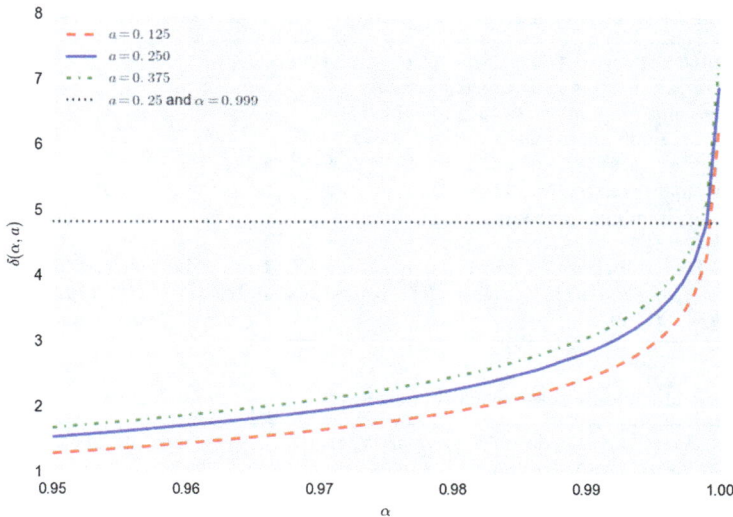

Fig. 6.11 *Choosing $\delta_\alpha(a)$*: This figure highlights the various values of $\delta_\alpha(a)$ for α levels in the range [0.95, 0.9999] for three possible choices of the a parameter: 0.125, 0.25, and 0.375. Currently, the recommended choice is to set a equal to 0.25.

Table 6.6 *CreditRisk+ parameter choices*: This table summarizes the key parameter choices found in our implementation of the one-factor CreditRisk+ model.

Parameter	Description	Value
a	Gamma parameter	0.25
ξ	LGD variance multiplier	0.25
ρ	ASRF systematic-factor loading	0.19
$\omega(\alpha)$	CreditRisk+ systematic-factor loading	[0.71, 0.88]
$\bar{\gamma}$	LGD mean	0.45
v_γ	LGD volatility	0.25
$B(a_1, b_1)$	LGD beta-distribution parameters	$B(1.35, 1.65)$
$\mathbb{E}(S)$	Systematic-factor mean	1.00
$\sigma(S)$	Systematic-factor volatility	2.00
$\delta_{0.999}(a)$	Regulatory constant	4.83

Algorithm 6.10 provides the Python implementation of the one-factor CreditRisk+ model to include stochastic loss-given-default. While it adds some computational expense—since a large number of additional random variables need to be generated—it only requires two additional lines of code. One line calls the function `calibrateBeta` to determine—by solving the system summarized in

Algorithm 6.10 Simulating the one-factor CreditRisk+ with stochastic loss-given-default

```
def crPlusOneFactorLGD (N,M,w,p,c,v,gBar,xi,alpha):
    a1,b1 = calibrateBeta(gBar,xi)
    LGD = np.random.beta(a1,b1,[M,N])
    S = np.random.gamma(v, 1/v, [M])
    wS = np.transpose(np.tile(1-w + w*S,[N,1]))
    pS = np.tile(p,[M,1])*wS
    H = np.random.poisson(pS,[M,N])
    lossIndicator = 1*np.greater_equal(H,1)
    lossDistribution = np.sort(np.dot(LGD*lossIndicator,c),axis=None)
    el,ul,var,es=util.computeRiskMeasures(M,lossDistribution,alpha)
    return el,ul,var,es
```

Algorithm 6.11 CreditRisk+ granularity adjustment

```
def granularityAdjustmentCR (myA,myW,gBar,xi,p,c,myAlpha,isApprox=0):
    myDelta = getDelta(myA,myAlpha)
    Cn = getC(gBar,xi)
    RKn = getRK(gBar,myA,myW,p,myAlpha)
    Kn = getK(gBar,myA,myW,p,myAlpha)
    KStar = np.dot(c,Kn)
    myRatio = myLGDRatio(gBar,xi)
    if isApprox==0:
        t1 = myDelta*(Cn*RKn+np.power(RKn,2)*myRatio)
        t2 = Kn*(Cn+2*RKn*myRatio)
    else:
        t1 = myDelta*Cn*RKn
        t2 = Kn*Cn
    return np.dot(np.power(c,2),t1-t2)/(2*KStar)
```

equation 6.69—the necessary beta-distribution parameters. The second line uses numpy's random number engine to generate the appropriate beta variates.[25]

The final step in this experiment involves the use of equation 6.67 to execute the one-factor CreditRisk+ granularity-adjustment computation. The Python code is summarized in Algorithm 6.11.

A suite of ancillary functions are used to compute the various components of equation 6.67, which simplifies the effort and enhances readability. The `isApprox` flag variable also allows the user to call either the full adjustment or the approximation summarized in equation 6.72. We can use this feature to examine, from an empirical perspective, the difference between these two approaches.

After the necessary mathematical drudgery, we are now in a position to examine the accuracy of the CreditRisk+ granularity adjustment. Figure 6.12 provides a visual perspective on the results. The left-hand graphic compares, for our 100-

---

[25]To put Algorithm 6.10 into context and for more information on the practical implementation of the CreditRisk+ model, please see Chap. 3.

6.3 The Granularity Adjustment 345

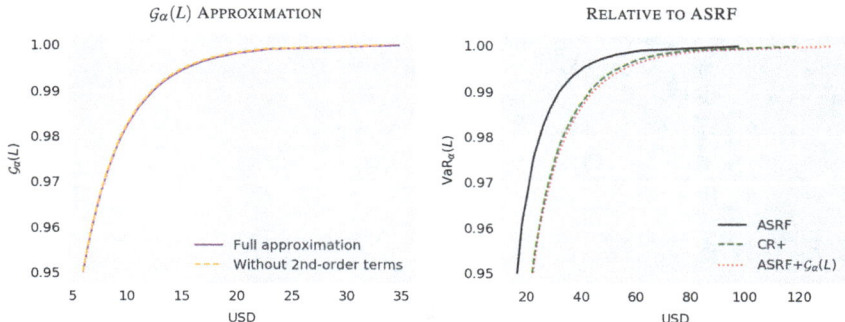

Fig. 6.12 $\mathcal{G}_\alpha(L)$ *accuracy*: The underlying figures focus on two questions. The first is: does the CreditRisk+ granularity adjustment provide a sensible approximation of concentration risk? Second, it asks: how important is the simplification in equation 6.72?

obligor portfolio summarized in Table 6.4, the full and abridged granularity adjustments from equations 6.67 and 6.72, respectively. For our portfolio, there is no visual difference between the two computations. This leads us to (cautiously) conclude that equation 6.72 is a justified approximation of equation 6.67.

The right-hand graphic in Fig. 6.12 illustrates three sets of tail probabilities: the ASRF, the full-blown one-factor CreditRisk+ model, and the ASRF plus the granularity adjustment. The distance between the ASRF and the one-factor CreditRisk+ model results can be attributed to concentration risk. Two points are worth making. First, the results are significantly smaller than in the one-factor Gaussian setting. The reason is that, for the CreditRisk+ model, the average loss-given-default is 0.45, whereas it is set to unity in the previous example. The second point is that, although we've taken great pains to ensure consistency between the ASRF and CreditRisk+ estimates, there may be a small amount of parametric uncertainty. We expect it to be modest, but feel obliged to mention some proportion of what we are treating as concentration risk might, in fact, be attributed to model differences.

When the granularity adjustment is added to the infinitely grained ASRF estimate, the result is visually quite consistent with the full-blown simulated one-factor CreditRisk+ outcomes. As we move further into the tail of the distribution, about around the 99.95th quantile, there starts to be some divergence. This is supported by the numerical values, for a selected set of quantiles, outlined in Table 6.7. The error, while not terribly large, is an increasing function of the confidence level, α. This is consistent with Gordy and Lütkebohmert (2007) who suggest that the granularity adjustment might have difficulty with smaller portfolios and that it "errs on the conservative (i.e., it overstates the effect of granularity)." We have seen that, although not dramatic, this tendency is present in both the Gaussian threshold and CreditRisk+ settings. It is somewhat larger in the latter case, but this is presumably complicated by the challenges in creating exact equivalency between the ASRF and CreditRisk+ approaches. Comparing a simple model to a more complex alternative

Table 6.7 *Testing the CreditRisk+ granularity adjustment*: This table provides a numerical assessment of the accuracy of the CreditRisk+ granularity adjustment. It compares the full-blown CreditRisk+ estimate, for our 100-obligor portfolio, to the infinitely granular ASRF estimate plus the granularity adjustment.

Quantile	Actual			Correction		Error
	ASRF	CR+	Δ	$\mathcal{G}_\alpha(L)$	ASRF + $\mathcal{G}_\alpha(L)$	
95.00th	16	22	6	6	22	−0
97.00th	21	28	7	8	29	−1
99.00th	32	43	11	12	44	−1
99.90th	62	80	18	23	85	−5
99.97th	90	112	22	29	119	−7
99.99th	98	121	24	35	132	−11

is a time-honoured sanity check, which can provide useful intuition and verification. Asset-pricing specialists, by way of analogy, often compare complex simulation or lattice models to simple analytic formulae for this reason. The Basel-IRB approach has the potential to play the same role in the credit-risk modelling environment.

Overall, however, we can see that, although mathematically complex, the granularity adjustment appears to be a powerful tool for regulators in the approximation of concentration risk. It quite successfully allows one to simply *add* an adjustment to one's infinitely grained, portfolio-invariant risk estimates. The result is surprisingly close to the true concentrated risk estimate associated with the full-blown underlying default-risk model. The granularity adjustment quite clearly finds important application in the regulatory sphere. It can also, in a less orthodox manner, be used to benchmark one's model outputs. Basel-IRB and granularity adjustments are readily computed and, based on this analysis, with a few refinements, should allow one to arrive close to one's base model estimates.

6.3.8 Final Thoughts

Regulatory guidance and internal-modelling frameworks can not only co-exist, they can effectively complement one another. The widespread use of regulatory models provides a common language and an effective benchmark for the discussion of credit-default risk. Internal models permit bespoke analysis of one's portfolio, allow for the relaxation of many restrictive regulatory assumptions, and provide enhanced insight into one's portfolio risks. We've seen that regulatory practice is informed by extant modelling approaches. This will certainly continue to be the case as the Basel guidance evolves. It is, however, also the case that the regulatory approach has an influence on default-risk models.

One fascinating and enormously practical model diagnostic virtually falls out of the Basel IRB framework. The assumption of portfolio invariance and the ability to simply sum one's risk contributions invite the analyst to extract and investigate the structure of one's risk contributions. The granularity adjustment, operating on the

6.3 The Granularity Adjustment

individual-obligor level, preserves this additivity. Combining these two elements would act to strengthen this useful model diagnostic.

Figure 6.13 thus performs an interesting computation. It takes the 100 ordered risk-capital contributions from our sample portfolio provided in Fig. 6.6 and adds the appropriate granularity adjustment using the approach detailed in the previous discussion. A total Basel IRB economic-capital estimate of roughly $160 is thus increased by about $38. This concentration risk brings the total to almost $200, which is quite close to the other internal models highlighted in Table 6.2. What makes Fig. 6.13 so compelling, however, is the decomposition of the *total* economic-capital contributions to their systematic and idiosyncratic components. The pattern of concentration risk is related, but not entirely consistent, with the associated systematic risk. A number of obligors with virtually no systematic contribution end up, via the granularity adjustment, with significant contributions to the total risk. This provides some awareness of those obligors with sizable levels of idiosyncratic risk; as we see clearly in Fig. 6.13, these need not necessarily coincide with the large contributors of systematic risk. Investigation of such issues has the potential to significantly augment one's understanding of one's portfolio.

The following chapter picks up on this discussion and investigates a variety of more general techniques for computing obligor contributions to our risk measures. While technically quite heavy, this is a worthwhile investment. The resulting model diagnostic enhances our ability to investigate, communicate, and trouble-shoot our model results. Given the complexity of the underlying models, this is a welcome tool and an important mitigant in the struggle to manage model risk.

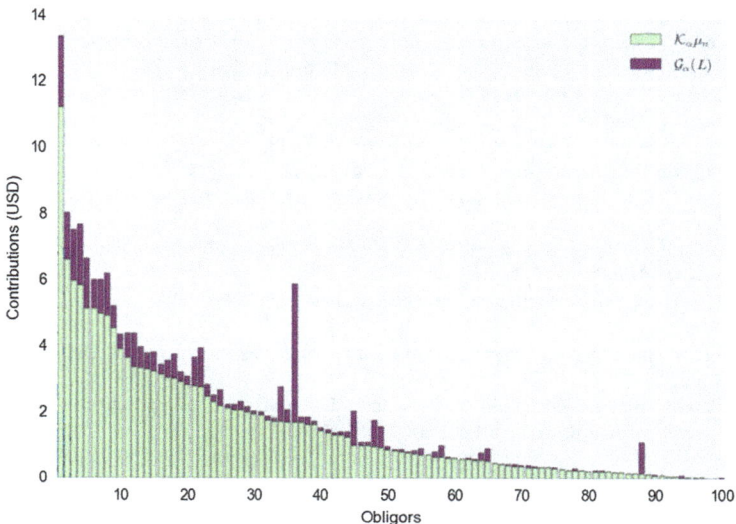

Fig. 6.13 *Revisiting economic-capital contributions*: This figure returns to the base Basel IRB contributions provided in Fig. 6.6 and, using the previously described granularity adjustment approach, adds the concentration-risk contributions.

References

Billingsley, P. (1995). *Probability and measure* (3rd edn.). Third Avenue, New York, NY: Wiley.
BIS. (2001). The internal ratings-based approach. *Technical report*. Bank for International Settlements.
BIS. (2004). International convergence of capital measurement and capital standards: A revised framework. *Technical report*. Bank for International Settlements.
BIS. (2005). An explanatory note on the Basel II IRB risk weight functions. *Technical report*. Bank for International Settlements.
BIS. (2006b). Studies on credit risk concentration. *Technical report*. Bank for International Settlements.
Bolder, D. J. (2015). *Fixed income portfolio analytics: A practical guide to implementing, monitoring and understanding fixed-income portfolios*. Heidelberg, Germany: Springer.
Casella, G., & Berger, R. L. (1990). *Statistical inference*. Belmont, CA: Duxbury Press.
Emmer, S., & Tasche, D. (2005). Calculating credit risk capital charges with the one-factor model. *Technical report*. Deutsche Bundesbank.
Gordy, M. B. (2002). A risk-factor model foundation for ratings-based bank capital rules. Board of Governors of the Federal Reserve System.
Gordy, M. B. (2003). A risk-factor model foundation for ratings-based bank capital rules. *Journal of Financial Intermediation, 12*, 199–232.
Gordy, M. B., & Lütkebohmert, E. (2007). Granularity Adjustment for Basel II. Deutsche Bundesbank, Banking and Financial Studies, No 01–2007.
Gordy, M. B., & Lütkebohmert, E. (2013). *Granularity adjustment for regulatory capital assessment*. University of Freiburg.
Gourieroux, C., Laurent, J., & Scaillet, O. (2000). Sensitivity analysis of values at risk. *Journal of Empirical Finance, 7*(3–4), 225–245.
Gundlach, M., & Lehrbass, F. (2004). *CreditRisk+ in the banking industry* (1st edn.). Berlin: Springer-Verlag.
Johnson, N. L., Kotz, S., & Balakrishnan, N. (1997). *Continuous univariate distributions: volume 2*. New York, NY: John Wiley & Sons. *Wiley series in probability and statistics*.
Lütkebohmert, E. (2008). *Concentration risk in credit portfolios*. Berlin: Springer-Verlag.
Martin, R. J., & Wilde, T. (2002). Unsystematic credit risk. *Risk*, 123–128.
Merton, R. (1974). On the pricing of corporate debt: The risk structure of interest rates. *Journal of Finance, 29*, 449–470.
Rau-Bredow, H. (2004). Value-at-risk, expected shortfall, and marginal risk contribution. In G. Szegö (Ed.), *Risk measures for the 21st century*. John Wiley & Sons.
Sharpe, W. F. (1963). A simplified model for portfolio analysis. *Management Science, 9*(2), 277–293.
Sharpe, W. F. (1964). Capital asset prices: A theory of market equilibrium under conditions of risk. *The Journal of Finance, 19*(3), 425–442.
S&P. (2010). Bank capital methodology and assumptions. *Technical report*. Standard & Poor's Global Inc.
S&P. (2011). General criteria: Principles of credit ratings. *Technical report*. Standard & Poor's Global Inc.
S&P. (2012). Multilateral lending institutions and other supranational institutions ratings methodology. *Technical report*. Standard & Poor's Global Inc.
Torell, B. (2013). Name concentration risk and pillar 2 compliance: The granularity adjustment. *Technical report*. Royal Institute of Technology, Stockholm, Sweden.
Vasicek, O. A. (1987). *Probability of loss on loan distribution*. KMV Corporation.
Vasicek, O. A. (1991). *Limiting loan loss probability distribution*. KMV Corporation.
Vasicek, O. A. (2002). The distribution of loan portfolio value. *Risk*, (12), 160–162.
Wilde, T. (1997). *CreditRisk+: A credit risk management framework*. Credit Suisse First Boston.

Yago, K. (2013). *The financial history of the Bank for international settlements*. London: Routledge, Taylor & Francis Group.
Zaccone, G. (2015). *Python parallel programming cookbook*. Birmingham, UK: Packt Publishing.

Chapter 7
Risk Attribution

> *Mathematicians are like Frenchmen: whatever you say to them they translate into their own language and forthwith it is something entirely different.*
> (Johann Wolfgang von Goethe)

In mathematical finance, computing the price of a financial security—ranging from simple bonds and equities to complex contingent claims—is an area of intense scrutiny. There are myriad articles and books dedicated to this important undertaking. In the practitioner setting, pricing is also paramount. Addition of new financial instruments to one's trading book requires more, however, than a pricing algorithm. One also needs to be able to compute the sensitivity of this price to various underlying risk factors. These hedge quantities—sometimes colourfully referred to as *Greeks* in an option setting—are often straightforward to compute, but become progressively more complex along with the intricacy of the underlying instrument. Their role is to assist in hedging unwanted exposures, but also to enhance understanding of one's positions.

In short, pricing algorithms and hedge ratios go hand in hand. In a similar vein, risk measures and attributions are intimately related. Risk metrics, by construction, collect information about outcomes and probabilities at the portfolio level and seek to quantify potential undesirable financial consequences. Such information is essential to take informed decisions about the structure of one's portfolios, to determine limits on risk exposures, and to aid one's reaction to adverse market conditions. Like our pricing analogy, we also seek insight into the structure of our risk measures. To this end, a number of risk-attribution techniques have been developed to help quantitative analysts. In many ways, they play a role similar to that of the hedge ratio in the pricing environment.

Risk attribution is, however, a non-trivial undertaking. Even in the simplest settings, a relative high level of complexity is present. It is nevertheless possible, under certain conditions, to decompose (or attribute) the contribution of individual risk factors to the overall risk measure; this is the standard operating procedure in the market-risk setting. In the credit-risk realm, however, we seek to attribute the total risk measure to the contribution from individual obligors or—when the necessary information is available and pertinent—regional or industry groupings.

Such decompositions are available in both analytic and numerical forms and can be applied to both VaR and expected-shortfall measures. In this chapter, we will consider the theoretical arguments and practical details that permit computation of risk-measure contributions in a simulation framework.[1] In brief, we turn our focus on the attribution of credit risk.

7.1 The Main Idea

In any portfolio setting, whether examining the situation from a market- or credit-risk perspective, we can typically sum our losses as follows

$$L = \sum_{n=1}^{N} X_n. \tag{7.1}$$

Each X_n can, depending on the context, be defined in terms of position, risk-factor weight, or individual credit obligor. The key point is that each X_n has some underlying meaning and importance for the managers of the portfolio. Allocating or attributing the amount of risk associated with each X_n thus has interpretative and policy value.

The general result for the decomposition of any risk measure, which follows from Euler's theorem for homogeneous functions, holds that the contribution of each X_n to VaR—a similar relationship holds for expected shortfall—is given uniquely as,

$$\text{VaR}_\alpha(L) = \sum_{n=1}^{N} \text{VaR}_\alpha(X_n), \tag{7.2}$$

$$= \sum_{n=1}^{N} X_n \frac{\partial \text{VaR}_\alpha(X)}{\partial X_n},$$

where $X = \begin{bmatrix} X_1 & \cdots & X_n \end{bmatrix}$. The partial derivative, $\dfrac{\partial \text{VaR}_\alpha(L)}{\partial X_n}$, is termed the marginal VaR. Intuitively, it is the change in the VaR measure for a vanishingly small change in the exposure of the nth portfolio weight or exposure. The consequence of equation 7.2 is that the total VaR can be written as the sum of the exposure-weighted marginal VaR estimates and the product of each exposure. Marginal VaR is a key element of any estimate of a portfolio constituent's contribution to the overall VaR. Other decompositions are possible, but this is essentially the only meaningful choice.

---

[1] A detailed description of the main conditions for the risk-factor decomposition of the parametric VaR measure are found in Bolder (2015).

7.1 The Main Idea

For this to hold, it is necessary that $\text{VaR}_\alpha(L)$ be a first-order homogeneous function of the exposure, weighting, or position vector, $X \in \mathbb{R}^N$. In general, if $\delta(x)$ is risk-metric, then it is considered a homogeneous function of the first order if,

$$\delta(\lambda x) = \lambda \delta(x), \tag{7.3}$$

for any choice of $\lambda \in \mathbb{R}$.[2] This is not at all a general attribute found outside of all but linear functions. Technicalities aside, however, this is a sensible and reasonable property for a risk measure. Simply put, it implies that if you double your position, you double your risk. Both the VaR and expected-shortfall risk measures are, quite happily, first-order homogeneous in their portfolio weights or exposures. This is also one of the properties of a coherent risk measure; see Artzner et al. (2001) for more on this notion.

Equation 7.2 represents both an enlightening and directly useful result. The decomposition of our risk metrics is thus intimately related to the computation of the partial derivatives of the underlying risk measure. In a market risk setting, when using analytical approximations of the market VaR, the risk contributions are readily computed from equation 7.1 and 7.2.

It is instructive to consider the ease of the market-risk computation. Consider a portfolio with a vector of K risk-factor weights $\zeta \in \mathbb{R}^K$ along with variance-covariance matrix, Ω. Assuming zero return and abstracting from questions of length of time horizon, the parametric VaR, at confidence level α for this portfolio, is

$$\text{VaR}_\alpha(L) = \Phi^{-1}(1-\alpha)\sqrt{\zeta^T \Omega \zeta}. \tag{7.4}$$

The VaR is, therefore, simply the appropriate quantile of the return distribution; given the assumption of Gaussianity, it reduces to a multiple of the portfolio-return variance, $\sqrt{\zeta^T \Omega \zeta}$. The first step in the risk-factor decomposition of $\text{VaR}_\alpha(L)$ is to compute the marginal value-at-risk. This is essentially the gradient vector of partial derivatives of the VaR measure to the risk-factor exposures summarized in ζ. Mathematically, it has the following form,

$$\frac{\partial \text{VaR}_\alpha(L)}{\partial \zeta} = \Phi^{-1}(1-\alpha)\frac{\partial \sqrt{\zeta^T \Omega \zeta}}{\partial \zeta}, \tag{7.5}$$

At this point, we note that computing $\frac{\partial \sqrt{\zeta^T \Omega \zeta}}{\partial \zeta}$ directly is not so straightforward. Let us define the portfolio volatility as,

$$\sigma_p(\zeta) = \sqrt{\zeta^T \Omega \zeta}. \tag{7.6}$$

---

[2] See Loomis and Sternberg (2014) for more discussion on homogeneous functions.

Working with the square-root sign is a bit annoying, so let's work with the square of the portfolio volatility (or, rather, variance),

$$\frac{\partial \left(\sigma_p(\zeta)^2\right)}{\partial \zeta} = 2\sigma_p(\zeta)\frac{\partial \sigma_p(\zeta)}{\partial \zeta}, \qquad (7.7)$$

$$\frac{\partial \sigma_p(\zeta)}{\partial \zeta} = \frac{1}{2\sigma_p(\zeta)}\frac{\partial \left(\sigma_p(\zeta)^2\right)}{\partial \zeta},$$

$$\frac{\partial \sqrt{\zeta^T \Omega \zeta}}{\partial \zeta} = \frac{1}{2\sigma_p(\zeta)}\frac{\partial \left(\sqrt{\zeta^T \Omega \zeta}^2\right)}{\partial \zeta},$$

$$= \frac{1}{2\sigma_p(\zeta)}\frac{\partial \left(\zeta^T \Omega \zeta\right)}{\partial \zeta},$$

$$= \frac{\Omega \zeta}{\sqrt{\zeta^T \Omega \zeta}},$$

which is a vector in $\mathbb{R}^K$. Plugging this result into equation 7.5, we have our desired result,

$$\frac{\partial \text{VaR}_\alpha(L)}{\partial \zeta} = \Phi^{-1}(1-\alpha)\underbrace{\frac{\Omega \zeta}{\sqrt{\zeta^T \Omega \zeta}}}_{\text{Equation 7.7}}. \qquad (7.8)$$

An element-by-element multiplication, however, of these marginal value-at-risk values and the sensitivity matrix, ζ, provides the contribution of each risk-factor to the overall risk. That is,

$$\text{Contribution of } n\text{th risk factor to VaR}_\alpha(L) = \zeta_n \cdot \frac{\partial \text{VaR}_\alpha(L)}{\partial \zeta_n}, \qquad (7.9)$$

where ζ_n refers to the nth element of the risk-weight vector, ζ. It is reasonable to enquire if the sum of these individual risk-factor contributions actually sums to the overall risk-measure value. This is easily verified. Consider the dot product of the sensitivity vector, ζ, and the marginal value-at-risk gradient,

$$\zeta^T \frac{\partial \text{VaR}_\alpha(L)}{\partial \zeta} = \zeta^T \underbrace{\left(\Phi^{-1}(1-\alpha)\frac{\Omega \zeta}{\sqrt{\zeta^T \Omega \zeta}}\right)}_{\text{Equation 7.8}}, \qquad (7.10)$$

$$= \Phi^{-1}(1-\alpha)\frac{\zeta^T \Omega \zeta}{\sqrt{\zeta^T \Omega \zeta}},$$

7.2 A Surprising Relationship

$$= \underbrace{\Phi^{-1}(1-\alpha)\sqrt{\zeta^T \Omega \zeta}}_{\text{Equation 7.4}},$$

$$= \text{VaR}_\alpha(L).$$

Thus, we have, in a few short lines, found a sensible and practical approach for the allocation of VaR to individual market-risk factors.

Sadly, in a credit risk setting, direct computation of the marginal VaR is generally difficult. Credit-risk analysts, therefore, face significant challenges in this area relative to their market-risk colleagues. There exists, as demonstrated by Tasche (2000b), Gourieroux et al. (2000), and Rau-Bredow (2004), a surprising relationship, which facilitates the computation of the marginal VaR figures. This relationship—derived in the following sections—implies that the nth VaR contribution is equivalent to the expected loss of counterparty n conditional on being at the αth quantile of the overall loss distribution. This leads to a direct Monte Carlo estimator for the marginal VaR and, as a consequence, the VaR contributions.

In a multi-factor model setting, the Monte Carlo approach is basically the only viable alternative for the computation of VaR contributions. In a one-factor setting, which applies to many of the models from the previous chapters, a powerful analytic approximation is available. This method, termed the *saddlepoint* approximation, is used in statistics and mathematical physics and can, in many cases, provide precise approximations of tail probabilities, the loss density, shortfall integrals, and risk-metric decompositions for a broad range of default-loss distributions. After our treatment of the general Monte-Carlo approach, we will work through the details of this useful tool and apply it to the class of one-factor threshold and mixture models.

7.2 A Surprising Relationship

The key object in the decomposition of a homogeneous risk measure is the marginal risk metric. There turns out to a surprising link between the marginal VaR and expected shortfall and conditional expectation. This link—introduced by Tasche (2000b), Gourieroux et al. (2000), and Rau-Bredow (2004)—leads to a useful approach for the decomposition of risk measures outside of the parametric setting. It is nonetheless significantly more complex.

We return to the usual jumping-off point for the credit-risk problem. The total loss is written as the following familiar sum over the N counterparties in the credit portfolio,

$$L = \sum_{n=1}^{N} c_n \underbrace{\mathbb{I}_{\mathcal{D}_n}}_{X_n}, \qquad (7.11)$$

where $\mathbb{I}_{\mathcal{D}_n}$ and c_n denote the nth default indicator and exposure at default, respectively. The unconditional default probability may, or may not, be a function of one or more underlying state variables. This provides us with a fairly generic definition.

We now take a new (i.e., marginal) credit position of $h \in \mathbb{R}$ units in a given obligor exposure, X_n. This new position is hX_n. That is, we marginally increase the portfolio by adding h units in the nth credit counterpart. The choice of n is arbitrary and may be applied to any of the $n = 1, \ldots, N$ entities in the portfolio.[3]

This implies that the risk of our new position, measured by the VaR, might be approximated by the quantity, $\text{VaR}_\alpha(L + hX_n)$. The idea is simple. We have a portfolio comprised of positions L and we add a new position, hX_n. Our task is to understand the sensitivity of our VaR measure to this new position: this amounts to the marginal VaR associated with our h units of obligor, X_n.

With the necessary notation, we may proceed to the main result. Under positive homogeneity conditions, the VaR measure for the portfolio loss, L, can be written as the sum of the marginal value-at-risk values. That is,

$$\text{Contribution of } n\text{th counterparty to } \text{VaR}_\alpha(L) = \left.\frac{\partial \text{VaR}_\alpha(L + hX_n)}{\partial h}\right|_{h=0}, \quad (7.12)$$

$$= \mathbb{E}\left(X_n \,\middle|\, L = \text{VaR}_\alpha(L)\right).$$

This is the key result. It is natural to enquire if the sum of these individual risk-factor contributions actually leads to the overall risk-measure value. Indeed, it holds trivially. Consider the following result,

$$\sum_{n=1}^{N} \left.\frac{\partial \text{VaR}_\alpha(L + hX_n)}{\partial h}\right|_{h=0} = \sum_{n=1}^{N} \mathbb{E}\left(X_n \,\middle|\, L = \text{VaR}_\alpha(L)\right), \quad (7.13)$$

$$= \mathbb{E}\left(\underbrace{\sum_{n=1}^{N} X_n}_{=L} \,\middle|\, L = \text{VaR}_\alpha(L)\right),$$

$$= \mathbb{E}\left(L \,\middle|\, L = \text{VaR}_\alpha(L)\right),$$

$$= \text{VaR}_\alpha(L).$$

---

[3] We further assume that both L and X_n are absolutely continuous random variables with associated density functions, $f_L(l)$ and $f_{X_n}(x)$, respectively.

7.2 A Surprising Relationship

This is not completely identical to the Euler theorem approach introduced in the previous section, but it is tantalizingly close.[4]

In the expected shortfall setting, the expression is quite similar. We have that,

$$\text{Contribution of } n\text{th counterparty to } \mathcal{E}_\alpha(L) = \left.\frac{\partial \mathcal{E}_\alpha(L + hX_n)}{\partial h}\right|_{h=0}, \quad (7.14)$$

$$= \mathbb{E}\left(X_n \mid L \geq \text{VaR}_\alpha(L)\right).$$

It is also easy to see, if somewhat repetitive, that the sum of the contributions yields the total expected shortfall,

$$\sum_{n=1}^{N} \left.\frac{\partial \mathcal{E}_\alpha(L + hX_n)}{\partial h}\right|_{h=0} = \sum_{n=1}^{N} \mathbb{E}\left(X_n \mid L \geq \text{VaR}_\alpha(L)\right), \quad (7.15)$$

$$= \mathbb{E}\left(\underbrace{\sum_{n=1}^{N} X_n}_{=L} \mid L \geq \text{VaR}_\alpha(L)\right),$$

$$= \mathbb{E}\left(L \mid L \geq \text{VaR}_\alpha(L)\right),$$

$$= \frac{1}{1-\alpha}\int_\alpha^1 \text{VaR}_u(L)du,$$

$$= \mathcal{E}_\alpha(L).$$

In brief, there exists a concrete representation of both the partial derivatives of the VaR and expected-shortfall measures.

7.2.1 The Justification

The claims in equations 7.12 and 7.14, however, need to be demonstrated. This requires some effort and a bit of pain. What we seek is the following quantity,

$$\frac{\partial \text{VaR}_\alpha(L + hX_n)}{\partial h}. \quad (7.16)$$

This is not so easy to identify. Equation 7.16 is also not a very useful place to begin. Let's instead begin with the actual definition of VaR and use this as our starting

---

[4] To reconcile with equation 7.9, merely note that the weight in front of each X_n is equal to one. This is because we are working in currency units and not return space.

point. $\text{VaR}_\alpha(L + hX_n)$ is, in fact, a function that solves the following equation,

$$\mathbb{P}\left(L + hX_n \geq \underbrace{\text{VaR}_\alpha(L + hX_n)}_{\ell_\alpha}\right) = 1 - \alpha, \tag{7.17}$$

where, going forward, we set

$$\ell_\alpha = \text{VaR}_\alpha(L + hX_n), \tag{7.18}$$

to ease the notation. Equation 7.17 is a working definition of VaR. It gives us something concrete to manipulate.

Let's now differentiate both sides with respect to h and simplify,

$$\frac{\partial \mathbb{P}\left(L + hX_n \geq \ell_\alpha\right)}{\partial h} = \frac{\partial (1 - \alpha)}{\partial h}, \tag{7.19}$$

$$\frac{\partial \mathbb{P}\left(L \geq \ell_\alpha - hX_n\right)}{\partial h} = 0,$$

$$\frac{\partial}{\partial h}\left(\int_{-\infty}^{\infty}\int_{\ell_\alpha - hx}^{\infty} f(l,x)dldx\right) = 0,$$

$$\int_{-\infty}^{\infty}\frac{\partial}{\partial h}\left(\int_{\ell_\alpha - hx}^{\infty} f(l,x)dl\right)dx = 0,$$

$$\int_{-\infty}^{\infty}\frac{\partial (\ell_\alpha - hx)}{\partial h} f(\ell_\alpha - hx, x)dx = 0,$$

$$\int_{-\infty}^{\infty}\left(\frac{\partial \ell_\alpha}{\partial h} - x\right) f(\ell_\alpha - hx, x)dx = 0,$$

$$\frac{\partial \ell_\alpha}{\partial h}\int_{-\infty}^{\infty} f(\ell_\alpha - hx, x)dx = \int_{-\infty}^{\infty} xf(\ell_\alpha - hx, x)dx,$$

$$\frac{\partial \ell_\alpha}{\partial h} = \frac{\int_{-\infty}^{\infty} xf(\ell_\alpha - hx, x)dx}{\int_{-\infty}^{\infty} f(\ell_\alpha - hx, x)dx},$$

$$\frac{\partial \ell_\alpha}{\partial h} = \mathbb{E}\left(X_n \bigg| L = \ell_\alpha - hX_n\right),$$

$$\frac{\partial \ell_\alpha}{\partial h} = \mathbb{E}\left(X_n \bigg| L + hX_n = \ell_\alpha\right).$$

7.2 A Surprising Relationship

Replacing our original definitions, we arrive at the following (fairly surprising) relationship,

$$\frac{\partial \text{VaR}_\alpha(L + hX_n)}{\partial h} = \mathbb{E}\bigg(X_n \bigg| L + hX_n = \text{VaR}_\alpha(L + hX_n)\bigg). \quad (7.20)$$

If we evaluate the preceding partial derivative at $h = 0$, then we obtain,

$$\frac{\partial \text{VaR}_\alpha(L + hX_n)}{\partial h}\bigg|_{h=0} = \mathbb{E}\bigg(X_n \bigg| L = \text{VaR}_\alpha(L)\bigg). \quad (7.21)$$

The risk contribution of the nth credit counterparty, therefore, can be considered its average value in precisely those situations where the portfolio value outcome is exactly equal to the VaR measure.

Similar, although more complex, computations are possible for the expected shortfall. Using Tasche (2000b) and Rau-Bredow (2004) as our inspiration, we provide the following derivation:

$$\frac{\partial \mathcal{E}_\alpha(L + hX_n)}{\partial h} = \frac{\partial}{\partial h}\left(\frac{1}{1-\alpha}\int_\alpha^1 \text{VaR}_u(L + hX_n)du\right), \quad (7.22)$$

$$= \frac{\partial}{\partial h}\left(\mathbb{E}\bigg(L + hX_n \bigg| L + hX_n \geq \ell_\alpha\bigg)\right),$$

$$= \frac{\partial}{\partial h}\left(\mathbb{E}\bigg(L \bigg| L + hX_n \geq \ell_\alpha\bigg) + h\,\mathbb{E}\bigg(X_n \bigg| L + hX_n \geq \ell_\alpha\bigg)\right),$$

$$= \frac{\partial}{\partial h}\left(\mathbb{E}\bigg(L \bigg| L + hX_n \geq \ell_\alpha\bigg)\right) + \frac{\partial}{\partial h}\left(h\,\mathbb{E}\bigg(X_n \bigg| L + hX_n \geq \ell_\alpha\bigg)\right),$$

$$= \frac{\partial \mathbb{E}\bigg(L \bigg| L + hX_n \geq \ell_\alpha\bigg)}{\partial h}$$

$$+ \underbrace{\mathbb{E}\bigg(X_n \bigg| L + hX_n \geq \ell_\alpha\bigg) + h\,\frac{\partial \mathbb{E}\bigg(X_n \bigg| L + hX_n \geq \ell_\alpha\bigg)}{\partial h}}_{\text{By product rule}}.$$

By a series of very detailed, non-trivial, and not particularly illuminating manipulations, Rau-Bredow (2003, 2004) and Tasche (2000a,b) demonstrate that the first and last terms in equation 7.22 are, when evaluated at $h = 0$, given as

$$\left.\frac{\partial\, \mathbb{E}\left(L \mid L + hX_n \geq \ell_\alpha\right)}{\partial h}\right|_{h=0} = \left.\frac{\mathrm{var}\left(L \mid L + hX_n = \ell_\alpha\right) f_L(\ell_\alpha)}{1 - \alpha}\right|_{h=0}, \quad (7.23)$$

$$= \frac{\mathrm{var}\left(L \mid L = \mathrm{VaR}_\alpha(L)\right) f_L(\mathrm{VaR}_\alpha(L))}{1 - \alpha},$$

and

$$\left.\frac{\partial\, \mathbb{E}\left(X_n \mid L + hX_n \geq \ell_\alpha\right)}{\partial h}\right|_{h=0} = \left.\frac{\mathrm{var}\left(X_n \mid L + hX_n = \ell_\alpha\right) f_{L+hX_n}(\ell_\alpha)}{1 - \alpha}\right|_{h=0} \quad (7.24)$$

$$= \frac{\mathrm{var}\left(X_n \mid L = \mathrm{VaR}_\alpha(L)\right) f_L(\mathrm{VaR}_\alpha(L))}{1 - \alpha},$$

As we'll see in moment, however, neither term makes much of a contribution to the result in equation 7.22 when we plug them in. The final result is,

$$\left.\frac{\partial\, \mathcal{E}_\alpha\,(L + hX_n)}{\partial h}\right|_{h=0} = \underbrace{\overbrace{\frac{\mathrm{var}\left(L \mid L = \mathrm{VaR}_\alpha(L)\right) f_L(\mathrm{VaR}_\alpha(L))}{1 - \alpha}}^{=0}}_{\text{Equation 7.23}}$$

$$+ \left.\mathbb{E}\left(X_n \mid L + hX_n \geq \mathrm{VaR}_\alpha(L)\right)\right|_{h=0} \quad (7.25)$$

$$+ 0 \cdot \underbrace{\frac{\mathrm{var}\left(X_n \mid L = \mathrm{VaR}_\alpha(L)\right) f_L(\mathrm{VaR}_\alpha(L))}{1 - \alpha}}_{\text{Equation 7.24}},$$

$$= \mathbb{E}\left(X_n \mid L \geq \mathrm{VaR}_\alpha(L)\right),$$

which is as claimed in equation 7.14. The term $\mathrm{var}\left(L \mid L = \mathrm{VaR}_\alpha(L)\right) = 0$ since, if we condition on the fact that $L = \mathrm{VaR}_\alpha(L)$, the variance operator is applied to a constant and, as a consequence, vanishes.

7.2.2 A Direct Algorithm

When an analytic solution is not available, then equation 7.12 may be exploited to numerically compute the marginal VaR and, subsequently, the contribution of each credit counterparty to the overall VaR. A similar approach can be employed, using equation 7.14, to decompose the expected-shortfall measure into its individual obligor contributions. In practice, however, this is a computationally intensive undertaking. A large number of simulations are required to determine the average set of outcomes, given that the profit-and-loss outcomes exactly coincide with the desired VaR_α or $\mathcal{E}_\alpha$ levels. As cogently stated in Glasserman (2006),

> each contribution depends on the probability of a rare event (a default) conditional on an even rarer event (an extreme loss for the portfolio).

Credit-risk is always about managing rare events, but computation of obligor risk contributions is unfortunately more extreme than usual.

Despite the computational burden, it is indeed possible. Moreover, since it represents the most direct and generally applicable approach to this problem, it is an excellent starting place. Other techniques are possible—and in many cases present dramatic improvements in efficiency—but they often apply only in certain settings and require additional assumptions. The brute-force simulation algorithm is, in contrast, relatively straightforward and can be applied to virtually any model; it need only be estimable by stochastic simulation, which is a fairly non-restrictive requirement. We thus begin with this basic, unadorned approach and offer, in the following sections, alternative techniques to reduce the computational costs.

To start, we denote ξ_n as the nth counterparty contribution to $\text{VaR}_\alpha(L)$, where by construction

$$\sum_{n=1}^{N} \xi_n = \text{VaR}_\alpha(L). \quad (7.26)$$

The algorithm proceeds in a number of steps. To begin, set $m = 1$ as the first iteration.

1. Generate the necessary state variable or variables. This will, depending on the choice of model, require the simulation of various random variates.
2. From these inputs, construct the default-indicator variable, $y_n^{(m)}$ for $n = 1, \ldots, N$. Again, this will have different forms depending on the choice of model. Let us use y to represent this combination of all underlying state variables to determine the default outcome. The actual default condition is some function of this state variable, which we denote as $f\left(y_n^{(m)}\right)$.
3. For each counterparty, construct the default indicator: $\mathbb{I}_{\mathcal{D}_n}^{(m)} = \mathbb{I}_{\left\{f\left(y_n^{(m)}\right)\right\}}$ for $n = 1, \ldots, N$.

4. Using the exposure and default indicator information, compute the simulated loss for each counterparty: $X_n^{(m)} = c_n \mathbb{I}_{\{f(y_n^{(m)})\}}$ for $n = 1, \ldots, N$. This leads the vector,

$$X^{(m)} = \begin{bmatrix} X_1^{(m)} & \cdots & X_N^{(m)} \end{bmatrix}, \quad (7.27)$$

and the portfolio loss outcome,

$$L^{(m)} = \sum_{n=1}^{N} X_n^{(m)}. \quad (7.28)$$

5. Repeat the previous steps for $m = 1, \ldots, M$.

This simulation provides us with all of the necessary ingredients for the estimation of the VaR (and expected-shortfall) contributions. To estimate the VaR$_\alpha(L)$, one merely places the collection of portfolio losses, $\{L^{(m)}; m = 1, \ldots, M\}$ in ascending order. If we denote this vector of simulated portfolio losses as $\tilde{L}$, then our VaR estimator is

$$\widehat{\text{VaR}_\alpha(L)} = \tilde{L}(\lceil \alpha \cdot M \rceil), \quad (7.29)$$

where $\lceil x \rceil$ denotes the smallest integer greater or equal to x.[5]

The VaR contributions require us to recall a fundamental property of conditional expectation. In particular, it is true that

$$\xi_n = \mathbb{E}(X_n | L = \text{VaR}_\alpha(L)), \quad (7.30)$$

$$= \frac{\mathbb{E}\left(X_n \mathbb{I}_{\{L = \text{VaR}_\alpha(L)\}}\right)}{\mathbb{P}(L = \text{VaR}_\alpha(L))},$$

where this manipulation is the average number of events occurring under the specified condition normalized by the probability of the conditioning set. This more convenient form leads to the following computation,

$$\hat{\xi}_n = \frac{\sum_{m=1}^{M} X_n^{(m)} \mathbb{I}_{\left\{L^{(m)} = \widehat{\text{VaR}_\alpha(L)}\right\}}}{\sum_{m=1}^{M} \mathbb{I}_{\left\{L^{(m)} = \widehat{\text{VaR}_\alpha(L)}\right\}}}, \quad (7.31)$$

---

[5] This is also referred to as the ceiling function, which maps a real number into the largest adjacent integer.

7.2 A Surprising Relationship

for $n = 1, \ldots, N$. In practice, the denominator should often be equal to one, since for each collection of simulations, one will typically only observe a single VaR measure at the α level. It is possible, in some cases, however, to have multiple observations at the VaR outcome.

An analogous expression—with the same basic form and derivation as equation 7.31—is available for the shortfall as

$$\hat{\zeta}_n = \frac{\sum_{m=1}^{M} X_n^{(m)} \mathbb{I}_{\left\{L^{(m)} \geq \widehat{\text{VaR}_\alpha(L)}\right\}}}{\sum_{m=1}^{M} \mathbb{I}_{\left\{L^{(m)} \geq \widehat{\text{VaR}_\alpha(L)}\right\}}}, \qquad (7.32)$$

where $\hat{\zeta}_n$ denotes the expected-shortfall contribution and $n = 1, \ldots, N$. In this case, however, there should always be, for $\alpha < 1$, a sizable number of observations in the denominator. In fact, it will be equal to $(1 - \alpha) \cdot M$. Equations 7.31 and 7.32 are quite expedient, because they allow us to estimate both measures with the same algorithm; they require, after all, the same essential inputs.

Even with a large value of simulations—say, for example, $M = 10,000,000$—the VaR and expected-shortfall contributions will exhibit some noise. As a general rule, however, since the expected shortfall is an average over a range—instead of a specific point—we should expect it to be more stable. To manage this inherent noisiness, the entire algorithm is repeatedly performed and the average contributions are determined. This modifies the final VaR computation as follows,

$$\hat{\xi}_n^{(s)} = \frac{\sum_{m=1}^{M} X_n^{(m,s)} \mathbb{I}_{\left\{L^{(m,s)} = \widehat{\text{VaR}_\alpha(L)}^{(s)}\right\}}}{\sum_{m=1}^{M} \mathbb{I}_{\left\{L^{(m,s)} = \widehat{\text{VaR}_\alpha(L)}^{(s)}\right\}}}, \qquad (7.33)$$

where $s = 1, \ldots, S$ and

$$\hat{\xi}_n = \frac{1}{S} \sum_{s=1}^{S} \hat{\xi}_n^{(s)}, \qquad (7.34)$$

for $n = 1, \ldots, N$. A similar calculation is performed for the expected shortfall. Monte Carlo estimation of the weights requires, therefore, repeated simulation of the α-level VaR. In the next section, we show examples with $S = 100$ and $M = 1,000,000$. Overall, therefore, the VaR and expected-shortfall decomposition involves 100 million model realizations, which is a lot of computation by virtually any standard. This suggests the desirability, as discussed in Chap. 8, for the use of variance-reduction techniques to reduce the computational burden.

Algorithm 7.1 Monte-Carlo-based VaR and expected-shortfall decompositions

```
import cmUtilities as util
import thresholdModels as th
def mcThresholdTDecomposition(N,M,S,p,c,rho,nu,isT,myAlpha):
    contributions = np.zeros([N,S,2])
    var = np.zeros(S)
    es = np.zeros(S)
    K = myT.ppf(p,nu)*np.ones((M,1))
    for s in range(0,S):
        Y = th.getY(N,M,p,rho,nu,isT)
        myD = 1*np.less(Y,K)
        myLoss = np.sort(np.dot(myD,c),axis=None)
        el,ul,var[s],es[s]=util.computeRiskMeasures(M,myLoss,np.array([myAlpha]))
        varVector = c*myD[np.dot(myD,c)==var[s],:]
        esVector = c*myD[np.dot(myD,c)>=var[s],:]
        contributions[:,s,0] = np.sum(varVector,0)/varVector.shape[0]
        contributions[:,s,1] = np.sum(esVector,0)/esVector.shape[0]
    return contributions,var,es
```

7.2.3 Some Illustrative Results

Algorithm 7.1 highlights a sample Python algorithm using the t-threshold model; the choice of model is irrelevant since this will apply to any simulation model. The implementation is not overly complex. N denotes the number of obligors, M are the simulation iterations, S is the number of total repetitions, c are the exposures, p are the unconditional default probabilities, and myAlpha is the desired degree of confidence. It also includes the t-threshold model parameters, rho and nu.

Algorithm 7.1 uses the functions from the thresholdModels and cmUtilities libraries to extract the simulated loss outcomes and construct the risk measures.[6] It then identifies the default outcomes both at and above the VaR threshold. The remainder is essentially an averaging exercise.

Figure 7.1 illustrates the application of Algorithm 7.1 to our usual 100-obligor portfolio. The input parameter M was set to one million and S was fixed at 100. rho and nu are set to about 0.17 and 10—as in Chap. 4 to target 5% and 1% default-correlation and tail-dependence coefficients, respectively. Thus, as previously indicated, a total of 100 million simulations was employed for the estimation of risk-metric contributions. This is a significant amount of effort. The right-hand graphic shows the VaR and expected shortfall for each individual credit

---

[6]The getY function is used to capture the state variables. As indicated in previous chapters, this is one of those situations where it is convenient to simulate state variables independent of the risk-metric calculation.

7.2 A Surprising Relationship

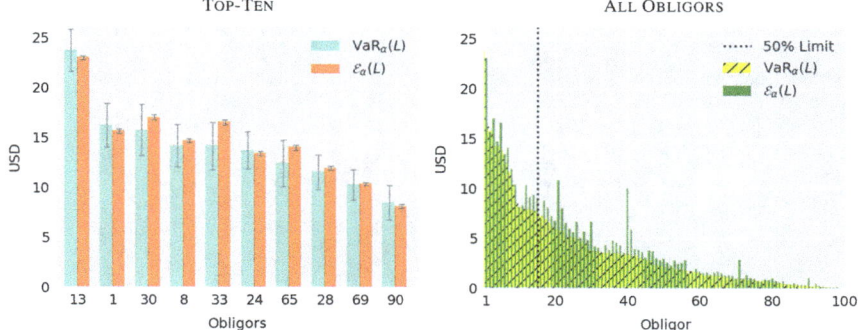

Fig. 7.1 *Monte-Carlo obligor contributions*: This figure illustrates the ten-largest and total obligor VaR and expected-shortfall contributions—ordered by their VaR contributions—for our running 100-obligor sample portfolio. All computations stem from a one-factor t-threshold model—a 95% confidence interval for the top-ten estimates is provided to underscore the noise in the estimates. About 50% of the total risk contribution arises from the 16 largest credit counterparts.

counterpart. Given the large number of entities, this is not an easy figure to read. It does, however, clearly indicate the concentration of the portfolio. One half of the total risk can, in fact, be allocated to the 16 largest obligors. This suggests that some additional focus would be useful.

The left-hand graphic of Fig. 7.1 follows this suggestion by examining the top-ten contributors to both VaR and expected shortfall; each contributor is identified by their original number ranging from 1 to 100. A number of observations can be drawn from the left-hand graphic of Fig. 7.1. First, the expected shortfall, in almost every case, exceeds the VaR contribution. This is virtually a logical necessity. If an obligor contributes to the VaR outcome, then by definition, it contributes to the expected shortfall. At the least, the VaR contribution should be equal to the expected-shortfall value. Second, Fig. 7.1 includes a constructed 99% confidence bound around our estimators.[7] There is, as we expected, significantly more uncertainty in the VaR estimates. The VaR contributions, even with 100 million simulations, demonstrate a 95% confidence of approximately $2-5 dollars. The expected-shortfall confidence interval, conversely, is barely visible and is certainly less than $1.

A final point relates to the ordering. The top-ten contributions were determined using the VaR measure. The results indicate, however, that the ordering is not preserved when we move to the expected-shortfall metric. While the differences are not dramatic, we nonetheless see evidence that the rankings would change somewhat if we were to have sorted on expected shortfall. This clearly suggests that the information in these two risk measures do *not* exactly coincide.

---

[7]In Chap. 8 we will discuss, in detail, how these interval estimates are constructed.

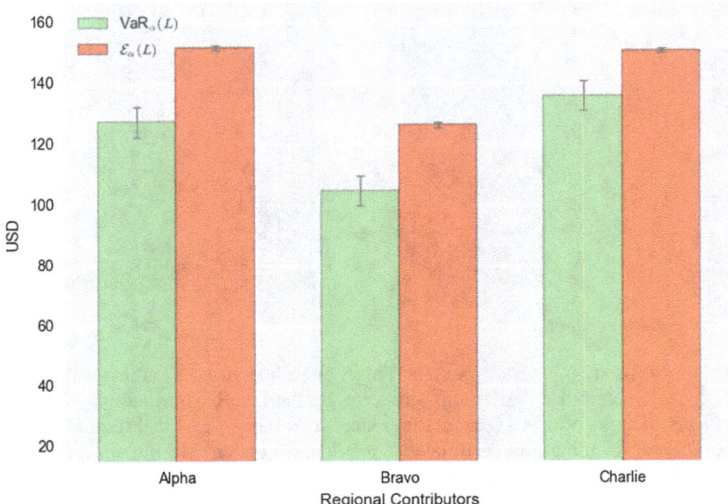

Fig. 7.2 *Regional contributions*: This figure illustrates the regional VaR and expected-shortfall contributions for our running 100-obligor sample portfolio; the computations stem from a one-factor Gaussian threshold model.

Figure 7.2 collects the data in an alternative manner. It sums each of the VaR and expected-shortfall contributions by the three pre-defined regions: Alpha, Bravo, and Charlie. Similar conclusions can be drawn about the relative size and uncertainty of the estimates. Figure 7.2 is, however, a powerful example of the benefits of decomposing one's risk measures. Although this is an invented example, it is clear that such information could prove extremely useful in interpreting and communicating one's risk estimates.

7.2.4 A Shrewd Suggestion

As clearly demonstrated in the previous section, attaining a reasonable degree of convergence for the Monte-Carlo-based VaR contributions requires significant computational expense. Interestingly, the same computation for expected shortfall is substantially more stable. The difference is that our expected-shortfall estimates are the *average* over the values beyond a certain quantile. The VaR figures, conversely, depend on the values *at* a specific quantile. This latter quantile is harder to pinpoint and subject to substantial simulation noise.

7.2 A Surprising Relationship

This would, on balance, argue for the use of expected shortfall as one's economic-capital measure.[8] VaR is a fairly common measure and it is understandable that firms might be reluctant to change their risk-management frameworks to immediately incorporate expected shortfall.

There is, numerically at least, a middle ground for the purposes of risk attribution. Bluhm et al. (2003, Chapter 5), keenly aware of the inherent instability of Monte-Carlo-based VaR risk attribution, offer a possible solution. The idea is to find the quantile that equates one's VaR-based estimate to the shortfall. They refer to this as *VaR-matched* expected shortfall. Conceptually, this is straightforward. One computes, using the predetermined level of confidence α, the usual VaR-based economic capital: we denote this value, as usual, as $\text{VaR}_\alpha(L)$. We seek, however, a new quantile, let's call it α^*, that equates expected shortfall with $\text{VaR}_\alpha(L)$. Mathematically, we seek the α^*-root of the following equation,

$$\text{VaR}_\alpha(L) - \mathcal{E}_{\alpha^*}(L) = 0. \tag{7.35}$$

In some specialized situations, equation 7.35 may have an analytic solution. More generally, however, this is readily solved numerically. Practically, we may write α^* as the solution to the following one-dimensional non-linear optimization problem

$$\min_{\alpha^*} \left\| \text{VaR}_\alpha(L) - \mathcal{E}_{\alpha^*}(L) \right\|_p, \tag{7.36}$$

where $\alpha^* \in (0, 1)$ and p describes the norm used to characterize distance. Given it is a one-dimensional problem, one can use a grid-search algorithm to resolve equation 7.36; the ultimate result is the conceptually identical.

Figure 7.3 replicates our previous analysis for the one-factor t-threshold model. 100 repetitions of one million iterations are employed to construct relatively broad VaR contributions estimates. By solving equation 7.36 for $p = 2$, we arrive at a value of approximately $\alpha^* = 0.9997$.[9] Using the same algorithm—with only 10 repetitions of one million iterations—we generate the so-called VaR-matched expected shortfall contributions. Some caution is required, because this is essentially a comparison of apples to pears. That is, although we've equalized the total risk, the VaR and expected-shortfall contributions are not measuring the same thing. Nevertheless, the basic results are not dramatically different and the VaR-matched contributions are dramatically more robust than the VaR equivalents. Bluhm et al. (2003, Chapter 5)'s pragmatic suggestion, therefore, might be worth considering for

---

[8]The coherence of expected shortfall—as described by Artzner et al. (2001)—is a further argument for moving to this risk measure.

[9]The choice of $p = 2$ leads to the L^2 notion of distance, which essentially amounts to a least-squares estimator. Other choices of p are, of course, possible, but this option is numerically robust and easily understood.

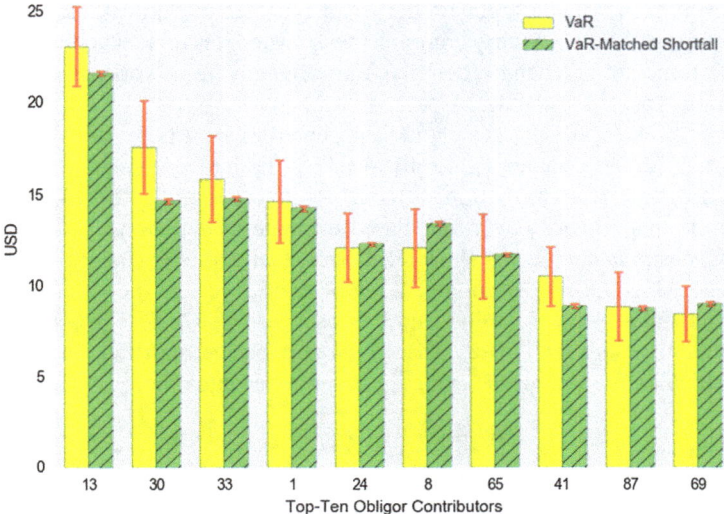

Fig. 7.3 *VaR-matched comparison*: This figure compares—for $\alpha^* \approx 0.9997$—the ten largest VaR and VaR-matched expected-shortfall contributions for the t-threshold one-factor also displayed in Fig. 7.1.

institutions using multivariate models that, as a consequence, are constrained to use of numerical methods for their risk attributions.

7.3 The Normal Approximation

Since the default loss is, in fact, the sum of independent—or conditionally independent—individual contributions for the various obligors, it is tempting to use the central-limit theorem to determine the contributions of the individual credit counterparties. Although this does not work terribly well, it is sometimes used in practice. More importantly, it does offer some insight into the nature of the computation. The idea is simple and it is best developed in the independent-default setting; from this point, it is readily generalized to the conditionally independent setting we've developed in the previous chapters.

Given our standard definition of the default loss as $L = \sum_{n=1}^{N} c_n \mathbb{I}_{\mathcal{D}_n}$ with $\mathbb{P}(\mathcal{D}_n) = p_n$, then we have already demonstrated in previous chapters that

$$\mathbb{E}(L) = \sum_{n=1}^{N} c_n p_n, \tag{7.37}$$

7.3 The Normal Approximation

and

$$\text{var}(L) = \sum_{n=1}^{N} c_n^2 p_n(1-p_n). \tag{7.38}$$

L is, by construction, the sum of independent, identically distributed random variables. By the central-limit theorem, therefore, it approximately follows that

$$L \sim \mathcal{N}\left(\underbrace{\sum_{n=1}^{N} c_n p_n}_{\mu_L}, \underbrace{\sum_{n=1}^{N} c_n^2 p_n(1-p_n)}_{\sigma_L^2}\right), \tag{7.39}$$

where we use μ_L and σ_L^2 to denote the default-loss expectation and variance, respectively. How good an approximation the Gaussian distribution is to our actual default loss is an open question. It will depend on the size and composition of the portfolio. In general, we have reason to be sceptical, since we have typically used rather different distributional choices to describe the skewed leptokurtotic form we expect from default losses.

Let's, however, set aside our scepticism and see where this assumption might take us. The implications for the tail probability—or rather the VaR—are quite clear,

$$\begin{aligned}
\mathbb{P}(L > \ell_\alpha) &= \int_{\ell_\alpha}^{\infty} f_L(u) du, \tag{7.40} \\
&= 1 - \int_{-\infty}^{\ell_\alpha} f_L(u) du, \\
&= 1 - \mathbb{P}(L \leq \ell_\alpha), \\
&= 1 - \Phi_L(\ell_\alpha), \\
&= \Phi_L(-\ell_\alpha), \\
&= \Phi\left(-\left(\frac{\ell_\alpha - \mu_L}{\sqrt{\sigma_L^2}}\right)\right), \\
&= \Phi\left(\frac{\mu_L - \ell_\alpha}{\sigma_L}\right),
\end{aligned}$$

where $\Phi_L(\cdot)$ denotes the normal inverse cumulative distribution function associated with our assumed Gaussian default-loss density in equation 7.39. A more concise description of the tail probability would be difficult to find.

The main idea behind this approach is the simple analytic expression for the upper-tail probability and thus, by extension, the VaR. As we've seen, the contribution of each obligor to the VaR is, by virtue of Euler's theorem, equal to the first partial derivative of the VaR with respect to the n exposure. In our case, we need to compute,

$$\frac{\partial \mathbb{P}(L > \ell_\alpha)}{\partial c_n} = \frac{\partial}{\partial c_n}\left(\Phi\left(\frac{\mu_L - \ell_\alpha}{\sigma_L}\right)\right), \tag{7.41}$$

$$= \underbrace{\frac{\partial}{\partial c_n}\left(\frac{\mu_L - \ell_\alpha}{\sigma_L}\right) \phi\left(\frac{\mu_L - \ell_\alpha}{\sigma_L}\right)}_{\text{By the chain rule}},$$

$$= \underbrace{\left(\frac{\frac{\partial}{\partial c_n}(\mu_L - \ell_\alpha)\sigma_L - (\mu_L - \ell_\alpha)\frac{\partial \sigma_L}{\partial c_n}}{\sigma_L^2}\right)}_{\text{By the quotient rule}} \phi\left(\frac{\mu_L - \ell_\alpha}{\sigma_L}\right).$$

To evaluate this further, we need to determine the form of the partial derivatives in the final term. Given equation 7.37, it follows naturally that

$$\frac{\partial}{\partial c_n}(\mu_L - \ell_\alpha) = \frac{\partial}{\partial c_n}\left(\sum_{k=1}^{N} c_k p_k - \ell_\alpha\right), \tag{7.42}$$

$$= p_n - \frac{\partial \ell_\alpha}{\partial c_n}.$$

The second term, while not terribly difficult, involves a bit more effort. We use the same trick as employed in equation 7.7 to isolate the partial derivative with respect to σ_L in terms of the known σ_L^2. The result is,

$$\frac{\partial \sigma_L^2}{\partial c_n} = \underbrace{2\sigma_L \frac{\partial \sigma_L}{\partial c_n}}_{\substack{\text{By the}\\\text{chain rule}}}, \tag{7.43}$$

$$\frac{\partial \sigma_L}{\partial c_n} = \frac{1}{2\sigma_L}\frac{\partial \sigma_L}{\partial c_n},$$

$$= \frac{1}{2\sigma_L}\frac{\partial}{\partial c_n}\left(\sum_{k=1}^{N} c_k^2 p_k(1-p_k)\right),$$

$$= \frac{c_n p_n(1-p_n)}{\sigma_L}.$$

7.3 The Normal Approximation

If we plug these two quantities into equation 7.41, we arrive at the following formula,

$$\underbrace{\frac{\partial \overbrace{\mathbb{P}(L > \ell_\alpha)}^{1-\alpha}}{\partial c_n}}_{=0} = \left(\frac{\left(p_n - \frac{\partial \ell_\alpha}{\partial c_n}\right)\sigma_L - (\mu_L - \ell_\alpha)\frac{c_n p_n(1-p_n)}{\sigma_L}}{\sigma_L^2}\right) \phi\left(\frac{\mu_L - \ell_\alpha}{\sigma_L}\right),$$

$$\left(\frac{\left(p_n - \frac{\partial \ell_\alpha}{\partial c_n}\right)\sigma_L}{\sigma_L^2}\right) \phi\left(\frac{\mu_L - \ell_\alpha}{\sigma_L}\right) = \left(\frac{(\mu_L - \ell_\alpha)\frac{c_n p_n(1-p_n)}{\sigma_L}}{\sigma_L^2}\right) \phi\left(\frac{\mu_L - \ell_\alpha}{\sigma_L}\right),$$

$$p_n - \frac{\partial \ell_\alpha}{\partial c_n} = (\mu_L - \ell_\alpha)\frac{c_n p_n(1 - p_n)}{\sigma_L^2},$$

$$\frac{\partial \ell_\alpha}{\partial c_n} = p_n - (\mu_L - \ell_\alpha)\frac{c_n p_n(1 - p_n)}{\sigma_L^2}. \tag{7.44}$$

This is, in fact, an elegant description of the VaR contribution in terms of quantities that are entirely known to us: the default probabilities, the exposures, the level of the VaR, and the loss expectation and volatility. Indeed, if we use Euler's theorem to sum the exposure-weighted partial derivatives, we easily recover the total VaR. This is easily shown as,

$$\sum_{n=1}^{N} c_n \frac{\partial \ell_\alpha}{\partial c_n} = \sum_{n=1}^{N} c_n \left(p_n - (\mu_L - \ell_\alpha)\frac{c_n p_n(1 - p_n)}{\sigma_L^2}\right), \tag{7.45}$$

$$= \underbrace{\sum_{n=1}^{N} c_n p_n}_{\mu_L} - \frac{(\mu_L - \ell_\alpha)}{\sigma_L^2} \underbrace{\sum_{n=1}^{N} c_n^2 p_n(1 - p_n)}_{\sigma_L^2},$$

$$= \mu_L - \frac{(\mu_L - \ell_\alpha)}{\sigma_L^2}\sigma_L^2,$$

$$= \ell_\alpha.$$

Thus, on the surface, we seem to have a sensible description of the VaR contribution from each of our individual obligors.

We should not, however, celebrate too soon. Figure 7.4 outlines the accuracy of the normal VaR approximation in equation 7.40 for two alternative portfolios: one diversified and the other concentrated. Both portfolios have the same total value, but the diversified portfolio has 10,000 small exposures, whereas the concentrated portfolio is our usual 100-obligor example. The diversified portfolio details are, in fact, the same as found in Chap. 6 in our investigation of the granularity adjustment.

Using these two portfolios, Fig. 7.4 constructs the full tail probabilities over the interval [0.9500, 0.9999] for both the Monte Carlo and normal-approximation. The results demonstrate clearly that the normal approximation is only reasonable in the

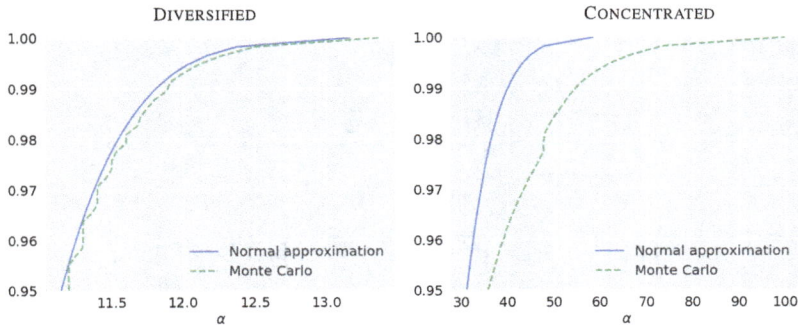

Fig. 7.4 *Accuracy of normal VaR approximation*: This figure outlines the accuracy of the normal VaR approximation in equation 7.40 for two alternative portfolios: one diversified and the other concentrated. Both portfolios have the same total value, but the diversified portfolio has 10,000 small exposures, whereas the concentrated portfolio is our usual 100-obligor example. The normal approximation is only reasonable in the diversified case.

diversified case. This is not terribly surprising given that the central-limit theorem is an asymptotic result that only holds as N gets very large. For a small or concentrated portfolio, it is not a great approximation.

What does this imply for the use of the elegant normal-approximation decomposition summarized in equations 7.44 and 7.45? If the overall VaR estimate is poor, then it does not make logical sense to depend on its decomposition. It may perform relatively well in a large, diversified portfolio setting, but this poses a logical conundrum. To have the required accuracy, one requires a large, finely grained portfolio. In such circumstances, however, the VaR decomposition is of limited value; indeed, the contributions are all small and roughly equal in the finely grained case. It is precisely in instances of concentrated exposure that it is useful and important to attribute one's risk measure to individual obligors. The unfortunate consequence is that, outside fairly restrictive circumstances, the normal approximation is *not* a legitimate solution to the determination of risk contributions. In the following sections we will consider a more complex, but accurate, approach.

7.4 Introducing the Saddlepoint Approximation

The determination of the contribution of each individual credit counterparty to one's overall credit VaR or expected-shortfall estimate provides enormous insight into these risk values. In modern credit-risk computations, counterparty level risk attribution is an indispensable model-diagnostic, communication, and policy tool. Their efficient computation is nonetheless a complex undertaking. The following sections will introduce an accurate and robust method for their computation. It is referred to as the saddlepoint approximation.

7.4 Introducing the Saddlepoint Approximation

Saddlepoint approximations were first introduced to finance by a series of articles from Martin and Thompson (2001a,b). Our non-rigorous exposition also profits significantly from a number of useful sources including Daniels (1954, 1987), Wilde (1997, Chapter 7), Goutis and Casella (1995), Antonov et al. (2005), Huzurbazar (1999), Martin (2011), Tasche (2000a,b), Martin (2004, Chapter 7), DasGupta (2008, Chapter 14), Hallerbach (2002), Muromachi (2007), Broda and Paolella (2010), and Huang et al. (2007, 2006).

7.4.1 The Intuition

A compelling and useful way to motivate this approach is to start in the deterministic setting—as in Goutis and Casella (1995)—and work toward the notion of function approximation. Imagine, for example, that we have a well-behaved and twice-differentiable function, $f(x)$, and we would like to approximate its value at x_0. This is readily accomplished with a second-order Taylor-series expansion around x_0, as

$$f(x) = \sum_{n=0}^{\infty} \frac{1}{n!} \left. f^{(n)}(x) \right|_{x=x_0} (x - x_0)^n, \qquad (7.46)$$

$$\approx \left. f^{(0)}(x) \right|_{x=x_0} (x-x_0)^0 + \left. f'(x) \right|_{x=x_0} (x-x_0)^1 + \frac{1}{2} \left. f''(x) \right|_{x=x_0} (x-x_0)^2,$$

$$\approx f(x_0) + f'(x_0)(x-x_0) + \frac{f''(x_0)}{2}(x-x_0)^2,$$

where $f^{(n)}(x)$ denotes the nth derivative of $f(x)$ with respect to x. Locally, close to x_0, this is often a good approximation of $f(x)$.[10] Now, let us repeat the previous exercise with $g(x) = \ln f(x)$. Again, we wish to approximate $g(x)$ around the arbitrary point, x_0. This may seem like an odd step, but with some cleverness and a bit of patience, it leads to a very useful result. Moreover, this result is at the heart of the saddlepoint approximation. Back to our approximation of $g(x)$:

$$\ln f(x) = g(x), \qquad (7.47)$$

$$e^{\ln f(x)} = e^{g(x)},$$

$$f(x) \approx e^{g(x_0) + g'(x_0)(x-x_0) + \frac{g''(x_0)}{2}(x-x_0)^2}.$$

Again, this may not seem very useful, but two clever tricks are now employed that change the situation. The first is to judiciously select $x_0 = \tilde{x}$ such that,

$$g'(\tilde{x}) = 0. \qquad (7.48)$$

---

[10]Naturally, for a quadratic function, the approximation will be perfect.

This leads to an important simplification of equation 7.47,

$$f(x) \approx e^{g(\tilde{x}) + \underbrace{g'(\tilde{x})(x-\tilde{x})}_{=0} + \frac{g''(\tilde{x})}{2}(x-\tilde{x})^2}, \qquad (7.49)$$

$$\approx e^{g(\tilde{x})} e^{\frac{g''(\tilde{x})}{2}(x-\tilde{x})^2},$$

$$\approx e^{g(\tilde{x})} e^{\left(\frac{(x-\tilde{x})^2}{2\left(\frac{1}{g''(\tilde{x})}\right)}\right)}.$$

The second trick is to recognize the kernel of the Gaussian density function, $e^{\frac{(x-\mu)^2}{2\sigma^2}}$, with $\mu = \tilde{x}$ and $\sigma^2 = -\frac{1}{g''(\tilde{x})}$. If we integrate both sides of equation 7.49, we arrive at a remarkable result,

$$\int f(x)dx \approx \int e^{g(\tilde{x})} e^{\left(\frac{(x-\tilde{x})^2}{2\left(-\frac{1}{g''(\tilde{x})}\right)}\right)} dx, \qquad (7.50)$$

$$\approx e^{g(\tilde{x})} \int \underbrace{\frac{\sqrt{2\pi\left(-\frac{1}{g''(\tilde{x})}\right)}}{\sqrt{2\pi\left(-\frac{1}{g''(\tilde{x})}\right)}}}_{=1} e^{-\frac{(x-\tilde{x})^2}{\frac{2}{g''(\tilde{x})}}} dx,$$

$$\approx e^{g(\tilde{x})} \sqrt{2\pi\left(-\frac{1}{g''(\tilde{x})}\right)} \underbrace{\int \frac{1}{\sqrt{2\pi\left(-\frac{1}{g''(\tilde{x})}\right)}} e^{-\frac{(x-\tilde{x})^2}{\frac{2}{g''(\tilde{x})}}} dx}_{=1},$$

$$\approx e^{g(\tilde{x})} \sqrt{\frac{-2\pi}{g''(\tilde{x})}}.$$

This is termed the Laplace approximation—alternatively referred to as Laplace's method—of the integral $\int f(x)dx$. It is a very old trick used to approximate difficult integrals.

7.4 Introducing the Saddlepoint Approximation

An extension of this idea is for the approximation of integral equations of the form,

$$f(x) = \int m(x,t)dt, \tag{7.51}$$

for some function, $m(x,t) > 0$. We will use the same basic tricks by setting $k(x,t) = \ln m(x,t)$.[11] This implies that, with a bit of manipulation, we have

$$f(x) = \int m(x,t)dt, \tag{7.52}$$

$$= \int e^{\ln m(x,t)} dt,$$

$$= \int e^{k(x,t)} dt,$$

$$\approx \int e^{k(x,t_0) + \left.\frac{\partial k(x,t)}{\partial t}\right|_{t=t_0}(t-t_0) + \frac{1}{2}\left.\frac{\partial^2 k(x,t)}{\partial t^2}\right|_{t=t_0}(t-t_0)^2} dt.$$

If we wish to use the previous approach again, then we need to select the largest $\tilde{t}_x$ such that,

$$\frac{\partial k(x,t)}{\partial t} = 0. \tag{7.53}$$

We use the notation $\tilde{t}_x$ to underscore the fact that this value depends explicitly on the choice of x. As x changes, the value of $\tilde{t}_x$ must be determined anew. This point has important computational implications for our credit-risk application, which will be addressed in latter discussion. Moreover, since equation 7.53 is a maximum, it follows that $\frac{\partial^2 k(x,\tilde{t}_x)}{\partial t} < 0$. This permits us to restate equation 7.52 as,

$$f(x) \approx \int \exp\left(k(x,\tilde{t}_x) + \overbrace{\left.\frac{\partial k(x,t)}{\partial t}\right|_{t=\tilde{t}_x}}^{=0}(t-\tilde{t}_x) - \frac{1}{2}\left.\frac{\partial^2 k(x,t)}{\partial t^2}\right|_{t=\tilde{t}_x}(t-\tilde{t}_x)^2 \right) dt, \tag{7.54}$$

$$\approx \int \exp\left(k(x,\tilde{t}_x) - \frac{1}{2}\frac{\partial^2 k(x,\tilde{t}_x)}{\partial t^2}(t-\tilde{t}_x)^2 \right) dt,$$

---

[11] This is the reason we require positivity of $m(x,t)$.

$$\approx e^{k(x,\tilde{t}_x)} \int e^{-\frac{(t-\tilde{t}_x)^2}{2\left(\frac{\partial^2 k(x,\tilde{t}_x)}{\partial t^2}\right)^{-1}}} dt,$$

$$\approx e^{k(x,\tilde{t}_x)} \sqrt{\frac{2\pi}{\frac{\partial^2 k(x,\tilde{t}_x)}{\partial t^2}}},$$

following the same logic as described in equation 7.50—that is, we have the kernel of the Gaussian density with mean and variance, $\tilde{t}_x$ and $\left(\frac{\partial^2 k(x,\tilde{t}_x)}{\partial t^2}\right)^{-1}$, respectively. This is referred to as the saddlepoint approximation of $f(x)$ and, not surprisingly, $\tilde{t}_x$ is called the saddlepoint.

A few points should be mentioned:

- To approximate the function $f(x)$ in this fashion, one needs a useful candidate for the function $k(x,t)$. This, in turn, requires some knowledge of $f(x)$. As we'll see in our particular application, the structure found in equation 7.51 occurs naturally.
- The usefulness of the approximation depends on the choice of x. If you seek to describe $f(s)$ with the saddlepoint approximation, $\tilde{t}_x$, where the distance between x and s is large, we should not expect accurate results.
- While equation 7.54 has a convenient analytic form, its computation is not without expense. A root to equation 7.53 must be found—moreover, it must often be determined numerically. In addition, the function $k(x,\tilde{t}_x)$ and its second partial derivative, $\frac{\partial^2 k(x,\tilde{t}_x)}{\partial t^2}$, must also be computed. This can, therefore, be considered a semi-analytic approximation.

7.4.2 The Density Approximation

As hinted at previously, in a statistical setting, integral equations of the form in equation 7.54 arise naturally. Let us consider a random variable, X, with density function, $f_X(x)$. The moment-generating function is defined as,

$$M_X(t) = \mathbb{E}\left(e^{tX}\right), \qquad (7.55)$$

$$= \int_{-\infty}^{\infty} e^{tx} f_X(x) dx,$$

for $x, t \in \mathbb{R}$.[12] The moment-generating function is thus a real-valued function of the probability density—we recover a function of x by integrating out the variable, t.

---

[12] See Casella and Berger (1990, Chapter 2) or Billingsley (1995) for more details.

7.4 Introducing the Saddlepoint Approximation

Equation 7.55 establishes a link between the density of the random variable and the moment-generating function. Indeed, a statistical distribution function is fully described by its moment-generating function. Not all distribution functions possess a moment-generating function, but we will only be working with functions that do. All distributions, however, have a characteristic function. For this reason, many of the proofs and results in this area, and the references provided, work with characteristic functions. To keep things as simple as possible, we will not consider this level of generality. Indeed, since our focus is on applications, we skip over many of the technical conditions and fine print.

Equation 7.55 can also be inverted to write the density function in terms of its moment-generating function.[13] The inversion of equation 7.55 yields,

$$f_X(x) = \frac{1}{2\pi i} \int_{-i\infty,(0+)}^{+i\infty} M_X(t) e^{-tx} dt, \qquad (7.56)$$

for $x \in \mathbb{R}$, $t \in \mathbb{C}$ and where i is the imaginary number satisfying $i = \sqrt{-1}$. The probability density function is thus a real-valued function of the moment-generating function, but involves integration over the complex plane—we recover a real-valued function of x by integrating over the complex variable, t. This is precisely the integral-equation form identified in equation 7.51. It occurs naturally through the moment generating function. Thus, we find ourselves with a convenient form for the density function, but are simultaneously obliged to deal with complex integration.

To make some progress on simplifying the form of equation 7.56, we introduce $K_X(t) = \ln M_X(t)$, which is termed the cumulant generating function. Using the usual steps, we can re-write equation 7.56 as,

$$f_X(x) = \frac{1}{2\pi i} \int_{-i\infty,(0+)}^{+i\infty} e^{\ln M_X(t)} e^{-tx} dt, \qquad (7.57)$$

$$= \frac{1}{2\pi i} \int_{-i\infty,(0+)}^{+i\infty} e^{K_X(t)-tx} dt.$$

To use the saddlepoint distribution, we need to consider a second-order Taylor series expansion of $K_X(t) - tx$ around the point $\tilde{t}_x$, that solves

$$\frac{\partial K_X(t) - tx}{\partial t} = 0, \qquad (7.58)$$

$$K'_X(t) - x = 0,$$

---

[13] The moment-generating function is the statistical analogue of the Laplace transform. The density function is recovered through the inverse Laplace transform. See Billingsley (1995, Section 26) for much more detail and rigour.

with $K''(\tilde{t}_x) < 0$. If we substitute our Taylor expansion into equation 7.57 at the saddlepoint, we have

$$f_X(x) \approx \frac{1}{2\pi i} \int_{-i\infty,(0+)}^{+i\infty} e^{K_X(\tilde{t}_x) - \tilde{t}_x x - \frac{K''(\tilde{t}_x)}{2}(t-\tilde{t}_x)^2} dt, \qquad (7.59)$$

$$\approx \frac{e^{K_X(\tilde{t}_x) - \tilde{t}_x x}}{2\pi i} \int_{-i\infty,(0+)}^{+i\infty} e^{-\left(\frac{(t-\tilde{t}_x)^2}{2(K''(\tilde{t}_x))^{-1}}\right)} dt,$$

$$\approx \frac{e^{K_X(\tilde{t}_x) - \tilde{t}_x x}}{2\pi i} \int_{-i\infty,(0+)}^{+i\infty} \underbrace{\frac{\sqrt{K_{X''}(\tilde{t}_x)^{-1}}}{\sqrt{K_{X''}(\tilde{t}_x)^{-1}}}}_{=1} e^{-\left(\frac{(t-\tilde{t}_x)^2}{2(K''(\tilde{t}_x))^{-1}}\right)} dt,$$

$$\approx \frac{\sqrt{K_{X''}(\tilde{t}_x)^{-1}} e^{K_X(\tilde{t}_x) - \tilde{t}_x x}}{\sqrt{2\pi}} \underbrace{\int_{-i\infty,(0+)}^{+i\infty} \frac{1}{\sqrt{2\pi i K_{X''}(\tilde{t}_x)^{-1}}} e^{-\left(\frac{(t-\tilde{t}_x)^2}{2(K''(\tilde{t}_x))^{-1}}\right)} dt}_{=1},$$

$$\approx \frac{e^{K_X(\tilde{t}_x) - \tilde{t}_x x}}{\sqrt{2\pi K_X''(\tilde{t}_x)}}.$$

The final step is essentially a line integral over the complex plane. While this outcome is analogous to the Laplace method result in equation 7.50, it is not easily shown in a rigorous manner. We quote DasGupta (2008), who notes *"the need to use delicate complex analysis in a rigorous proof of these steps."* This is not a book on complex analysis, so the reader is referred to Daniels (1954) for such a rigorous proof. Tricky technical details aside, this is referred to as the saddlepoint approximation to a density function. In practice, it turns out to be a fairly accurate approximation; we will investigate its precision by applying it to our usual example.

7.4.3 The Tail Probability Approximation

Our interest, in a risk-management setting, is not directly with the density. Instead, we are much more interested in tail probabilities. In particularly, for a generic random variable X, whose density is approximated in equation 7.59, we would like to know,

$$\mathbb{P}(X > x) = \int_x^\infty f_X(u) du, \qquad (7.60)$$

7.4 Introducing the Saddlepoint Approximation

for a rather extreme value x in the support of the distribution of X. In words, we seek the upper tail probability. Equation 7.60 could be solved by numerical integration. It turns out, however, that the saddlepoint approximation can also be used in this context. The inverse-Laplace transform for the tail probability may be, with a bit of manipulation inspired by Gil-Pelaez (1951), given the following form

$$\mathbb{P}(X > x) = \int_x^\infty \underbrace{\frac{1}{2\pi i} \int_{-i\infty,(0+)}^{+i\infty} e^{K_X(t) - tu} dt \, du}_{\text{Equation 7.57}}, \tag{7.61}$$

$$= \frac{1}{2\pi i} \int_{-i\infty,(0+)}^{+i\infty} \int_x^\infty e^{K_X(t) - tu} du \, dt,$$

$$= \frac{1}{2\pi i} \int_{-i\infty,(0+)}^{+i\infty} e^{K_X(t)} \left(\int_x^\infty e^{-tu} du \right) dt,$$

$$= \frac{1}{2\pi i} \int_{-i\infty,(0+)}^{+i\infty} e^{K_X(t)} \left[-\frac{e^{-tu}}{t} \right]_x^\infty dt,$$

$$= \frac{1}{2\pi i} \int_{-i\infty,(0+)}^{+i\infty} e^{K_X(t)} \left[0 - \left(-\frac{e^{-tx}}{t} \right) \right] dt,$$

$$= \frac{1}{2\pi i} \int_{-i\infty,(0+)}^{+i\infty} \frac{e^{K_X(t) - tx}}{t} dt.$$

This is important progress, but the final representation for the upper tail probability, $\mathbb{P}(X > x)$, is not immediately obvious. Again, it involves a complex (or contour) integral, which is computed to the right of the origin. Moreover, although it is also not particularly easy, an analytic expression for the final solution to equation 7.61 is possible.[14] It has the following form,

$$\mathbb{P}(X > x) \approx \begin{cases} e^{K_X(\tilde{t}_x) - \tilde{t}_x x} e^{\frac{\tilde{t}_x^2 K_X''(\tilde{t}_x)}{2}} \Phi\left(-\sqrt{\tilde{t}_x^2 K_X''(\tilde{t}_x)}\right) & : x > \mathbb{E}(X) \\ 1/2 & : x = \mathbb{E}(X) \\ 1 - e^{K_X(\tilde{t}_x) - \tilde{t}_x x} e^{\frac{\tilde{t}_x^2 K_X''(\tilde{t}_x)}{2}} \Phi\left(-\sqrt{\tilde{t}_x^2 K_X''(\tilde{t}_x)}\right) & : x < \mathbb{E}(X) \end{cases},$$

(7.62)

which depends on the choice of x relative to the expected value of X, $\mathbb{E}(X)$. This is closely related to the sign of the saddlepoint estimate. When it takes a negative sign, then a correction is necessary. Given our focus on far out in the upper tail for credit-risk applications, where the saddlepoint is positive, we will principally be using the

---

[14] See Daniels (1987) for a useful exposition of the derivation.

first term in equation 7.62. As we generalize to the conditionally independent case, we will need to make a few adjustments to accommodate this issue.

In other words, for our purposes, since generally $x \gg \mathbb{E}(X)$, we can (reasonably) safely use the following simplification of equation 7.62,

$$\mathbb{P}(X > x) \approx e^{K_X(\tilde{t}_x) - \tilde{t}_x x} e^{\frac{\tilde{t}_x^2 K_X''(\tilde{t}_x)}{2}} \Phi\left(-\sqrt{\tilde{t}_x^2 K_X''(\tilde{t}_x)}\right). \quad (7.63)$$

Figure 7.5 provides some visual support for this statement by computing the saddlepoint values for a range of confidence levels $\alpha \in [0.9500, 0.9999]$. While we will discuss the numerical aspect of this computation in the following sections, we first wish to establish that it generally takes a positive value. From Fig. 7.5, considering two alternative models in the context of our typical 100-obligor example, this is clearly the case. This allows us to avoid a number of complicated technical precisions related to forcing the contour integral to pass through the saddlepoint.

It turns out that this is not the only possible solution. Another analytic expression—which has the advantage of not requiring a split over the range of possible x values—is suggested by Lugannani and Rice (1980). Equation 7.62 is nevertheless perfectly sufficient for our purposes. Moreover, to again quote from an extremely useful reference DasGupta (2008), *"no practically important difference in quality is seen among the various approximations."*

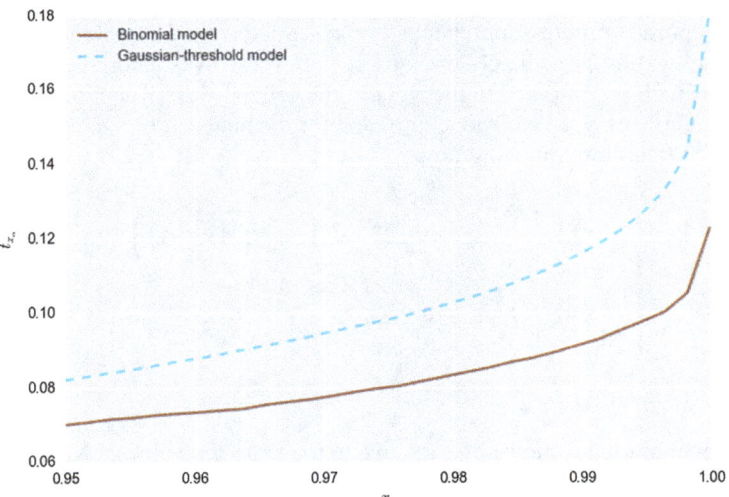

Fig. 7.5 *Saddlepoint positivity*: This figure computes the saddlepoint values for a range of confidence levels $\alpha \in [0.9500, 0.9999]$. We will discuss precisely how this is implemented in the following sections, but we first wish to establish that it generally takes a positive value.

7.4 Introducing the Saddlepoint Approximation

7.4.4 Expected Shortfall

The VaR is an upper-tail probability, which implies that we are well on our way to constructing a VaR saddlepoint approximation. We are, however, also interested in the expected shortfall. It has a rather more complex form, but also lends itself to the saddlepoint approximation. Ultimately, we proceed in the same manner as with equation 7.61. We have,

$$\mathbb{E}(X|X > x) = \frac{1}{\mathbb{P}(X > x)} \int_x^\infty u f_X(u) du, \tag{7.64}$$

$$= \frac{1}{\mathbb{P}(X > x)} \int_x^\infty u \underbrace{\frac{1}{2\pi i} \int_{-i\infty,(0+)}^{+i\infty} e^{K_X(t)-tu} dt}_{\text{Equation 7.57}} du,$$

$$= \frac{1}{\mathbb{P}(X > x)} \frac{1}{2\pi i} \int_{-i\infty,(0+)}^{+i\infty} e^{K_X(t)} \left(\int_x^\infty u e^{-tu} du \right) dt,$$

$$= \frac{1}{\mathbb{P}(X > x)} \frac{1}{2\pi i} \int_{-i\infty,(0+)}^{+i\infty} e^{K_X(t)} \left[\left(-\frac{u}{t} - \frac{1}{t^2} \right) e^{-tu} \right]_x^\infty dt,$$

$$= \frac{1}{\mathbb{P}(X > x)} \frac{1}{2\pi i} \int_{-i\infty,(0+)}^{+i\infty} e^{K_X(t)} \left[-\left(-\frac{x}{t} - \frac{1}{t^2} \right) e^{-tx} \right] dt,$$

$$= \frac{1}{\mathbb{P}(X > x)} \frac{1}{2\pi i} \int_{-i\infty,(0+)}^{+i\infty} e^{K_X(t)} \left[\frac{xe^{-tx}}{t} + \frac{e^{-tx}}{t^2} \right] dt,$$

$$= \frac{1}{\mathbb{P}(X > x)} \left(\frac{1}{2\pi i} \int_{-i\infty,(0+)}^{+i\infty} \frac{e^{K_X(t)-tx}}{t^2} dt + x \underbrace{\frac{1}{2\pi i} \int_{-i\infty,(0+)}^{+i\infty} \frac{e^{K_X(t)-tx}}{t} dt}_{\mathbb{P}(X>x)} \right),$$

$$= \frac{1}{\mathbb{P}(X > x)} \left(\frac{1}{2\pi i} \int_{-i\infty,(0+)}^{+i\infty} \frac{e^{K_X(t)-tx}}{t^2} dt + x\mathbb{P}(X > x) \right),$$

$$= x + \frac{1}{\mathbb{P}(X > x)} \left(\frac{1}{2\pi i} \int_{-i\infty,(0+)}^{+i\infty} \frac{e^{K_X(t)-tx}}{t^2} dt \right).$$

Using the saddlepoint approximation, which again requires the evaluation of a non-trivial contour integral, we have

$$\mathbb{E}(X|X>x) \approx x + \frac{1}{\mathbb{P}(X > x)} \left(e^{K_X(\tilde{t}_x) - \tilde{t}_x x} \left(\sqrt{\frac{K_X''(\tilde{t}_x)}{2\pi}} - K_X''(\tilde{t}_x)|\tilde{t}_x| e^{\frac{K_X''(\tilde{t}_x)\tilde{t}_x^2}{2}} \Phi\left(-\sqrt{\tilde{t}_x^2 K_X''(\tilde{t}_x)} \right) \right) \right).$$

$$(7.65)$$

Although this expression a bit nasty to look at, it is actually a fairly nice result. The expected-shortfall integral reduces to an, admittedly elaborate, add-on to the VaR expression, x.

7.4.5 A Bit of Organization

We can, however, simplify things somewhat. There is a common pattern in the saddlepoint density, tail probability, and shortfall integral that can help us to simultaneously reduce our notational burden and organize our thinking. Antonov et al. (2005) offer a clever and useful suggestion, which we will adopt in our discussion. We have, over the preceding pages, examined *three* integrals of the form,

$$I_k(x) = \frac{1}{2\pi i} \int_{-i\infty,(0+)}^{+i\infty} \frac{e^{K_X(t)-tx}}{t^k} dt. \tag{7.66}$$

If we set $k = 0$, we have the density function as in equation 7.57. A value of $k = 1$ leads us to the tail-probability as summarized in equation 7.61. And, finally, the kernel of the shortfall integral in equation 7.64 stems from setting $k = 2$.

If we now perform the saddlepoint approximation of equation 7.66, we can simplify the integral somewhat further as,

$$I_k(x) \approx \frac{1}{2\pi i} \int_{-i\infty,(0+)}^{+i\infty} \frac{\exp\left((K_X(\tilde{t}_x) - \tilde{t}_x x)(t - \tilde{t}_x)^0 + \overbrace{(K'_X(\tilde{t}_x)-x)}^{=0}(t - \tilde{t}_x)^1 + \frac{K_X''(\tilde{t}_x)(t-\tilde{t}_x)^2}{2}\right)}{t^k} dt,$$

$$\approx e^{K_X(\tilde{t}_x)-\tilde{t}_x x} \left(\frac{1}{2\pi i} \int_{-i\infty,(0+)}^{+i\infty} \frac{e^{\frac{K_X''(\tilde{t}_x)(t-\tilde{t}_x)^2}{2}}}{t^k} dt\right),$$

$$\approx e^{K_X(\tilde{t}_x)-\tilde{t}_x x} \mathcal{I}_k\left(K_X''(\tilde{t}_x), \tilde{t}_x\right), \tag{7.67}$$

where

$$\mathcal{I}_0(\kappa, \tilde{t}) = \frac{1}{\sqrt{2\pi\kappa}}, \tag{7.68}$$

$$\mathcal{I}_1(\kappa, \tilde{t}) = e^{\frac{\kappa \tilde{t}^2}{2}} \Phi\left(-\sqrt{\kappa \tilde{t}^2}\right), \tag{7.69}$$

and

$$\mathcal{I}_2(\kappa, \tilde{t}) = \sqrt{\frac{\kappa}{2\pi}} - \kappa |\tilde{t}| e^{\frac{\kappa \tilde{t}^2}{2}} \Phi\left(-\sqrt{\kappa \tilde{t}^2}\right). \tag{7.70}$$

7.4 Introducing the Saddlepoint Approximation 383

Table 7.1 *Key saddlepoint approximations*: This table uses the definitions in the previous section to provide (relatively) succinct expressions for the loss density, upper tail probability, and shortfall integrals.

Object	Description	Approximation
$f_X(x)$	Loss density	$\exp\left(K_X(\tilde{t}_x) - \tilde{t}_x x\right) \mathcal{J}_0\left(K_X''(\tilde{t}_x), \tilde{t}_x\right)$
$\mathbb{P}(X > x)$	Upper-tail probability	$\exp\left(K_X(\tilde{t}_x) - \tilde{t}_x x\right) \mathcal{J}_1\left(K_X''(\tilde{t}_x), \tilde{t}_x\right)$
$\mathbb{E}(X\|X > x)$	Shortfall integral	$x + \dfrac{\exp\left(K_X(\tilde{t}_x) - \tilde{t}_x x\right) \mathcal{J}_2\left(K_X''(\tilde{t}_x), \tilde{t}_x\right)}{\mathbb{P}(X > x)}$

These definitions allow us to re-state each of these quantities in a (relatively) succinct fashion. Table 7.1 summarizes our revised expressions; with these formulae, we can readily and conveniently construct analytic approximations of our three risk measures of interest.

A useful fact, that will prove essential in our determination of the attribution of expected shortfall into individual obligor contributions, is

$$\kappa\left(\mathcal{J}_0(\kappa, \tilde{t}) - \tilde{t}\mathcal{J}_1(\kappa, \tilde{t})\right) = \kappa\left(\left(\frac{1}{\sqrt{2\pi\kappa}}\right) - \tilde{t}\left(e^{\frac{\kappa\tilde{t}^2}{2}}\Phi\left(-\sqrt{\kappa\tilde{t}^2}\right)\right)\right), \quad (7.71)$$

$$= \frac{\kappa}{\sqrt{2\pi\kappa}} - \kappa\tilde{t}e^{\frac{\kappa\tilde{t}^2}{2}}\Phi\left(-\sqrt{\kappa\tilde{t}^2}\right),$$

$$= \sqrt{\frac{\kappa}{2\pi}} - \kappa\tilde{t}e^{\frac{\kappa\tilde{t}^2}{2}}\Phi\left(-\sqrt{\kappa\tilde{t}^2}\right),$$

$$= \mathcal{J}_2(\kappa, \tilde{t}).$$

Thus, as long as the saddlepoint is positive, which, for the moment, holds clearly in the current setting, we have a clean relationship between these three integrals. There are a number of additional technical elements that we need to consider, but Table 7.1 summarizes the vast majority of the intuition required to implement the saddlepoint approximation.

Algorithm 7.2 illustrates the implementation of the $\mathcal{J}$ function. The `myOrder` argument determines whether we seek the loss density, tail probability, or shortfall integral. It should be relatively easy to follow, although we make use of the not-yet-introduced `computeCGF_2` function. This is the second partial derivative of the cumulant generating function evaluated at the saddlepoint. This important object will be derived along with its implementation code in the next section.

Algorithm 7.2 The $\mathcal{J}_k$ function for $j = 0, 1, 2$

```
def getJ(l,p,c,t_1,myOrder):
    K2 = computeCGF_2(t_1,p,c)
    if myOrder==0:
        return np.sqrt(np.divide(1,2*math.pi*K2))
    if myOrder==1:
        t0 = K2*(t_1**2)
        return np.exp(0.5*t0)*norm.cdf(-np.sqrt(t0))
    if myOrder==2:
        return K2*(getJ(l,p,c,t_1,0)-t_1*getJ(l,p,c,t_1,1))
```

7.5 Concrete Saddlepoint Details

Up to this point, we have examined a collection of relatively abstract results. We have, however, virtually all of the important ideas in place. In this section, we will make the analysis slightly more concrete by specializing our results to the credit-risk case. We will begin with the independent-default binomial model since saddlepoint approximations only work well with sums of independent random variables. This is nevertheless not an issue, since by exploiting conditional independence we can, and will, recycle all of the main results and generalize the approach to all one-factor threshold and mixture credit-risk models.

The total default-loss, L, on an arbitrary credit portfolio is concisely written as $\sum_{n=1}^{N} X_n$ where each independent X_n is defined as $c_n \mathbb{I}_{\mathcal{D}_n}$. The default-loss is thus, to repeat *ad nauseum*, the sum of a collection of independent random variables: $\{X_n; n = 1, \ldots, N\}$. Each of these random variables has its own moment-generating function,

$$M_{X_n}(t) = \mathbb{E}\left(e^{tX_n}\right), \tag{7.72}$$

$$= \mathbb{E}\left(e^{tc_n \mathbb{I}_{\mathcal{D}_n}}\right),$$

$$= \mathbb{P}\left(\mathbb{I}_{\mathcal{D}_n} = 1\right) e^{tc_n(1)} + \mathbb{P}\left(\mathbb{I}_{\mathcal{D}_n} = 0\right) e^{tc_n(0)},$$

$$= \underbrace{\mathbb{P}\left(\mathbb{I}_{\mathcal{D}_n} = 1\right)}_{p_n} e^{tc_n} + \left(1 - \underbrace{\mathbb{P}\left(\mathbb{I}_{\mathcal{D}_n} = 1\right)}_{p_n}\right),$$

$$= 1 - p_n + p_n e^{tc_n},$$

which yields a concrete and practical form. Indeed, this represents the foundation for all of our subsequent computations.

7.5 Concrete Saddlepoint Details

One of the useful properties of the moment-generating function is that, if L is the sum of i.i.d. random variables, $\{X_n; n = 1, \ldots, N\}$, then its moment-generating function is given as,

$$M_L(t) = \prod_{n=1}^{N} M_{X_n}(t), \tag{7.73}$$

$$= \underbrace{\prod_{n=1}^{N} \mathbb{E}\left(e^{tX_n}\right)}_{\text{Equation 7.72}},$$

$$= \prod_{n=1}^{N} \left(1 - p_n + p_n e^{tC_n}\right).$$

In words, the moment-generating function of the sum of a collection of independent random variables reduces to the product of the individual moment-generating functions of the members of the sum. This property is, in fact, the reason we require independence. In general, the density function of the sum of a collection of random variables is a convolution of their individual density functions. One can think of a convolution as being a complicated combination of the individual density functions into a new function. This property of the moment-generating function along with the inverse-Laplace transform greatly simplify the identification of the density of a sum.

Equation 7.73 can be made even more convenient. Recall that the cumulant generating function, $K_X(t) = \ln M_X(t)$, and we may determine that,

$$\ln M_L(t) = \ln\left(\prod_{n=1}^{N} M_{X_n}(t)\right), \tag{7.74}$$

$$K_L(t) = \sum_{n=1}^{N} \underbrace{\overbrace{\ln M_{X_n}(t)}^{K_{X_n}(t)}}_{\text{Equation 7.72}},$$

$$= \sum_{n=1}^{N} \ln\left(1 - p_n + p_n e^{tC_n}\right).$$

Both the moment and cumulant-generating functions are thus readily computed. Algorithm 7.3 provides sample code for their computation. It takes three inputs arguments: the variable of integration, t, a vector of default probabilities, p, and

Algorithm 7.3 Moment and cumulant generating functions

```
def computeMGF(t,p,c):
    return 1-p+p*np.exp(c*t)
def computeCGF(t,p,c):
    return np.sum(np.log(computeMGF(t,p,c)))
```

a vector of counterparty exposures, c. To enhance efficiency and readability, the cumulant generating function, computeCGF, makes direct use of the moment-generating function, computeMGF. This simple code—each requiring a single line—can also be used irrespective of whether unconditional or conditional default probabilities are employed. This will become important as we generalize to threshold and mixture models.

We now have a precise form for the cumulant generating function of L, which will allow us to use the saddlepoint approximations for the loss density and tail probabilities. To use these approximations, however, we require a number of derivatives of $K_L(t)$. Thankfully, these are readily calculated. The first derivative of $K_L(t)$ is,

$$K'_L(t) = \frac{\partial}{\partial t}\underbrace{\left(\sum_{n=1}^{N}\ln\left(1-p_n+p_n e^{tc_n}\right)\right)}_{\text{Equation 7.74}}, \qquad (7.75)$$

$$= \sum_{n=1}^{N}\frac{\partial}{\partial t}\left(\ln\left(1-p_n+p_n e^{tc_n}\right)\right),$$

$$= \sum_{n=1}^{N}\frac{c_n p_n e^{tc_n}}{(1-p_n+p_n e^{tc_n})},$$

while the second derivative—using the quotient rule—is given as,

$$K''_L(t) = \frac{\partial}{\partial t}\underbrace{\left(\sum_{n=1}^{N}\frac{c_n p_n e^{tc_n}}{(1-p_n+p_n e^{tc_n})}\right)}_{\text{Equation 7.75}}, \qquad (7.76)$$

$$= \sum_{n=1}^{N}\frac{\partial}{\partial t}\left(\frac{c_n p_n e^{tc_n}}{(1-p_n+p_n e^{tc_n})}\right),$$

7.5 Concrete Saddlepoint Details

Algorithm 7.4 First and second derivatives of $K_L(t)$

```
def computeCGF_1(t,p,c):
    num = c*p*np.exp(c*t)
    den = computeMGF(t,p,c)
    return np.sum(np.divide(num,den))
def computeCGF_2(t,p,c,asVector=0):
    num = (1-p)*(c**2)*p*np.exp(c*t)
    den = np.power(computeMGF(t,p,c),2)
    if asVector==1:
        return np.divide(num,den)
    else:
        return np.sum(np.divide(num,den))
```

$$= \sum_{n=1}^{N} \frac{c_n^2 p_n e^{tc_n} \left(1 - p_n + p_n e^{tc_n}\right) - c_n p_n e^{tc_n} \left(c_n p_n e^{tc_n}\right)}{(1 - p_n + p_n e^{tc_n})^2},$$

$$= \sum_{n=1}^{N} \frac{c_n^2 p_n e^{tc_n} \left(1 - p_n + \cancel{p_n e^{tc_n}} - \cancel{p_n e^{tc_n}}\right)}{(1 - p_n + p_n e^{tc_n})^2},$$

$$= \sum_{n=1}^{N} \frac{c_n^2 p_n e^{tc_n} (1 - p_n)}{(1 - p_n + p_n e^{tc_n})^2}.$$

These expressions are simple manipulations of the variable, t, the default probabilities and the counterparty exposures. Algorithm 7.4 provides some sample Python code implementing equations 7.75 and 7.76. Again, the implementation is straightforward and general. Indeed, both make liberal use of computeMGF, from Algorithm 7.3, to describe the denominators thereby facilitating the computations and making the code more readable.

The second derivative function, computeCGF_2, allows the user to return the results as a single value or as a vector of outcomes indexed to the individual obligors. At this point, the default is to return the sum as summarized in equation 7.76. Later in the development, when we seek to determine the contributions to the expected shortfall, we will require this element-by-element information.

7.5.1 The Saddlepoint Density

This information permits us to find the saddlepoint approximation of the density; it follows directly from equation 7.59 that

$$f_L(\ell) \approx \frac{e^{K_L(\tilde{t}_\ell) - \tilde{t}_\ell \ell}}{\sqrt{2\pi K_L''(\tilde{t}_\ell)}}, \tag{7.77}$$

$$\approx e^{K_L(\tilde{t}_\ell) - \tilde{t}_\ell \ell} \mathcal{J}_0\left(K_L''(\tilde{t}_\ell), \tilde{t}_\ell\right),$$

$$\approx \exp\left(\sum_{n=1}^{N} \ln\left(1 - p_n + p_n e^{\tilde{t}_\ell c_n}\right) - \tilde{t}_\ell \ell\right) \mathcal{J}_0\left(\sum_{n=1}^{N} \frac{c_n^2 p_n e^{\tilde{t}_\ell c_n}(1-p_n)}{\left(1 - p_n + p_n e^{\tilde{t}_\ell c_n}\right)^2}, \tilde{t}_\ell\right),$$

where the saddlepoint, $\tilde{t}_\ell$, is the solution to the following optimization problem,

$$\frac{\partial \left(K_L(t) - t\ell\right)}{\partial t} = 0, \tag{7.78}$$

$$K_L'(t) - \ell = 0,$$

$$\underbrace{\sum_{n=1}^{N} \frac{c_n p_n e^{t c_n}}{(1 - p_n + p_n e^{t c_n})}}_{\text{Equation 7.75}} - \ell = 0.$$

This root-solving problem merits a bit of additional attention. We seek to find a root in t, the saddlepoint $\tilde{t}_\ell$, given the values of the exposures and default probabilities. Let's consider the range of values of the default probabilities. If all default probabilities take the value of zero, then irrespective of the exposures, $K_L'(t)$ vanishes. If conversely, the probability of default for all counterparties is unity, then $K_L'(t) = \sum_{n=1}^{N} c_n$. This implies that, as a function of the default probabilities, $K_L'(t) \in \left[0, \sum_{n=1}^{N} c_n\right]$. The important conclusion is that $K_L'(t)$ is a bounded function.

The first derivative of the cumulant generating function, $K_L'(t)$, is also a monotonically increasing function of t.[15] Figure 7.6 provides visual evidence of the monotone form of $K_L'(t)$ in t. The combination of its monotonicity and boundedness

---

[15] Both the numerator and denominator of $K_L'(t)$ have a e^{tc_n} term, but the coefficient on the numerator, $c_n p_n$, will, for positive exposures, dominate the p_n coefficient in the denominator.

7.5 Concrete Saddlepoint Details

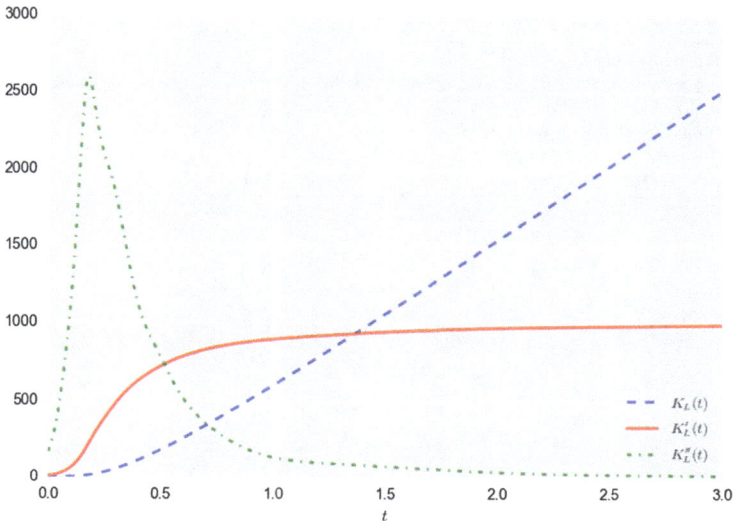

Fig. 7.6 *Cumulant generating function and derivatives*: This figure highlights, for a portfolio of 100 randomly generated heterogeneous exposures and default probabilities, the cumulant generating function and its first two derivatives. Note that $K'_L(t)$ is a monotonically increasing function of t.

implies that there exists a unique solution to equation 7.78 for any choice of
$$\ell \in \left[0, \sum_{n=1}^{N} c_n\right].$$

Given a choice of ℓ, the computation of $\tilde{t}_\ell$ is thus readily performed. Algorithm 7.5 provides the sample Python code. It takes the value ℓ, default probabilities, and exposures as input arguments and solves equation 7.78 for the saddlepoint, $\tilde{t}_\ell$, or t_1 in the underlying code. This requires use of a root-finding algorithm similar to that employed in the calibration of the mixture models in previous chapters. root is a one-dimensional version of the previously used fsolve Python function.[16] Other than the root-solving exercise, this is a straightforward computation.

The previous development and a practical sub-routine for the determination of the saddlepoint allow us to actually execute our saddlepoint density approximation. The result is surprisingly pleasant; it requires merely two lines of code. This is possible by employing the getSaddlePoint, computeCGF, and getJ functions introduced in algorithms 7.5, 7.3, and 7.2, respectively.

How well does our density approximation work? Using our example portfolio of 100 randomly generated heterogeneous exposures and default probabilities, Fig. 7.7 compares our density approximation from equation 7.77 and Algorithm 7.6

---

[16] In both cases, Python is merely a wrapper around MINPACK's hybrd algorithms.

Algorithm 7.5 Finding the saddlepoint: $\tilde{t}_\ell$

```
def getSaddlePoint(p,c,l):
    r = scipy.optimize.root(computeCGFRoot,0.00025,args=(p,c,l),method='hybr')
    return r.x
def computeCGFRoot(t,p,c,l):
    cgfDerivative = computeCGF_1(t,p,c)
    return cgfDerivative-l
```

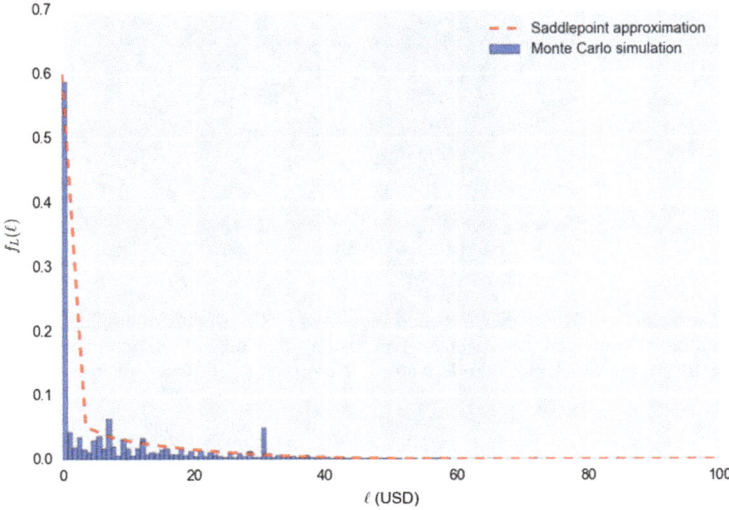

Fig. 7.7 *Independent-default density approximation*: This figure illustrates, for our running portfolio of 100 randomly generated heterogeneous exposures and default probabilities, density approximation. The density approximation, $f_L(\ell)$, is compared with a Monte Carlo binomial independent-default simulation employing 10 million iterations.

Algorithm 7.6 The saddlepoint density

```
def saddlePointDensity(l,p,c):
    t_l = getSaddlePoint(p,c,l)
    return np.exp(computeCGF(t_l,p,c)-t_l*l)*getJ(l,p,c,t_l,0)
```

with a Monte Carlo simulation employing 10 million iterations. The overall fit is generally quite good, despite being a continuous approximation to a discrete function. The approximation does certainly appear to capture the general form of the loss distribution. It is hard, given the scale, to say much about its performance as we move out into the tail. This will be addressed in the next section when we consider the tail probabilities.

7.5.2 Tail Probabilities and Shortfall Integralls

While the saddle density is interesting, our principal interest is in the tail probabilities. If we return to our general expression from Equation 7.63 and Table 7.1, with our precise definitions of the cumulant generating function and its derivatives, we arrive at the following result,

$$\mathbb{P}(L > \ell) \approx e^{K_L(\tilde{t}_\ell) - \tilde{t}_\ell \ell} \mathcal{J}_1\left(K_L''(\tilde{t}_\ell), \tilde{t}_\ell\right), \tag{7.79}$$

$$\approx \exp\left(\sum_{n=1}^{N} \ln\left(1 - p_n + p_n e^{\tilde{t}_\ell c_n}\right) - \tilde{t}_\ell \ell\right) \mathcal{J}_1\left(\sum_{n=1}^{N} \frac{c_n^2 p_n e^{\tilde{t}_\ell c_n}(1 - p_n)}{\left(1 - p_n + p_n e^{\tilde{t}_\ell c_n}\right)^2}, \tilde{t}_\ell\right).$$

The shortfall integral is also now readily expressed as,

$$\mathbb{E}(L|L > \ell) \approx \ell + \frac{e^{K_L(\tilde{t}_\ell) - \tilde{t}_\ell \ell} \mathcal{J}_2\left(K_L''(\tilde{t}_\ell), \tilde{t}_\ell\right)}{\mathbb{P}(L > \ell)}, \tag{7.80}$$

$$\approx \ell + \frac{\exp\left(\sum_{n=1}^{N} \ln\left(1 - p_n + p_n e^{\tilde{t}_\ell c_n}\right) - \tilde{t}_\ell \ell\right) \mathcal{J}_2\left(\sum_{n=1}^{N} \frac{c_n^2 p_n e^{\tilde{t}_\ell c_n}(1 - p_n)}{\left(1 - p_n + p_n e^{\tilde{t}_\ell c_n}\right)^2}, \tilde{t}_\ell\right)}{\exp\left(\sum_{n=1}^{N} \ln\left(1 - p_n + p_n e^{\tilde{t}_\ell c_n}\right) - \tilde{t}_\ell \ell\right) \mathcal{J}_1\left(\sum_{n=1}^{N} \frac{c_n^2 p_n e^{\tilde{t}_\ell c_n}(1 - p_n)}{\left(1 - p_n + p_n e^{\tilde{t}_\ell c_n}\right)^2}, \tilde{t}_\ell\right)},$$

$$\approx \ell + \frac{\mathcal{J}_2\left(\sum_{n=1}^{N} \frac{c_n^2 p_n e^{\tilde{t}_\ell c_n}(1 - p_n)}{\left(1 - p_n + p_n e^{\tilde{t}_\ell c_n}\right)^2}, \tilde{t}_\ell\right)}{\mathcal{J}_1\left(\sum_{n=1}^{N} \frac{c_n^2 p_n e^{\tilde{t}_\ell c_n}(1 - p_n)}{\left(1 - p_n + p_n e^{\tilde{t}_\ell c_n}\right)^2}, \tilde{t}_\ell\right)}.$$

The expected shortfall integral has some nice cancellation leading to a quite succinct representation. When we will move to the conditionally independent setting, however, this will no longer be possible and the original form is required. Once again, however, the computations involve only standard manipulations of the saddlepoint, the default probabilities, and the counterparty exposures.

Algorithm 7.7 highlights the Python implementation of equations 7.79 and 7.80. As before, each solves for $\tilde{t}_\ell$ using a root-solving algorithm and employs previously defined `getJ` to great effect, reducing each implementation to a few short lines of code.

Algorithm 7.7 Computing the saddle-point tail probabilities and shortfall integrals

```
def saddlePointTailProbability(l,p,c):
    t_l = getSaddlePoint(p,c,l)
    return np.exp(computeCGF(t_l,p,c)-t_l*l)*getJ(l,p,c,t_l,1)
def saddlePointShortfallIntegral(l,p,c):
    den = saddlePointTailProbability(l,p,c)
    t_l = getSaddlePoint(p,c,l)
    return l + np.exp(computeCGF(t_l,p,c)-t_l*l)*getJ(l,p,c,t_l,2)/den
```

7.5.3 A Quick Aside

Before turning to consider the performance of these two estimates in the context of our usual sample portfolio, let us first address an additional practical issue. Equation 7.79 is, for the computation of VaR estimates, of significant usefulness if ℓ is known. It operates, however, in the opposite direction from our usual approach. That is, Monte-Carlo simulations and saddlepoint approximations work in contrasting ways. In the Monte-Carlo setting, we typically select our level of confidence, α, and compute $\text{VaR}_\alpha(L)$. With the saddlepoint approximation, it runs the opposite direction. We define the level of loss, ℓ, and compute the level of confidence or tail probability, $\mathbb{P}(L > \ell)$. Table 7.2 summarizes this idea.

This is inconvenient, but can be comfortably resolved. In particular, we recall that if L is the overall loss of the portfolio, then by definition

$$\mathbb{P}\left(L > \ell_\alpha\right) = 1 - \alpha. \tag{7.81}$$

We now merely use a numerical method to identify the specific value of ℓ_α that solves equation 7.81. In particular, we need only find the root of

$$\underbrace{\left(e^{K_L(\tilde{t}_\ell) - \tilde{t}_\ell \ell_\alpha} \mathcal{J}_1\left(K_L''(\tilde{t}_\ell), \tilde{t}_\ell\right)\right)}_{\mathbb{P}(L > \ell_\alpha)} - (1 - \alpha) = 0. \tag{7.82}$$

Our objective function to solve this optimization problem is summarized in Algorithm 7.8. We have elected to minimize the squared deviation between the

Table 7.2 *Computational direction*: In the Monte-Carlo setting, we typically select our level of confidence, α, and compute $\text{VaR}_\alpha(L)$. With the saddlepoint approximation, it runs the opposite direction. We define the level of loss, ℓ, and compute the level of confidence, $\mathbb{P}(L > \ell)$. This table summarizes this important detail.

Technique	Input	Output
Monte-Carlo simulation	α	$\text{VaR}_\alpha(L)$
Saddlepoint approximation	ℓ	$\mathbb{P}(L > \ell)$

7.5 Concrete Saddlepoint Details

Algorithm 7.8 Identifying a specific saddlepoint VaR estimate

```
def identifyVaRInd(x,p,c,myAlpha):
    tpY = saddlePointTailProbability(x,p,c)
    return 1e4*np.power((1-tpY)-myAlpha,2)
```

saddlepoint tail probability and our desired quantile. Since these figures are quite small, which can introduce rounding errors, we further multiply the squared distance by a large positive factor.

The actual implementation of Algorithm 7.8 merely requires an optimization. We have further used, as in previous chapters, Powell's method available in `scipy`'s minimization library. The actual function call is,

$$r = \text{scipy.optimize.minimize(identifyVaRInd,x0,args=(p,c,myAlpha),} \\ \text{method='Powell')}, \quad (7.83)$$

where `x0` is our starting point and `myAlpha` is the target quantile. We will use this general approach—which is incredibly fast in the independent-default setting—to identify our VaR estimates. These values then serve as the reference point for our expected-shortfall computations. We continue to use this idea in the conditionally independent setting, although it becomes a bit more computationally expensive.

7.5.4 Illustrative Results

Again, we should naturally ask: how well do these approximations perform? An intuitive point of comparison is the Monte Carlo estimator. Using our classical example, Fig. 7.8 compares the saddlepoint tail-probability approximations to their Monte Carlo equivalents.

Generally speaking, the fit is quite close. Figure 7.8 also highlights an important point. The Monte Carlo estimators, by their very nature, are subject to simulation noise. At 10 million iterations, a significant amount of computation was employed to create a smooth associated tail-probability and expected-shortfall profile. The saddlepoint approximation, in contrast, is predominately analytic computation and, as such, exhibits significantly less noise. Although it is naturally only an approximation, and as such has some error, it appears to give very similar results.

At the lower quantiles, the saddlepoint tail-probability estimates do exhibit a higher degree of error relative to their Monte Carlo equivalents. Table 7.3, which provides some numerical values for selected tail-probability and shortfall estimates, suggests that these can approach $2 at the 95th quantile. As we move further into the tail, however, the accuracy increases. Moreover, the expected-shortfall estimates are similarly accurate at all quantiles.

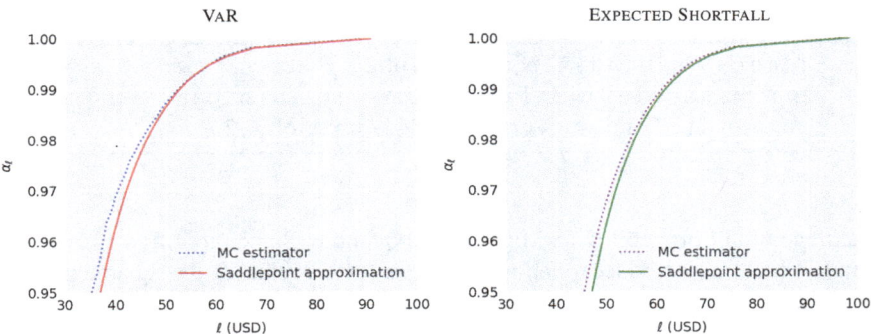

Fig. 7.8 *Independent-default tail-probability and shortfall-integral approximations*: This figure outlines—again for our usual 100-obligor example portfolio—the saddlepoint tail-probability and shortfall-integral approximations compared with a Monte Carlo simulation employing the usual 10 million iterations. Both approximations are extremely fast and quite accurate.

Table 7.3 *Numerical saddlepoint results*: For a selected number of quantiles, this figure summarizes the tail-probability and shortfall results from Fig. 7.8. The level of accuracy is generally quite acceptable.

Quantile	$\text{VaR}_\alpha(L)$			$\mathcal{E}_\alpha(L)$		
	MC	SP	Δ	MC	SP	Δ
95.00th	$35.3	$37.0	$1.7	$45.6	$47.1	$1.6
97.00th	$40.5	$42.2	$1.7	$50.8	$52.0	$1.2
99.00th	$52.5	$52.8	$0.3	$61.2	$61.9	$0.7
99.50th	$59.0	$59.1	$0.1	$66.9	$67.9	$1.0
99.90th	$77.9	$78.8	$0.9	$85.9	$86.7	$0.8
99.97th	$87.2	$88.1	$1.0	$94.9	$95.8	$0.9
99.99th	$89.8	$90.8	$1.0	$97.4	$98.4	$0.9

It is possible, through the use of higher-order approximations, to obtain a greater degree of accuracy in these estimates.[17] The cost is, however, a higher degree of implementation complexity. In this analysis, we have opted to avoid these higher-order terms, but depending on one's application, they may prove necessary.

7.6 Obligor-Level Risk Contributions

After much heavy lifting, we are finally in a position to consider our ultimate goal: the estimation of obligor-level contributions to our risk metrics using the saddlepoint method. Again, this requires some involved computations, but the final results are intuitive and readily implemented.

---

[17] These require, as one would expect, higher-order derivatives of the cumulant generating function.

7.6 Obligor-Level Risk Contributions

7.6.1 The VaR Contributions

As we established at the beginning of this chapter, determination of risk-metric contributions requires computation of the marginal VaR or expected shortfall. Our initial focus will be VaR. It turns out that we have already developed the vast majority of ingredients required for the determination of the marginal VaR. In this section, we will derive the final results—again, in the independent-default case—and demonstrate how one can use the saddlepoint approximation to accurately estimate the risk-metric contributions.

The tail probability is, as we've seen many times, given a particular choice of α, a VaR computation. We begin, therefore, our development with the partial derivative of a tail probability, evaluated at $\ell_\alpha = \text{VaR}_\alpha(L)$, taken with respect to the nth counterparty exposure, c_n. After some extensive manipulation, we arrive at a useful and practical result. Computing the marginal VaR directly yields

$$\frac{\partial \mathbb{P}(L > \ell_\alpha)}{\partial c_n} = \frac{\partial}{\partial c_n}\left(\frac{1}{2\pi i}\int_{-i\infty,(0+)}^{+i\infty}\frac{e^{K_L(t)-t\ell_\alpha}}{t}dt\right), \quad (7.84)$$

$$\frac{\partial \overbrace{\mathbb{P}(L > \ell_\alpha)}^{1-\alpha}}{\partial c_n} = \frac{1}{2\pi i}\int_{-i\infty,(0+)}^{+i\infty}\frac{\partial}{\partial c_n}\left(\frac{e^{K_L(t)-t\ell_\alpha}}{t}\right)dt,$$

$$\underbrace{\frac{\partial(1-\alpha)}{\partial c_n}}_{=0} = \frac{1}{2\pi i}\int_{-i\infty,(0+)}^{+i\infty}\frac{1}{t}\left(\frac{\partial K_L(t)}{\partial c_n} - t\frac{\partial \ell_\alpha}{\partial c_n}\right)e^{K_L(t)-t\ell_\alpha}dt,$$

$$\frac{1}{2\pi i}\int_{-i\infty,(0+)}^{+i\infty}\frac{\partial \ell_\alpha}{\partial c_n}e^{K_L(t)-t\ell_\alpha}dt = \frac{1}{2\pi i}\int_{-i\infty,(0+)}^{+i\infty}\frac{\partial K_L(t)}{\partial c_n}\frac{e^{K_L(t)-t\ell_\alpha}}{t}dt,$$

$$\frac{\partial \ell_\alpha}{\partial c_n} = \frac{\dfrac{1}{2\pi i}\displaystyle\int_{-i\infty,(0+)}^{+i\infty}\dfrac{\partial K_L(t)}{\partial c_n}\dfrac{e^{K_L(t)-t\ell_\alpha}}{t}dt}{\underbrace{\dfrac{1}{2\pi i}\displaystyle\int_{-i\infty,(0+)}^{+i\infty}e^{K_L(t)-t\ell_\alpha}dt}_{f_L(\ell_\alpha)}},$$

$$\frac{\partial \ell_\alpha}{\partial c_n} \approx \frac{\dfrac{1}{2\pi i}\displaystyle\int_{-i\infty,(0+)}^{+i\infty}\left(\dfrac{1}{t}\dfrac{\partial K_L(t)}{\partial c_n}\right)e^{K_L(t)-t\ell_\alpha}dt}{e^{K_L(\tilde{t}_\ell) - \tilde{t}_\ell \ell_\alpha} \mathcal{J}_0(K_L''(\tilde{t}_\ell), \tilde{t}_\ell)}.$$

This is significant progress. We have identified the marginal VaR and represented it as a fraction of two integrals. The denominator can furthermore be proxied, following equation 7.77, as a saddlepoint density approximation with the saddlepoint, $\tilde{t}_\ell$, associated with the VaR estimate.

The complicating element is the numerator of equation 7.84. At this point, there are *two* alternative ways to proceed. The first, and most straightforward, is to follow the cleverly suggested approach by Martin and Thompson (2001b). The idea is to use the same saddlepoint as found in the denominator, $\tilde{t}_\ell$, to approximate the partial derivative term—this amounts to a zero-order Taylor-series expansion. In this case, we have the following,

$$\frac{\partial \ell_\alpha}{\partial c_n} \approx \frac{\frac{1}{2\pi i} \int_{-i\infty,(0+)}^{+i\infty} \left(\frac{1}{t}\frac{\partial K_L(t)}{\partial c_n}\right) e^{K_L(t) - t\ell_\alpha} dt}{e^{K_L(\tilde{t}_\ell) - \tilde{t}_\ell \ell_\alpha} \mathcal{J}_0(K_L''(\tilde{t}_\ell), \tilde{t}_\ell)}, \qquad (7.85)$$

$$\approx \frac{\left(\frac{1}{\tilde{t}_\ell}\frac{\partial K_L(\tilde{t}_\ell)}{\partial c_n}\right) \frac{1}{2\pi i} \int_{-i\infty,(0+)}^{+i\infty} e^{K_L(t) - t\ell_\alpha} dt}{e^{K_L(\tilde{t}_\ell) - \tilde{t}_\ell \ell_\alpha} \mathcal{J}_0(K_L''(\tilde{t}_\ell), \tilde{t}_\ell)},$$

$$\approx \frac{1}{\tilde{t}_\ell}\frac{\partial K_L(\tilde{t}_\ell)}{\partial c_n} \left(\frac{\cancel{e^{K_L(\tilde{t}_\ell) - \tilde{t}_\ell \ell_\alpha} \mathcal{J}_0(K_L''(\tilde{t}_\ell), \tilde{t}_\ell)}}{\cancel{e^{K_L(\tilde{t}_\ell) - \tilde{t}_\ell \ell_\alpha} \mathcal{J}_0(K_L''(\tilde{t}_\ell), \tilde{t}_\ell)}}\right),$$

$$\approx \frac{1}{\tilde{t}_\ell}\frac{\partial K_L(\tilde{t}_\ell)}{\partial c_n}.$$

This yields a concrete expression for the marginal VaR and allows us to use directly Euler's theorem for homogeneous functions. The price is the time to solve for a single saddlepoint. In our experience, this is a reasonably accurate and useful way to approach the problem and, despite being something of an approximation, it works well across a variety of portfolio structures.

An alternative approach is suggested by Huang et al. (2007, 2006). Ultimately it provides similar results, albeit with a higher degree of complexity. The reader is referred to the references for a more detailed discussion of this alternative. The former approach works quite well and generalizes more easily, in our opinion, to the expected shortfall case; as such, we will restrict our attention to this method.

To practically move forward with equation 7.85, we need to determine the specific form of the partial derivative,

$$\frac{\partial K_L(t)}{\partial c_n} = \frac{\partial}{\partial c_n} \underbrace{\left(\sum_{k=1}^{N} \ln\left(1 - p_k + p_k e^{tc_k}\right)\right)}_{\text{Equation 7.74}}, \qquad (7.86)$$

$$= \sum_{k=1}^{N} \frac{\partial}{\partial c_n}\left(\ln\left(1 - p_k + p_k e^{tc_k}\right)\right),$$

$$= \frac{tp_n e^{tc_n}}{(1 - p_n + p_n e^{tc_n})},$$

7.6 Obligor-Level Risk Contributions

where the sum reduces to a single term since all other partial derivatives, save the one with respect to c_n, vanish.

Adding the specific partial derivative from equation 7.86 into 7.85 evaluated at the saddlepoint $\tilde{t}_\ell$, we arrive at a rather simple and pleasant expression for the VaR contribution

$$\frac{\partial \ell_\alpha}{\partial c_n} \approx \frac{1}{\tilde{t}_\ell} \frac{\partial K_L(\tilde{t}_\ell)}{\partial c_n}, \qquad (7.87)$$

$$\approx \frac{1}{\tilde{t}_\ell} \underbrace{\frac{\tilde{t}_\ell p_n e^{\tilde{t}_\ell c_n}}{(1 - p_n + p_n e^{c_n \tilde{t}_\ell})}}_{\text{Equation 7.86}},$$

$$\approx \frac{p_n e^{\tilde{t}_\ell c_n}}{1 - p_n + p_n e^{\tilde{t}_\ell c_n}}.$$

It is hard to imagine a more concise and straightforward result. Equation 7.87 is merely a simple combination of the obligor's exposure, its unconditional default probability, and the saddlepoint.

If we now apply Euler's theorem for homogeneous functions to these partial derivatives and sum the individual contributions, we have

$$\sum_{n=1}^{N} c_n \frac{\partial \ell_\alpha}{\partial c_n} = \underbrace{\sum_{n=1}^{N} c_n \frac{p_n e^{\tilde{t}_\ell c_n}}{\left(1 - p_n + p_n e^{\tilde{t}_\ell c_n}\right)}}_{\text{Equation 7.75}}, \qquad (7.88)$$

$$= \underbrace{K'_L(\tilde{t}_\ell)}_{\ell_\alpha},$$

$$= \text{VaR}_\alpha(L),$$

where the last step holds, by the very definition of the saddlepoint, from equation 7.78. The sum of the individual saddlepoint approximated VaR contributions thus add up, as one would expect, to the total VaR.

Algorithm 7.9 demonstrates how, in two lines of code, we can compute the VaR contributions for the independent-default model. Since the denominator is, in fact, the moment-generating function, we can even re-use the computeMGF sub-routine.

Algorithm 7.9 Computing the independent-default VaR contributions

```
def getVaRC(l,p,c):
    t_l = getSaddlePoint(p,c,l)
    return np.divide(p*np.exp(c*t_l),computeMGF(t_l,p,c))
```

Table 7.4 *VaR contribution results*: This table uses the approach described in equation 7.88 to illustrate the ten largest 99.99th quantile VaR contributions associated with our 100 counterparty example. It also includes the default probability, exposure and ranking as well as the saddlepoint and Monte Carlo VaR contribution estimates.

Rank	ID	p	c	SP	MC
1	13	6.8%	$30.4	$23.0	$25.2
2	30	0.7%	$30.9	$7.8	$8.0
3	33	1.1%	$28.4	$7.4	$4.8
4	1	1.2%	$27.1	$6.9	$6.0
5	8	1.1%	$25.8	$5.4	$5.9
6	24	1.4%	$22.9	$4.3	$3.9
7	65	0.5%	$28.3	$4.0	$3.1
8	35	0.0%	$47.8	$3.7	$2.9
9	28	1.2%	$21.0	$2.9	$2.3
10	69	1.4%	$18.4	$2.2	$2.6
Other			$719.0	$23.2	$25.5
Total			$1000.0	$90.8	$90.2
Time (seconds)				0.6	885.6

The best way to appreciate the usefulness of the saddlepoint approximation is to consider a practical example. Table 7.4 presents the ten largest 99.99th quantile VaR contributions for our ongoing 100-obligor portfolio. To provide some context, the default probabilities and exposures are also included. This is important, because the VaR contribution is logically an increasing function of these two variables. A coincidence of large default probabilities and exposure should, all else equal, lead to a significant contribution to the overall VaR figure.

Table 7.4 also indicates the computational time associated with the VaR contribution approximations. The Monte Carlo estimator requires several minutes to arrive at its approximation. This is because it necessitates an enormous amount of iterations to compute the expectation of each counterparty loss conditional on being at the 99.99th tail of the distribution. This is essentially a combination of two extremely rare events—to obtain a precise value for this expectation is thus computationally very intensive. Indeed, we used 300 million iterations for these approximations.

The Monte Carlo VaR-contribution estimator, in addition to being relatively slow, is also somewhat noisy. Figure 7.9 graphically illustrates the results of Table 7.4. What is interesting about this graphic is that it provides a 99% confidence interval for the Monte Carlo estimates along with the saddlepoint approximations. Generally speaking the results agree quite closely. The punchline, however, is that there is a substantial amount of uncertainty associated with the Monte Carlo estimates. In a multi-factor setting, this is inevitable since the saddlepoint approximation is not available. In the one-factor world, however, the saddlepoint, by virtue of its accuracy and speed, seems to be the better choice. Figure 7.9 also offers another argument, when forced to use numerical methods for one's risk attribution, to explore Bluhm et al. (2003)'s VaR-matched expected shortfall approach.

Overall, VaR contribution analysis is a powerful and useful tool. Figure 7.10 is a case in point. It compares, in percentage terms, the unconditional default probabilities, counterparty exposures, and VaR contributions for the 99.99th percentile

7.6 Obligor-Level Risk Contributions

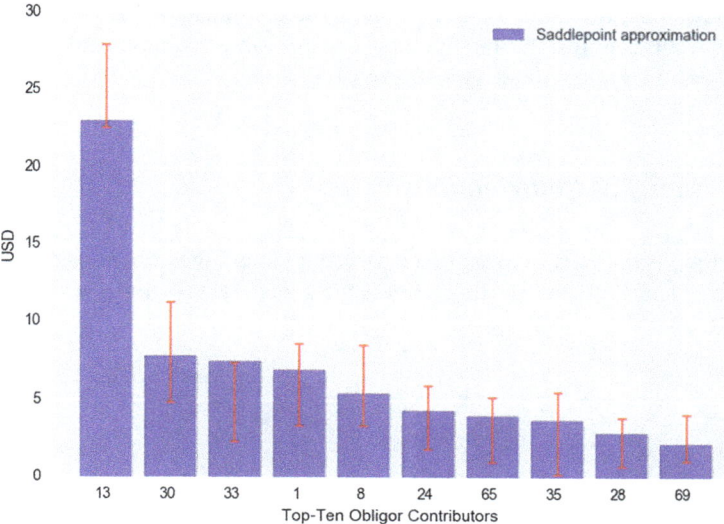

Fig. 7.9 *Independent-default VaR-contribution approximation*: This figure outlines—for our usual portfolio of randomly generated heterogeneous exposures and default probabilities—the saddlepoint VaR-contribution approximations. They are compared with a Monte-Carlo estimate employing 100 million iterations. Recall that the total portfolio values $1,000 and the red bars denote a 99% confidence interval around the Monte Carlo approximations.

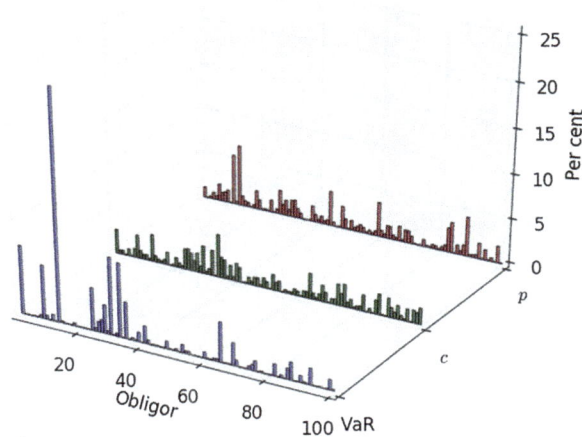

Fig. 7.10 *VaR contribution perspective*: This figure compares, in percentage terms, the unconditional default probabilities, obligor exposures, and VaR contributions for the 99.99th percentile in the independent-default binomial model. The combination of these three factors provided significant insight into the model output.

in the independent-default binomial model. The combination of these three factors provides significant insight into the model output, eases communication of results, and acts as a helpful model diagnostic tool.

7.6.2 Shortfall Contributions

Our next job is to determine, using the preceding ideas, the individual contributions to expected shortfall. Our starting point is the general expression,

$$\underbrace{\mathbb{E}(L|L > \ell_\alpha)}_{\mathcal{E}_\alpha(L)} = \frac{1}{1-\alpha} \left(\frac{1}{2\pi i} \int_{-i\infty, (0+)}^{+i\infty} \frac{e^{K_L(t) - t\ell_\alpha}}{t^2} dt \right), \qquad (7.89)$$

$$\approx \frac{\exp\left(K_L(\tilde{t}_{\ell_\alpha}) - \tilde{t}_{\ell_\alpha} \ell_\alpha\right) \mathcal{I}_2\left(K_L''(\tilde{t}_{\ell_\alpha}), \tilde{t}_{\ell_\alpha}\right)}{\mathbb{P}(L > \ell_\alpha)} + \ell_\alpha,$$

where ℓ_α is, as usual, a short-hand representation for $\text{VaR}_\alpha(L)$. By definition, therefore, we have that $\mathbb{P}(L > \ell_\alpha) = 1 - \alpha$, which will eventually permit us to slightly simplify the denominator.

To begin a rather daunting and strenuous derivation, we compute the partial derivative of equation 7.89 with respect to c_n. This leads us to

$$\frac{\partial \mathcal{E}_\alpha(L)}{\partial c_n} = \frac{\partial}{\partial c_n} \left(\frac{1}{1-\alpha} \left(\frac{1}{2\pi i} \int_{-i\infty, (0+)}^{+i\infty} \frac{e^{K_L(t) - t\ell_\alpha}}{t^2} dt \right) \right), \qquad (7.90)$$

$$= \frac{1}{1-\alpha} \left(\frac{1}{2\pi i} \int_{-i\infty, (0+)}^{+i\infty} \frac{\partial}{\partial c_n} \left(\frac{e^{K_L(t) - t\ell_\alpha}}{t^2} \right) dt \right),$$

$$= \frac{1}{1-\alpha} \left(\frac{1}{2\pi i} \int_{-i\infty, (0+)}^{+i\infty} \frac{\partial (K_L(t) - t\ell_\alpha)}{\partial c_n} \frac{e^{K_L(t) - t\ell_\alpha}}{t^2} dt \right),$$

$$= \frac{1}{1-\alpha} \left(\frac{1}{2\pi i} \int_{-i\infty, (0+)}^{+i\infty} \frac{\partial K_L(t)}{\partial c_n} \frac{e^{K_L(t) - t\ell_\alpha}}{t^2} dt \right),$$

$$= \frac{1}{1-\alpha} \left(\frac{1}{2\pi i} \int_{-i\infty, (0+)}^{+i\infty} \underbrace{\frac{t p_n e^{tc_n}}{(1 - p_n + p_n e^{tc_n})}}_{\text{Equation 7.86}} \frac{e^{K_L(t) - t\ell_\alpha}}{t^2} dt \right),$$

$$= \frac{1}{1-\alpha} \left(\frac{1}{2\pi i} \int_{-i\infty, (0+)}^{+i\infty} \frac{p_n e^{tc_n}}{(1 - p_n + p_n e^{tc_n})} \frac{e^{K_L(t) - t\ell_\alpha}}{t} dt \right).$$

7.6 Obligor-Level Risk Contributions

Note that, in this case, we have assumed that ℓ_α is a constant and, as such, $\frac{\partial \ell_\alpha}{\partial c_n} = 0$. Not surprisingly, we will again make use of the saddlepoint approximation to evaluate this expression. Unlike the VaR setting, however, we will need to perform a first-order—and not zero-order—Taylor series expansion of the c_n partial-derivative coefficient in equation 7.90. It is a bit messy, although not difficult, and the result is

$$\frac{p_n e^{tc_n}}{(1-p_n+p_n e^{tc_n})} \approx \frac{p_n e^{\tilde{t}_\ell c_n}}{(1-p_n+p_n e^{\tilde{t}_\ell c_n})}(t-\tilde{t}_\ell)^0 + \frac{\partial}{\partial t}\left(\frac{p_n e^{tc_n}}{(1-p_n+p_n e^{tc_n})}\right)\bigg|_{t=\tilde{t}_\ell}(t-\tilde{t}_\ell)^1, \quad (7.91)$$

$$\approx \frac{p_n e^{\tilde{t}_\ell c_n}}{(1-p_n+p_n e^{\tilde{t}_\ell c_n})} + \underbrace{\left(\frac{c_n p_n e^{tc_n}(1-p_n+p_n e^{tc_n}) - c_n p_n^2 e^{2tc_n}}{(1-p_n+p_n e^{tc_n})^2}\right)\bigg|_{t=\tilde{t}_\ell}}_{\text{Quotient rule}}(t-\tilde{t}_\ell),$$

$$\approx \frac{p_n e^{\tilde{t}_\ell c_n}}{(1-p_n+p_n e^{\tilde{t}_\ell c_n})} + \left(\frac{c_n p_n e^{tc_n}(1-p_n)}{(1-p_n+p_n e^{tc_n})^2}\right)\bigg|_{t=\tilde{t}_\ell}(t-\tilde{t}_\ell),$$

$$\approx \underbrace{\frac{p_n e^{c_n \tilde{t}_\ell}}{(1-p_n+p_n e^{c_n \tilde{t}_\ell})}}_{\mathcal{A}(\tilde{t}_\ell)} + \underbrace{\frac{c_n p_n e^{c_n \tilde{t}_\ell}(1-p_n)}{(1-p_n+p_n e^{c_n \tilde{t}_\ell})^2}(t-\tilde{t}_\ell)}_{\mathcal{B}(\tilde{t}_\ell)},$$

$$\approx \mathcal{A}(\tilde{t}_\ell) + \mathcal{B}(\tilde{t}_\ell).$$

We may combine this with our usual saddlepoint approximation of the integral in equation 7.90. The introduction of the abbreviations $\mathcal{A}(\tilde{t}_\ell)$ and $\mathcal{B}(\tilde{t}_\ell)$ is to keep the expressions to a manageable length. We will identify two distinct terms that we will address separately and then recombine at the end. While a long process, the final pay-off is a worthwhile and intuitive expression.

$$\frac{\partial \mathcal{E}_\alpha(L)}{\partial c_n} \approx \frac{1}{1-\alpha}\left(\frac{1}{2\pi i}\int_{-i\infty,(0+)}^{+i\infty}\left(\mathcal{A}(\tilde{t}_\ell)+\mathcal{B}(\tilde{t}_\ell)\right)\frac{e^{K_L(\tilde{t}_\ell)-\tilde{t}_\ell \ell_\alpha}e^{K_L''(\tilde{t}_\ell)(\tilde{t}_\ell-t)^2}}{t}dt\right), \quad (7.92)$$

$$\approx \underbrace{\frac{e^{K_L(\tilde{t}_\ell)-\tilde{t}_\ell \ell_\alpha}}{(1-\alpha)2\pi i}\int_{-i\infty,(0+)}^{+i\infty}\mathcal{A}(\tilde{t}_\ell)\frac{e^{K_L''(\tilde{t}_\ell)(\tilde{t}_\ell-t)^2}}{t}dt}_{\mathcal{T}_1}$$

$$+\underbrace{\frac{e^{K_L(\tilde{t}_\ell)-\tilde{t}_\ell \ell_\alpha}}{(1-\alpha)2\pi i}\int_{-i\infty,(0+)}^{+i\infty}\mathcal{B}(\tilde{t}_\ell)\frac{e^{K_L''(\tilde{t}_\ell)(\tilde{t}_\ell-t)^2}}{t}dt}_{\mathcal{T}_2},$$

In the spirit of dividing and conquering, we will first simplify $\mathcal{T}_1$ and then proceed to consider $\mathcal{T}_2$.

The first term is actually, once we make a few manipulations, quite familiar.

$$\mathcal{T}_1 \approx \frac{e^{K_L(\tilde{t}_\ell) - \tilde{t}_\ell \ell_\alpha}}{(1-\alpha)2\pi i} \int_{-i\infty,(0+)}^{+i\infty} \mathcal{A}(\tilde{t}_\ell) \frac{e^{K_L''(\tilde{t}_\ell)(\tilde{t}_\ell - t)^2}}{t} dt, \quad (7.93)$$

$$\approx \underbrace{\left(\frac{p_n e^{c_n \tilde{t}_\ell}}{(1-p_n + p_n e^{c_n \tilde{t}_\ell})}\right)}_{\mathcal{A}(\tilde{t}_\ell)} \frac{e^{K_L(\tilde{t}_\ell) - \tilde{t}_\ell \ell_\alpha}}{(1-\alpha)2\pi i} \int_{-i\infty,(0+)}^{+i\infty} \frac{e^{K_L''(\tilde{t}_\ell)(\tilde{t}_\ell - t)^2}}{t} dt,$$

$$\approx \left(\frac{p_n e^{c_n \tilde{t}_\ell}}{1-p_n + p_n e^{c_n \tilde{t}_\ell}}\right) \frac{1}{(1-\alpha)} e^{K_L(\tilde{t}_\ell) - \tilde{t}_\ell \ell_\alpha} \underbrace{\int_{-i\infty,(0+)}^{+i\infty} \frac{1}{2\pi i} \frac{e^{K_L''(\tilde{t}_\ell)(\tilde{t}_\ell - t)^2}}{t} dt}_{\mathcal{J}_1(K_L''(\tilde{t}_\ell), \tilde{t}_\ell)},$$

$$\approx \left(\frac{p_n e^{c_n \tilde{t}_\ell}}{1-p_n + p_n e^{c_n \tilde{t}_\ell}}\right) \frac{1}{(1-\alpha)} \underbrace{e^{K_L(\tilde{t}_\ell) - \tilde{t}_\ell \ell_\alpha} \mathcal{J}_1(K_L''(\tilde{t}_\ell), \tilde{t}_\ell)}_{\mathbb{P}(L > \ell_\alpha)},$$

$$\approx \left(\frac{p_n e^{c_n \tilde{t}_\ell}}{1-p_n + p_n e^{c_n \tilde{t}_\ell}}\right) \frac{\overbrace{\mathbb{P}(L > \ell_\alpha)}^{1-\alpha}}{(1-\alpha)},$$

$$\approx \frac{p_n e^{c_n \tilde{t}_\ell}}{1-p_n + p_n e^{c_n \tilde{t}_\ell}}.$$

This, as the reader will recall from equation 7.87, is actually equivalent to the $\frac{\partial \text{VaR}_\alpha(L)}{\partial c_n}$. This makes sense as a starting point. The evaluation of the second term, $\mathcal{T}_2$, represents the additional contribution, associated with each obligor, stemming from the expected shortfall.

The intimidating second term also reduces, with some effort, to a manageable and insightful expression.

$$\mathcal{T}_2 \approx \frac{e^{K_L(\tilde{t}_\ell) - \tilde{t}_\ell \ell_\alpha}}{(1-\alpha)2\pi i} \int_{-i\infty,(0+)}^{+i\infty} \mathcal{B}(\tilde{t}_\ell) \frac{e^{K_L''(\tilde{t}_\ell)(\tilde{t}_\ell - t)^2}}{t} dt, \quad (7.94)$$

$$\approx \frac{e^{K_L(\tilde{t}_\ell) - \tilde{t}_\ell \ell_\alpha}}{(1-\alpha)2\pi i} \int_{-i\infty,(0+)}^{+i\infty} \underbrace{\frac{c_n p_n e^{c_n \tilde{t}_\ell}(1-p_n)}{\left(1-p_n + p_n e^{c_n \tilde{t}_\ell}\right)^2}(t - \tilde{t}_\ell)}_{\mathcal{B}(\tilde{t}_\ell)} \frac{e^{K_L''(\tilde{t}_\ell)(\tilde{t}_\ell - t)^2}}{t} dt,$$

$$\approx \left(\frac{c_n p_n e^{c_n \tilde{t}_\ell}(1-p_n)}{\left(1-p_n + p_n e^{c_n \tilde{t}_\ell}\right)^2}\right) \frac{e^{K_L(\tilde{t}_\ell) - \tilde{t}_\ell \ell_\alpha}}{(1-\alpha)2\pi i} \int_{-i\infty,(0+)}^{+i\infty}$$

$$\left(e^{K_L''(\tilde{t}_\ell)(\tilde{t}_\ell - t)^2} - \tilde{t}_\ell \frac{e^{K_L''(\tilde{t}_\ell)(\tilde{t}_\ell - t)^2}}{t}\right) dt,$$

7.6 Obligor-Level Risk Contributions

$$\approx \left(\frac{c_n p_n e^{c_n \tilde{t}_\ell} (1 - p_n)}{(1 - p_n + p_n e^{c_n \tilde{t}_\ell})^2} \right) \frac{e^{K_L(\tilde{t}_\ell) - \tilde{t}_\ell \ell_\alpha}}{(1 - \alpha)}$$

$$\left(\underbrace{\int_{-i\infty,(0+)}^{+i\infty} \frac{1}{2\pi i} e^{K_L''(\tilde{t}_\ell)(\tilde{t}_\ell - t)^2} dt}_{\mathcal{J}_0(K_L''(\tilde{t}_\ell), \tilde{t}_\ell)} - \tilde{t}_\ell \underbrace{\int_{-i\infty,(0+)}^{+i\infty} \frac{1}{2\pi i} \frac{e^{K_L''(\tilde{t}_\ell)(\tilde{t}_\ell - t)^2}}{t} dt}_{\mathcal{J}_1(K_L''(\tilde{t}_\ell), \tilde{t}_\ell)} \right),$$

$$\approx \left(\frac{c_n p_n e^{c_n \tilde{t}_\ell} (1 - p_n)}{(1 - p_n + p_n e^{c_n \tilde{t}_\ell})^2} \right) \frac{e^{K_L(\tilde{t}_\ell) - \tilde{t}_\ell \ell_\alpha}}{(1 - \alpha)} \underbrace{\left(\mathcal{J}_0(K_L''(\tilde{t}_\ell), \tilde{t}_\ell) - \tilde{t}_\ell \mathcal{J}_1(K_L''(\tilde{t}_\ell), \tilde{t}_\ell) \right)}_{\frac{\mathcal{J}_2(K_L''(\tilde{t}_\ell), \tilde{t}_\ell)}{K''(\tilde{t}_\ell)} \text{ from Equation 7.71}},$$

$$\approx \frac{c_n p_n e^{c_n \tilde{t}_\ell} (1 - p_n)}{(1 - p_n + p_n e^{c_n \tilde{t}_\ell})^2} \frac{e^{K_L(\tilde{t}_\ell) - \tilde{t}_\ell \ell_\alpha}}{(1 - \alpha)} \frac{\mathcal{J}_2(K_L''(\tilde{t}_\ell), \tilde{t}_\ell)}{K''(\tilde{t}_\ell)}.$$

We now have concrete expression for both $\mathcal{T}_1$ and $\mathcal{T}_2$. Let's combine them together into their final form,

$$\frac{\partial \mathcal{E}_\alpha(L)}{\partial c_n} \approx \underbrace{\frac{p_n e^{c_n \tilde{t}_\ell}}{(1 - p_n + p_n e^{c_n \tilde{t}_\ell})}}_{\text{Equation 7.93}}$$

$$+ \underbrace{\frac{c_n p_n e^{c_n \tilde{t}_\ell} (1 - p_n)}{(1 - p_n + p_n e^{c_n \tilde{t}_\ell})^2} \frac{e^{K_L(\tilde{t}_\ell) - \tilde{t}_\ell \ell_\alpha}}{(1 - \alpha)} \frac{\mathcal{J}_2(K_L''(\tilde{t}_\ell), \tilde{t}_\ell)}{K''(\tilde{t}_\ell)}}_{\text{Equation 7.94}}, \quad (7.95)$$

where, as promised, we made use of the identity from equation 7.71 to simplify our final expression for $\mathcal{T}_2$. While this might still seem a bit unwieldy, a pleasant thing happens when we use Euler's theorem to compute the total contribution to expected shortfall, which corroborates the correctness of the derivation. The ultimate result is,

$$\sum_{n=1}^{N} c_n \frac{\partial \mathcal{E}_\alpha(L)}{\partial c_n} \approx \underbrace{\sum_{n=1}^{N} \frac{c_n p_n e^{c_n \tilde{t}_\ell}}{(1 - p_n + p_n e^{c_n \tilde{t}_\ell})}}_{K_L'(\tilde{t}_\ell)} + \frac{e^{K_L(\tilde{t}_\ell) - \tilde{t}_\ell \ell_\alpha}}{(1 - \alpha)} \frac{\mathcal{J}_2(K_L''(\tilde{t}_\ell), \tilde{t}_\ell)}{K''(\tilde{t}_\ell)} \underbrace{\sum_{n=1}^{N} \frac{c_n^2 p_n e^{c_n \tilde{t}_\ell} (1 - p_n)}{(1 - p_n + p_n e^{c_n \tilde{t}_\ell})^2}}_{K_L''(\tilde{t}_\ell)}, \quad (7.96)$$

$$\approx \underbrace{K_L'(\tilde{t}_\ell)}_{\text{VaR}_\alpha(L)} + \frac{e^{K_L(\tilde{t}_\ell) - \tilde{t}_\ell \ell_\alpha}}{(1 - \alpha)} \frac{\mathcal{J}_2(K_L''(\tilde{t}_\ell), \tilde{t}_\ell)}{\cancel{K''(\tilde{t}_\ell)}} \cancel{K_L''(\tilde{t}_\ell)},$$

$$\approx \text{VaR}_\alpha(L) + \underbrace{\frac{e^{K_L(\tilde{t}_\ell) - \tilde{t}_\ell \ell_\alpha} \mathcal{J}_2(K_L''(\tilde{t}_\ell), \tilde{t}_\ell)}{(1 - \alpha)}}_{\mathcal{E}_\alpha(L) - \text{VaR}_\alpha(L)},$$

Algorithm 7.10 Computing the independent-default shortfall contributions

```
def getESC(l,p,c):
    varPart = getVaRC(l,p,c)
    myAlpha = saddlePointTailProbability(l,p,c)
    t_1 = getSaddlePoint(p,c,l)
    K2 = computeCGF_2(t_1,p,c)
    myW = computeCGF_2(t_1,p,c,1)
    t0 = np.exp(computeCGF(t_1,p,c)-t_1*1)*getJ(l,p,c,t_1,2)
    return varPart + np.divide(t0*np.divide(myW,K2),myAlpha)
```

$$\approx \cancel{\text{VaR}_\alpha(L)} + \mathcal{E}_\alpha(L) - \cancel{\text{VaR}_\alpha(L)},$$

$$= \mathcal{E}_\alpha(L),$$

where the final manipulation is a direct consequence of the final row in Table 7.1. The conclusion is that the first term estimates the contribution of each obligor to the VaR, whereas the second term represents the additional counterparty contribution associated with expected shortfall. In this way, the two computations are intimately linked. Indeed, computing VaR contributions is a sub-set of the expected shortfall calculation. If calculating both elements, which seems a sensible approach, one need only add the second term to VaR contributions to arrive at the expected-shortfall decomposition.

The actual implementation, summarized in Algorithm 7.10, is not terribly heavy. We first employ algorithms 7.5, 7.7, and 7.9 to obtain the saddlepoint, the tail probability, and the VaR contributions. We then need to call Algorithm 7.4, with the asVector flag set to unity, to obtain the element-wise representation of the second partial derivatives of the cumulant generating function with respect to c_n. The remaining manipulations follow directly from equation 7.96.

The results of application of Algorithm 7.10 to our sample portfolio are summarized in Table 7.5. As in Table 7.4, the top-ten contributors are provided with the mid-point of the Monte-Carlo estimators.

The estimates in Table 7.5 relate to the 99.99th quantile. Although the Monte-Carlo estimates are relatively close to the saddlepoint values, there is no overlap between the two for a number of obligors. This is clearly visible in Fig. 7.11. Since we are dealing with a quantile so incredibly far out into the tail of the default-loss distribution, we might be relatively sceptical of the Monte-Carlo results. Nevertheless, generating millions of iterations, there are still hundreds of observations upon which to base the estimates. The more likely culprit of the differences relates to the fact that the saddlepoint approximation assumes continuity of the underlying default loss distribution. In reality, it is, of course, discrete. For large diversified portfolios, this is a very good assumption. For smaller concentrated portfolios, it might smooth out the analysis somewhat. There is, however, no economic difference between the estimates and the total expected-shortfall figures agree almost perfectly.

7.6 Obligor-Level Risk Contributions

Table 7.5 *Shortfall contribution results*: This table uses the approach described in equation 7.96 to illustrate the ten largest 99.99th quantile expected-shortfall contributions associated with our 100 counterparty example. It also includes the default probability, exposure, and ranking as well as the saddlepoint and Monte Carlo VaR contribution estimates.

Rank	ID	p	c	SP	MC
1	13	6.8%	$30.4	$23.8	$25.2
2	30	0.7%	$30.9	$8.7	$8.0
3	33	1.1%	$28.4	$8.2	$4.8
4	1	1.2%	$27.1	$7.6	$6.0
5	8	1.1%	$25.8	$6.0	$5.9
6	24	1.4%	$22.9	$4.7	$3.9
7	65	0.5%	$28.3	$4.5	$3.1
8	35	0.0%	$47.8	$4.5	$2.9
9	28	1.2%	$21.0	$3.2	$2.3
10	69	1.4%	$18.4	$2.4	$2.6
Other			$719.0	$24.8	$33.0
Total			$1000.0	$98.4	$97.7
Time (seconds)				0.6	885.6

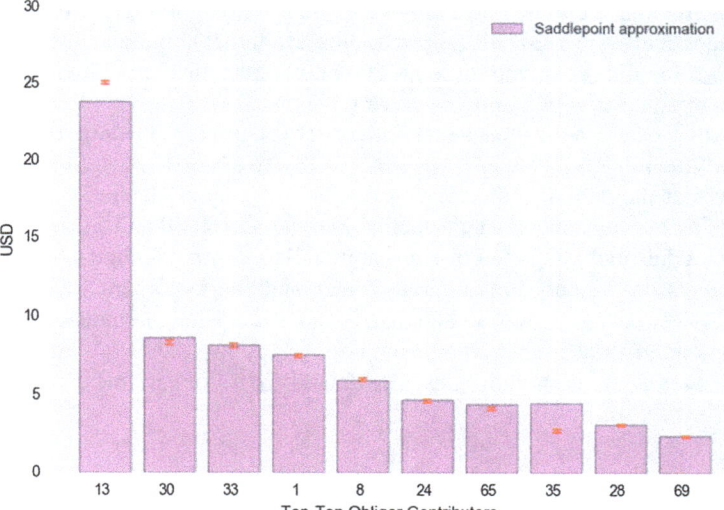

Fig. 7.11 *Independent-default shortfall-contribution approximation*: This figure outlines the saddlepoint expected-shortfall contribution approximations. They are compared with a Monte-Carlo estimate employing 500 million iterations. The red bars, as before, denote a 99% confidence interval around the Monte Carlo approximations.

These important points notwithstanding, the saddlepoint approximation appears, in the independent-default setting, to be a quite reasonable approximation. It is certainly computationally efficient. We see in Table 7.5 that it requires less than a second of computer time. The Monte-Carlo estimates, by contrast, require almost 15 minutes. This is a strong advantage. The current disadvantage, however, is that our development applies only to the independent-default setting. In the next section, we will consider how this might—at the price of some additional computational expense—be resolved.

7.7 The Conditionally Independent Saddlepoint Approximation

As discussed extensively in previous sections, the independent-default binomial model, by virtue of its thin tails and lack of default dependence, is not a very realistic credit-risk model. Its exclusive use will lead to an underestimate of default risk in one's portfolio. As a consequence, the independent-default approach is rarely used in practice.[18] This means that our approach needs to be extended to the more realistic models described in the previous chapters. The bad news is that our saddlepoint-approximation development cannot be used directly using the results from the previous sections. The good news is that the overall logic and structure are almost identical—we need only perform some additional numerical integration. The final result involves relatively little additional mathematical complexity, but rather some incremental computational overhead.

We will proceed in the most general manner to encompass the largest possible set of underlying models. We will, however, identify two broad cases. In the first case, we have that the default indicator is a function of a single latent systematic state variable, Y. This covers the single-factor Gaussian threshold and a large number of mixture models including the Poisson-gamma, Beta-binomial, logit-normal, probit-normal, Poisson-Weibull, Poisson log-normal, and one-factor CreditRisk+ models. In each of these cases, the default indicators are no longer independent in the unconditional default probabilities, $\{p_n; n = 1, \ldots n\}$, but they are independent in the *conditional* default probabilities. The default-indicator variable can be rewritten as,

$$\mathbb{I}_{\mathcal{D}_n}(y) = \begin{cases} 1 : \text{default with probability } p_n(y) \\ 0 : \text{survival with probability } 1 - p_n(y) \end{cases}. \quad (7.97)$$

where,

$$p_n(y) = \mathbb{P}\left(\mathbb{I}_{\mathcal{D}_n}(y)\big| Y = y\right), \quad (7.98)$$

for $n = 1, \ldots, N$ and $Y = y$. The conditional portfolio is a function of the state-variable outcome, the unconditional default probability and the model parameters. This quantity is available in closed-form for all of the previously described models.

---

[18]It is, however, an extremely useful model, given its simplicity, for the purposes of comparison, benchmarking, and model diagnostics.

7.7 The Conditionally Independent Saddlepoint Approximation

The normal-variance-mixture threshold model is the second case. This includes the t, variance-gamma, and generalized-hyperbolic threshold models. Each of these choices depend upon two state variables, the usual systematic global state variable, Y, and a second, mixing variable to generate the desired mixture distribution. The default-indicator variable is thus represented as,

$$\mathbb{I}_{\mathcal{D}_n}(y, v) = \begin{cases} 1 : \text{default with probability } p_n(y, v) \\ 0 : \text{survival with probability } 1 - p_n(y, v) \end{cases}. \quad (7.99)$$

where,

$$p_n(y, v) = \mathbb{P}\left(\mathbb{I}_{\mathcal{D}_n}(y, v) \middle| Y = y, V = v\right), \quad (7.100)$$

for $n = 1, \ldots, N$. The only difference between these two cases is the additional state variable.

Conditional independence is the key feature, which permits us to employ the saddlepoint approximation technique for these two cases. We will work with the first case since the approach is identical and the only distinction is some additional computation. The idea is quite simple. We may use, following from the development in the previous sections, a saddlepoint approximation to the *conditional* loss density as,

$$f_L(\ell | Y = y) = e^{K_L(\tilde{t}_\ell, y) - \tilde{t}_\ell \ell} \mathcal{J}_0\left(K_L''(\tilde{t}_\ell, y), \tilde{t}_\ell\right), \quad (7.101)$$

where the cumulant generating function and its derivatives are all defined in terms of the conditional default probability, $p_n(y)$, rather than p_n. To keep the notation under control, we let $K_L''(\tilde{t}_\ell, y) \equiv \frac{\partial K_L(t, y)}{\partial t}\Big|_{t=\tilde{t}_\ell}$. This is a slight abuse of notation, but given the already cluttered state of our expressions, we hope the reader will forgive this transgression. The structure of all the previously defined mathematical objects is identical save for the additional dependence on the value of the latent state variable, y. Given y, we conditionally preserve independence and maintain the overall structure of the saddlepoint approximation.

While this might seem uninteresting, since we are focused on the unconditional loss density, it is actually an important step in the right direction. Indeed, the unconditional loss expectation can be directly extracted from equation 7.101 by merely integrating out—or averaging over—the variable, y, as follows,

$$f_L(\ell) = \mathbb{E}\left(e^{K_L(\tilde{t}_\ell, y) - \tilde{t}_\ell \ell} \mathcal{J}_0\left(K_L''(\tilde{t}_\ell, y), \tilde{t}_\ell\right) \middle| Y = y\right), \quad (7.102)$$

$$= \int_{-\infty}^{\infty} e^{K_L(\tilde{t}_\ell, y) - \tilde{t}_\ell \ell} \mathcal{J}_0\left(K_L''(\tilde{t}_\ell, y), \tilde{t}_\ell\right) f_Y(y) dy,$$

where $f_Y(y)$ denotes the density function of the latent variable, Y, assuming it has infinite support. The only difference, therefore, is the necessity of averaging over the possible values of y.

There is, however, a small fly in the ointment. To evaluate these integrals, it is necessary to compute a separate saddlepoint approximation using the conditional-default probabilities for each individual choice of y—and v as well in the normal-variance mixture setting. In many cases, this leads to negative saddlepoint values. Figure 7.12 illustrates this fact by demonstrating the interplay between the level of confidence (α), the value of the state variable (y), and the saddlepoint approximation ($\tilde{t}_\ell$). Unlike the independent-default case, the saddlepoint is negative for many values of y. As a consequence, we need to include some adjustments to our contour integrals to force them back through the origin. This will leads to some correction terms in our saddlepoint approximations.

The same basic trick is employed for all of the remaining elements. It is useful to review them quickly, because in some cases certain terms no longer cancel out and the final result is somewhat more complex. The tail-probability, however, has a similar form,

$$\mathbb{P}(L > \ell_\alpha | Y = y) = \mathcal{H}_{-\tilde{t}_\ell} + e^{K_L(\tilde{t}_\ell, y) - \tilde{t}_\ell \ell_\alpha} \mathcal{I}_1\left(K_L''(\tilde{t}_\ell, y), \tilde{t}_\ell\right), \quad (7.103)$$

$$\mathbb{P}(L > \ell_\alpha) = \mathbb{E}\left(\mathcal{H}_{-\tilde{t}_\ell} + e^{K_L(\tilde{t}_\ell, y) - \tilde{t}_\ell \ell_\alpha} \mathcal{I}_1\left(K_L''(\tilde{t}_\ell, y), \tilde{t}_\ell\right) \middle| Y = y\right),$$

$$= \int_{-\infty}^{\infty} \left(\mathcal{H}_{-\tilde{t}_\ell} + e^{K_L(\tilde{t}_\ell, y) - \tilde{t}_\ell \ell_\alpha} \mathcal{I}_1\left(K_L''(\tilde{t}_\ell, y), \tilde{t}_\ell\right)\right) f_Y(y) dy,$$

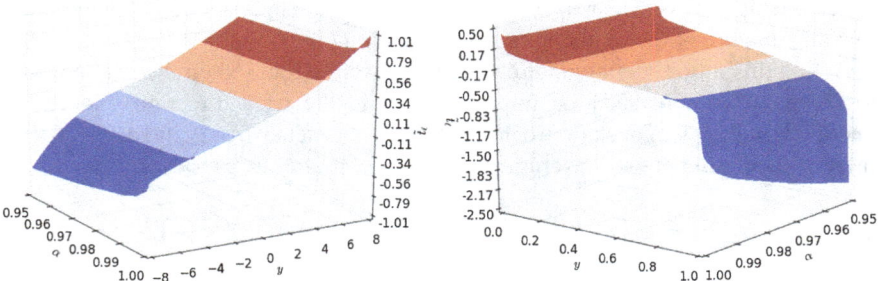

Fig. 7.12 *Integrated saddlepoint density approximation*: This figure illustrates, for two alternative models, the interplay between the level of confidence (α), the value of the state variable (y), and the saddlepoint approximation ($\tilde{t}_\ell$). Unlike the independent-default case, the saddlepoint is negative for many values of y. As a consequence, we need to include some adjustments to our contour integrals to force them back through the origin.

7.7 The Conditionally Independent Saddlepoint Approximation

where,

$$\mathcal{H}_x = \begin{cases} 1 : x \geq 0 \\ 0 : x < 0 \end{cases}, \tag{7.104}$$

is the Heavyside step function. It identifies the need for the correction term in the event of a negative saddlepoint. In the tail probability, it is merely necessary to add one to bring the integral back to the origin. These terms, of course, should have been present all along, but we have ignored them due to the positivity of our saddlepoint estimates.[19] In this case, with the exception of the requisite correction term, the final structure is almost identical to the independent case.

The shortfall integral is not as fortunate. It can be written as,

$$\mathbb{E}(L|L > \ell_\alpha, Y = y) \approx \ell_\alpha + \frac{e^{K_L(\tilde{t}_\ell, y) - \tilde{t}_\ell \ell_\alpha} \mathcal{I}_2\left(K_L''(\tilde{t}_\ell, y), \tilde{t}_\ell\right)}{\mathbb{P}(L > \ell_\alpha | Y = y)}, \tag{7.105}$$

$$\mathbb{E}(L|L > \ell_\alpha) \approx \ell_\alpha + \frac{\mathbb{E}\left(\mathcal{H}_{-\tilde{t}_\ell} C(y) + e^{K_L(\tilde{t}_\ell, y) - \tilde{t}_\ell \ell_\alpha} \mathcal{I}_2\left(K_L''(\tilde{t}_\ell, y), \tilde{t}_\ell\right) \bigg| Y = y\right)}{\mathbb{E}\left(\mathcal{H}_{-\tilde{t}_\ell} + e^{K_L(\tilde{t}_\ell, y) - \tilde{t}_\ell \ell_\alpha} \mathcal{I}_1\left(K_L''(\tilde{t}_\ell, y), \tilde{t}_\ell\right) \bigg| Y = y\right)},$$

$$\approx \ell_\alpha + \frac{\int_{-\infty}^{\infty} \left(\mathcal{H}_{-\tilde{t}_\ell} C(y) + e^{K_L(\tilde{t}_\ell, y) - \tilde{t}_\ell \ell_\alpha} \mathcal{I}_2\left(K_L''(\tilde{t}_\ell, y), \tilde{t}_\ell\right)\right) f_Y(y) dy}{\int_{-\infty}^{\infty} \left(\mathcal{H}_{-\tilde{t}_\ell} + e^{K_L(\tilde{t}_\ell, y) - \tilde{t}_\ell \ell_\alpha} \mathcal{I}_1\left(K_L''(\tilde{t}_\ell, y), \tilde{t}_\ell\right)\right) f_Y(y) dy},$$

where the correction term in the numerator is,

$$C(y) = \sum_{n=1}^{N} c_n \left(p_n(y) + \frac{p_n(y) e^{c_n \tilde{t}_\ell}}{1 - p_n(y) + p_n(y) e^{c_n \tilde{t}_\ell}}\right), \tag{7.106}$$

$$= \sum_{n=1}^{N} c_n p_n(y) + \underbrace{\frac{c_n p_n(y) e^{c_n \tilde{t}_\ell}}{1 - p_n(y) + p_n(y) e^{c_n \tilde{t}_\ell}}}_{\text{Equation 7.75}},$$

$$= \mathbb{E}(L|Y = y) + K'\left(\tilde{t}_\ell, y\right),$$

$$= \mathbb{E}(L|Y = y) + \ell_\alpha.$$

The correction term, therefore, makes an adjustment based on the conditional mean of the default loss given the systematic state variable, y.

---

[19] This complexity is part of the price one must pay to use this technique. The reader is referred to Daniels (1954, 1987), DasGupta (2008, Chapter 14), Huzurbazar (1999), and Lugannani and Rice (1980) for much more rigour on the evaluation of the contour integrals in the saddlepoint approximation.

Unlike the independent case in equation 7.80, there is no cancellation between the numerator and denominator. Instead, we are required to evaluate a separate numerical integral for both terms in the ratio. The consequence is more computational effort. The intuition of the previous sections, however, is not lost. The shortfall integral remains an add-on to the VaR computation.

The VaR contributions also lose some of their simplicity since certain terms no longer cancel out one another. To actually see the result, it is necessary to work through, a second time, some of the development originally found in equation 7.84. Some of the shortcuts used in the derivation are no longer available to us. The consequence is,

$$\frac{\partial \mathbb{P}(L > \ell_\alpha | Y = y)}{\partial c_n} = \frac{\partial}{\partial c_n}\left(\frac{1}{2\pi i}\int_{-i\infty,(0+)}^{+i\infty}\frac{e^{K_L(t,y)-t\ell_\alpha}}{t}dt\right), \quad (7.107)$$

$$\underbrace{\frac{\partial(1-\alpha)}{\partial c_n}}_{=0} = \frac{1}{2\pi i}\int_{-i\infty,(0+)}^{+i\infty}\frac{1}{t}\left(\frac{\partial K_L(t,y)}{\partial c_n} - t\frac{\partial \ell_\alpha}{\partial c_n}\right)e^{K_L(t,y)-t\ell_\alpha}dt,$$

$$\frac{\partial \ell_\alpha}{\partial c_n}\frac{1}{2\pi i}\int_{-i\infty,(0+)}^{+i\infty}e^{K_L(t,y)-t\ell_\alpha}dt = \frac{1}{2\pi i}\int_{-i\infty,(0+)}^{+i\infty}\frac{\partial K_L(t,y)}{\partial c_n}\frac{e^{K_L(t,y)-t\ell_\alpha}}{t}dt,$$

$$\frac{\partial \ell_\alpha}{\partial c_n}\left(e^{K_L(\tilde{t}_\ell,y)-\tilde{t}_\ell\ell_\alpha}\mathcal{I}_0\left(K_L''(\tilde{t}_\ell,y),\tilde{t}_\ell\right)\right) \approx \frac{p_n(y)e^{c_n\tilde{t}_\ell}}{1-p_n(y)+p_n(y)e^{c_n\tilde{t}_\ell}}e^{K_L(\tilde{t}_\ell,y)-\tilde{t}_\ell\ell_\alpha}\mathcal{I}_0\left(K_L''(\tilde{t}_\ell,y),\tilde{t}_\ell\right),$$

$$\mathbb{E}\left[\frac{\partial \ell_\alpha}{\partial c_n}\left(e^{K_L(\tilde{t}_\ell,y)-\tilde{t}_\ell\ell_\alpha}\mathcal{I}_0\left(K_L''(\tilde{t}_\ell,y),\tilde{t}_\ell\right)\right)\right] \approx \mathbb{E}\left[\frac{p_n(y)e^{c_n\tilde{t}_\ell}}{1-p_n(y)+p_n(y)e^{c_n\tilde{t}_\ell}}e^{K_L(\tilde{t}_\ell,y)-\tilde{t}_\ell\ell_\alpha}\mathcal{I}_0\left(K_L''(\tilde{t}_\ell,y),\tilde{t}_\ell\right)\right],$$

$$\frac{\partial \ell_\alpha}{\partial c_n} \approx \frac{\mathbb{E}\left[\frac{p_n(y)e^{c_n\tilde{t}_\ell}}{1-p_n(y)+p_n(y)e^{c_n\tilde{t}_\ell}}e^{K_L(\tilde{t}_\ell,y)-\tilde{t}_\ell\ell_\alpha}\mathcal{I}_0\left(K_L''(\tilde{t}_\ell,y),\tilde{t}_\ell\right)\right]}{\mathbb{E}\left[e^{K_L(\tilde{t}_\ell,y)-\tilde{t}_\ell\ell_\alpha}\mathcal{I}_0\left(K_L''(\tilde{t}_\ell,y),\tilde{t}_\ell\right)\right]},$$

$$\approx \frac{\int_{-\infty}^{\infty}\frac{p_n(y)e^{c_n\tilde{t}_\ell}}{1-p_n(y)+p_n(y)e^{c_n\tilde{t}_\ell}}e^{K_L(\tilde{t}_\ell,y)-\tilde{t}_\ell\ell_\alpha}\mathcal{I}_0\left(K_L''(\tilde{t}_\ell,y),\tilde{t}_\ell\right)f_Y(y)dy}{\int_{-\infty}^{\infty}e^{K_L(\tilde{t}_\ell,y)-\tilde{t}_\ell\ell_\alpha}\mathcal{I}_0\left(K_L''(\tilde{t}_\ell,y),\tilde{t}_\ell\right)f_Y(y)dy}.$$

We can see the independent-default VaR contribution to the marginal VaR term in the numerator of equation 7.107. It now has to be weighted by the appropriate saddlepoint and state-variable densities to incorporate the role of default dependence. Moreover, since the saddlepoint contributions require only the use of the saddlepoint densities, no correction term is required.

Equation 7.95 provides us with the final marginal independent-default expected-shortfall expression required to determine the associated obligor contributions. Its generalization to the dependent setting is, fortunately, not much more complex. There was relatively little cancellation in the original expression and, thus, the final result is a fairly straightforward numerical integration. The marginal expected-

7.7 The Conditionally Independent Saddlepoint Approximation

shortfall is summarized as,

$$\frac{\partial \mathcal{E}_\alpha(L)}{\partial c_n} \approx \underbrace{\frac{\partial \ell_\alpha}{\partial c_n}}_{\text{Eqn 7.107}} + \frac{1}{1-\alpha} \mathbb{E}\left[\mathcal{H}_{-\tilde{t}_\ell} C_n(y)\right.$$

$$\left. + \left(\frac{c_n p_n(y) e^{c_n \tilde{t}_\ell}(1 - p_n(y))}{(1 - p_n(y) + p_n(y) e^{c_n \tilde{t}_\ell})^2}\right) \frac{e^{K_L(\tilde{t}_\ell, y) - \tilde{t}_\ell \ell_\alpha}}{K''(\tilde{t}_\ell, y)} \mathcal{I}_2(K_L''(\tilde{t}_\ell, y), \tilde{t}_\ell)\right], \qquad (7.108)$$

$$\approx \underbrace{\frac{\partial \ell_\alpha}{\partial c_n}}_{\substack{\text{VaR} \\ \text{term}}} + \underbrace{\frac{1}{\mathbb{P}(L > \ell_\alpha)}}_{\substack{\text{Eqn} \\ 7.103}} \overbrace{\int_\infty^\infty \mathcal{H}_{-\tilde{t}_\ell} C_n(y) f_Y(y) dy}^{\text{Correction term}}$$

$$+ \underbrace{\int_\infty^\infty \left(\frac{c_n p_n(y) e^{c_n \tilde{t}_\ell}(1 - p_n(y))}{(1 - p_n(y) + p_n(y) e^{c_n \tilde{t}_\ell})^2}\right) \frac{e^{K_L(\tilde{t}_\ell, y) - \tilde{t}_\ell \ell_\alpha}}{K''(\tilde{t}_\ell, y)} \mathcal{I}_2(K_L''(\tilde{t}_\ell, y), \tilde{t}_\ell) f_Y(y) dy}_{\text{Expected-shortfall term}},$$

where the correction term is now,

$$C_n(y) = p_n(y) + \frac{p_n(y) e^{c_n \tilde{t}_\ell}}{1 - p_n(y) + p_n(y) e^{c_n \tilde{t}_\ell}}. \qquad (7.109)$$

It should be clear that if we sum equation 7.109 over each obligor and multiply each term by the appropriate c_n, we recover the total correction $C(y)$ described in equation 7.106. Thus, when we use the integrated partial expected-shortfall derivatives along with Euler's theorem to compute the individual contributions, they will, by definition, sum to the total shortfall integral. Again, a pleasant and intuitive form is maintained. There are, however, now three aspects: the first term for the marginal VaR, the second term for the corrections associated with a negative saddlepoint, and the final piece that describes the shortfall contributions.

Equations 7.102 to 7.109, relying on the hard-earned lessons and derivations from the previous sections, summarize the approach required to tease out key quantities in the one-factor dependent-default setting. If we move to the normal-variance mixture environment, nothing changes. The only difference is that each one-dimensional integral with respect to Y becomes a two-dimensional integral over the Y and V densities. This leads to significant increases in computational time, reducing somewhat the appeal of the saddlepoint approximation. This is, in fact, the main reason for our inability to apply the saddlepoint technique in the multivariate domain. The numerical integration becomes simply too heavy.

7.7.1 Implementation

The implementation in the dependent-default case is more intricate. Much of the existing code, fortunately, can be reused. Our first step, however, is some consolidation. Algorithm 7.11 outlines a general Python sub-routine, termed `saddlePointApprox`, which can be used to compute our three main saddlepoint integrals. It takes the usual arguments `l`, `c`, and `p` to describe the target loss, the exposures, and the default probabilities. It also takes an additional argument, `myDegree`, which can takes the values 0, 1, and 2. This allows one to evaluate integrals of the form summarized in equation 7.66. This simplifies our execution and makes the code easier to read. It also permits, if we wish, to send conditional default probabilities—$p_n(y)$ instead of p_n—to construct the expressions required for the dependent-default setting. Finally, the term `constant` allows for an incorporation of the correction term required for negative saddlepoint estimates; the `step` function is merely a one-line implementation of equation 7.104.

The next step is the performance of our numerical integration in a general form. We have a variety of integrals that need to be evaluated with a broad range of choices for the conditional default probabilities and their associated state-variable density. This can quickly become overwhelming. To organize ourselves, therefore, we introduce a general integral function in Algorithm 7.12. This Python function, called `computeYIntegral`, performs three tasks. Given a choice of model—summarized by the `whichModel` flag—it

1. uses the provided parameters, `p1` and `p2`, to obtain the conditional default probabilities from `getPy` and stores them in the array, `pY`;
2. uses the parameters, choice of model, and value for the state-variable, `y`, to call the function `getYDensity` to return the appropriate density value; and
3. returns the previous two values as an input to the numerical-integration engine.

The saddlepoint, `t_1`, is also computed separately so as to avoid solving the optimization problem more often than strictly necessary.

Algorithm 7.11 A general saddlepoint function

```
def saddlePointApprox(l,p,c,t_1,myDegree,constant=0):
    if myDegree==1:
        constant = step(-t_1)
    elif myDegree==2:
        constant = step(-t_1)*(np.dot(p,c)-1)
    coefficient = np.exp(computeCGF(t_1,p,c)-t_1*l)
    return constant + coefficient*getJ(l,p,c,t_1,myDegree)
```

7.7 The Conditionally Independent Saddlepoint Approximation

Algorithm 7.12 A general integral function

```
def computeYIntegral(y,l,p,c,p1,p2,whichModel,myDegree):
    pY = getPy(p,y,p1,p2,whichModel)
    d = getYDensity(y,p1,p2,whichModel)
    t_1 = getSaddlePoint(pY,c,1)
    return saddlePointApprox(l,pY,c,t_1,myDegree)*d
```

Algorithm 7.13 Integrated densities, tail probabilities and shortfall integrals

```
import scipy.integrate as nInt
def myApprox(l,p,c,p1,p2,whichModel,myDegree,constant=0,den=1):
    lB,uB = getIntegrationBounds(whichModel)
    if myDegree==2:
        constant = 1
        den,e = nInt.quad(computeYIntegral,lB,uB,\
                args=(l,p,c,p1,p2,whichModel,1))
    num,e = nInt.quad(computeYIntegral,lB,uB,\
            args=(l,p,c,p1,p2,whichModel,myDegree))
    return constant + np.divide(num,den)
```

You can think of `getPy` and `getYDensity` as two look-up functions that, depending on the value of `whichModel`, scroll through a list of alternatives and return the desired values. This code is quite repetitive and, as such, it has not been displayed. As a final point, not all models require two parameters, but we've created two placeholders to ensure that all two-parameter cases can be accommodated.

With these two helper functions, we are now in a position to be able to estimate the loss density, tail probabilities, and the shortfall integrals within a single function. Using the `scipy` integrate function, Algorithm 7.13 calls `computeYIntegral` for the desired model and required choice of `myDegree` flag. The generic form is a constant plus a ratio of two values. In the case of the loss density and the tail probability, the constant and denominator take the values of zero and unity, respectively. For the expected-shortfall integral, however, the constant is the VaR estimate and the denominator is the tail probability.

An additional support function, named `getIntegrationBounds`, ensures that the range of integration is consistent with the choice of model. If the state-variable is Gaussian, for example, the limits of integration are $[-10, 10]$ to proxy its infinite support. Conversely, should the state-variable follow a beta distribution, then the integration is confined to the unit interval. The benefit of this approach is that one need not write a separate Python function for each separate model. One need only determine its conditional default probabilities, the state-variable density, and the appropriate bounds of integration.

Algorithm 7.14 Computing integrated VaR risk contributions

```
def myVaRCY(l,p,c,p1,p2,whichModel):
    lB,uB = getIntegrationBounds(whichModel)
    den = myApprox(l,p,c,p1,p2,whichModel,0)
    num = np.zeros(len(p))
    for n in range(0,len(p)):
        num[n],err = nInt.quad(varCNumerator,lB,uB,\
                        args=(l,n,p,c,p1,p2,whichModel))
    return c*np.divide(num,den)
```

Algorithm 7.15 Determining the numerator for integrated VaR contributions

```
def varCNumerator(y,l,myN,p,c,p1,p2,whichModel,v=0):
    pY = getPy(p,y,p1,p2,whichModel,v)
    d = getYDensity(y,p1,p2,whichModel,v)
    t_1 = getSaddlePoint(pY,c,l)
    num = pY[myN]*np.exp(c[myN]*t_1)
    den = computeMGF(t_1,pY[myN],c[myN])
    return np.divide(num,den)*saddlePointApprox(l,pY,c,t_1,0)*d
```

The final step is to compute the contributions from VaR and expected-shortfall. As we've seen, these are the ratio of multiple integrals. Both approaches are conceptually very similar thus, to avoid making this lengthy chapter even longer, we will focus solely on the VaR case.[20] Algorithm 7.14 describes the base function, getVARCY, used to estimate the VaR contributions for a given level of loss. It performs the same job as Algorithm 7.9, but in a rather less straightforward manner. The whichModel identifier uses the function getIntegrationBounds to identify the appropriate integration bounds; these depend, of course, on the support of y in the underlying models.

Algorithm 7.14 starts by computing the denominator associated with each of the N contribution terms. This requires a single call to the myApprox function in Algorithm 7.13. It then loops over each of the N numerator terms and passes off the work to the varCNumerator function summarized in Algorithm 7.15. Each iteration of the numerical integration, quad, requires new conditional default-probabilities, a new density function evaluation, and a new saddlepoint. Indeed, virtually every step in the computation of the numerator relies on previously defined Python functions.

---

[20] The expected-shortfall code is, however, found in the varContributions library. See Appendix D for more details on the library structure.

7.7 The Conditionally Independent Saddlepoint Approximation

This same basic approach is repeated for the expected-shortfall computation. The main difference is the three separate integrands: the VaR component, the correction for negative saddlepoints, and the expected-shortfall element. The incremental complexity of the integrand implies a bit more computation effort for its valuation. The actual implementation is, however, conceptually identical.

If we move to the normal-variance mixture models, then the only difference is a second variable of integration. Examination of Algorithm 7.15 reveals the optional argument, v, whose default value is zero. This allows us to recycle this code for the two-dimensional case. This is performed using `scipy`'s `nquad` function. As we'll see in the next section, this is a much slower proposition.

7.7.2 A Multi-Model Example

Saddlepoint approximations involve some rather complicated mathematics and a variety of numerical techniques such as root-solving and numerical integration. Is this additional effort worthwhile? This is an important question that merits a serious answer. Complex techniques are more easily misunderstood and incorrectly applied than simpler, more robust methods. Complexity should not be used, therefore, only for complexity's sake, but to improve the speed, accuracy, and interpretation of model results.

A useful approach to assessing this question is to examine a range of saddlepoint and Monte Carlo computations in the context of multiple models. This will allow us to assess the accuracy and speed of the saddlepoint approximations in a variety of contexts. We will, of course, use our usual 100-obligor, heterogeneous and concentrated portfolio. Each of the models are calibrated to the same average level of default probability, 1%, and, where necessary, target default correlation of roughly 0.05. Using these common inputs, four distinct one-period default credit-risk models will be considered—the Gaussian-threshold, the beta-binomial-mixture, the one-factor CreditRisk+, and the Poisson-Weibull-mixture models. We have many other models that might be considered, but this sample allows us to avoid clutter and still cover a broad range of approaches.

Table 7.6 compares the USD difference—for a range of possible quantiles—between the Monte-Carlo and saddlepoint VaR and expected-shortfall estimates. The usual 10 million iterations were used perform the simulation-based estimates. A number of conclusions may be drawn from Table 7.6. First, the saddlepoint approximations demonstrate an impressive amount of accuracy for each of the models under consideration. Differences between the Monte Carlo and saddlepoint estimates do not exceed more than a few units of currency, or percentage points, on a $1,000 portfolio.[21] This result suggests that the saddlepoint approximation is

---

[21] The differences are even smaller if we were to construct confidence intervals for the Monte Carlo estimates. This important question is addressed in the following chapter.

Table 7.6 *Multiple-model saddlepoint comparison*: This table uses our saddlepoint approximation to estimate the VaR and expected-shortfall for four competing models: the Gaussian threshold ($\mathcal{N}$), the beta-binomial (β-B), the one-factor CreditRisk+ (CR+), and the Poisson-Weibull ($\mathcal{P}$-W) approaches. The percentage error of each saddlepoint estimate is compared to its Monte Carlo equivalent computed using 10 million iterations.

Quantile	$\text{VaR}_\alpha(L)$				$\mathcal{E}_\alpha(L)$			
	$\mathcal{N}$	β-B	CR+	$\mathcal{P}$-W	$\mathcal{N}$	β-B	CR+	$\mathcal{P}$-W
95.00th	4.3	4.9	3.4	5.4	1.1	0.6	0.1	0.7
97.00th	2.3	3.3	2.4	3.5	0.9	0.4	0.2	0.5
99.00th	1.9	1.6	1.3	1.8	0.8	0.3	0.1	0.2
99.50th	1.6	1.2	0.8	1.1	0.6	0.3	−0.1	0.0
99.90th	0.8	0.6	0.3	0.3	0.3	0.1	−0.2	−0.2
99.97th	0.4	−0.1	−0.1	−0.3	0.2	0.2	−0.5	−0.0
99.99th	0.3	0.2	−0.7	−0.0	0.1	0.5	−0.5	0.3

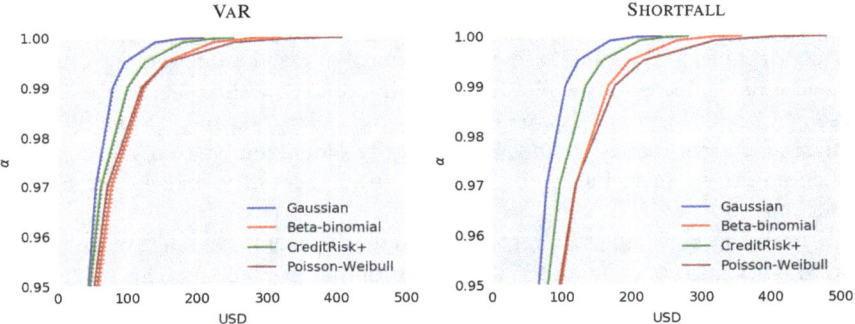

Fig. 7.13 *Multiple-model visual risk-metric comparison*: This figure graphically depicts the application of the saddlepoint approximation—for both VaR and expected-shortfall estimates—across four alternative models. The saddlepoint estimates are quite accurate and improve as we move further out into the tail.

robust to the choice of model and appears to work well in both the independent and conditionally independent settings.

The second point is that, for all models, the accuracy actually improves as we move further into the tail of the default-loss distribution. This pattern holds for all four models. The saddlepoint estimator's reputation—as particularly good for extreme outcomes—thus appears to be well founded. A third observation is that the overall closeness of fit is somewhat better for the expected-shortfall measures relative to the VaR outcomes. This may be related to the superiority of the Monte-Carlo expected-shortfall estimates. Estimating an integral over a given interval should be a bit more stable than the approximation of a specific quantile.

Figure 7.13 graphically depicts the results from Table 7.7 by comparing visually the saddlepoint and Monte Carlo VaR estimates. The dotted lines, of the same colour, denote the Monte Carlo estimates—the results indicate a robust fit across a range of quantiles ranging from 0.95 to 0.9999. Indeed, the dotted lines are virtually

7.7 The Conditionally Independent Saddlepoint Approximation

Table 7.7 *Multiple-model top-ten VaR contributions*: This table highlights—along with the rank, unconditional default probability, and exposure—the ten largest 99.99th quantile VaR contributions associated with our randomly generated heterogeneous portfolio. In all cases, the figures come from the saddlepoint approximation.

Obligor details				$\text{VaR}_\alpha(L)$				$\mathcal{E}_\alpha(L)$			
Rank	ID	p	c	GA	BE	CR+	PW	GA	BE	CR+	PW
1	13	6.8%	$30.4	$22.1	$11.7	$28.4	$13.8	$23.4	$13.1	$29.1	$15.9
2	33	1.1%	$28.4	$10.3	$10.6	$12.6	$12.6	$11.6	$11.8	$14.1	$14.6
3	1	1.2%	$27.1	$10.1	$10.0	$12.6	$12.0	$11.3	$11.1	$14.1	$13.8
4	30	0.7%	$30.9	$9.9	$12.0	$11.3	$14.1	$11.2	$13.4	$12.7	$16.2
5	8	1.1%	$25.8	$9.0	$9.3	$11.1	$11.3	$10.1	$10.4	$12.5	$13.1
6	24	1.4%	$22.9	$8.3	$7.9	$10.8	$9.8	$9.4	$8.8	$12.1	$11.3
7	65	0.5%	$28.3	$6.9	$10.6	$7.2	$12.6	$8.0	$11.8	$8.3	$14.6
8	28	1.2%	$21.0	$6.9	$7.1	$8.8	$8.8	$7.8	$7.9	$9.8	$10.3
9	69	1.4%	$18.4	$6.2	$5.9	$8.2	$7.6	$7.1	$6.7	$9.2	$8.8
10	93	1.8%	$15.3	$5.5	$4.7	$7.6	$6.2	$6.2	$5.3	$8.4	$7.2
Other				$751.5	$116.6	$234.1	$135.3	$302.9	$138.0	$262.3	$151.6	$354.3
Total				$1000.0	$211.8	$323.9	$253.9	$411.7	$244.1	$362.6	$281.9	$480.1

invisible. For the purposes of VaR and expected-shortfall computation, therefore, the saddlepoint appears to be a viable alternative in terms of accuracy.

The key incentive for the use of the saddlepoint approximation, however, does not relate to the computation of VaR and expected shortfall, but rather the attribution of risk to each counterparty in one's portfolio. We have, therefore, also used the saddlepoint approximation to compute the 99.99th quantile VaR and expected-shortfall contributions for each model. Figure 7.14 graphically illustrates the top-ten risk contributions in USD terms. The ordering, however, was determined by the Gaussian-threshold model.

Figure 7.14 reveals rather important differences in the structure and magnitude of the risk contributions associated with our four models. There are similarities between the Gaussian-threshold and CreditRisk+ models on the one hand, and the beta-binomial and Poisson-Weibull models, on the other. The reason is simple. Both the Gaussian-threshold and CreditRisk+ models rely on the individual unconditional default probabilities as a central input. Consequently, all else equal, high-risk obligors will have larger risk contributions. The other mixture models randomize default probability for all obligors enforcing only an overall average level. The result is that the size of the exposure plays a more important role. Instead of attempting to determine which approach is superior, we rather underscore the value of having two competing approaches to address the same question.

Table 7.7, in a busy array of numbers, displays the values summarized graphically in Fig. 7.14. It also highlights the rank, unconditional default probability, and exposures. The combination of VaR and expected-shortfall contributions along with unconditional default and counterparty exposure data provide us more insight into the final figures. A coincidence of large exposure and default probabilities generates

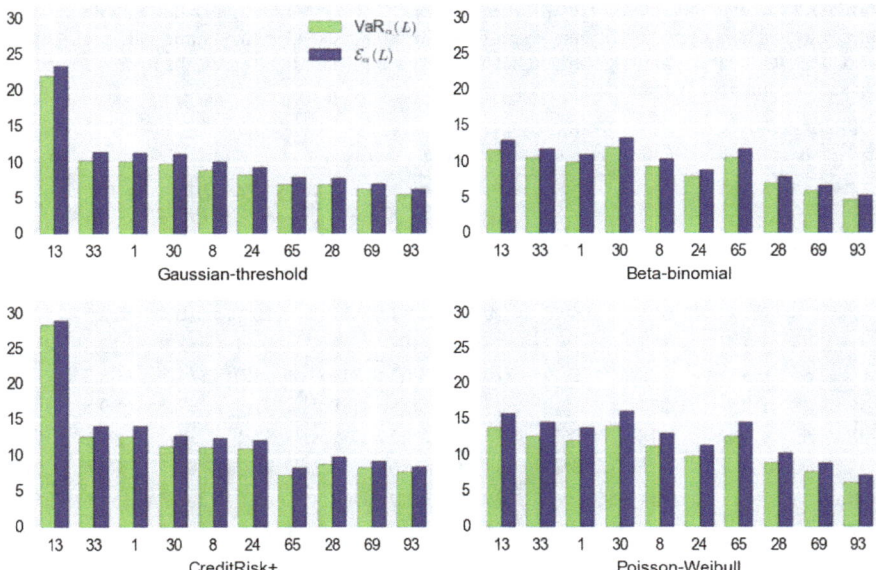

Fig. 7.14 *Saddlepoint risk-measure contributions*: This figure graphically shows the top-ten 99.99th quantile VaR and expected-shortfall contributions for each of our four models. The various approaches exhibit rather different implications for the individual contributions of specific obligors. All values are in USD.

the largest overall exposure. Nevertheless, large exposures with modest default probabilities or large default probabilities with modest exposures can also generate large VaR contributions. The relationship is, however, relatively complex, which demonstrates the important need for risk attribution.

The differences in model treatment of individual obligors is clearly visible in Fig. 7.15. All of the risk contributions for the 99.99th quantile expected-shortfall estimates are ordered—again by the Gaussian-threshold model—for each model. All values are represented in percentage terms relative to the total portfolio value. The largest risk contributions, therefore, rise to approximately 2.5% of the $1,000 portfolio. The ordering, if not the magnitude, of the Gaussian-threshold and CreditRisk+ models appear to agree fairly closely.

We also see, however, rather more closely the alternative behaviour of the beta-binomial and Poisson-Weibull models. They do not attribute as sizable an amount of risk to the high-risk, high-exposure obligors. They do, however, assign significantly higher risk to obligors who, in the other approaches, receive a very small allocation. The reason relates to the size of the exposure. In the Gaussian-threshold and CreditRisk+ setting, a large exposure with a vanishingly small unconditional default probability is deemed low risk. While entirely reasonable, the other models approach the problem differently. Since all obligors share a common

7.7 The Conditionally Independent Saddlepoint Approximation 419

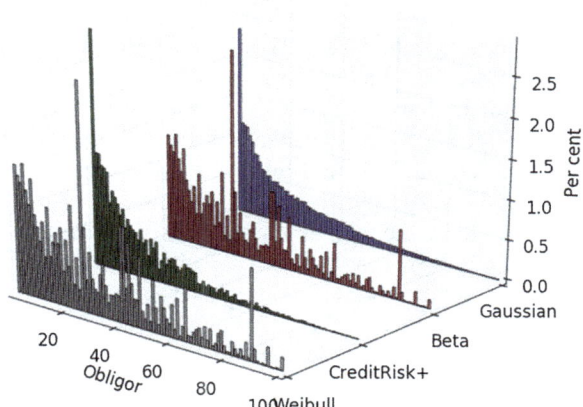

Fig. 7.15 *Proportional saddlepoint shortfall contributions*: This figure highlights ordered—following the Gaussian threshold model—expected-shortfall contributions for all models and all obligors. It is interesting to examine how the various models have different implications for the structure of credit-counterparty risk.

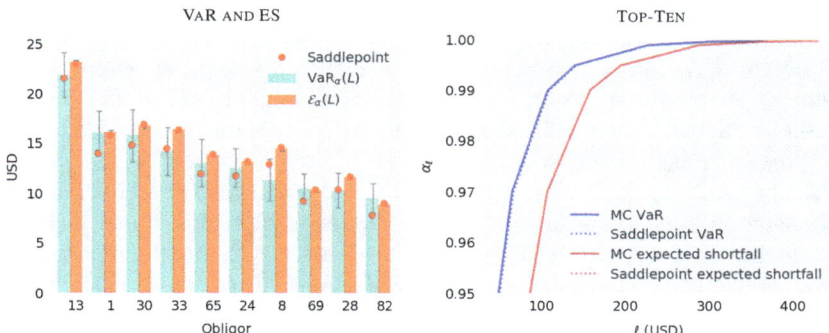

Fig. 7.16 *Back to the beginning: The t-threshold model*: This figure returns to the Monte-Carlo estimates of the ten largest one-factor t-threshold VaR and expected-shortfall contributions. For comparison purposes, it shows the associated saddlepoint approximations computed using two-dimensional numerical integration. It also demonstrates the accuracy of the saddlepoint tail-probability and shortfall-integral estimates.

default probability, the size of the underlying exposure plays a more important role. This perspective can help identify potential risks, which are ignored in other settings.

Before addressing the important issue of computational speed, we return to our original t-threshold model Monte-Carlo-based risk-attribution exercise. Figure 7.16 performs two tasks. First, it reproduces the VaR and expected-shortfall contribution estimates from Fig. 7.1, but also includes the saddlepoint approximation. The latter values are represented as round points. These figures are computed using two-dimensional integration. The only difference with the previous four models is that we integrate with respect to $f_Y(y)$ and $f_V(v)$, where these are the standard-normal Gaussian and χ^2 densities, respectively. The agreement with the expected-shortfall estimates is quite close. The VaR contributions also fall into the Monte-Carlo estimators 95% confidence band, but typically on the lower end.

The second aspect of Fig. 7.16 is to demonstrate the closeness of fit associated with the saddlepoint VaR and shortfall-integral estimates in the two-dimensional environment. The right-hand graphic in Fig. 7.16 demonstrates that the accuracy of the integrated t-threshold saddlepoint approximation is quite high. It is nonetheless rather slow. The reason is the necessity of numerically evaluating a two-dimensional integral. Although the accuracy does not suffer, as we see in the next section, it is a time-consuming computational exercise.

Having now examined this technique across six alternative models with positive results, we may cautiously conclude that it appears to be a viable alternative.

7.7.3 Computational Burden

Before making our final conclusion with regard to the saddlepoint method, there remains a final critical dimension: computational time. Table 7.8 compares the time, in seconds, to perform a number of different saddlepoint approximations for each of our *six* alternative approaches: the independent-default binomial, the Gaussian-threshold, the beta-binomial, the one-factor CreditRisk+, the Poisson-Weibull, and the t-threshold models.

The first task, determination of the loss threshold for a given level of confidence, is the solution to the optimization problem in equation 7.82. This should, in general, be considered a computationally intensive task. It seeks to find the root to a non-linear equation and thus requires numerous function evaluations for its derivatives. While it is almost instantaneous in the independent-default setting, it takes almost 20 seconds for the one-dimensional integrated models. For the t-threshold model, this

Table 7.8 *Comparing computational times*: This table compares and contrasts the computational time, in seconds, for a variety of tasks across different models. The various tasks include computation of a loss-density value, a given tail-probability, a specific shortfall integral, a set of VaR contributions, and a collection of expected-shortfall contributions.

| Description | Notation | Models ||||||
		Ind.	$\mathcal{N}$	β-B	CR+	$\mathcal{P}$-W	$\mathcal{T}$
Loss threshold	ℓ_α	0.0	18.0	18.0	18.0	18.0	305.2
Loss density	$f_L(\ell_\alpha)$	0.0	0.1	0.1	0.1	0.1	5.9
Tail probability	$\text{VaR}_\alpha(L)$	0.0	0.1	0.1	0.1	0.1	7.2
Shortfall integral	$\mathcal{E}_\alpha(L)$	0.0	0.2	0.2	0.2	0.2	25.9
VaR contributions	$c_n \frac{\partial \text{VaR}_\alpha(L)}{\partial c_n}$ for $n = 1, \ldots, N$	0.0	7.1	7.1	7.1	7.1	341.2
ES contributions	$c_n \frac{\partial \mathcal{E}_\alpha(L)}{\partial c_n}$ for $n = 1, \ldots, N$	0.0	13.9	13.9	13.9	13.9	1343.0

requires almost five minutes.[22] Clearly, the movement from one- to two-dimensional integration requires a large increase in computational complexity.[23]

The computation of loss densities, tail probabilities, and shortfall integrals are, as we've seen and demonstrated, related problems. The one-dimensional integrated models require less than one quarter of a second for their resolution. This is quite fast for the degree of accuracy achieved. Even in the t-threshold setting, this requires between 10 to 30 seconds.

VaR and expected-shortfall contributions require the evaluation of $N + 1$ numerical integrals. This should require some effort and this is clearly indicated in Table 7.8. Determination of the set of VaR contributions for each of the one-dimensional integrated models necessitates about 7 seconds of computation. Roughly twice as much is required to evaluate the expected-shortfall risk attributions. Again, the accuracy and time trade-off appears to be quite favourable. The two-dimensional t-threshold model is another story. The VaR contributions require about six minutes, while the collection of expected-shortfall attributions for a given level of confidence involves in excess of 20 minutes of computer time. Given the accuracy, it is still not a bad solution, but the trade-off is growing less favourable. 100 million Monte-Carlo iterations of the one-dimensional t-threshold model has computational requirements in the same basic ballpark. The decision is, in this two-dimensional setting, less clear.

We may, however, conclude that the saddlepoint is both accurate and computationally efficient. While its ability to replicate VaR and expected-shortfall estimates to a very reasonable degree of accuracy across many models is impressive, the real benefit is its fast and robust approximation for individual obligor risk contributions. Risk attribution is useful as a model diagnostic, but also has important applications of its own. One may use them to determine individual counterparty limits, determine the pricing associated with incremental business with a given counterparty, or assess the profitability of transactions with a given counterparty. Any robust, fast, and accurate technique for risk attribution is thus both welcome and extremely useful.

7.8 An Interesting Connection

Before we move to the next chapter, it is helpful to identify a final characteristic of the saddlepoint method arising from an illuminating and motivational derivation found in Martin (2011). We will borrow it here and work through it in some detail, not because it was used directly in the previous development of the saddlepoint approximation, but rather due to the link it establishes with a key technique used in

---

[22] One could construct a simple grid to find ℓ_α in the t-threshold model and estimate one's desired loss threshold with interpolation. A 100-point grid could be constructed in approximately five minutes. If one seeks a single loss threshold, it is still probably better to use the optimization approach.

[23] This is precisely what makes the saddlepoint approximation inappropriate for a multi-factor model.

Chap. 8. The idea is that we would like to create an approximation for the sum of a collection of independent and identically distributed random variables: $X_1,\ldots,X_N$. We define the sum as,

$$Y = \sum_{n=1}^{N} X_n. \tag{7.110}$$

The sum of our default events has precisely this form. The central-limit-theorem—see, for example, Durrett (1996)—is typically used to approximate Y with the Gaussian distribution. Another common approach is the so-called Edgeworth expansion—see DasGupta (2008, Chapter 13)—to approximate the distribution of Y in terms of its cumulants. While both choices are widely used, neither is particularly efficient as one moves further out into the tail; they are, however, good approximations for the centre of Y's distribution.

An alternative idea is—based on an underlying assumption of Y being approximately Gaussian—to construct an alternative adjusted density function for the variable, $\tilde{Y}$. It has the following form,

$$f_{\tilde{Y}}(y) = \frac{e^{\lambda y} f_Y(y)}{M_Y(\lambda)}, \tag{7.111}$$

where $f_Y(y)$ is the standard-normal distribution and $M_Y(\lambda)$ is the moment-generating function of Y. This is a familiar character from our previous development. At this point, we will, given our knowledge of $f_Y(y)$, compute its actual form. By definition, we have

$$M_Y(\lambda) = \mathbb{E}\left(e^{\lambda Y}\right), \tag{7.112}$$

$$= \int_{-\infty}^{\infty} e^{\lambda y} f_Y(y) dy,$$

$$= \int_{-\infty}^{\infty} e^{\lambda y} \frac{1}{\sqrt{2\pi}} e^{-\frac{y^2}{2}} dy,$$

$$= \int_{-\infty}^{\infty} \frac{1}{\sqrt{2\pi}} e^{\frac{-y^2+2\lambda y}{2}} dy,$$

$$= \int_{-\infty}^{\infty} \frac{1}{\sqrt{2\pi}} e^{\frac{-y^2+2\lambda y+\overbrace{\lambda^2-\lambda^2}^{=0}}{2}} dy,$$

$$= e^{\frac{\lambda^2}{2}} \int_{-\infty}^{\infty} \frac{1}{\sqrt{2\pi}} e^{\frac{-(y^2-2\lambda y+\lambda^2)}{2}} dy,$$

7.8 An Interesting Connection

$$= e^{\frac{\lambda^2}{2}} \underbrace{\int_{-\infty}^{\infty} \frac{1}{\sqrt{2\pi}} e^{-\frac{(y-\lambda)^2}{2}} dy}_{=1},$$

$$= e^{\frac{\lambda^2}{2}}.$$

By completing the square, we find that the moment-generating function of Y in λ has a compact form. If we return to equation 7.111 and use the result from equation 7.112, we see more clearly the role of $M_Y(\lambda)$,

$$f_{\tilde{Y}}(y) = \frac{e^{\lambda y} f_Y(y)}{\underbrace{e^{\frac{\lambda^2}{2}}}_{\text{Equation 7.112}}}, \qquad (7.113)$$

$$\int_{-\infty}^{\infty} f_{\tilde{Y}}(y) dy = \int_{-\infty}^{\infty} \frac{e^{\lambda y} f_Y(y)}{e^{\frac{\lambda^2}{2}}} dy,$$

$$= \frac{1}{e^{\frac{\lambda^2}{2}}} \underbrace{\int_{-\infty}^{\infty} e^{\lambda y} f_Y(y) dy}_{=M_Y(\lambda) = e^{\frac{\lambda^2}{2}}},$$

$$= 1.$$

The role of $M_Y(\lambda)$ in the denominator is to normalize the density function of $\tilde{Y}$ so that it integrates to one.

Our new random variable, $\tilde{Y}$, has some useful properties. Its expectation is,

$$\mathbb{E}\left(\tilde{Y}\right) = \int_{-\infty}^{\infty} y f_{\tilde{Y}}(y) dy, \qquad (7.114)$$

$$= \int_{-\infty}^{\infty} y \frac{e^{\lambda y} f_Y(y)}{e^{\frac{\lambda^2}{2}}} dy,$$

$$= \frac{e^{\frac{\lambda^2}{2}}}{e^{\frac{\lambda^2}{2}}} \underbrace{\int_{-\infty}^{\infty} y \frac{1}{\sqrt{2\pi}} e^{-\frac{(y-\lambda)^2}{2}} dy}_{=\lambda},$$

$$= \lambda,$$

where we used the complete-the-square trick a second time and remarked that the associated integral is the expectation of a $\mathcal{N}(0, \lambda)$ random variable. This is quite pertinent, since the transformation of Y into $\tilde{Y}$ involves a shift of its mean from zero to λ. The variance, however, is unaffected. This is easily verified by a third application of completing the square,

$$\text{var}\left(\tilde{Y}\right) = \mathbb{E}(\tilde{Y}^2) - \mathbb{E}(\tilde{(Y)})^2, \qquad (7.115)$$

$$= \int_{-\infty}^{\infty} y^2 f_{\tilde{Y}}(y) dy - \lambda^2,$$

$$= \underbrace{\int_{-\infty}^{\infty} y^2 \frac{1}{\sqrt{2\pi}} e^{\frac{-(y-\lambda)^2}{2}} dy}_{1+\lambda^2} - \lambda^2,$$

$$= 1.$$

In short, therefore, we have that, as an approximation at least, $\tilde{Y} \sim (\lambda, 1)$.

We now introduce the cumulant generating function—which has also played a headlining role in the previous discussion—as

$$K_Y(\lambda) = \ln M_Y(\lambda), \qquad (7.116)$$

$$= \ln\left(e^{\frac{\lambda^2}{2}}\right),$$

$$= \frac{\lambda^2}{2}.$$

If we compute the first two derivatives of $K_Y(\lambda)$, we make a remarkable observation. The first derivative is,

$$K'_Y(\lambda) = \lambda = \mathbb{E}\left(\tilde{Y}\right), \qquad (7.117)$$

whereas the second derivative is,

$$K''_Y(\lambda) = 1 = \text{var}\left(\tilde{Y}\right). \qquad (7.118)$$

In other words, we can restate our approximated distributional assumption as $\tilde{Y} \sim \left(K'_Y(\lambda), K''_Y(\lambda)\right)$.

This might not seem immediately useful, but it helps us get to the essence of the transformation in equation 7.111. We have the ability to determine the mean of the transformed distribution through our choice of λ. Imagine, for example, that we select $\tilde{\lambda}_y$ such that

$$\underbrace{\mathbb{E}(\tilde{Y})}_{K'_Y(\lambda)} = y. \qquad (7.119)$$

7.8 An Interesting Connection

The choice of $\tilde{\lambda}_y$ implies that we have shifted the middle of the new transformed distribution, $f_{\tilde{Y}}(y)$, to y. We write $\tilde{\lambda}_y$ to explicitly denote its dependence on the specific choice of y. A transformation of this nature is quite helpful, when we seek to examine tail probabilities, because we can move the centre of the transformed distribution to our area of interest.

Moreover, since equation 7.111 incorporates both the original and transformed distribution, we can *invert* the expression to recover $f_Y(y)$. A bit of manipulation with the previous definitions yields

$$f_{\tilde{Y}}(y) = \frac{e^{\lambda y} f_Y(y)}{M_Y(\lambda)}, \tag{7.120}$$

$$f_Y(y) = \frac{M_Y(\lambda) f_{\tilde{Y}}(y)}{e^{\lambda y}},$$

$$= M_Y(\lambda) e^{-\lambda y} f_{\tilde{Y}}(y),$$

$$= e^{\ln M_Y(\lambda)} e^{-\lambda y} f_{\tilde{Y}}(y),$$

$$= e^{K_Y(\lambda) - \lambda y} f_{\tilde{Y}}(y).$$

Now, since we have that $\tilde{Y} \sim \left(K'_Y(\lambda), K''_Y(\lambda) \right)$, we can re-write this as,

$$f_{\tilde{Y}}(y) = e^{K_Y(\lambda) - \lambda y} \frac{1}{\sqrt{2\pi \operatorname{var}(\tilde{Y})}} e^{-\frac{(y - \mathbb{E}(\tilde{Y}))^2}{2}}, \tag{7.121}$$

$$= e^{K_Y(\lambda) - \lambda y} \frac{1}{\sqrt{2\pi K''_Y(\lambda)}} e^{-\frac{(y - K'_Y(\lambda))^2}{2}}.$$

If we proceed to select $\tilde{\lambda}_y$ so that it satisfies equation 7.119, then we may further simplify our density approximation as,

$$f_{\tilde{Y}}(y) = e^{K_Y(\tilde{\lambda}_y) - \tilde{\lambda}_y y} \frac{1}{\sqrt{2\pi K''_Y(\tilde{\lambda}_y)}} e^{-\overbrace{\frac{(y - K'_Y(\tilde{\lambda}_y))^2}{2}}^{=0}}, \tag{7.122}$$

$$= \frac{e^{K_Y(\tilde{\lambda}_y) - \tilde{\lambda}_y y}}{\sqrt{2\pi K''_Y(\tilde{\lambda}_y)}},$$

$$= e^{K_Y(\tilde{\lambda}_y) - \tilde{\lambda}_y y} \mathfrak{J}_0 \left(K''_Y(\tilde{\lambda}_y), \tilde{\lambda}_y \right).$$

As we've already seen, $\tilde{\lambda}_y$ is none other than our saddlepoint and this is the saddlepoint approximation of the density of Y. The transformation used in equation 7.111, permitting us to arrive at this point, is referred to as the Esscher transform or, more colloquially, the exponential twist. The driving idea is to move the centre of the transformed distribution to our location of interest. In this case, an arbitrarily selected outcome y; in a practical setting, y will be informed by the structure of one's problem.

While it turns out that it is more efficient to develop the saddlepoint approximation through the use of the Laplace transform technique, the Esscher transform is a useful approach. Indeed, it forms the backbone of the importance-sampling variance-reduction technique used in Chap. 8 to reduce the computational expense associated with Monte-Carlo estimation of our risk and risk-decomposition metrics. Although it has required some effort, we feel that it is worthwhile to establish a conceptual link between these two important areas of quantitative credit-risk study.

7.9 Final Thoughts

The saddlepoint method is a viable, accurate, and robust approach to perform risk attribution in the independent-default, one-dimensional integrated, and even in the two-dimensional integrated setting. Although it is rather heavy to perform the necessary two-dimensional integration required for the class of normal-variance mixture models, one is probably still better off using the saddlepoint approximation. As we move to the three- and four-dimensional cases, however, the balance tips strongly in favour of Monte-Carlo techniques. If moving from one- to two-dimensional integration adds a few orders of magnitude to one's computational burden, it is clear that higher dimensions are simply not viable.

Monte-Carlo methods also offer a number of inefficiencies. The final link between the Esscher transform and the saddlepoint approximation is critical. It offers us—as cleverly developed in a series of papers Glasserman (2004b, 2006); Glasserman and Li (2005)—a set of approaches to reduce the variance in our Monte-Carlo estimators. This permits us to simultaneously moderate the computational expense and improve the accuracy of our simulation estimates. In doing so, it opens a number of additional possibilities for the performance of these important computations.

References

Antonov, A., Mechkov, S., & Misirpashaev, T. (2005). Analytical techniques for synthetic CDOs and credit default risk measures. NumeriX LLC Working Paper.

Artzner, P., Delbaen, F., Eber, J.-M., & Heath, D. (2001). Coherent measures of risk. *Mathematical Finance, 9*(3), 203–228.

References

Billingsley, P. (1995). *Probability and measure* (3rd edn.). Third Avenue, New York, NY: Wiley.
Bluhm, C., Overbeck, L., & Wagner, C. (2003). *An introduction to credit risk modelling* (1st edn.). Boca Raton: Chapman & Hall, CRC Press.
Bolder, D. J. (2015). *Fixed income portfolio analytics: A practical guide to implementing, monitoring and understanding fixed-income portfolios*. Heidelberg, Germany: Springer.
Broda, S. A., & Paolella, M. S. (2010). *Saddlepoint approximation of expected shortfall for transformed means*. Amsterdam School of Economics.
Casella, G., & Berger, R. L. (1990). *Statistical inference*. Belmont, CA: Duxbury Press.
Daniels, H. E. (1954). Saddlepoint approximations in statistics. *The Annals of Mathematical Statistics, 25*(4), 631–650.
Daniels, H. E. (1987). Tail probability approximations. *International Statistical Review, 55*(1), 37–48.
DasGupta, A. (2008). *Asymptotic theory of statistics and probability*. New York: Springer-Verlag.
Durrett, R. (1996). *Probability: Theory and examples* (2nd edn.). Belmont, CA: Duxbury Press.
Gil-Pelaez, J. (1951). Note on the inversion theorem. *Biometrika, 38*(3/4), 481–482.
Glasserman, P. (2004b). *Tail approximations for portfolio credit risk*. Columbia University.
Glasserman, P. (2006). Measuring marginal risk contributions in credit portfolios. Risk Measurement Research Program of the FDIC Center for Financial Research.
Glasserman, P., & Li, J. (2005). Importance sampling for portfolio credit risk. *Management Science, 51*(11), 1643–1656.
Gourieroux, C., Laurent, J., & Scaillet, O. (2000). Sensitivity analysis of values at risk. *Journal of Empirical Finance, 7*(3–4), 225–245.
Goutis, C., & Casella, G. (1995). Explaining the saddlepoint approximation. Universidad Carlos III de Madrid: Working Paper 95–63.
Hallerbach, W. G. (2002). *Decomposing portfolio value-at-risk: A general analysis*. Erasmus University Rotterdam.
Huang, X., Oosterlee, C. W., & Mesters, M. (2007). Computation of var and var contribution in the vasicek portfolio credit loss model: A comparative study. *Journal of Credit Risk, 3*(3), 75–96.
Huang, X., Oosterlee, C. W., & van der Weide, J. (2006). Higher-order saddlepoint approximations in the vasicek portfolio credit loss model. Delft University of Technology, Department of Applied Mathematical Analysis.
Huzurbazar, S. (1999). Practical saddlepoint approximations. *The American Statistician, 53*(3), 225–232.
Loomis, L. H., & Sternberg, S. (2014). *Advanced calculus*. Hackensack, NJ: World Scientific Publishing, Inc.
Lugannani, R., & Rice, S. (1980). Saddlepoint approximation for the distribution of the sum of independent random variables. *Advanced Applied Probability, 12*(2), 475–490.
Martin, R. J. (2004). *Credit portfolio modeling handbook*. Credit Suisse First Boston.
Martin, R. J. (2011). Saddlepoint methods in portfolio theory. In A. Rennie, & A. Lipton (Eds.), *Handbook of credit derivatives*. Oxford: Oxford University Press.
Martin, R. J., & Thompson, K. (2001a). Taking to the saddle. *Risk*, 91–94.
Martin, R. J., & Thompson, K. (2001b). VaR: Who contributes and how much? *Risk*, 91–94.
Muromachi, Y. (2007). *Decomposing the total risk of a portfolio in to the contribution of individual assets*. Tokyo Metropolitan University.
Rau-Bredow, H. (2003). *Derivatives of value at risk and expected shortfall*. University of Würzburg.
Rau-Bredow, H. (2004). Value-at-risk, expected shortfall, and marginal risk contribution. In G. Szegö (Ed.), *Risk measures for the 21st century*. John Wiley & Sons.
Tasche, D. (2000a). *Conditional expectation as quantile derivative*. Zentrum Mathematik, Technische Universität München.
Tasche, D. (2000b). *Risk contributions and performance measurement*. Zentrum Mathematik, Technische Universität München.
Wilde, T. (1997). *CreditRisk+: A credit risk management framework*. Credit Suisse First Boston.

Chapter 8
Monte Carlo Methods

An algorithm must be seen to be believed.

(Donald Knuth)

Since the beginning of this work, we have made generous use of the Monte-Carlo method. Indeed, with the notable exception of the Basel IRB approach—addressed in Chap. 6 and BIS (2001, 2004, 2005)—it has been used to estimate every one of the models considered. The independent-default, the range of mixture, the various threshold, and the Merton (1974) models have all relied, to one extent or another, upon stochastic simulation for their solution. Clearly, therefore, it is a powerful and flexible tool. It is, however, like most tools, an approach that requires some care and caution. Incorrectly implemented, Monte-Carlo methods can generate misleading or incorrect results, fail to converge, or consume excessive amounts of one's computational resources.

Care and caution, when referring to a quantitative tool, are equivalent to understanding. The better the analyst understands a method or technique, the lower the probability of its misapplication. This makes complete sense. Learning about the details of a particular approach reveals its strengths and weaknesses. Using a mathematical method as a black box, conversely, makes it easier to walk directly into pitfalls that could be avoided with a bit more background. This chapter, therefore, is dedicated to reviewing and exploring the foundations of the Monte-Carlo method.

The experienced practitioner may already have been exposed to, or at least be aware of, much of the following discussion. The objective, however, is to structure the discussion so that it has some value for all readers. Indeed, we start from the very basics of Monte-Carlo integration to motivate the key idea behind the method. We then consider some of the underlying theory and examine the important role of standard errors and confidence bounds. These latter quantities form essential communication and model-diagnostic devices for an implementation of the Monte-Carlo method. We then proceed to address some advanced tools available to the quantitative analyst. All of these aspects are investigated in the context of our usual 100-obligor example. This discussion raises and wrestles with the central weakness of this approach: computational expense. In other words, Monte-Carlo simulation is *slow*.

This core feature cannot be changed, but it can be mitigated. There are a number of tricks—referred to as variance-reduction techniques—that can be usefully employed to improve the speed of our estimators. This is not, however, a free lunch. The cost takes the form of additional mathematical complexity. We will cover one popular and useful variance-reduction approach— suggested by Glasserman and Li (2005) and Glasserman (2006)—related to the saddlepoint approximation considered in Chap. 7. Hopefully, the examination of a specific variance-reduction approach will create a meaningful link to developments in previous chapters and prove thought-provoking and useful even for experienced practitioners.

8.1 Brains or Brawn?

In the previous chapter, we explored a complicated, but immensely useful, risk-measure estimation technique. While powerful, its development is quite complicated. Given the cumulant-generating function, K_L, and its derivatives we can approximate the default-loss density as,

$$f_L(\ell) \approx \frac{e^{K_L(\tilde{t}_\ell) - \tilde{t}_\ell \ell}}{\sqrt{2\pi K_L''(\tilde{t}_\ell)}}, \qquad (8.1)$$

where $\tilde{t}_\ell$, referred to as the saddlepoint, solves the equation,

$$K_L(t_\ell) - \ell = 0. \qquad (8.2)$$

With some additional effort, the saddlepoint method equation 8.1 can be extended to provide efficient and accurate approximations of the default-loss density, tail probabilities, shortfall integrals and for determining obligor contributions to VaR and expected shortfall.

This raises a legitimate question: why do we need, in the context of the credit-risk problem, to employ the Monte-Carlo method at all? Stochastic simulation is a brute-force technique where we essentially replace mathematical sophistication with raw computer power. All of the applications in the previous chapters have used at least 10 million iterations to derive our underlying risk measures. This is a tremendous amount of effort.

The short answer to the previous question is: in many cases, we do not need the Monte-Carlo method. In the one-factor setting—even with more complex normal-variance mixture models—one might reasonably rely upon the saddlepoint approximation for all of one's estimation needs. Under certain simplifying assumptions, we've also seen other analytic approximations in the previous chapters. The CreditRisk+ model, for example, is traditionally estimated using the so-called Panjer

8.2   A Silly, But Informative Problem   431

(1981) recursion.[1] To the extent that these methods are available, well understood, and apply to one's specific problem, they should be employed.

The Monte-Carlo method is thus a kind of last resort. That is, when other more efficient and clever approaches are not available, then one grudgingly turns to this technique. The long answer to the previously posed question, however, is that we do absolutely need it. Realistic models are often complex models. One-factor models provide useful intuition and can be helpful in many applications, but many organizations require multivariate structures to incorporate regional, industrial, or sectoral effects. The saddlepoint method—and indeed most other analytic approximations—break down in this setting. If we are going to use Monte Carlo in more complex situations, then it is essential to gain experience and confidence with it in the univariate case. This is the ultimate justification for its use among one-factor models presented in the previous chapters.

After making these rather disparaging comments about the Monte-Carlo method, we feel obliged to defend it. The next section, therefore, uses a very simple example to tease out the strengths and weaknesses of stochastic simulation. While elementary, this exercise helps underscore our argumentation.

8.2   A Silly, But Informative Problem

We're going to do something a bit frivolous, which involves a return to secondary-school calculus. The plan is to concoct a perfectly ridiculous, but completely legitimate, mathematical function. We will then try to integrate this function over an interval on the real line using a variety of different approaches. In fact, the mapping is a very simple one-dimensional function,

$$f : \mathbb{R} \to \mathbb{R}. \tag{8.3}$$

where its form is,

$$f(x) = 7\sin(5x) + 5x^3 + 3e^{\frac{x}{2}}. \tag{8.4}$$

f is thus a linear combination of a polynomial and two transcendental functions. It is neither complicated nor completely trivial. It has, of course, no particular significance, but was rather invented to perform a simple experiment. Figure 8.1 illustrates the behaviour of our function in equation 8.4 over the unit interval. The sine function gives it an oscillatory form.

Although it might seem like overkill, Algorithm 8.1 summarizes the one-line sub-routine required to implement equation 8.4. This will, believe it or not, actually prove useful in the following discussion.

---

[1] Wilde (1997) and Gundlach and Lehrbass (2004) provide more detail on this approach.

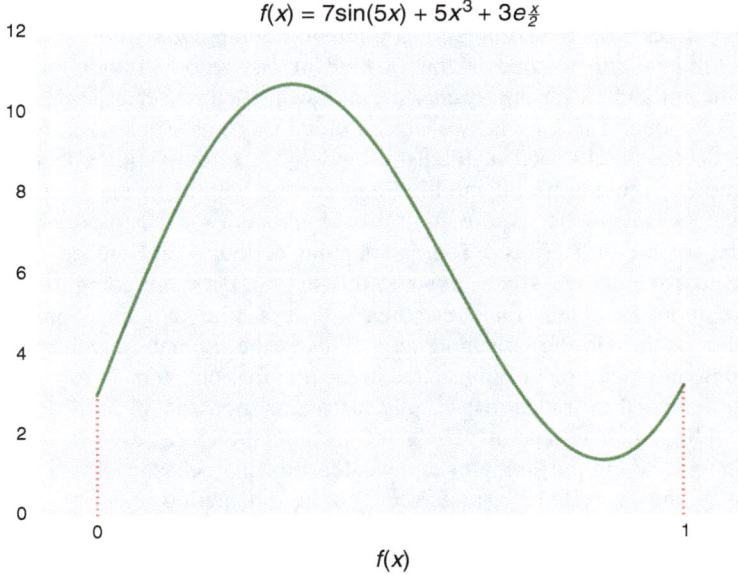

Fig. 8.1 *A silly function*: The underlying figure summarizes the behaviour of the function summarized in equation 8.4 over the unit interval.

Algorithm 8.1 Evaluating our silly function

```
def myFunction(x):
    return 7*np.sin(5*x) + 5*np.power(x,3) + 3*np.exp(x/2)
```

We have set ourselves a relatively trivial task. We want to integrate this function on the unit interval, [0, 1]. There is nothing particularly specific about the range of integration; it can be adjusted to any desired range without loss of generality, but it simplifies our treatment. In short, therefore, we seek to solve the following definite integral:

$$\int_0^1 f(x)dx = \int_0^1 \left(7\sin(5x) + 5x^3 + 3e^{\frac{x}{2}}\right) dx, \tag{8.5}$$

$$= \left[-\frac{7\cos(5x)}{5} + \frac{5x^4}{4} + 6e^{\frac{x}{2}}\right]_0^1,$$

$$= \left(-\frac{7\cos(5)}{5} + \frac{5}{4} + 6e^{\frac{1}{2}}\right) - \left(-\frac{7}{5} + 6\right),$$

$$= 6.14520.$$

8.2 A Silly, But Informative Problem

This elementary calculus problem is, of course, readily solved by any secondary-school student since our definite integral easily admits an analytic solution. We may compute its value to virtually any desired precision. In particular, the area under this curve on [0, 1] is about 6.15 units.

This is hardly an interesting question. It gets worse, because we now come to the silly part. We are going to try to solve this problem numerically. In a one-dimensional real-valued setting, a definite integral is, of course, geometrically equivalent to computing the area under the curve. All numerical integration techniques, therefore, essentially try to break down the entire region into sub-intervals and approximate each sub-region with a simpler function such as a rectangle or a trapezoid. The area is then approximated by the sum of the area of these sub-intervals.

Let's begin with the naive approach. Our plan is to break the space under the curve in Fig. 8.1 into a collection of non-overlapping rectangles; the area of each sub-rectangle is easily computed. Since the rectangle does not perfectly describe the area of the sub-interval, there will be some error. Figure 8.2 provides a schematic illustration of this approach. Interesting questions arise as to whether one should set the height of the rectangle at the function's value at the beginning or end of the interval. Naturally, as the number of rectangles increases and their size gets correspondingly smaller, the accuracy of this estimate will increase.

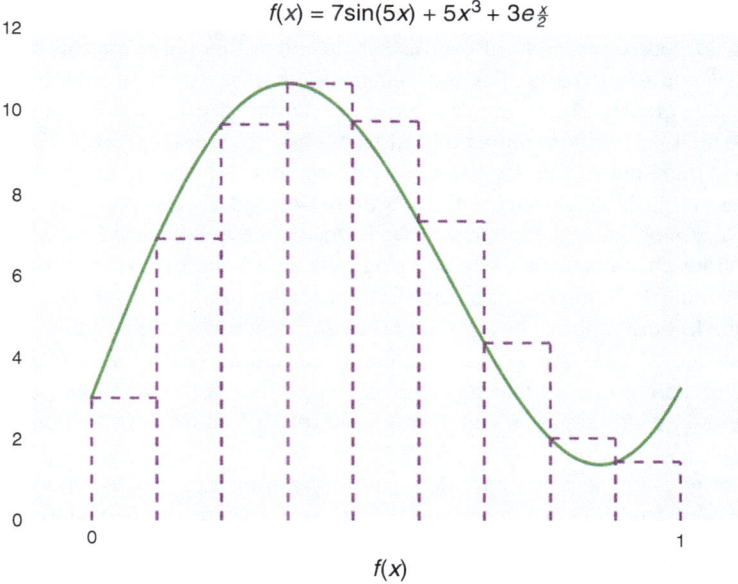

Fig. 8.2 *A simple, but effective, approach*: The underlying figure summarizes a naive, but effective, approach to numerical integration. The interval is partitioned into a number of non-overlapping sub-intervals; the area of each sub-interval is approximated by a rectangle.

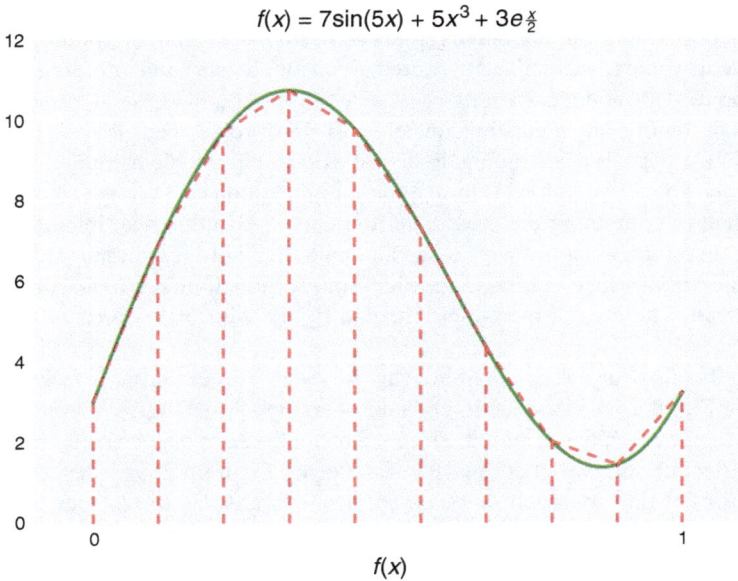

Fig. 8.3 *A clever technique*: The underlying figure summarizes the geometric logic behind the Gaussian quadrature technique for numerical integration of our simple function.

It is, of course, possible to do much better than this naive approach. A better method is to use both rectangles and triangles. This case, summarized schematically in Fig. 8.3, creates clever trapezoid approximation of each sub-interval, thereby minimizing the required number of partitions and the associated error. This method is termed quadrature. See, for example, Ralston and Rabinowitz (1978, Chapter 4) and Press et al. (1992, Chapter 4) for a more detailed discussion of its derivation, technical conditions and application. We have, in fact, already used it extensively in the previous chapters through Python scipy's quad and nquad routines.

Algorithm 8.2 summarizes the Python code used to estimate the naive and quadrature approaches to numerical integration. The first function, naiveNumericalIntegration, creates a simple grid of K equally spaced rectangles and uses myFunction from Algorithm 8.1 to compute their area. The second quadrature approach borrows the quad routine from Python's scipy library.

Although elementary, we have done this for a reason. Let's try to understand what these two approaches have in common. To be very specific, both techniques

1. construct a grid—either evenly spaced or not—to partition the space; and
2. allow the number of sub-intervals to become very small—that is, they use a limiting argument.

8.2 A Silly, But Informative Problem

Algorithm 8.2 Two numerical integration options

```
def naiveNumericalIntegration(K,a,b):
    myGrid = np.linspace(a,b,K)
    myIntegral = 0
    for n in range(0,K-1):
        dx = myGrid[n+1]-myGrid[n]
        myIntegral += myFunction(myGrid[n+1])*dx
    return myIntegral
# Quadrature approach
import scipy.integrate as nInt
myIntegral,err = nInt.quad(myFunction,0,1)
```

Can either of these assumptions be avoided when we numerically integrate a function? The limiting argument—if one wishes convergence to the true answer—cannot really be avoided. The quadrature approach is clever in using function information to determine and optimize its choice of grid points, but it is still ultimately based on a type of limiting argument. We might have a bit more luck in relaxing the first aspect. What if instead of a creating a grid in a *deterministic* way, we opted to partition the space in an alternative fashion? The most obvious choice would be to try to randomly select the points. This might seem a bit odd—and ultimately even a waste of time—but let's see where it takes us.

Random generation of our grid is a bit vague; we need to specify exactly how, or under what assumptions, we perform this task. The simplest possible approach would be to uniformly select points along the unit interval. Figure 8.4 performs this task by drawing 100 points in the unit interval from the standard uniform distribution. The range for our definite integral is not as cleanly defined, but it does appear to cover much of the range of integration.

One might be wondering what precisely is the point of this exercise. In reality, this turns out to be a fairly clever, and completely legitimate, approach to numerically integrate almost any function. We can demonstrate this with a bit of basic probability. Assume that we have a random variable $X \in \mathcal{U}[0, 1]$. It is well known that the density function of X is straightforwardly written as $g_X(x) = 1$. That is, each observation in the unit interval is equally likely. Imagine further that we have some relatively well-behaved function, f. The quantity $f(X)$ is itself a random variable. We can deduce its density and moments, but imagine that we seek its expected value. In this case, from first principles, we have

$$\mathbb{E}\left(f(X)\right) = \int_0^1 f(x) \underbrace{g_X(x)}_{=1} dx, \qquad (8.6)$$

$$= \int_0^1 f(x) dx.$$

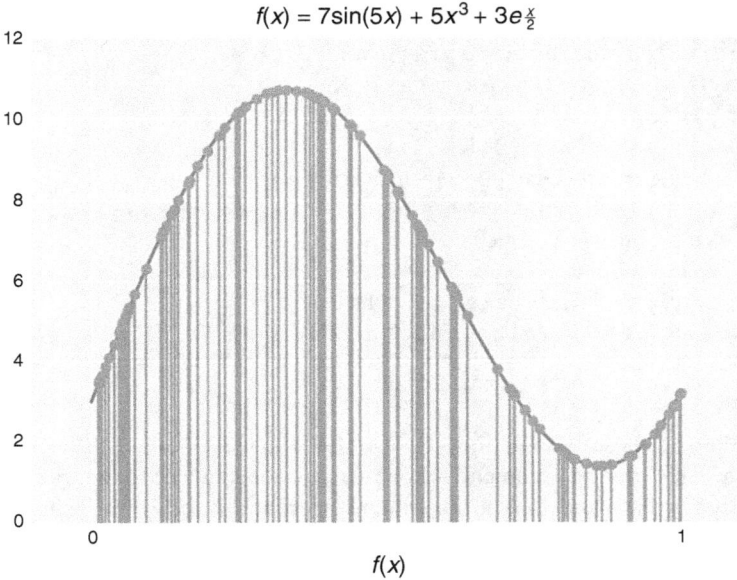

Fig. 8.4 *A randomized grid*: The underlying figure randomly draws 100 points for our silly function within the unit interval—our range of integration—from the standard uniform distribution. Although somewhat uneven, it does appear to cover much of the range.

This is quite elegant. Moving backwards, we may transform a deterministic problem into a stochastic one. X need not, in fact, be a random variable at all. By assuming that it is, however, we arrive at a method to evaluate its integral.

We thus need only approximate the expectation in equation 8.6 to resolve our integral. To accomplish this, we merely select a sequence of uniform-random variates,

$$U_m \sim \mathcal{U}[0, 1], \tag{8.7}$$

for $m = 1, \ldots, M$. The expectation from equation 8.6 is readily approximated as,

$$\mathbb{E}(f(X)) \equiv \int_0^1 f(x)dx, \tag{8.8}$$

$$\approx \frac{1}{M} \sum_{m=1}^{M} f(U_m).$$

Equality occurs in the limit as M becomes infinitely large. That is,

$$\int_0^1 f(x)dx = \lim_{M \to \infty} \frac{1}{M} \sum_{m=1}^{M} f(U_m). \tag{8.9}$$

8.2 A Silly, But Informative Problem

Algorithm 8.3 Monte-Carlo integration

```
def mcIntegration(M):
    U = np.random.uniform(0,1,M)
    return np.mean(vr.myFunction(U))
```

The efficiency and accuracy of this estimator, of course, depends entirely on how quickly it converges to the true integral as we increase the value of M.

The Python code in Algorithm 8.3, used to actually estimate the integral of our function in equation 8.4 on the unit interval follows directly from equation 8.8. As we've seen in previous applications, it is easy to follow and does not require a tremendous amount of code. The hard work is performed by numpy's uniform random number generator.

This may seem like a mathematical parlour trick, but moving back and forth between deterministic and random settings has a long tradition in mathematics. In Chap. 7, we identified the kernel of the Gaussian density to approximate an otherwise thorny integral. Recall $f(x) = \ln g(x)$, we approximated the integral as,

$$f(x) \approx e^{g(\tilde{x})} \exp\left(-\frac{(x-\tilde{x})^2}{2\left(-\frac{1}{g''(\tilde{x})}\right)}\right), \tag{8.10}$$

$$\int f(x)dx \approx e^{g(\tilde{x})} \sqrt{\frac{-2\pi}{g''(\tilde{x})}}.$$

where $\tilde{x}$ solves $g'(x) = 0$. This is, if you recall, termed the Laplace approximation. In this case, we introduced a uniform density—which takes the value of unity—to evaluate an integral. It also shows up in more complex settings. The Feyman-Kac formula establishes, for example, a rather unexpected link between deterministic partial-differential equations and stochastic processes. It is used in mathematical finance to price complex contingent claims.[2]

Let us now actually compare these techniques to estimate the definite integral of our function in equation 8.2. We know the answer already, so there is relatively little suspense in terms of the final answer. Our objective is to understand the differences in speed and accuracy between our three main approaches: naive integration, quadrature, and the Monte-Carlo integral. The results—including the number of function evaluations, the integral estimate, the error, and the computational time—are summarized in Table 8.1.

---

[2]The interested reader is referred to Heunis (2011).

Table 8.1 *Integration results*: This table aids us to understand the differences in speed and accuracy between our *three* main approaches to numerically integrating equation 8.4: naive integration, quadrature, and the Monte-Carlo integral. The results include the number of function evaluations, the integral estimate, the error, and the computational time.

Estimation technique	Function evaluations	Estimated value	Error term	Time (microseconds)
Gaussian quadrature	21	6.145201	6.8e−14	314
Naive quadrature	100,000	6.145202	−1.2e−06	419,583
Monte-Carlo integration	100,000	6.147121	−1.9e−03	11,091
Monte-Carlo integration	1,000,000	6.142752	2.4e−03	113,351
Monte-Carlo integration	10,000,000	6.146728	−1.5e−03	1,121,457
Monte-Carlo integration	100,000,000	6.145075	1.3e−04	11,103,411
True value	0	6.145201	0.0e+00	92

The results are fascinating. Computation of the true, known value requires about 100 microseconds.[3] The quadrature method requires only about twice as long to arrive at an estimate with error roughly equivalent to machine accuracy. We should probably stop at this point. Numerical evaluation of a one-dimensional integral—at least in this case—seems terribly well addressed by Python's quad algorithm. It is hard to imagine we can do much better.

Indeed, our naive approach to integration requires about half of a second—several orders of magnitude of error slower than the quadrature approach—with error creeping into the sixth decimal place. To accomplish this, it required partitioning the unit interval into 100,000 equally spaced rectangles. It is clearly a sub-optimal approach. The Monte-Carlo integral, however, is dramatically worse. 100 million function evaluations—requiring an enormous 11 seconds of computational effort—are required to obtain roughly the same level of accuracy as the naive interpolation method. We have shown four different choices of M—increasing from 100,000 to 100 million—in an effort to demonstrate the slow, but steady convergence to the true value.

This may be a somewhat silly example, but we have learned a number of interesting things. Gaussian quadrature is both fantastically accurate and fast. The naive approach is not particularly great, but basically does the job. Naive integration, despite its shortcomings, is still significantly better than the Monte Carlo approach. Even with 100 million iterations, the Monte Carlo technique is far from convergence. Although we have seen that Monte-Carlo integration is based on a very clever—even elegant—mathematical trick, the bad news is that it does not appear to work very well.

The conclusion is uncontroversial and entirely consistent with our claim in the introduction: in a one-dimensional setting, there is *no* compelling reason to use Monte Carlo to approximate integrals. Grid-based methods are exceptionally

---

[3] One microsecond is one millionth of a second.

suited to the low-dimensional environment. This is clear in numerous areas of mathematical finance: lattice and grid-based techniques for evaluation of integrals and differential equations of many kinds are used to price many types of securities. This is an important lesson, but it is also useful to recall that we do not always find ourselves in low-dimensional settings. As the dimension gets to about three or more, for instance, these approaches begin to flounder. At this point, the brute-force Monte-Carlo method offers a legitimate alternative.

8.3 The Monte Carlo Method

The idea behind the Monte-Carlo method is apparently quite old. The idea of statistical sampling—see, for example, Buffon's needle—precedes the formal technique.[4] The idea, as clever as it might be, was not very practical without computers to do the heavy lifting. It first came to life during WWII as large-scale computing became a reality. It was used, by John von Neumann and others, to explore fission reactions in the Los Alamos project. Apparently, Nick Metropolis coined the name after a colleague's uncle who borrowed money from relatives to *go to Monte Carlo*. This flashy name to describe the notion of randomization—albeit in the context of a gambling strategy—apparently stuck. Figure 8.5 provides a picture of the eponymous location in Monaco.

The origins of the technique and its use for military applications are fascinating. The interested reader is referred, as a useful starting point, to Eckhardt (1987) and Metropolis (1987) for more on its historical background. It genesis aside, it has become an essential tool for engineers, statisticians, physicists, geneticists, and finance professionals. Indeed, it is a ubiquitous approach found in myriad applications in virtually all areas of science and computing.

What is the reason for its popularity when our previous example so clearly highlighted its shortcomings? The answer is simple: dimensionality. Numerical integration techniques operate under tight dimensional limits. In low dimensions, the regular grid used by numerical integration techniques is, as we've seen, a decided advantage. As we move to higher dimensions, however, it quickly becomes a problem. This general phenomenon was described by Bellman (1961) in the context of optimal control. He referred to the challenges of solving high-dimensional problems as "the curse of dimensionality." Specifically, Bellman (1961) was referring to the exponential increase in the number of points required to create

---

[4]In the late 1880, the Comte de Buffon offered an interesting probability problem that led to a surprising geometric result involving π. The renowned mathematician, Pierre-Simon Laplace later remarked that this problem could be used, employing the Monte-Carlo method, to experimentally approximate π. The interested reader is referred to Arnow (1994) and Badger (1994) for more detail.

Fig. 8.5 *The eponymous location*: This figure illustrates the Monte-Carlo Casino in the city-state of Monaco on the French Riviera. It feels a bit more like James Bond than mathematics.

a hypergrid. A simple example vividly illustrates this point. Consider the following sequence of *three* grids in increasing dimensionality:

1. a one-dimension integral with a 100-point grid requires only 100 sampling points;
2. a similar grid for a two-dimensional integral comprises 10,000 ($100 \times 100 = 100^2$) sampling points; and
3. the same treatment for a 10-dimensional integral leads to $100^{10} = 100,000,000,000,000,000,000$ sample points.

It is easy to see that even the two-dimensional setting is quite heavy. As we move to three dimensions or beyond, the number of function evaluations to evaluate even a modest grid often exceeds our computational capacity. Enter the Monte-Carlo method. Even though its overall performance is slow, the speed of convergence is independent of the dimension. This makes stochastic simulation particularly appealing for high-dimensional problems. Since many realistic scientific applications occur in high-dimensional settings, the Monte-Carlo method has become a standard scientific tool.

8.3.1 Monte Carlo in Finance

Stochastic simulation has been a tool in mathematical finance for roughly 40 years. The Monte-Carlo method was introduced into finance by Boyle (1977), a trained physicist who recognized the enormous benefits of this technique for complex asset-pricing problems. It has, over the years, become an indispensable tool in pricing and risk management. The reason is quite easy to see. Perhaps the most important problem in mathematical finance is the computation of the prices of various assets. When we examine this problem more closely, we will see how the Monte-Carlo

8.3 The Monte Carlo Method

approach is a natural fit. Imagine, for example, an underlying market-risk factor, X_t, defined on $(\Omega, \mathcal{F}, \mathbb{P})$. The price of a contingent claim, $\varphi(X_t)$, can, under assumption of lack of arbitrage and market completeness, be written as,

$$\varphi(X_t) = \mathbb{E}^{\mathbb{Q}}\left(e^{-r(T-t)}\varphi(X_T)\bigg| \mathcal{F}_t\right), \quad (8.11)$$

$$= e^{-r(T-t)} \int_{A \in \mathcal{F}_t} \varphi(X_T) d\mathbb{Q}.$$

where T is the tenor, r is the risk-free rate, and $\mathbb{Q}$ is the equivalent martingale measure induced through Girsanov's change-of-measure technique given a particular choice of numeraire asset.[5] The most common choice for the numeraire asset, perhaps, is the money market account. This gives rise to the so-called risk-neutral probability measure. Loosely speaking, therefore, the price of any security can be represented as its discounted, *risk-neutral* expectation. Expectations are, of course, integrals. Thus, in principle, every asset-pricing problem reduces to an integration problem. Many can be resolved using grid-based techniques such as those explored in the previous section. More complex, high-dimensional securities, however, are readily evaluated with the Monte-Carlo method.

Let's now introduce a bit of formalism into our discussion. Define the definite integral of an arbitrary function, f, as,

$$\lambda = \int_0^1 f(x)dx. \quad (8.12)$$

We can write our Monte-Carlo estimate of equation 8.12 as

$$\hat{\lambda}_M = \frac{1}{M} \sum_{m=1}^{M} f(u_m), \quad (8.13)$$

$$= \frac{\overbrace{f(u_1) + f(u_2) + \cdots + f(u_M)}^{S_M}}{M},$$

$$= \frac{S_M}{M}.$$

Each function evaluation, $f(u_m)$, is, by its very construction, an independent, identically distributed random variable. S_M is thus, by simple extension, the sum of M i.i.d. random variables.

Monte Carlo is, at its heart, the averaging of collections of independent random numbers. Such sequences, which form the foundation for a variety of flavours of the

---

[5] See Harrison and Kreps (1979) and Harrison and Pliska (1981) for the original references with respect to this fundamental concept. Duffie (1996) also offers an excellent alternative reference.

central-limit theorem, have been extensively studied and are very well understood.[6] We have convergence, we know its speed, and we can estimate the error. To be more precise, by the weak law of large numbers, for all $\epsilon > 0$, we have

$$\lim_{M \to \infty} \mathbb{P}\left(\left|\frac{S_M}{M} - \lambda\right| \geq \epsilon\right) = 0. \tag{8.14}$$

In other words, S_M converges to λM in probability. By the central-limit theorem, we may conclude that

$$\mathbb{P}\left(\frac{S_M - M\lambda}{\sqrt{M}}\right) \xrightarrow{d} \mathcal{N}(0, \sigma^2). \tag{8.15}$$

where

$$\sigma^2 = \mathbb{E}\left((f(U) - \lambda)^2\right), \tag{8.16}$$

is the population variance of the estimator. The normalized distance between our estimate and the true function converges, in distribution, to a Gaussian distribution. The error of the Monte-Carlo estimator goes to zero as $\sqrt{M}$ goes to infinity. This is termed $O\left(\frac{1}{\sqrt{M}}\right)$ convergence. To cut the error in half—to make this a bit more practical—one needs to increase the iterations by four times.

Let's return to the finance setting and see how this fits into our specific credit-risk environment. Consider again a random variable, X, defined on the probability space, $(\Omega, \mathcal{F}, \mathbb{Q})$. Further assume that X is absolutely continuous and has the density, $f_X(x)$. Given some measurable function, g, the expectation of $g(X)$ is defined as follows,

$$\mathbb{E}(g(X)) = \int_{-\infty}^{\infty} g(x) f_X(x) dx. \tag{8.17}$$

Evaluation of equation 8.17 may be straightforward, but can be analytically difficult or even impossible to solve. In this case, one need only use the density, $f_X(x)$, to generate M independent realizations of X. This leads to the set, $\{X_1, \ldots, X_M\}$. Then, we may estimate equation 8.17 as,

$$\widehat{\mathbb{E}(g(X))} = \frac{1}{M} \sum_{m=1}^{M} g(X_m). \tag{8.18}$$

---

[6]See Billingsley (1995) and Durrett (1996) for a rigorous and detailed discussion of the results associated with the central-limit theorem.

8.3 The Monte Carlo Method

As we've seen from equations 8.14 to 8.16, our solution converges to the true value as M tends to infinity.

Equations 8.17 and 8.18 are just a more general representation of our original problem. At first glance, it does not appear that this technique can be directly applied to our credit-risk problem. After all, we are not attempting to compute expectations, but trying to identify quantiles of the credit-default loss distribution. As we've seen many times, the default-loss expression is given as,

$$L = \sum_{n=1}^{N} c_n \mathbb{I}_{\mathcal{D}_n}, \tag{8.19}$$

where N denotes the total number of credit counterparties, c_n is the nth exposure, and $\mathbb{I}_{\mathcal{D}_n}$ is the default-indicator variable. Our interest is in the following probability,

$$\mathbb{P}(L \geq \ell_\alpha) = \mathbb{P}\left(\underbrace{\sum_{n=1}^{N} c_n \mathbb{I}_{\mathcal{D}_n}}_{\text{Equation 8.19}} \geq \ell_\alpha\right), \tag{8.20}$$

where ℓ_α is the loss associated with a given statistical level of confidence. This is an upper tail probability or, as it is more commonly referred to, a VaR computation.

There is, happily, a well-known solution. The probability of an event $\mathcal{A}$ can be transformed into an expectation if we recall that $\mathbb{P}(\mathcal{A}) = \mathbb{E}(\mathbb{I}_\mathcal{A})$. Equation 8.20 is thus, following this logic, readily transformed into the following expectation,

$$\begin{aligned} \mathbb{P}(L \geq \ell_\alpha) &= \mathbb{E}\left(\mathbb{I}_{\left\{\sum_{n=1}^{N} c_n \mathbb{I}_{\mathcal{D}_n} \geq \ell_\alpha\right\}}\right), \\ &= \int_{-\infty}^{\infty} \cdots \int_{-\infty}^{\infty} \mathbb{I}_{\left\{\sum_{n=1}^{N} c_n \mathbb{I}_{\mathcal{D}_n} \geq \ell_\alpha\right\}} f(\mathbb{I}_{\mathcal{D}_1}, \ldots, \mathbb{I}_{\mathcal{D}_N}) d\mathbb{I}_{\mathcal{D}_1} \cdots d\mathbb{I}_{\mathcal{D}_N}, \end{aligned} \tag{8.21}$$

where $f(\mathbb{I}_{\mathcal{D}_1}, \ldots, \mathbb{I}_{\mathcal{D}_N})$ denotes the joint-default density. Quite simply, the tail probability is a multidimensional integral, which depends on the joint-default distribution of the collection of credit counterparties in one's portfolio. This places our VaR computation into the same structure as described in equation 8.17. This implies that we may employ the estimator in equation 8.18.

The expected shortfall, conversely, is already an expectation so we have

$$\mathbb{E}\left(L \,\bigg|\, L \geq \ell_\alpha\right) = \frac{\mathbb{E}\left(L \mathbb{I}_{\{L \geq \ell_\alpha\}}\right)}{\mathbb{P}(L \geq \ell_\alpha)}, \tag{8.22}$$

$$= \underbrace{\frac{\int_{-\infty}^{\infty} \cdots \int_{-\infty}^{\infty} L \mathbb{I}_{\{L \geq \ell_\alpha\}} f(\mathbb{I}_{\mathcal{D}_1}, \ldots, \mathbb{I}_{\mathcal{D}_N}) d\mathbb{I}_{\mathcal{D}_1} \cdots d\mathbb{I}_{\mathcal{D}_N}}{\int_{-\infty}^{\infty} \cdots \int_{-\infty}^{\infty} \mathbb{I}_{\{L \geq \ell_\alpha\}} f(\mathbb{I}_{\mathcal{D}_1}, \ldots, \mathbb{I}_{\mathcal{D}_N}) d\mathbb{I}_{\mathcal{D}_1} \cdots d\mathbb{I}_{\mathcal{D}_N}}}_{\text{Equation 8.21}}.$$

The expected shortfall is, therefore, actually the ratio of two multidimensional integrals. We now know—officially, that is—that Monte-Carlo applies to our situation. The remaining challenge is that, since we are considering rather extreme outcomes for ℓ_α, stochastic simulation is both slow and computationally expensive.

8.3.2 Dealing with Slowness

Mathematics technicalities aside, the Monte-Carlo method is on sound theoretical ground; it is a legitimate approach. It has been extensively studied with many useful references; Fishman (1995) and Glasserman (2004a) are two excellent choices for technical details, underlying theory, and practical ideas. There is, however, one important drawback: convergence is slow.

Although slow, this method also, in multidimensional settings, of great importance. We rely heavily on this technique—we've already used it extensively in the previous chapters—but simultaneously wish to find ways to manage its slow convergence. Fortunately, there are, at least, *two* mitigating strategies that one can follow when using simulation techniques. These include:

1. the computation of standard errors and construction of confidence intervals for one's estimates to understand their behaviour and the lower-bound on the amount of computational effort required; and
2. the employment of so-called *variance-reduction* techniques to improve efficiency and, hopefully, the speed of one's estimates.

In the remaining discussion in this chapter, we will consider both strategies. The first is essential to understanding the inherent trade-off between speed and accuracy in one's specific problem. This does not reduce speed, but it ensures one is able to make intelligent and defensible decisions regarding one's choice of M. The second strategy involves cleverness; as usual, the cost is mathematical complexity.

8.4 Interval Estimation

Let us start our assessment of mitigating approaches to the slowness of Monte-Carlo estimates with a formal description of the inherent error and its uses. We begin our discussion with a *strong* and definitive statement:

> It is basically irresponsible to compute Monte Carlo estimates without also assessing, to the extent possible, the uncertainty associated with these values.

Despite the fact that, in general, Monte-Carlo methods have $O\left(\frac{1}{\sqrt{M}}\right)$ convergence, not all problems exhibit the same levels of practical estimation noise. Different settings exhibit varying levels of underlying uncertainty. This impacts convergence. This argues strongly for ongoing analysis of the behaviour of each individual problem. Often, however, Monte-Carlo methods are used without thorough understanding of a problem's convergence properties. This is unfortunate and unprofessional. The safe and sensible way to use this approach, therefore, is to consistently provide an assessment of the uncertainty embedded in one's estimates.

8.4.1 A Rough, But Workable Solution

Unfortunately, it turns out to be relatively involved to compute formal interval estimates for our Monte-Carlo approximations. Questions of bias and consistency arise and can complicate matters.[7] In our approach, however, we will consider rough approximations that are a logical corollary of the central-limit theorem. Let's consider the most general setting. Imagine that we seek to estimate the average of a sequence of S i.i.d. simulated random variables, $X_1, X_2, \ldots, X_S$ with true, but unknown, expected value μ. The sum of these random variables is denoted as,

$$X_S = \sum_{s=1}^{S} X_s, \tag{8.23}$$

while our mean estimate is simply,

$$\bar{X}_S = \frac{X_S}{S}, \tag{8.24}$$

which is the sample average. For our purposes, we can think of $\bar{X}_S$ as the mean Monte-Carlo approximation of a particular quantile of the default-loss distribution.

---

[7] See Fishman (1995, Chapter 1) and Glasserman (2004a, Chapter 1) for more technical details.

What is the uncertainty in this estimate? That is our current preoccupation. By the central-limit theorem, as we saw in equation 8.15, we have that

$$\mathbb{P}\left(\frac{X_S - S\mu}{\sqrt{S}}\right) \xrightarrow{d} \mathcal{N}(0, \sigma_X^2). \tag{8.25}$$

where,

$$\sigma_X^2 = \mathbb{E}\left((X - \mu)^2\right), \tag{8.26}$$

$$= \lim_{S \to \infty} \sum_{s=1}^{S} \frac{(X_s - \mu)^2}{S - 1}.$$

While technically correct, this is practically rather unhelpful, since its computation requires the unobserved and unknown value of the population parameter, μ. To arrive at a workable solution, we need to replace it with the sample mean, $\bar{X}_S$, from equation 8.24. The standard error, or sample standard deviation is thus,

$$\hat{\sigma}_S = \sqrt{\sum_{s=1}^{S} \frac{(\hat{X}_s - \bar{X}_S)^2}{S - 1}}. \tag{8.27}$$

With a sample mean and variance, we can formally assess the uncertainty of our estimate. It is, however, for an arbitrary choice of S, no longer the case that our estimator follows a Gaussian distribution. In practice, our estimate has the following distribution

$$\hat{\mu} \sim \mathcal{T}_{S-1}\left(\bar{X}_S, \hat{\sigma}_S^2\right). \tag{8.28}$$

In other words, it follows a univariate t-distribution with $S - 1$ degrees of freedom. The t-distribution arises because we employ the sample variance rather than the true, unknown, variance.[8] The corresponding two-sided γ confidence interval is thus merely,

$$\bar{X}_S \pm \mathcal{T}_{S-1}^{-1}\left(\frac{1 - \gamma}{2}\right) \frac{\hat{\sigma}_S}{\sqrt{S - 1}}. \tag{8.29}$$

As a practical matter, for $S > 30$ or so, there is very little difference between $\mathcal{T}_{S-1}^{-1}(x)$ and $\Phi^{-1}(x)$, where $\Phi^{-1}(\cdot)$ denotes the inverse Gaussian distribution, or quantile, function. The use of the t-distribution might, therefore, be considered a bit

---

[8]See Box (1981, 1987) for the interesting history of this distribution's inventor, William Sealy Gosset. We will return to these questions of statistical inference in the following two chapters.

8.4 Interval Estimation

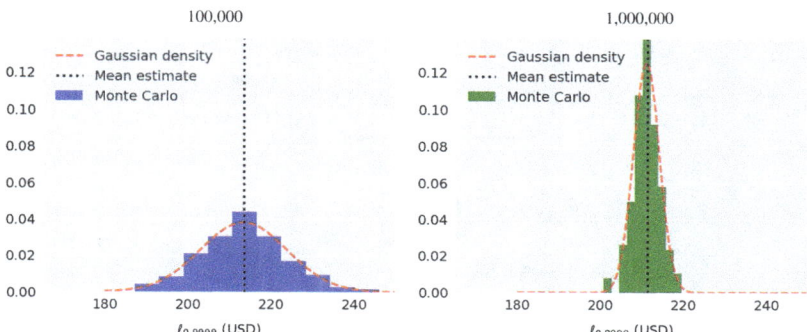

Fig. 8.6 *The role of M*: This figure illustrates the histogram of 200 separate estimations of the 99.99th quantile VaR of our sample portfolio using a one-factor Gaussian threshold model. The first estimate uses 100,000 iterations for each estimate, whereas the second employs 1 million. In each case, the distribution of Monte Carlo estimates is overlaid with a Gaussian distribution using the sample mean and sample error—the fit is quite reasonable.

pedantic; it is nonetheless an attempt to add a bit more conservatism to small-sample confidence intervals.

Although this approach is not perfect, it is practically fairly reasonable. Figure 8.6 provides some useful motivation. It provides histograms of 200 separate estimations of the 99.99th quantile VaR of our sample portfolio using a one-factor Gaussian threshold model. The first estimate uses 100,000 iterations for each estimate, whereas the second employs 1 million. In each case, the empirical distribution of Monte Carlo estimates is overlaid with a Gaussian distribution using the sample mean and sample error. The fit, if not perfect, is quite reasonable and provides a sensible first-order approximation of the simulation uncertainty.

8.4.2 An Example of Convergence Analysis

It is useful to perform a more formal examination of the convergence properties of our Monte-Carlo estimators. Using our rough approximation, we can readily compute interval estimates for any risk-metric approximation. Let's first compute a confidence interval by hand. We will then examine a broader range of quantiles, iterations, and interval estimates.

Consider again the 99.99th quantile VaR estimate with one million iterations. In this case, unlike the illustration in Fig. 8.6, we employ a one-factor t-threshold model; the common ρ parameter is set to 0.20 and the degrees of freedom, ν, are fixed at 12. The analysis reveals that, for 100 repetitions of 1 million iterations, the mean value is $\bar{X}_{100} = \$361.8$ and the standard error is $\hat{\sigma}_{100} = \$6.3$. We thus have an estimate—computed with a total of 100 million iterations—and a rough assessment of its uncertainty. The two-sided 99% confidence interval is,

$$\bar{X}_{100} \pm \underbrace{\mathcal{T}_{99}^{-1}\left(1 - \frac{0.01}{2}\right)}_{2.62} \cdot \left(\frac{\hat{\sigma}_{100}}{\sqrt{100-1}}\right), \tag{8.30}$$

$$\$361.8 \pm 2.62 \cdot \left(\frac{\$6.3}{\sqrt{99}}\right),$$

$$\$361.8 \pm \$1.65,$$

$$[\$360.2 \; , \; \$363.5].$$

This suggests that, with a 99% level of confidence, the true value is practically between about $360 and $364. This is useful information about the relative accuracy of your estimate. The quantitative analyst may find this entirely appropriate for her desired application. Conversely, a higher degree of precision might be desired. In the latter case, it is necessary to increase the number of iterations (or repetitions).

Let us now proceed to examine this question in a more formal manner. We have performed 100 repetitions for a range of alternative iteration counts—ranging from 100,000 to 10 million—of the t-threshold model applied to our 100-obligor example. Figure 8.7 gives us a first look at the results. In particular, it provides two alternative, but complementary, perspectives on the convergence properties of our Monte-Carlo estimator. The left-hand graphic illustrates the reduction in standard error for a variety of levels of confidence as we increase the number of iterations. Two useful observations can be made. First, although the shapes of the curves are similar for the various quantiles, the uncertainty and computational effort are higher for the more extreme quantiles. This makes complete sense, since only a small

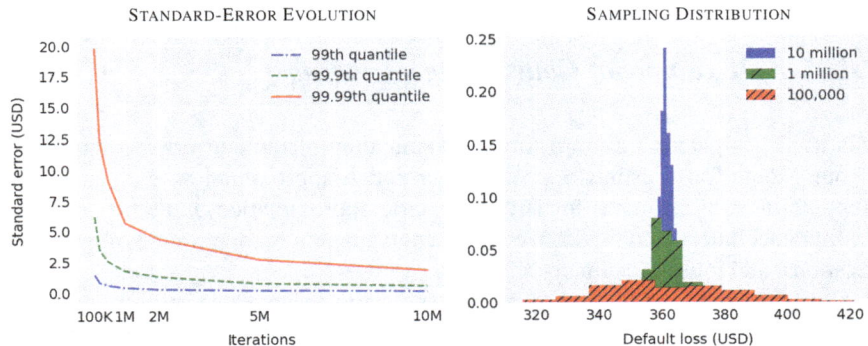

Fig. 8.7 *Understanding convergence*: The underlying figures provide two perspectives on the convergence properties of our Monte-Carlo estimator of the one-factor t-threshold model applied to our usual example. The left-hand graphic illustrates the reduction in standard error for a variety of levels of confidence as we increase the number of iterations, while the right-hand graphic highlights the sampling distributions for a fixed 99.99th quantile by various iteration settings.

8.4 Interval Estimation

Table 8.2 *Convergence analysis*: This table supplements the data presented in Fig. 8.7. For three alternative quantiles, the mean, standard error, and a 99% confidence interval are presented. Each of these estimates are considered in the context of 100,000, one million, and 10 million iterations.

Value	Statistic	Results (USD millions)		
		100,000	1,000,000	10,000,000
99.00	Mean estimate	$105.9	$105.9	$106.0
	Standard error	$1.3	$0.4	$0.1
	99% C.I.	[$105.6, $106.3]	[$105.8, $106.0]	[$106.0, $106.0]
99.90	Mean estimate	$223.3	$222.1	$222.3
	Standard error	$5.2	$2.0	$0.6
	99% C.I.	[$222.0, $224.7]	[$221.5, $222.6]	[$222.1, $222.4]
99.99	Mean estimate	$362.9	$361.8	$361.1
	Standard error	$18.6	$6.3	$1.9
	99% C.I.	[$358.0, $367.8]	[$360.2, $363.5]	[$360.6, $361.6]
Time	Average	0.6 s	5.6 s	55.8 s
	Total	0.9 min	9.3 min	93.0 min
Iterations	Total	10,000,000	100,000,000	1,000,000,000

fraction of the iterations actually inform these outcomes. The second point is that convergence is painfully slow. Even for 10 million iterations, the standard error remains appreciably different from zero.

The right-hand graphic of Fig. 8.7 highlights the sampling distributions for a fixed 99.99th quantile level associated with various iteration settings. The Monte-Carlo estimates are, when using 10 million iterations, much tighter than their 100,000 iteration equivalents. Nonetheless, as we saw in the right-hand graphic, a reasonable amount of uncertainty still remains. The acceptableness of this variation, of course, depends on one's needs and applications.

Table 8.2 provides some actual numerical values to accompany the graphic display in Fig. 8.7. For three alternative quantiles, the mean, standard error, and a 99% confidence interval are presented. Each of these estimates are considered in the context of 100,000, one million, and 10 million iterations.[9] The interplay between the iterations and the quantile is clear. One million iterations at the 99th quantile give a very tight estimate; indeed, the uncertainty for 100,000 iterations is already quite good for this quantile. 10 million iterations in this setting feel like overkill. At the 99.99th quantile, conversely, 100,000 iterations lead to a rather loose estimate of the VaR and something closer to 10 million feel more appropriate. This discussion also explains why we have extensively used 10 million iterations for all of our previous analysis, despite not providing any interval estimates. The intention was to keep the simulation noise to manageable levels until we could formally introduce and examine the notion of confidence intervals.

---

[9] We can easily identify, from the third row and fourth column, the computation from equation 8.30 in Table 8.2.

An additional, and ultimately quite important, element of Table 8.2 is an assessment of the computational time and effort. Each repetition of the Monte-Carlo estimation technique requires less than a second for 100,000 iterations, but almost a minute for 10 million. This is a roughly linear increase in computational time. A single repetition is thus not overly onerous. For 100 repetitions, the amount of work quickly adds up. At 10 million iterations, this leads to roughly 1.5 hours of computational effort and a total of one billion iterations! At 100,000 iterations, the entire affair requires less than a minute, but the accuracy is not particularly good at the higher quantiles. All told, however, the improvements in accuracy when moving from 100,000 to 10 million iterations are not terribly impressive when one considers that we have increased our effort by two orders of magnitude. In a nutshell, this is the fundamental challenge of the Monte-Carlo method. It also explains quite starkly the general interest in the area of variance-reduction techniques.

8.4.3 Taking Stock

Model risk is a real issue for the quantitative analyst. When estimating a model with Monte Carlo methods, there are, at least, *two* main sources of risk. These include:

1. the use of the wrong model; and
2. the estimate of our current model is wrong.

We mitigate the first issue by using a range of benchmark models; this explains why we examined so many different approaches in the first chapters of this work. The second can be dealt with by using convergence analysis similar to what we have just performed in the context of our usual example.

The computation of rough, first-order interval estimates for our Monte-Carlo approximations is conceptually quite straightforward. Given the general slowness of this method—with its $O\left(\frac{1}{\sqrt{M}}\right)$ convergence rate—we strongly argue that this should become a habit. It is certainly more computational effort, but the alternative is basically the equivalent to flying blind. In the remainder of this chapter, we will turn our attention to a methodology that might help to reduce this burden somewhat.

8.5 Variance-Reduction Techniques

Although Monte-Carlo methods are destined to be relatively slow, there are other ways to improve their efficiency. A collection of ideas and tricks—referred to as variance-reduction techniques—are used to enhance the precision of simulation-based estimates. As the name suggests, these approaches seek, quite simply, to reduce sample variance and thus improve the sharpness of one's estimate for a given amount of computational effort. Figure 8.8 provides a list of five alternative

8.5 Variance-Reduction Techniques

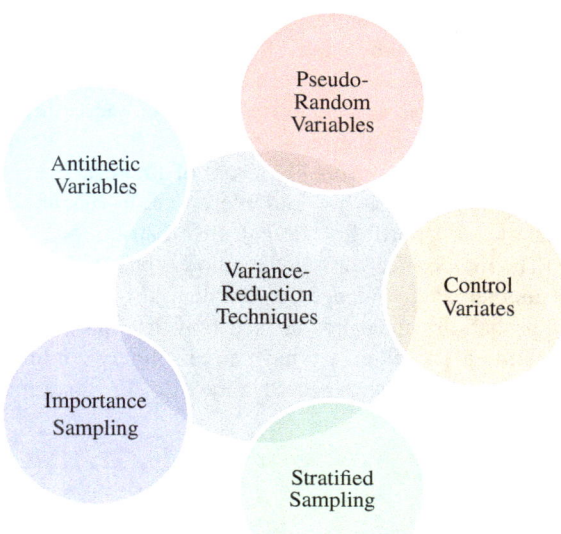

Fig. 8.8 *Variance-reduction techniques*: This figure illustrates schematically a range of alternative variance-reduction techniques. Not all methods perform equally well for every problem. The greatest improvements generally arise from exploiting knowledge of one's specific problem.

categories of variance-reduction approaches. All of them are potentially useful. McLeish (2005) is an excellent reference for more detail on the specifics of these alternatives.

There is, of course, one important catch. Variance-reduction techniques are often quite complicated. Usually, it is necessary to perform reasonably complicated mathematical and statistical manipulations to improve the situation. Even worse, while many approaches exist, the success of a given technique often depends on the specifics of one's problem. There are, of course, some generic approaches that can help in virtually all situations, but the greatest improvements generally arise from exploiting knowledge of one's specific problem. Our situation is no exception. We will, in the following pages, make use of the importance-sampling technique and rely heavily on the work from Glasserman (2004a, 2006); Glasserman et al. (1999); Glasserman and Li (2005), Brereton et al. (2012), Kang and Shahabuddin (2005) and McNeil et al. (2015) in this area.

8.5.1 Introducing Importance Sampling

Importance sampling, as one can deduce from its name, is about drawing random realizations from the most important part of one's distribution. What important means, of course, depends entirely on one's specific problem. If one seeks the centre of a distribution, then standard techniques are already essentially performing a type of importance sampling. If one is interested in events quite far from the distribution's centre, however, then this can be a useful approach. It is used extensively in mathematical finance to value far out-of-the-money contingent claims whose exercise is highly unlikely. Standard Monte-Carlo will, in these cases, generate a large number of relatively uninteresting sample paths containing little information about the pay-off function. This technique permits a greater focus on the, relatively less probable, event of interest. It is not particularly difficult to see how such an approach would interest credit-risk analysts seeking to describe the far quantiles of a portfolio's default-loss distribution.

The importance-sampling method is based on a clever approach used extensively in mathematical finance for a rather different application involving the change of probability measure. To see how it works, let's return to the example from equation 8.18, where we have an absolutely continuous random variable, X, with density, $f_X(x)$. The expected value of $g(X)$, for some measurable function, g is,

$$\mathbb{E}(g(X)) = \int_{-\infty}^{\infty} g(x) f_X(x) dx. \tag{8.31}$$

and our Monte-Carlo estimator is merely,

$$\widehat{\mathbb{E}(g(X))} = \frac{1}{M} \sum_{m=1}^{M} g(X_m), \tag{8.32}$$

where each independent X_m is randomly drawn from the distribution of X. To this point, we have done nothing special; this is the standard approach introduced in equation 8.17 and 8.18. We will now do something a bit odd, which has a slightly surprising result. Let us introduce, therefore, a new density function, h. At this point, other than meeting the definition of a statistical density function, no additional structure is required. Returning to our original expectation in equation 8.31, we now proceed to perform a few subtle manipulations,

$$\mathbb{E}(g(X)) = \int_{-\infty}^{\infty} g(x) \underbrace{\frac{h(x)}{h(x)}}_{=1} f_X(x) dx, \tag{8.33}$$

$$= \int_{-\infty}^{\infty} g(x) \frac{f_X(x)}{h(x)} h(x) dx,$$

$$= \mathbb{E}^{\mathcal{H}}\left(g(X) \frac{f_X(X)}{h(X)}\right).$$

Since the ratio of $h(x)$ with itself is obviously equal to unity, we have not changed the final value of the integral. The expectation, as reflected by the $\mathbb{E}^{\mathcal{H}}(\cdot)$ operator, is now taken with respect to a different density function. In other words, the integrand is a bit more complicated, with an additional term, and we are no longer taking the expectation with respect to the original density $f_X(x)$, but instead with respect to $h(x)$. This probably feels a bit weird and not immediately beneficial.

8.5 Variance-Reduction Techniques

What exactly have we done? We have performed a mathematical sleight of hand to change the density of integration from f_X to h. The ratio, $\frac{f_X(X)}{h(X)}$—often referred to as either the *Radon-Nikodym* derivative or the likelihood ratio—basically acts to rescale g to take into account the adjustment in the density function. This is closely related to the Girsanov theorem used extensively in mathematical finance and probability theory. This so-called change-of-measure technique can be used to handle risk preferences—such as moving from $\mathbb{P}$ to $\mathbb{Q}$—or simply to make a complex integral easier to solve.[10] The Radon-Nikodym derivative is essentially the translation between the two density functions. We may now update equation 8.32 and restate our Monte Carlo estimator as:

$$\widehat{\mathbb{E}(g(X))} = \frac{1}{M} \sum_{m=1}^{M} g(X_m) \frac{f_X(X_m)}{h(X_m)}. \tag{8.34}$$

The effectiveness of this technique stems, in large part, from identifying a judicious choice of $h(x)$—a good selection will improve the accuracy of our estimate, whereas a poor choice can make it worse. Generally, $h(x)$ is selected to minimize the variance of the estimator in equation 8.34; this is entirely consistent with the fundamental notion behind all variance-reduction techniques.

The basic idea is, at its heart, nothing short of brilliant. In the credit-risk setting, as we've learned through discussion and practical examples, we routinely use simulation techniques to measure very *rare* events.[11] Enormous numbers of stochastic-simulation iterations are required to generate sufficient observations to describe these extreme outcomes. Importance sampling amounts to a change of probability measure. The objective, therefore, is to select h in such a way that we make extreme tail outcomes more likely. In this case, we will, by construction, have many more observations to estimate our integrals in those areas of interest to us. A common example is a far out-of-the-money barrier option; the centre of the underlying distribution provides relatively little information about this option's value. If done effectively, this will increase the accuracy of our estimates for a given amount of computational effort.

8.5.2 Setting Up the Problem

The next step is to see how this might be applied to our specific problem. Glasserman and Li (2005) begin with the independent-default setting. This turns out to be very informative, because the general case takes advantage of the conditional

---

[10] See Musiela and Rutkowski (1998, Appendix B), Karatzas and Shreve (1991, Section 3.5), or Heunis (2011, Section 5.7) for more background on the Girsanov result.

[11] Importance sampling has interesting links to risk and the notion of entropy. See Reesor and McLeish (2001) for a discussion of these useful concepts.

independence of the default outcomes when one conditions on the latent factors. We will follow their lead and also begin with the independent-default case.

The starting point is to write out explicitly our tail probability—or VaR—using the indicator variable trick from equation 8.21. The consequence is the following expression,

$$\mathbb{P}(L \geq \ell_\alpha) = \mathbb{P}\left(\sum_{n=1}^{N} c_n \mathbb{I}_{\mathcal{D}_n} \geq \ell_\alpha\right), \tag{8.35}$$

$$= \int_{-\infty}^{\infty} \cdots \int_{-\infty}^{\infty} \mathbb{I}_{\left\{\sum_{n=1}^{N} c_n \mathbb{I}_{\mathcal{D}_n} \geq \ell_\alpha\right\}} f(\mathbb{I}_{\mathcal{D}_1}, \ldots, \mathbb{I}_{\mathcal{D}_N}) d\mathbb{I}_{\mathcal{D}_1} \cdots d\mathbb{I}_{\mathcal{D}_N},$$

where the joint density of the individual default events follows a binomial law and has the following form,

$$f(\mathbb{I}_{\mathcal{D}_1}, \ldots, \mathbb{I}_{\mathcal{D}_N}) = \prod_{n=1}^{N} p_n^{\mathbb{I}_{\mathcal{D}_n}} (1 - p_n)^{1-\mathbb{I}_{\mathcal{D}_n}}. \tag{8.36}$$

As usual, we use ℓ_α to denote the default loss associated with a particular quantile, α.

While useful, this expression requires default independence. The independent-default setting basically undermines the overall purpose of the threshold and mixture structures introduced in the previous chapters. We can, happily, continue to exploit the conditional independence of the default events, by re-arranging the joint default-indicator density as,

$$f(\mathbb{I}_{\mathcal{D}_1}, \ldots, \mathbb{I}_{\mathcal{D}_N} | Y = y) = \prod_{n=1}^{N} p_n(y)^{\mathbb{I}_{\mathcal{D}_n}} \left(1 - p_n(y)\right)^{1-\mathbb{I}_{\mathcal{D}_n}}, \tag{8.37}$$

where $p_n(y)$ is the conditional default probability computed from a realization $\{Y = y\}$ of the global state variable. Equation 8.37 is a conditional joint default density. In full generality, Y can be a random vector to permit the accommodation of multi-factor models. To recall our previous treatment, we write this conditional-default probability more formally as,

$$p_n(y) = \mathbb{P}\left(\mathbb{I}_{\mathcal{D}_n} \big| Y = y\right). \tag{8.38}$$

The key point is that all of the models considered in the previous chapters possess the property of conditional independence. That is, given a realization of the state variable (or variables), each obligor's default is independent. The result is that equation 8.37 consequently represents a very general representation of the credit-risk problem.

8.5 Variance-Reduction Techniques

Bringing it all together, we now have a practical, concrete expression for the tail probability, or credit VaR, of the default-loss distribution for a generic credit risk model. It is summarized as,

$$\mathbb{P}(L \geq \ell_\alpha | Y = y)$$

$$= \int_{-\infty}^{\infty} \cdots \int_{-\infty}^{\infty} \mathbb{I}_{\left\{\sum_{n=1}^{N} c_n \mathbb{I}_{\mathcal{D}_n} \geq \ell_\alpha\right\}} \underbrace{\prod_{n=1}^{N} p_n(y)^{\mathbb{I}_{\mathcal{D}_n}} (1 - p_n(y))^{1-\mathbb{I}_{\mathcal{D}_n}} d\mathbb{I}_{\mathcal{D}_1} \cdots d\mathbb{I}_{\mathcal{D}_N}}_{f(\mathbb{I}_{\mathcal{D}_1},\ldots,\mathbb{I}_{\mathcal{D}_N}|Y=y)}. \quad (8.39)$$

This allows us to include the dependence structure of our model and, simultaneously, use the conditional independence to write out a straightforward and manageable expression for the probability of the tail default loss for our portfolio. We have used conditional VaR (and expected-shortfall) expressions like this extensively in our past analysis; generally, some form of numerical integration is involved to arrive at an unconditional risk-measure estimate. The good news in this setting is that the value of the state variable, $Y = y$, is naturally integrated out through the use of the Monte-Carlo method. In other words, the numerical integration is performed by stochastic simulation.

We are also interested, of course, in the shortfall integral as a risk metric. This suggests that we also need to examine its form. Fortunately, it is readily determined using Bayes theorem as we've seen in previous chapters. It is described as,

$$\mathbb{E}(L | L \geq \ell_\alpha, Y = y)$$

$$= \frac{\mathbb{E}\left(L \mathbb{I}_{L \geq \ell_\alpha} | Y = y\right)}{\mathbb{P}(L \geq \ell_\alpha) | Y = y}, \quad (8.40)$$

$$= \frac{\int_{-\infty}^{\infty} \cdots \int_{-\infty}^{\infty} \left(\sum_{n=1}^{N} c_n \mathbb{I}_{\mathcal{D}_n}\right) \mathbb{I}_{\left\{\sum_{n=1}^{N} c_n \mathbb{I}_{\mathcal{D}_n} \geq \ell_\alpha\right\}} \prod_{n=1}^{N} p_n(y)^{\mathbb{I}_{\mathcal{D}_n}} (1-p_n(y))^{1-\mathbb{I}_{\mathcal{D}_n}} d\mathbb{I}_{\mathcal{D}_1} \cdots d\mathbb{I}_{\mathcal{D}_N}}{\int_{-\infty}^{\infty} \cdots \int_{-\infty}^{\infty} \mathbb{I}_{\left\{\sum_{n=1}^{N} c_n \mathbb{I}_{\mathcal{D}_n} \geq \ell_\alpha\right\}} \prod_{n=1}^{N} p_n(y)^{\mathbb{I}_{\mathcal{D}_n}} (1-p_n(y))^{1-\mathbb{I}_{\mathcal{D}_n}} d\mathbb{I}_{\mathcal{D}_1} \cdots d\mathbb{I}_{\mathcal{D}_N}}.$$

The result is relatively predictable. The shortfall integral is, itself, the ratio of two multidimensional integrals. Indeed, since the VaR problem summarized in equation 8.39 has the same basic form—and, in fact, arises in the denominator of the shortfall integral—we will focus on it. The expected shortfall treatment is virtually the same, if slightly more effort.

At this point, we are ready to derive what Glasserman and Li (2005) refer to as the basic importance sampling identity by changing the conditional default probabilities,

$$\left\{p_n(y); n = 1, \ldots, N\right\}, \quad (8.41)$$

to a new set of conditional probabilities, which we will refer to as,

$$\left\{q_n(y); n = 1, \ldots, N\right\}. \tag{8.42}$$

Changing the probabilities associated with each individual event is, as we'll see in a moment, equivalent to adjusting the underlying probability density function.

This is accomplished by a simple manipulation inspired by equation 8.33 as follows,

$$\mathbb{P}(L \geq \ell_\alpha | Y = y) = \int_{-\infty}^{\infty} \cdots \int_{-\infty}^{\infty} \mathbb{I}_{\{L \geq \ell_\alpha\}} f_Y(\mathbb{I}_{\mathcal{D}_1}, \ldots, \mathbb{I}_{\mathcal{D}_N}) d\mathbb{I}_{\mathcal{D}_1} \cdots d\mathbb{I}_{\mathcal{D}_N}, \tag{8.43}$$

$$= \int_{-\infty}^{\infty} \cdots \int_{-\infty}^{\infty} \mathbb{I}_{\{L \geq \ell_\alpha\}} \underbrace{\prod_{n=1}^{N} p_n(y)^{\mathbb{I}_{\mathcal{D}_n}} (1 - p_n(y))^{1 - \mathbb{I}_{\mathcal{D}_n}}}_{\text{Equation 8.37}} d\mathbb{I}_{\mathcal{D}_1} \cdots d\mathbb{I}_{\mathcal{D}_N},$$

$$= \int_{-\infty}^{\infty} \cdots \int_{-\infty}^{\infty} \mathbb{I}_{\{L \geq \ell_\alpha\}} \underbrace{\left(\frac{\prod_{n=1}^{N} q_n(y)^{\mathbb{I}_{\mathcal{D}_n}} (1 - q_n(y))^{1 - \mathbb{I}_{\mathcal{D}_n}}}{\prod_{n=1}^{N} q_n(y)^{\mathbb{I}_{\mathcal{D}_n}} (1 - q_n(y))^{1 - \mathbb{I}_{\mathcal{D}_n}}}\right)}_{=1}$$

$$\times \prod_{n=1}^{N} p_n(y)^{\mathbb{I}_{\mathcal{D}_n}} (1 - p_n(y))^{1 - \mathbb{I}_{\mathcal{D}_n}} d\mathbb{I}_{\mathcal{D}_1} \cdots d\mathbb{I}_{\mathcal{D}_N},$$

$$= \int_{-\infty}^{\infty} \cdots \int_{-\infty}^{\infty} \mathbb{I}_{\{L \geq \ell_\alpha\}} \prod_{n=1}^{N} \left(\frac{p_n(y)}{q_n(y)}\right)^{\mathbb{I}_{\mathcal{D}_n}} \left(\frac{1 - p_n(y)}{1 - q_n(y)}\right)^{1 - \mathbb{I}_{\mathcal{D}_n}}$$

$$\times \underbrace{\prod_{n=1}^{N} q_n(y)^{\mathbb{I}_{\mathcal{D}_n}} (1 - q_n(y))^{1 - \mathbb{I}_{\mathcal{D}_n}} d\mathbb{I}_{\mathcal{D}_1} \cdots d\mathbb{I}_{\mathcal{D}_N}}_{f_Q(\mathbb{I}_{\mathcal{D}_1}, \ldots, \mathbb{I}_{\mathcal{D}_N} | Y = y)},$$

$$= \mathbb{E}^Q \left(\mathbb{I}_{\{L \geq \ell_\alpha\}} \prod_{n=1}^{N} \left(\frac{p_n(y)}{q_n(y)}\right)^{\mathbb{I}_{\mathcal{D}_n}} \left(\frac{1 - p_n(y)}{1 - q_n(y)}\right)^{1 - \mathbb{I}_{\mathcal{D}_n}} \bigg| Y = y\right),$$

where $\mathbb{E}^Q(\cdot)$ denotes the conditional expectation with respect to the new measure induced by the change of default probabilities given the value of the latent state variable and $f_Q(\mathbb{I}_{\mathcal{D}_1}, \ldots, \mathbb{I}_{\mathcal{D}_N} | Y = y)$ is the new conditional joint-default density function. Most importantly, equation 8.43 represents an unbiased estimator of the tail probability.[12] What remains is the determination of an appropriate form for the adjusted conditional default probabilities introduced in equation 8.42.

---

[12] Important discussion and results regarding the unbiasedness of this estimator along with upper and lower bounds are found in Glasserman and Li (2005).

8.5 Variance-Reduction Techniques

8.5.3 The Esscher Transform

The logic behind Glasserman and Li (2005) is to increase the default probabilities thus making extreme events more probable and providing a more efficient, and lower variance, estimate of the tail probability. An arbitrary increase in the default probabilities, however, is undesirable. They propose the definition of a parametrized set of default probabilities of the following form,

$$q_n(\theta, y) = \frac{p_n(y)e^{c_n\theta}}{1 + p_n(y)(e^{c_n\theta} - 1)}, \qquad (8.44)$$

where $\theta \in \mathbb{R}$ is a parameter and, as per our usual treatment, c_n denotes the exposure at default of the nth credit counterparty. It is easy to see that if $\theta = 0$, then $q_n(\theta, y) = p_n(y)$. For any positive value of θ, therefore, the new set of default probabilities will be larger than the initial unconditional default probabilities, $\{p_n(y); n = 1, \ldots, N\}$.

Equation 8.44 is, of course, not an entirely unfamiliar commodity. It looks suspiciously like many of the cumulant-generating-function derivative objects introduced and used extensively in the development of the saddlepoint approximation in Chap. 7. This, as we saw at the tail end of Chap. 7, is not a coincidence. Once we have developed the basic structure, we will return and establish the important and intuitive link to the saddlepoint method.

Furnished with the definition in equation 8.44, we may now revise the general definition of the Radon-Nikodym derivative in equation 8.43 to determine the new form of our integrand. This tedious, but illuminating exercise, leads to

$$\prod_{n=1}^{N} \left(\frac{p_n(y)}{q_n(y)}\right)^{\mathbb{I}_{\mathcal{D}_n}} \left(\frac{1 - p_n(y)}{1 - q_n(y)}\right)^{1-\mathbb{I}_{\mathcal{D}_n}}$$

$$= \prod_{n=1}^{N} \left(\frac{p_n(y)}{q_n(\theta, y)}\right)^{\mathbb{I}_{\mathcal{D}_n}} \left(\frac{1 - p_n(y)}{1 - q_n(\theta, y)}\right)^{1-\mathbb{I}_{\mathcal{D}_n}}, \qquad (8.45)$$

$$= \prod_{n=1}^{N} \left(\frac{p_n(y)}{\frac{p_n(y)e^{c_n\theta}}{1+p_n(y)(e^{c_n\theta}-1)}}\right)^{\mathbb{I}_{\mathcal{D}_n}} \left(\frac{1 - p_n(y)}{1 - \left(\frac{p_n(y)e^{c_n\theta}}{1+p_n(y)(e^{c_n\theta}-1)}\right)}\right)^{1-\mathbb{I}_{\mathcal{D}_n}},$$

$$= \prod_{n=1}^{N} \left(\frac{\cancel{p_n(y)}(1 + p_n(y)(e^{c_n\theta} - 1))}{\cancel{p_n(y)}e^{c_n\theta}}\right)^{\mathbb{I}_{\mathcal{D}_n}} \left(\frac{1 - p_n(y)}{\frac{1+p_n(y)(e^{c_n\theta}-1)-p_n(y)e^{c_n\theta}}{1+p_n(y)(e^{c_n\theta}-1)}}\right)^{1-\mathbb{I}_{\mathcal{D}_n}},$$

$$= \prod_{n=1}^{N} \left(\frac{1 + p_n(y)(e^{c_n\theta} - 1)}{e^{c_n\theta}}\right)^{\mathbb{I}_{\mathcal{D}_n}} \left(\frac{1 - \cancel{p_n(y)}}{\frac{1-\cancel{p_n(y)}}{1+p_n(y)(e^{c_n\theta}-1)}}\right)^{1-\mathbb{I}_{\mathcal{D}_n}},$$

$$= \prod_{n=1}^{N} \left(\frac{1 + p_n(y)(e^{c_n\theta} - 1)}{e^{c_n\theta}} \right)^{\mathbb{I}_{\mathcal{D}_n}} \left(1 + p_n(y)(e^{c_n\theta} - 1)\right)^{1-\mathbb{I}_{\mathcal{D}_n}},$$

$$= \prod_{n=1}^{N} \left(e^{c_n\theta}\right)^{-\mathbb{I}_{\mathcal{D}_n}} \left(1 + p_n(y)(e^{c_n\theta} - 1)\right).$$

This is already a dramatic improvement, but applying natural logarithms allows us to achieve further simplification and reveals a few pleasant surprises. In particular, we have

$$\ln\left(\prod_{n=1}^{N}\left(\frac{p_n(y)}{q_n(y)}\right)^{\mathbb{I}_{\mathcal{D}_n}}\left(\frac{1-p_n(y)}{1-q_n(y)}\right)^{1-\mathbb{I}_{\mathcal{D}_n}}\right) = \ln\left(\prod_{n=1}^{N}\left(e^{c_n\theta}\right)^{-\mathbb{I}_{\mathcal{D}_n}}\left(1+p_n(y)(e^{c_n\theta}-1)\right)\right), \quad (8.46)$$

$$= \sum_{n=1}^{N} \ln\left(\left(e^{c_n\theta}\right)^{-\mathbb{I}_{\mathcal{D}_n}}\left(1+p_n(y)(e^{c_n\theta}-1)\right)\right),$$

$$= \sum_{n=1}^{N} \left[\ln\left(\left(e^{c_n\theta}\right)^{-\mathbb{I}_{\mathcal{D}_n}}\right) + \ln\left(1+p_n(y)(e^{c_n\theta}-1)\right)\right],$$

$$= \underbrace{-\theta \sum_{n=1}^{N} c_n \mathbb{I}_{\mathcal{D}_n}}_{L} + \underbrace{\sum_{n=1}^{N} \ln\left(1+p_n(y)(e^{c_n\theta}-1)\right)}_{K_L(\theta,y)},$$

$$= -\theta L + K_L(\theta, y),$$

$$\prod_{n=1}^{N}\left(\frac{p_n(y)}{q_n(y)}\right)^{\mathbb{I}_{\mathcal{D}_n}}\left(\frac{1-p_n(y)}{1-q_n(y)}\right)^{1-\mathbb{I}_{\mathcal{D}_n}} = e^{K_L(\theta,y)-\theta L}.$$

The consequence is a concise and intuitive form for the Radon-Nikodym derivative.

As an interesting, and non-coincidental, aside, we recall that the moment generating function of a random variable, X, is given as $\mathbb{E}(e^{tX})$.[13] If we define our random variable as $X_n = c_n\theta$, replace the usual variable of integration, t, with the parameter, θ, then

$$\mathbb{E}\left(e^{c_n \mathbb{I}_{\mathcal{D}_n} \theta} \,\Big|\, Y = y\right) = \underbrace{\mathbb{P}\left(\mathbb{I}_{\mathcal{D}_n} = 1 \,\Big|\, Y = y\right)}_{p_n(y)} e^{c_n(1)\theta} + \underbrace{\mathbb{P}\left(\mathbb{I}_{\mathcal{D}_n} = 0 \,\Big|\, Y = y\right)}_{1-p_n(y)} e^{c_n(0)\theta}, \quad (8.47)$$

$$= p_n(y)e^{c_n\theta} + (1 - p_n(y)).$$

---

[13] When X is continuous, the moment-generating function coincides with the Laplace transform of the density function. For a more detailed discussion of moment-generating functions, see Casella and Berger (1990, Chapter 2) or the relevant discussion in Chap. 7.

8.5 Variance-Reduction Techniques

This allows us to work backwards to verify our definition in equation 8.46. Specifically, recalling that each default event $\mathbb{I}_{\mathcal{D}_n}$ is independent,

$$\sum_{n=1}^{N} \ln\left(1 + p_n(y)(e^{c_n\theta} - 1)\right) = \sum_{n=1}^{N} \ln\left(\underbrace{p_n(y)e^{c_n\theta} + (1 - p_n(y))}_{\text{Equation 8.47}}\right), \quad (8.48)$$

$$= \sum_{n=1}^{N} \ln\left(\mathbb{E}(e^{c_n \mathbb{I}_{\mathcal{D}_n}\theta})\right),$$

$$= \ln\left(\prod_{n=1}^{N} \mathbb{E}(e^{c_n \mathbb{I}_{\mathcal{D}_n}\theta})\right),$$

$$= \ln\left(\mathbb{E}\left(\prod_{n=1}^{N} e^{c_n \theta \mathbb{I}_{\mathcal{D}_n}}\right)\right),$$

$$= \ln\left(\mathbb{E}\left(\exp\left(\theta \underbrace{\sum_{n=1}^{N} c_n \mathbb{I}_{\mathcal{D}_n}}_{L}\right)\right)\right),$$

$$= \ln\left(\mathbb{E}\left(e^{\theta L}\right)\right),$$

$$= K_L(\theta, y).$$

which is the cumulant generating function of L.[14] This indicates that the Radon-Nikodym derivative may be defined as $e^{K_L(\theta,y) - \theta L}$.[15] In other words, the new default probabilities have vanished. The punchline, therefore, is that, as promised, the transformation of the default probabilities in equation 8.44 is equivalent to directly adjusting the default-loss density.

Ultimately, therefore, we have that

$$\mathbb{P}(L \geq \ell_\alpha | Y = y) = \mathbb{E}^Q\left(\mathbb{I}_{\{L \geq \ell_\alpha\}} e^{K_L(\theta,y) - \theta L} \middle| Y = y\right), \quad (8.49)$$

is an unbiased estimator for the tail probability. This specific change-of-measure technique is termed an *exponential twist*, an exponential tilt, or more formally, it is referred to as an Esscher transform.

---

[14] Recall that the cumulant generating function, or cumulant, is simply the natural logarithm of the moment-generating function.

[15] This is, in fact, the integrand from the inverse Laplace transform used to identify the loss density in Chap. 7.

8.5.4 Finding θ

A sensible value for the probability distorting parameter $\theta \in \mathbb{R}$ is required. Glasserman and Li (2005) suggest selecting θ to minimize the second moment of our estimator in equation 8.49; this is basically the same idea as minimizing the variance of our estimator. If we compute the second moment, we have

$$\mathbb{E}^Q \left(\mathbb{P}(L \geq \ell_\alpha)^2 \bigg| Y = y \right) = \mathbb{E}^Q \left(\bigg(\underbrace{\mathbb{I}_{\{L \geq \ell_\alpha\}} e^{K_L(\theta,y) - \theta L}}_{\text{Equation 8.49}} \bigg)^2 \bigg| Y = y \right), \quad (8.50)$$

$$= \mathbb{E}^Q \left(\mathbb{I}_{\{L \geq \ell_\alpha\}} e^{2K_L(\theta,y) - 2\theta L} \bigg| Y = y \right).$$

Minimization of this second moment is relatively difficult. As a consequence, Glasserman and Li (2005) propose an upper bound of the following form,

$$\mathbb{E}^Q \left(\mathbb{I}_{\{L \geq \ell_\alpha\}} e^{2K_L(\theta,y) - 2\theta L} \bigg| Y = y \right) \leq e^{2K_L(\theta,y) - 2\theta \ell_\alpha}. \quad (8.51)$$

Minimization of this upper bound is not as difficult. Indeed, it leads to a sequence of equivalent optimization problems,

$$\min_\theta e^{2K_L(\theta,y) - 2\theta \ell_\alpha} = \min_\theta -2 \left(\theta \ell_\alpha - K_L(\theta, y) \right), \quad (8.52)$$

$$= \max_\theta \theta \ell_\alpha - K_L(\theta, y),$$

all indexed to a specific choice of $Y = y$ and the overall level of tail loss one is targeting (i.e., ℓ_α). The indexation is critical, because it implies that each choice of θ relates to a particular outcome of y. This will have important consequences for the implementation of this technique.

Computing the first-order condition associated with the optimization problem in equation 8.52, we have

$$\frac{\partial}{\partial \theta} \left(\theta \ell_\alpha - K_L(\theta, y) \right) = 0, \quad (8.53)$$

$$\ell_\alpha - \frac{\partial K_L(\theta, y)}{\partial \theta} = 0,$$

where,

$$\frac{\partial K_L(\theta, y)}{\partial \theta} = \frac{\partial}{\partial \theta} \left(\sum_{n=1}^N \ln \left(1 + p_n(y)(e^{c_n \theta} - 1) \right) \right), \quad (8.54)$$

$$= \sum_{n=1}^N \frac{\partial}{\partial \theta} \left(\ln \left(1 + p_n(y)(e^{c_n \theta} - 1) \right) \right),$$

8.5 Variance-Reduction Techniques

$$= \sum_{n=1}^{N} \frac{c_n p_n(y) e^{c_n \theta}}{1 + p_n(y)(e^{c_n \theta} - 1)},$$

$$= \sum_{n=1}^{N} c_n q_n(\theta, y).$$

A good choice of θ, therefore, is the root of the following equation,

$$\ell_\alpha - \sum_{n=1}^{N} c_n q_n(\theta, y) = 0, \tag{8.55}$$

which we will denote as $\tilde{\theta}_\ell(y)$ to indicate its dependence on *both* the level of default loss and the outcome of the latent state variable. This is easily solved numerically with a root-finding algorithm.[16]

The reader will, of course, recognize from Chap. 7 that $\tilde{\theta}_\ell(y)$ is none other than the saddlepoint approximation, which we previously denoted as $\tilde{t}_\ell$. To understand more clearly what is going on, let's compute the conditional expected return under our two density assumptions. In the usual case, we have

$$\mathbb{E}(L|Y=y) = \mathbb{E}\left(\sum_{n=1}^{N} c_n \mathbb{I}_{\mathcal{D}_n} \middle| Y=y\right), \tag{8.56}$$

$$= \sum_{n=1}^{N} c_n \underbrace{\mathbb{E}\left(\mathbb{I}_{\mathcal{D}_n} \middle| Y=y\right)}_{p_n(y)},$$

$$= \sum_{n=1}^{N} c_n p_n(y).$$

This is not particularly exciting. The expected default loss depends on the realization of the state variable (or variables), Y. This is the usual case and the specific conditional default probability will depend on the choice of model.

Now, let us turn to compute the expected return under the Esscher transform. The result, if we set the parameter θ to the solution to equation 8.55, is

$$\mathbb{E}^Q(L|Y=y) = \mathbb{E}^Q\left(\sum_{n=1}^{N} c_n \mathbb{I}_{\mathcal{D}_n} \middle| Y=y\right), \tag{8.57}$$

$$= \sum_{n=1}^{N} c_n \underbrace{\mathbb{E}^Q\left(\mathbb{I}_{\mathcal{D}_n} \middle| Y=y\right)}_{q_n(y)},$$

---

[16] We employed the modified Powell method from the Python `scipy` library, which requires only a fraction of a second to converge to a good solution.

$$= \sum_{n=1}^{N} c_n q_n(y),$$

$$= \underbrace{\frac{K_L(\theta, y)}{\partial \theta}\bigg|_{\theta=\tilde{\theta}_\ell(y)}}_{\substack{\text{Root of} \\ \text{Equation 8.54}}},$$

$$= \underbrace{\text{VaR}_\alpha(L|Y = y)}_{\ell_\alpha}.$$

Under the twisted default probabilities, the expected value of the loss distribution has been moved to the α quantile or VaR measure associated with the specific realization of y. The centre of the distribution has thus been shifted in precisely the way we would have hoped. The loss distribution is adjusted such that the *rare* VaR event becomes the average, or expected, result.

8.5.5 Implementing the Twist

It is also relatively straightforward, having worked through the saddlepoint approximation in Chap. 7, to practically implement the Esscher transform. The first job is the determination of the θ parameter. This essentially reduces to finding the saddlepoint, which allows us, as summarized in Algorithm 8.4, to recycle the getSaddlepoint function from the varContributions library.[17] One requires, as we see from equation 8.53, the default probabilities p, the exposures c, and the targeted amount of loss l. We've also set the starting point for the optimization to zero, which appears to yield more efficient results.

The second step is to use the parameter, theta, to compute the twisted probabilities. This is accomplished in a single line in Algorithm 8.5 following from equation 8.44. Again, we re-use code from Chap. 7 since the denominator is nothing other than the moment-generating function. Algorithm 8.5 may be used in the independent and conditionally independent settings, since the input argument, p, may be either the unconditional or conditional default probabilities.

Algorithm 8.4 Identifying θ

```
import varContributions as vc
theta0 = 0.0
theta = vc.getSaddlePoint(p,c,l,theta0)
```

---

[17] See Appendix D for more details on the overall Python library structure and organization.

8.5 Variance-Reduction Techniques

Algorithm 8.5 Computing twisted probabilities

```
def getQ(theta,c,p):
    return np.divide(p*np.exp(c*theta),vc.computeMGF(theta,p,c))
```

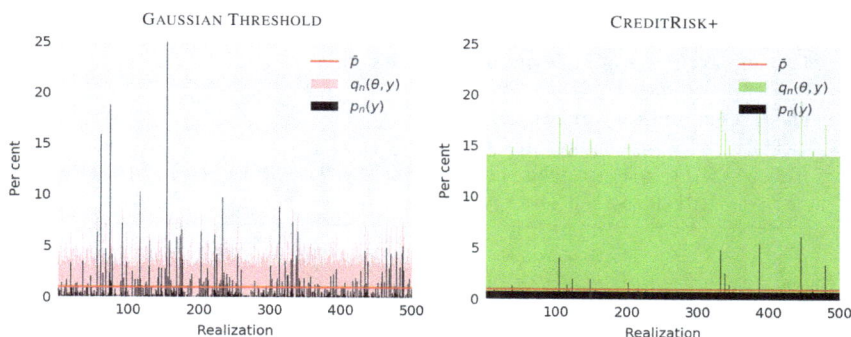

Fig. 8.9 *Twisting probabilities*: This figure illustrates the average twisted and non-twisted of conditional default probabilities for two competing models: the one-factor Gaussian threshold and one-factor CreditRisk+ approaches. In both cases, the θ parameter was selected to force the mean of the default distribution to be equal to the 99.99th VaR quantile.

Algorithm 8.6 Computing the Esscher-transform Radon-Nikodym derivative

```
def computeRND(theta,L,cgf):
    return np.exp(-np.multiply(theta,L)+cgf)
```

Figure 8.9 demonstrates the impact of twisting probabilities on two familiar models: the one-factor Gaussian threshold and CreditRisk+ approaches. In each case, the average conditional, untransformed conditional, and twisted conditional default probabilities are displayed for 500 iterations of the simulation algorithm; the θ parameters are determined based on a 99.99th quantile default loss. It is very clear how the twisted default probabilities have essentially been shifted upwards in both cases. Visually, this is the impact of adjusting the density to sample from those areas of greater default loss. The likelihood ratio, or Radon-Nikodym derivative, then pulls back the resulting estimates to the real world.

The actual likelihood ratio computation is also reasonably undemanding to compute. The Python code involves another one-line function found in Algorithm 8.6. It is summarized in the form described in equation 8.46 and takes the twist parameter `theta`, the total default loss `L`, and the cumulant generating function, `cgf`, as input arguments. In the independent-default case, this is a constant for all iterations. In the conditionally independent environment, it is a vector with a different value for each realization of the global state variable. Indeed, this makes logical sense since there is a vector of twisting parameters, $\theta \in \mathbb{R}^M$. A different likelihood ratio is required for each of these cases.

Algorithm 8.7 Importance sampling in the independent default case

```
def isIndDefault(N,M,p,c,l):
    U = np.random.uniform(0,1,[M,N])
    theta = vc.getSaddlePoint(p,c,l,0.0)
    qZ = getQ(theta,c,p)
    cgf = vc.computeCGF(theta,p,c)
    I = np.transpose(1*np.less(U,qZ))
    L = np.dot(c,I)
    rn = computeRND(theta,L,cgf)
    tailProb = np.mean(np.multiply(L>l,rn))
    eShortfall = np.mean(np.multiply(L*(L>l),rn))/tailProb
    return tailProb,eShortfall
```

While not terribly realistic, the independent-default setting is always a useful starting point due to its simplicity. Algorithm 8.7 provides the various steps in the importance-sampling approach described in the previous sections for this easily accessible situation. It is worth examining in some detail, because it forms the blueprint for the more involved conditionally independent cases. The first step is common with the raw Monte-Carlo approach from Chap. 2; we generate a large number of uniform random variables. The `getSaddlepoint` function from Algorithm 8.5 is then called to determine the (single) twisting parameter, `theta`. Algorithm 8.6 follows with the calculation of the new twisted default probabilities; we salvage the `computeCGF` sub-routine from our discussion in Chap. 7 to determine the cumulant generating function associated with `theta`. `I` is the collection of simulated default indicator functions estimated using the transformed default probabilities. `L` is the default loss associated with `I`. If we were to compute `np.mean(L>l)`, then we would have an estimate of the upper default-tail probability associated with a loss of magnitude `l` under the Esscher transformed probability measure. This is not what we want; instead, we seek the true estimate of the tail probability. Computing the Radon-Nikodyn derivative with Algorithm 8.6, we first calculate the product of the tail-loss and the likelihood ratio as `np.multiply(L>l,rn)`—the mean of this quantity is our true tail-probability estimate.

Algorithm 8.7 is not only a useful overview of the method, it also highlights an important point. Similar to the saddlepoint approach, the importance-sampling technique works backwards relative to the raw Monte-Carlo method. In raw Monte Carlo, we select our level of confidence, α, and compute $\text{VaR}_\alpha(L)$. With the saddlepoint approximation, we define the level of loss, ℓ, and compute the level of confidence associated with it. This yields ℓ_α. With the Esscher transform, it works the same way. We specify our desired level of VaR, $\ell_\alpha = \text{VaR}_\alpha(L)$, we determine the optimal twist, and the algorithm returns $\mathbb{P}(L \geq \ell_\alpha)$. This idea is summarized, with comparison to the raw Monte-Carlo and saddlepoint settings, in Table 8.3.

8.5 Variance-Reduction Techniques

Table 8.3 *A different perspective*: This table illustrates the directionality of the computations in the raw Monte-Carlo, saddlepoint, and importance-sampling approaches. Use of the Esscher-transform technique to reduce simulation variance is, in this respect, conceptually very similar to the employment of the saddlepoint approximation.

Technique	Input	Parameter	Output
Monte-Carlo simulation	α	None	$\text{VaR}_\alpha(L)$
Saddlepoint approximation	ℓ	$\tilde{t}_\ell$	$\mathbb{P}(L > \ell)$
Esscher transform	ℓ	θ_ℓ	$\mathbb{P}(L > \ell)$

Practically, therefore, one needs to select a number of different levels of ℓ, compute the associated parameters θ_ℓ, and receive the desired tail probabilities from the algorithm. This is a bit troublesome since one requires the actual loss distribution to determine the desired default-loss levels. In the saddlepoint setting, as we saw in Chap. 7, it was possible to numerically solve for the default associated with a given tail probability. In this case, numerical optimization is not a practically feasible solution. While it is conceptually possible, realistically it is a slow and noisy procedure. The best one can do is to run the raw Monte-Carlo estimator with a small number of iterations to trace out a reasonable domain for the default losses and use the importance-sampling estimator to narrow in and sharpen up these estimates.

Table 8.4 applies Algorithm 8.7 to our usual example for the independent-default binomial model. We also provide for comparison—although we have not yet examined all of the algorithmic details—the results for the one-factor Gaussian threshold model. Each model employs three estimators. The approaches include *two* raw Monte-Carlo estimates using 10 million and 100,000 iterations, respectively. The third method is the importance-sampling (IS) technique with 3,000 iterations.

While there are a dizzying array of numbers in Table 8.4, a few lessons can be drawn. The first point is that the Esscher transform works remarkably well in the independent-default setting. At vanishingly small levels of computational expense, it generates VaR estimates that legitimately compete with the raw simulations involving multiple repetitions of 10 million iterations. The principle behind the importance-sampling technique, therefore, has significant appeal in the credit-risk setting. As a second point, to curb our enthusiasm, we move to the bad news. The one-factor Gaussian threshold results are, quite simply, underwhelming. It is outperformed, at every presented quantile, by raw simulation involving 100,000 iterations.

What is responsible for this difference? Clearly, the deviation relates to the fundamental distinction between the independent-default and threshold models: the latent state variable. We may shift the mean of the loss distribution, but the threshold and mixture model loss outcomes depend importantly on the outcome of the latent state variables. To improve the situation, therefore, we need to reach into our bag of variance-reduction techniques to solve the disturbing behaviour described in Table 8.4.

Table 8.4 *A first try:* This table summarizes a number of summary statistics for three different possible Monte-Carlo estimators of the independent-default and one-factor Gaussian threshold model VaR applied to our usual example. The approaches include two raw Monte-Carlo estimates using 10 million and 100,000 iterations, respectively. The third method is the importance-sampling (IS) technique with 3,000 iterations.

		Independent default			Gaussian threshold		
Value	Statistic	Raw (10M)	IS (3,000)	Raw (100,000)	Raw (10M)	IS (3,000)	Raw (100,000)
95.00	Mean	$35.3	$35.3	$35.3	$48.5	$51.0	$48.5
	SE	$0.0	$0.3	$0.1	$0.1	$5.2	$0.4
	99% C.I.	[$35.3, $35.3]	[$35.2, $35.4]	[$35.2, $35.3]	[$48.5, $48.6]	[$49.0, $53.0]	[$48.3, $48.6]
97.00	Mean	$40.4	$40.9	$40.4	$63.5	$66.4	$63.6
	SE	$0.0	$0.3	$0.2	$0.1	$5.6	$0.5
	99% C.I.	[$40.4, $40.5]	[$40.8, $41.0]	[$40.3, $40.5]	[$63.5, $63.6]	[$64.2, $68.5]	[$63.5, $63.8]
99.00	Mean	$52.6	$52.5	$52.6	$102.0	$102.4	$102.2
	SE	$0.0	$0.5	$0.4	$0.1	$9.1	$1.1
	99% C.I.	[$52.6, $52.7]	[$52.3, $52.7]	[$52.5, $52.8]	[$102.0, $102.1]	[$98.9, $105.9]	[$101.8, $102.6]
99.90	Mean	$72.3	$72.4	$72.3	$204.3	$202.2	$204.4
	SE	$0.1	$0.4	$0.8	$0.6	$24.9	$4.0
	99% C.I.	[$72.2, $72.3]	[$72.2, $72.5]	[$72.0, $72.6]	[$203.9, $204.6]	[$192.7, $211.8]	[$202.8, $205.9]
99.99	Mean	$90.2	$91.0	$90.8	$324.1	$287.3	$327.6
	SE	$0.2	$0.2	$2.3	$1.8	$51.2	$19.7
	99% C.I.	[$90.1, $90.3]	[$90.9, $91.0]	[$89.9, $91.7]	[$323.1, $325.1]	[$267.7, $306.9]	[$320.0, $335.1]
Time	Mean	15.6 s	0.0 s	0.2 s	50.2 s	0.8 s	0.5 s
	Total	6.5 min	0.0 min	0.1 min	20.9 min	0.7 min	0.4 min
	Per m	2 μs	2 μs	2 μs	5 μs	262 μs	5 μs
Iterations	Total	250,000,000	150,000	5,000,000	250,000,000	150,000	5,000,000

8.5 Variance-Reduction Techniques

The Esscher transform is a good start and, for some models as we'll see, it is the only technique at our disposition. In the threshold setting, particularly with Gaussian or conditionally Gaussian state variables, we can further reduce the variance of our Monte-Carlo estimates. The techniques used to accomplish this are discussed in the following sections and, fortunately, can be used in conjunction with the Esscher transform.

8.5.6 Shifting the Mean

Imagine that Y is $\mathcal{N}(0, 1)$ and you wish to use the Monte-Carlo method to estimate the integral, $\mathbb{E}(g(Y))$, for some measurable function, g. This is easily accommodated with equation 8.31. Imagine further, however, that $g(Y)$ put particular emphasis on extreme positive outcomes of Y. Stochastic-simulation may prove even slower than usual in such cases, simply because large positive outcomes of Y are quite rare; consequently, many simulations are required to generate a sufficiently large number of extreme realizations to reasonably estimate the expectation of $g(Y)$. This seems like a natural situation to apply importance sampling. In fact, there is a quite elegant transformation that can be employed in this context.

The approach starts from the general integration problem,

$$\mathbb{E}(g(Y)) = \int_{-\infty}^{\infty} g(y)\phi(y)dy, \tag{8.58}$$

$$= \int_{-\infty}^{\infty} g(y)\frac{e^{-\frac{y^2}{2}}}{\sqrt{2\pi}}dy,$$

$$= \int_{-\infty}^{\infty} g(y)\frac{e^{-\frac{y^2}{2}}}{\sqrt{2\pi}}\underbrace{\left(\frac{\exp\left(\frac{2\mu y-\mu^2}{2}\right)}{\exp\left(\frac{2\mu y-\mu^2}{2}\right)}\right)}_{=1}dy,$$

$$= \int_{-\infty}^{\infty} g(y)e^{\frac{-2\mu y+\mu^2}{2}}\left(\frac{e^{\frac{-(y^2-2\mu y+\mu^2)}{2}}}{\sqrt{2\pi}}\right)dy,$$

$$= \int_{-\infty}^{\infty} g(y)e^{\frac{-2\mu y+\mu^2}{2}}\underbrace{\left(\frac{e^{\frac{-(y-\mu)^2}{2}}}{\sqrt{2\pi}}\right)}_{h(y)}dy,$$

$$= \mathbb{E}^{\mathcal{H}}\left(g(y)e^{\frac{-2\mu y+\mu^2}{2}}\right).$$

The new density—$\mathcal{N}(\mu, 1)$—is a Gaussian distribution with a shifted mean, μ, and the Radon Nikodym derivative is $e^{\frac{-2\mu y + \mu^2}{2}}$. If one carefully selects the choice of μ such that more positive extreme outcomes are observed, then the Monte-Carlo estimator will become more efficient. Note also that, although we've derived this quantity in a univariate setting, it can be applied equally to a multivariate Gaussian distribution.

This is particularly pertinent in our case, because although the Esscher transform does a great job of shifting the conditional loss distribution to our area of interest, the conditionality still poses a challenge. The saddlepoint used to determine the twisted probabilities is a function of the worst-case loss given a particular outcome for the state variable, Y. Recall that default losses are at their highest, in the Gaussian and normal-variance mixture settings, when Y is large and negative. Figure 8.10 illustrates the evolution of the 99.99th quantile of the loss distribution conditioning each time on a different, but fixed, level of the state variable, $Y = y$. The results, presented for the one-factor Gaussian model, are compared to the unconditional VaR, which is basically the average across all state-variable outcomes. Similar patterns are evident—albeit with different signs in some cases—in all of our previously examined threshold and mixture models.

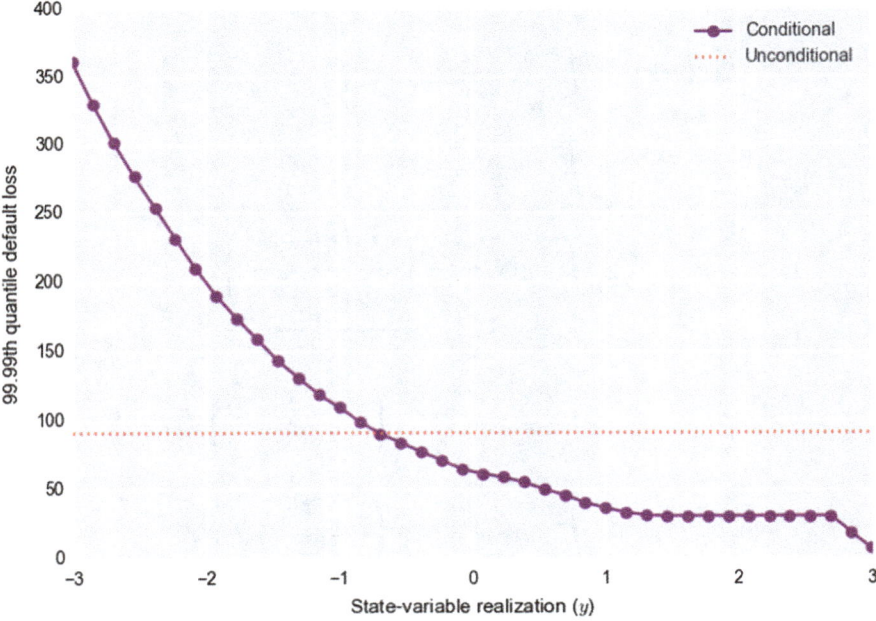

Fig. 8.10 *Conditional VaR*: This figure illustrates the evolution of the 99.99th quantile of the loss distribution conditioning each time on a different, but fixed, level of the state variable, $Y = y$. The results, presented for the one-factor Gaussian model, are compared to the unconditional VaR, which is basically the average across all state-variable outcomes. Similar patterns are evident in all previously examined threshold and mixture models.

8.5 Variance-Reduction Techniques

The unfortunate consequence of Fig. 8.10 is that the Esscher transform provides improvement, but does not—outside of the independent-default setting—provide enormous gains in computational efficiency. In some cases, however, there is a solution. In particular, in the Gaussian and normal-mixture threshold settings, we can use an approach similar to that described in equation 8.58. The difficult aspect of this approach, however, is the determination of the appropriate level of μ.

Selecting μ requires us to identify a sensible object, related to our problem, to optimize. Glasserman (2006), Glasserman et al. (2007), and Glasserman and Li (2005) suggest that we use the law of total variance as our starting point.[18] For any random variable, X, we can decompose its variance as follows,

$$\text{var}(X) = \mathbb{E}\Big(\text{var}(X|Y)\Big) + \text{var}\Big(\mathbb{E}(X|Y)\Big). \tag{8.59}$$

In our case, the random variable is $\widehat{\mathbb{P}(L \geq \ell_\alpha)}$, which is our estimator of $\mathbb{P}(L \geq \ell_\alpha)$. The conditioning random variable is the state variable Y, which could be in either scalar or vector format. This allows us to restate equation 8.59 as follows,

$$\text{var}\Big(\underbrace{\widehat{\mathbb{P}(L \geq \ell_\alpha)}}_{X}\Big) = \mathbb{E}\Big(\text{var}\Big(\underbrace{\widehat{\mathbb{P}(L \geq \ell_\alpha)}}_{X}\Big| Y = y\Big)\Big) + \text{var}\Big(\mathbb{E}\Big(\underbrace{\widehat{\mathbb{P}(L \geq \ell_\alpha)}}_{X}\Big| Y = y\Big)\Big). \tag{8.60}$$

Since we seek to estimate the upper-tail probability, it makes sense to minimize the variability of this estimate. Equation 8.60 provides a bit more insight into the structure of the variance of our estimator. Assuming that we have already performed the Esscher transform, then given a realization of the state variable, we expect, on average, that $\text{var}\Big(\widehat{\mathbb{P}(L \geq \ell_\alpha)}\Big| Y = y\Big)$ will be quite small. This was, in fact, the whole point of the previous exercise. The consequence is that we can ignore the first term. Our attention, therefore, is directed to the second term.

If we use a clever choice of importance-sampling density, we can actually reduce the variance of our estimator to zero. Consider the following manipulation of the integrand of the second term in equation 8.60,

$$\mathbb{E}\left(\mathbb{P}(L \geq \ell_\alpha)| Y = y\right) = \int_{-\infty}^{\infty} \mathbb{P}(L \geq \ell_\alpha) \frac{e^{-\frac{y^2}{2}}}{\sqrt{2\pi}} dy, \tag{8.61}$$

$$= \int_{-\infty}^{\infty} \mathbb{P}(L \geq \ell_\alpha) \frac{e^{-\frac{y^2}{2}}}{\sqrt{2\pi}} \underbrace{\left(\frac{e^{-\frac{y^2}{2}}\mathbb{P}(L \geq \ell_\alpha)}{e^{-\frac{y^2}{2}}\mathbb{P}(L \geq \ell_\alpha)}\right)}_{=1} dy,$$

---

[18]See Held and Bové (2014, Appendix A) for more information.

$$= \int_{-\infty}^{\infty} \mathbb{P}(L \geq \ell_\alpha) \frac{e^{-\frac{y^2}{2}}(2\pi)^{-\frac{1}{2}}}{e^{-\frac{y^2}{2}}\mathbb{P}(L \geq \ell_\alpha)} \underbrace{e^{-\frac{y^2}{2}}\mathbb{P}(L \geq \ell_\alpha)}_{h(y)} dy,$$

$$= \mathbb{E}^{\mathcal{H}}\left(\cancel{\mathbb{P}(L \geq \ell_\alpha)}\frac{\cancel{e^{-\frac{y^2}{2}}}(2\pi)^{-\frac{1}{2}}}{\cancel{e^{-\frac{y^2}{2}}}\cancel{\mathbb{P}(L \geq \ell_\alpha)}}\right),$$

$$= \mathbb{E}^{\mathcal{H}}\left((2\pi)^{-\frac{1}{2}}\right),$$

where, of course,

$$\text{var}\left(\mathbb{E}^{\mathcal{H}}\left((2\pi)^{-\frac{1}{2}}\right)\right) = 0. \tag{8.62}$$

This is a common trick where the importance-sampling density is selected to dramatically simplify the integral of interest. In this case, the optimal choice reduces the variance of our estimator to zero. The catch, however, is that this choice is not available to us, because we do not observe $\mathbb{P}(L \geq \ell_\alpha)$. Indeed, this is the quantity that we seek to estimate; if we had it, then we would no longer require the use of the Monte-Carlo estimator.

The development in equation 8.61 is not, however, a waste of time. Glasserman et al. (1999) use it as a starting point and propose an alternative, more workable choice. In particular, they suggest

$$\mu = \arg\max_{y} \mathbb{P}(L \geq \ell_\alpha | Y = y) e^{\frac{-y^2}{2}}. \tag{8.63}$$

While this problem is not terribly easy to solve, it does suggest a few approximations that are, in fact, readily determined. The first, which is referred to as the constant approximation, exchanges the default loss, L, for its conditional expectation, $\mathbb{E}(L|Y = y)$. This leads us to restate equation 8.63 as,

$$\mu = \arg\max_{y} \mathbb{P}\left(\mathbb{E}(L|Y=y) \geq \ell_\alpha | Y = y\right) e^{\frac{-y^2}{2}}, \tag{8.64}$$

$$= \arg\max_{y} \mathbb{I}_{\{\mathbb{E}(L|Y=y) \geq \ell_\alpha\}} e^{\frac{-y^2}{2}},$$

which is basically the maximization of $e^{\frac{-y^2}{2}}$ subject to the conditional default loss exceeding ℓ_α.

Glasserman (2006), Glasserman et al. (2007), and Glasserman and Li (2005) also put forward another alternative closely related to the saddlepoint estimator from Chap. 7. In our development of the saddlepoint method, we found an explicit approximation for the conditional upper-tail probability. In particular, it can be

8.5 Variance-Reduction Techniques

written as

$$\mathbb{P}(L \geq \ell_\alpha | Y = y) = \mathbb{E}\left(\frac{1}{2\pi i}\int_{-i\infty,(0+)}^{+i\infty}\frac{e^{K_L(t,y)-t\ell_\alpha}}{t}dt \bigg| Y = y\right), \quad (8.65)$$

$$\approx \mathbb{E}\left(e^{K_L(\tilde{t}_\ell,y) - \tilde{t}_\ell \ell_\alpha} \underbrace{\mathcal{J}_1\left(\frac{\partial^2 K_L(t,y)}{\partial t^2}\bigg|_{t=\tilde{t}_\ell}, \tilde{t}_\ell\right)}_{\text{Assume} \approx 1} \bigg| Y = y\right),$$

$$\leq \mathbb{E}\left(e^{K_L(\tilde{t}_\ell,y) - \tilde{t}_\ell \ell_\alpha} \bigg| Y = y\right),$$

$$\approx e^{K_L(\tilde{t}_\ell,y) - \tilde{t}_\ell \ell_\alpha},$$

where, once again, the saddlepoint, $\tilde{t}_\ell$, makes an appearance. Figure 8.11—in an attempt to justify the replacement of $\mathcal{J}_1 = 1$ in equation 8.65—illustrates the evolution of the $\mathcal{J}_1$ function for the 99.99th quantile of the Gaussian and t-threshold

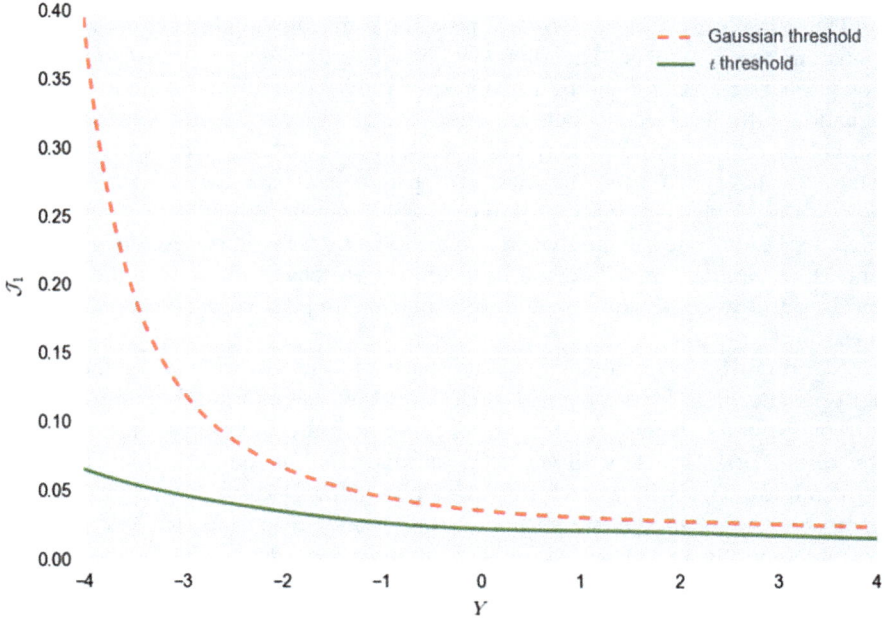

Fig. 8.11 $\mathcal{J}_1$ *behaviour*: This figure illustrates the evolution of $\mathcal{J}_1$ function for the 99.99th quantile of the Gaussian and t-threshold models across a broad range of state variable values. In both cases, we observe that it remains significantly below unity supporting the upper bound employed in equation 8.65.

models across a broad range of state variable values. In both cases, we observe that it remains significantly below unity supporting the upper bound.

Using this approximation, we may return to the original problem represented in equation 8.63 and use the following approximation,

$$\mu = \arg\max_{y} \underbrace{e^{K_L(\tilde{t}_\ell, y) - \tilde{t}_\ell \ell_\alpha}}_{\approx \mathbb{P}(L \geq \ell_\alpha | Y = y)} e^{\frac{-y^2}{2}}, \tag{8.66}$$

$$= \arg\max_{y} K_L(\tilde{t}_\ell, y) - \tilde{t}_\ell \ell_\alpha - \frac{y^2}{2}.$$

Either one of these approaches will allow us, for Gaussian models, to shift the state-variable mean so that, on average, many more extreme, negative realizations occur. The Radon-Nikodym derivative will then perform the necessary adjustment to return to the true upper-tail probability estimate. For many of the non-Gaussian-based mixture models, however, this is not a possibility. The mean of the beta distribution involves two parameters and is not easily shifted. The CreditRisk+ model forces the mean of the gamma-distributed state variable to take a value of unity. Shifting the mean would change the structure of the model. In these cases, this approach is not possible and we are left with using only the Esscher transform.

The consequence is a two-step procedure for the importance-sampling of Gaussian-based models. The first step involves the Esscher transform of the conditional-default probabilities; given θ, this gives rise to the likelihood ratio in equation 8.46. The second step requires shifting the mean of the Gaussian state variable. Armed with μ, we may add a second likelihood ratio from equation 8.58. Thus, we have two new densities, two parameters, and two Radon-Nikodym derivatives. Using both together provides a bit more flexibility over our problem. Figure 8.12 illustrates, in the context of the one-factor Gaussian threshold model, the impact on the transformed default probabilities associated solely with the Esscher transform. It also provides some insight into the impact arising from adding the mean shift.

Algorithm 8.8 illustrates the Python code for the discernment of the mean-shift parameter, μ, for a threshold model. It makes use of a number of previously introduced code examples. As a first step, it uses the `computeP` routine from the `thresholdModels` library to determine the Gaussian conditional default probabilities. This will work for all threshold models—even the normal-mixture approaches—due to the presence of a Gaussian global state variable. Clearly we are ignoring the mixing variable, but it will be dealt with in the following section. The `getSaddlePoint` function appears again to determine the approximation of the conditional tail probability. The final step in the construction of the objective function, `meanShiftOF`, is to replicate equation 8.66. The `getOptimalMeanShift` sub-routine is merely a wrapper around the optimization of this objective function. Observe that, because we seek to maximize equation 8.66, and most optimization packages only minimize, we return the negative value of our objective function in `meanShiftOF`. The effect is, of

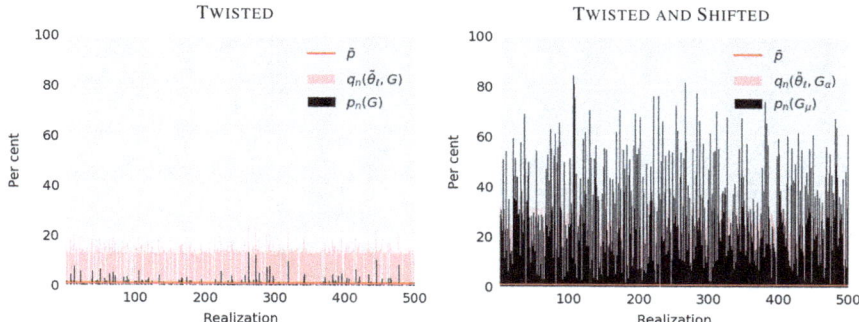

Fig. 8.12 *Twisting and shifting*: This figure illustrates the average unconditional, untransformed conditional, and transformed conditional default probabilities in the context of a one-factor Gaussian threshold model for two cases. The first involves solely an Esscher transform of the conditional default probabilities. The second is a simultaneous twist and shift of the same probabilities. Note the dramatically more extreme average conditional probability of default in the latter case.

Algorithm 8.8 Finding the mean-shift parameter

```
import thresholdModels as th
def meanShiftOF(mu,c,p,l,myRho):
    pZ = th.computeP(p,myRho,mu)
    theta = vc.getSaddlePoint(pZ,c,l,0.0)
    f_l = -theta*l + vc.computeCGF(theta,pZ,c)
    return -(f_l - 0.5*np.dot(mu,mu))
def getOptimalMeanShift(c,p,l,myRho):
    r = scipy.optimize.minimize(meanShiftOF,-1.0,args=(c,p,l,myRho),
                                method='SLSQP',jac=None,bounds=[(-4.0,4.0)])
    return r.x
```

course, maximization. As a final point, we have used Python's `SLSQP` method—which is a sequential non-linear least-squares method—because it permits the introduction of concrete bounds.[19] As the domain of a Gaussian random variable is the entire real line, it is useful, to avoid numerical problems, to restrict the domain somewhat. In the context of our simple example, the optimization requires about 8,000 microseconds for its solution.

Figure 8.13 provides a quick view—once again in the context of the one-factor Gaussian threshold model—of the relationship between the choice of ℓ and the new shifted mean, μ. As ℓ increases in value, Algorithm 8.8 selects a μ value with a larger negative value. As ℓ gets arbitrarily large, the mean tends to be less than -3, which is close to the practical limits of the Gaussian distribution. This is consistent with the negative relationship between the global state variable and the conditional default probability in the threshold-model milieu.

---

[19] See Beck (2014, Chapter 3) for a bit of context on the non-linear least-squares problem.

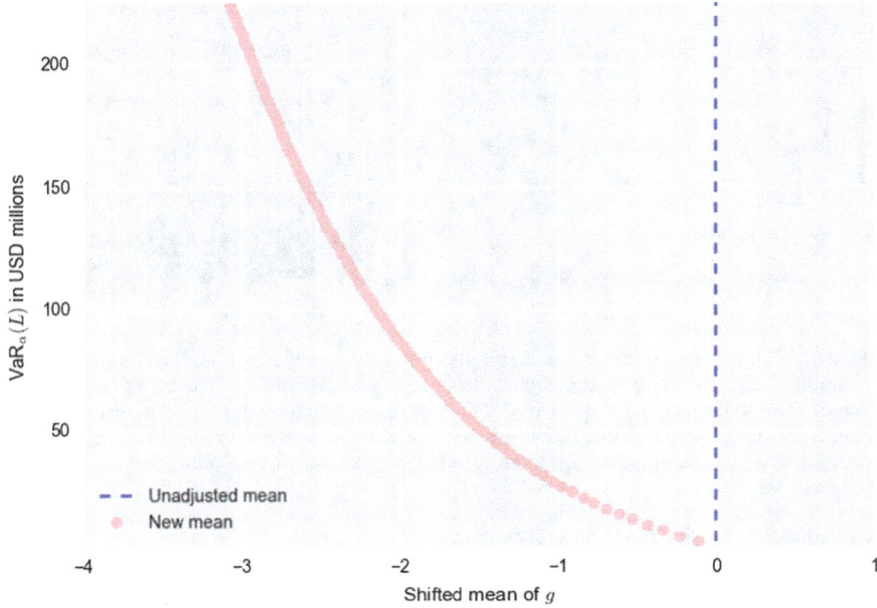

Fig. 8.13 *The shifted mean*: This figure illustrates, in the context of the one-factor Gaussian threshold model, the relationship between the choice of ℓ and the shifted mean, μ.

The preceding technique, for the normal-variance mixture models, assumes a fixed level of the mixing variable. Practically, if not theoretically, this is not particularly problematic. It turns out, however, that this is not optimal for all models. Indeed, as we'll see in the next section, taking it a step further has some benefits for the t-threshold model.

8.5.7 Yet Another Twist

As we'll learn in the latter sections, when we return to our practical example, the Esscher transform of the conditional-default probabilities and the shifting of the mean of the global-state variable (or variables)—when this is possible—actually do a good job of reducing the variance of our Monte-Carlo estimators. This continues to be true in the normal-variance mixture setting when, in fact, the conditional default probabilities depend on two quantities: the global variables and a mixing variable. In the case of the t-threshold model, the results are a bit unstable due to the heaviness of the tails. Kang and Shahabuddin (2005) suggest, therefore, a three-step procedure to improve the situation.

8.5 Variance-Reduction Techniques

The basic idea is to perform an additional twist of the mixing variable, W. In the t-threshold model this is a chi-squared variable with ν degrees of freedom. The general integration problem is,

$$\mathbb{P}(L > \ell_\alpha) = \int_{-\infty}^{\infty} \underbrace{\left(\int_0^\infty \mathbb{I}_{L > \ell_\alpha} f_W(w) dw \right)}_{\text{Our focus}} f_Y(y) dy. \qquad (8.67)$$

Our interest is on the inner integral, which is conditionally Gaussian. That is, given W, the result of this integral follows a Gaussian law. If we work with the distribution of W, we can shift the values of W so that more negative events occur. Before we start, we should recall that if $W \sim \chi^2(\nu)$, then it is also $\Gamma\left(\frac{\nu}{2}, 2\right)$.[20] With this fact in mind, we may now manipulate the inner integral of equation 8.67 as follows,

$$\int_0^\infty \mathbb{I}_{L > \ell_\alpha} f_W(w) dw = \int_0^\infty \mathbb{I}_{L > \ell_\alpha} \frac{w^{\frac{\nu}{2}-1} e^{-\frac{w}{2}}}{2^{\frac{\nu}{2}} \Gamma\left(\frac{\nu}{2}\right)} dw, \qquad (8.68)$$

$$= \int_0^\infty \mathbb{I}_{L > \ell_\alpha} \frac{w^{\frac{\nu}{2}-1} e^{-\frac{w}{2}}}{2^{\frac{\nu}{2}} \Gamma\left(\frac{\nu}{2}\right)} \underbrace{\left(\frac{e^{\gamma w} \left(\frac{1}{1-2\gamma}\right)^{\frac{\nu}{2}}}{e^{\gamma w} \left(\frac{1}{1-2\gamma}\right)^{\frac{\nu}{2}}} \right)}_{=1} dw,$$

$$= \int_0^\infty \mathbb{I}_{L > \ell_\alpha} \frac{w^{\frac{\nu}{2}-1} e^{-\frac{w}{2}+\gamma w}}{2^{\frac{\nu}{2}} \left(\frac{1}{1-2\gamma}\right)^{\frac{\nu}{2}} \Gamma\left(\frac{\nu}{2}\right)} \left(\frac{\left(\frac{1}{1-2\gamma}\right)^{\frac{\nu}{2}}}{e^{\gamma w}} \right) dw,$$

$$= \int_0^\infty \mathbb{I}_{L > \ell_\alpha} \frac{w^{\frac{\nu}{2}-1} e^{-\frac{(1-2\gamma)w}{2}}}{\left(\frac{2}{1-2\gamma}\right)^{\frac{\nu}{2}} \Gamma\left(\frac{\nu}{2}\right)} \left(e^{-\gamma w} (1-2\gamma)^{-\frac{\nu}{2}} \right) dw,$$

$$= \int_0^\infty \mathbb{I}_{L > \ell_\alpha} \underbrace{\frac{w^{\frac{\nu}{2}-1} e^{-\frac{w}{\left(\frac{2}{1-2\gamma}\right)}}}{\left(\frac{2}{1-2\gamma}\right)^{\frac{\nu}{2}} \Gamma\left(\frac{\nu}{2}\right)}}_{\Gamma\left(\frac{\nu}{2}, \frac{2}{1-2\gamma}\right)} \left(e^{-\gamma w} (1-2\gamma)^{-\frac{\nu}{2}} \right) dw,$$

$$= \mathbb{E}^\mathcal{V} \left(\mathbb{I}_{L > \ell_\alpha} \left(e^{-\gamma w} (1-2\gamma)^{-\frac{\nu}{2}} \right) \right).$$

Thus, the integral is now the expectation with respect to the density of a new twisted random variable, $V \sim \Gamma\left(\frac{\nu}{2}, \frac{2}{1-2\gamma}\right)$. If we manipulate the Radon-Nikodym

---

[20] See Johnson et al. (1994, Chapters 17–18) for more information on the chi-squared and gamma distributions and their properties.

derivative somewhat, we will find a familiar form,

$$e^{-\gamma w}(1-2\gamma)^{-\frac{\nu}{2}} = e^{-\gamma w - \frac{\nu}{2}\ln(1-2\gamma)}, \qquad (8.69)$$
$$= e^{K_w(\gamma) - \gamma w},$$

where $K_w(\gamma)$ is the cumulant generating function of the chi-squared distribution. This is the same, and well-known, form found in the Esscher transform of the binomially distributed conditional default events in equation 8.46. It is the usual integrand used in the inverse Laplace transform of the moment-generating function. This is, therefore, a general approach that can be applied to a range of alternative distributions. Asmussen and Glynn (2007, Chapters V and VI) and Fuh et al. (2013) are an excellent entry point into this area of study.

Kang and Shahabuddin (2005) offer an algorithm for the determination of γ and then make the discernment of μ conditional on γ. This requires a separate optimization problem for each realization of $V \sim \Gamma\left(\frac{\nu}{2}, \frac{2}{1-2\gamma}\right)$. This is computationally quite heavy, which essentially defeats the purpose of variance reduction. Kang and Shahabuddin (2005) propose stratified sampling to resolve this issue, but our experience suggests that importance sampling works relatively well for the untransformed use of $W \sim \chi^2(\nu)$. The use of the previous algorithm is helpful, however, to ensure a greater degree of stability. For this purpose, we have set $\gamma \approx -2$, and employ a three-step approach.

8.5.8 Tying Up Loose Ends

The typical stochastic-simulation algorithm, as we've discussed, generates a collection of losses and allows one to estimate a particular quantile from these ordered losses. This algorithm works in the opposite direction: one specifies a range of discrete losses, say over the interval, $[\ell_a, \ell_b]$, and one determines the quantile (or tail probability) associated with each of these losses. Ultimately, the results are equivalent.

Let us work through, before we turn to examine the practical results of imposing this approach on our usual portfolio, the key elements required to implement the two and three-step conditionally independent importance-sampling algorithms for a generic one-factor threshold model. We begin by setting $m = 1$ as the first iteration of the simulation algorithm and then select a particular choice of $\ell \in [\ell_a, \ell_b]$. With these preliminaries, we then proceed to:

1. Generate the global, mixing, and idiosyncratic factors: $G^{(m)}$, $W^{(m)}$, and $\left\{\epsilon_n^{(m)}; n = 1, \ldots, N\right\}$. In the case of the Gaussian-threshold model we might set $W^{(m)} \equiv 1$ for all $m = 1, \ldots, M$. These M latent state-variable outcomes are precisely what allow us to integrate out the conditionality from our VaR and expected-shortfall expressions.

8.5 Variance-Reduction Techniques

2. If it is a t-threshold model, set the twist parameter for the $W \sim \chi^2(\nu)$ random mixing variable to $\gamma = -2$. In the other cases, we set $\gamma \equiv 0$.
3. Define $Y^{(m)}$ as the indicator state variable. It has the following general form for the class of one-factor threshold models,

$$Y_n^{(m)} = \sqrt{f(W)}\left(\sqrt{\rho}G^{(m)} + \sqrt{1-\rho}\epsilon_n^{(m)}\right), \quad (8.70)$$

for $n = 1, \ldots, N$. $f(W)$, as in Chap. 4, represents any necessary transformation of the mixing variable.

4. Compute the conditional default probabilities, $\{p_n\left(G^{(m)}, W^{(m)}\right), n = 1, \ldots, N\}$. They can, as we've already seen, be written as

$$p_n\left(G^{(m)}, W^{(m)}\right) = \frac{\sqrt{\frac{1}{f(W^{(m)})}} F_Y^{-1}(p_n) - \sqrt{\rho}G^{(m)}}{\sqrt{1-\rho}}, \quad (8.71)$$

where $F_Y^{-1}(\cdot)$ is the appropriate inverse cumulative distribution function for the indicator state variable, Y.

5. Solve equation 8.55 for $\theta_\ell\left(G^{(m)}, W^{(m)}\right)$, which for the sake of brevity, we will simply denote as $\theta_\ell^{(m)}$. This is the saddlepoint used to compute the exponentially twisted conditional default probabilities, $\{q_n\left(\theta_\ell^{(m)}, G^{(m)}, W^{(m)}\right); n=1, \ldots, N\}$ from equation 8.44. That is, for posterity, it is written in full for each $n = 1, \ldots, N$ as,

$$q_n\left(\theta_\ell^{(m)}, G^{(m)}, W^{(m)}\right) = \frac{p_n\left(G^{(m)}, W^{(m)}\right) e^{c_n \theta_\ell^{(m)}}}{1 + p_n\left(G^{(m)}, W^{(m)}\right)(e^{c_n \theta_\ell^{(m)}} - 1)}. \quad (8.72)$$

6. For each counterparty, construct the default indicator function for each individual obligor,

$$\mathbb{I}_{\mathcal{D}_n}^{(m)} = \mathbb{I}_{\left\{Y_n^{(m)} \leq F_Y^{-1}(p_n)\right\}}, \quad (8.73)$$

$$= \mathbb{I}_{\left\{\underbrace{\sqrt{f(W)}\left(\sqrt{\rho}G^{(m)} + \sqrt{1-\rho}\epsilon_n^{(m)}\right)}_{\text{Equation 8.70}} \leq F_Y^{-1}(p_n)\right\}},$$

$$= \mathbb{I}_{\left\{\epsilon_n^{(m)} \leq \frac{\sqrt{\frac{1}{f(W^{(m)})}} F_Y^{-1}(p_n) - \sqrt{\rho} G^{(m)}}{\sqrt{1-\rho}}\right\}},$$

$$= \mathbb{I}_{\left\{\epsilon_n^{(m)} \leq \Phi^{-1}\left(p_n(G^{(m)}, W^{(m)})\right)\right\}}.$$

With the default event written in terms of the conditional-default probabilities, movement to the new set of twisted probabilities merely requires the following restatement of equation 8.73 as,

$$\mathbb{I}_{\mathcal{D}_n}^{(m)} = \mathbb{I}_{\left\{\epsilon_n^{(m)} \leq \Phi^{-1}\left(q_n(G^{(m)}, W^{(m)})\right)\right\}}. \tag{8.74}$$

7. Using the exposure and default indicator information, compute the simulated portfolio loss as,

$$L^{(m)} = \sum_{n=1}^{N} c_n \mathbb{I}_{\mathcal{D}_n}^{(m)}. \tag{8.75}$$

8. Using the previous inputs and generalizing equation 8.49, we may construct the following estimator,

$$\mathbb{I}_{\{L^{(m)} \geq \ell\}} e^{-\theta_\ell^{(m)} L^{(m)} + K_L\left(\theta_\ell^{(m)}, G^{(m)}, W^{(m)}\right) - \mu G^{(m)} + \frac{\mu^2}{2} - \gamma W^{(m)} - \frac{\nu}{2} \ln(1-2\gamma)}. \tag{8.76}$$

The likelihood ratio, or Radon-Nikodyn derivative, becomes quite sizable with the three-step approach. If we set the parameters θ_ℓ, μ, and γ to zero, however, it reduces to unity and we return to the raw Monte Carlo situation.

9. We now need to repeat the previous steps for $m = 1, \ldots, M$.

To estimate the tail probability associated with x, we now merely use

$$\widehat{\mathbb{P}(L \geq \ell_\alpha)} \approx \int_{-\infty}^{\infty} \int_{0}^{\infty} \mathbb{P}(L > \ell_\alpha | Y = y, W = w) f_W(w) f_Y(y) dw dy, \tag{8.77}$$

$$\approx \frac{1}{M} \sum_{m=1}^{M} \mathbb{I}_{\{L^{(m)} \geq \ell\}} e^{-\theta_\ell^{(m)} L^{(m)} + K_L\left(\theta_\ell^{(m)}, G^{(m)}, W^{(m)}\right) - \mu G^{(m)} + \frac{\mu^2}{2} - \gamma W^{(m)} - \frac{\nu}{2} \ln(1-2\gamma)},$$

which amounts to averaging over the loss estimates constructed in equation 8.76; we are also simultaneously integrating out the latent state variables thereby removing the conditionality. This is then repeated for a range of ℓ values to map out the tail of the loss distribution.

If one wishes to compute the expected shortfall, then we merely adjust equation 8.77 using the logic from equation 8.23. The result is,

$$
\begin{aligned}
&\mathbb{E}(\widehat{L|L \geq \ell_\alpha}) \\
&\approx \int_\infty^\infty \int_0^\infty \mathbb{E}(L|L > \ell_\alpha, Y = y, W = w) f_W(w) f_Y(y) dw dy, \qquad (8.78) \\
&\approx \frac{\frac{1}{M} \sum_{m=1}^M L^{(m)} \mathbb{I}_{\{L^{(m)} \geq \ell\}} e^{-\theta_\ell^{(m)} L^{(m)} + K_L\left(\theta_\ell^{(m)}, G^{(m)}, W^{(m)}\right) - \mu G^{(m)} + \frac{\mu^2}{2} - \gamma W^{(m)} - \frac{\nu}{2} \ln(1 - 2\gamma)}}{\widehat{\mathbb{P}(L \geq \ell_\alpha)}}.
\end{aligned}
$$

While the form appears, at first glance, quite complex, it is easily constructed from the ingredients of the simulation algorithm.

Algorithm 8.9 provides the Python code used to generate the VaR and expected-shortfall estimates with the importance-sampling approach for the threshold models. It is, quite likely, the longest algorithm demonstrated in this book, but this is due to its attempt to handle multiple elements, rather than any particular complexity. It has two flag arguments: `shiftMean` and `isT`. The former allows one to turn on or off the mean shift to the Gaussian global state variable. The latter determines the type of threshold model: 0, 1, and 2 denote the Gaussian, t, and variance-gamma models, respectively.[21] Incorporation of these multiple cases adds a substantial number of lines to Algorithm 8.9.

The structure of the code is essentially identical to the independent-default case summarized in Algorithm 8.7; it merely has a number of additional *bells and whistles*. When `isT=1`, for example, Algorithm 8.9 incorporates the exponential twist of the chi-squared mixing variable. This leads to the creation of an additional contribution, `rnChi`, to the Radon-Nikodym derivative. Another key aspect is the need to compute a different `theta` parameter, set of twisted probabilities, `qZ`, and the cumulant-generating function, `cgf` for each state-variable outcome. This requires a loop over the M iterations. The consequence is the addition, in the conditionally independent case, of a significant amount of overhead for each iteration. Equipped with these elements, the tail probability and expected shortfall are computed in precisely the same manner as described in equations 8.77 and 8.78.

8.5.9 Does It Work?

To close out this chapter and to examine the effectiveness of the importance-sampling technique, let's return to our running 100-obligor sample portfolio. Using the calibration parameters from the previous chapters, we will consider five

---

[21] The `invVector` argument passes the numerically computed inverse cumulative distribution function values for the variance-gamma and generalized-hyperbolic models. This is purely to speed the code by performing this computationally expensive step once rather than many times.

Algorithm 8.9 The *Full Monte* threshold-model importance-sampling algorithm

```
def isThreshold(N,M,p,c,l,myRho,nu,shiftMean,isT,invVector=0):
    mu = 0.0
    gamma = 0.0
    if shiftMean==1:
        mu = getOptimalMeanShift(c,p,l,myRho)
    theta = np.zeros(M)
    cgf = np.zeros(M)
    qZ = np.zeros([M,N])
    G = np.transpose(np.tile(np.random.normal(mu,1,M),(N,1)))
    e = np.random.normal(0,1,[M,N])
    if isT==1:
        gamma = -2
        W = np.random.chisquare(nu,M)
        myV = W/(1-2*gamma)
        V = np.transpose(np.sqrt(np.tile(myV,(N,1))/nu))
        num = (1/V)*myT.ppf(p,nu)*np.ones((M,1))-np.multiply(np.sqrt(myRho),G)
        pZ = norm.cdf(np.divide(num,np.sqrt(1-myRho)))
    elif isT==2:
        V = np.transpose(np.sqrt(np.tile(np.random.gamma(nu,1/nu,M),(N,1))))
        num = (1/V)*invVector*np.ones((M,1))-np.multiply(np.sqrt(myRho),G)
        pZ = norm.cdf(np.divide(num,np.sqrt(1-myRho)))
    else:
        pZ = th.computeP(p,myRho,G)
    for n in range(0,M):
        theta[n] = vc.getSaddlePoint(pZ[n,:],c,1,0.0)
        qZ[n,:] = getQ(theta[n],c,pZ[n,:])
        cgf[n] = vc.computeCGF(theta[n],pZ[n,:],c)
    I = np.transpose(1*np.less(e,norm.ppf(qZ)))
    L = np.dot(c,I)
    if isT==1:
        rnChi = np.exp(-gamma*myV-(nu/2)*np.log(1-2*gamma))
    else:
        rnChi = np.ones(M)
    if shiftMean==1:
        rn = computeRND(theta,L,cgf)*np.exp(-mu*G[:,0]+0.5*(mu**2))*rnChi
    else:
        rn = computeRND(theta,L,cgf)*rnChi
    tailProb = np.mean(np.multiply(L>l,rn))
    eShortfall = np.mean(np.multiply(L*(L>l),rn))/tailProb
    return tailProb,eShortfall
```

alternative models: the independent-default binomial, the Gaussian threshold, the CreditRisk+, the *t*-threshold, and the variance-gamma threshold models. All are examined in the one-factor setting. The plan is to use the aforementioned methods to produce relatively efficient and fast risk-metric estimates for each of these modelling choices.

To assess the accuracy of the importance-sampling estimators, however, we need to have a solid estimate of the true VaR and expected-shortfall risk metrics for a broad range of confidence levels. As initially shown in Table 8.4, each set of model results was averaged over separate runs of 25 repetitions of 10 million iterations. Our estimate of each true risk-metric value is thus based on 250 million iterations of the

8.5 Variance-Reduction Techniques

raw stochastic-simulation algorithms. There will, of course, be some residual noise, but these will nonetheless be quite sharp—albeit highly computationally intensive—estimates and form a very good proxy for the true underlying values.

Figure 8.14 starts the analysis with a comparison of our raw and importance-sampling VaR estimates for a range of alternative confidence levels. The importance-sampling values are computed as the average of 50 repetitions of 3,000 iterations. We are, in essence, comparing raw simulation of 250 millions iterations to variance-reduced estimation with 150,000 random draws. This is, quite frankly, a fairly cheeky thing to do. There is, therefore, to be very explicit, a difference of three orders of magnitude in the overall number of iterations. The results are quite remarkable. In four of five cases, there is relatively little visual difference between the two competing estimates.

The exception is the one-factor CreditRisk+ mixture model. The importance-sampling estimator appears to overestimate the tail probabilities associated with various levels of default loss. Indeed, the tail-probability profile of the CreditRisk+ importance-sampling estimates compare much more closely with the independent-default binomial model than with the raw Monte-Carlo estimates. Twisting the conditional-default probabilities—as with the one-factor Gaussian threshold model in Table 8.4—does not seem to help much in this setting. The reason appears to stem from the role of the single global state variable impacting all obligors. Twisting the

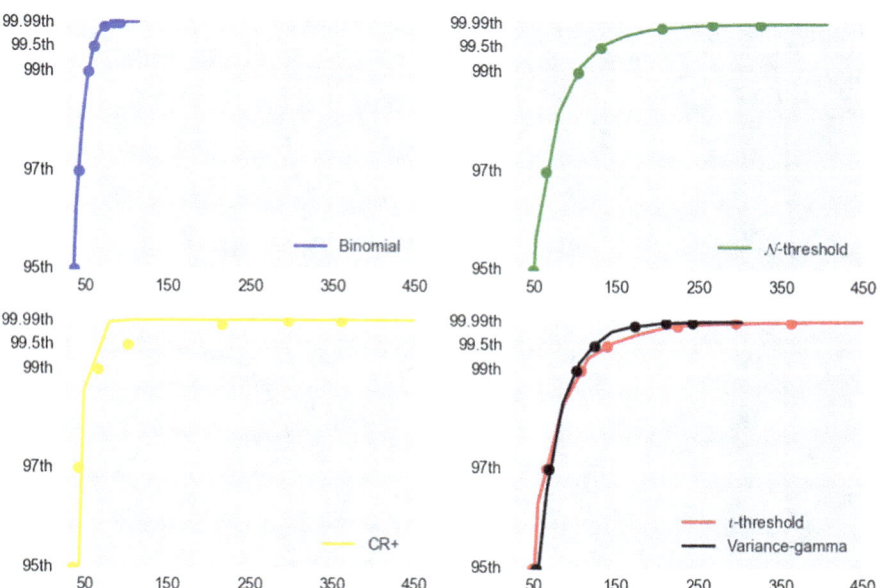

Fig. 8.14 *Raw vs. IS VaR comparison*: This figure compares the raw and importance-sampling VaR estimates and associated tail probabilities across five alternative models using our usual 100-obligor example. The IS values are computed with 50 repetitions of 3,000 iterations, whereas the raw simulation values involve 250 repetitions of 10 million iterations.

482   8 Monte Carlo Methods

default probabilities does not capture this dimension. The situation might potentially be improved by shifting this global gamma-distributed state variable, but it has already been restricted to have a unit mean. Shifting it, therefore, would upset the balance of the model.

Figure 8.15 repeats the analysis from Fig. 8.14, but focuses now on the expected-shortfall measure. A very similar pattern is evident. The raw and variance-reduced estimates exhibit a close match despite a dramatically lower number of total simulation iterations. Once again, our importance-sampling technique does not appear to perform particularly well in the mixture-model setting.

We might be inclined to conclude, on the basis of these two figures, that the importance-sampling approach is an unmitigated success—the notable exception being the CreditRisk+ model—and that it should be used exclusively in the threshold setting. As is often the case, reality is a bit more nuanced. We have demonstrated that a reasonable fit is possible using importance sampling with dramatically fewer iterations. The number of iterations alone, however, does not completely determine the overall computational expense. Each iteration, in the importance-sampling setting, requires significantly more time.

To assess this important dimension, we perform one final experiment. We examine the mean, standard-error, and computational time associated with seven distinct confidence levels for three distinct estimators: the brute-force 10 million iterations Monte-Carlo, our 3,000 iteration importance-sampling, and raw Monte-

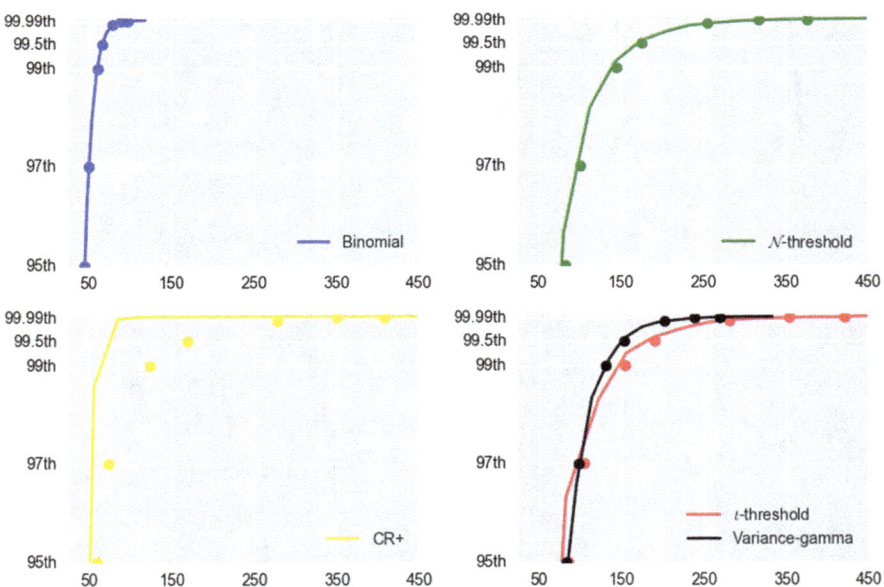

Fig. 8.15 *A tighter comparison*: This figure compares the raw and importance-sampling expected-shortfall estimates and associated tail probabilities across five alternative models using our usual 100-obligor example. As in Fig. 8.14, the IS values are computed with 50 repetitions of 3,000 iterations, whereas the raw simulation values involve 250 repetitions of 10 million iterations.

8.5 Variance-Reduction Techniques

Carlo with 100,000 iterations. To keep the sheer number of figures to a minimum, we present only the one-factor Gaussian threshold model. The results are similar for the other approaches—with the notable exception of CreditRisk+—although, in the interests of full disclosure, it does exhibit the best behaviour.

Table 8.5 displays the results. Each repetition of the 10 million iteration approach requires roughly 50 seconds; at 25 repetitions, a total of almost 21 minutes of computer time is utilized. The standard error for this brute-force estimator is extremely small for lower levels of confidence, but increases significantly as we reach the extreme part of the tail of the default-loss distribution. The same is true for the importance-sampling and 100,000 iteration raw-simulation estimates. As the level of confidence increases, so does the standard error. The difference, however, is the rate of speed of the increase. The importance-sampling standard error grows much more slowly. At the 95th percentile, the variance-reduced standard error is numerous multiples of the 10 million and 100,000 raw-simulation estimates. At the 99.99th quantile, however, it is roughly 50% greater than the 10 million iteration result and only about 15% of the 100,000 iteration standard error. The consequence

Table 8.5 *Variance-reduction report card*: This table summarizes a number of summary statistics for three different possible Monte-Carlo estimators of the one-factor Gaussian threshold model VaR applied to our usual example. The approaches include two raw Monte-Carlo estimates using 10 million and 100,000 iterations, respectively. The third method is the importance-sampling technique with 3,000 iterations.

Value	Statistic	Results (USD millions)		
		Raw (10M)	IS (3,000)	Raw (100,000)
95.00	Mean	$48.5	$49.6	$48.5
	SE	$0.1	$1.4	$0.4
	99% C.I.	[$48.5, $48.6]	[$49.0, $50.1]	[$48.3, $48.6]
97.00	Mean	$63.5	$65.8	$63.6
	SE	$0.1	$1.6	$0.5
	99% C.I.	[$63.5, $63.6]	[$65.2, $66.4]	[$63.5, $63.8]
99.00	Mean	$102.0	$102.5	$102.2
	SE	$0.1	$1.6	$1.1
	99% C.I.	[$102.0, $102.1]	[$101.9, $103.2]	[$101.8, $102.6]
99.90	Mean	$204.3	$205.4	$204.4
	SE	$0.6	$3.0	$4.0
	99% C.I.	[$203.9, $204.6]	[$204.2, $206.7]	[$202.8, $205.9]
99.99	Mean	$324.1	$324.5	$327.6
	SE	$1.8	$2.8	$19.7
	99% C.I.	[$323.1, $325.1]	[$323.4, $325.6]	[$320.0, $335.1]
Time	Average	50.2 s	0.7 s	0.5 s
	Total	20.9 min	0.5 min	0.4 min
	Per m	5 μs	220 μs	5 μs
Iterations	Total	250,000,000	150,000	5,000,000

is rather tight importance-sampling VaR estimates at the most extreme quantiles. It is particularly useful to compare the outcomes to the unshifted state-variable mean results in Table 8.4.

At the lower quantiles, therefore, the 100,000 raw estimator outperforms the importance-sampling technique. At the 99.99th quantile, however, the variance-reduction technique succeeds in competing with the 10 million iteration raw Monte-Carlo approach. This is consistent with the underlying notion of importance sampling: transforming the default-loss density so that raw events become more likely. This transformation comes at a cost. It requires the solution of a number of sub-problems at each iteration. We see clearly from Table 8.5 that each iteration in the standard simulation setting requires about 5 microseconds (μs).[22] The equivalent computation for the importance-sampling estimator is more than 40 times slower at 220 microseconds. This implies that 3,000 variance-reduced iterations require roughly the same time as our 100,000 iteration approach.

It is nevertheless important to add that the true amount of computational effort for the variance-reduced approach can be practically larger than implied by this analysis. The results in Table 8.5 summarize the computational time for a single run of the importance-sampling method. A separate run is, in principle, required for each level of default loss. This means the actual estimates could, depending on one's knowledge of the loss distribution, require rather more effort to isolate the actual tail probabilities. In contrast to the saddlepoint approximation in Chap. 7, this cannot be solved easily with numerical methods. Instead, it requires a bit of trial and error. This is not necessarily a huge issue, and can be resolved with a simple grid, but it is worth consideration and adds to the overall computational burden.

The importance-sampling technique is not a panacea. It is mathematically complex and, as we've seen in the case of the CreditRisk+ model, does not perform admirably in all settings. It is also, with its reliance on the solution of non-linear optimization sub-problems, somewhat fragile. It is possible, in some cases, for the optimization algorithm in some states to diverge thereby actually leading to noisier estimates. Importance-sampling, working from default loss to tail probability, also has some hidden grid-search costs. It does also offer some advantages. Raw Monte-Carlo seems like a sensible alternative in our 100-obligor example, but the true power of this approach becomes more evident when examining large portfolios with thousands of obligors. In this case, millions of raw Monte-Carlo iterations become extremely computationally expensive whereas the importance-sampling approach can work to reduce the burden. It is also worth stressing that the importance-sampling technique also readily computes expected-shortfall estimates and, using the simulation techniques from Chap. 7, the risk-metric obligor contributions. Despite its shortcomings, therefore, importance-sampling is an effective tool that merits inclusion in the quantitative analyst's toolkit. Since it requires close analyst attention, the trick is to know when it can and it should be employed.

---

[22] Recall that one microsecond is a millionth of a second.

On a closing note, effective use of any model requires sensitivity analysis. The challenge with sensitivity analysis in the simulation setting is that significant computation is necessary. There is a very real danger that, faced with time constraints, the analyst will dispense with sensitivity analysis altogether. This unfortunate decision, while understandable, will generally lead to lower levels of model understanding and troubleshooting; the corollary, of course, is an attendant increase in model risk. To the extent that importance-sampling techniques, when judiciously employed, can reduce the overall computational cost and relax time constraints, it will act as a positive contribution to one's overall model diagnostics. One can, in fact, consider it to be a model diagnostic in its own right.

8.6 Final Thoughts

This is the final chapter in a sequence of tools available to the quantitative analyst for the enhancement of model understanding. What have we learned? Stochastic simulation, which at its heart is basically a numerical integration technique, is a key tool with particular importance in the credit-risk modelling setting. The reasons for its usefulness are twofold: dimensionality and rarity. Monte Carlo numerical integration, as we saw in the first sections of this chapter, is an inefficient approach. It is easily outperformed by grid-based methods. Grid-based approaches, however, fail spectacularly as we increase the dimensionality. This weakness is referred to, by Bellman (1961), as the curse of dimensionality. Like the slow-and-steady tortoise in Aesop's famous parable, the Monte-Carlo technique, while still slow, exhibits dimension-independent convergence, which makes it literally the only practical contender for dealing with high-dimensional integrals.

Although extremely useful, Monte-Carlo simulation is not to be used without caution. A close eye should always be kept on convergence properties of one's estimators. Practically, this implies that it is good practice to always compute standard errors and confidence intervals for one's estimates. Instead of running 10 million iterations, it is preferable to compute 10 repetitions of one million iterations and use this resulting information to construct both point and interval estimates. No additional computation effort is required and one gains significant insight into the accuracy of one's estimates. The method nevertheless, with its $O\left(\frac{1}{\sqrt{M}}\right)$ convergence, remains stubbornly slow. Variance-reduction techniques are potentially helpful, but neither offer a panacea nor are they particularly easy to use and implement. Another area, not discussed in this work, is to harness one's computational infrastructure through the use of parallel-computing techniques to bring more processing power to bear on the raw Monte-Carlo simulation method. Zaccone (2015) is a good starting point. To summarize, simulation is a critical tool, which is easy to learn, but offers numerous hidden pitfalls and is difficult to master. As with all high-powered tools, it requires caution, skill, and a careful touch.

References

Arnow, B. J. (1994). On laplace's extension of the buffon needle problem. *The College Mathematics Journal, 25*(1), 40–43.

Asmussen, S., & Glynn, P. W. (2007). *Stochastic simulation: Algorithms and analysis* (1st edn.). Heidelberg, Germany: Springer-Verlag. *Stochastic modelling and applied probability.*

Badger, L. (1994). Lazzarini's lucky approximation of π. *Mathematics Magazine, 67*(2), 83–91.

Beck, A. (2014). *Introduction to nonlinear optimization: Theory, algorithms, and applications with MATLAB* (1st edn.). Philadelphia, USA: Society for Industrial and Applied Mathematics.

Bellman, R. (1961). *Adaptive control processes: A guided tour*. Princeton: Princeton University Press.

Billingsley, P. (1995). *Probability and measure* (3rd edn.). Third Avenue, New York, NY: Wiley.

BIS. (2001). The internal ratings-based approach. *Technical report*. Bank for International Settlements.

BIS. (2004). International convergence of capital measurement and capital standards: A revised framework. *Technical report*. Bank for International Settlements.

BIS. (2005). An explanatory note on the Basel II IRB risk weight functions. *Technical report*. Bank for International Settlements.

Box, J. F. (1981). Gosset, Fisher, and the t-distribution. *The American Statistician, 35*(2), 61–66.

Box, J. F. (1987). Guiness, Gosset, Fisher, and small samples. *Statistical Science, 2*(1), 45–52.

Boyle, P. (1977). Options: A Monte Carlo approach. *Journal of Financial Economics, 4*(3), 323–338.

Brereton, T. J., Kroese, D. P., & Chan, J. C. (2012). *Monte Carlo methods for portfolio credit risk*. Australia: The University of Queensland.

Casella, G., & Berger, R. L. (1990). *Statistical inference*. Belmont, CA: Duxbury Press.

Duffie, D. (1996). *Dynamic asset pricing theory* (2nd edn.). Princeton, NJ: Princeton University Press.

Durrett, R. (1996). *Probability: Theory and examples* (2nd edn.). Belmont, CA: Duxbury Press.

Eckhardt, R. (1987). Stan Ulam, John Von Neumann and the Monte Carlo method. *Los Alamos Science*, 131–141.

Fishman, G. S. (1995). *Monte Carlo: Concepts, algorithms, and applications*. 175 Fifth Avenue, New York, NY: Springer-Verlag. *Springer series in operations research.*

Fuh, C.-D., Teng, H.-W., & Wang, R.-H. (2013). *Efficient importance sampling for rare event simulation with applications*. Taiwan: National Central University.

Glasserman, P. (2004a). *Monte Carlo methods in financial engineering* (1st edn.). Berlin: Springer.

Glasserman, P. (2006). Measuring marginal risk contributions in credit portfolios. Risk Measurement Research Program of the FDIC Center for Financial Research.

Glasserman, P., Heidelberger, P., & Shahabuddin, P. (1999). Asymptotically optimal importance sampling and stratification for pricing path-dependent options. *Mathematical Finance, 9*(2), 203–228.

Glasserman, P., Kang, W., & Shahabuddin, P. (2007). *Fast simulation for multifactor portfolio credit risk*. Graduate School of Business, Columbia University.

Glasserman, P., & Li, J. (2005). Importance sampling for portfolio credit risk. *Management Science, 51*(11), 1643–1656.

Gundlach, M., & Lehrbass, F. (2004). *CreditRisk+ in the banking industry* (1st edn.). Berlin: Springer-Verlag.

Harrison, J., & Kreps, D. (1979). Martingales and arbitrage in multiperiod security markets. *Journal of Economic Theory, 20*, 381–408.

Harrison, J., & Pliska, S. (1981). Martingales and stochastic integrals in the theory of continuous trading. *Stochastic Processes and Their Applications, 11*, 215–260.

Held, L., & Bové, D. S. (2014). *Applied statistical inference*. Berlin, Germany: Springer-Verlag.

Heunis, A. J. (2011). *Notes on stochastic calculus*. University of Waterloo.

References

Johnson, N. L., Kotz, S., & Balakrishnan, N. (1994). *Continuous univariate distributions: volume I* (2nd edn.). New York, NY: John Wiley & Sons.

Kang, W., & Shahabuddin, P. (2005). Fast simulation for multifactor portfolio credit risk in the t-copula model. Proceedings of the 2005 Winter Simulation Conference.

Karatzas, I., & Shreve, S. E. (1991). *Brownian motion and stochastic calculus* (2nd edn.). Berlin: Springer-Verlag.

McLeish, D. (2005). *Monte Carlo simulation and finance* (1st edn.). Wiley.

McNeil, A. J., Frey, R., & Embrechts, P. (2015). *Quantitative risk management: Concepts, tools and techniques*. Princeton, NJ: Princeton University Press.

Merton, R. (1974). On the pricing of corporate debt: The risk structure of interest rates. *Journal of Finance, 29*, 449–470.

Metropolis, N. (1987). The beginning of the Monte Carlo method. *Los Alamos Science*, 125–130.

Musiela, M., & Rutkowski, M. (1998). *Martingale methods in financial modelling* (1st edn.). Berlin: Springer-Verlag.

Panjer, H. H. (1981). Recursive evaluation of a family of compound distributions. *Astin Bulletin*, (12), 22–26.

Press, W. H., Teukolsky, S. A., Vetterling, W. T., & Flannery, B. P. (1992). *Numerical recipes in C: The art of scientific computing* (2nd edn.). Trumpington Street, Cambridge: Cambridge University Press.

Ralston, A., & Rabinowitz, P. (1978). *A first course in numerical analysis* (2nd edn.). Mineola, NY: Dover Publications.

Reesor, R. M., & McLeish, D. L. (2001). Risk, entropy and the transformation of distributions, *Working paper*, Department of Statistics and Actuarial Science, University of Waterloo, Waterloo, Ontario, Canada.

Wilde, T. (1997). *CreditRisk+: A credit risk management framework*. Credit Suisse First Boston.

Zaccone, G. (2015). *Python parallel programming cookbook*. Birmingham, UK: Packt Publishing.

Part III
Parameter Estimation

Models require parameters, but these values sadly do *not* fall from the sky. Optimally, one seeks model parameters that are consistent with observed reality. Although this is a reasonable and defensible statement, it raises some practical challenges. The first question, of course, is which observed reality? What is, for example, the appropriate composition and time period to use in the compilation of one's datasets? While an important and difficult question, we leave this point for other authors. A second question consumes our attention: what is a reasonable way to extract parameter estimates from observed default and transition outcomes? Chapter 9 tackles this question for the unconditional default probabilities associated with individual obligors. The maximum-likelihood framework is a useful tool for the production of both point and interval default-probability estimates. The latter notion of interval estimate—often concretely represented by confidence intervals—is essential given the noisiness of typical datasets. Rarity of default implies few observations, which in turn makes statistical estimation challenging. A more general method, referred to as statistical bootstrapping, is presented to help assess estimation uncertainty. Finally, an alternative calibration approach—using credit-default swap data—is investigated for the identification of default probabilities. While these two approaches are *not* equivalent, there is value and insight in their joint examination. Chapter 10 concludes with a difficult task. It seeks to consider methods for determining parameters that govern default dependence between obligors. Unlike default probabilities, these parameters are model specific, which enhances the complexity of their estimation. A variety of techniques, of varying degrees of sophistication and robustness, are considered. Equally importantly, a concerted effort is consistently made to determine the informativeness of the data and the associated certainty of one's parameter estimates. For quantitative analysts seeking to manage the risk of model usage, this is a critical undertaking.

Chapter 9
Default Probabilities

> *A map is not the territory it represents, but, if correct, it has a similar structure to the territory, which accounts for its usefulness.*
>
> (Alfred Korzybski)

In credit-risk modelling, perhaps the single most important modelling input is an assessment of the relative creditworthiness of the individual obligors in one's portfolio. This assessment typically takes the form of an estimate of the unconditional probability of default for each individual credit counterparty, in one's portfolio, over a predetermined period of time; the usual starting point is one year. While there are a variety of sources for this information—indeed one can even purchase credit-risk analysis from a broad range of third parties vendors—it is essential for the quantitative analyst to develop a fundamental understanding of how these quantities are determined and, equally importantly, how their relative accuracy or reliability might be appraised. This chapter is thus dedicated to precisely this task.

There are, to simplify things somewhat, two broad approaches toward the determination of default probabilities: estimation and calibration. The former approach uses transition and default history for similarly classified obligors over time and employs statistical techniques to estimate the desired values. It involves a variety of assumptions. The first and perhaps most important point is that there is a relatively small and finite number of credit states and their dynamics are well approximated by a mathematical object termed a Markov chain. The pay-off for these assumptions is access to the full apparatus associated with the field of statistical estimation and inference.

The second approach takes an alternative tact. It examines market instruments—such as bond obligations or credit-default swaps—and seeks to extract implied default probabilities from their observed prices. Since the market prices are related to the individual obligors, the corresponding estimates relate directly to specific obligors rather than groups with similar characteristics. The actual computation, which is a bit involved, reduces to a non-linear optimization problem. In many ways, the exercise is very similar to extracting zero-coupon rates from observed bond prices. Statistical inference, however, is not possible.

While tempting to consider these two approaches as alternatives, this would be a mistake. They operate in two rather distinct settings and require a significant effort and not a few assumptions to link them explicitly. More specifically, parameter estimation operates under the physical probability measure, whereas calibration occurs under the equivalent-martingale measure. The distance between these two risk measures relates to aggregate attitudes toward risk. Reconciling these alternative estimates, therefore, involves grappling with risk preferences. Perhaps more importantly, these alternative estimates naturally relate to different applications. Estimation, incorporating market risk premia, is suited to risk management. Calibration, conversely, is more consistent with asset pricing. We will, in the following sections, consider both approaches separately. In the final part of this chapter, we will further illustrate how they might be compared and, to a certain extent, reconciled.

9.1 Some Preliminary Motivation

Over the course of the previous eight chapters, we have worked with a common 100-obligor portfolio example. Embedded in that portfolio are distinct, and randomly generated default-probability estimates. This, as many readers have certainly noted, is not terribly realistic. This shortcoming will, however, be rectified in the following discussion. Before we get to this critical improvement, however, we pose an interesting question. How important are the default probabilities? Fairly important, we suspect. If all default probabilities are arbitrarily set to zero, the credit VaR and expected-shortfall estimates will be identically zero. Similarly, setting these values to unity will create a 100% loss of the portfolio. Our risk measures are thus an increasing function of default probabilities. Certainly, therefore, they are important.

It is still fair to ask our question. Yes, we agree that default probabilities are important: but, what is the nature of the relationship? Is the increase linear or non-linear? How strongly does it depend on the model? To test this, in a quick—but hopefully illuminating manner—we scale the default probabilities by a linear factor, γ. This leads us to re-state the VaR measure as,

$$\text{VaR}_\alpha(L) \equiv \text{VaR}_\alpha \left(\sum_{n=1}^{N} p_n(\gamma) c_n \right). \tag{9.1}$$

The default probabilities, as well as the VaR measure, are a direct function of the γ parameter. We can scale up, or down, the value of γ and see what is the impact of a linear movement in the unconditional default probabilities on the VaR measure. Practically, we accomplish this by selecting a range of γ values in the interval, $(0, 2)$. For each value, we estimate the 99.9th quantile VaR employing raw simulation of the Gaussian and t-distributed threshold models with two million iterations. The results are summarized in Fig. 9.1.

9.1 Some Preliminary Motivation

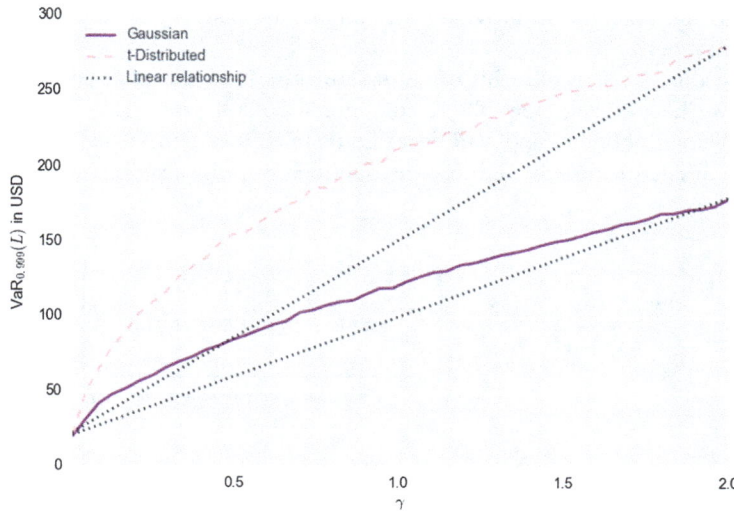

Fig. 9.1 *VaR sensitivity to default probabilities*: The underlying figure highlights the impact on our 99.99th quantile VaR measure, applied to our usual 100-obligor example, for a linear increase in all default probabilities.

As the value of γ approaches zero, our VaR measure also tends, as we would expect, to zero. As the default probabilities are doubled, we observe significant increases in the magnitude of the VaR estimates. The size of the increase is larger for the t-distributed relative to the Gaussian model. In the former case, the risk almost doubles, where in the latter situation it increases by roughly 50%. In both cases, however, we clearly perceive a non-linear movement in the VaR measure associated with a linear increment in the value of γ. We conclude, therefore, that not only are default probabilities important, but that they have non-linear effects on our risk measures that vary importantly among alternative models. While suggestive, with a bit more effort, we can gain even more insight into the relationship between our risk measures and the assessment of our obligors' creditworthiness.

9.1.1 A More Nuanced Perspective

Another way to appreciate the importance of default probabilities involves a more direct, and classic, assessment of the sensitivity of a risk-metric. In Chap. 7, we performed risk attribution with, among other things, the partial derivatives of our risk measures with respect to individual exposures. We can, of course, recycle some of this analysis to compute the partial derivatives of a given risk metric with respect to each individual default probability. This will not help us perform risk attribution, but it has the potential to provide some interesting insight into the role of these important inputs.

To do this, we'll need a bit of set-up. Imagine that, for a general one-factor model, we have a state variable y, conditional default probability $p_n(y)$ for the nth of N total obligors, and VaR measure denoted as ℓ_α. As usual, α represents the appropriate quantile of the loss distribution. We will also need the following quantity,

$$\frac{\partial K_L(t,y)}{\partial p_n(y)} = \frac{\partial}{\partial p_n(y)} \underbrace{\left(\sum_{k=1}^{N} \ln\left(1 - p_k(y) + p_k(y)e^{tc_k}\right) \right)}_{\text{Cumulant generating function}}, \qquad (9.2)$$

$$= \sum_{k=1}^{N} \frac{\partial}{\partial p_n(y)} \left(\ln\left(1 - p_k(y) + p_k(y)e^{tc_k}\right) \right),$$

$$= \frac{e^{tc_n} - 1}{(1 - p_n(y) + p_n(y)e^{tc_n})},$$

where, as with the exposures, the sum collapses to a single term. All other terms, except those involving $p_n(y)$, vanish.

The partial derivative of ℓ_α—or our αth quantile VaR—can be approximated with the following saddlepoint technique borrowed from Chap. 7,

$$\frac{\partial \mathbb{P}(L > \ell_\alpha | Y = y)}{\partial p_n(y)} \qquad (9.3)$$

$$= \frac{\partial}{\partial p_n(y)} \left(\frac{1}{2\pi i} \int_{-i\infty,(0+)}^{+i\infty} \frac{e^{K_L(t,y) - t\ell_\alpha}}{t} dt \right),$$

$$\underbrace{\frac{\partial(1-\alpha)}{\partial p_n(y)}}_{=0}$$

$$= \frac{1}{2\pi i} \int_{-i\infty,(0+)}^{+i\infty} \frac{1}{t} \left(\frac{\partial K_L(t,y)}{\partial p_n(y)} - t \frac{\partial \ell_\alpha}{\partial p_n(y)} \right) e^{K_L(t,y) - t\ell_\alpha} dt,$$

$$\frac{\partial \ell_\alpha}{\partial p_n(y)} \frac{1}{2\pi i} \int_{-i\infty,(0+)}^{+i\infty} e^{K_L(t,y) - t\ell_\alpha} dt$$

$$= \frac{1}{2\pi i} \int_{-i\infty,(0+)}^{+i\infty} \frac{\partial K_L(t,y)}{\partial p_n(y)} \frac{e^{K_L(t,y) - t\ell_\alpha}}{t} dt,$$

$$\frac{\partial \ell_\alpha}{\partial p_n(y)} \left(e^{K_L(\tilde{t}_\ell, y) - \tilde{t}_\ell \ell_\alpha} \mathcal{J}_0 \left(K_L''(\tilde{t}_\ell, y), \tilde{t}_\ell \right) \right)$$

9.1 Some Preliminary Motivation

$$\approx \underbrace{\left(\frac{1}{\tilde{t}_\ell} \cdot \frac{e^{c_n\tilde{t}_\ell}-1}{1-p_n(y)+p_n(y)e^{c_n\tilde{t}_\ell}}\right)}_{\text{Equation 9.2}} e^{K_L(\tilde{t}_\ell, y) - \tilde{t}_\ell \ell_\alpha} \mathcal{J}_0\left(K_L''(\tilde{t}_\ell, y), \tilde{t}_\ell\right),$$

$$\underbrace{\mathbb{E}\left[\frac{\partial \ell_\alpha}{\partial p_n(y)}\right] \mathbb{E}\left[\left(e^{K_L(\tilde{t}_\ell, y) - \tilde{t}_\ell \ell_\alpha} \mathcal{J}_0\left(K_L''(\tilde{t}_\ell, y), \tilde{t}_\ell\right)\right)\right]}_{\text{A perhaps questionable assumption}}$$

$$\approx \mathbb{E}\left[\left(\frac{1}{\tilde{t}_\ell} \cdot \frac{e^{c_n\tilde{t}_\ell}-1}{1-p_n(y)+p_n(y)e^{c_n\tilde{t}_\ell}}\right) e^{K_L(\tilde{t}_\ell, y) - \tilde{t}_\ell \ell_\alpha} \mathcal{J}_0\left(K_L''(\tilde{t}_\ell, y), \tilde{t}_\ell\right)\right],$$

$$\mathbb{E}\left[\frac{\partial \ell_\alpha}{\partial p_n(y)}\right]$$

$$\approx \frac{\mathbb{E}\left[\left(\frac{1}{\tilde{t}_\ell} \cdot \frac{e^{c_n\tilde{t}_\ell}-1}{1-p_n(y)+p_n(y)e^{c_n\tilde{t}_\ell}}\right) e^{K_L(\tilde{t}_\ell, y) - \tilde{t}_\ell \ell_\alpha} \mathcal{J}_0\left(K_L''(\tilde{t}_\ell, y), \tilde{t}_\ell\right)\right]}{\mathbb{E}\left[e^{K_L(\tilde{t}_\ell, y) - \tilde{t}_\ell \ell_\alpha} \mathcal{J}_0\left(K_L''(\tilde{t}_\ell, y), \tilde{t}_\ell\right)\right]},$$

$$\frac{\partial \ell_\alpha}{\partial p_n}$$

$$\approx \frac{\int_{-\infty}^{\infty} \left(\frac{1}{\tilde{t}_\ell} \cdot \frac{e^{c_n\tilde{t}_\ell}-1}{1-p_n(y)+p_n(y)e^{c_n\tilde{t}_\ell}}\right) e^{K_L(\tilde{t}_\ell, y) - \tilde{t}_\ell \ell_\alpha} \mathcal{J}_0\left(K_L''(\tilde{t}_\ell, y), \tilde{t}_\ell\right) f_Y(y) dy}{\int_{-\infty}^{\infty} e^{K_L(\tilde{t}_\ell, y) - \tilde{t}_\ell \ell_\alpha} \mathcal{J}_0\left(K_L''(\tilde{t}_\ell, y), \tilde{t}_\ell\right) f_Y(y) dy}.$$

While a bit involved, this looks virtually identical to the $\frac{\partial \ell_\alpha}{\partial c_n}$ quantity derived in Chap. 7.[1] One key difference, however, is that each $p_n(y)$, unlike the exposures, depends on the state-variable outcome. This implies that we needed to make a fairly questionable assumption to isolate our partial derivative of interest. Since this is intended to assess—in the form of sensitivity analysis—the relative importance of the default probabilities, the reader will hopefully not view this as a executable offence.

Given a form for the partial derivative, let us now define the following quantity,

$$D_{p_n} = \frac{1}{\ell_\alpha} \cdot \frac{\partial \ell_\alpha}{\partial p_n}. \tag{9.4}$$

In words, we can interpret D_{p_n} as the normalized sensitivity of a given VaR measure to an infinitesimal change in the nth obligor's unconditional default probability.

---

[1] Although more effort is involved, we can use equation 9.3 and the results in Chap. 7 to derive a similar expression for expected shortfall.

Table 9.1 *Ranking VaR default-probability sensitivities*: The underlying table highlights the five largest and smallest default-probability sensitivities—defined in equation 9.4—for our ongoing, fictitious 100-obligor example. The original obligor number, default probability, and exposure are also provided for context and interpretation.

Rank	#	p_n	c_n	D_{p_n}
1	35	0.02%	$47.8	29.0
2	30	0.74%	$30.9	6.3
3	65	0.50%	$28.3	5.4
4	33	1.07%	$28.4	5.0
5	76	0.19%	$25.2	4.4
96	4	0.76%	$0.3	0.0
97	43	0.53%	$0.3	0.0
98	56	0.14%	$0.2	0.0
99	89	4.43%	$0.1	0.0
100	83	2.74%	$0.1	0.0

Since it is a partial derivative, naturally all other inputs and parameters are held constant.

How do we interpret this value? Consider the following approximation,

$$D_{p_n} \approx \frac{1}{\ell_\alpha} \cdot \frac{\Delta \ell_\alpha}{\Delta p_n}, \qquad (9.5)$$

$$D_{p_n} \Delta p_n \approx \frac{\Delta \ell_\alpha}{\ell_\alpha},$$

where Δx denotes a small finite difference in the variable, x. Simply put, the product of D_{p_n} and Δp_n describes the percentage movement in the VaR metric.[2] This normalization is particularly useful in this setting since the units of the partial derivative are not immediately helpful.

The final step involves the application of this sensitivity measure to our ongoing, fictitious portfolio example. Table 9.1 outlines the five largest, and smallest, default-probability sensitivities as defined in equation 9.4. The results are enlightening and quite intuitive. The largest outcome, with a value of roughly 30, is associated with obligor 35. Although this counterparty has a significant exposure—indeed, it is the largest in the portfolio—its default probability is only about 2 basis points. Our measure suggests that a one percent increase in obligor 35's unconditional default probability would lead to a roughly 30% increase in the overall VaR measure. By any standard, that is a high degree of sensitivity. On the other hand, the lowest sensitivity stems from a counterpart with a substantial default probability—close to

---

[2] The reader familiar with fixed-income markets will, of course, see the strong similarities with the modified duration measure.

9.1 Some Preliminary Motivation

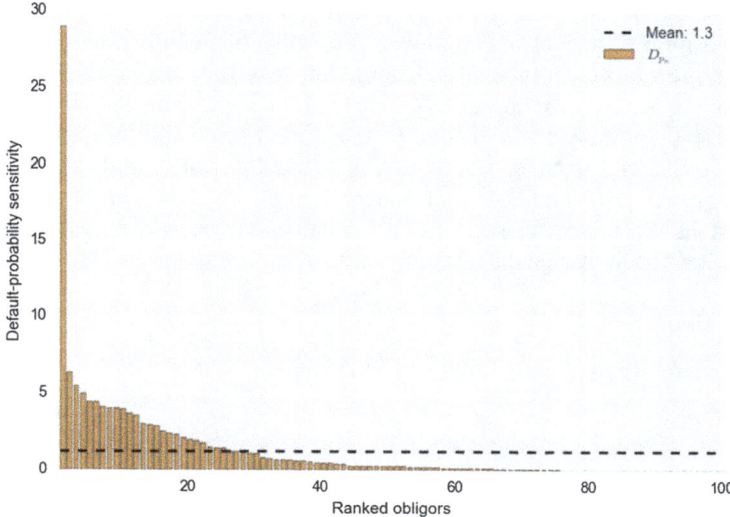

Fig. 9.2 *Obligor differentiation*: The underlying figure provides a visual perspective on the D_{p_n} estimates, from equation 9.4, for all 100 obligors in our sample portfolio. The average value is slightly greater than unity, although there are important differences across counterparties.

3%—but a minuscule exposure; a deterioration in this obligor's credit quality has virtually no impact on the VaR outcome.

The clear takeaway from this analysis is that unconditional default probability estimates do *not* generally have equal importance to one's portfolio. Depending on the structure of one's exposures, modest changes to one's estimates of creditworthiness can have dramatic (or underwhelming) effects on one's risk measurements. Figure 9.2 provides a high-level perspective on all 100 obligors in our sample portfolio. Although the average D_{p_n} value is slightly greater than unity—suggesting a one-to-one relationship between default-probability and VaR changes—there are important differences across individual counterparties.

The objective of the preceding motivational examples is to demonstrate that default probabilities matter. There is no simple relationship between our risk estimates and default probabilities; small changes in the latter can lead to sizable movements in the former. Moreover, default probabilities are *not* observable. They need to be determined by the quantitative analyst. Sadly, there are essentially only *three* alternative sources of information. These include:

1. *estimation* from historical default and transition data; or
2. *calibration* from the value of market instruments; or
3. *determination* by using personal or expert judgement.

The last option, given its subjectivity, is not very appealing as the sole source of default-probability input. It may nonetheless prove useful to adjust or verify more objective methods. The other two approaches, while outwardly objective, also involve some level of subjectivity through the selection of empirical data and imposition of model assumptions. The rarity of default events reduces the information in our datasets and makes this process particularly challenging. The consequence is that this important input requires significant thought and effort. Nevertheless, the analyst is well advised to use every piece of information available, be it subjective or objective. Thankfully, substantial practical guidance exists. This is a rich area of academic and practitioner research, which we will try to carefully represent in the following sections. We begin with the estimation approach to determining unconditional default probabilities.

9.2 Estimation

Estimating default probabilities requires a bit of structure. To start, we require a useful data source. A natural point of reference is the credit-rating agencies. Since the early part of the 20th century, there have existed a number of well-organized, well-regarded, and well-financed credit-rating agencies. Their role has been, and remains, the collection of obligor-level data and its transformation into a credit rating for various entities or individual securities. The role of these agencies has come into question in the years following the 2008 mortgage-lending crisis; see, for example, White (2010), for a more detailed discussion of these issues. While there may be scope to optimize their function within the global regulatory framework, it is indisputable that these organizations serve a useful role for credit analysts. Specifically, they categorize, on a regular basis, a broad range of obligors according to their credit quality. This has led to a rich data history of rating and default observations, which from a risk-management perspective, is an essentially priceless resource.

Many credit-risk professionals rely heavily on these agencies. Others construct their own individual assessments. This is generally a useful, if resource-intensive, undertaking. Even in this case, however, the agencies form an essential information source—and benchmark—for an internal credit-rating system. Rating agencies also typically produce, on a periodic basis using their ratings histories, unconditional default-probability estimates. It is certainly tempting to blindly use these as one's model inputs. This is not advisable, because black boxes are dangerous. If one is going to use an external party's assessment of default probabilities, then one, at the very least, requires a basic understanding of what they are doing. Anything else would be irresponsible, because without knowledge of the key assumptions, the risk of misapplication is simply too high. This requires, of course, a bit of mathematics and patience.

9.2.1 A Useful Mathematical Object

Rating agencies do not provide individual default-probability assessments for distinct obligors. Instead, they assign obligors to a small and finite number of credit states. These states are typically designated with an alphanumeric code: AAA, for example, denotes the highest level of credit quality, while D represents default status.[3] 15 to 20 intermediate categories provide a reasonable degree of granularity for the description of obligor's creditworthiness.

Clearly, some information is lost by assigning obligors to a small set of credit categories. Conceptually, we would still expect a degree of heterogeneity among the counterparties assigned to a single credit class. On the other hand, this discrete creditworthiness classification offers some advantages. The most important of which is the ability to estimate a compact set of default probabilities rather than a distinct value for each obligor. It forms a kind of system where each obligor finds itself in a particular state at a particular time. As time evolves, the obligor may remain in its current state or transition to a new one. The consensus approach—dating back to at least Jarrow et al. (1997) and Gupton et al. (1997)—used to model this system is a discrete-state, discrete-time Markov-chain process.

Although the Markov chain is a well-known mathematical object, this book does not assume one has in-depth knowledge of this literature. We will, therefore, briefly motivate it to help the reader understand it better.[4] This will also serve as a warm-up for our derivation of our default-probability estimators. Although the study of Markov chains can be very deep, only its basic properties are required for this exercise.

To begin, therefore, let us introduce a stochastic process, $\{X_t; t \geq 0\}$ with an associated sample space defined as,

$$\Omega = \{1, 2\}. \tag{9.6}$$

This is a very simple object; each X_t can only take one of two values: 1 or 2. Time evolves discretely—that is, $t = 0, 1, 2, 3, \cdots$. Each individual X_t is a random variable. It is the collection, or sequence, of X_t's that forms a stochastic process.

In this sense, a sequence of coin flips forms a—not particularly interesting—stochastic process. Our example is a bit more complex. Unlike tossing a coin, each X_t outcome in this example is *not* independent, but instead depends upon the path of previous outcomes. This feature, referred to as time dependence, simultaneously makes things more realistic, interesting, and complicated.

---

[3]Different agencies use slightly different classifications, but there is generally a fairly straightforward one-to-one mapping between these groupings.

[4]Appendix C also summarizes a number of useful technical aspects related to the Markov chain.

There are, of course, different degrees of time dependence. It is thus important to ask: what type are we talking about? Consider the following conditional probability,

$$\mathbb{P}(X_t = j \mid \underbrace{X_{t-1} = i_{t-1}, X_{t-2} = i_{t-2}, X_{t-3} = i_{t-3}, \cdots}_{\text{History of the process}}). \tag{9.7}$$

Equation 9.7 attempts to assess the probability that our process, X_t, takes a particular value in its sample space conditional on its history. In general, of course, the value of X_t depends on *all* of its previous values. We may characterize the relative importance of the entire history as the process' memory. In the case of a Markov-chain process, it has only a (small) amount of memory. In fact, it has the minimal amount of memory that one might possess. The value of X_t depends only on the previous value, X_{t-1}. That is,

$$\mathbb{P}(X_t = j \mid X_{t-1} = i). \tag{9.8}$$

All of the previous history is immaterial; to be blunt, the only information required to determine the process' current outcome is its previous value. The next outcome depends only on the current value and so on. This feature is termed the *Markov property*.

While the Markov property leads to a number of interesting mathematical results, more practically it dramatically simplifies our analysis. It implies that we can ignore all but the most recent history and, consequently, the dynamics of our process are easier and more succinctly described. Indeed, in our two-state example, there are precisely *four* possible outcomes from one period to the next. These include:

1. two possibilities involve no change in the process; and
2. another two possibilities involve transition.

These events are succinctly summarized in Table 9.2. A bit of reflection reveals that, if we are told the value from the previous period, then there are only two possible outcomes: staying put or moving to the other state.

If we are given the relative probabilities associated with the four outcomes in Table 9.2, then we have a complete description of the process' intertemporal

Table 9.2 *Counting Markov-Chain outcomes*: This table illustrates the four possible outcomes associated with a two-state Markov-chain's dynamics over a one-period interval.

Process		Description	What happens?
X_{t-1}	X_t		
1	1	No change	Stays in state 1
1	2	Transition	Moves from state 1 to 2
2	1	Transition	Moves from state 2 to 1
2	2	No change	Stays in state 2

9.2 Estimation

dynamics. Let us collect these probabilities into a single matrix,

$$P = \text{from} \left\{ \begin{bmatrix} \overbrace{\underbrace{\mathbb{P}(X_t = 1 | X_{t-1} = 1)}_{p_{11}} \; \underbrace{\mathbb{P}(X_t = 2 | X_{t-1} = 1)}_{p_{12}}}^{\text{to}} \\ \underbrace{\mathbb{P}(X_t = 1 | X_{t-1} = 2)}_{p_{21}} \; \underbrace{\mathbb{P}(X_t = 2 | X_{t-1} = 2)}_{p_{22}} \end{bmatrix} \right. = \begin{bmatrix} p_{11} & p_{12} \\ p_{21} & p_{22} \end{bmatrix}. \quad (9.9)$$

This immensely useful object, P, is termed a *transition matrix*. A more mnemonic way of considering this transition probability is as $p_{\text{from,to}}$. That is, given the state of the process at time $t - 1$—where you are coming *from*—it will tell you the probability of X_t, where you are going *to*. A bit of caution is advised, since the indices on the elements of the transition matrix are contrary to those used in the conditional probability from equation 9.8. Moreover, in some textbooks, the opposite order is employed.

A stochastic process whose dynamics are described by a transition matrix is termed—after the Russian mathematician, A.A. Markov, who popularized it—a *Markov chain*. Although we have, and will continue, to examine rather simple implementations of this object, it is used to efficiently model a rather complex range of behaviour. Markov chains are used, for example, to describe a wide diversity of phenomena covering the areas of genetics, wave mechanics, speech recognition, and even Google's web-search engine. In the finance and macroeconomic setting, they are used to describe business-cycle and high and low volatility state mechanics.[5] Apparently, Markov's first application of the Markov chain was to model the arrival of letters in a famous Pushkin poem. For more history and context on the Markov chain, von Hilgers and Langville (2006) is a fascinating read.

To effectively use and understand its basic properties, it is important to examine the transition matrix in a bit more detail. Although transition matrices provide a highly parsimonious description of a Markov chain, there are a number of important restrictions on the form of P. More specifically, a transition matrix has two key properties:

1. each entry, p_{ij}, lies in the unit interval; and
2. the entries of each *row* must sum to unity.[6]

Mathematically, this implies that for an N-state Markov chain,

$$p_{ij} \in [0, 1], \quad (9.10)$$

---

[5]See Hamilton (1989) and Filardo (1993) for use of the so-called hidden Markov model in these settings as well as Bolder (2002) for an application of this approach to yield-curve modelling.

[6]It is possible to force the columns of P to sum to unity, but this changes the interpretation of the matrix. Although either choice is correct, it is important to select a perspective and stay with it.

for $i, j = 1, \ldots, N$, which is rather sensible, since each entry is, after all, a probability. The second constraint holds that,

$$\sum_{j=1}^{N} p_{ij} = 1, \qquad (9.11)$$

for $i = 1, \ldots, N$. This is basically a conservation-of-mass type constraint. If you start in state i, then you are logically required to move to one of a set of disjoint outcomes—including, of course, i. The total probability of moving to these states has to sum to unity. If not, then some probability mass is lost and the integrity of the transition matrix is compromised.

The transition matrix allows one to move the process, X, through time. It also permits determination of the probability of a given outcome for any number of arbitrary periods in the future. This is a powerful tool that can be efficiently employed, but a practical example can help us understand it better. For ease of exposition, and without any loss of generality, we'll stay in two dimensions. Let's define the following concrete transition matrix for our two-state Markov chain:

$$P = \begin{bmatrix} p_{11} & p_{12} \\ p_{21} & p_{22} \end{bmatrix}, \qquad (9.12)$$

$$= \begin{bmatrix} 0.75 & 0.25 \\ 0.90 & 0.10 \end{bmatrix}.$$

Is this a proper transition matrix? It certainly seems to meet our criteria. Each element falls in the unit interval and the two rows sum to unity.

Imagine that we are standing at time t, but X_t has not yet been revealed. We know that $X_{t-1} = 1$ and we wish to determine the probability that X_{t+1} is also equal to one. More mathematically, we seek the conditional probability, $\mathbb{P}(X_{t+1} = 1 \mid X_{t-1} = 1)$. This is thus a two-step forward probability and it will depend on the outcomes of X_t and X_{t+1} and their associated probabilities. All the required information is found in P and it is essentially a book-keeping exercise to find our desired probability. It is a bit messy to describe in words, but is quite efficiently described in a graphical format. Figure 9.3 maps out, over two periods, the path of a Markov chain in the form of a tree diagram.

What does Fig. 9.3 tell us? Starting from $X_{t-1} = 1$, there are three forks in the road and four final destinations at X_{t+1} Only two of them are of interest to us, because their terminal value for X_{t+1} is unity. By collecting the individual probabilities from the tree diagram, we may conclude that $\mathbb{P}(X_{t+1} = 1 \mid X_{t-1} = 1) = 0.5625 + 0.225 = 0.7875$. This result can be written out in more detail as,

$$\mathbb{P}(X_{t+1} = 1 \mid X_{t-1} = 1) = p_{11} \cdot p_{11} + p_{12} \cdot p_{21}. \qquad (9.13)$$

9.2 Estimation

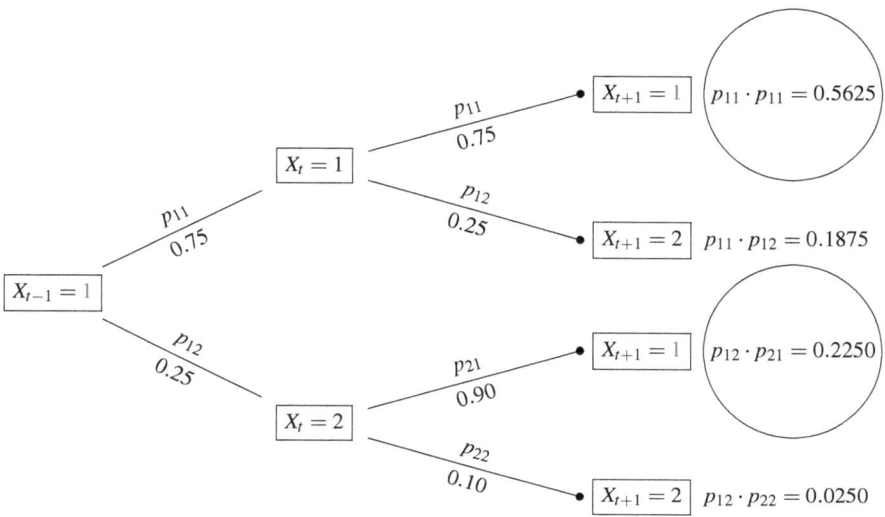

Fig. 9.3 *Calculating forward Markov-Chain probabilities*: The underlying schematic diagram constructs a two-step tree diagram outlining the various outcomes of a two-state Markov chain. This allows us, with some book-keeping, to determine $\mathbb{P}(X_{t+1} = 1 | X_{t-1} = 1)$.

That is, there are basic two paths to $X_{t+1} = 1$: it stays in state one the entire time or it makes one-period pit-stop in state 2 and then returns to one.

While this is fairly obvious, as we add states and time periods, it will become quite painful to perform the necessary book-keeping. It turns out, quite happily, that a simple operation can handle all the necessary details. Consider the following,

$$P \cdot P \equiv P^2 = \begin{bmatrix} p_{11} & p_{12} \\ p_{21} & p_{22} \end{bmatrix} \begin{bmatrix} p_{11} & p_{12} \\ p_{21} & p_{22} \end{bmatrix}, \quad (9.14)$$

$$= \begin{bmatrix} \overbrace{(p_{11} \cdot p_{11} + p_{12} \cdot p_{21})}^{\text{Equation 9.13}} & (p_{11} \cdot p_{12} + p_{12} \cdot p_{22}) \\ (p_{21} \cdot p_{11} + p_{22} \cdot p_{21}) & (p_{21} \cdot p_{12} + p_{22} \cdot p_{22}) \end{bmatrix}.$$

We have multiplied the transition matrix by itself. Observe that the p_{11}-th element of P^2 is the probability of staying in first state over *two* periods. If one computes, P^3, the equivalent element will yield the probability of remaining in the first state for three periods. This result generalizes to the n-step probability coming from the appropriate element of P^n. P^n is a forecast, or extrapolation, of the one-period P's. More generally stated, the p_{ij}th element of the matrix P^n contains the conditional probability of starting in state i at time $t - 1$ and arriving in state j after n periods.

This convenient property is frequently used to transform one-period transition probabilities into their n period equivalents. If one uses the Markov-chain approach to construct transition probabilities for credit obligors between a distinct set of credit states, this can be quite useful. One might estimate one-year default probabilities, but for practical reasons, require three-year default outcomes. It is possible to arrange one's dataset into three-year increments, but one loses a large number of observations. An alternative would be to use equation 9.14 to forecast the three-year default probabilities. These two figures are unlikely to be statistically equivalent—unless the underlying process is truly a Markov-chain—but, in some cases, one may simply have no choice.

Mathematics is often concerned with the characteristics of objects when we move to extremes. From this perspective, therefore, we might inquire: what happens if we use equation 9.14 and set $n = 1,000$? That is, we forecast 1,000 steps into the future. Let's do the computation:

$$P^{1,000} = \begin{bmatrix} 0.75 & 0.25 \\ 0.90 & 0.10 \end{bmatrix}^{1,000}, \qquad (9.15)$$

$$= \begin{bmatrix} 0.7826087 & 0.2173913 \\ 0.7826087 & 0.2173913 \end{bmatrix}.$$

Interestingly, each row and column are identical. If we define the following vector as,

$$\pi = \lim_{n \to \infty} \text{diag}\left(P^n\right), \qquad (9.16)$$

$$= \begin{bmatrix} \mathbb{P}(X_t = 1) \\ \mathbb{P}(X_t = 2) \end{bmatrix},$$

$$\approx \begin{bmatrix} 0.78 \\ 0.22 \end{bmatrix},$$

we see that the probability of being in states one and two converges to 0.78 and 0.22, respectively. The idea is that as we move very far into the future, the conditioning information loses its value. Conceptually, this makes sense. If we know that $X_{t-1} = 1$, this has a significant impact on X_{t+1} or X_{t+2}. But if we seek $X_{t+1,000}$, then it is no longer particularly helpful. The probability, therefore, converges to an unconditional estimate; the unimportance of conditioning information is why the vector, π, has only two elements. In the context of Markov-chains, these are also occasionally referred to as ergodic probabilities; this term can be a bit misleading, since not all Markov chains are ergodic.

There exists, in fact, an entire language and extensive literature associated with Markov chains. The reader is referred to Brémaud (1999), Taylor and Karlin (1994) or Meyn and Tweedie (1993) for excellent overviews and the technical details.[7]

---

[7] A few important technical features of Markov chains, and elaboration of some of these ideas, are summarized in Appendix C.

9.2 Estimation

A number of terms are used to describe individual features and types of Markov chains. The following list highlights just a few of the terms used in this setting.

Accessibility: Describes the ability of a Markov chain to get from state i to state j.
Communication: Indicates that states i and j are both mutually accessible.
Absorption: A state i is termed an absorbing state if it is impossible to leave it.
Transience: This alludes to a non-zero probability you will never return to this state.
Recurrence: The opposite of transient—a chain is either recurrent or transient.
Irreducibility: Summarizes a Markov chain where you can get from one state to any other state.

This is a lot to digest and, for the unfamiliar reader, merits independent study.

The Markov chains used in credit-risk applications require the notion of absorption. As a consequence, we will examine it in more detail. Absorption arises naturally in the definition of a default outcome. We can imagine that a given obligor may freely communicate between different levels of credit risk. It may move from a very high credit-quality state to lower levels and back again. If it defaults, however, this is the end of the road.[8] To see this more clearly, consider the following transition matrix,

$$P = \begin{bmatrix} 0.75 & 0.25 \\ 0 & 1 \end{bmatrix}. \tag{9.17}$$

This is the mathematical form of absorption in a Markov-chain environment. If the process starts in the first state, it has a 0.75 probability of remaining in it. Should it transition to the second state, which occurs with probability 0.25, there is no exit. The probability of transitioning back to the first state, given the process finds itself in the second state, is identically zero.

If we compute the unconditional probabilities for equation 9.17, we arrive at the following vector,

$$\pi = \begin{bmatrix} \mathbb{P}(X_t = 1) \\ \mathbb{P}(X_t = 2) \end{bmatrix} = \begin{bmatrix} 0 \\ 1 \end{bmatrix}. \tag{9.18}$$

This is not particularly interesting, since absent any information you would bet that the process is in the second state. Credit-risk processes, although invariably involving higher dimensionality, also have this property. The bottom line is that if you wait long enough, everyone defaults. The ergodic, or unconditional probabilities put all of the probability mass on the default outcome. Economically, this may seem to be a dubious assumption. Pragmatically, however, it usually takes a very long time—typically hundreds of years—to converge to such an outcome. Since the time

---

[8] Facetiously, one might refer to this as the *Hotel California* state: "you can check-out any time you like, but you can never leave."

horizon associated with credit-risk analysis rarely exceeds three to five years, this seemingly extreme assumption does not typically pose any practical problems.

9.2.2 Applying This Idea

Ratings transitions are, as previously indicated, typically modelled as a Markov chain process, $\{X_t, t = 1, \ldots T\}$, where each X_t is an N-dimensional vector, one for each obligor in one's portfolio, of K distinct credit-quality states. In general, $K \ll N$. That is, one might have 1,000 credit obligors in one's portfolio, but they will typically all fall into one of, approximately, 10 to 20 different credit states. There is a rich literature on the estimation of these models and the following development relies quite heavily on the results described, in much more detail, in Jafry and Schuermann (2004), Schuermann (2008), and Lando and Skødeberg (2002).

To make our computations a bit more general, we will consider an N-dimensional transition matrix, $P \in \mathbb{R}^{K \times K}$. This implies that X_t may take only the values $1, 2, \ldots, K$. In the earlier discussion, we treated X_t as a theoretical object. X_t is actually performing double duty in the following discussion. On the one hand, it represents the theoretical Markov-chain process, which we assume to describe the dynamics of counterparties credit-state evolution. On the other hand, it represents the observed transition of our N obligors among the K disparate credit states over a finite period of time steps, T. In other words, we also use $\{X_t, t = 1, \ldots, T\}$ to describe our dataset.

The transition matrix is defined generically as,

$$P = \begin{bmatrix} p_{11} & p_{12} & \cdots & p_{1K} \\ p_{21} & p_{22} & \cdots & p_{2K} \\ \vdots & \vdots & \ddots & \vdots \\ p_{K1} & p_{K2} & \cdots & p_{KK} \end{bmatrix}, \tag{9.19}$$

where none of the individual entries is known. They need to be inferred from the data. As usual,

$$p_{ij} = \mathbb{P}(X_t = j | X_{t-1} = i), \tag{9.20}$$

for $i, j = 1, \ldots, K$ and the rows of P sum to unity as,

$$\sum_{j=1}^{K} \underbrace{\mathbb{P}(X_t = j | X_{t-1} = i)}_{p_{ij}} = 1. \tag{9.21}$$

Equations 9.20 and 9.21 are constraints, or restrictions, that need to be respected in our identification of the entries in P.

9.2 Estimation

To this point, nothing new has been introduced. If we wish to estimate the individual transition probabilities from observed transition data, then we require a bit more structure. We will employ the common method of maximum likelihood. The conceptual structure of this method, merits a brief aside. Consider an independent, identically distributed sample $y_1, \ldots, y_n$ from a population with probability density (or mass) function, $f(y|\theta)$, where $\theta \in \mathbb{R}^m$ denotes the m parameters of the associated probability density function. The joint density of the sample given θ is, by our assumption of independence, described as,

$$f(y_1, \ldots, y_n | \theta) = \prod_{i=1}^{n} f(y_i | \theta), \tag{9.22}$$

where $f : \mathbb{R}^n \to \mathbb{R}$. This mapping of n random-variable outcomes to the real line describes, as one would expect, the probability of this collection of events given the density parameters, θ. Since the observed are i.i.d., it has an easily derived multiplicative form.

This is not directly useful for parameter estimation, but it gets more interesting as one twists it around as,

$$\mathcal{L}(\theta | y_1, \ldots, y_n) = \prod_{i=1}^{n} f(y_i | \theta). \tag{9.23}$$

where $\mathcal{L} : \mathbb{R}^m \to \mathbb{R}$. All we have done is change the order of the conditioning information on the left-hand side of equation 9.22. It is now a mapping of m parameter values to the real line. Indeed, equation 9.23 is *no* longer a proper probability density function, but is instead referred to as the likelihood function. The likelihood function provides the probability, for a given dataset, of observing particular parameters.

In equation 9.22, the parameters are fixed and the data varies. Equation 9.23, conversely, fixes the data, but treats the parameters as the input value. Maximum-likelihood exploits this latter form to find a reasonable estimate of θ. Those parameters that *maximize* the probability of having observed one's data are termed maximum-likelihood estimators. This reduces a relative difficult conceptual task to an optimization problem.

Since optima are preserved under any logarithmic transformation, and we seek to maximize equation 9.23, it is more convenient to use the log-likelihood function:

$$\ell(\theta | y_1, \ldots, y_n) = \ln\left(\mathcal{L}(\theta | y_1, \ldots, y_n)\right), \tag{9.24}$$

$$= \ln\left(\prod_{i=1}^{n} f(y_i | \theta)\right),$$

$$= \sum_{i=1}^{n} \ln\left(f(y_i | \theta)\right).$$

The vector maximum-likelihood estimates for this very general problem are thus simply,

$$\hat{\theta} = \arg\max_{\theta} \ell(\theta|y_1, \ldots, y_n). \quad (9.25)$$

$\hat{\theta}$ represents the most likely parameter set associated with the observed data and it possesses many useful properties—see Casella and Berger (1990), Pawitan (2001) or Judge et al. (1985) for more details.

With this in mind, the likelihood function for our Markov chain is written as,

$$\mathcal{L}(P|X_1, \ldots, X_T) = \underbrace{\mathbb{P}(X_1 = x_1)}_{\text{Starting point}} \prod_{i=1}^{T} \prod_{j=1}^{T} p_{ij}^{n_{ij}}, \quad (9.26)$$

where T is the total number of periods in one's dataset, $\{X_1 = x_2\}$ is the arbitrary starting point of the process, and n_{ij} denotes the number of occurrences of the state value i followed by the state value j. The joint density, as seen in the right-hand-side of equation 9.26, is thus simply the possible outcomes—which are all disjoint—adjusted for the number of times that they occur—each value is anchored by its initial value. The form of the likelihood function can also be inferred by examining, and generalizing, the tree structure summarized in Fig. 9.3.

Broadly speaking, there are two alternative approaches to the estimation of Markov-chain transition probabilities within the context of maximum-likelihood: the so-called cohort and hazard-rate approaches. We will consider each in turn.

9.2.3 Cohort Approach

Using the ideas from the previous section, we seek to identify the choice of parameters, $\{p_{ij}; i, j = 1, \ldots, K\}$, by maximizing the value of the likelihood function in equation 9.26. That is, we seek the transition probabilities that are most consistent with the observed data. More concretely, this is accomplished by solving the following optimization problem,

$$\max_{P} \mathcal{L}(P|X_t, \ldots, X_T), \quad (9.27)$$

$$\text{subject to:} \quad \sum_{j=1}^{K} p_{ij} = 1 \text{ for } i = 1, \ldots, K.$$

This is a simple problem in calculus. We will use the method of Lagrange multipliers to determine the first-order conditions, set them zero, and solve for the appropriate parameter values. First, as previously suggested, to facilitate the computations, we

9.2 Estimation

will work with the log-likelihood function.

$$\ln(\mathcal{L}(P|X_t,\ldots,X_T)) = \ln\underbrace{\left(\mathbb{P}(X_1 = x_1)\prod_{i=1}^{T}\prod_{j=1}^{T}p_{ij}^{n_{ij}}\right)}_{\text{Equation 9.26}}, \quad (9.28)$$

$$\ell(P|X_t,\ldots,X_T) = \ln\left(\mathbb{P}(X_1 = x_1)\right) + \ln\left(\prod_{i=1}^{T}\prod_{j=1}^{T}p_{ij}^{n_{ij}}\right),$$

$$= \ln\left(\mathbb{P}(X_1 = x_1)\right) + \sum_{i=1}^{T}\sum_{j=1}^{T}\ln\left(p_{ij}^{n_{ij}}\right),$$

$$= \ln\left(\mathbb{P}(X_1 = x_1)\right) + \sum_{i=1}^{T}\sum_{j=1}^{T}n_{ij}\ln\left(p_{ij}\right).$$

The log-transform simplifies things somewhat by moving from a multiplicative to an additive setting. The first-order conditions are, in this case, simplified when certain terms—such as the starting point in our case—fall out of the computation.

We may now restate the maximization problem in equation 9.27 more concretely as,

$$\max_{\{p_{ij}\}i,j=1,\ldots,K} \ln\left(\mathbb{P}(X_1 = x_1)\right) + \sum_{i=1}^{T}\sum_{j=1}^{T}n_{ij}\ln\left(p_{ij}\right), \quad (9.29)$$

$$\text{subject to: } \sum_{j=1}^{K}p_{ij} = 1 \text{ for } i = 1,\ldots,K.$$

The method of Lagrange multipliers proceeds by bringing the constraint in the objective function as follows,

$$\max_{\{p_{ij}\}i,j=1,\ldots,K} \underbrace{\ln\left(\mathbb{P}(X_1 = x_1)\right) + \sum_{i=1}^{T}\sum_{j=1}^{T}n_{ij}\ln\left(p_{ij}\right)}_{\ell(P|X_t,\ldots,X_T)} - \underbrace{\sum_{i=1}^{K}\lambda_i\left(\sum_{j=1}^{K}p_{ij} - 1\right)}_{\text{Constraint}}, \quad (9.30)$$

where $\{\lambda_i; i = 1, \ldots, K\}$ denote the eponymous Lagrange multipliers. We now proceed to differentiate the revised objective function with respect to p_{ij}, set it equal to zero, and simplify as follows,

$$\frac{\partial}{\partial p_{ij}} \left(\mathbb{P}(X_1 = x_1) + \sum_{i=1}^{T} \sum_{j=1}^{T} n_{ij} \ln \left(p_{ij} \right) - \sum_{i=1}^{K} \lambda_i \left(\sum_{j=1}^{K} p_{ij} - 1 \right) \right) = 0, \quad (9.31)$$

$$\frac{n_{ij}}{p_{ij}} - \lambda_i = 0,$$

$$p_{ij} = \frac{n_{ij}}{\lambda_i},$$

for $i, j = 1, \ldots, K$. This amounts to K^2 first-order conditions associated with an equal number of transition-probability parameters.

This does not look terribly promising, until we observe that we may recover the constraint by summing both sides over j as follows,

$$\underbrace{\sum_{j=1}^{K} p_{ij}}_{=1} = \sum_{j=1}^{K} \frac{n_{ij}}{\lambda_i}, \quad (9.32)$$

$$1 = \frac{1}{\lambda_i} \sum_{j=1}^{K} n_{ij},$$

$$\lambda_i = \sum_{j=1}^{K} n_{ij},$$

for $i, j = 1, \ldots, K$. The first-order conditions in equation 9.31 imply, given the Lagrange multipliers in equation 9.32, that the maximum-likelihood estimator of the transition probability is,

$$\hat{p}_{ij} = \frac{n_{ij}}{\sum_{j=1}^{K} n_{ij}}, \quad (9.33)$$

for $i, j = 1, \ldots, K$. In words, therefore, the maximum-likelihood estimator for the ijth transition probability is the ratio of occurrences of state i followed by state j, normalized by the total number of observations within that rating class or *cohort*. For this reason, transition matrices estimated using the result in equation 9.33 are referred to as using the cohort approach. This is, as of the time this chapter was prepared, the industry standard.

9.2.4 Hazard-Rate Approach

The cohort approach is not, however, the only technique that one might use to compute a transition matrix. An alternative, called the hazard-rate approach, has been offered as a viable alternative to the previously described cohort technique. This appears to be particularly the case, when a rich amount of intra-year credit transition information is available. While data richness is not considered in our setting, the technique can still be used. Gaining some awareness of how it works is a good idea.

The hazard-rate approach is a bit more complex. We will dispense with the derivation and provide a brief description of the technique. The basic idea is that transition probabilities for a K-dimensional time-homogeneous Markov chain model are determined through an intensity matrix, $\Lambda \in \mathbb{R}^{K \times K}$. With knowledge of the intensity or—as it is also often referred to—generator matrix, the transition matrix has the following form,

$$P_t = e^{\Lambda t}, \tag{9.34}$$

for all $t \geq 0$ and where equation 9.34 refers to the matrix exponential.[9] The transition matrix is, from this alternative perspective, indexed to time. Similar to the previous development, one may find maximum-likelihood estimators for the intensity matrix. The individual elements of Λ are described as,

$$\lambda_{ij} = \frac{n_{ij}(t, T)}{\int_t^T Y_i(s) ds}, \tag{9.36}$$

for all $i \neq j$. $Y_i(t)$ denotes the total number of counterparties in state i at time t and $n_{ij}(t, T)$ represents the number of transitions from state i to j over the period, $[t, T]$. In this manner, it permits a more granular approach to aggregating weekly or even daily transition data into annual transition matrix estimates. In our setting, we have only annual observations. As a consequence, we can rewrite equation 9.36 as follows,

$$\lambda_{ij} = \frac{n_{ij}}{\sum_{t=1}^{T} n_i}. \tag{9.37}$$

---

[9]The multivariate analogue of the exponential function, the matrix exponential, is defined as,

$$e^{\Lambda t} = \sum_{k=0}^{\infty} \frac{(\Lambda t)^k}{k!}. \tag{9.35}$$

See Golub and Loan (2012, Chapter 9) for a discussion of various algorithms for its computation.

The off-diagonal elements of Λ are thus the number of occurrences of state i followed by state j divided by the total number of occurrences of state i. The diagonal elements, however, are given as,

$$\lambda_{ij} = -\sum_{i \neq j} \lambda_{ij}. \tag{9.38}$$

This technique can become significantly more involved, particularly if one relaxes the assumption of time homogeneity. See Lando and Skødeberg (2002) for more details.

9.2.5 Getting More Practical

The previous analysis is somewhat abstract. To make it easier to understand and more concrete, we will, as usual, consider a practical example. The idea is to categorize and model the credit quality of 100 distinct obligors—eventually, we will link this analysis with our usual, running example. To keep things manageable and permit exposition of all results, we define only *four* credit states in our Markov Chain: A, B, C, and D. State D, with its suggestive letter designation, is considered to be an *absorbing* default state. This is significantly less than the 20-odd credit states used in industrial settings, but works well as an illustrative example.[10]

Schematically, the transition matrix for our simplified example is summarized as follows,

$$P = \begin{bmatrix} p_{AA} & p_{AB} & p_{AC} & p_{AD} \\ p_{BA} & p_{BB} & p_{BC} & p_{BD} \\ p_{CA} & p_{CB} & p_{CC} & p_{CD} \\ p_{DA} & p_{DB} & p_{DC} & p_{DD} \end{bmatrix}. \tag{9.39}$$

A is the highest possible credit state, followed by categories B and C, respectively. Given that D is an absorbing state, however, we have that $p_{DA} = p_{DB} = p_{DC} = 0$ and $p_{DD} = 1$. That is, the final row of P is determined in advance. This implies that, in fact, we need only estimate 12 separate parameters. Our interest, of course, is principally in the first *three* elements of the final column, since these describe the default probabilities. To be clear, therefore, the default-probability estimates are simply a subset of the transition matrix. They are not, however, estimated in isolation, but rather in conjunction with the full set of (non-absorbing) transition probabilities.

---

[10]Since credit ratings are often associated with not just a current value, but also a positive or negative outlook, the effective number of credit states can become quite a bit more sizable than one might expect.

9.2 Estimation

The transition matrix, and most particularly, the default probabilities need to be estimated using transition data. We require, therefore, a dataset. While not difficult to obtain, using an external dataset offers some drawbacks. The largest shortcoming is that we do *not* know, nor will we ever, the true population parameters. This is a simple fact of life, but it makes it difficult, pedagogically at least, to assess the relative accuracy of our estimation approach. To make this possible, the following analysis is founded on an unreasonable, and entirely unrealistic, assumption: we assume that we know the true form of our transition matrix. In particular, we define the true P as

$$P = \begin{bmatrix} 0.9600 & 0.0290 & 0.0100 & 0.0010 \\ 0.1000 & 0.7750 & 0.1200 & 0.0050 \\ 0.1200 & 0.2200 & 0.6500 & 0.0100 \\ 0.0000 & 0.0000 & 0.0000 & 1.0000 \end{bmatrix}. \tag{9.40}$$

This is a bit silly. If we did know the true transition matrix, we would hardly be concerned with its estimation or with confidence bands. Lack of realism notwithstanding, this assumption simply serves the purpose of assisting our understanding of the estimation procedure.

The values in equation 9.40 are broadly consistent—albeit involving far fewer credit categories—with the types of transition probabilities observed in real applications. We can refer to these as empirical stylized facts. There are, at least, *three* different points of interest associated with our transition matrix in equation 9.40. In particular,

- P is diagonally dominant;
- default is an absorbing state; and
- the probability of default (entering state D) is very small, but increases monotonically as credit quality deteriorates.

Diagonal dominance implies that the probability of staying in the current state is high relative to the probability of transition to a new credit-quality category. There is, therefore, a relatively high level of inertia in credit ratings. This is a common, and conceptually reasonable, property for a credit-rating system. The default probabilities also decline monotonically as credit quality deteriorates. Empirically, one does not always observe perfect monotonicity—due to a paucity of default observations and associated estimation noise—but economically it is a reasonable and logical property.

9.2.6 Generating Markov-Chain Outcomes

Although we know the true transition matrix, we cannot use it directly to construct point and interval estimates. Instead, we will use equation 9.40 to simulate a collection of obligor credit-state transitions over a predefined period of time. The idea is to generate a dataset derived from a true Markov-chain process. This dataset will then be employed, along with the previously outlined algorithm, to estimate the transition probabilities. To generate a reasonable dataset of typical and realistic size, we have elected to consider $N = 100$ credit counterparties over a 30-year time period, which we will denote as T. Most industrial datasets cover a similar time horizon, but generally include a much larger number of individual obligors.[11]

Our first estimation task, therefore, is to simulate a four-state Markov chain describing a system of 100 credit counterparties over a 30-year horizon. Furnished with this data, we will employ it to construct point and interval estimates of the elements of P. Since we know the true parameter values, we will find ourselves in the privileged position of being able to truly see how well these approaches perform. This is a common trick used in statistics to assess the efficacy of a particular technique—it is often referred to as a simulation study.

While useful, some caution is nevertheless warranted in the interpretation. We have, quite explicitly, assumed that our credit-rating system follows a Markov-chain process in this analysis. We may find that our estimation approach works very well. This is, however, not a global verification of the efficacy of the estimation algorithm. If the true data-generating process for rating transitions does not follow a Markov chain, then we have introduced some structural bias into our examination. Indeed, it is quite likely that credit states are not a *true* Markov chain. Our hope, however, is that it represents a reasonable approximation.

Simulating a Markov-chain process is, thankfully, a relatively straightforward and somewhat familiar exercise. Each obligor will be, for this analysis, simulated independently.[12] To start, however, one requires knowledge of the initial state associated with each credit counterparty. The Markov property ensures that these initial conditions, and the transition matrix, are all that are required to simulate Markov-chain outcomes. The starting points, for our analysis, will also be simulated. This is not particularly problematic since, as we've seen in the basic algorithm, they play no role in the final estimates.

Imagine that the initial state, at $t = 1$, for counterparty n is A. This implies that the transition from $t = 1$ to $t = 2$ is governed by the first row in P. We thus require an efficient algorithm to randomly assign the transition to A with 0.96 probability, to B with 0.029 probability, to C with 0.01 probability, and with the small remaining probability mass, to assign default. Statistically, this is equivalent to a multinomial trial. Simulation of a multinomial trial is, however, a logical extension of binomial-trial simulation. Having simulated binomial trials in Chaps. 2 and 3, the reader will be, to a certain extent, acquainted with this procedure.

---

[11] Some datasets extend back as far as the 1930s. While the extra observations are certainly welcome, they do create other issues. In particular, it creates worries about the applicability of observations from very different economic settings to the current environment.

[12] We will relax this assumption in Chap. 10.

9.2 Estimation

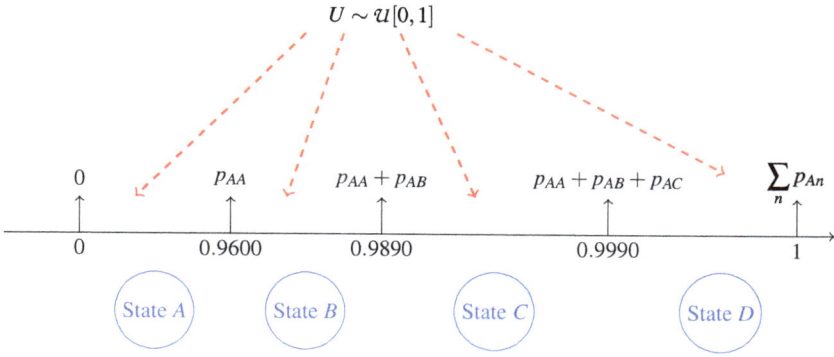

Fig. 9.4 *Simulating a Markov chain*: The simulation of a Markov chain is conceptually similar to the generation of a Bernoulli trial, albeit with more possible outcomes. First, one draws a standard uniform random variate, $U \sim \mathcal{U}[0, 1]$. Then, we use the row of the transition matrix corresponding to the current state—in this case, state A—to partition the unit interval. The outcome of U, as described in the underlying figure, determine the new state.

The typical approach is to draw a standard uniform-random variate, $U \sim \mathcal{U}[0, 1]$, from the unit interval. U falls, by definition, uniformly across $[0, 1]$. Imagine one defines a sub-interval, $[a, b]$, where $0 < a < b < 1$, and one draws repeatedly from the standard uniform distribution. With sufficiently many draws, the proportion of outcomes will tend to $\frac{b-a}{1-0} = b - a$. We can, therefore, think of $b - a$ as representing the probability of a given U falling in the interval, $[a, b]$. We exploit this idea in simulating Markov chain outcomes.[13]

The simple—and somewhat poorly scaled—schematic in Fig. 9.4 displays the basic idea. Using the first row of P, meaning that we are currently in state A, we partition the unit interval into the following *four* sub-intervals:

$$[0, p_{AA}], \tag{9.41}$$

$$(p_{AA}, p_{AA} + p_{AB}],$$

$$(p_{AA} + p_{AB}, p_{AA} + p_{AB} + p_{AC}],$$

$$(p_{AA} + p_{AB} + p_{AC}, 1].$$

The union of these sub-intervals covers the entire unit interval, but each sub-interval is disjoint. Most importantly, the size of each sub-interval is p_{AA}, p_{AB}, p_{AC} and p_{AD}, respectively. Given this partition and an outcome U, one merely determines which sub-interval the value of U falls. If U arrives in the first sub-interval, the system remains in state A. If it falls in the second sub-interval, the credit obligor

---

[13] There are faster and more complex ways to simulate a Markov chain—see, for example, Fishman (1983)—but this simple approach fully meets our needs.

transitions to state B and so on. Once this is determined, the current state is determined and one merely repeats the process with the appropriate row of P.

More formally, this is termed the inverse-transform method—see Fishman (1995, Section 3.2) for more details. In full generality, it is a bit more difficult to decipher, but we will work through it for completeness. We define our sample space—or rather, the set of possible credit states—as

$$S = \{1, \ldots, K\}. \tag{9.42}$$

We now denote the credit state of counterparty n at time t as $X_{n,t}^{(m)}$. m refers to the specific iteration of our simulation algorithm—in the first case, when we desire a single dataset, $m = 1$. This will be relaxed later when using these ideas to construct interval estimates.

We wish to write down the approach to simulate $X_{n,t+1}^{(m)}$ given $X_{n,t}^{(m)}$ and P. The possible transition probabilities for $X_{n,t+1}^{(m)}$, given that $X_{n,t}^{(m)} = i$, are written as,

$$\mathbb{P}\left(X_{n,t+1}^{(m)} = j \mid X_{n,t}^{(m)} = i\right) \in \{p_{ij}\}_{i \in S}, \tag{9.43}$$

for $n = 1, \ldots, N$, which amounts to mathematical shorthand for the ith row of P.

We now draw a standard uniform random variate,

$$U_{n,t+1}^{(m)} \sim \mathcal{U}[0, 1], \tag{9.44}$$

for $n = 1, \ldots, N$. It is naturally indexed to the appropriate time, $t + 1$, and credit counterparty, n. We also index it to m, the mth iteration of our Monte Carlo simulation. Using this simulated outcome, we may assign a value to $X_{n,t+1}^{(m)}$ as follows,

$$X_{n,t+1}^{(m)} = \begin{cases} 1 & : \text{if } U_{n,t+1}^{(m)} \in [0, p_{i1}] \\ 2 & : \text{if } U_{n,t+1}^{(m)} \in (p_{i1}, p_{i1} + p_{i2}] \\ \vdots & \\ K & : \text{if } U_{n,t+1}^{(m)} \in \left(\sum_{j=1}^{K-1} p_{ij}, 1\right] \end{cases}, \tag{9.45}$$

This need only be repeated for each of the $n = 1, \ldots, N$ credit counterparties and each $t = 1, \ldots, T$ periods. At this point, as indicated, we set $m = 1$.

9.2 Estimation

Algorithm 9.1 Simulating a four-state Markov chain

```
def simulateRatingData(N,T,P,wStart):
    tStart = initalizeCounterparties(N,wStart)
    D = np.zeros([N,T])
    for n in range(0,N):
        D[n,0] = tStart[n]
        for t in range(1,T):
            D[n,t] = transitionStep(D[n,t-1].astype(int),P)
    return D
def transitionStep(currentState,P):
    myP = np.cumsum(P[currentState-1,:]) # Determine current row
    u = np.random.uniform(0,1) # Draw uniform variate
    if ((u>0) & (u<=myP[0])): # in [0,pAj]
        return 1
    elif ((u>myP[0]) & (u<=myP[1])): # in [pAj,pAj+pBj]
        return 2
    elif ((u>myP[1]) & (u<=myP[2])): # in [pAj+pBj,pAj+pBj+pCj]
        return 3
    elif ((u>myP[2]) & (u<=myP[3])): # in [pAj+pBj+pCj,1]
        return 4
```

Algorithm 9.1 takes this mathematical discussion a step further toward practical implementation by illustrating two Python functions: simulateRatingData and transitionStep. The first sub-routine, simulateRatingData, randomly generates the initial values from a set of starting weights, wStart. These can, of course, take any values. For this exercise, we assign 25% to state A and C, respectively, with the remaining 50% in state B. This does not have a tremendously important impact on the results; the only key point is to ensure a reasonable coverage of the various credit categories.[14]

At inception, it is assumed that no credit obligors are in the default state. It then enters a double loop over the credit counterparties and time horizon—N and T—using the preceding logic at each step. P is the 4×4 transition matrix, while wStart is a vector of weights to assign, at inception, to each credit state. The hard work, performed in the sub-routine, transitionStep, is thus a direct translation of equation 9.45 for a four-state Markov chain. This implementation is specific to the four-state case and must be generalized to handle a different number of states. Each step requires the current state and the transition matrix—along with a uniform-random variable—to simulate its next outcome.

---

[14] Starting all of the obligors in credit state A will reduce the observations in states B and C, which might increase the noise of their transition-probability estimates.

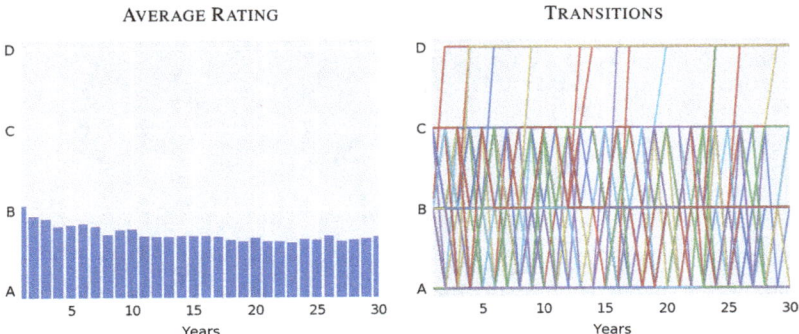

Fig. 9.5 *Transition overview*: The two graphics in this figure provide some insight into the characteristics of our simulated data. The left-hand graphic describes the average credit rating—it appears to be around the level of state B. The right-hand graphic, conversely, highlights the individual transitions for all of our 100 obligors. Observe that only a handful of default events occur.

9.2.7 Point Estimates and Transition Statistics

Having described how transition outcomes are simulated from P, we may now proceed to examine the results of a single set of realizations. More simply, we now have a dataset and are able to test our estimation algorithm. With our simulated dataset of 100 counterparties over 30 years, this yields 3,000 credit outcomes, but only 2,900 transition observations.[15] This is a reasonable amount of data and, as previously mentioned, is broadly consistent with the typical time horizon used in industrial settings.

The general rule with analysis is, before applying any formal techniques, to carefully survey one's data. Figure 9.5, in an effort to follow this important advice, summarizes some of the characteristics of the simulated data. The left-hand graphic highlights the annual evolution of the average credit rating. While there is some variance over the period, it does remain relatively close to the state B. There does, however, appear to be a slight improvement in average credit quality over the period.

The right-hand graphic of Fig. 9.5, conversely, highlights the individual transitions for all of our 100 obligors. This is reasonably confusing to read and is not intended to provide information about a specific obligor. Instead, it is illustrative of the general pattern of transition. We observe quite clearly, for example, that there has been only a handful of default events. Transition between the other states, however, seems to have occurred with significantly greater frequency. This is consistent with our expectations for a Markov-chain associated with the parametrization in equation 9.40.

---

[15]Unlike some real-life datasets, we assume that each credit obligor is rated in every period.

9.2 Estimation

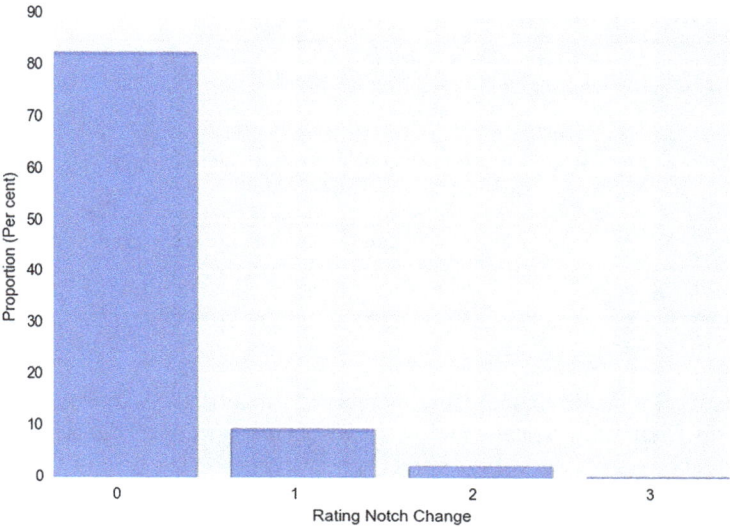

Fig. 9.6 *Type of transitions*: This figure displays a histogram of the transitions from our simulated data. Almost 80% of the time, an obligor remains in its current state. About 10% of the cases involve a one-state transition. Very few two-state transitions and a vanishingly small number of three-state transitions occur.

Figure 9.6 takes our analysis a step further and provides a histogram of the number of transition steps across our collection of transition outcomes. Given the diagonal dominance of the true transition matrix, it should be no surprise that credit-state inertia—or no change in credit state—is the most popular outcome. Indeed, more than 80% of the time, an obligor can be expected to stay in its current state. In slightly more than 10% of the transitions, a single step is observed. The remainder are predominately two-step transitions, but a vanishingly small proportion of three-step transitions—the largest possible movement—are evident.

Following from the maximum-likelihood-based cohort approach, outlined in the previous section, we may now proceed to estimate the elements of the transition matrix from this (simulated) data. The general expression, to repeat the key result, for p_{ij} is:

$$\hat{p}_{ij} = \frac{n_{ij}}{\sum_{j=1}^{K} n_{ij}}, \quad (9.46)$$

for $i, j = 1, \ldots, K$. The central point is that the maximum-likelihood estimator involves nothing more than counting transitions, which is an intuitive and practical

Algorithm 9.2 Counting transitions

```
def getTransitionCount(K,T,N,data):
    n_ij = np.zeros([K,K])
    # Count the n_ij or n_{to,from}
    for i in range(0,K): # to
        for j in range(0,K): # from
            for n in range(0,N): # obligor
                for tau in range(1,T): # time
                    if ((data[n,tau]==j+1) & (data[n,tau-1]==i+1)):
                        n_ij[i,j] += 1
    return n_ij
```

result. Indeed, it makes logical sense that the information about transition probabilities stems from the physical count of state transitions in one's dataset. Moreover, other methods—such as the hazard-rate approach—also make use of the transition count, although this information is employed in a slightly different way.

While easily described, computing the $K \times K$ terms from equation 9.46 requires a bit of effort and organization. Algorithm 9.2 highlights the Python code used to count our ratings transitions. It takes the number of states, time periods, and obligors in the form of input arguments K, T, and N, respectively. The final requirement is the $N \times T$ array of data observations, data.

Counting transitions is not a particularly elegant task, but instead rather hard work. It involves *four* layers of for loops to determine the transitions from each state to another: two loops for the credit states, one for all obligors, and another for the time steps. The results, summarized in the $K \times K$ array, n_ij, have the same dimensions and location as the transition matrix to avoid confusion. Each entry thus represents the number of observed transitions from state i outcomes followed by arrival in j.

The sub-routine, getTransitionCount, should work quite generally to determine the number of transition counts associated with any dataset. The identification of the maximum-likelihood transition probability estimates, armed with a matrix of $\{n_{ij}; i, j = 1, \ldots K\}$ transition counts, is a comparatively leisurely job. Algorithm 9.3 uses the standard cohort technique, from equation 9.46, to return the matrix of transition-probability estimates. The only twist in Algorithm 9.3 is the argument, myPeriod. Its default value is unity, which implies one-year values consistent with our underlying dataset. Any value of myPeriod> 1 will generate forecast transition probabilities for a correspondingly longer horizon. There are also a few manipulations of the final row relating to the absorbing default state. This is merely for security. If no defaults occur in the—real or simulated—dataset, you will

9.2 Estimation

Algorithm 9.3 Cohort transition-matrix estimation

```
def estimateTransitionMatrix(K,T,N,data,myPeriod=1):
    n_ij = getTransitionCount(K,T,N,data)
    H = np.zeros([K,K])
    for i in range(0,K):
        for j in range(0,K):
            H[i,j] = np.divide(n_ij[i,j],np.sum(n_ij[i,:]))
    H[-1,:] = np.zeros(K)
    H[-1,-1] = 1
    return anp.matrix_power(H,myPeriod)
```

find yourself in the uncomfortable position of dividing zero by itself. The last two lines ensure that default is treated appropriately and no infinite values are returned.[16]

An application of Algorithm 9.2 to our dataset yields a $K \times K$ matrix of transitions. The actual transition counts are summarized in the underlying matrix,

$$\{n_{ij}\}_{i,j=1,\ldots,4} = \begin{bmatrix} 1535 & 60 & 16 & 4 \\ 75 & 586 & 76 & 7 \\ 47 & 67 & 212 & 3 \\ 0 & 0 & 0 & 212 \end{bmatrix}. \tag{9.47}$$

The sum of all elements in equation 9.47 is equal to the total number of transitions; in our case, this is 2,900. A total of 14 defaults was observed and, as expected, the vast majority of the observations fall along the diagonal. This starkly illustrates the typical data scarcity problem faced by all credit-risk practitioners. We have almost 3,000 data points, but only 14 of them are informative about the default event. The final row provides information on how long the defaulting entities have been in that state. $\frac{212}{14} \approx 15$ thus represents the average number of years each defaulted entity, over our 30-year period, remained in the absorbing state; this suggests, as we would expect, that default occurs roughly uniformly over the period.

Figure 9.7 provides a three-dimensional view of the counted transitions organized in transition-matrix format. The diagonal dominance is clearly visible. While this data is fictitious and involves significantly too few credit states and obligors, its basic structure is quite similar to that observed within actual datasets.

Equation 9.47 and Fig. 9.7 provide all the information necessary to construct maximum-likelihood transition-probability estimates. Indeed, one could readily perform the necessary computations with a hand-held calculator. Saving the reader the headache, we use equation 9.46 and Algorithm 9.3 to provide us with our desired

---

[16] A lack of observations can, in principle, occur for any credit state. In our setting, this is highly improbable, but a more general implementation would take this possibility into consideration.

Fig. 9.7 *Simulated transition counts*: Using Algorithm 9.2, we count the number of possible transitions from our simulated credit-quality data. The underlying three-dimensional figure highlights the distribution of these counts.

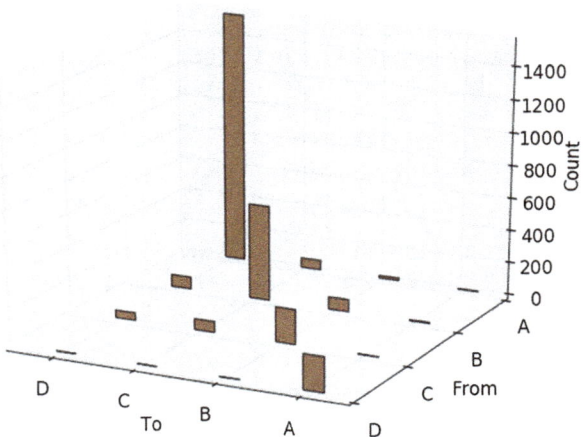

values. We denote the estimated transition matrix as $\hat{P}$. It has the following form,

$$\hat{P} = \begin{bmatrix} 0.9505 & 0.0372 & 0.0099 & 0.0025 \\ 0.1008 & 0.7876 & 0.1022 & 0.0094 \\ 0.1429 & 0.2036 & 0.6444 & 0.0091 \\ 0.0000 & 0.0000 & 0.0000 & 1.0000 \end{bmatrix}. \tag{9.48}$$

Our estimates are not, of course, exact. Comparison with the *true* values in equation 9.40, however, reveals that our estimates are reasonably precise. It is useful to examine the absolute value of the differences between the true, known matrix elements and our estimates. The result is,

$$\left| P - \hat{P} \right| = \begin{bmatrix} 0.0095 & 0.0082 & 0.0001 & 0.0015 \\ 0.0008 & 0.0126 & 0.0178 & 0.0044 \\ 0.0229 & 0.0164 & 0.0056 & 0.0009 \\ 0.0000 & 0.0000 & 0.0000 & 0.0000 \end{bmatrix}. \tag{9.49}$$

Since we know the true transition matrix, this comparison is easily performed. In reality, the true values are unknown. Moreover, even with knowledge of the truth, the average transition-probability error, across the non-absorbing states, is approximately 80 basis points. This is probably acceptable for a diagonal element with large absolute value, but rather less so for an already quite small default probability.

Equation 9.49 provides an *ad hoc* comparison of the estimation error. While useful, even this basic analysis is impossible in a real-life setting. Faced with this situation, we desire a formal assessment of estimation uncertainty, which may be employed to construct interval estimates for our point estimates highlighted in equation 9.48.

9.2.8 Describing Uncertainty

Estimating model parameters using a statistical technique without some notion of the uncertainty of these estimates is neither terribly responsible nor particularly useful. Data provides information, but not all datasets are equally informative. It is the responsibility of the analyst to investigate and communicate the degree of confidence in one's estimates. These parameters find their way into models and ultimately inform financial decisions. A point estimate alone can create a false sense of security. Interval estimates, which are formulated through the measurement of statistical uncertainty, provide some insight into how much one might trust various parameter estimates. From a model-risk perspective, it is one of the few effective tools at our disposition to help us judiciously use and communicate model results.

Statistically, assessing the accuracy of an estimate requires a description of its distribution. Such a description is *not* always easy to obtain. Often, we only know, or may infer, its asymptotic behaviour—that is, the distribution of our estimate as the sample size becomes very large. In some cases, the distribution is available analytically, but often it must also be determined numerically. We will consider *two* alternatives:

1. use of the maximum-likelihood framework; and
2. the application of a technique known as *bootstrapping*.

We will apply both of these approaches to our known example credit-transition matrix. Since we know the true values, we can also assess the efficiency of the corresponding error bounds and compare the relative success of the two approaches. We feel obliged to indicate that within the vast array of statistical theory and practice, we will only be scratching the surface. Much more rigour and context may be found in Casella and Berger (1990), Held and Bové (2014), Stuart and Ord (1987), and Judge et al. (1985).

9.2.8.1 Likelihood Theory

The notion of the likelihood function was introduced to statistics by R.A. Fisher in the early part of the 20th century. Fisher's original papers are both relatively difficult to read and no longer available in general circulation.[17] Consequently, we make use of the excellent reference Pawitan (2001) as the foundation for the discussion in this section. This text provides a broad range of practical insights into likelihood theory that we hope to apply to this specific problem.

Tremendous insight can be gained into the method of maximum likelihood by graphing the likelihood function. In a multivariate setting, however, this can be quite challenging since, quite clearly, a function of more than two variables cannot be

---

[17] The interested reader is referred to Savage (1976) for an overview of Fisher's voluminous contributions to statistics.

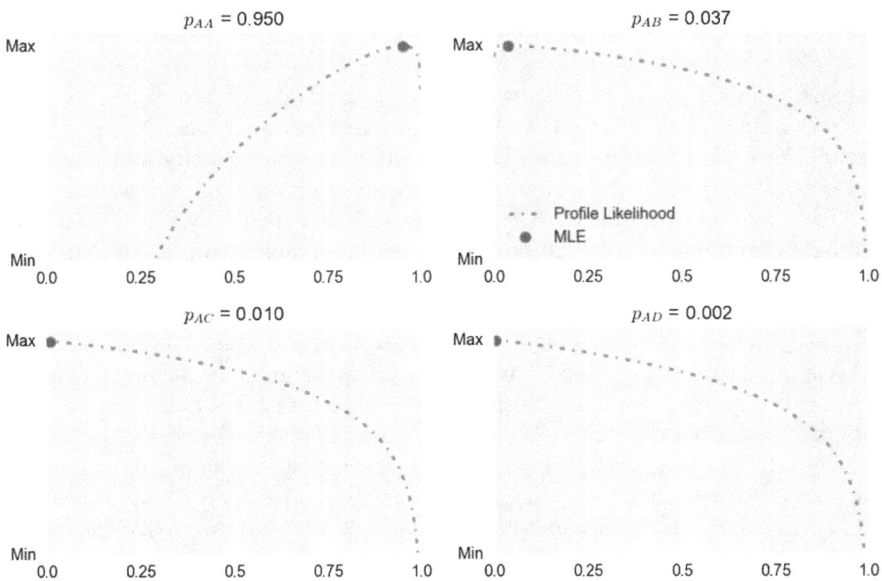

Fig. 9.8 *The profile likelihood*: Graphing the likelihood function provides important insight into the certainty of one's parameter estimates. In a multivariate setting, this is challenging. The profile likelihood addresses this issue by examining one parameter at a time—this figure examines the four parameters of the first row of P.

plotted by conventional methods. For this reason, the notion of a *profile likelihood* was introduced. In brief, the idea is to examine the evolution of the likelihood as a function of a single parameter, where the other parameters are held fixed.[18]

In our case, we have a transition matrix with 16 parameters, although only nine are of interest given the existence of an absorbing state. We can further reduce this dimension somewhat by noting that, conditional on X_t, the value of the K-dimensional Markov chain, X_{t+1}, follows a multinomial distribution with K possible outcomes. The consequence is that we may work with each row in isolation. Figure 9.8 provides the profile likelihoods associated with the elements of the first row of $\hat{P}$. In each case, we observe the value of likelihood function for each choice of the parameter and, at its highest value, the maximum-likelihood estimator.[19]

We can see from Fig. 9.8 that, for large parts of the unit interval, different parameter values provide relatively poor likelihood estimates. Visually, the

---

[18] Typically, one fixes the other parameters at their maximum-likelihood value.

[19] Construction of Fig. 9.8 requires a bit of caution. One cannot simply fix the other parameters at their maximum-likelihood values, because of the constraint that all elements in a given row of $\hat{P}$ must sum to unity. As we vary the value of a given parameter of interest, the remaining weight is thus allocated equally to each of the other fixed parameters. There may exist a more clever solution, but this potentially imperfect approach meets our purposes.

9.2 Estimation

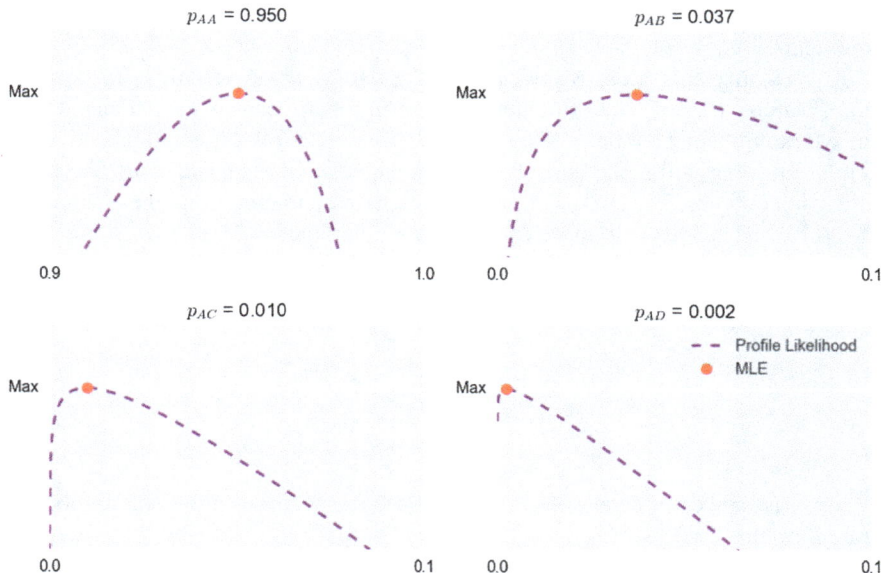

Fig. 9.9 *Likelihood curvature*: As highlighted in Pawitan (2001), the larger the curvature around the maximum of the likelihood function, the higher the implied level of parameter certainty. The figure magnifies the results from Fig. 9.8 to help us visualize the degree of MLE-curvature for each of the four estimates in the first row of $\hat{P}$.

maximum-likelihood estimator occurs, by definition, at the function's maximum. A natural question, however, is how good are other possible parameter values in the neighbourhood of the maximum? Logically, if the maximum-likelihood value falls at the top of a narrow or acute peak, it would inspire greater confidence. In other words, other parameters in the region of the maximum exhibit, in relative terms at least, significantly lower likelihood outcomes.

This is not easy to see from Fig. 9.8. As a consequence, Fig. 9.9 magnifies, or zooms, our perspective on each of the profile likelihoods to the vicinity of the maximum. In three of four cases, we observe that the log-likelihood function has an approximately quadratic form in the neighbourhood of its maximum. Such likelihood functions are termed *regular*. Conceptually, this is important, because the second derivative of a quadratic function, which has a parabolic form, is a constant. The larger this constant, the greater the curvature around the focus of the parabola—this is also, in our case, the maximum of the likelihood function.

The likelihood curvature in the neighbourhood of its maximum value, therefore, provides useful information about the level of parametric certainty in our maximum-likelihood estimate. The sharper, tighter, or pointier the peak of the likelihood function, intuitively, the more information we possess about the parameter. Conversely, if the likelihood is flat, or approximately linear, around its maximum point, then we consider ourselves to have less information. In the following development, we will formalize these ideas (slightly), but the intuition flows from these geometric arguments.[20]

To formalize this idea, we need to make use of some calculus. To describe the curvature, we will need to determine the first two derivatives associated with our Markov-chain parameter set. This will involve a bit of notation and some terminology, but will make our job easier as we develop the basic ideas. We begin with some notation. For a K-state Markov chain, our parameters are the elements of $P \in \mathbb{R}^{K \times K}$. The idea is to place each of the K^2 parameters, or elements of P, into a single vector $\theta \in \mathbb{R}^{K^2 \times 1}$. This form is more convenient. The result is,

$$\theta = \begin{bmatrix} p_{11} \\ p_{12} \\ \vdots \\ p_{KK} \end{bmatrix}. \tag{9.50}$$

As a consequence, the likelihood function maps the vector θ into a real number or, more formally, $\ell : \theta \to \mathbb{R}$.

The *first* derivative of $\ell(\theta)$—also known as the *score* function, $\nabla \ell(\theta) \in \mathbb{R}^{K^2 \times 1}$—is a gradient vector with the following form,

$$\frac{\partial \ell(\theta)}{\partial \theta} = \nabla \ell(\theta) = \begin{bmatrix} \frac{\partial \ell(\theta)}{\partial \theta_1} \\ \frac{\partial \ell(\theta)}{\partial \theta_2} \\ \vdots \\ \frac{\partial \ell(\theta)}{\partial \theta_{K^2}} \end{bmatrix}, \tag{9.51}$$

$$\nabla \ell(P) \equiv \begin{bmatrix} \frac{\partial \ell(\theta)}{\partial p_{11}} \\ \frac{\partial \ell(\theta)}{\partial p_{12}} \\ \vdots \\ \frac{\partial \ell(\theta)}{\partial p_{KK}} \end{bmatrix}.$$

The maximum-likelihood estimator is, of course, determined by setting the score function to zero—that is, $\frac{\partial \ell(\theta)}{\partial \theta} = 0$—and solving for θ. We have dispensed with the constraint in this development, because it is zero at the maximum and it does not show up in our objective of interest, the collection of second derivatives of $\ell(\theta)$.

---

[20] For a rigorous treatment of these topics, please see Held and Bové (2014).

9.2 Estimation

The *second* derivative of $\ell(\theta)$ leads to the so-called Hessian matrix, $H(\theta) \in \mathbb{R}^{K^2 \times K^2}$, with the following form

$$\frac{\partial^2 \ell(\theta)}{\partial \theta^2} = H(\theta) = \begin{bmatrix} \frac{\partial^2 \ell(\theta)}{\partial \theta_1^2} & \cdots & \frac{\partial^2 \ell(\theta)}{\partial \theta_1 \partial \theta_K} \\ \vdots & \ddots & \vdots \\ \frac{\partial^2 \ell(\theta)}{\partial \theta_K \partial \theta_1} & \cdots & \frac{\partial^2 \ell(\theta)}{\partial \theta_K^2} \end{bmatrix}, \quad (9.52)$$

$$H(P) \equiv \begin{bmatrix} \frac{\partial^2 \ell(\theta)}{\partial p_{11}^2} & \cdots & \frac{\partial^2 \ell(\theta)}{\partial p_{11} \partial p_{KK}} \\ \vdots & \ddots & \vdots \\ \frac{\partial^2 \ell(\theta)}{\partial p_{KK} \partial p_{11}} & \cdots & \frac{\partial^2 \ell(\theta)}{\partial p_{KK}^2} \end{bmatrix}.$$

Constructing the Hessian matrix is basically a book-keeping exercise; it is important to place the appropriate entry in the right place. It also plays an important role, since second-derivative information provides insight into the curvature of $\ell(\theta)$ around its maximum.

Without a formal development, let us motivate the role of the Hessian matrix by performing a multivariate second-order Taylor series expansion of the log-likelihood function around the vector of maximum-likelihood estimates, $\hat{\theta}$. This leads to the following expression,

$$\ln(\ell(\theta)) \approx \ln\left(\ell(\hat{\theta})\right) + (\theta - \hat{\theta})^T \underbrace{\nabla(\hat{\theta})}_{=0} + \frac{1}{2}(\theta - \hat{\theta})^T H(\hat{\theta})(\theta - \hat{\theta}), \quad (9.53)$$

$$\ln\left(\frac{\ell(\theta)}{\ell(\hat{\theta})}\right) \approx \frac{1}{2}(\theta - \hat{\theta})^T H(\hat{\theta})(\theta - \hat{\theta}).$$

Conceptually, this implies that, if we normalize this representation of the log-likelihood function, it has an approximately quadratic form. If this is a good approximation—or, equivalently, the likelihood function is regular—then the Hessian matrix is the key source of information about its curvature and, by extension, our degree of certainty about the maximum-likelihood estimates.

In practice, we do not work directly with the Hessian matrix, but instead adjust it somewhat. In particular, the negative of the Hessian matrix,

$$I(\theta) = -H(\theta), \quad (9.54)$$

$$= -\frac{\partial^2 \ell(\theta)}{\partial \theta^2},$$

is termed the Fisher information. The negative sign handles the fact that, at the maximum, all second derivatives should be negative. It is, from a calculus perspective, an assessment of the curvature of the likelihood function around a particular choice of θ. At $\hat{\theta}$, it measures the curvature of one's likelihood around its maximum. The larger this curvature, or rather the greater $I(\hat{\theta})$, the less uncertainty, or the more information we possess about our maximum-likelihood estimate.

While this can be made much more precise, the basic idea is that the quadratic expansion in equation 9.53 amounts to a normal approximation of $\hat{\theta}$—this provides a working definition of regularity. As stated by Pawitan (2001), *"regular problems are those where we can approximate the log-likelihood around the MLE by a quadratic function."* Given a regular likelihood function, one can show that, asymptotically at least, the estimation error is described as,

$$\hat{\theta} - \theta \sim \mathcal{N}\left(\underbrace{\mathbb{E}\left(\frac{\partial \ell(\theta)}{\partial \theta}\right)}_{=0}, \underbrace{\mathbb{E}\left(\frac{\partial^2 \ell(\theta)}{\partial \theta^2}\right)^{-1}}_{I^{-1}(\theta)}\right) \equiv \mathcal{N}\left(0, I^{-1}(\hat{\theta})\right). \quad (9.55)$$

That is, the estimation error is asymptotically normal. Very practically, therefore, the negative of the inverse of the Hessian matrix of the likelihood function evaluated at its maximum, $\hat{\theta}$, is a sensible estimate of the asymptotic, estimation-error covariance matrix. Moreover, the so-called Cramér-Rao bound, holds that $I^{-1}(\hat{\theta})$ is a lower bound for the (unknown) variance of our maximum-likelihood estimator,

$$\text{var}(\hat{\theta}) \geq I^{-1}(\hat{\theta}). \quad (9.56)$$

The interested reader is referred to DasGupta (2008, Chapter 16) or Pawitan (2001, Chapter 8) for additional generality, much more rigour, and technical details. The geometrical intuition found in Fig. 9.9 provides, formal derivation notwithstanding, significant conceptual support for the use of the second-derivative information regarding our likelihood function in the construction of a measure of parametric uncertainty.

All that remains is to determine, analytically if possible, the precise form of these aforementioned quantities in our specific problem. Let's use the log-likelihood function from equation 9.28 to evaluate the necessary ingredients. The ijth first derivative is,

$$\frac{\partial \ell(\theta)}{p_{ij}} = \frac{\partial}{\partial p_{ij}}\left(\ln\left(\mathbb{P}(X_1 = x_1)\right) + \sum_{i=1}^{T}\sum_{j=1}^{T} n_{ij} \ln\left(p_{ij}\right)\right), \quad (9.57)$$

$$= \frac{n_{ij}}{p_{ij}},$$

9.2 Estimation

for $i, j = 1, \ldots, K$. The ijth element of the Hessian matrix is simply,

$$\frac{\partial^2 \ell(\theta)}{p_{ij}^2} = \frac{\partial}{\partial p_{ij}} \left(\frac{n_{ij}}{p_{ij}} \right), \qquad (9.58)$$

$$= -\frac{n_{ij}}{p_{ij}^2},$$

when $i = j$ and zero otherwise. These quantities may then be readily employed in constructing our key quantities. We can write the score function, or gradient vector, as

$$\nabla \ell(P) = \begin{bmatrix} \frac{n_{11}}{p_{11}} \\ \frac{n_{11}}{p_{12}} \\ \vdots \\ \frac{n_{44}}{p_{44}} \end{bmatrix}. \qquad (9.59)$$

The Fisher information, which is merely a function of the Hessian matrix of $\ell(\theta)$ has a convenient diagonal form,

$$I(\theta) = -H(\theta), \qquad (9.60)$$

$$= \begin{bmatrix} \frac{n_{11}}{p_{11}^2} & 0 & \cdots & 0 \\ 0 & \frac{n_{12}}{p_{12}^2} & \cdots & 0 \\ \vdots & \vdots & \ddots & \vdots \\ 0 & 0 & \cdots & \frac{n_{44}}{p_{44}^2} \end{bmatrix}.$$

An important distinction is made between the true Fisher information, $I(\theta)$, and the *observed* Fisher information, $I(\hat{\theta})$. The former is a function, whereas the latter is a scalar value, or matrix, depending on the dimensionality of the parameter space. We should think of the observed Fisher information as an estimator of the true Fisher information, computed using the observed dataset.

The diagonal form of equation 9.60—which is incidentally not a general property of Hessian matrices—makes the inversion of the Fisher information significantly easier. Specifically, our standard-error (SE) estimate is simply:

$$\text{SE}(\hat{p}_{ij}) = \sqrt{I^{-1}(\hat{p}_{ij})}, \qquad (9.61)$$

$$= \sqrt{\frac{\hat{p}_{ij}^2}{n_{ij}}},$$

$$= \frac{\hat{p}_{ij}}{\sqrt{n_{ij}}}.$$

This suggests that the standard error tends to zero—relatively slowly due to the square root—as n_{ij} tends toward infinity. An alternative form involves elimination of the $\hat{p}_{ij}$ term, but plugging in our explicit maximum-likelihood estimate from equation 9.46. This permits us to restate equation 9.61 as,

$$\text{SE}\left(\hat{p}_{ij}\right) = \underbrace{\left(\frac{n_{ij}}{\sum_{j=1}^{K} n_{ij}}\right)}_{\text{Equation 9.46}} \frac{1}{\sqrt{n_{ij}}}, \qquad (9.62)$$

$$= \frac{\sqrt{n_{ij}}}{\sum_{j=1}^{K} n_{ij}}.$$

Here the interpretation is a bit more subtle. As before, the error in our parameter estimate tends toward zero as the sample size increases, but the relative number of transitions within a given state also plays a role. Indeed, the uncertainty of a given transition probability is the square root of the observed number of transitions normalized by the total transition count into that credit state. This is an intuitive and, by virtue of its simplicity, an immensely useful result.

One of the drawbacks of this approach, however, is that in the event that no transitions are observed into or out of a given credit category, the denominator will be zero and the corresponding standard error estimate will be undefined. While inconvenient, this result is nonetheless realistic. No information is available on the transition, so no meaningful standard-error estimate can be constructed. A more annoying consequence, however, is that in the event that $n_{ij} = 0$, but $\sum_{j=1}^{K} n_{ij} > 0$, the standard-error estimate will be identically zero. While a complete lack transitions to a given state nonetheless provides us with some information, it may not be entirely appropriate to assign a high degree of certainty to our estimate.

Before we apply this result to our specific estimation example, we examine an alternative, somewhat more practically motivated, approach to the assessment of transition-probability standard errors.

9.2.8.2 The Bootstrap Technique

Although standard-error bounds arising from likelihood theory are useful and convenient, they rely on the regularity of the likelihood function around its maximum and, by construction, are inherently asymptotic results. That is, accuracy improves as our

9.2 Estimation

sample size gets very large; full confidence in our standard-error values occurs only in the limit. In practice, as with our example industry typical dataset, one is obliged to work with relatively small finite samples. This argues for the investigation of an alternative technique exhibiting less dependence on the asymptotic behaviour of our estimates.

A useful non-parametric technique for application in precisely this type of situation was introduced by Efron (1979). It is a resampling technique termed bootstrapping.[21] Other useful references used to motivate the following discussion include Kulperger and Rao (1989), Fuh (1993)—who work with bootstrapping in the Markov-chain setting—and the fantastic general reference, Chernick (1999). Stuart and Ord (1987, Chapter 10) also provide some useful perspective and Efron and Tibshirani (1986) serve admirably as a helpful introduction to the main ideas.

The basic idea behind the bootstrap technique is, in the words of Chernick (1999), *"to replace the unknown population distribution with the known empirical distribution."* Consider, for example, a collection of independent, identically distributed observations from an unknown distribution, which we will denote $\{X_1, \ldots, X_N\}$. The empirical distribution, which is often represented as F_N, assigns a $\frac{1}{N}$ probability to each observed outcome. Imagine further that we seek to estimate the unknown population parameter vector, θ. Let us denote, as usual, this estimate as $\hat{\theta}$. If we make some assumptions about the distribution of our observations, we may proceed—using, for example, the observed Fisher information—to formulate a standard error for our estimate.

What if, however, we do not wish to make any assumptions regarding the $X's$? This may occur because the likelihood function is irregular, implying the normal approximation fails, or due to the lack of a reasonable analytic expression. Another motivation is a small sample size, which makes the analyst reluctant to rely on asymptotic results. Whatever our reason, Efron (1979)'s original idea was to use the empirical distribution to assist us in our efforts. It requires the repetition of the following *two* steps:

1. using the empirical distribution, F_N, simulate with replacement a random sample of size N,

$$X^*(m) = \{X_1^*, \ldots, X_N^*\}; \text{ and} \qquad (9.63)$$

2. using this simulated data, estimate the population parameter—denote this as $\theta^*(m)$.

One repeats these two steps for $m = 1, \ldots, M$ to construct an empirical, or bootstrap, distribution for θ. As again clearly stated in Chernick (1999), this yields the distribution of $\theta^* - \hat{\theta}$, when we, in fact, seek the distribution of $\theta - \hat{\theta}$. Intuitively,

---

[21] For finance professionals, this is not to be confused with the recursive technique used for extracting zero-coupon rates from bond prices. As we'll see in this discussion, it refers to an entirely different procedure.

one might expect that these two distributions converge for sufficiently large N. In this case, the intuition is indeed correct. While the technical details are wisely left to the quoted references, it is worth noting that the empirical distribution, F_N, is the nonparametric maximum-likelihood estimate of the unknown true distribution, F. This implies that any sample statistic computed from F_N is also, conceptually at least, a nonparametric maximum-likelihood estimate. This fact in no way amounts to a proof of convergence or unbiasedness, but it does suggest that bootstrap estimates are operating on relatively solid conceptual ground.

Instead of reviewing the mathematical details, let's instead focus on a simple example. Although not a proof of the methodology, it nonetheless provides practical motivation of its usefulness. Consider a random variable taking the following two values:

$$Y = \begin{cases} 1 : \text{with } \textit{unknown} \text{ probability, } p \\ 0 : \text{with } \textit{unknown} \text{ probability, } 1 - p \end{cases}. \tag{9.64}$$

In other words, Y, is thought to be a Bernoulli trial with population parameter, p. Given a dataset, we would like to formulate an estimate of p along with an assessment of our estimation uncertainty.

Imagine further that we observed the following dataset,

$$Y_N = \{1, 0, 0, 1, 0, 0, 0, 1, 0, 1\}. \tag{9.65}$$

Quite simply, $N = 10$, and the number of observations of $\{Y = 1\}$ is 4—we will denote this quantity as $\hat{k}_N$. A collection of independent Bernoulli trials follows, of course, a binomial distribution. The log-likelihood function has the following form,

$$\ell(p|N, k) = \ln\binom{N}{k} + k\ln(p) + (N - k)\ln(1 - p). \tag{9.66}$$

The first and second derivatives of this log-likelihood function are readily determined as,

$$\ell'(p|N, k) = \frac{k}{p} - \frac{N - k}{1 - p}, \tag{9.67}$$

and

$$\ell''(p|N, k) = \frac{\partial}{\partial p}\left(\frac{k}{p} - \frac{(N - k)}{1 - p}\right), \tag{9.68}$$

$$= -\frac{k}{p^2} - \frac{(N - k)}{(1 - p)^2}.$$

9.2 Estimation

Equation 9.67 is also the score function; by setting it to zero and solving for p, we easily find the maximum-likelihood estimator,

$$s(\hat{p}|N, k) = 0, \tag{9.69}$$

$$\frac{k}{\hat{p}} - \frac{(N-k)}{1-\hat{p}} = 0,$$

$$\hat{p} = \frac{k}{N}.$$

Moreover, the observed Fisher information, is also readily, albeit somewhat tediously, computed as,

$$I(\hat{p}|N, k) = -\ell''(\hat{p}|N, k), \tag{9.70}$$

$$= \frac{k}{\hat{p}^2} + \frac{(N-k)}{(1-\hat{p})^2},$$

$$= \frac{k}{\left(\frac{k}{N}\right)^2} + \frac{(N-k)}{(1-\left(\frac{k}{N}\right))^2},$$

$$= \frac{N^2}{k} + \frac{N^2}{(N-k)},$$

$$= \frac{(N-k)N^2 + kN^2}{k(N-k)},$$

$$= \frac{N^3}{k(N-k)}.$$

With a bit of manipulation, one can easily show that

$$I^{-1}(\hat{p}) = \frac{k(N-k)}{N^3}, \tag{9.71}$$

$$= \frac{k}{N^2} - \frac{k^2}{N^3},$$

$$= \frac{k}{N^2}\left(1 - \frac{k}{N}\right).$$

This flurry of simple algebra provides us with our maximum-likelihood estimate and its observed Fisher information.

Figure 9.10 plots, as previously counselled, our binomial log-likelihood function—conditional on the data in equation 9.65—across the range of possible population parameter values in the unit interval. The likelihood visually displays a clear quadratic form, implying that this is a regular problem, but it is fairly flat around its maximum. The relative flatness is due to the small sample size—this suggests a substantial degree of uncertainty regarding the true value of p.

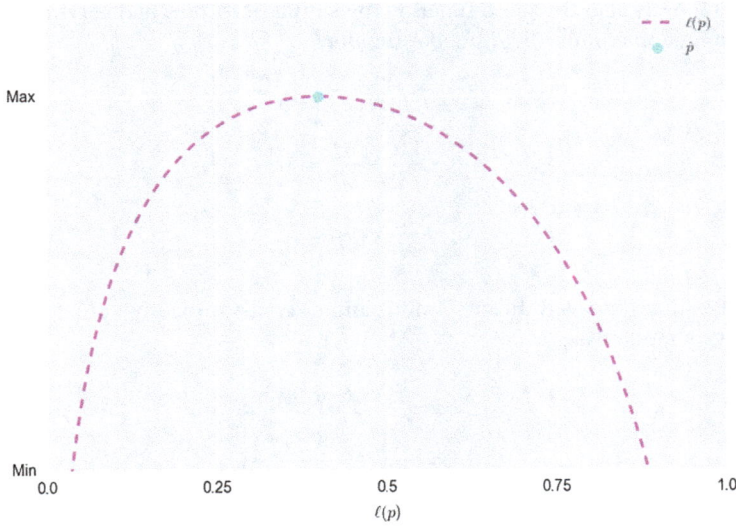

Fig. 9.10 *Binomial likelihood*: This figure plots the binomial log-likelihood function—conditional on the data in equation 9.65—across the range of possible population parameter values. While the likelihood clearly has a quadratic form, which implies regularity, it is relatively flat around its maximum. This is due to the small sample size.

Since we have established that the likelihood function is regular around its maximum, we may reasonably approximate the distribution of our estimation error as,

$$p - \hat{p} \sim \mathcal{N}\left(0, \frac{k}{N^2}\left(1 - \frac{k}{N}\right)\right). \tag{9.72}$$

An additional advantage of this example is that it permits us to review the likelihood framework in some detail. Given the dataset in equation 9.65 along with our hard-earned analytic formulae, our maximum-likelihood estimator is now simply

$$\begin{aligned}\hat{p} &= \frac{\hat{k}_N}{N}, \\ &= \frac{4}{10}.\end{aligned} \tag{9.73}$$

9.2 Estimation

Using the maximum-likelihood framework, assuming the Gaussianity of the error distribution, and equation 9.72, a 95% confidence interval for $\hat{p}$ is given as,

$$\hat{p} \pm 1.96\sqrt{I^{-1}(\hat{p})} = \hat{p} \pm 1.96\sqrt{\frac{\hat{k}_N}{N^2}\left(1 - \frac{\hat{k}_N}{N}\right)}, \quad (9.74)$$

$$= \frac{4}{10} \pm 1.96\sqrt{\frac{4}{10^2}\left(1 - \frac{4}{10}\right)},$$

$$= \frac{4}{10} \pm 1.96\sqrt{\frac{24}{1,000}},$$

$$= 0.4 \pm 0.304,$$

$$= [0.096, 0.704].$$

Our interval estimate thus suggests that, with a 95% level of confidence, the true (and unknown) parameter value, p, lies between 0.10 and 0.70. While this is a surprisingly wide interval and may prove disappointing to the analyst—or her boss—it is nonetheless a realistic and prudent estimate.

At this point, we would like to employ the bootstrapping technique to this problem. The idea is relatively simple. We use $\hat{p}$ to generate a sequence of random samples—each of size N. For each simulated sample, we estimate an alternative population parameter estimate, p^*.[22] This yields an empirical distribution for $\hat{p}$, which we employ to construct an associated interval estimate.

Figure 9.11 summarizes the bootstrap distribution arising from a sequence of 10,000 simulated samples of $N = 10$ observations.[23] To provide some additional perspective, Fig. 9.11 also highlights the maximum-likelihood estimator and normal approximation, computed using the MLE and observed Fisher-information-motivated standard-error estimates. Visually, there is strong agreement among the results.

How do the results compare numerically? The average parameter is, as one would expect, virtually indistinguishable from 0.4. The confidence intervals may be constructed in two ways. The first is to estimate the standard error of $\hat{p}$ from the empirical distribution of p^*'s. In this case, we have

$$\bar{p}^* \pm 1.96 \cdot \text{SE}(p^*) = 0.4 \pm 1.96 \cdot 0.156, \quad (9.75)$$

$$= [0.095, 0.705],$$

---

[22] It appears to be common practice in the bootstrapping literature to refer to the bootstrap data, distribution or estimator, X, as X^*. We will not deviate from this tradition.

[23] Each Bernoulli trial is simulated using $\hat{p}$ and the uniform-random-variate trick summarized—albeit in the multinomial setting—in Fig. 9.4.

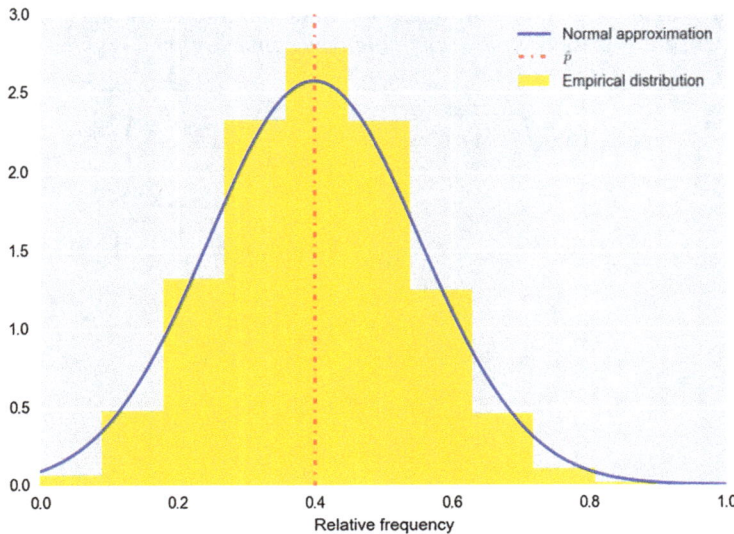

Fig. 9.11 *Binomial bootstrap distribution*: This figure summarizes the bootstrap distribution of $\hat{p}^*$ associated with 10,000 simulated samples of 10 observations. For perspective, the maximum-likelihood estimator and normal approximation are also illustrated. There is strong agreement among the results.

which agrees almost perfectly with the resulting 95% confidence interval from equation 9.74.[24] The alternative is to use the quantiles of the bootstrap distribution summarized in Fig. 9.11. That is,

$$\left[F_N^{-1}(0.025), F_N^{-1}(0.975)\right] = [0.10, 0.70]. \qquad (9.78)$$

---

[24] $\bar{p}^*$ and $\text{SE}(p^*)$ are computed precisely as one would expect. Given M simulations, they are defined as,

$$\bar{p}^* = \sum_{m=1}^{M} \frac{p^*(m)}{M}, \qquad (9.76)$$

and

$$\text{SE}(p^*) = \sqrt{\sum_{m=1}^{M} \frac{(p^*(m) - \bar{p}^*)^2}{M-1}}, \qquad (9.77)$$

where $p^*(m)$ denotes the maximum-likelihood estimate from the mth simulated dataset in our bootstrap computation.

9.2 Estimation

Again, the agreement is very close, but the quantiles are, in this case, round figures. The reason is simple: the empirical distribution is constructed of a collection of simulated samples, each possessing only 10 observations. More granular estimates are not possible.

While instructive, this example may be of limited use since we already have asymptotic estimates of parametric uncertainty. The agreement of the two approaches is nonetheless reassuring. Moreover, we see clearly that bootstrapping is a more general approach and requires both fewer assumptions and analytic complexity. As Efron and Tibshirani (1986) state, bootstrapping often amounts to *"the substitution of raw computing power for theoretical analysis."* Its generality and ability to handle relatively small samples make it a sensible alternative approach for the construction of default-probability interval estimates.

To complete this section, let us quickly describe the approach used to construct bootstrap estimates for our transition-probability estimates. We start from the observed data $\{X_{n,t}; t = 1, \ldots, T \text{ and } n = 1, \ldots, N\}$ and use it to estimate, in the previously described manner, $\hat{P}$. We then perform the following *two* steps,

- generate a sample, $\{X_{n,t}^{*,(m)}; t = 1, \ldots, T \text{ and } n = 1, \ldots, N\}$, from the empirical distribution using $\hat{P}$—this amounts to precisely the same approach used to simulate our data and described in Algorithm 9.1 on page 517;
- Using this information, we compute $\hat{P}^{*,(m)}$ using the cohort estimate.

The entire procedure is repeated $m = 1, \ldots, M$ times yielding a bootstrap distribution of $\hat{P}^*$. In short, the steps required to compute the bootstrap distribution are identical to those followed, earlier in the chapter, to simulate a Markov-chain process from a given transition matrix. In this case, however, we employ $\hat{P}$ as the starting point for the simulation.

The implementation, described in Algorithm 9.4, is also a pleasant surprise. It simply involves a recycling of a number of previous sub-routines. We define the total number of runs as S. For each iteration, we simulate the data

Algorithm 9.4 Generating our bootstrap distribution

```
def bootstrapDistribution(K,N,T,PEstimate,wStart,S):
    PBootstrap = np.zeros([K,K,S])
    for s in range(0,S):
        DBootstrap = simulateRatingData(N,T,PEstimate,wStart)
        NBootstrap = getTransitionCount(K,T,N,DBootstrap)
        PBootstrap[:,:,s] = estimateCohortTransitionMatrix(K,NBootstrap,1)
    return PBootstrap
```

using `simulateRatingData`, then count the number of transitions with `getTransitionCount`, and finally employ `estimateCohortTransitionMatrix` to construct the associated maximum-likelihood estimates. The return value, termed `PBootstrap`, is a $K \times K \times S$ array collecting the full bootstrap distribution of transition-probability estimates. Each element of P, therefore, has S bootstrap estimates that can be ordered and used to construct both point and interval estimates.

9.2.9 Interval Estimates

At this point, we have introduced all of the machinery required to construct interval estimates for our default probabilities. Let's, therefore, immediately jump to the final results. The underlying matrix provides, using the maximum-likelihood framework, a 95% confidence interval for each of the transition-matrix entries,

$$\hat{P} \pm \mathcal{T}_{T-1}^{-1}\left(1 - \frac{0.05}{2}\right)\sqrt{I^{-1}\left(\hat{P}\right)} = \begin{bmatrix} [0.90, 1.00] & [0.03, 0.05] & [0.00, 0.01] & [0.00, 0.01] \\ [0.08, 0.12] & [0.72, 0.85] & [0.08, 0.13] & [0.00, 0.02] \\ [0.10, 0.19] & [0.15, 0.25] & [0.55, 0.73] & [0.00, 0.02] \\ [0.00, 0.00] & [0.00, 0.00] & [0.00, 0.00] & [1.00, 1.00] \end{bmatrix}.$$

(9.79)

Conversely, this second matrix uses bootstrap distribution—constructed with 1,000 simulated draws of $N \times T$ samples from the empirical distribution summarized by $\hat{P}$—to construct a 95% quantile-based set of confidence intervals,

$$\left[F_{\hat{P},N}^{-1}(0.025), F_{\hat{P},N}^{-1}(0.975)\right] = \begin{bmatrix} [0.94, 0.96] & [0.03, 0.05] & [0.01, 0.01] & [0.00, 0.01] \\ [0.08, 0.12] & [0.76, 0.82] & [0.08, 0.12] & [0.00, 0.02] \\ [0.11, 0.19] & [0.16, 0.25] & [0.59, 0.69] & [0.00, 0.02] \\ [0.00, 0.00] & [0.00, 0.00] & [0.00, 0.00] & [1.00, 1.00] \end{bmatrix}.$$

(9.80)

Neither equation 9.79 nor equation 9.80 is, with its proliferation of numbers, particularly easy to read or interpret. Figure 9.12 helps somewhat by displaying the *nine* most important transition probabilities—ignoring the absorbing state—in graphic form.

These numerical and graphical results permit a few cautious observations. While the two approaches provide confidence bounds, which do not agree exactly, they provide qualitatively similar results. The bootstrap intervals, on the whole, do seem to be tighter. Second, the true values—found in equation 9.40—appear, in virtually

9.2 Estimation

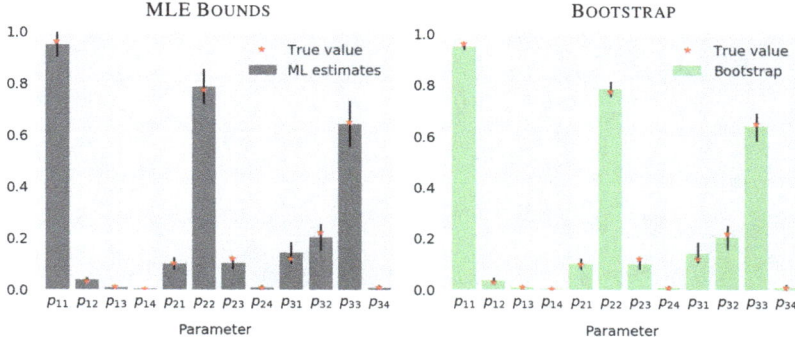

Fig. 9.12 *Comparing approaches*: The two underlying graphics display the estimates of each transition probability along with true values and a 95% confidence interval for the MLE and bootstrap approaches, respectively. The results are, visually at least, in broad agreement.

all cases, to fall comfortably within the reported confidence bound. This is, at least, true to the reported number of significant digits. This naturally need not, in principle, always be the case, but it is generally reassuring. As a final point, the default probabilities, by reason of their small magnitude, are particularly difficult to interpret. Since these quantities are our principal object of interest, we will spend the rest of this section, through more in-depth analysis and focused use of graphics, on these default quantities.

Let us begin by magnifying our perspective on the three default probabilities. Figure 9.13 highlights the MLE and bootstrap point estimates and their associated 95% confidence intervals—for reference, the true values are also provided. Interestingly, in contrast to our conclusion from examining the full transition matrices, the bootstrap method appears to assign slightly greater, or at least equal, uncertainty to default-probability estimates from the analytic approach. This appears to suggest that, for default probability computations, these are not entirely equivalent approaches and they interpret the data in different ways. This result, given the divergence of results, argues for the coincident use of both techniques as a mutual sanity check.

The larger issue, however, is the relatively high degree of uncertainty in the default-probability estimates. The true default probability of state C is twice that for obligors in credit category, B. Our point and interval estimates, however, indicate that there is essentially no difference between these two default probabilities. We do not need to look far for the culprit: a small amount of data. Our simple ten-observation binomial example clearly supports this conclusion. More technically, the flatness of the profile log-likelihood functions speaks volumes about the challenges of obtaining precise default-probability values. This is, however, life. We can neither manufacture data nor can we pretend that our estimates have a high degree of certainty. The appropriate solution is to communicate this fact explicitly in one's credit-risk computations.

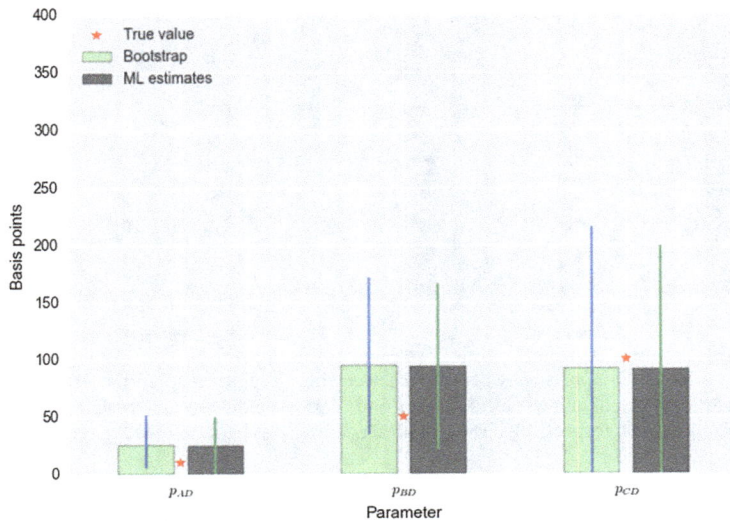

Fig. 9.13 *Default-probability confidence bounds*: This figure takes the default probability results from Fig. 9.12 and magnifies them providing a closer look. We may conclude that, in this instance at least, the bootstrap provides a more accurate and conservative assessment of parameter uncertainty.

Table 9.3 *Numerical parameter uncertainty*: This table highlights, using the bootstrap with the true and estimated transition matrix, the point estimate, standard error and 95% confidence interval. As one would expect, there is noise involved when using the estimated transition matrix. The lesson is that interval estimates are an essential tool in a small-sample setting. All figures are reported in percentage terms.

Quantity	True value	Bootstrap with P			Bootstrap with $\hat{P}$			$\max \left| p^*_{Dn} - \hat{p}^*_{Dn} \right|$
		$\bar{X}$	SE	CI	$\bar{X}$	SE	CI	
p_{AD}	0.10	0.10	0.08	[0.00, 0.27]	0.25	0.12	[0.06, 0.52]	0.72
p_{BD}	0.50	0.50	0.28	[0.00, 1.13]	0.95	0.36	[0.30, 1.71]	2.26
p_{CD}	1.00	1.01	0.55	[0.00, 2.24]	0.92	0.55	[0.00, 2.18]	2.93

To complete this section, we will examine the bootstrap in somewhat more detail both numerically and graphically. Table 9.3 highlights a number of summary statistics from the bootstrap distribution. The bootstrap results, however, are computed twice. The first computation involves the *true* known data parameters—this essentially amounts to drawing from the population parameters.[25] The second computation, much more realistically, uses the maximum-likelihood estimates as the basis for the bootstrap simulations.

---

[25]This is, of course, conceptually quite odd. If we really knew the true parameters, we would not require any estimate of their uncertainty. At the same time, we can view it as another practical verification of the bootstrap approach.

Fig. 9.14 *Bootstrap results*: The underlying graphic displays the bootstrapped distributions associated with the default probabilities for states A, B, and C. These distributions are used to determine our desired error bounds.

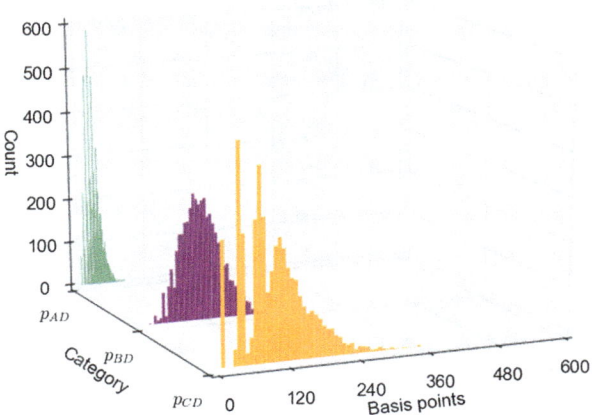

As the sample size, $N \times T$, grows sufficiently large, we would expect to observe a convergence of these results. The evidence in Table 9.3, however, suggests that, when using $\hat{P}$, there is greater uncertainty in the simulations. The larger point, however, is that with a finite sample comprised of approximately 2,900 transitions, there is a significant amount of uncertainty about the actual default probabilities. An attempt at summarizing this point is found in the final column of Table 9.3. It summarizes—across all the simulations—the maximum absolute distance between simulated estimates under the true and estimated transition probabilities.

Independent small-sample draws from the same distribution can lead to parameter estimates that differ significantly from the true population parameters. While this is intellectually well understood, in practice, point estimates continue to be used without any assessment of uncertainty. The results of this analysis indicate that standard-error and interval estimates are, in fact, an essential tool in this setting.

A powerful way to appreciate this fact is to visually inspect the bootstrap distributions. Figure 9.14 provides, in a three-dimensional manner, the bootstrap distributions associated with the three default-probability estimates. The clearly symmetric form and significant deviation of point estimates are clearly visible. The probability of defaulting, given you find yourself in state C, ranges from close to zero to almost 4%.

9.2.10 Risk-Metric Implications

In this final section, we attempt to gain some insight into the implications of default-probability uncertainty on an important application of these figures. As we've seen in the previous chapters, these are critical inputs into the computation of VaR and expected-shortfall measures. We have seen that, with a sample similar to that employed in practical applications, there is a substantial amount of uncertainty

Table 9.4 *Categorizing obligors*: The underlying table describes the logic and numerical outcomes associated with the assignment of the 100 individual obligors, from our running example, to our three non-default credit categories: A, B, and C. We use q_x—where $x \in [0, 1]$—to denote the xth quantile of the original unconditional default probabilities. All numerical values are denoted in basis points.

Treatment	Category	Interval	Values (bps.)
Top third	A	$[q_0, q_{\frac{1}{3}})$	$[0.1, 34.6)$
Middle third	B	$[q_{\frac{1}{3}}, q_{\frac{2}{3}})$	$[34.6, 106.8)$
Bottom third	C	$[q_{\frac{2}{3}}, q_1]$	$[106.8, 682.1]$

regarding the default probabilities. Our question is simply: does this translate into material uncertainty with respect to our credit-risk-measure estimates?

While this question cannot be answered definitively, we can nonetheless perform some simple analysis to reach some tentative conclusions. To do this, however, we require a portfolio and a credit-risk model. For our portfolio, we will naturally employ our running 100-obligor example. A few changes are nonetheless required. In particular, we originally assigned separate default probabilities to each individual credit counterpart. In this example, there are only three distinct, non-default credit states; within each credit category, all obligors share the same default probability.

To assign each obligor one of these default probabilities, we use a very simple approach. The obligor are ordered by their unconditional default-probability value and the top third, in terms of quality, are assigned to credit category A. The second third are placed in category B, while the remaining obligors fall into the final group, C. This is a bit arbitrary, but not conceptually inconsistent with the work performed by credit-rating agencies. Everyone is clearly aware that there are differences between all individual credit counterparts within a given rating class. They are nevertheless, for analytic convenience, partitioned into disjoint groups of obligors sharing similar characteristics.

Table 9.4 highlights the specifics of this approach along with the numerical outcomes. Each third of the 100 individual obligors, from our running example, is simply allocated based on its original default probability to our three non-default credit categories: A, B, and C. By construction, therefore an (approximately) equal number of credit counterparties find their way into each of our three credit states.

Our second requirement is a model. We will employ two different, albeit related, approaches: the one-factor Gaussian and t-distributed threshold models.[26] Both models are discussed in detail in Chap. 4. To be fair, however, the choice of model is not particularly important. Similar results are observed with other underlying credit-risk models.

Table 9.5 summarizes, for a few illustrative quantiles, the VaR estimates associated with both models for our sample portfolio. Both risk measures were

---

[26]The state-variable correlation parameter, ρ, is set to 0.3 for both models, while the degrees-of-freedom parameter in the t-distributed implementation is 9.

9.2 Estimation

Table 9.5 *Parameter uncertainty and VaR*: Using a simple (simulated) portfolio, we can use our default-probability interval estimates to assess the corresponding VaR uncertainty at different quantiles; very similar results are observed for expected shortfall. As expected, the outcomes are quite sensitive to the underlying default parameters.

Quantile	$1 - \mathcal{N}$		$1 - \mathcal{T}$	
	PE	CI	PE	CI
Bootstrap				
99.00th	$72	[$20, $112]	$102	[$21, $155]
99.90th	$136	[$37, $198]	$234	[$56, $310]
99.99th	$213	[$59, $294]	$388	[$103, $473]
ML estimates				
99.00th	$72	[$27, $114]	$102	[$31, $157]
99.90th	$136	[$51, $202]	$234	[$85, $313]
99.99th	$214	[$82, $301]	$391	[$169, $477]

numerically estimated using raw Monte-Carlo with 10 million iterations. To keep the clutter to a minimum, we do not report the residual simulation noise. It is, we hasten to mention, dramatically smaller than the variation caused by default-probability uncertainty. As one would expect, given its leptokurtotic form and tail dependence, the t-distributed model yields higher point estimates (PE) across all quantiles with the difference increasing as we move further out into the tail of the default-loss distribution.

Each of the VaR estimates found in Table 9.5 are computed using the point-estimate default probabilities extracted from $\hat{P}$. Since we have also taken considerable pains to construct interval estimates for these default probabilities, we can use them to re-estimate the commensurate VaR amounts. For each quantile, model, and interval-estimate technique, a 95% confidence interval is presented. The potential differences from the base-line values are important. Small distinctions between the maximum-likelihood and bootstrap techniques are evident, but economically the results are consistent. Relatively small changes in one's default probability inputs clearly lead to significant differences in one's risk computations; most importantly, our risk measures inherit the uncertainty embedded in our default-probability inputs.

While it is typically not widespread practice to use interval estimates to gauge the uncertainty in one's risk estimates, this analysis suggests that this would be a prudent approach. Parametric uncertainty in default-probability estimates is non-trivial and this can have important spillover effects on interpretation of risk measurement. At the very least, it makes sense to perform sensitivity analysis similar to what is summarized in Table 9.5 and to communicate the implications of this parametric uncertainty to senior decision makers. This information will certainly colour their interpretation of the results and permit better decision making.

9.3 Risk-Neutral Default Probabilities

Estimation of Markov-chain parameters from credit-category transition data is *not* the only approach to determining default-probability inputs. The alternative approach—which we refer to broadly as calibration—involves extracting risk-neutral default probabilities from the observed prices of appropriate market instruments. The most common choices are bonds or credit-default swaps. While both can prove quite useful, we will focus our attention on the somewhat more general and broadly used credit-default swap approach.

The calibration technique—since it works directly with market prices—will require some review of the fundamentals of credit-risk pricing. Our investigation will remain relatively elementary and provide only the basic ideas required to competently implement our calibration exercise. Credit-risk pricing is, in fact, a separate discipline and we will not have time to do it justice. The interested reader is referred to Bielecki and Rutkowski (2002) for a comprehensive and useful overview.

9.3.1 Basic Cash-Flow Analysis

Let us introduce some basic notation to describe a generic fixed-income financial claim. Let us first use β to denote the total number of payment dates.[27] This implies that, at inception, there is the following sequence of cash-flow times,

$$\{T_1, T_2, \cdots, T_\beta\}. \tag{9.81}$$

We define N as the notional amount and let c_i represent the ith coupon rate, which may be fixed or floating. The discounted coupon stream is then written as,

$$\text{Coupon Stream} = \sum_{i=1}^{\beta} c_i N \delta(t, T_i), \tag{9.82}$$

where $\delta(t, T_i)$ represents the risk-free, zero-coupon bond price—or discount rate—between the current time, t, and the ith payment date, T_i. To ease the notation burden, we will summarize this as δ_i.[28] The discounted principal cash-flow stream

---

[27] We have, in the interests of full disclosure, pinched some convenient notation from Brigo and Mercurio (2001).

[28] That is, to be explicit

$$\delta_i \equiv \delta(t, T_i), \tag{9.83}$$

for $i = 1, \ldots, \beta$.

9.3 Risk-Neutral Default Probabilities

is similarly written as,

$$\text{Principal Repayment} = N\delta_\beta. \tag{9.84}$$

The set of discounted cash-flows is simply the combination of equations 9.82 and 9.84,

$$P = \underbrace{\sum_{i=1}^{\beta} c_i N \delta_i}_{\text{Coupon stream}} + \underbrace{N\delta_\beta}_{\text{Principal repayment}}. \tag{9.85}$$

This all seems quite straightforward and obvious, but we are not quite done. Indeed, we have merely set the stage. The price of any financial security, given an absence of arbitrage and complete financial markets, can be written as an expectation with respect to the equivalent-martingale measure, $\mathbb{Q}$.[29] We can thus restate equation 9.85 as,

$$\mathbb{E}^{\mathbb{Q}}\left(P \mid \mathcal{F}_t\right) = \mathbb{E}^{\mathbb{Q}}\left(\sum_{i=1}^{\beta} c_i N \delta_i + N\delta_\beta \,\bigg|\, \mathcal{F}_t\right) \tag{9.86}$$

$$= \sum_{i=1}^{\beta} \mathbb{E}^{\mathbb{Q}}\left(c_i \,\bigg|\, \mathcal{F}_t\right) N \, \mathbb{E}^{\mathbb{Q}}\left(\delta_i \,\bigg|\, \mathcal{F}_t\right) + \sum_{i=1}^{\beta} N \, \mathbb{E}^{\mathbb{Q}}\left(\delta_\beta \,\bigg|\, \mathcal{F}_t\right),$$

on the probability space, $(\Omega, \mathcal{F}, \mathbb{Q})$. At first glance, this may not seem terribly meaningful. We have, in fact, quietly made an important assumption. That is, we have assumed that the discount factor and the coupon payment are statistically independent, which permits us to write their joint expectation as the product of their individual expectations.[30]

While interest rates evolve stochastically, the discount rates are $\mathcal{F}_t$-measurable and can generally treated as constants. One can, however, be much more precise. Technically, we have that

$$\delta(t, T) = \mathbb{E}^{\mathbb{Q}}\left(\exp\left(-\int_t^T r(u)du\right)\right), \tag{9.87}$$

---

[29] See Harrison and Kreps (1979) and Harrison and Pliska (1981) for the original references in this regard. Duffie (1996) also offers an excellent alternative reference.
[30] Of course, if c_i is constant, then independence follows trivially.

where $r(t)$ is the instantaneous short rate. When simply discounting cash-flows, we typically do not worry much about this structure. It is nevertheless always lurking in the background and will be explicitly required in the following discussion.

The only possible, remaining, random term in this expression is the coupon rate. We need, therefore, to investigate the structure of the ith coupon payment, c_i. In general, we write it as,

$$c_i = \Delta(T_{i-1}, T_i) K_i, \tag{9.88}$$

where $\Delta_i \equiv \Delta(T_{i-1}, T_i)$ is a measure of the distance, in units of time, between the $(i-1)$th and ith payment dates.[31] The K_i term is important. It may either be fixed—say $K_i \equiv K$—or floating. The former case, as indicated, is simple. The latter case is more involved. Typically, the floating-rate coupon is indexed to a particular LIBOR setting. Let us, therefore, denote the floating coupon as,

$$K_i = \underbrace{L(T_{i-1}, T_i)}_{L_i}, \tag{9.89}$$

where L is a particular LIBOR tenor, T_{i-1} is the reset date, and T_i is the payment date.[32] Again, to keep our expressions sufficiently short, we will often denote the LIBOR rate as L_i.

Let us now, armed with this structure, return to determine the expectation of the coupon value from equation 9.86,

$$\mathbb{E}^{\mathbb{Q}}\left(K_i | \mathcal{F}_t\right) = \mathbb{E}^{\mathbb{Q}}\left(L(T_{i-1}, T_i) | \mathcal{F}_t\right), \tag{9.90}$$

$$\mathbb{E}_t^{\mathbb{Q}}(K_i) = \mathbb{E}_t^{\mathbb{Q}}(L(T_{i-1}, T_i)),$$

$$= F(t, T_{i-1}, T_i),$$

where $F(t, T_{i-1}, T_i)$ is the implied LIBOR-based forward rate at time t for a zero-coupon bond contract starting at time T_{i-1} and maturing at time T_i. This so-called break-even rate is, in fact, as stated in Brigo and Mercurio (2001, Chapter 1) "the expectation of $L(T_{i-1}, T_i)$ at time t under a suitable probability measure." The *suitable* measure is the equivalent martingale measure induced by the money-market account as the numeraire asset; this is commonly referred to as the risk-neutral measure. As a final point, again in an attempt to ease the notation, we write $\mathbb{E}_t^{\mathbb{Q}}(\cdot)$, as is common practice in economics, to denote the conditionality on the filtration, $\mathcal{F}_t$.

---

[31] This quantity is often referred to as the day-count convention. While the specific choice is certainly important in practice, it is neither complex nor stochastic. As such, we do not allocate it any particular attention.

[32] Often a fixed spread, s, is added to reflect the creditworthiness of the individual counterparty. We will set $s = 0$ without any loss of generality.

9.3 Risk-Neutral Default Probabilities

To summarize, therefore, our pricing relation in the absence of credit risk is given as,

$$\mathbb{E}_t^{\mathbb{Q}}(P) = N\delta_\beta + \sum_{i=1}^{\beta} K \Delta_i N \delta_i, \qquad (9.91)$$

for fixed rate bonds and,

$$\mathbb{E}_t^{\mathbb{Q}}(P) = N\delta_\beta + \sum_{i=1}^{\beta} F(t, T_{i-1}, T_i) \Delta_i N \delta_i, \qquad (9.92)$$

for floating-rate bonds. In the former case, the effort amounts to identifying cash-flows dates, magnitudes, and discount rates. The latter case involves the same basic effort, but additionally requires the construction of a forward interest-rate curve consistent with the LIBOR tenor used to construct the cash-flow index. The curve used to determine the forward rates and subsequently the forecast future cash-flows is referred to as the projection curve. Historically, both the discount and projection curves were assumed to be equivalent, which led to some simplification in the underlying mathematics. In modern applications, however, separate curves are used to generate the cash-flow stream and discount them. This approach, known broadly as overnight-index swap (OIS) discounting, is now widely considered to be best practice.

The reasons for the previous development are twofold. First, it helps us introduce a common and flexible notation to use in the following development. This will pay dividends when we price individual credit-default swap contracts. The second, and more important, objective is to explicitly remind the reader that even simple cash-flow discounting occurs within a broader and deeper asset-pricing framework. We may use implied forward rates to proxy floating-rate payments and discount future cash-flows using observed discount rates, but these quantities are risk-neutral expectations of stochastic variables. This reality becomes even starker as we introduce the notion of default risk.

9.3.2 Introducing Default Risk

The previous section introduced a simple structure for the pricing of a coupon-bearing, risk-free fixed-income security. We now seek to add the notion of default. Default is modelled as a random event and to incorporate it into a pricing framework, it requires some structure. Following Jeanblanc (2002), we introduce the notion of a default time τ as a positive-valued, random-variable defined the probability space

$(\Omega, \mathcal{F}, \mathbb{Q})$.[33] Having introduced the default time as a random event, we can, at any given point in time or state of the world, concisely describe its occurrence with the use of the underlying indicator variable,

$$\mathbb{I}_{\tau > T} = \begin{cases} 1 : \tau > T \text{ (or survival to } T) \\ 0 : \tau \leq T \text{ (or default prior to } T) \end{cases}. \quad (9.93)$$

The plan is to integrate this random description of default into our pricing formulae. The easiest place to start is with zero-coupon instruments. Following from Duffie and Singleton (2003), equation 9.93 can be used to gracefully describe a zero-coupon bond price subject to default risk (at time t) with a tenor of T years as,

$$\tilde{\delta}(t, T) = \underbrace{\mathbb{E}_t^{\mathbb{Q}} \left(e^{-\int_t^T r(u) du} \mathbb{I}_{\tau > T} \right)}_{\text{Survival}} + \underbrace{\mathbb{E}_t^{\mathbb{Q}} \left(e^{-\int_t^T r(u) du} R \mathbb{I}_{\tau \leq T} \right)}_{\text{Default}}, \quad (9.94)$$

where R denotes the recovery rate and $r(t)$ is the instantaneous short-rate process. Logically, this is very appealing. There is a single unit cash-flow at time T. If the obligor survives and does not default, one receives one unit of currency. If they default, the pay-off is $R \in [0, 1]$. If the recovery is assumed to be zero, then the second term disappears entirely.

To develop pricing formulae, we will exploit the main idea from equation 9.94. To simplify matters, and follow common pricing practice, we will assume that $r(t)$ and τ are independent.[34] In this case, equation 9.94 reduces to,

$$\tilde{\delta}(t, T) = \mathbb{E}_t^{\mathbb{Q}} \left(e^{-\int_t^T r(u) du} \right) \mathbb{E}_t^{\mathbb{Q}} \left(\mathbb{I}_{\tau > T} \right) + \mathbb{E}_t^{\mathbb{Q}} \left(e^{-\int_t^T r(u) du} \right) R \mathbb{E}_t^{\mathbb{Q}} \left(\mathbb{I}_{\tau \leq T} \right), \quad (9.95)$$

$$= \delta(t, T) \mathbb{Q}(\tau > T) + \delta(t, T) R \mathbb{Q}(\tau \leq T),$$

$$= \delta(t, T) \Big(\mathbb{Q}(\tau > T) + R \mathbb{Q}(\tau \leq T) \Big),$$

where $\delta(t, T)$ and $\tilde{\delta}(t, T)$ represent risk-free and default-risky T period zero-coupon bonds, respectively. The terms $\mathbb{Q}(\tau > T)$ and $\mathbb{Q}(\tau \leq T)$ are, in fact, the survival and default probabilities over the time interval $[0, T]$, respectively. The survival

---

[33] Other approaches are, of course, possible. Merton (1974) treats τ as the first-passage time of a firm's assets falling below its liabilities. This leads to the so-called structural approach to default risk addressed in Chaps. 4 and 5.

[34] Duffie and Singleton (2003) do not make this assumption and develop an entire framework incorporating uncertainty with regard to both quantities. While both elegant and useful, we do not require this level of sophistication for our application.

9.3 Risk-Neutral Default Probabilities

probability, up to a given time, plays an important role in the pricing of default-risky securities. We define it, using the fairly common notation from Bluhm et al. (2003), as

$$S(t) = \mathbb{Q}(\tau > t), \tag{9.96}$$

where $S(t)$ is typically referred to as the survival function. If we set $R = 0$ and use equation 9.96 to restate our zero-coupon bond price from equation 9.95, we have

$$\tilde{\delta}(t, T) = \delta(t, T)s(T). \tag{9.97}$$

This is an interesting and insightful expression. Since both the risk-free discount factor and survival probability are restricted to the unit interval, they both act to deflate the risky bond's cash-flow. $\delta(t, T)$ discounts the final payment for the time value of money, whereas $s(T)$ deflates it for the risk of default. One can think, therefore, of the survival function as being analogous to the pure-discount-bond-price function.

Survival probabilities thus, acting as a kind of discount factor, play an important role. We also require information about future (or forward) default probabilities. As we'll see, these operate quite similar to forward interest rates. Moreover, since forward interest rates can be derived from zero-coupon bond prices, forward default probabilities can be deduced from survival probabilities. To see this, we first introduce a few basic ideas. First of all, consider two points in time, $t_1 \leq t_2$. In this case, we have that

$$\mathbb{Q}\bigg(\tau \in (t_1, t_2)\bigg) = \mathbb{Q}(\tau > t_1) - \mathbb{Q}(\tau > t_2), \tag{9.98}$$

$$= S(t_1) - S(t_2),$$

for $t_1 \leq t_2$. That is, the default probability over a given interval can be determined by the difference between the survival probabilities. Since, by definition, $S(t)$ is monotonically decreasing and restricted to the unit interval, the definition in equation 9.98 is entirely consistent with a default probability.

It also follows, from basic principles of probability theory, that the cumulative distribution function of the default time also depends on the survival function,

$$F_\tau(t) = 1 - S(t), \tag{9.99}$$

$$= \mathbb{Q}(\tau \leq t).$$

More simply, if one knows the survival function, then one also has the default distribution. If one seeks the probability density function, then—assuming that

$F_\tau(t)$ is continuous and differentiable—it has the following form,

$$f_\tau(t) = \frac{\partial F_\tau(t)}{\partial t}, \tag{9.100}$$

$$= \frac{\partial}{\partial t}\Big(1 - S(t)\Big),$$

$$= -S'(t).$$

Thus, both the density and distribution functions depend on the form of the survival function.

Now, we may introduce the notion of a conditional survival probability. Consider the following object,

$$S(t_2|t_1) = \mathbb{Q}(\tau > t_2 | \tau > t_1), \tag{9.101}$$

for $t_1 \leq t_2$. This is the probability of an obligor surviving, without default, until time t_2, given that they have already survived to time t_1. Using Bayes theorem, we can manipulate equation 9.101 further as,

$$S(t_2|t_1) = \mathbb{Q}(\tau > t_2 | \tau > t_1), \tag{9.102}$$

$$= \underbrace{\frac{\mathbb{Q}\Big(\{\tau > t_2\} \cap \{\tau > t_1\}\Big)}{\mathbb{Q}(\tau > t_1)}}_{\text{By Bayes' theorem}},$$

$$= \frac{\mathbb{Q}(\tau > t_2)}{\mathbb{Q}(\tau > t_1)},$$

$$= \frac{S(t_2)}{S(t_1)}.$$

The previous development is mostly definitional. The joint probability of survival to times t_1 and t_2, however, collapses to survival to time t_2, because survival to t_2 requires survival to t_1. The larger point is that the conditional survival probability is the ratio of unconditional survival function values. The conditional *default* probability is thus simply $1 - S(t_2|t_1)$ for all $t_1 \leq t_2$.

The final step is the introduction of the so-called hazard function, which is also commonly referred to as the hazard rate or failure rate in operational research and actuarial science applications.[35] It is the relative frequency of default in a given period $(t, t + \Delta t]$, conditional on survival to the beginning of the time interval (i.e.,

---

[35] See, for example, Taylor and Karlin (1994, Chapter 1) or Stuart and Ord (1987, Section 5.34) for more background information on this concept.

9.3 Risk-Neutral Default Probabilities

t). One can think of it as the default risk one faces over a small period starting at some arbitrary point in the future. It has a convenient form, but requires some work for its derivation; we have, however, already explored the necessary background facts. Starting with the formal definition,

$$\mathbb{Q}\left(\tau \in (t, t+\Delta t] \,\middle|\, \tau > t\right) = \underbrace{\frac{\mathbb{Q}\left(\{\tau \in (t, t+\Delta t]\} \cap \{\tau > t\}\right)}{\mathbb{Q}(\tau > t)}}_{\text{By Bayes' theorem}}, \quad (9.103)$$

$$= \frac{\mathbb{Q}\left(\tau \in (t, t+\Delta t]\right)}{S(t)},$$

$$= \frac{\overbrace{\mathbb{Q}(\tau \le t+\Delta t) - \mathbb{Q}(\tau \le t)}^{\text{Equation 9.98}}}{S(t)},$$

$$= \frac{F_\tau(t+\Delta t) - F_\tau(t)}{S(t)}.$$

The hazard function is now defined as the following limit of a slight modification of equation 9.103 as $\Delta t \to 0$, then we recover the typical form of the hazard function,

$$h(t) = \lim_{\Delta t \to 0} \frac{\mathbb{Q}\left(\tau \in (t, t+\Delta t] \,\middle|\, \tau > t\right)}{\Delta t} = \lim_{\Delta t \to 0} \frac{1}{S(t)} \underbrace{\frac{F_\tau(t+\Delta t) - F_\tau(t)}{\Delta t}}_{\substack{\text{Forward finite-difference} \\ \text{approximation to } F_\tau'(t)}}, \quad (9.104)$$

$$= \frac{F_\tau'(t)}{S(t)},$$

$$= \frac{f_\tau(t)}{S(t)},$$

$$h(t) = -\frac{S'(t)}{S(t)}.$$

We can see, therefore, that we have identified the forward finite-difference approximation to the default-time density, $f_\tau(t) = F_\tau'(t)$. This is not only a convenient and concise form, but it is also a first-order homogeneous differential equation that is readily solved to determine $S(t)$ in terms of $h(t)$.

This is a simple calculus exercise. By integrating both sides of equation 9.104, we have

$$\int_0^t h(u)\,du = -\int_0^t \frac{S'(u)}{S(u)}\,du. \quad (9.105)$$

Substituting $v = S(u)$ leads to $dv = S'(u)du$ and

$$\int_0^t h(u)du = -\int_0^t \frac{1}{v}dv, \qquad (9.106)$$

$$\left[\ln(v)\right]_{S(0)}^{S(t)} = -\int_0^t h(u)du,$$

$$\ln S(t) - \underbrace{\ln S(0)}_{=0} = -\int_0^t h(u)du,$$

$$S(t) = e^{-\int_0^t h(u)du},$$

where $\ln S(0)$, because $S(0) \equiv 1$; that is, the survival probability is equal to unity at inception.

Having an explicit form for our survival function, in terms of the hazard rate, allows us to restate the default-time distribution and density functions. The cumulative distribution function is simply restated as,

$$F_\tau(t) = 1 - S(t), \qquad (9.107)$$

$$= 1 - e^{-\int_0^t h(u)du},$$

while the density function is,

$$f_\tau(t) = F'_\tau(t), \qquad (9.108)$$

$$= \frac{\partial\left(1 - e^{-\int_0^t h(u)du}\right)}{\partial t},$$

$$= \frac{\partial\left(-\int_0^t h(u)du\right)}{\partial t} e^{-\int_0^t h(u)du},$$

$$= -\left(-h(t) - \underbrace{h(0)}_{=0}\right) S(t),$$

$$= h(t)S(t).$$

This is an elegant result. The default density is merely the product of the hazard and survival functions; moreover, both depend on the specification of the hazard rate.

Different choices of hazard rate are, of course, possible and each have different implications for the distribution of the default event. Imagine, for example, the simplest possible choice: setting $h(t) = \lambda \in \mathbb{R}_+$. In this case, the default density

9.3 Risk-Neutral Default Probabilities

reduces to,

$$f_\tau(t) = h(t)e^{-\int_0^t h(u)du}, \quad (9.109)$$
$$= \lambda e^{-\int_0^t \lambda du},$$
$$= \lambda e^{-\lambda t},$$

which is, of course, the density of the well-known exponential distribution.[36]

This is the jumping-off point for the entire field of reduced-form default-risk modelling and the introduction of the Poisson process. This is basically similar in spirit to the independent-increment Brownian motions, save that the increments are Poisson distributed.[37] Thus one speaks of the exponentially distributed parameter, λ, as the *intensity* of default arrival. As λ is made stochastic and given its own dynamics, it is common to refer to these as stochastic-intensity models.[38]

In many practical applications—and this work will not be an exception—the hazard rate is assumed to be a piecewise linear function of time. The individual values are inferred from an obligor-associated credit-default swap. This requires additional attention, but will ultimately, as can be seen from the previous development, allows one to infer survival and default-probability values. First, however, we return briefly to our generic pricing formulae to see how the introduction of default risk changes the development.

9.3.3 Incorporating Default Risk

Let us quickly return to our general expression for a floating-rate fixed-income security. In the absence of credit risk, recall that the price is described as,

$$P = \underbrace{\sum_{i=1}^\beta L_i \Delta_i N\delta_i}_{\text{Interest}} + \underbrace{N\delta_\beta}_{\text{Principal}}. \quad (9.110)$$

We have not yet applied the expectation operator to these cash-flows, because our first task is to incorporate the default event. Conceptually, all payments are received

---

[36] See Casella and Berger (1990) or Johnson et al. (1994, Chapter 19) for much more information on the exponential distribution.

[37] See Andersen et al. (1993) for a deep dive into the underlying theory of Poisson and Cox processes.

[38] Duffie and Singleton (2003) use stochastic models—such as the Ornstein-Uhlenbeck and CIR processes—to describe the infinitesimal dynamics of $\lambda(t)$, which leads to a sophisticated theory combining interest-rate and default elements into a single framework.

in the event of survival. A final payment of a proportion of the remaining principal—determined by the recovery rate assumption—is made in the event of default. We know the individual payment data, so we can add an indicator variable to each payment to reflect its conditionality. The recovery payment, in the event of default, depends on the location of τ. We will just leave it as a random occurrence and determine how it might be handled once we take risk-neutral expectations.

Incorporating default risk thus leads to the following revision of equation 9.110,

$$P = \sum_{i=1}^{\beta} \underbrace{\mathbb{I}_{\tau > T_i} L_i \Delta_i N \delta_i}_{\text{Conditional interest payments}} + \underbrace{\mathbb{I}_{\tau > T_\beta} N \delta_\beta}_{\text{Conditional notional payment}} + \underbrace{R(\tau) N \delta(t, \tau)}_{\text{Conditional default recovery}}, \quad (9.111)$$

where $R(\tau)$ denotes the—very generally defined—recovery rate associated with default at time τ. The consequence is three distinct terms: the conditional interest payments, the conditional notional payments, and a conditional recovery payment in the event of a default outcome. The final term will require the most effort; the good news, however, is that we have already developed the basic structure required for its treatment.

Applying risk-neutral expectations to both sides of equation 9.111 leads to the following result,

$$\mathbb{E}_t^{\mathbb{Q}}(P) = \mathbb{E}_t^{\mathbb{Q}}\left(\sum_{i=1}^{\beta} \mathbb{I}_{\tau > T_i} L_i \Delta_i N \delta_i + \mathbb{I}_{\tau > T_\beta} N \delta_\beta + R(\tau) N \delta(t, \tau)\right), \quad (9.112)$$

$$= N \sum_{i=1}^{\beta} \mathbb{E}_t^{\mathbb{Q}}\left(\mathbb{I}_{\tau > T_i}\right) \mathbb{E}_t^{\mathbb{Q}}(L_i) \Delta_i \delta_i + \mathbb{E}_t^{\mathbb{Q}}\left(\mathbb{I}_{\tau > T_\beta}\right) N \delta_\beta + \mathbb{E}_t^{\mathbb{Q}}\left(R(\tau) N \delta(t, \tau)\right),$$

$$= N \sum_{i=1}^{\beta} \mathbb{Q}(\tau > T_i) F(t, T_{i-1}, T_i) \Delta_i \delta_i + \mathbb{Q}(\tau > T_\beta) N \delta_\beta + N \int_t^T R(u) \delta(t, u) f_\tau(u) du,$$

$$= N \sum_{i=1}^{\beta} S(T_i) F(t, T_{i-1}, T_i) \Delta_i \delta_i + S(T_\beta) N \delta_\beta + N \int_t^T R(u) \delta(t, u) \underbrace{h(u) S(u)}_{\text{Equation 9.108}} du.$$

With the exception of the final term, the individual cash-flows are merely deflated by the appropriate risk-neutral survival probabilities and discount factors. The final term requires some assumptions and additional structure to arrive at a workable and implementable expression.

Two assumptions will prove helpful in this regard. The first is that the recovery function, $R(t)$, is assumed to be constant; that is, $R \in \mathbb{R}$. This allows us to pull it out of the integrand. The second, and perhaps more helpful, choice is to give $h(t)$ a piecewise linear form. We discretize the full time interval from inception to final

9.3 Risk-Neutral Default Probabilities

maturity, $[t, T]$, in K sub-intervals as

$$\left\{ t = t_0, t_1, t_2, \cdots, t_{K-1}, t_K = T \right\}. \tag{9.113}$$

This grid need not coincide with the individual payment dates; instead, it will depend on the available tenors of the credit-default-swap contracts. It serves to calculate the requisite survival probabilities used to calibrate our piecewise linear hazard function.

We then proceed to define a constant hazard rate value, h_k, on each sub-interval. That is,

$$h(t) = \sum_{k=1}^{K} h_k \mathbb{I}_{t \in [t_{k-1}, t_k)}. \tag{9.114}$$

Since the sub-intervals are disjoint, only one of the elements in the sum in equation 9.114 is non-zero. Thus each $h(t) = h_k$ for some $t \in [t_{k-1}, t_k)$. This allows us to determine the form of the survival function,

$$S(t) = e^{-\int_0^t h(u)du}, \tag{9.115}$$

$$= \exp\left(-\int_0^t \sum_{k=1}^{K} h_k \mathbb{I}_{t \in [t_{k-1}, t_k)} du\right),$$

$$= \exp\left(-\sum_{k=1}^{K} h_k \mathbb{I}_{t \in [t_{k-1}, t_k)} t\right).$$

The corresponding density function is thus simply,

$$h(t)S(t) = \underbrace{\sum_{k=1}^{K} h_k \mathbb{I}_{t \in [t_{k-1}, t_k)}}_{\text{Equation 9.114}} \underbrace{\exp\left(-\sum_{k=1}^{K} h_k \mathbb{I}_{t \in [t_{k-1}, t_k)} t\right)}_{\text{Equation 9.115}}, \tag{9.116}$$

$$f_\tau(t) = \sum_{k=1}^{K} \underbrace{h_k e^{-h_k t}}_{\substack{\text{Exponential} \\ \text{density}}} \mathbb{I}_{t \in [t_{k-1}, t_k)},$$

where some of the clutter disappears since logically we require neither two indicator functions nor two sums. As hinted at in equation 9.109, the default time is exponentially distributed on each sub-interval $t \in [t_{k-1}, t_k)$ with constant parameter, h_k.

This allows us to re-write the integral at the end of equation 9.112 as the following sum,

$$N \int_t^T \underbrace{R(u)}_{R} \delta(t,u) f_\tau(u) du = NR \sum_{i=1}^{\beta} \int_{T_{i-1}}^{T_i} \delta(t,u) f_\tau(u) du, \qquad (9.117)$$

$$\approx NR \sum_{i=1}^{\beta} \delta(t, T_i) \int_{T_{i-1}}^{T_i} f_\tau(u) du,$$

$$\approx NR \sum_{i=1}^{\beta} \delta_i \int_{T_{i-1}}^{T_i} F'_\tau(u) du,$$

$$\approx NR \sum_{i=1}^{\beta} \delta_i \left(F_\tau(T_i) - F_\tau(T_{i-1}) \right),$$

$$\approx NR \sum_{i=1}^{\beta} \delta_i \left((1 - S(T_i)) - (1 - S(T_{i-1})) \right),$$

$$\approx NR \sum_{i=1}^{\beta} \delta_i \left(S(T_{i-1}) - S(T_i) \right).$$

The consequence is two alternative grids: one to define the payment schedule and the other for the piecewise linear hazard rate. It would, of course, be convenient if they were to overlap. This is unrealistic and undesirable given the large number of possible payment schedules and our complete lack of control over the individual tenors in the credit-default-swap market.

The construction in equation 9.117 does, it should noted, assume that default occurs at the upper time point of each sub-interval, T_i. This might be a bit unrealistic. There are alternatives. One could force the default event to occur anywhere in the payment interval. The mid-point is probably the least controversial and defensible approach. This implementation only requires replacing Δ_i with $\frac{\Delta_i}{2}$ and $e^{-h_i T_i}$ as $e^{-h_i \left(T_i - \frac{\Delta_i}{2} \right)}$. For short payment intervals, this effect may not be too important.

Combining all of these elements together, we arrive at a final expression for the default-risk equivalent-martingale-measure expectation of generic coupon-bearing

9.3 Risk-Neutral Default Probabilities

fixed-income security. It has the following form,

$$\mathbb{E}_t^{\mathbb{Q}}(P) \approx N \left(\underbrace{\sum_{i=1}^{\beta} \delta_i S(T_i) \Delta_i F(t, T_{i-1}, T_i)}_{\text{Interest}} + \underbrace{\delta_\beta S(T_\beta)}_{\text{Principal}} + R \underbrace{\sum_{i=1}^{\beta} \delta_i \left(S(T_{i-1}) - S(T_i) \right)}_{\text{Recovery}} \right).$$

(9.118)

Ultimately, therefore, our pricing relation reduces to a function of the cash-flows, a recovery rate assumption, the risk-free discount factors, and the survival probabilities. The next, and final, step involves developing an approach to identify the stepwise hazard function from market credit-default swap contracts.

9.3.4 Inferring Hazard Rates

We are finally ready to extract our object of interest: risk-free default probabilities. The ideas from the preceding sections can be used to price a broad range of plain-vanilla credit instruments. Credit-default-swap contracts are particularly useful, however, because they are frequently priced, typically cover a broad range of tenors, and are available for numerous individual obligors. By construct pricing formulae for an arbitrary credit-default-swap contract, we can use the observed prices to infer the levels of the hazard rates—and thus indirectly survival and default probabilities—associated with the underlying obligor. Conceptually, this is similar to extracting zero-coupon rates from liquid government-bond prices.

A credit-default swap is structurally very similar to an insurance agreement. There are, in fact, three main elements:

1. a stream of premium payments in the form of a periodic coupon;
2. a protection payment conditional on the default of the underlying obligor; and
3. a second conditional payment of the accrued protection premium from the last payment date to the actual default.

The final element is market convention. The idea is that the insurance purchaser is required to pay premium right up until the default event. The first and last payments are made by the party seeking protection, whereas the second payment is made by the protection seller. When the expected value of the protection seller and buyer cash-flows are equal, a credit-default-swap contract is said to trade at par. This is analogous to a fixed-floating interest-rate swap; by convention, market quotes are at par. Over time, of course, relative changes in credit quality will push a given contract—like an interest-rate swap—in or out of the money.

Finding a reasonable description of current market assessments of risk-neutral survival and default probabilities reduces to a two-step problem. The first step involves developing reasonable mathematical descriptions of the value associated with the *three* key elements of a credit-default swap in terms of hazard rates. The second step is to find a functional form for the hazard rate function that ensures all theoretical credit-default-swap price estimates are maximally consistent with market quotes. This is, from a conceptual perspective, analogous to the extraction of zero-coupon rates from a collection of bond prices or swap rates.[39]

Having built the foundation in previous sections, we can move relatively quickly to construct the necessary pricing relations. Let us begin with the coupon payment. With the establishment of a K-step piecewise linear hazard function, for each obligor, we observe K distinct credit-default-swap tenors. Operating one obligor at a time, we denote the *fixed* credit-default swap coupon of the kth tenor, at time t, as c_k. We further describe the sequence of cash-flow times for the kth contract as

$$\left\{ T_1, \ldots, T_{\beta_k} \equiv T_k \right\}. \quad (9.119)$$

The stream of premium coupon payments is given as,

$$\mathbb{E}_t^{\mathbb{Q}}\left(P_C(t)\right) = \mathbb{E}_t^{\mathbb{Q}}\left(\sum_{i=1}^{\beta_k} e^{-\int_t^{T_i} r(u)du} \Delta(T_{i-1}, T_i) \mathbb{I}_{\tau > T_i} c_k N \right), \quad (9.120)$$

$$V_C(t) = c_k N \sum_{i=1}^{\beta_k} \delta(t, T_i) \Delta_i S(T_i),$$

$$= c_k N \sum_{i=1}^{\beta_k} \delta_i \Delta_i \underbrace{\sum_{k=1}^{K} e^{-h_k(T_i - t)} \mathbb{I}_{T_i \in [t_{k-1}, t_k)}}_{S(T_i)},$$

where N denotes the notional amount.[40] We also use $V_C(t)$, in the interest of brevity, to represent the risk-neutral expectation of the premium-coupon cash-flows.

The conditional accrued premium-coupon payment, in the event of default, requires a bit more effort. Like the conditional premium payment, its occurrence is uncertain and, as a consequence, we need to make use of the integral form. The

---

[39] See Bolder and Stréliski (1999) and Bolder and Gusba (2002) for a detailed description and practical examination of algorithms used to extract zero-coupon rates from government bond prices.
[40] This can be set to unity without any loss of generality.

9.3 Risk-Neutral Default Probabilities

basic structure is,

$$\mathbb{E}_t^{\mathbb{Q}}\left(P_A(t)\right) = \mathbb{E}_t^{\mathbb{Q}}\left(e^{-\int_t^{\tau} r(u)du} \Delta(T_{i-1}, \tau) c_k N\right), \tag{9.121}$$

$$V_A(t) = c_k N \mathbb{E}_t^{\mathbb{Q}}\left(\delta(t, \tau)\Delta(T_{i-1}, \tau)\right),$$

$$= c_k N \int_t^{T_k} \delta(t, u)\Delta(T_{i-1}, u) f_\tau(u) du,$$

$$= c_k N \sum_{i=1}^{\beta_k} \int_{T_{i-1}}^{T_i} \delta(T_{i-1}, u)\Delta(T_{i-1}, u) f_\tau(u) du.$$

The final step essentially breaks up the integral, as done previously, over our collection of disjoint sub-intervals. We can further simplify equation 9.121 if we make an assumption about when, in each sub-interval, default is assumed to occur. Forcing default to occur in the mid-point of each sub-interval is, while arbitrary, the most defensible choice. This leads to,

$$V_A(t) \approx c_k N \sum_{i=1}^{\beta_k} \int_{T_{i-1}}^{T_i} \delta\left(T_{i-1}, T_{i-1} + \frac{\Delta_i}{2}\right) \frac{\Delta_i}{2} f_\tau(u) du, \tag{9.122}$$

$$\approx c_k N \sum_{i=1}^{\beta_k} \delta\left(T_{i-1}, T_{i-1} + \frac{\Delta_i}{2}\right) \frac{\Delta_i}{2} \int_{T_{i-1}}^{T_i} F_\tau'(u) du,$$

$$\approx c_k N \sum_{i=1}^{\beta_k} \delta\left(T_{i-1}, T_{i-1} + \frac{\Delta_i}{2}\right) \frac{\Delta_i}{2} \left(F_\tau(T_{i-1}) - F_\tau(T_i)\right),$$

$$\approx c_k N \sum_{i=1}^{\beta_k} \delta\left(T_{i-1}, T_{i-1} + \frac{\Delta_i}{2}\right) \frac{\Delta_i}{2} \left((1 - S(T_{i-1})) - (1 - S(T_i))\right),$$

$$\approx c_k N \sum_{i=1}^{\beta_k} \delta\left(T_{i-1}, T_{i-1} + \frac{\Delta_i}{2}\right) \frac{\Delta_i}{2} \left(S(T_i) - S(T_{i-1})\right),$$

$$\approx c_k N \sum_{i=1}^{\beta_k} \delta\left(T_{i-1}, T_{i-1} + \frac{\Delta_i}{2}\right) \frac{\Delta_i}{2}$$

$$\times \left(\underbrace{\sum_{k=1}^{K} e^{-h_k(T_i - \frac{\Delta_i}{2} - t)} \mathbb{I}_{T_i \in [t_{k-1}, t_k)} - e^{-h_k(T_{i-1} - \frac{\Delta_i}{2} - t)} \mathbb{I}_{T_{i-1} \in [t_{k-1}, t_k)}}_{S(T_i) - S(T_{i-1})} \right).$$

The integral has thus been reduced to a relatively straightforward sum.

The final element is the conditional protection payment. It has the following (by now familiar) form,

$$\mathbb{E}_t^{\mathbb{Q}}\left(P_X(t)\right) = \mathbb{E}_t^{\mathbb{Q}}\left(e^{-\int_t^\tau r(u)du} N(1-R)\right), \tag{9.123}$$

$$V_X(t) = N(1-R) \int_t^{T_k} \delta(t, u) f_\tau(u) du,$$

$$= N(1-R) \sum_{i=1}^{\beta_k} \int_{T_{i-1}}^{T_i} \delta(T_{i-1}, u) f_\tau(u) du,$$

$$= N(1-R) \sum_{i=1}^{\beta_k} \delta\left(T_{i-1}, T_{i-1} + \frac{\Delta_i}{2}\right) \int_{T_{i-1}}^{T_i} F_\tau'(u) du,$$

$$= N(1-R) \sum_{i=1}^{\beta_k} \delta\left(T_{i-1}, T_{i-1} + \frac{\Delta_i}{2}\right)$$

$$\times \left(\underbrace{\sum_{k=1}^{K} e^{-h_k(T_i - \frac{\Delta_i}{2} - t)} \mathbb{I}_{T_i \in [t_{k-1}, t_k)} - e^{-h_k(T_{i-1} - \frac{\Delta_i}{2} - t)} \mathbb{I}_{T_{i-1} \in [t_{k-1}, t_k)}}_{S(T_i) - S(T_{i-1})} \right).$$

This uses the same trick and result from equations 9.117 and 9.122. Once again, absent a better assumption, we assume that default occurs at the mid-point of each interval.

We have three detailed expressions—$V_C(t)$, $V_A(t)$, and $V_X(t)$—to describe the risk-neutral expectation of the premium-coupon, conditional premium-coupon accrual, and conditional protection payment, respectively. We observe in the market a set of credit-default-swap par coupon-rate quotes, $\{c_1, \ldots, c_K\}$ and our task is

9.3 Risk-Neutral Default Probabilities

to find a set of fixed hazard rates, $\{h_1, \ldots, h_K\}$ that, using our previously derived expressions, replicates what is observed in the market.

This leads to a system of K non-linear equations in K unknowns. Indeed, we seek the collect of hazard rates that solve this set of equations,

$$V_X^{(1)}(t) - \left(V_C^{(1)}(t) + V_A^{(1)}(t)\right) = 0 \tag{9.124}$$

$$V_X^{(2)}(t) - \left(V_C^{(2)}(t) + V_A^{(2)}(t)\right) = 0$$

$$\vdots$$

$$V_X^{(K)}(t) - \left(V_C^{(K)}(t) + V_A^{(K)}(t)\right) = 0.$$

In plain English, we seek the set of hazard rates that, given current market coupon rates, reproduces the par conditions present in the market.

Like linear systems, there do exist fast and robust methods to solve such problems. Their reliability—unlike the linear setting—depends upon the stability of the underlying equations. One can try to solve it in a single step with a numerical optimization algorithm, but there is a chance it could run into difficulty or fail to converge. An alternative approach is to solve the system recursively. That is, the first equation, by construction, has a single market credit-default-swap coupon, c_1, and a single hazard rate, h_1. h_1 cannot be determined algebraically, but it reduces to a one-dimensional root-solving problem. Given a fixed and determined, h_1, one then proceeds to the second equation. In this case, we have a given c_2, h_1 has already been determined, and we have another one-dimensional root-finding exercise to determine, h_2. Armed with h_1 and h_2, one then proceeds to the third equation to identify h_3. The algorithm proceeds in this recursive manner until all K hazard rates are identified.

9.3.5 A Concrete Example

To demonstrate how piecewise linear hazard rates might be extracted from a collection of credit-default-swap fixed-rate coupon quotes, we will consider a practical example. We have already derived the requisite expressions. Now we require some actual data inputs. We will examine, therefore, *nine* distinct credit-default-swap coupon tenors ranging from one to 20 years. Each coupon is paid semi-annually and the notional value of each contract is assumed to be $100. The consequence is that we seek to identify $K = 9$ separate h_k parameters from nine credit-default swaps.

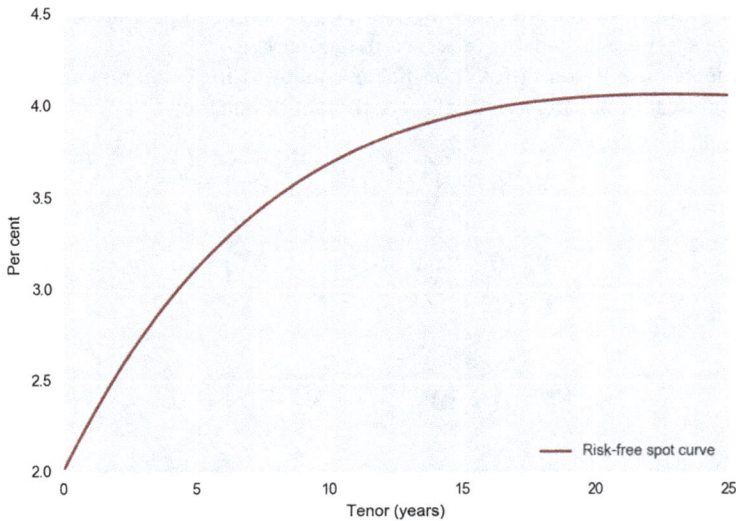

Fig. 9.15 *An assumed risk-free spot curve*: The underlying figure illustrates an assumed zero-coupon curve for our bootstrapping example. It is constructed using the Nelson and Siegel (1987) model with arbitrarily selected level, slope, and curvature parameters of 0.03, -0.01, and 0.05, respectively.

A particular important input is the risk-free spot curve used to discount the individual cash-flows. To simplify our life, we introduce the zero-coupon curve from a Nelson-Siegel model with arbitrarily selected parameters; a graph of this function is summarized in Fig. 9.15.[41] In a practical setting, however, these spot rates would more likely be bootstrapped from swap or government securities.[42]

We have K contracts which, as we saw in equation 9.124, give rise to K non-linear equations in K unknowns. These equations must be resolved numerically, which means instead of forcing each to hold precisely, we seek to force each equation as close as possible to a value of zero. For this purpose, it is useful to restate equation 9.124 as follows,

$$V_X^{(1)}(t) - \left(V_C^{(1)}(t) + V_A^{(1)}(t)\right) = \epsilon^{(1)}(t) \qquad (9.125)$$

$$V_X^{(2)}(t) - \left(V_C^{(2)}(t) + V_A^{(2)}(t)\right) = \epsilon^{(2)}(t)$$

---

[41] See Nelson and Siegel (1987) for more background on this commonly used model. Bolder (2015, Chapter 5) also provides some colour on the curve-fitting exercise.

[42] The swap rates would be used to form the projection curve, whereas government securities (or overnight-indexed swaps) would be employed for the risk-free discount curve.

9.3 Risk-Neutral Default Probabilities

$$\vdots$$

$$V_X^{(K)}(t) - \left(V_C^{(K)}(t) + V_A^{(K)}(t) \right) = \epsilon^{(K)}(t)$$

Our objective, therefore, is to simultaneously minimize the absolute value (i.e., each $\left| \epsilon^{(k)}(t) \right|$ term).

To accomplish this task, we need to implement the pricing equations. The first job is to write a brief function to return the appropriate h_k value for an arbitrary cash-flow time. Algorithm 9.5, part of our basic Python programming language implementation, takes a vector of arbitrary piecewise linear hazard rates, h, along with their associated tenors, tenor, and returns the hazard-rate value for the requested tenor, myT.

Algorithm 9.5 is extremely simple, but it allows us to build the survival-probability function required for the valuation of the three components in the credit-default swap price. Algorithm 9.6 illustrates the computation of the survival probability, for a given time period, using the getH function from Algorithm 9.6.

The survival probabilities play an important role in the valuation of credit-default swaps. Algorithm 9.7 provides an overview of the determination of the credit-default-swap price. As inputs, it takes a given set of hazard rates, h, credit-default-swap fixed coupons, cds, the associated tenors, tenor, the length of each payment period, Delta, the discount factors, d, and the recovery rate, R. It then proceeds to loop over the number of payments for each of the K contracts and determines a

Algorithm 9.5 Hazard-function implementation

```
def getH(myT, tenor, h):
    myRange = np.insert(tenor, 0, 0)
    for n in range(0, len(tenor)):
        if myT==0:
            loc = 0
        elif (myT>myRange[n]) & (myT<=myRange[n+1]):
            loc = n
    try: return h[loc]
    except:
        print("No value found!")
```

Algorithm 9.6 The survival probability

```
def survivalProb(myT, tenor, h):
    myH = getH(myT, tenor, h)
    return np.exp(-myH*myT)
```

Algorithm 9.7 Theoretical CDS prices

```
def cdsPrice(h,cds,tenor,Delta,d,R):
    K = len(h)
    cStream = np.zeros(K)
    aStream = np.zeros(K)
    xStream = np.zeros(K)
    beta = np.zeros(K)
    for n in range(0,K):
        beta = (tenor[n]/Delta).astype(int)
        cStream[n] = getCouponStream(beta,cds[n],tenor,h,Delta,d)
        aStream[n] = getPremiumAccrual(beta,cds[n],tenor,h,Delta,d)
        xStream[n] = getProtectionPayment(beta,cds[n],tenor,h,Delta,d,R)
    return cStream+aStream-xStream
```

price for each of the three components—the coupon stream, the conditional accrual of coupon premium, and the conditional premium payment. These calculations are performed by three additional Python sub-routines: getCouponStream, getPremiumAccrual, and getProtectionPayment. The result is a vector of theoretical prices—given h—for the sell and buy sides of the contract. It returns the vector of the sum of these two aspects; or, in other words, the ϵ's from equation 9.125.

Equation 9.126 below demonstrates how Algorithm 9.7 is used to determine the set of optimal hazard rates most consistent with observed credit-default-swap coupons at a given point in time.

```
r = scipy.optimize.root(cdsPrice,h0,args=(cds,tenor,Delta,d,R),method='hybr')
```
(9.126)

root—like fsolve—is a Python function, which we have already employed on numerous occasions in previous chapters.[43] It is particularly useful for finding the roots of a vector-valued function. Moreover, it finds the solution quite efficiently in a very short period of time. Indeed, experimentation with this approach reveals that it is equally accurate and significantly more efficient than a recursive approach.

Before examining the results, let us look at the structure of one of the pricing functions embedded in Algorithm 9.5. Since they are all quite similar, it suffices to examine just one. Algorithm 9.7 summarizes the Python code used to price

---

[43]This sub-routine, as the reader will recall, is essentially a wrapper around MINPACK's hybrd algorithm.

9.3 Risk-Neutral Default Probabilities

the credit-default-swap protection payment. It is actually quite compact and easy to follow. It simply loops over each of the payment dates, given by the integer `beta`, and reproduces each term in equation 9.123. `sIncrement` is the difference in survival probabilities—both evaluated with help of Algorithm 9.6—and `myProtection` is the return value containing an estimate of the integral used to price the protection payment. As a final but not incidental point, the recovery rate was set to 0.70 for exercise.

Algorithm 9.8 Pricing the protection payment

```
def getProtectionPayment(beta,fRate,tenor,h,Delta,d,R):
    myProtection = 0
    for i in range(0,beta):
        sIncrement = survivalProb(i*Delta,tenor,h)-\
                    survivalProb((i+1)*Delta,tenor,h)
        myProtection += (1-R)*np.interp(i*Delta+(Delta/2),tenor,d)*sIncrement
    return myProtection
```

Table 9.6 illustrates the results of applying the previous algorithms and the discount function in Fig. 9.15 to our set of nine, slightly fictional, but not unreasonable credit-default-swap coupons. For each tenor, it displays the credit-default-swap coupon, hazard rates, the price of each component (for a $100 notional), and the error term. It is clear that the hazard rates—determined using equation 9.126—do an admirable job of solving our system of simultaneous non-linear equations. It is also evident that most of the value, as one might expect, comes from the coupon stream and the conditional protection payment. The conditional accrual premium is quite small; indeed, it is only visible at the longest tenor.

Figure 9.16 illustrates the key inputs and outputs of the hazard-function bootstrap. The left-hand graphic displays the credit-default-swap coupon and optimal hazard rates in the same space. Both are piecewise linear step functions with the hazard rates dominating the associated coupons. The difference is an increasing function of tenor. An important advantage of this approach is the simplicity with which it generates risk-neutral probability estimates for a broad range of tenors. They rise, as seen in the final column of Table 9.6, from roughly 20 basis points at a one-year horizon to about 16% at 20 years.

The right-hand graphic illustrates, over the unit interval, the discount and survival-probability functions. Recall that these two functions work together to determine the price of a credit-risky security. The discount function captures the time value of money, whereas the survival-probability incorporates default risk. In this case, given the relatively low level of credit risk associated with the underlying obligor, the discount factor has a more rapid rate of decrease. The product of the two functions is also displayed to indicate the impact of applying both functions to

Table 9.6 *Bootstrapping results*: The underlying table illustrates the results of applying the previous algorithms and the discount function in Fig. 9.15 to our set of nine, slightly fictional, but representative credit-default-swap coupons.

Tenor (yrs.)	Values (bps.)		Contract (USD)				Probabilities	
	c_k	h_k	$V_C^{(k)}(t)$	$V_A^{(k)}(t)$	$V_X^{(k)}(t)$	$\epsilon^{(k)}(t)$	$\mathbb{Q}(\tau > T_k)$	$\mathbb{Q}(\tau < T_k)$
1	5	16	0.05	0.00	0.05	−0.000000	99.8%	0.2%
2	7	23	0.14	0.00	0.14	−0.000000	99.5%	0.5%
3	8	26	0.23	0.00	0.23	−0.000000	99.2%	0.8%
4	11	36	0.41	0.00	0.41	−0.000000	98.5%	1.5%
5	11	40	0.55	0.00	0.55	0.000000	98.0%	2.0%
7	15	50	0.92	0.00	0.93	−0.000000	96.5%	3.5%
10	20	69	1.65	0.00	1.65	0.000000	93.3%	6.7%
15	23	80	2.53	0.00	2.54	0.000000	88.7%	11.3%
20	25	88	3.28	0.01	3.29	−0.000000	83.8%	16.2%

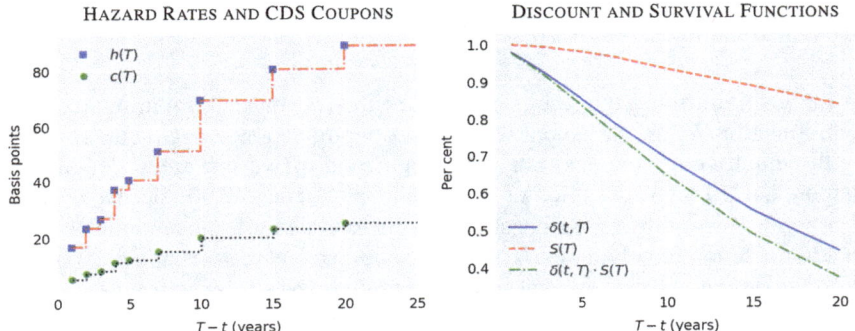

Fig. 9.16 *Key ingredients and outputs*: This figure outlines visually the piecewise linear credit-default-swap coupons, hazard rate functions as well as the relationship between the discount and survival functions for our simple illustrative example.

a given cash-flow. If one was to compute the zero-coupon rates associated with this *full-discount* factor, it would produce the implied credit curve for the underlying third-party obligor in the credit-default-swap contract.

When market participants buy and sell instruments with inherent credit risk, they are effectively assessing the obligor's creditworthiness. It is precisely this implied default-probability information that is being extracted from the credit-default swap. These figures do indeed provide useful insight into credit risk, but the results are rather better adapted to pricing securities than assessing risk. Given the richness and theoretical appeal of these implied default probabilities, one might nevertheless be inclined to conclude that they are close enough for risk-management analysis. There are, in fact, rather dramatic differences between estimation and calibration default-probability results. In the next, and final, section of this chapter, we will examine and motivate the sources of these deviations.

9.4 Back to Our ℙ's and ℚ's

We have now examined two alternative approaches for determining a key input into our default-risk models: the unconditional default probabilities. Estimation uses credit transition data from credit-rating agencies or internal assessments. The use of historical, time-series, transition data implies, quite importantly, that the corresponding default-probability estimates are formed under the physical-probability measure. This is the appropriate perspective for risk-management analysis. As a consequence, this is a reasonable and defensible approach to source difficult-to-obtain unconditional default probability data.

The principal alternative, which we refer to as calibration, involves inference of the *risk-neutral* default probability from an entity's credit-default-swap spread. In this final element of analysis in this chapter, we'd like to investigate the nature and magnitude of the difference between the physical and risk-neutral default probabilities. This will require a comparison, for a given set of obligors, of these competing estimates across a reasonable period of time.

We start this process by examining five-year-tenor, credit-default-swap spread data for four sizable international banks. The evolution of these spreads is summarized in Fig. 9.17. Unlike the slow-moving ratings-based historical default data, the credit-default-swap spreads move rapidly from one period to the next. They are, after all, market prices and exhibit much of the same behaviour as one would expect from market-based variables. Over an 18-month period, for example, spreads for our small collection of similarly rated counterparts have ranged from ten to 50 basis points. Moreover, despite their comparable credit quality, one can visually differentiate between the various entities.

We've already examined a formal and correct approach to the extraction of hazard rates and associated information. The premium is also often summarized in the form of a so-called credit-default swap spread.[44] Following a rough approximation offered by Hull et al. (2005), we can approximate the risk-neutral default intensity, or what we've been referring to as the hazard rate h_n, from its spread with the following expression

$$h_n \approx \frac{S_n}{1 - \mathcal{R}_n}, \qquad (9.127)$$

---

[44]There are a number of complex details, but roughly speaking, it is useful to think of the credit-default-swap spread as being loosely equivalent to the difference between the obligor's bond yield and the comparative risk-free rate in the economy. At par, we would expect agreement between credit-default-swaps spreads and coupons.

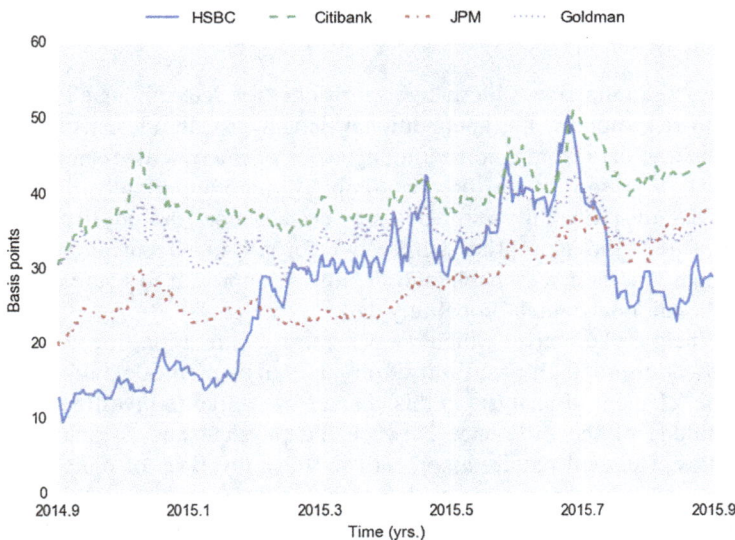

Fig. 9.17 *Key CDS spreads*: This figure summarizes the evolution of credit-default swap spreads, for four important international banks, over an approximately 18 months ending October 2016.

where $\mathcal{R}_n$ denotes the assumed recovery rate associated with the nth credit counterparty.[45] The associated risk-neutral default probability—from equation 9.107 over the time interval, $[t, T]$—is thus approximated as,

$$q_n = \mathbb{Q}(\tau \leq T), \tag{9.128}$$
$$\approx 1 - S(T),$$
$$\approx 1 - e^{-h_n(T-t)},$$

where, in this example, $T - t = 5$. Taking this a step further, actual and risk-neutral default probabilities are linked—see, for example, Jarrow (2012) for more detail—by the following identity:

$$q_n = \lambda_n p_n, \tag{9.129}$$

where $\lambda_n > 0$ is called the risk premium. Determining the appropriate value of λ_n, at any given point of time, is a very difficult empirical task. Equation 9.129, however, suggests that if we compute the physical and risk-neutral probabilities, that their ratio will provide some insight into the level of the risk premium (i.e., $\lambda_n = \frac{q_n}{p_n}$). Since we can infer risk-neutral default probabilities at a very high frequency and

---

[45] One can use this trick, with $R = 0.7$ in Table 9.6, to verify its reasonableness.

9.4 Back to Our ℙ's and ℚ's

Table 9.7 *Comparing* ℙ *and* ℚ *estimates*: The underlying table highlights our estimated and calibrated default-probability estimates under the two measures. It also uses this information to infer the risk-premium coefficient value.

Obligor	Rating	Estimate (bps.)		
		p_n	q_n	λ
HSBC	AA−	3.0	194.6	64.9
Citibank	A+	9.0	280.1	31.1
JPM	AA−	3.0	199.1	66.4
Goldman	A	9.0	246.1	27.3

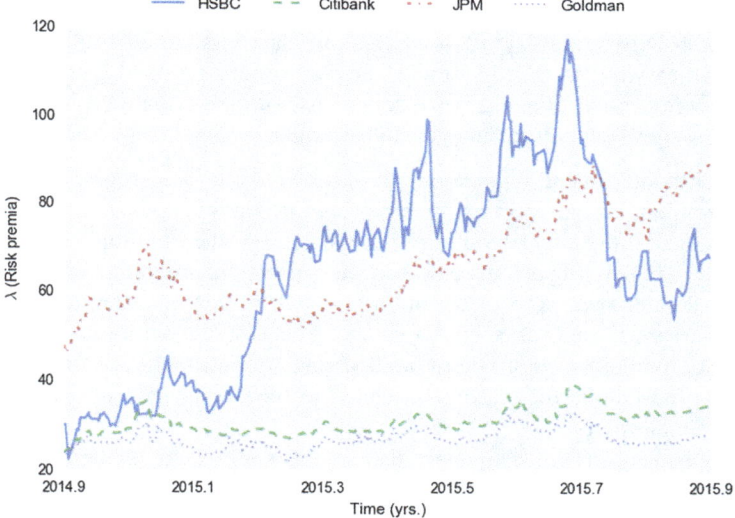

Fig. 9.18 *Approximate risk-premium evolution*: This figure highlights, for illustrative purposes, the approximate evolution of the risk premium over our sample period. The risk-neutral default probabilities change with each observation, while the values under the physical measure are held fixed.

their actual equivalents relatively rarely, direct comparison is a challenge. We can, therefore, approximate the average risk premium over a time interval, but it is very difficult to know it, with much precision, at a given point in time.

Table 9.7 takes the information in Fig. 9.17 and proceeds to compute—using equations 9.127 and 9.128—the average neutral default probabilities over our data period. This also allows us—since we've also looked up the long-term default probabilities offered by a well-known credit-rating agency—to estimate the risk premium. On average, the risk-neutral estimates are a factor of 30 to almost 70 times larger than the physical-measure equivalents. Figure 9.18 highlights, for illustrative purposes, the approximate evolution of risk premium over our sample period. The risk-neutral default probabilities change with each observation, while the values under the physical measure are held fixed.

Although this might appear surprising, it is consistent with values presented in the associated academic literature. Murthy (2011) is a useful survey of recent work in this area. This is admittedly not a terribly rigorous nor formal investigation of the differences between these two quantities. It does nonetheless underscore the basic point; the deviations between these two estimations are dramatic and structural. A quantitative risk analyst wishing to infer default probabilities from credit-default-swap data will have to wrestle with (time-varying) risk premia before being able to employ her estimates in her risk models.

Despite this shortcoming, the use of credit-default swap spreads—or, which is also possible, bond credit spreads—to infer default probabilities offers a number of conceptual advantages. It is forward looking and also incorporates the most recent information. Differentiation, as we've seen, is also possible among various credit counterparties who, from a historical perspective, are grouped together by their common credit rating. Practically, of course, use of market-based information also bears a few significant disadvantages. In particular,

- For risk-management applications, it is necessary to use an assessment of the risk premium to transform risk-neutral estimates into the physical world. This is difficult and ultimately becomes rather subjective—the situation is not improved by the overwhelming evidence of the time-varying nature of market-risk premia.
- Credit-default-swap spreads are not a pure description of market risk. They are financial prices and, as a consequence, reflect other factors such as relative liquidity, market microstructure, and market frictions. The result is that these spreads may move for reasons unrelated to the default risk of the underlying credit counterparty.
- The noisiness of the estimates also presents practical challenges for risk-management reporting and governance. It may be less problematic, if we could claim that the vast majority of credit-default-swap spread movements stem from changes in the market perception of default risk. This is not necessarily true; or, at least, it is difficult to prove. Changes in market conditions, market frictions, or movement in risk attitudes can, and do, lead to movements in market-based default probability estimates. This is difficult, as a risk manager, to explain and manage.

Use of risk-neutral default probabilities is not particularly easy and is fraught with practical challenges. Jarrow (2012), in fact, offers a number of compelling conceptual arguments against their use and strongly recommends the use of historically determined values. Although this argumentation is compelling, it is painful, as a practitioner, to ignore an important and potentially useful source of information. There is quite likely a role to be played from using both estimation and calibration techniques. At the very least, one might wish to estimate physical-measure probabilities, but closely monitor credit-default-swap spreads to catch any early market signals about deterioration of the creditworthiness of our swap counterparties. This inevitably requires some subjective analysis regarding the behaviour of the risk premia, but avoids ignoring possibly beneficial guidance about a critical, difficult-to-estimate quantity.

9.5 Final Thoughts

This lengthy, and highly detailed, chapter has focused on a critical input into one's credit-risk computation: the unconditional probability of default for each of the individual obligors in one's portfolio. Considered broadly, there are *two* main ways to approximate these important values. The first approach, referred to as estimation, seeks to infer default-probability values from time-series credit-state transition data. Typically assuming that distinct credit counterparts are allocated to a modest number of credit states—treated generally as a Markov-chain process—the estimates are made under the physical, or actual, probability measure. Use of statistical inference techniques—in this case, predominately the method of maximum likelihood—permits the analyst to produce confidence intervals to help assess underlying uncertainty arising of one's estimates. Given the relative paucity of default outcomes, this can be a powerful tool for the mitigation of model risk.

We have termed the second approach as calibration; it proceeds in a rather different fashion. It relies on the fact that pricing of market instruments exposed to default risk—such as bonds or credit-default swaps—necessarily involves a market assessment of the underlying obligor's ability, and willingness, to meet its obligations. Understanding the structure of the pricing formulae, one can proceed to extract market-implied estimates of obligor-specific default probabilities. Credit-default swaps are particularly useful in this regard, since they offer contracts covering a wide range of obligors and tenors. There is, of course, a catch. The resultant unconditional default probabilities are under the so-called equivalent-market measure. This is congruent for pricing applications, but is not generally appropriate for risk-management applications.

Theoretically, the difference between actual and risk-neutral probabilities relates to aggregate risk preferences held by market participants. Practically, the distance between the two is both large in magnitude and subject to variation over time. There is certainly information about counterparty credit quality in both quantities, although the actual estimates are, without manipulation, of most direct use for the risk analyst. Risk-neutral probabilities can also, in principle, be employed in the risk-management setting. We would, in fact, be remiss to ignore an additional source of information. Caution and conscientious analysis is nonetheless required to avoid misapplication of the results. To summarize, computation and application of default probabilities are of paramount importance, but it is *not* a trivial undertaking and requires a careful hand.

References

Andersen, P. K., Borgan, Ø., Gill, R. D., & Keiding, N. (1993). *Statistical models based on counting processes*. Heidelberg, Germany: Springer.

Bielecki, T. R., & Rutkowski, M. (2002). *Credit risk: Modeling, valuaton and hedging* (1st edn.). Berlin: Springer-Verlag.

Bluhm, C., Overbeck, L., & Wagner, C. (2003). *An introduction to credit risk modelling* (1st edn.). Boca Raton: Chapman & Hall, CRC Press.

Bolder, D. J. (2002). Towards a more complete debt strategy simulation framework. Bank of Canada: Working Paper 2002–13.

Bolder, D. J. (2015). *Fixed income portfolio analytics: A practical guide to implementing, monitoring and understanding fixed-income portfolios.* Heidelberg, Germany: Springer.

Bolder, D. J., & Gusba, S. (2002). Exponentials, polynomials, and fourier series: More yield curve modelling at the Bank of Canada. Bank of Canada: Working Paper 2002–29.

Bolder, D. J., & Stréliski, D. (1999). *Yield curve modelling at the Bank of Canada.* Technical Report No. 84. Ottawa: Bank of Canada.

Brémaud, P. (1999). *Markov chains: Gibbs fields, Monte Carlo simulation and queues.* 175 Fifth Avenue, New York, NY: Springer-Verlag. *Texts in applied mathematics.*

Brigo, D., & Mercurio, F. (2001). *Interest-rate models: Theory and practice* (1st edn.). Berlin, Germany: Springer-Verlag.

Casella, G., & Berger, R. L. (1990). *Statistical inference.* Belmont, CA: Duxbury Press.

Chernick, M. R. (1999). *Bootstrap methods: A practitioner's guide.* New York: John Wiley & Sons.

DasGupta, A. (2008). *Asymptotic theory of statistics and probability.* New York: Springer-Verlag.

Duffie, D. (1996). *Dynamic asset pricing theory* (2nd edn.). Princeton, NJ: Princeton University Press.

Duffie, D., & Singleton, K. (2003). *Credit risk* (1st edn.). Princeton: Princeton University Press.

Efron, B. (1979). Bootstrap methods: Another look at the jackknife. *The Annals of Statistics, 7*(1), 1–26.

Efron, B., & Tibshirani, R. (1986). Bootstrap methods for standard errors, confidence intervals, and other measures of statistical accuracy. *Statistical Science, 1*(1), 54–77.

Filardo, A. J. (1993). Business cycle phases and their transitional dynamics. Federal Reserve Bank of Kansas City: Working Paper No. 93-14.

Fishman, G. S. (1983). Accelerated accuracy in the simulation of Markov chains. *Operations Research, 31*(3), 466–487.

Fishman, G. S. (1995). *Monte Carlo: Concepts, algorithms, and applications.* 175 Fifth Avenue, New York, NY: Springer-Verlag. *Springer series in operations research.*

Fuh, C. D. (1993). Statistical inquiry for Markov chains by bootstrap method. *Statistica Sinica, 3*(1), 53–66.

Golub, G. H., & Loan, C. F. V. (2012). *Matrix computations.* Baltimore, Maryland: The John Hopkins University Press.

Gupton, G. M., Coughlan, G., & Wilson, M. (1997). *CreditMetrics — technical document.* New York: Morgan Guaranty Trust Company.

Hamilton, J. D. (1989). A new approach to the economic analysis of nonstationary time series and the business cycle. *Econometrica, 57,* 357–384.

Harrison, J., & Kreps, D. (1979). Martingales and arbitrage in multiperiod security markets. *Journal of Economic Theory, 20,* 381–408.

Harrison, J., & Pliska, S. (1981). Martingales and stochastic integrals in the theory of continuous trading. *Stochastic Processes and Their Applications, 11,* 215–260.

Held, L., & Bové, D. S. (2014). *Applied statistical inference.* Berlin, Germany: Springer-Verlag.

Hull, J., Predescu, M., & White, A. (2005). Bond prices, default probabilities, and risk premiums. *Technical report.* University of Toronto.

Jafry, Y., & Schuermann, T. (2004). Measurement, estimation, and comparison of credit migration matrices. *Journal of Banking & Finance, 28*(11), 2603–2639.

Jarrow, R. A. (2012). Problems with using CDS to infer default probabilities. *Technical report.* Johnson Graduate School of Management, Cornell University.

Jarrow, R. A., Lando, D., & Turnbull, S. M. (1997). A markov model for the term structure of credit risk spreads. *Review of Financial Studies, 10*(2), 481–523.

Jeanblanc, M. (2002). *Credit risk.* Université d'Evry.

Johnson, N. L., Kotz, S., & Balakrishnan, N. (1994). *Continuous univariate distributions: volume I* (2nd edn.). New York, NY: John Wiley & Sons.

References

Judge, G. G., Griffiths, W., Hill, R. C., Lütkepohl, H., & Lee, T.-C. (1985). *The theory and practice of econometrics* (2nd edn.). New York, NY: John Wiley & Sons.

Kulperger, R. J., & Rao, B. L. S. P. (1989). Bootstrapping a finite state Markov chain. *The Indian Journal of Statistics, 51*(2), 178–191.

Lando, D., & Skødeberg, T. M. (2002). Analyzing rating transitions and rating drift with continuous observations. *Journal of Banking & Finance, 26*(2–3), 423–444.

Merton, R. (1974). On the pricing of corporate debt: The risk structure of interest rates. *Journal of Finance, 29*, 449–470.

Meyn, S. P., & Tweedie, R. L. (1993). *Markov chains and stochastic stability*. 175 Fifth Avenue, New York, NY: Springer-Verlag.

Murthy, S. (2011). Market-implied risk-neutral probabilities, actual probabilities, credit risk and news. *Finance and Control, Indian Institute of Management Bangalore Management Review, 23*, 140–150.

Nelson, C. R., & Siegel, A. F. (1987). Parsimonious modeling of yield curves. *Journal of Business, 60*, 473–489.

Pawitan, Y. (2001). *In all likelihood: Statistical modelling and inference using likelihood*. Oxford: Clarendon Press.

Savage, L. J. (1976). On rereading R.A. Fisher. *The Annals of Statistics, 4*(3), 441–500.

Schuermann, T. (2008). Credit migration matrices. In E. Melnick, & B. Everitt (Eds.), *Encyclopedia of quantitative risk assessment*. John Wiley & Sons.

Stuart, A., & Ord, J. K. (1987). *Kendall's advanced theory of statistics*. New York: Oxford University Press.

Taylor, H. M., & Karlin, S. (1994). *An introduction to stochastic modeling*. San Diego, CA: Academic Press.

von Hilgers, P., & Langville, A. N. (2006). The five greatest applications of Markov chains. *Technical report*. Berlin, Germany: Max Planck Institute for History of Science.

White, L. J. (2010). Markets: The credit rating agencies. *The Journal of Economic Perspectives, 24*(2), 211–226.

Chapter 10
Default and Asset Correlation

> *What I cannot create, I do not understand.*
>
> (Richard Feynman)

As indicated on many occasions in the previous chapters, the heart of credit-risk modelling relates to one's assessment of the dependence between the default outcomes of the various obligors in one's portfolio. If one assumes independence of default, then default risk is entirely idiosyncratic. This assumption certainly eases the analysis; the correct modelling framework is thus well described by the binomial or Poisson independent-default approaches addressed in Chap. 2. Relaxing this unrealistic choice, however, is the source of most of the complexity in this field.

A variety of structures—in the threshold and mixture environments—is employed to induce dependence between the default outcomes of the various obligors in one's portfolio. In the threshold setting, it reduces to the notion of asset correlation, whereas in the mixture setting it is more directly related to default correlation. In other words, different models have differing implications for the structure of default dependence. All models nevertheless need to be calibrated to observed levels of default correlation.

We have already, to a certain extent, done some work in this area. In Chaps. 3 and 4, we arbitrarily specified a particular level of default correlation—and occasionally also average default probability or tail dependence—and calibrated the model parameters to be consistent with this choice. In the original Merton (1974) setting, addressed in Chap. 5, default correlation is typically inferred from observed levels of equity correlation. This is based on the assumption that asset correlation is roughly similar to its equity equivalent. While this is a conceptually reasonable approach, there are many settings where asset and equity notions—for example, with sovereign obligors—are not immediately helpful. Moreover, in general, one would perhaps prefer to surmise default correlations, and any associated asset-correlation parameters, directly from default or credit-transition data.

Inferring parameters from default and transition data makes this a statistical-estimation problem. This has conceptual appeal, because it permits empirical assessment of default dependence and is consistent with the typical approach used to approximate unconditional default probabilities. Indeed, it is both possible and desirable to use the same dataset. There are a number of extant approaches, which can be profitably employed for identification of the asset and default correlations. This final chapter thus seeks to examine, from the usual practical perspective, a few alternatives to answering this important question. Consistent with previous discussion, our focus is on practical, real-life examples and the illustration of the advantages and shortcomings of the various approaches.

The following discussion is divided into four distinct elements. We begin with a re-visitation of the notion of default correlation to recall some important relationships and set the stage for the subsequent approaches. We then consider *two* alternative techniques commonly used in statistical inference: the method of moments and maximum-likelihood estimation. These require a bit of adjustment to handle mixture or threshold models, but appear to work quite generally. Using the four credit-state structure from Chap. 9, we proceed to examine a number of alternative examples of varying complexity. The final example turns its attention to a method, which exploits the associated copula function, employed solely in the context of threshold models. Less commonly used due to its relative fragility, its examination, again in the maximum-likelihood framework, nonetheless provides useful insight into the underlying problem.

10.1 Revisiting Default Correlation

Ultimately, our interest is with the default correlation between two individual credit obligors. This can be common across any two pairs of credit counterparts or vary within and in between different groups. The nature of default-correlation differentiation can, in principle, be defined in any way that is meaningful to the quantitative analyst: some common examples include creditworthiness, geographic region, type of industry, or size of firm. In its most atomistic form, however, it relates to the dependence structure between two arbitrary obligors n and m. To make this more concrete and create a foundation for future discussion, let us revisit the form of the default correlation. It is sensible to begin with the default covariance between obligors n and m,

$$\operatorname{cov}\left(\mathbb{I}_{\mathcal{D}_n}, \mathbb{I}_{\mathcal{D}_m}\right) = \mathbb{E}\left(\left(\mathbb{I}_{\mathcal{D}_n} - \underbrace{\mathbb{E}\left(\mathbb{I}_{\mathcal{D}_n}\right)}_{p_n}\right)\left(\mathbb{I}_{\mathcal{D}_m} - \underbrace{\mathbb{E}\left(\mathbb{I}_{\mathcal{D}_m}\right)}_{p_m}\right)\right), \quad (10.1)$$

$$= \mathbb{E}\left(\left(\mathbb{I}_{\mathcal{D}_n} - p_n\right)\left(\mathbb{I}_{\mathcal{D}_m} - p_m\right)\right),$$

$$= \mathbb{E}\left(\mathbb{I}_{\mathcal{D}_n}\mathbb{I}_{\mathcal{D}_m}\right) - p_m \underbrace{\mathbb{E}\left(\mathbb{I}_{\mathcal{D}_n}\right)}_{p_n} - p_n \underbrace{\mathbb{E}\left(\mathbb{I}_{\mathcal{D}_m}\right)}_{p_m} + \cancel{p_n p_m},$$

$$= \mathbb{E}\left(\mathbb{I}_{\mathcal{D}_n}\mathbb{I}_{\mathcal{D}_m}\right) - p_n p_m.$$

10.1 Revisiting Default Correlation

While this notion applies quite generally, covariance is a difficult quantity to interpret and compare. By both convenience and convention, it is much more common to work with normalized covariance, or what is more commonly referred to as correlation. It has the following form,

$$\rho\left(\mathbb{I}_{\mathcal{D}_n}, \mathbb{I}_{\mathcal{D}_m}\right) = \frac{\text{cov}\left(\mathbb{I}_{\mathcal{D}_n}, \mathbb{I}_{\mathcal{D}_m}\right)}{\sigma\left(\mathbb{I}_{\mathcal{D}_n}\right) \sigma\left(\mathbb{I}_{\mathcal{D}_m}\right)}, \qquad (10.2)$$

$$= \frac{\mathbb{E}\left(\mathbb{I}_{\mathcal{D}_n}\mathbb{I}_{\mathcal{D}_m}\right) - p_n p_m}{\sqrt{p_n(1-p_n)}\sqrt{p_m(1-p_m)}}.$$

The default correlation, therefore, through examination of equation 10.2, depends on *three* separate elements:

1. the default probability of obligor m;
2. the default probability of obligor n; and
3. the term $\mathbb{E}\left(\mathbb{I}_{\mathcal{D}_n}\mathbb{I}_{\mathcal{D}_m}\right)$.

The first two quantities are comprehensively addressed in Chap. 9 and, while important, can be identified in a variety of ways. These elements, however, are obligor-specific and do not directly address the question of dependence. The interesting interaction element, and the focus of this chapter, arises from the final aspect: $\mathbb{E}\left(\mathbb{I}_{\mathcal{D}_n}\mathbb{I}_{\mathcal{D}_m}\right)$. It is nevertheless not particularly common to view this quantity as an expectation. More often it is described as,

$$\mathbb{E}\left(\mathbb{I}_{\mathcal{D}_n}\mathbb{I}_{\mathcal{D}_m}\right) = \mathbb{P}\left(\mathcal{D}_n \cap \mathcal{D}_m\right). \qquad (10.3)$$

This expectation is, of course, really just the joint probability of the two default events, which will depend upon the specific model choice. This is understandable since, in general, a distributional choice is required to evaluate this probability. One could avoid this choice, however, by assuming that the two default events are independent. The joint probability then reduces to the product of the two marginal default-event probabilities.

$$\mathbb{E}\left(\mathbb{I}_{\mathcal{D}_n}\mathbb{I}_{\mathcal{D}_m}\right) = \mathbb{E}\left(\mathbb{I}_{\mathcal{D}_n}\right)\mathbb{E}\left(\mathbb{I}_{\mathcal{D}_m}\right), \qquad (10.4)$$

$$= p_n p_m.$$

This seems like a great decision until we plug this quantity back into equation 10.2 and realize that—irrespective of the individual default probability choices—the default correlation collapses to zero. This is not terribly surprising, since this is basically a defining characteristic of the independent-default model introduced in Chap. 2, but is nonetheless useful to keep in mind.

Absent this extreme and unrealistic assumption, we need to grapple with the joint default probability. Let us start in the mixture setting where the default probability is randomized and common among all obligors. Consider a generic mixture variable, S, and recall that in this setting

$$\mathbb{E}(\mathbb{I}_{D_n}|S) = p(S) \tag{10.5}$$

for all $n = 1, \ldots, N$, where $p(S)$ denotes the conditional default probability. The joint-default probability has the following form,

$$\mathbb{P}(\mathcal{D}_n \cap \mathcal{D}_m) = \mathbb{E}\left(\mathbb{I}_{\mathcal{D}_n}\mathbb{I}_{\mathcal{D}_m}\right), \tag{10.6}$$

$$= \underbrace{\mathbb{E}\left(\mathbb{E}\left(\mathbb{I}_{\mathcal{D}_n}\mathbb{I}_{\mathcal{D}_m}\mid S\right)\right)}_{\text{By iterated expectations}},$$

$$= \underbrace{\mathbb{E}\left(\mathbb{E}\left(\mathbb{I}_{\mathcal{D}_n}\mid S\right)\mathbb{E}\left(\mathbb{I}_{\mathcal{D}_m}\mid S\right)\right)}_{\text{By conditional independence}},$$

$$= \underbrace{\mathbb{E}\left(p(S)p(S)\right)}_{\text{By equation 10.5}},$$

$$= \mathbb{E}\left(p(S)^2\right),$$

$$= \text{var}(p(S)) + \mathbb{E}(p(S))^2,$$

where the last step follows from the definition of variance. The joint-probability of default between any two obligors in the generic mixture-model milieu is thus a function of the mean and variance of the conditional default-probability specification. This seems quite reasonable and will, in the following sections, provide the basis for estimating model parameters consistent with a desired level of default correlation.

In the one-factor CreditRisk+ setting—which is, of course, also a mixture model—the result is similar. As shown in Chap. 3, the joint default probability is written as,

$$\mathbb{P}(\mathcal{D}_n \cap \mathcal{D}_m) = \mathbb{E}\left(\mathbb{I}_{\mathcal{D}_n}\mathbb{I}_{\mathcal{D}_m}\right), \tag{10.7}$$

$$= p_n p_m \left(1 + \frac{\omega_1^2}{a}\right),$$

where ω_1 is the factor-loading on the common, gamma-distributed, mixture variable, S, and a is the underlying parameter. Although equation 10.7's form is slightly different, its structure agrees closely with equation 10.6. Given the parameterization of the conditional-default probability, however, the individual unconditional-default probabilities also enter into the expression. This simultaneously enhances the realism of the model and, since few things in life are free, makes things more complicated.

In the threshold setting, however, an additional step is taken. This family of approaches, discussed extensively in Chaps. 4 and 5, has an explicit asset-correlation parameter that acts to indirectly drive the default correlation. To see this more clearly, let's briefly examine the one-factor Gaussian threshold model with a latent state variable, $y_n \sim \mathcal{N}(0, 1)$ for the nth obligor and where default event is defined as,

$$\{y_n \leq \Phi^{-1}(p_n)\}, \tag{10.8}$$

for $n = 1, \ldots, N$. If we re-introduce this specific definition into the context of the joint-default probability, we gain some insight into the notion of default correlation in the threshold setting. Indeed, it has the following form,

$$\begin{aligned}\mathbb{P}\left(\mathcal{D}_n \cap \mathcal{D}_m\right) &= \mathbb{E}\left(\mathbb{I}_{\mathcal{D}_n} \mathbb{I}_{\mathcal{D}_m}\right), \\ &= \mathbb{P}\left(y_n \leq \Phi^{-1}(p_n) \cap y_m \leq \Phi^{-1}(p_m)\right), \\ &= F_{n,m}\left(\Phi^{-1}(p_n), \Phi^{-1}(p_m)\right),\end{aligned} \tag{10.9}$$

where $F_{n,m}$ is the bivariate cumulative distribution function of the latent state variables, y_n and y_m, evaluated at their respective default thresholds. In the threshold model environment, the individual and joint default-state distribution need not be Gaussian. As identified in Chap. 4, the specific marginal and joint distribution is an analyst's choice. The joint-default probability, therefore, possesses a rather more general form than its mixture-model equivalents. Nevertheless, it ultimately boils down to a function of the underlying model parameters.

There appears to exist, therefore, a link between the fundamental assumption associated with each model, the joint default probability, and the model (or distributional) parameters. This chapter will seek to exploit these relationships to target an empirically consistent degree of default correlation. There are, in principle, three main avenues that one might use to accomplish this task. These include:

1. use of the method-of-moments approach, which, as we'll soon see, is strongly reminiscent of the calibration efforts in Chap. 3;
2. reliance on conditional independence and the binomial structure of default to construct a likelihood function; or
3. generalization of the bivariate-distribution function—in the form of the copula—to build an associated likelihood function.

The first approach, which we will examine next, is broadly applicable and generally straightforward to implement. It nevertheless has a few shortcomings and tends to be most useful in the generic mixture-model setting. The second, perhaps most popular, technique returns us to the likelihood framework. This offers both point and interval estimates and is relatively well-suited to the multivariate case. The final, most complex, method is quite general through its consideration of the full dependence structure, but is typically restricted to the threshold setting. Difficult to implement and somewhat fragile, it will be considered last. Before moving on to the algorithms, however, we require a dataset to ensure that our analysis remains both practical and concrete. This is our next task.

10.2 Simulating a Dataset

It is extremely helpful to examine any estimation approach in the context of a practical example. This is particularly the case when the example involves simulated data where the true underlying parameters are known. This permits application of the individual algorithms and some insight into their efficacy. This general idea was used, quite successfully, in the previous chapter to motivate and compare alternative default-probability estimation techniques.

The situation is, however, somewhat more complex in the context of model parameters governing default correlation. Default-probability estimates are obligor specific and, as such, are usefully independent of one's model choice. Default-correlation parameters, conversely, are intimately related to the underlying modelling choice. Mixture and threshold models—as evident in the preceding analysis—handle this aspect in alternative ways; even within each class of models, differences are evident. This complicates the task of creating a common dataset.

It is nonetheless clear that when simulating credit-state transition data, it is necessary to incorporate some notion of default correlation. The base Markov-chain framework, without any adjustment, assumes that each individual default is independent. The only true dependence in the class Markov-chain setting is that each outcome is dependent only on its immediately preceding value. If rating transitions actually followed such a process then, in principle, we could restrict our attention to the independent-default binomial and Poisson models from Chap. 2. In other words, all risk would be idiosyncratic.

For fundamental and defensible economic reasons, we expect to observe default dependence. If we wish to create a dataset, therefore, we will need to incorporate this element. This is not a particularly difficult task, but it requires some assumptions about the nature of this interaction. In short, it requires a model. We could generate a separate dataset for each individual model and use it to estimate that particular model's parameters. This is, since we have a large number of models, fairly cumbersome. The alternative is to use a single model to generate a set of correlated credit-state transitions. We will then employ this dataset in all—or rather most—of the subsequent estimation efforts.

This may seem unreasonable, but it is not dramatically different than the real-world setting. There is some true, physical model of rating transitions. We do not, nor will we ever, know it. We then consider alternative models and try to fit them to an incomplete set of observed outcomes from the real-world process.

10.2 Simulating a Dataset

In the following discussion, therefore, we will use the one-factor Gaussian threshold model to represent a simulated version of the true world. We will then seek to use a collection of techniques and models—including, of course, the true model—to extract estimates of the underlying level of default correlation. While not perfect, it feels like a reasonable compromise permitting a detailed look at the main estimation techniques—this is, after all, the objective of this chapter.

10.2.1 A Familiar Setting

To be explicit, we will simulate a one-factor, one-parameter, Gaussian threshold model with a single asset-correlation parameter, ρ_G. This will help us better understand the specific steps in the implementation and allow visual inspection of some key estimation objects. We can also use the general simulation methodology to bootstrap interval estimates and, in the latter parts of the chapters, slightly change the simulation approach to incorporate other threshold models.

Since we will be simulating transitions, we do not stray very far from the Markov-chain framework; we need only tweak it to incorporate a particular level of default dependence. One motivating principle is also to keep our treatment consistent with the previous chapter. We will thus further assume the existence of only *four* credit states: A, B, C, and D, where, as usual, D corresponds to an *absorbing* default state. Schematically, the transition matrix has the following format:

$$P = \begin{bmatrix} p_{AA} & p_{BA} & p_{CA} & p_{DA} \\ p_{AB} & p_{BB} & p_{CB} & p_{DB} \\ p_{AC} & p_{BC} & p_{CC} & p_{DC} \\ p_{AD} & p_{BD} & p_{CD} & p_{DD} \end{bmatrix}. \tag{10.10}$$

Underscoring our commitment to consistency, we use precisely the same structure and transition matrix as introduced in Chap. 9, which we repeat here for convenience,

$$P = \begin{bmatrix} 0.9600 & 0.0290 & 0.0100 & 0.0010 \\ 0.1000 & 0.7750 & 0.1200 & 0.0050 \\ 0.1200 & 0.2200 & 0.6500 & 0.0100 \\ 0.0000 & 0.0000 & 0.0000 & 1.0000 \end{bmatrix}. \tag{10.11}$$

As before, we will continue to examine 100 initial counterparties over a 30-year time horizon. This may appear to be a relatively small amount of data, but it is consistent with actual available data in practice. We could, for example, simulate out 100 years of data for thousands of counterparties, but realistically, we will rarely, if ever, have this luxury. Even if we did find such a comprehensive database, we would be plagued by questions of the structural relevance of transition and default outcomes from very different, past macroeconomic regimes. The consequence, therefore, is a relatively small dataset.

To simulate the credit-state outcomes over time, as in Chap. 9, it is necessary to make use of the entire transition matrix. Moreover, one needs to make assumptions about the composition of the portfolio at the beginning of the horizon. We assume, therefore, that, at inception, 40% of the 100 counterparties are in category A, with the remaining weight allocated equally to credit classes B and C.[1] No obligor is assumed to be in default at inception.

The next order of business is to introduce the underlying Gaussian default variable. Although this was already discussed extensively in Chap. 4, it involves a slight, albeit easily manageable, twist. Since transitions occur over many periods, we first need to add the time dimension to our latent default variable. This implies updating our usual Gaussian variable as,

$$y_{n,t} = \sqrt{\rho_G} G_t + \sqrt{1 - \rho_G} \epsilon_{n,t}, \qquad (10.12)$$

As usual, the global variables, $\{G_t\}_{t=1,\ldots,T}$, are independent of the idiosyncratic variables, $\{\epsilon_{n,t}; t = 1, \ldots, T; n = 1, \ldots, N\}$. The additional assumption is that the G and ϵ_n random variables are also serially independent. That is, more specifically, G_t and G_{t+1} are independent, for all possible choices of t—the same applies to the $\epsilon_{n,t}$'s. Systematic and idiosyncratic variables are thus both serially and cross-sectionally independent.

If we sought to simulate credit-state transition data—in the usual way highlighted in Chap. 9—we would condition each obligor transition probabilities on its starting value. We would then treat each set of outcomes as a multinomial trial using the appropriate row of the transition matrix to inform the probabilities. The actual outcome would be determined by a draw from an independent uniform distribution. We proceed in a similar manner, but dispense with the uniform random variable. Its role is undertaken—for obligor n in period t—by the latent default variable, $y_{n,t}$.

Figure 10.1 illustrates, in a schematic manner, the basic intuition. Practically, the support of this random, latent state variable is partitioned into *four* disjoint categories—one for each credit state. The particular interval where $y_{n,t}$ falls decides its outcome. The actual partition is determined by the transition probabilities from the appropriate row of the transition matrix. If each $y_{n,t}$ was statistically independent, then this would be equivalent to the uniform random-variable approach. Each $y_{n,t}$, however, all share an exposure to the systematic risk factor, G_t. Moreover, the relative importance of this factor is governed by the parameter, ρ_G.

The actual simulated data, therefore, is generated, sequentially, by the use of equation 10.12 to construct the latent state variables. The heavy lifting involves determining the appropriate partition of the domain of the standard normally distributed state variables. This turns out to be a bit involved. The first step requires cumulating, or summing, the transition-matrix entries across each row from right

---

[1] Practically speaking, we use a uniform distribution to assign counterparties to these categories using these weights. This implies that, given the randomness, the final portfolio weights can deviate slightly.

10.2 Simulating a Dataset

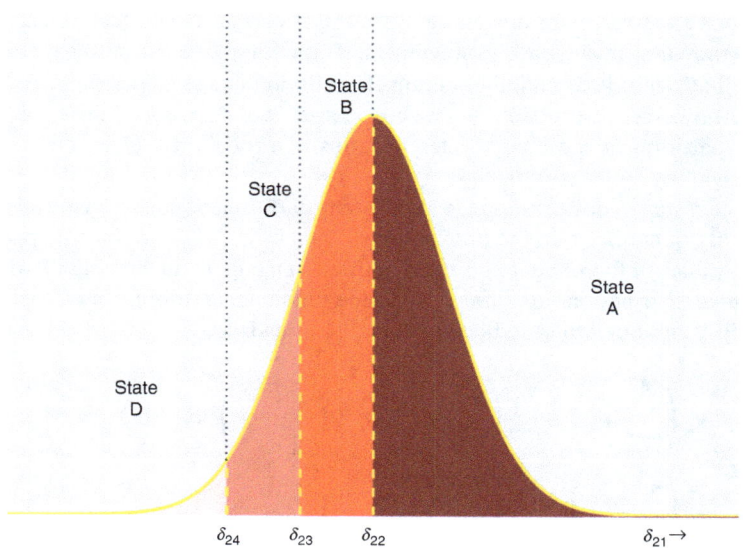

Fig. 10.1 *Creating transition data*: This figure visually illustrates the creation of the transition dataset. The transitions are delineated by a partition of the domain of the standard normal variable following the appropriate row in the transition matrix. The underlying schematic relates to a fictitious row of an arbitrary four-dimensional Markov chain with transition matrix, P.

to left. This leads to the so-called cumulative transition matrix,

$$Q = \begin{bmatrix} (p_{11} + p_{12} + p_{13} + p_{14}) & (p_{12} + p_{13} + p_{14}) & (p_{13} + p_{14}) & p_{14} \\ (p_{21} + p_{22} + p_{23} + p_{24}) & (p_{22} + p_{23} + p_{24}) & (p_{23} + p_{24}) & p_{24} \\ (p_{31} + p_{32} + p_{33} + p_{34}) & (p_{32} + p_{33} + p_{34}) & (p_{33} + p_{34}) & p_{34} \\ (p_{41} + p_{42} + p_{43} + p_{44}) & (p_{42} + p_{43} + p_{44}) & (p_{43} + p_{44}) & p_{44} \end{bmatrix}, \quad (10.13)$$

$$= \begin{bmatrix} \sum_{i=1}^{4} p_{1i} & \sum_{i=2}^{4} p_{1i} & \sum_{i=3}^{4} p_{1i} & \sum_{i=4}^{4} p_{1i} \\ \sum_{i=1}^{4} p_{2i} & \sum_{i=2}^{4} p_{2i} & \sum_{i=3}^{4} p_{2i} & \sum_{i=4}^{4} p_{2i} \\ \sum_{i=1}^{4} p_{3i} & \sum_{i=2}^{4} p_{3i} & \sum_{i=3}^{4} p_{3i} & \sum_{i=4}^{4} p_{3i} \\ \sum_{i=1}^{4} p_{4i} & \sum_{i=2}^{4} p_{4i} & \sum_{i=3}^{4} p_{4i} & \sum_{i=4}^{4} p_{4i} \end{bmatrix},$$

$$= \begin{bmatrix} 1.0000 & 0.0400 & 0.0110 & 0.0010 \\ 1.0000 & 0.9000 & 0.1250 & 0.0050 \\ 1.0000 & 0.8800 & 0.6600 & 0.0100 \\ 1.0000 & 1.0000 & 1.0000 & 1.0000 \end{bmatrix}.$$

Each one of the values in this cumulative transition matrix, Q, lies in the interval, $[0, 1]$—they are, in the end, probabilities. The cumulative transition matrix is an intermediate step. It essentially amounts to a slight change of perspective. We still remain, however, in probability space. In other words, we can think of Q as a generic partition of the credit states in terms of probability. Q is thus a general object awaiting a distributional assumption.

The next step in determining the appropriate partition requires transforming these probability quantities into the domain of y. The entries in Q are thus mapped back into the values of the standard normal distribution using the inverse standard normal cumulative distribution function.[2] This is the fundamental tool for moving between probability and random-variable outcomes. This leads to,

$$\Delta = \begin{bmatrix} \Phi^{-1}\left(\sum_{i=1}^{4} p_{1i}\right) & \Phi^{-1}\left(\sum_{i=2}^{4} p_{1i}\right) & \Phi^{-1}\left(\sum_{i=3}^{4} p_{1i}\right) & \Phi^{-1}\left(\sum_{i=4}^{4} p_{1i}\right) \\ \Phi^{-1}\left(\sum_{i=1}^{4} p_{2i}\right) & \Phi^{-1}\left(\sum_{i=2}^{4} p_{2i}\right) & \Phi^{-1}\left(\sum_{i=3}^{4} p_{2i}\right) & \Phi^{-1}\left(\sum_{i=4}^{4} p_{2i}\right) \\ \Phi^{-1}\left(\sum_{i=1}^{4} p_{3i}\right) & \Phi^{-1}\left(\sum_{i=2}^{4} p_{3i}\right) & \Phi^{-1}\left(\sum_{i=3}^{4} p_{3i}\right) & \Phi^{-1}\left(\sum_{i=4}^{4} p_{3i}\right) \\ \Phi^{-1}\left(\sum_{i=1}^{4} p_{4i}\right) & \Phi^{-1}\left(\sum_{i=2}^{4} p_{4i}\right) & \Phi^{-1}\left(\sum_{i=3}^{4} p_{4i}\right) & \Phi^{-1}\left(\sum_{i=4}^{4} p_{4i}\right) \end{bmatrix}, \quad (10.14)$$

$$= \begin{bmatrix} \Phi^{-1}(q_{11}) & \Phi^{-1}(q_{12}) & \Phi^{-1}(q_{13}) & \Phi^{-1}(q_{14}) \\ \Phi^{-1}(q_{23}) & \Phi^{-1}(q_{22}) & \Phi^{-1}(q_{23}) & \Phi^{-1}(q_{24}) \\ \Phi^{-1}(q_{31}) & \Phi^{-1}(q_{32}) & \Phi^{-1}(q_{33}) & \Phi^{-1}(q_{34}) \\ \Phi^{-1}(q_{41}) & \Phi^{-1}(q_{42}) & \Phi^{-1}(q_{43}) & \Phi^{-1}(q_{44}) \end{bmatrix},$$

$$= \begin{bmatrix} 5.0000 & -1.7507 & -2.2904 & -3.0902 \\ 5.0000 & 1.2816 & -1.1503 & -2.5758 \\ 5.0000 & 1.1750 & 0.4125 & -2.3263 \\ 5.0000 & 5.0000 & 5.0000 & 5.0000 \end{bmatrix}.$$

A bit of caution is required. Since the sum of each row of the transition matrix is unity, the first column of Q all take the value of one. That the support of the normal distribution is $(-\infty, \infty)$ implies, however, that one would assign a value of ∞ to the elements in the first column of Q. This will, of course, inevitably lead to numerical problems. Instead, we assign a value of 5, because the probability of observing an outcome from a standard normal distribution behind these values is vanishingly small.[3]

---

[2] The inverse normal distribution is employed since we find ourselves in the Gaussian threshold setting. The inverse cumulative distribution function is, however, determined by the latent state variable's underlying characteristics.

[3] Any sufficiently large positive number will, of course, work.

10.2 Simulating a Dataset

Algorithm 10.1 Computing Q and Δ

```
def cumulateTransitionMatrix(K,M):
    H = np.zeros([K,K])
    for n in range(0,K):
        for m in range(0,K):
            H[m,(K-1)-n] = np.sum(M[m,(K-1)-n:K])
    return H
def transformCumulativeTransitionMatrix(K,M_c):
    H = np.zeros([K,K])
    for n in range(0,K):
        for m in range(0,K):
            if M_c[n,m]>=0.9999999:
                H[n,m]=5
            elif M_c[n,m]<=0.0000001:
                H[n,m] = -5
            else:
                H[n,m] = norm.ppf(M_c[n,m])
    return H
```

Algorithm 10.1 summarizes two Python sub-routines for the computation of the Q and Δ matrices, respectively. Cumulating the transition matrix—denoted as the input argument, M—requires a double loop over the number of credit states, K. Computing Δ could be performed in a single line—by simply applying the function norm.ppf to M_c or Q—but this would yield some infinite outcomes. Algorithm 10.1 includes some logic to resolve this problem.

The matrix, Δ, is used to assign values to each of the rating outcomes. In other words, it is the key object into determining the appropriate partition of the latent-variable outcomes. Consider, at time t, a given counterparty, n, with a latent state variable outcomes of $y_{n,t}$. Furthermore, assume that this counterparty was in credit state, A, at time $t - 1$. The rating at time $t - 1$ determines the appropriate row of Δ to be used. In this case, because counterparty n is in credit category A, we make use of the first row of Δ. The rating category is assigned depending on where $y_{n,t}$ falls into the range of values from the appropriate row of Δ. Since $y_{n,t}$ is a standard normal variate, this makes complete sense.

The allocation rule is now simple. For our four-state example, we assign the time t rating to counterparty n, denoted $S_{n,t}$, using the following decision rule,

$$S_{n,t} = \begin{cases} A : \text{if } y_{n,t} \in (\infty, -1.7507] \\ B : \text{if } y_{n,t} \in (-1.7507, -2.2904] \\ C : \text{if } y_{n,t} \in (-2.2904, -3.0902] \\ D : \text{if } y_{n,t} \in (-3.0902, -\infty) \end{cases}. \quad (10.15)$$

Algorithm 10.2 Simulating correlated transition data

```
import markovChain as mc
import thresholdModels as th
def simulateCorrelatedTransitionData(K,N,T,Pin,wStart,myRho):
    Q = cumulateTransitionMatrix(K,Pin)
    Delta = transformCumulativeTransitionMatrix(K,Q)
    Y = np.zeros([N,T]) # latent variables
    S = np.zeros([N,T]) # credit states
    allP = np.zeros([N,T]) # default probabilities
    Slast = mc.initializeCounterparties(N,wStart) # initial states
    S0 = Slast
    Plast = Pin[(Slast-1).astype(int),-1]
    for t in range(0,T):
        Y[:,t] = th.getY(N,1,Plast,myRho)
        for n in range(0,N):
            if Slast[n] == 4:
                S[n,t] = 4
                continue
            else:
                S[n,t] = migrateRating(Slast[n],Delta,Y[n,t])
        allP[:,t] = Pin[(Slast-1).astype(int),-1]
        Plast = allP[:,t]
        Slast = S[:,t]
    return S,Y,Delta,allP,S0
```

Again, for the purposes of practical implementation in Algorithm 10.1, we brazenly replace ∞ and $-\infty$ with 5 and -5, respectively.

We have, as one would expect with a structural model, defined default as the value of $y_{n,t}$ falling below a particular threshold. The approach merely generalizes this idea by assigning different regions of the domain of $y_{n,t}$ to various rating outcomes. The size and location of these regions are informed—mechanically, through equations 10.13 and 10.14—by the transition matrix.

Algorithm 10.2 illustrates the function that, following the preceding description, simulates correlated transition data. While it is fairly lengthy, much of the effort is book-keeping. It also makes use of some functions from the `markovChain` and `thresholdModels` libraries. The first step is generating the Q and Delta matrices from Algorithm 10.1. This is followed by generation of the starting conditions with recycled code from the `markovChain` library introduced in Chap. 9. It then loops over the T periods and N obligors and uses the `migrateRating` function to determine the partition and allocate the appropriate credit class. A bit of caution is required to ensure that the conditioning information is preserved at each time step. Finally, the average unconditional default probabilities—for non-defaulted obligors—are computed for each time step. This information, stored in `allP`, will be useful in our estimation efforts.

10.2.2 The Actual Results

Having specified the transition probabilities, the only remaining choice is the asset correlation parameter, ρ_G. Equation 10.12 demonstrates clearly that, as we scale up ρ_G, we generate greater joint default probability and correspondingly larger amounts of default correlation. Figure 10.2 illustrates the impact of this parameter value—over the range of $\rho_G \in [0, 1]$—on both quantities.

Computation of pairwise default correlation requires, of course, an assessment of the unconditional default probability of each obligor. Figure 10.2 uses the average default probability of default at inception—using the initial weights of 0.4, 0.3, and 0.3 for states A, B, and C, respectively—for each obligor. This amounts to slightly less than 50 basis points. Based on this figure, and consistent with values used in the regulatory world, we set $\rho_G = 0.20$. The consequence of this choice is an average level of default correlation of about 0.016.[4] This feels like a reasonable choice and is generally consistent with the calibration targets established in Chaps. 3 and 4.

Armed with our parametric assumptions, we need only invoke Algorithm 10.2 to create a time series of rating transitions for each of our credit obligors. Our dataset is thus a 30-period collection of credit ratings for 100 counterparties. Figure 10.3 highlights, a single realization of this simulated dataset. It indicates the evolution of the number of credit counterparties (N_t), the number of defaults (k_t), and the default rate ($\hat{p}_t$). It also provides a view of the average default probabilities ($\bar{p}_t$) associated with the remaining non-default obligors in each period. There is an important difference between these final two quantities. The default rate is the observed proportion of defaults, whereas the default probability is the expected incidence of default assigned to each obligor based on its credit state.

Investigation of Fig. 10.3 is quite interesting. As default is a relatively rare event, its incidence is not particularly frequent. Indeed, only 10 defaults occur over the 30-year period. Interestingly, when it does occur, it appears to occur in clusters. In one third of the seven years where an event does occur, it involves multiple counterparties. This would argue for some level of default correlation in the dataset.

Table 10.1 drills deeper into our simulated dataset to examine some summary statistics across the entire 30-year period. It includes the mean, standard deviation, minimum, and maximum of the four quantities highlighted in Fig. 10.3. The sample default rate is about 35 basis points with a volatility of roughly 72 basis points. If one examines the unconditional default probabilities of the obligors—assigned by their credit rating—the average is about 30 basis points. The sample default rate is much more volatile, however, than the default probability. This stems from the infrequency and lumpiness of default arrival.

---

[4]It will, of course, vary depending on the actual default probability of the individual credit counterparties.

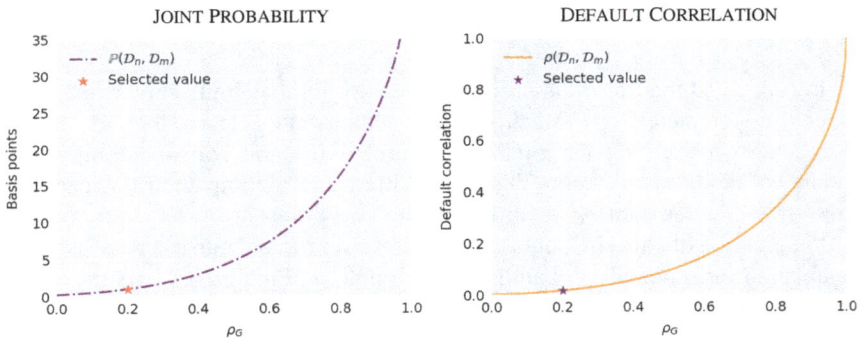

Fig. 10.2 *Choosing* ρ_G: This figure highlights the joint-default probability and default correlation for a variety of choices of the parameter ρ_G with an average starting default probability of approximately 50 basis points.

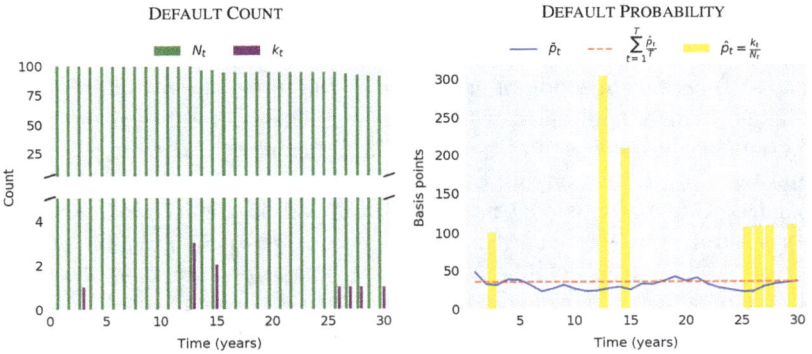

Fig. 10.3 *Aggregate default data*: These two figures summarize the intertemporal evolution of the number of credit counterparties, the number of defaults, the default rate, and the default probabilities. As default is a relatively rare event, its incidence is unsurprisingly infrequent.

The quantities in Fig. 10.3 and Table 10.1 are all determined in Algorithm 10.3. This may seem like a straightforward task, but it requires a bit of care. The input arguments are the number of periods T, the rating observations S, and the default probabilities for each obligor in each period, allP. A loop over the time dimension allows us to count the number of credit counterparties and defaults in each period: these are denoted nVec and kVec, respectively. The average default rate in each period is found in the array, pVec. These serve as critical inputs into our subsequent computations.[5]

---

[5] Although we do not provide the figures, if we estimate the default probabilities, the values are equivalent to those found in Chap. 9. Transition probabilities are thus preserved under the imposition of default correlation.

10.3 The Method of Moments

Table 10.1 *Simulated data at a glance*: This table provides some numerical support to the set of simulated, default-correlated, one-factor Gaussian model transition data results presented in Fig. 10.3. σ, min, and max, as one might expect, refer to standard deviation, minimum, and maximum, respectively.

Quantity		Statistics			
Symbol	Dimension	Mean	σ	min	max
N_t	Sample number of obligors	96.1	2.9	91.0	100.0
k_t	Sample number of defaults	0.3	0.7	0.0	3.0
$\hat{p}_t$	Sample default rate (bps.)	34.8	72.0	0.0	303.0
$\bar{p}_t$	Assigned average default probability (bps.)	30.5	6.3	21.9	47.7

Algorithm 10.3 Accounting for counterparties, defaults, and probabilities

```
def getSimpleEstimationData(T,S,allP):
    N,T = S.shape
    kVec = np.zeros(T)
    nVec = np.zeros(T)
    pVec = np.zeros(T)
    kVec[0] = np.sum(S[:,0]==4)
    nVec[0] = N
    pVec[0] = np.mean(allP[:,0])
    for t in range(1,T):
        kVec[t] = np.sum(S[:,t]==4)-np.sum(S[:,t-1]==4)
        nVec[t] = nVec[t-1] - kVec[t-1]
        pVec[t] = np.mean(allP[S[:,t-1]!=4,t])
    return pVec,nVec,kVec
```

10.3 The Method of Moments

The first estimation approach, termed the method of moments, is certainly the most intuitive. The basic idea behind this classic estimation technique—introduced by Karl Pearson in the late 1800's—is quite intuitive.[6] It essentially matches theoretical population moments of a distribution with the observed sample moments, thereby creating a system of simultaneous equations.

The theoretical moments are functions of the distribution's parameters, whereas the sample moments are observed real numbers. If one can identify k equations for k unknown parameters, then one can, in principle, solve the system to arrive at a set of parameter estimates. This approach is no longer as commonly used, because in the words of Casella and Berger (1990) "in many cases, unfortunately, this method yields estimators that may be improved on." The principal shortcomings

---

[6] See Magnello (2009) for an illuminating review of Karl Pearson's path-breaking contributions to the field of statistics.

are difficulty in performing statistical inference and unreasonable estimates. It is nonetheless a useful starting point for our analysis.[7]

This approach was first introduced for the determination of credit-risk model parameters by Koyluoglu and Hickman (1998) and Gordy (1998). The targeted moments include the expectation and the variance of the unconditional default rate. In the generic mixture models examined in Chap. 3, we can write the unconditional default probability as,

$$\mu_p = \int_{\mathcal{A}} p(s) f_S(s) ds, \qquad (10.16)$$

where $f_S(s)$ denotes the density function of the positive-valued common mixture variable, S. $\mathcal{A}$ denotes the domain of S; for the beta-binomial mixture model, the range of integration would be the unit interval, [0, 1]. For other models, it could be the entire or positive real line. Equation 10.16 simply averages over, or integrates out, the variable S from the (known) conditional default probability. Similarly, the unconditional default variance is

$$\sigma_p^2 = \int_{\mathcal{A}} \left(p(s) - \mu_p \right)^2 f_S(s) ds. \qquad (10.17)$$

Strictly speaking, the variance is not precisely the second moment, but rather the second *central* moment. This poses no problems and, in fact, makes it easier to compare to the sample data.

We now have two simultaneous equations, which both depend on the conditional default probability, $p(s)$, and its density $f_S(s)$. To the extent that these two quantities involve *two* parameters, we should be able to estimate them. First, however, we require a link to our dataset. We have simulated T periods of cross-sectional data and, in each period, we observe the credit state of N_t distinct obligors in each of these periods. We use k_t to denote the number of them—quite possibly zero—that actually fall into default in period t. The default rate in period t is, of course, defined as,

$$\hat{p}_t = \frac{k_t}{N_t}. \qquad (10.18)$$

A reasonable estimator of the empirical unconditional default probability is simply the sample average or,

$$\hat{\mu}_p = \frac{1}{T} \sum_{t=1}^{T} \hat{p}_t, \qquad (10.19)$$

---

[7] Hansen (1982) offers a full-fledged framework, referred to as the generalized method of moments, where the moment equations are replaced with orthogonality conditions inferred from one's model. While a useful and powerful approach, it seems a bit excessive in this context.

10.3 The Method of Moments

while the sample unconditional default-variance estimator is,

$$\hat{\sigma}_p = \frac{1}{T-1} \sum_{t=1}^{T} \left(\hat{p}_t - \hat{\mu}_p\right)^2. \tag{10.20}$$

This is all that we require to construct our two non-linear simultaneous equations,

$$\underbrace{\int_{\mathcal{A}} p(s) f_S(s) ds}_{\mu_p} - \underbrace{\frac{1}{T} \sum_{t=1}^{T} \hat{p}_t}_{\hat{\mu}_p} = 0, \tag{10.21}$$

$$\underbrace{\int_{\mathcal{A}} \left(p(s) - \mu_p\right)^2 f_S(s) ds}_{\sigma_p} - \underbrace{\frac{1}{T-1} \sum_{t=1}^{T} \left(\hat{p}_t - \hat{\mu}_p\right)^2}_{\hat{\sigma}_p} = 0.$$

Each of the two-parameter, generic, mixture models are readily incorporated into this estimation framework.

Solution of equation 10.21 thus requires only two data inputs: the mean and variance of observed default volatility associated with a given dataset. The actual implementation, however, requires a reasonable amount of computation. In general, the moments need to be determined through numerical integration, whereas the system is also solved numerically. The good news, however, is that most of this effort was already performed in Chap. 3 while examining calibration techniques for the mixture models.

Borrowing heavily from the mixtureModels library, Algorithm 10.4 highlights the objective function for the method-of-moments estimation of the generic mixture-model parameters. It incorporates *six* distinct implementations highlighted in Chap. 3. Although lengthy, the structure is, in fact, quite straightforward. Given the model choice, governed by the myModel flag, the first and second moments of the default probability distribution are computed; each is creatively referred to as M1 and M2. Finally the two equations, f1 and f2, are computed using these quantities as indicated in Equation 10.21. The input arguments, myP and myV, are the sample moments. The optimally consistent set of mixture-model parameters associated with this system of equations are identified by the following function call:

x = scipy.optimize.fsolve(mixtureMethodOfMoment,x0,args=(myP,myV,myModel)), (10.22)

where x0 are one's starting values. The optimization is performed by scipy's now-familiar fsolve algorithm. In each case, it requires only a few seconds to arrive at a solution.

How well does it work? This can readily be assessed, but first we require some empirical estimates of the mean and variance of the unconditional default probability. Enter the dataset provided in the previous section. Table 10.1 indicates

Algorithm 10.4 Simultaneous system of mixture model method of moments equations

```
import mixtureModels as mx
def mixtureMethodOfMoment(x,myP,myV,myModel):
    if myModel==0: # Beta-binomial
        M1 = mx.betaMoment(x[0],x[1],1)
        M2 = mx.betaMoment(x[0],x[1],2)
    elif myModel==1: # Logit
        M1,err = nInt.quad(mx.logitProbitMoment,-8,8,args=(x[0],x[1],1,1))
        M2,err = nInt.quad(mx.logitProbitMoment,-8,8,args=(x[0],x[1],2,1))
    elif myModel==2: # Probit
        M1,err = nInt.quad(mx.logitProbitMoment,-8,8,args=(x[0],x[1],1,0))
        M2,err= nInt.quad(mix.logitProbitMoment,-8,8,args=(x[0],x[1],2,0))
    elif myModel==3: # Poisson-gamma
        M1 = mx.poissonGammaMoment(x[0],x[1],1)
        M2 = mx.poissonGammaMoment(x[0],x[1],2)
    elif myModel==4: # Poisson-lognormal
        M1,err = nInt.quad(mx.poissonMixtureMoment,0.0001,0.9999,args=(x[0],x[1],1,0))
        M2,err = nInt.quad(mx.poissonMixtureMoment,0.0001,0.9999,args=(x[0],x[1],2,0))
    elif myModel==5: # Poisson-Weibull
        M1,err=nInt.quad(mix.poissonMixtureMoment,0.0001,0.9999,args=(x[0],x[1],1,1))
        M2,err = nInt.quad(mix.poissonMixtureMoment,0.0001,0.9999,args=(x[0],x[1],2,1))
    f1 = M1 - myP
    f2 =(M2-M1**2) - myV
    return [1e4*f1, 1e4*f2]
```

Table 10.2 *Mixture method of moments results*: This table highlights the theoretical and sample moments along with the parameter estimates for six of the one-factor, two-parameter mixture models introduced in Chap. 3.

Model	Mean			Volatility			Parameters		
	$\hat{\mu}_p$	μ_p	Δ	$\hat{\sigma}_p$	σ_p	Δ	p_1	p_2	$\rho_\mathcal{D}$
Beta-binomial	34.80	34.80	0.00	72.04	72.04	0.00	0.23	65.59	0.015
Logit-normal	34.80	34.80	−0.00	72.04	72.04	−0.00	−6.54	1.34	0.015
Probit-normal	34.80	34.80	−0.00	72.04	72.04	−0.00	−3.05	0.53	0.015
Poisson-gamma	34.80	34.80	0.00	72.04	72.04	0.00	0.23	65.21	0.015
Poisson-lognormal	34.80	34.81	−0.01	72.04	72.03	0.01	−6.52	1.32	0.015
Poisson-Weibull	34.80	34.86	−0.06	72.04	72.01	0.03	0.52	0.00	0.015

that the mean unconditional default probability, or rate, is equal to about 35 basis points and its associated variance to roughly 72 basis points. These values were computed from our simulated dataset constructed from a proper transition matrix with a given magnitude of default correlation; as previously indicated, the true underlying model is a one-factor, Gaussian threshold approach.

Table 10.2 summarizes the results of the application of Algorithm 10.4 to our two sample moments across *six* distinct mixture-model instances. The theoretical and sample moments, along with their differences, are provided. The fitted values are quite close and, in all cases, yield a default correlation coefficient of approximately 0.015. This quantity is computed by plugging the result from equation 10.6 into equation 10.2 and simplifying.[8] The theoretical level of default correlation—using

---

[8] The mixture-model default correlation expression can also be found in Chap. 3.

the initial credit states—is about 0.016. If we use the sample default rate of 35 basis points with the true ρ_G parameter of 0.2, this leads to roughly consistent value of about 0.013. In short, therefore, this approach appears to work reasonably well.

10.3.1 The Threshold Case

The pair of model parameters for each model—which are the stars of this show—form our first estimates in the area of default correlation. It is precisely these values that, in conjunction with the distributional assumptions of the model, yield an empirically consistent level of default probability, variance, and correlation. Estimates are also possible for the set of threshold models. A few adjustments are nonetheless required.

One key difference in the threshold setting is the reliance of the conditional default probability, $p(G)$, on the individual obligor's unconditional default probability. For this reason, it is typically referred to as $p_n(G)$. To resolve this, we will need to use the average default probability and stay with the $p(G)$ form. The second issue is that $p(G)$ refers to the one-factor, Gaussian threshold setting.[9] In the normal-variance-mixture cases, which include some interesting alternatives, a second mixing variable arises. Thus, we have conditional default probabilities of two variables; let's write this as $p(G, V)$ where again we use an average default probability. The implication is that our sample moments involve double integrals of the form,

$$\mu_p = \int_{\mathbb{R}_+} \int_{\mathbb{R}} p(g, v) f_G(g) f_V(v) dg dv, \quad (10.23)$$

where $f_S(s)$ denotes the density function of the positive-valued common mixture variable, S. We simply average over, or "integrate out", the variable S from the (known) conditional default probability. Similarly, the unconditional default variance is

$$\sigma_p^2 = \int_{\mathbb{R}_+} \int_{\mathbb{R}} \left(p(g, v) - \mu_p \right)^2 f_G(g) f_V(v) dg. \quad (10.24)$$

In all cases, the integral with respect to the Gaussian density covers the real line, whereas the mixing variable is integrated over the positive reals.

Handling these double integrals augments the amount of computation expense and makes implementation a bit heavier. Algorithm 10.5 gets started with the Python `thresholdMoment` function for the normal-variance mixture models

---

[9]This is, incidentally, the only model in this group with a single parameter value.

Algorithm 10.5 Computing threshold model moments

```
import thresholdModels as th
def thresholdMoment(g,v,p1,p2,myP,whichModel,myMoment,invCdf=0):
    d1 = util.gaussianDensity(g,0,1)
    if whichModel==1: # t
        d2 = util.chi2Density(v,p2)
        integrand = np.power(th.computeP_t(myP,p1,g,v,p2),myMoment)
    if whichModel==2: # Variance-gamma
        d2 = util.gammaDensity(v,p2,p2)
        integrand = np.power(th.computeP_NVM(myP,p1,g,v,p2,invCdf),myMoment)
    if whichModel==3: # Generalized hyperbolic
        d2 = util.gigDensity(w,p2)
        integrand = np.power(th.computeP_NVM(myP,p1,g,v,p2,invCdf),myMoment)
    return integrand*d1*d2
```

Algorithm 10.6 Simultaneous system of threshold model method of moments equations

```
def thresholdMethodOfMoment(x,myP,myV,whichModel):
    if (x[0]<=0) | (x[0]>1):
        return 100
    M1,M2 = getThresholdMoments(x,myP,whichModel)
    f1 = M1 - myP
    f2 =(M2-M1**2) - myV
    return [1e4*f1,1e4*f2]
```

covered in Chap. 4.[10] It takes a number of inputs arguments: these include the two latent state variables, g and v, the type input parameters, p1 and p2, the average default probability, myP, and two additional flags. The first flag, whichModel, allows a single implementation for the three model choices. The second flag, myMoment, determines the appropriate moment to compute. The final input argument, invCdf, is the inverse cumulative distribution function—representing the default threshold—required for the conditional default probability. Since it is computationally expensive to compute, and fixed for each set of model parameters, it is passed as an argument to increase efficiency.

Algorithm 10.6 is used to actually handle the optimization exercise. A second function, getThresholdMoments, manages the numerical integration required for each of the moments. It is not shown here, because it has basically the same form as the nested if statement in Algorithm 10.4. In short, the treatment of the threshold models is conceptually very similar to the mixture-model setting. The key differences relate to the form of the conditional default probabilities and the mixture densities.

Table 10.3 highlights—following the same structure as the mixture results in Table 10.2—the threshold estimates. We again observe, for the optimal parameters,

---

[10] The Gaussian model, which takes only a single state variable, is performed separately.

10.4 Likelihood Approach

Table 10.3 *Threshold method of moments results*: This table highlights the theoretical and sample moments along with the parameter estimates for each of the four generic, one-factor, two-parameter threshold models introduced in Chap. 4.

Model	Mean			Volatility			Parameters		
	$\hat{\mu}_p$	μ_p	Δ	$\hat{\sigma}_p$	σ_p	Δ	p_1	p_2	$\rho_\mathcal{D}$
Gaussian	34.80	34.79	0.00	72.04	72.04	0.00	0.22	0.00	0.015
t	34.80	34.80	0.00	72.04	72.04	−0.00	0.11	26.18	0.015
Variance-gamma	34.80	34.80	−0.00	72.04	72.04	0.00	0.10	7.17	0.015
Generalized hyperbolic	34.80	34.80	0.00	72.04	72.04	0.00	0.16	19.73	0.015

a numerically close fit of the sample and theoretical moments. Since the true underlying model used to generate the dataset stems from a one-factor, Gaussian threshold model with an asset correlation of 0.20, it is comforting to observe that the method-of-moments estimate is 0.22 for this model. Equally comforting is the 0.015 implied default correlation coefficient across all of the threshold models. This agrees with the common mixture-model estimate and is an underlying feature of the dataset. The various frameworks and individual models, when forced to match the first two sample moments, quite naturally replicate this empirical result.

The method of moments technique is essentially an exercise in calibration of one's model parameters to a selection of sample moments. It should feel very much like the efforts performed in Chaps. 3 and 4 for the identification and comparison of model parameters. The key difference, in this setting, is that the calibrating values are informed by empirical data rather than arbitrarily selected targets. It has the important advantages of being conceptually straightforward and quite generally applicable. We have been able to use it to estimate ten different mixture and threshold models. A shortcoming of this approach, however, is the difficulty of statistical inference.[11] The more formal likelihood framework, addressed in the following section, addresses this issue.

10.4 Likelihood Approach

The conditional-default probability was essential in the implementation of the method-of-moments technique. It represents an important link between default outcomes, the systematic state variable (or variables), and the distributional parameters governing asset correlation. Most importantly, by exploiting conditional independence, we can obtain usable results by integrating out the systematic factor. This quantity can, however, also be employed in an alternative manner. Indeed, it is the foundation of the use of maximum-likelihood estimation in this setting.

---

[11] It is, of course, possible to use Efron (1979)'s bootstrap technique to generate standard errors.

10.4.1 The Basic Insight

The key insight from the work in this area—Gordy and Heitfield (2002), Hamerle et al. (1986), and Demey et al. (2004)—is that the number of defaults in a given year, conditional on the systematic variable, follows a binomial law. Let's begin with a general form, before specializing to a few concrete examples. We define Z_t as the generic systematic latent variable at time t and θ as the vector—which may collapse to a scalar—containing the model parameters. The probability of observing k_t defaults with N_t counterparties has the following distribution function,

$$f_t(k_t, N_t | \theta, Z_t = z) = \binom{N_t}{k_t} p(\bar{p}_t, z)^{k_t} (1 - p(\bar{p}_t, z))^{N_t - k_t}, \quad (10.25)$$

where $p(\bar{p}_t, z)$ is the conditional default probability and,

$$\bar{p}_t = \frac{1}{N_t} \sum_{n=1}^{N_t} p_{n,t}. \quad (10.26)$$

$p_{n,t}$ represents the unconditional default probability of counterparty n at time t. We use, at this point, the average default probability across all obligors to evaluate the conditional default probability. For most mixture models, this is standard procedure. The exception is the CreditRisk+ framework. The threshold models, conversely, each make explicit use of individual obligor creditworthiness. To derive a manageable likelihood function, we average across all credit counterparts. This is a bit unrealistic, but it simplifies the development somewhat. It is not particularly criminal, because it will be relaxed shortly.

To slightly simplify the notational burden, let us write the conditional default probability as,

$$p(\bar{p}_t, z) \equiv p_t(z). \quad (10.27)$$

In the Gaussian threshold model, this is written as,

$$p_t(z) = \Phi\left(\frac{\Phi^{-1}(\bar{p}_t) - \sqrt{\rho_G} z}{\sqrt{1 - \rho_G}}\right), \quad (10.28)$$

with $z \sim \mathcal{N}(0, 1)$ and $\theta = \rho_G$ while the one-factor CreditRisk+ model is,

$$p_t(z) = \bar{p}_t \left((1 - \omega) + \omega z\right). \quad (10.29)$$

In this latter case, $z \sim \Gamma(a, a)$ implying that $\theta \in \mathbb{R}^2$. This implies that the likelihood approach can be applied to all models since each has a conditional default probability. Both ρ_G and ω play a similar role; they are the loading on the systematic

10.4 Likelihood Approach

factor. Alternative expressions are available for all mixture and threshold models. To keep the exposition manageable, however, we will focus on these two models.

The driving idea behind this method is that if we observe default outcomes over multiple periods, then we should be able to make sensible inferences about the model parameters in θ influencing default correlation. We first, however, need to eliminate the systematic variable using the same trick as employed in previous sections and chapters. Integrating out the systematic state variable, we have

$$f_t(k_t, N_t | \theta) = \int_{\mathcal{A}} \binom{N_t}{k_t} p_t(z)^{k_t} (1 - p_t(z))^{N_t - k_t} f_Z(z) dz, \quad (10.30)$$

where $\mathcal{A} \equiv \mathbb{R}$ for choices of Z with infinite support and $\mathcal{A} \equiv \mathbb{R}_+$ for systematic variables confined to the positive half line.

Equation 10.30, while interesting and useful, only describes the joint default density at a single point in time. We need to consider the full data period. Given the assumption of serial independence, this is a straightforward exercise. We may collect each of the t-period densities together over time as,

$$f_{t,\ldots,T}(\mathbf{k}, \mathbf{N} | \theta) = \prod_{t=1}^{T} \underbrace{\int_{\mathcal{A}} \binom{N_t}{k_t} p_t(z)^{k_t} (1 - p_t(z))^{N_t - k_t} f_Z(z) dz}_{\text{Equation 10.30}}, \quad (10.31)$$

where for notational convenience,

$$\mathbf{k} \stackrel{\Delta}{=} \{k_1, \ldots, k_T\}, \quad (10.32)$$

and

$$\mathbf{N} \stackrel{\Delta}{=} \{N_1, \ldots, N_T\}, \quad (10.33)$$

represent the time series of observed defaults and number of counterparties, respectively. The T period joint default density, by serial independence of the systematic variables, is merely the product of the individual t period densities.

Having developed the joint default density, the desired likelihood function is merely a slight change of perspective. Specifically, we have that

$$\mathcal{L}(\theta | \mathbf{k}, \mathbf{N}) = \prod_{t=1}^{T} \int_{\mathcal{A}} \binom{N_t}{k_t} p_t(z)^{k_t} (1 - p_t(z))^{N_t - k_t} f_Z(z) dz. \quad (10.34)$$

As introduced in Chap. 9, the joint density is the probability of a constellation of outcomes given a particular parameter set. The likelihood function, conversely, is the probability of observing a given set of parameters conditional on one's observed data. This is more than a mathematical sleight of hand; it is a fundamental change in the frame of reference.

The derivation of the much easier-to-handle, log-likelihood kernel is merely a bit of algebraic manipulation,

$$\ell(\theta|\mathbf{k}, \mathbf{N}) = \ln \left(\prod_{t=1}^{T} \int_{\mathcal{A}} \binom{N_t}{k_t} p_t(z)^{k_t} (1 - p_t(z))^{N_t - k_t} f_Z(z) dz \right), \quad (10.35)$$

$$= \sum_{t=1}^{T} \ln \left(\int_{\mathcal{A}} \binom{N_t}{k_t} p_t(z)^{k_t} (1 - p_t(z))^{N_t - k_t} f_Z(z) dz \right).$$

Given, therefore, a sequence of counterparties and default observations, one can then use equation 10.35 to determine the maximum-likelihood estimate of θ. While it cannot be solved analytically, it is readily handled numerically. It nonetheless reasonably computationally expensive since it will require the evaluation of T numerical integrals.

10.4.2 A One-Parameter Example

Our starting point is the one-factor, Gaussian threshold model. This approach has the important advantage of possessing a single parameter, ρ_G. Moreover, the underlying dataset was also produced using this modelling approach. All that is required, having derived the log-likelihood function and armed with a proper dataset, is the maximization of equation 10.35. The Python code used to perform this task is found in Algorithm 10.7. It requires *two* functions: `getCMF` and `logLSimple`. The former is essentially the integrand found in equation 10.30, whereas the latter is the log-likelihood function. It loops over every period, numerically computing the necessary integral at each step, and then sums the results and applies the log operator. Given the necessity of numerically integrating the likelihood function, it is relatively slow. A single function evaluation requires almost two thirds of a second. While the clever analyst might identify a few tricks to speed it up somewhat, T separate one-dimensional numerical integrations cannot be avoided and will require some time.

The first step in solving this one-dimensional optimization problem, following the good practice underscored in the previous chapter, is to graph the normalized

10.4 Likelihood Approach

Algorithm 10.7 Single-parameter log-likelihood functions

```
def getCMF(g,myRho,myP,myN,myK):
    pg = th.computeP(myP,myRho,g)
    f=util.getBC(myN,myK)*np.power(pg,myK)*np.power(1-pg,myN-myK)
    cmf = f*util.gaussianDensity(g,0,1)
    return cmf
def logLSimple(x,T,pVec,nVec,kVec):
    L = np.zeros(T)
    for t in range(0,T):
        L[t],err = nInt.quad(getCMF,-5,5,
                        args=(x,pVec[t],nVec[t],kVec[t]))
    logL = np.sum(np.log(L))
    return -logL
```

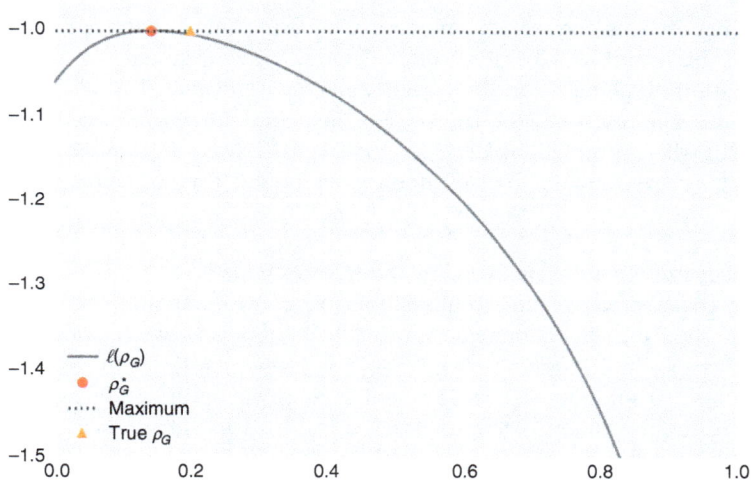

Fig. 10.4 *Single-parameter likelihood function*: Using the data from Fig. 10.3, this graphic traces—over the permissible domain of the parameter, $\rho \in [0,1]$—the shape of the log-likelihood function. It also indicates the estimated and true population parameters. The log-likelihood appears to be regular.

log-likelihood function.[12] Figure 10.4 highlights the log-likelihood function, evaluated across a fine grid of values over the unit interval. The maximum occurs at approximately 0.15, which is somewhat below the true parameter value of 0.2. The shape of the likelihood function is, however, reassuring parabolic and, as such, we may relatively safely conclude that this problem is regular. This will allow us to compute standard errors using the observed Fisher information.

---

[12] Normalized, in this context, means dividing all reported log-likelihood values by its maximum.

Table 10.4 *Single-dataset results*: This table summarizes the key results from use of the maximum-likelihood estimate of ρ_G from the dataset and likelihood functions described in Figs. 10.3 and 10.4.

Quantity	Symbol	Value
True parameter value	ρ_G	0.20000
Grid search optimum	$\rho_{G,0}$	0.14213
Hill-climber optimum	ρ_G^*	0.14438
Score function	$s(\rho_G^*)$	−0.00000
Observed Fisher information	$I(\rho_G^*)$	79.22
Standard error	$\sqrt{I^{-1}(\rho_G^*)}$	0.11
95% confidence interval	$\rho_G^* \pm 1.96\sqrt{I^{-1}(\rho_G^*)}]$	[0.00,0.36]

In a one-dimensional setting, numerical optimization is not really necessary. We are able, from Fig. 10.4, to visually determine the optimum. We nevertheless employ a hill-climbing algorithm to determine the true value. This also permits us to compute other useful quantities, such as the value of the score function and the observed Fisher information. Using Algorithm 10.7, we simply make use of the following `scipy` function to determine the maximum of our function,[13]

```
rHat = scipy.optimize.minimize(logLSimple,x0,args=(T,pVec,nVec,kVec),method='TNC')
```
(10.36)

This makes use of a truncated version of the Newton method.[14]

Table 10.4 summarizes the results of this exercise for our simulated dataset. As expected, the estimated parameter is roughly 0.2, but now we also have that the standard error is quite significant at 0.11. This leads to a 95% confidence interval ranging from 0 to about 0.4. This is not a complete surprise given the relative flatness of the likelihood function observed in Fig. 10.4. In summary, we have a reasonable, if somewhat downward biased, parameter estimate with a sizable degree of uncertainty. The downward bias is highlighted in Gordy and Heitfield (2002), Hamerle et al. (1986), and Demey et al. (2004). Moreover, they demonstrate how, as the length of the sample increases, it dissipates. While useful to be aware of this fact, the restricted size of existing datasets suggests that we need to be aware of this bias. We will attempt to explore this, if not answer it, in the future sections.

To this point, we've performed a single estimate on one (simulated) dataset. A more comprehensive and formal analysis is nonetheless warranted. To this end, therefore, we simulate multiple, credit-state time series—using the same basic approach—and estimate the known parameters across each simulation. This simulation-study approach allows us to consider how the estimation algorithm behaves for different possible parameter values and across a broad range of possible data outcomes. We can also verify, since the true parameter values are known, if there exists any material bias in the estimates.

---

[13] In this case, of course, we are minimizing the negative of the log-likelihood function, which amounts to maximization.

[14] See Press et al. (1992, Chapter 10) for more detail on non-linear optimization algorithms.

10.4 Likelihood Approach

Table 10.5 *Single-parameter simulation study results*: This table illustrates the results of estimating the *three* alternative ρ_G parameters—0.15, 0.30, and 0.45—using the maximum-likelihood approach with 100 simulated datasets for each parameter value.

Parameter						
ρ_G	$\hat{\rho}_G^*$	$\sigma(\hat{\rho}_G^*)$	Score function	Fisher information	Standard error	Success ratio
0.15	0.19	0.16	0.64051	62.9	0.17	100.0%
0.30	0.30	0.18	0.24813	45.4	0.18	98.0%
0.45	0.40	0.20	0.21013	34.5	0.19	95.0%

The simulation study structure is quite simple. We consider *three* different values of ρ_G: 0.15, 0.30, and 0.45. For each of these asset-correlation parameter values, we then proceed to simulate 100 datasets using the aforementioned approach. As in the previous section, each dataset has $N \times T$ credit-rating observations with $N = 100$ and $T = 30$. With this large collection of datasets and estimates under three distinct parameters—this yields 300 separate asset-correlation estimates—we may address a number of interesting questions:

1. Are, on average, the estimates of ρ_G close to the true population parameters?
2. How much volatility are there in these estimates?
3. Does the optimization algorithm always converge? If not, how often does it fail?
4. What is the standard error of these estimates?

Table 10.5 summarizes the results of this simulation study and provides some insight into these queries. Our first observation is that, on average, the maximum-likelihood estimates do indeed appear to reasonably closely match the true population parameters. The difference between the true and estimated parameters is relatively modest and, quite encouragingly, a clear difference in average estimate is evident for the different values of ρ_G. There is, however, substantial variation in the estimates. The third column of Table 10.5 outlines the variation in the estimates across the 100 datasets for each ρ_G value. Indeed, the confidence bands overlap suggesting that, for any given estimate, one could mistake a true ρ_G parameter of 0.15 for a value of 0.45. This fact is also graphically illustrated in Fig. 10.5.

This does not, however, appear to be a problem of biasedness, but rather due to the relative flatness of the maximum-likelihood functions associated with the rarity of default events. The uncertainty in the estimates will fall asymptotically as we increase the breadth and depth of the dataset. Practically, of course, this does not help us much. We may increase the size of our simulated dataset at our leisure, but true default data is typically available only for periods of 20 to 30 years.

The careful reader may have noticed that the standard errors derived from the Fisher information are extremely close to the uncertainty estimates derived from the simulation experiment. This is not an accident. Our simulation experiment basically amounts to a bootstrapping exercise, which is a non-parametric resampling

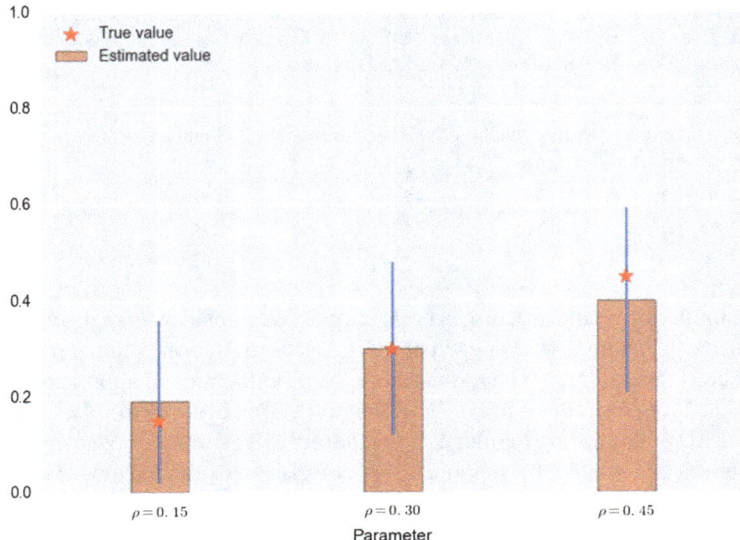

Fig. 10.5 *Single-parameter estimates*: Here we observe the true and estimated parameters for our single-parameter simulation study. The blue lines indicate, using observed noise in optimization results, an indication of the uncertainty in the maximum-likelihood estimates.

technique introduced by Efron (1979).[15] This approach is also considered in significant detail in Chap. 9.

The final column of Table 10.5 provides some insight into our question regarding model convergence. The model does not always converge. On occasion the optimizer fails to find a solution. This may be because it has hit a boundary or the data is so uninformative and the likelihood function so flat that a numerical solution cannot be identified. It does not, however, appear to happen very frequently. Only a handful of the 100 estimations for each value of ρ_G failed to converge and, as such, was excluded from our computations. This also leads us to cautiously conclude that this estimation approach is relatively stable.

10.4.3 Another Example

Although, in principle, any mixture or threshold model permits use of the maximum-likelihood approach, we will only consider one other example: the one-factor

---

[15] Chernick (1999) provides an excellent overview of this technique with a wealth of useful references.

10.4 Likelihood Approach

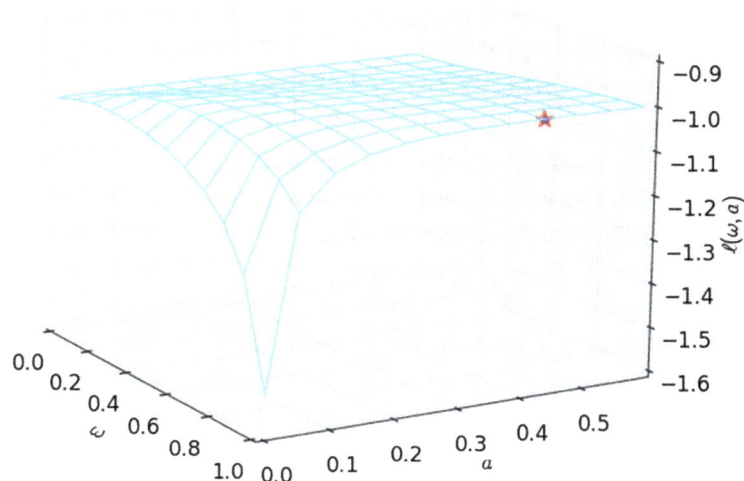

Fig. 10.6 *The CreditRisk+ likelihood surface*: The presence of two CreditRisk+ model parameters directly implies a two-dimensional likelihood function. This allows us to visually determine the regularity of the log-likelihood.

CreditRisk+ implementation. Not only is this a popular industry model, but it also highlights many of the practical issues faced by the quantitative analyst.

The first point is that the one-factor CreditRisk+ model requires estimation of two parameters. The first parameter is ω, which determines the loading on the systematic factor.[16] Broadly speaking, therefore, ω is roughly equivalent to the asset-correlation parameter found in the one-factor, Gaussian model. The second parameter, a, describes both the location and scale of the gamma-distributed systematic risk factor.[17]

As a second point, other than the presence of two parameters and a correspondingly different form for the conditional default probability, the maximum-likelihood function is constructed in precisely the same manner as in the previous example. Figure 10.6 skips right to the chase by examining the two-dimensional, normalized, log-likelihood function. The optimal solution involves a value of a roughly equal to about 0.46, with a ω value approaching unity. It is represented by a bold star in Fig. 10.6.

The optimal solution notwithstanding, Fig. 10.6 demonstrates the usefulness of this diagnostic tool. The likelihood surface is incredibly flat for all values of the a parameter beyond about 0.3. In other words, there is very little information

---

[16]By extension, $1 - \omega$ describes the relative importance of the idiosyncratic element.

[17]Typically, the gamma distribution requires two parameters, but a parameter is lost due to the necessity of constraining the expectation of this random variable to unity. See Chap. 3 for much more detail.

about ω for specific choices of a in this region of the likelihood surface. A similar pattern is evident for most simulated datasets. This lack of regularity suggests that the maximum-likelihood estimator is not terribly informative and that we should probably be somewhat suspicious of our estimates.

It is very possible that there is simply insufficient information in the data to determine both parameters. That is, it is difficult to distinguish between default correlation—as summarized by the parameter ω—and volatility in the default rate, which is captured by the parameter, a. The method-of-moments estimator distinguished between these two aspects, but these ideas are mixed together in the CreditRisk+ likelihood set-up. This is not a general feature of maximum-likelihood estimation, but is rather model specific. In the CreditRisk+ model, implications of the movements in both parameters appear to be observationally equivalent and, as a consequence, the data is not terribly informative about their relative values. Indeed, the one point that the likelihood surface in Fig. 10.6 tells us emphatically is that simultaneously high ω and low a values is sub-optimal. All other parameter combinations seem to be approximately equally probable.

This depressing revelation is, thankfully, not a completely unique situation. On the contrary, it is a common problem. Parameter identification restrictions are a critical element of statistical estimation.[18] There are a variety of techniques, but perhaps the most common and powerful alternative involves fixing one of the parameters. Conceptually, this is like having n equations for $n + 1$ unknowns. If we eliminate a variable, by fixing its values, we have a solvable system of n equations in n unknowns.[19] This reduction of parameter-space dimensionality allows the data to inform the remaining parameters. Fixing parameters is naturally not without some danger and is best done after careful reflection.

There is evidence that practitioner's have also found themselves in this situation and followed a similar approach. Gordy and Lütkebohmert (2007, 2013) discuss the CreditRisk+ model in the context of the regulatory granularity adjustment computations discussed in Chap. 6. This requires an assessment of the parameter, a. They suggest that, although it can be estimated by maximum likelihood, "it is most common to impose an arbitrary round value such as $a = 1$ or to fix the factor loading as $\omega = 1$ and estimate a using the variance of the default frequency." In the same discussion, they suggest the use of the concrete values $a = 0.125$ or $a = 0.25$ for regulatory purposes. Their analysis thus provides some support for the idea of fixing a and indicates a few possible choices.

Figure 10.7 takes the next natural step and illustrates *four* alternative one-dimensional log-likelihood functions for distinct choices of a. We've selected the values of 0.125 and 0.25 suggested by Gordy and Lütkebohmert (2007, 2013) as well as the overall unconstrained optimum of 0.46. Finally, to add an additional perspective a smaller setting of $a = 0.05$ was included. The first, and perhaps most interesting, observation is that all four likelihood functions appear to be quite

---

[18]See, for example, Judge et al. (1985, Chapter 17) or Dhrymes (1994, Chapter 6) for a more rigorous discussion of identification.

[19]Not all systems of n equations and n unknowns have solutions, of course, but this is the basic idea.

10.4 Likelihood Approach

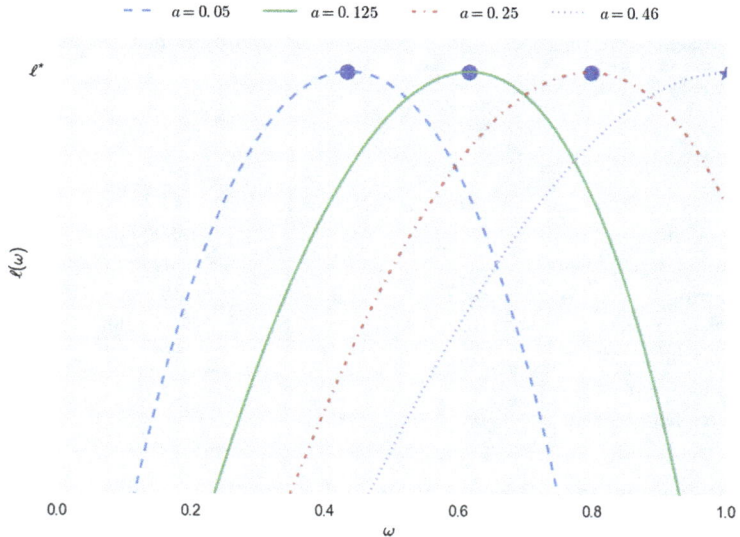

Fig. 10.7 *Fixed a CreditRisk+ likelihood functions*: This graphic illustrates four alternative, normalized, one-dimensional log-likelihood functions for distinct fixed choices of a. This permits us to assess their regularity and degree of informativeness.

regular. The unconstrained optimum likelihood function—that is, $a \approx 0.46$—is truncated by the constraint of $\omega \in [0, 1]$. This is a fairly common situation where modelling constraints conflict with empirical reality.[20]

If you are going to fix a parameter, then it is important to understand the implications of your choice. Table 10.6 performs this task. For each of the four fixed choices of a, it highlights a range of summary statistics permitting a useful comparison. The results are quite enlightening. As we reduce the fixed setting of a, we simultaneously reduce the overall maximum-likelihood function value and the standard error of our ω estimate. Moreover, the amount of weight on the systematic factor falls in conjunction with the value of a. Based on the goodness of fit to the sample moments and implicit default correlation and the degree of statistical certainty, a setting of $a = 0.05$ seems sensible.

This is, however, a bit misleading. As we arbitrarily reduce the value of a, the performance of the model deteriorates. At $a = 0.01$, for example, the optimizer no longer converges. The solution, therefore, is not as simple as it might appear in Table 10.6. Selecting a very low level of a seems as unsatisfying as using the relatively unconstrained value. This explains the, by most standards quite defensible, regulatory choice of setting $a = 0.125$ or $a = 0.25$. This admittedly

---

[20]The reader might be thinking that, since this is a simulated dataset, there is no empirical reality. While true, the actual dataset was generated from an alternative model, which is not unlike the situation faced by real-life statisticians.

Table 10.6 *CreditRisk+ MLE results*: This table summarizes the key results from use of the maximum-likelihood estimate of ω from our simulated data using the four fixed value of a highlighted in Fig. 10.7.

Quantity	Fixed a value			
	0.05	0.125	0.25	0.455
ω^*	0.44	0.62	0.80	1.00
$\ell(\omega^*)$	−23.6	−23.3	−23.2	−23.2
Score function	0.00008	0.00042	−0.00061	−0.00151
Observed Fisher information	2130.2	79.2	11.1	9.0
Standard error	0.02	0.11	0.30	0.33
95% confidence interval	[0.40,0.48]	[0.40,0.84]	[0.21,1.00]	[0.35,1.00]
Average default probability	34.7	34.8	34.8	34.8
Average default volatility	66.6	60.8	55.4	51.5
Default correlation coefficient	0.013	0.011	0.009	0.008

feels unsatisfying and subjective. It is, unfortunately, a reality for quantitative analyst working with short datasets involving rare events. Complete objectivity is not available and consequently, the quantitative analyst is forced to employ good judgement. It also argues for transparency in the parameter selection process to ensure that all stakeholders have some insight—and, optimally, a certain degree of input—into the steps involved.

10.4.4 A More Complicated Situation

There are other ways that additional parameters can enter into the analysis. Returning to the one-factor Gaussian setting, it may very well be the case that we expect to have more than a single asset-correlation parameter. This could arise in a few different ways. First, in the most general case, each individual counterparty could be assigned its own parameter, $\rho_{G,n}$. Given the difficulties of identifying a single parameter, it is rather improbable that we would have much success with such a parametrization. One could, however, easily imagine a more modest, restricted parametric implementation. This could involve assigning each obligor to different groups with varying degrees of asset correlation. A second reason for a multivariate parameter space naturally arises in a multi-factor setting. Demey et al. (2004) offer a number of clever modifications and parametric restrictions to handle this situation.

In this final section pertaining to the classic maximum-likelihood approach, we will remain in the one-factor Gaussian setting, but relax the single asset-correlation parameter assumption. This will provide some insight into the additional steps required in a more general, multi-factor setting without adding too much complexity. These ideas can be generalized to handle different regions, industries, countries, or even firm size.

Gordy and Heitfield (2002) examine the implications of assigning a different asset-correlation parameter to each distinct credit rating. While plausible, we would

10.4 Likelihood Approach

argue that other asset-correlation dimensions are also interesting and, perhaps, more practically pertinent. In our example, therefore, we will permit different asset-correlations between *two* distinct geographical regions. In our simple four-state Markov-chain example, this amounts to two asset-correlation parameters: $\rho_{G,1}$ and $\rho_{G,2}$. Naturally, within practical limits, one can have as many regions as one would like. This is the analyst's choice and will involve trade-offs between realism and complexity of implementation. We keep the dimensionality low to permit graphical visualization and a manageable amount of computational time.

Let us first develop the necessary results in generality. Consider R regions, each possessing its own asset-correlation coefficient. This requires a few adjustments to our framework. As a start, we need to modify the dynamics of our latent state variable. It depends not only on the counterparty, but also on the geographical region, r. This leads to the following restatement of equation 10.12 as follows

$$y_{n,t}(r) = \sqrt{\rho_{G,r}} G_t + \sqrt{1 - \rho_{G,r}} \epsilon_{n,t}. \tag{10.37}$$

Quite simply, therefore, the latent state variable also depends on the obligor's regional status. There are thankfully no changes to the structure of the G_t and $\epsilon_{n,t}$ random variables.

We also need to perform our counting by region, rather than in aggregate. This merely requires a slight adjustment to our notation. In each period, for each credit class, we observe $N_{r,t}$ credit counterparties and $k_{r,t}$ defaults in that category. The default rate, for the rth category, is simply

$$\hat{p}_{r,t} = \frac{k_{r,t}}{N_{r,t}}, \tag{10.38}$$

for $r = 1, \ldots, R$. If we sum over the credit counterparties, we retrieve the previously N_t and k_t quantities. That is,

$$k_t = \sum_{r=1}^{R} k_{r,t}, \tag{10.39}$$

and

$$N_t = \sum_{r=1}^{R} N_{r,t}. \tag{10.40}$$

These are essentially just book-keeping items, but it is important that these details are correctly handled.

Implicit in our previous development, we assumed that the default probability, by each counterparty, was a constant over time. Since we used only a single default probability, which was an average across all counterparties at time t, we need to compute a weighted average for each period. This average is defined in equation 10.26. Given that we now treat each region as a distinct group, we need to generalize this approach.[21] The unconditional default probability of region r at time t is approximated by,

$$\bar{p}_{r,t} = \frac{1}{N_{r,t}} \sum_{n=1}^{N_{r,t}} p_{n,r,t}, \qquad (10.41)$$

where $p_{n,r,t}$ is the default probability of the nth obligor in the rth region at time t. The conditional default probability for region r at time t, given a particular value of $G = g$, is now simply,

$$p_{r,t}(g) = \Phi\left(\frac{\Phi^{-1}(\bar{p}_{r,t}) - \sqrt{\rho_{G,r}}g}{\sqrt{1-\rho_{G,r}}}\right). \qquad (10.42)$$

The conditional probability of default, therefore, now depends on the systematic variable *and* one's region. Each region is simply loading onto the global risk factor in different ways.

This leads to the following (conditional) binomial cumulative mass function for the rth region at time t,

$$f_{r,t}(k_{r,t}, N_{r,t}; \rho_r|g) = \binom{N_{r,t}}{k_{r,t}} \Phi\left(\frac{\Phi^{-1}(\bar{p}_{r,t}) - \sqrt{\rho_{G,r}}g}{\sqrt{1-\rho_{G,r}}}\right)^{k_{r,t}} \qquad (10.43)$$

$$\times \left(1 - \Phi\left(\frac{\Phi^{-1}(\bar{p}_{r,t}) - \sqrt{\rho_r}g}{\sqrt{1-\rho_r}}\right)\right)^{N_{r,t}-k_{r,t}},$$

$$= \binom{N_{r,t}}{k_{r,t}} p_{r,t}(g)^{k_{r,t}} \left(1 - p_{r,t}(g)\right)^{N_{r,t}-k_{r,t}}.$$

Given the value of g, the default outcomes are (conditionally) independent across all counterparties—this includes regions. As a consequence, the full joint cumulative mass function, at time t, is the product of equation 10.43 over all geographical regions. More specifically,

$$f_t(k_t, N_t; \boldsymbol{\rho}|g) = \prod_{r=1}^{R} \binom{N_{r,t}}{k_{r,t}} p_{r,t}(g)^{k_{r,t}} \left(1 - p_{r,t}(g)\right)^{N_{r,t}-k_{r,t}}, \qquad (10.44)$$

where $\boldsymbol{\rho} = \begin{bmatrix} \rho_{G,1} & \cdots & \rho_{G,R} \end{bmatrix}$ denotes the parameter space.

We can recover the unconditional joint cumulative mass function by merely integrating out, as usual, the systematic variable. This remains a one-dimensional

---

[21] If, as in Gordy and Heitfield (2002), we used the credit-rating, this would not be necessary.

10.4 Likelihood Approach

problem, because all credit categories depend on a single global factor. The result is,

$$f_t(k_t, N_t; \rho) = \underbrace{\int_{\mathbb{R}} \left(\prod_{r=1}^{R} \binom{N_{r,t}}{k_{r,t}} p_{r,t}(g)^{k_{r,t}} \left(1 - p_{r,t}(g)\right)^{N_{r,t}-k_{r,t}} \right) \phi(g) dg}_{\text{Equation 10.44}}. \quad (10.45)$$

The full joint default density function—collecting all of the terms over time—is thus simply,

$$f(\mathbf{k}, \mathbf{N}; \rho) = \prod_{t=1}^{T} \underbrace{\int_{\mathbb{R}} \left(\prod_{r=1}^{R} \binom{N_{r,t}}{k_{r,t}} p_{r,t}(g)^{k_{r,t}} \left(1 - p_{r,t}(g)\right)^{N_{r,t}-k_{r,t}} \right) \phi(g) dg}_{\text{Equation 10.45}}. \quad (10.46)$$

The associated log-likelihood kernel—computed by eliminating multiplicative constants from the Gaussian density—follows directly as,

$$\ell(\rho|\mathbf{k}, \mathbf{N}) = \sum_{t=1}^{T} \ln \left(\int_{\mathbb{R}} \left(\prod_{r=1}^{R} \binom{N_{r,t}}{k_{r,t}} p_{r,t}(g)^{k_{r,t}} \left(1 - p_{r,t}(g)\right)^{N_{r,t}-k_{r,t}} \right) e^{\frac{-g^2}{2}} dg \right). \quad (10.47)$$

The good news is that the product remains *inside* the integral, which does not add much numerical complexity. Indeed, it is not much more computationally expensive to evaluate than the single ρ parameter equivalent found in equation 10.35. Finding the vector, $\rho \in \mathbb{R}^R$, that maximizes equation 10.47 will be somewhat more difficult, but still computationally manageable.[22]

The biggest issue, however, is that by partitioning the credit obligors into R distinct regional groups, the number of default observations will be, by construction, even smaller in each sub-category. Defaults are, in aggregate, a rarity. As we partition them into sub-categories—be they regions, industries, or credit classes—the already meagre amount of default observations are spread more thinly among our parameters. This is in stark contrast to market-risk analysis, where the richness of our data permits extreme focus on a broad range of market segments. Again, the quantitative analyst faces a trade-off between realism and complexity and data informativeness. A rich high-dimensional model may have enormous conceptual appeal, but if one cannot reliably or accurately determine its parameters, then its usefulness is dramatically hampered.

---

[22] As R increases, however, it will become increasingly difficult and time-consuming to maximize the likelihood kernel.

From an algorithmic perspective, nothing particularly new is required. We do, however, need to simulate a new dataset, since our last collection of simulated data involved only a single parameter. The only change required to Algorithm 10.2 is to allocate the obligors to a given region at inception and generate a two-dimensional latent state variable. With these slight changes, we generate a second dataset with two regions: the first and second regions are assigned a 0.15 and 0.45 asset-correlation coefficient, respectively. We also place roughly 70% of our initial 100-obligor portfolio into the first region with the remainder in the second. This implies that, by virtue of its larger dataset, we have more information about the first region.

Figure 10.8 displays both the number of obligors and the number of defaults across the simulated 30-year period. Technically, the first region experiences more defaults, but also has more obligors. This has an important influence on the sample moments. Both regions have a sample default rate of about 50 basis points; the default-rate volatility in the first region is about 90 basis points, but more than 160 basis points in the second region. There is also some empirical evidence of higher default correlation in the second region. One of three periods with default events involves multiple defaults. In the first region, conversely, only one of eight periods exhibiting defaults has more than a single default.

The construction of the log-likelihood function—following from equation 10.47—requires a small amount of extra organization and effort. Algorithm 10.8 summarizes three Python functions that handle this task. The first, getProdCMF, takes vector input arguments and constructs the distinct elements in the sum inside the integral. getCMF2r collects these terms, takes their product, and inserts the Gaussian density function. Finally, logL2r, extracts this quantity for each period, $t = 1, \ldots, T$, and builds the log-likelihood function. Again, since

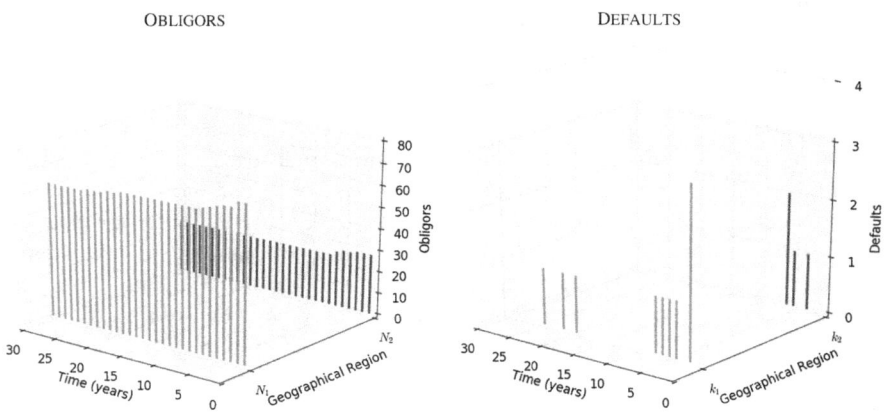

Fig. 10.8 *Regional default data*: These two figures illustrate the number of obligors, across our 30-year simulated dataset, and the incidence of default by counterparty and time period. In the vast majority of periods, no default event occurs.

10.4 Likelihood Approach

Algorithm 10.8 Two-parameter regional log-likelihood functions

```
def getProdCMF(g,myRho,myP,myN,myK):
    pg = th.computeP(myP,myRho,g)
    return np.multiply(util.getBC(myN,myK),np.power(pg,myK)*np.power(1-pg,myN-myK))
def getCMF2r(g,myRho,pVec3,nVec3,kVec3):
    myF=getProdCMF(g,myRho,pVec3,nVec3,kVec3)
    return np.prod(myF)*util.gaussianDensity(g,0,1)
def logL2r(x,T,pMat,nMat,kMat):
    L = np.zeros(T)
    for t in range(0,T):
        L[t],err = nInt.quad(getCMF2r,-5,5,args=(x,pMat[t,:],nMat[t,:],kMat[t,:]))
    return -np.sum(np.log(L))
```

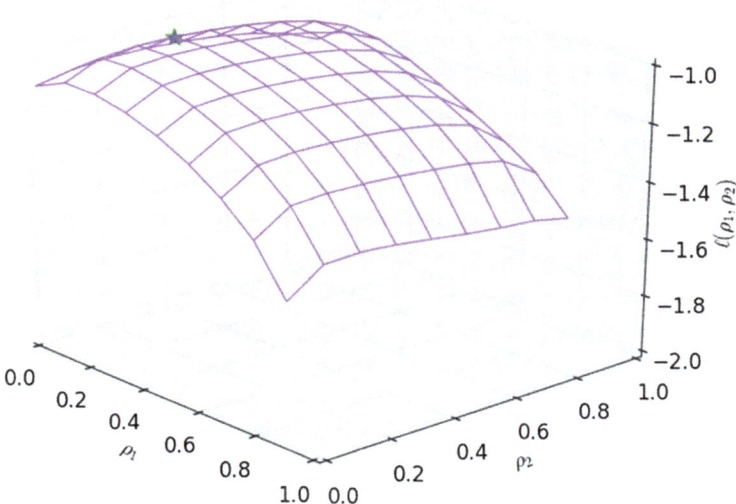

Fig. 10.9 *The regional likelihood surface*: Given that we have a two-dimensional likelihood function, we may plot the overall likelihood surface. This allows us to visually determine the regularity of the log-likelihood.

we will be using `scipy` minimizers to maximize our log-likelihood, we apply a negative sign to the objective function.

As counselled on many occasions, the first order of business is to visually examine the likelihood function. Let's begin with the two-dimensional view. Figure 10.9 illustrates the regional, normalized, log-likelihood surface. While the neighbourhood of its maximum appears relatively flat, the function does slope away from this point in all directions. In other words, it appears to be reasonably regular.

Although we find ourselves in the fortunate position of being able to examine our objective function in three dimensions, additional information is also provided by the more generally applicable profile likelihoods. This notion can be used irrespective of the problem's dimensionality. Figure 10.10 summarizes these values

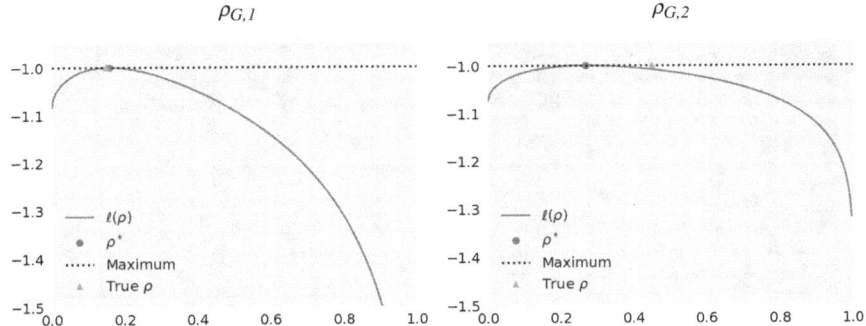

Fig. 10.10 *Regional profile likelihoods*: As a complement to the full two-dimensional likelihood, illustrated in Fig. 10.9, this figure highlights the profile likelihoods for each parameter. Profile likelihoods are computed by tracing out the likelihood function for all values of one parameter, while holding other parameters fixed at their maximum-likelihood estimator.

Table 10.7 *Regional single-dataset results*: This table summarizes the key results from use of the maximum-likelihood estimate of ρ_1 and ρ_2 from the geometric results summarized in Figs. 10.8 and 10.9.

Quantity	$\rho_{G,1}$	$\rho_{G,2}$
True parameter value	0.15000	0.45000
Grid search optimum	0.11189	0.33367
Hill-climber optimum	0.16047	0.26856
Score function	0.00005	0.00008
Observed Fisher information	113.57	27.77
Standard error	0.09	0.19
95% confidence interval	[0.00,0.33]	[0.08,0.82]

for both regional parameters. While both appear to have an approximately parabolic form—and can thus be considered regular problems—their flatness around the maxima differs importantly. In particular, the second region's asset-correlation parameter appears much less informative than the first region's equivalent. This suggests that not only do we expect a less accurate estimate of $\rho_{G,2}$, but that it will receive a much broader confidence interval.

Table 10.7 verifies this suspicion with a summary of the estimation results. The maximum-likelihood estimate of 0.16 is reassuringly close to the true asset-correlation parameter of 0.15. The same cannot be said for the second region. The optimal estimate is about 0.27, which is quite far from the true value. If one re-examines Fig. 10.10, this is not terribly surprising. The profile likelihood is quite flat between these two points. This is also reflected in the Fisher information and standard-error estimates. The second region has less than one third of the observed Fisher information and twice the standard error relative to the first region. The culprit is the smaller overall amount of data and the correspondingly smaller number of defaults.

How does this address the inherent default correlation in this model? This is a difficult question to answer definitively, but we can make a reasonable approximation.

10.4 Likelihood Approach

To compute the joint default probability—and hence the default correlation—one requires both default probabilities and asset-correlation coefficients. The true and estimated ρ values are clear. To estimate the default probabilities, we take the average values for each region at inception—in both cases, these are in the neighbourhood of about 50 basis points. Using the true, unobserved, asset-correlation figures, the default-correlation matrix is,

$$\begin{bmatrix} \rho\left(\mathbb{I}_{\mathcal{D}_{r_1}}, \mathbb{I}_{\mathcal{D}_{r_1}}\right) & \rho\left(\mathbb{I}_{\mathcal{D}_{r_1}}, \mathbb{I}_{\mathcal{D}_{r_2}}\right) \\ \rho\left(\mathbb{I}_{\mathcal{D}_{r_2}}, \mathbb{I}_{\mathcal{D}_{r_1}}\right) & \rho\left(\mathbb{I}_{\mathcal{D}_{r_2}}, \mathbb{I}_{\mathcal{D}_{r_2}}\right) \end{bmatrix} = \begin{bmatrix} 0.010 & 0.027 \\ 0.027 & 0.081 \end{bmatrix}, \quad (10.48)$$

where $\rho\left(\mathbb{I}_{\mathcal{D}_{r_i}}, \mathbb{I}_{\mathcal{D}_{r_j}}\right)$ denotes the default correlation between obligors i and j.[23] The matrix of results is quite intuitive. If both obligors are in the first region, they share a 0.01 default correlation. Should they both fall in the second region, this rises significantly to about 0.08. If two obligors are in different regions, the default correlation coefficient is roughly 0.03. This diagnostic tool provides us with a better understanding of the implications of these parameter values for our real quantity of interest: default correlation.

We can now repeat this exercise for the estimated parameters in Table 10.7. The results are,

$$\begin{bmatrix} \hat{\rho}\left(\mathbb{I}_{\mathcal{D}_{r_1}}, \mathbb{I}_{\mathcal{D}_{r_1}}\right) & \hat{\rho}\left(\mathbb{I}_{\mathcal{D}_{r_1}}, \mathbb{I}_{\mathcal{D}_{r_2}}\right) \\ \hat{\rho}\left(\mathbb{I}_{\mathcal{D}_{r_2}}, \mathbb{I}_{\mathcal{D}_{r_1}}\right) & \hat{\rho}\left(\mathbb{I}_{\mathcal{D}_{r_2}}, \mathbb{I}_{\mathcal{D}_{r_2}}\right) \end{bmatrix} = \begin{bmatrix} 0.011 & 0.018 \\ 0.018 & 0.030 \end{bmatrix}. \quad (10.49)$$

Since our estimate of $\rho_{G,1}$ is virtually the same as the true population parameter, the first region's dependence is extremely close to equation 10.48. The low estimate of $\rho_{G,2}$, conversely, implies inter- and intraregional default correlation involving the second region is substantially lower.

We can, using the 95% confidence intervals in Table 10.7, take this analysis a step further by computing corresponding interval estimates for the regional default-correlation coefficients. The results are,

$$\begin{bmatrix} \hat{\rho}\left(\mathbb{I}_{\mathcal{D}_{r_1}}, \mathbb{I}_{\mathcal{D}_{r_1}}\right) & \hat{\rho}\left(\mathbb{I}_{\mathcal{D}_{r_1}}, \mathbb{I}_{\mathcal{D}_{r_2}}\right) \\ \hat{\rho}\left(\mathbb{I}_{\mathcal{D}_{r_2}}, \mathbb{I}_{\mathcal{D}_{r_1}}\right) & \hat{\rho}\left(\mathbb{I}_{\mathcal{D}_{r_2}}, \mathbb{I}_{\mathcal{D}_{r_2}}\right) \end{bmatrix} \in \begin{bmatrix} [0.000, 0.040] & [0.000, 0.140] \\ [0.000, 0.140] & [0.005, 1.000] \end{bmatrix}. \quad (10.50)$$

The first region's default-correlation coefficient is expected, with a high degree of probability, to range from independent to 0.04. While this is not fantastic, it is a tight range in comparison to the second region. In that case, it is basically expected to fall in the unit interval. The punchline is that while asset-correlation coefficients are somewhat difficult to interpret, their uncertainty spills over to the default correlation.

---

[23] This follows directly from equation 10.9.

Table 10.8 *Regional simulation study results*: This table illustrates the results of estimating our two regional ρ parameters—$\rho_{G,1} = 0.15$ and $\rho_{G,2} = 0.45$—using the maximum-likelihood approach with 100 simulated datasets.

Geographical region	Parameter			Score function	Fisher information	Standard error	Success ratio
	$\rho_{G,r}$	$\hat{\rho}^*_{G,r}$	$\sigma(\rho^*_{G,r})$				
1	0.15	0.31	0.22	0.00079	71.3	0.18	76.0%
2	0.45	0.44	0.23	0.00286	35.2	0.29	71.0%

Confidence bands on these quantities are meaningful and should be well understood. In this case, the dataset is simply too small to be informative about the second region's asset-correlation parameter.

The final question addressed in this section is: how representative is this dataset? Our dataset can be viewed as a single realization; perhaps it is particularly uninformative. Another simulation study is thus warranted to confront this reasonable question. Table 10.8 summarizes the results of 100 simulated datasets with the same structure and parameters as the single realization examined in Table 10.7. The results are interesting, informative, and ultimately a bit disappointing. The second parameter, $\rho_{G,2}$, has an average value that coincides almost exactly with the true population parameter. The average estimate of $\rho_{G,1}$, however, is almost twice its true value. There thus appears to be a bias in the estimation of the first parameter. Interestingly, the average Fisher observed information is significantly higher for the first parameter. This is consistent with the previous analysis and the consequence of a higher number of obligors and thus data for this region. It is also noteworthy that the standard-deviation of the 100 simulation-study estimates agrees quite closely with the likelihood framework standard-error figures. The former are essentially bootstrap standard-error estimates from Efron (1979)'s algorithm. This is a form of confirmation of the regularity of the underlying likelihood functions.

The ultimate lesson from this simulation study is unsurprising. When working with a limited dataset, the likelihood functions are quite flat and there is a correspondingly high degree of parameter uncertainty. Figure 10.11 underscores this point by illustrating our average estimates along with a 95% confidence interval. These problems dissipate as we increase the amount of available data, but this is unfortunately hardly realistic. The objective of this chapter is not only to examine a variety of estimation techniques for different models, but to do so under honest and realistic circumstances.

This maximum-likelihood estimation technique exploits conditional independence of the default events by integrating out the systematic factor and counting the number of default events. Not only does it work relatively well, but it generalizes effectively to multiple parameters. As we've seen, one can easily distinguish between obligor size, country, region, credit-state, or industry. Although it can be employed in a multi-factor setting, there are, at least, *two* practical issues. First, we run into serious computational problems as the number of state variables increases; as we move to three and four dimension numerical integration, the

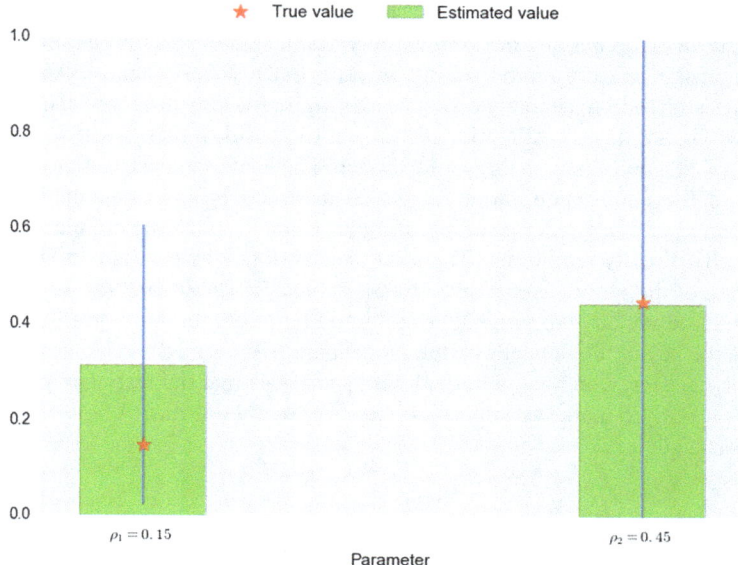

Fig. 10.11 *Regional estimates*: This figure illustrates graphically the main results of the regional simulation study found in Table 10.8.

problem becomes virtually insoluble. The second challenge, as we experienced in the previous example, is that adding parameters leads to spreading out an already small number of defaults. This makes it harder to identify each individual parameter. This creates trade-offs.[24] In the next section, we will investigate another approach that offers potential solutions to these problems. We will stay in the likelihood setting, but seek to expand our perspective with respect to our dataset.

10.5 Transition Likelihood Approach

In the previous section, we used default outcomes to inform the asset-correlation parameters. One of the drawbacks of this approach is the rarity of default events. Although the maximum-likelihood estimators are asymptotically consistent, we operate in small-sample settings, where a real possibility of biasedness exists. Another approach attempts to address this shortcoming by using transition data—including defaults—to determine the asset-correlation parameters. In many ways, this is a generalization of the previous method, but it only really applies within the

---

[24]Some useful studies that address these issues, to a certain extent, include Hashimoto (2009) and Martin (2013).

context of structural threshold models. The reason is that we require a link between the transition outcomes and the systematic-variable outcome. This can, in principle, be done in the mixture-model setting as well, but it is less natural. As we'll see shortly, it is already reasonably difficult and fragile for threshold models.

We will examine this estimation technique, to provide a bit of variety, within the context of the one-factor t-threshold model. The method makes use of the joint density of the latent state variables, which naturally brings us to the associated copula function. For implementation of the threshold model, the copula function is not actually directly required.[25] The joint distribution, however, is pertinent for the estimation of the model asset-correlation parameters. The idea is to use the copula function to derive an alternative log-likelihood function; the joint density function, which is the key input, depends on the model parameters and the response variables. The tricky aspect, which we will deal with shortly, is that the response variables—our global or systematic variables—are *not* observable. It has to be inferred from the transition data.

10.5.1 The Elliptical Copula

The construction of the log-likelihood function makes use of the copula density and not the copula function itself. Obtaining the copula density is not completely straightforward. The first step, therefore, is to map out its derivation. To derive the copula density, we begin by introducing a set of random variables, $X_1, \ldots, X_N$, each with marginal distributions,

$$\mathbb{P}(X_n \leq x_n) = F_n(x_n), \tag{10.51}$$

for $n = 1, \ldots, N$. Let F be the N-dimensional joint cumulative distribution function for this collection of random variables. It has the following general form,

$$\mathbb{P}(X_1 \leq x_1, \ldots, X_N \leq x_N) = F(x_1, \ldots, x_N). \tag{10.52}$$

In this setting, a famous result from Sklar (1959), holds that there exists an N dimensional mapping from the unit cube to the unit interval, $C : [0, 1]^N \to [0, 1]$, such that for all x in the domain of F,

$$F(x_1, \ldots, x_N) = C\bigg(F_1(x_1), \ldots, F_N(x_N)\bigg). \tag{10.53}$$

C is termed a copula function and it forms the fundamental link between the marginal and joint distributions. The original source, Sklar (1959), is quite brief. It has been extensively discussed and explored in the statistical and, also increasingly, in the financial literature. See Sklar (1996), McNeil et al. (2015, Chapter 7), Bouyé

---

[25] It is, of course, implicit in the calculations.

10.5 Transition Likelihood Approach

et al. (2000), Embrechts et al. (2001), or Yan (2007) for a small subset of the work in this area.

Let's continue to remain fairly abstract. In the threshold setting, the marginals of the model-response variables obey standard, univariate distributions as follows,

$$F_n(x_n) = \int_{-\infty}^{x} dF_n(u)du, \qquad (10.54)$$

$$= \int_{-\infty}^{x} f_n(u)du,$$

for $n = 1, \ldots, N$. Following our discussion in Chap. 4, the form of each F_n and f_n may be Gaussian, t, variance-gamma, or generalized hyperbolic. By definition, irrespective of its ultimate form, each $F_n(x_n) \in [0, 1]$ maps the response variable, x_n into the unit interval. This is, as we saw earlier, how the threshold model determines default, or migration, of a given obligor over a particular period. The specific choice of C depends on the desired choice of joint distribution.

In this brief discussion, we will focus on the family of elliptical distributions—this will permit us to freely consider both the Gaussian and the t-distributed cases. In this case, our copula function may be, following from Fang et al. (1990), written as,

$$C\bigg(F_1(x_1), \ldots, F_N(x_N)\bigg) = \check{F}\bigg(\check{F}_1^{-1}(F_1(x_1)), \ldots, \check{F}_N^{-1}(F_N(x_N))\bigg), \quad (10.55)$$

where $\check{F}$ denotes the multivariate elliptical distribution function, $\{\check{F}_n; n = 1, \ldots, N\}$ represent the marginals of the elliptical distribution, and $\{\check{F}_n^{-1}; n = 1, \ldots, N\}$ are its inverse-cumulative-distribution (or quantile) functions.[26]

The relationship between a density and cumulative distribution functions is well known. In particular, for a random variable X with an absolutely continuous probability measure,[27]

$$\mathbb{P}(X \leq x) = F(x), \qquad (10.56)$$

$$F(x) = \int_{-\infty}^{x} f(u)du,$$

$$\frac{\partial F(x)}{\partial x} = \frac{\partial}{\partial x}\left(\int_{-\infty}^{x} f(u)du\right),$$

$$dF(x) = f(x).$$

---

[26] It is clear that there is some simplification if $\check{F}_n \equiv F_n$ for $n = 1, \ldots, N$; we will address this point shortly.

[27] For much more technical rigour on the notion of absolute continuity, see Billingsley (1995)

In our multivariate setting, this generalizes to the following expression,

$$c(x_1,\ldots,x_N) = \frac{\partial C\left(F_1(x_1),\ldots,F_N(x_N)\right)}{\partial x_1 \cdots \partial x_N}, \tag{10.57}$$

$$= \frac{\partial \check{F}\left(\check{F}_1^{-1}(F_1(x_1)),\ldots,\check{F}_N^{-1}(F_N(x_N))\right)}{\partial x_1 \cdots \partial x_N},$$

$$= \frac{\partial}{\partial x_1 \cdots \partial x_N}\left(\int_{-\infty}^{\check{F}_1^{-1}(F_1(x_1))} \cdots \int_{-\infty}^{\check{F}_N^{-1}(F_N(x_N))} \check{f}(y_1,\ldots,y_N)dy_1 \cdots dy_N\right),$$

$$= \check{f}\left(\check{F}_1^{-1}(F_1(x_1)),\ldots,\check{F}_N^{-1}(F_N(x_N))\right)\left(\prod_{n=1}^{N} \frac{\partial \check{F}_n^{-1}(F_n(x_n))}{\partial F_n(x_n)} \frac{\partial F_n(x_n)}{\partial x_n}\right),$$

$$= \check{f}\left(\check{F}_1^{-1}(F_1(x_1)),\ldots,\check{F}_N^{-1}(F_N(x_N))\right)\left(\prod_{n=1}^{N} \frac{\partial \check{F}_i^{-1}(F_n(x_n))}{\partial F_n(x_n)} f_n(x_n)\right),$$

where the result follows from a liberal application of the fundamental theorem of calculus, the chain rule, and the properties of distribution and density functions. We are, however, not quite complete. The derivative of the inverse cumulative distribution function ($\check{F}_i^{-1}$) requires some more work. It may be solved as follows,

$$y = \check{F}_n^{-1}(F_n(x_n)), \tag{10.58}$$

$$\check{F}_n(y) = \check{F}_n\left(\check{F}_n^{-1}(F_n(x_n))\right),$$

$$\check{F}_n(y) = F_n(x_n),$$

$$\check{F}_n'(y) = \frac{dF_n(x_n)}{dy},$$

$$\frac{1}{\check{F}_n'(y)} = \frac{dy}{dF_n(x_n)},$$

$$\frac{d\check{F}_n^{-1}(F_n(x_n))}{dF_n(x_n)} = \frac{1}{\check{F}_n'\left(\check{F}_n^{-1}(F_n(x_n))\right)},$$

$$\frac{\partial \check{F}_n^{-1}(F_n(x_n))}{\partial F_n(x_n)} = \frac{1}{\check{f}_n\left(\check{F}_n^{-1}(F_n(x_n))\right)}.$$

This allows us to update equation 10.57 to arrive at a generic expression for the elliptical copula density with, as yet unspecified, one-dimensional elliptical

10.5 Transition Likelihood Approach

distributions for the marginals:

$$c(x_1,\ldots,x_N) = \check{f}\left(\check{F}_1^{-1}(F_1(x_1)),\ldots,\check{F}_N^{-1}(F_N(x_N))\right)\left(\prod_{n=1}^{N}\frac{f_n(x_n)}{\check{f}_n\left(\check{F}_n^{-1}(F_n(x_n))\right)}\right). \quad (10.59)$$

At first glance, the elliptical copula density may appear somewhat intimidating. It contains an extensive number of mathematical objects. In particular, it includes:

1. the multivariate elliptical cumulative density function, $\check{f}$;
2. the inverse elliptical marginal cumulative distribution functions, $\{\check{F}_n^{-1}; n = 1,\ldots,N\}$;
3. the elliptical marginal density functions, $\{\check{f}_n; n = 1,\ldots,N\}$;
4. the marginal cumulative distribution functions, $\{F_n; n = 1,\ldots,N\}$; and
5. the marginal density functions, $\{f_n; n = 1,\ldots,N\}$.

The elliptical copula density thus clearly includes a broad range of abstract ingredients. At this point, therefore, it is constructive to make the discussion somewhat more tangible. Motivated by an interesting exposition in Bluhm et al. (2003, Chapter 2), we will consider three cases: joint and marginal Gaussian distributions, joint and marginal t distributions, and a joint t but marginal Gaussian density. This provides a reasonable and practical overview of a range of possible model implementations.

Let's examine the first case. As we'll see, the complexity of equation 10.59 may be maintained, or dramatically reduced, depending on one's choice of elliptical copula function. One of the key benefits of the elliptical family of distributions is that they are all described by a correlation matrix. We set $\check{f}$ to the multivariate Gaussian distribution with correlation matrix, R_θ—we'll denote it as $f_{\mathcal{N},R}$. For notational simplicity, we will also refer to $R_\theta \equiv R$, but it is important to underscore that this is where the parameter vector, θ, enters into the problem. Finally, each $\check{F}_n$ and F_n distribution—and their associated densities $\check{f}_n$ and f_n—also follow standard normal distributions for each $n = 1,\ldots,N$. This yields the following result

$$c(y_1,\ldots,y_N) = \check{f}\left(\check{F}_1^{-1}(F_1(y_1)),\ldots,\check{F}_N^{-1}(F_N(y_N))\right)\left(\prod_{n=1}^{N}\frac{f_n(y_n)}{\check{f}_n\left(\check{F}_n^{-1}(F_n(y_n))\right)}\right), \quad (10.60)$$

$$= f_{\mathcal{N},R}\left(\Phi^{-1}(\Phi(y_1)),\ldots,\Phi^{-1}(\Phi(y_N))\right)\left(\prod_{n=1}^{N}\frac{\phi(y_n)}{\phi(\Phi^{-1}(\Phi(y_n)))}\right),$$

$$= f_{\mathcal{N},R}(y_1,\ldots,y_N)\left(\prod_{n=1}^{N}\frac{\phi(y_n)}{\phi(y_n)}\right),$$

$$= f_{\mathcal{N},R}(y_1,\ldots,y_N),$$

$$= \frac{1}{\sqrt{(2\pi)^N|R|}}\exp\left(-\frac{y^T R^{-1} y}{2}\right),$$

where $y = \begin{bmatrix} y_1 & \cdots & y_N \end{bmatrix}^T$ is the vector of latent Gaussian state (or model-response) variables and $|\cdot|$ denotes the determinant of a matrix. This result is rather less intimidating since, as one would expect, the Gaussian copula density essentially amounts to the multivariate normal density function. Note we have replaced the generic x_n notation to y_n to be consistent with the usual symbols used thus far to describe the threshold state variables.

The second case is not much more complex. We specify a multivariate t distribution, $f_{\mathcal{T}_\nu, R}$, with ν degrees of freedom as the joint distribution. As a member of the elliptical distribution, it also has a correlation matrix, R. Analogous to the previous case, the marginal distributions, and their densities, each obey univariate t laws; they are written as $F_{\mathcal{T}_\nu}$ and $f_{\mathcal{T}_\nu}$, respectively. This leads to a similar pattern of cancellation,

$$c(y_1, \ldots, y_N) = \breve{f}\left(\breve{F}_1^{-1}(F_1(y_1)), \ldots, \breve{F}_N^{-1}(F_N(y_N))\right) \left(\prod_{n=1}^{N} \frac{f_n(y_n)}{\breve{f}_n\left(\breve{F}_n^{-1}(F_n(y_n))\right)}\right), \quad (10.61)$$

$$= f_{\mathcal{T}_\nu, R}\left(F_{\mathcal{T}_\nu}^{-1}(F_{\mathcal{T}_\nu}(y_1)), \ldots, F_{\mathcal{T}_\nu}^{-1}(F_{\mathcal{T}_\nu}(y_N))\right) \left(\prod_{n=1}^{N} \frac{f_{\mathcal{T}_\nu}(y_n)}{f_{\mathcal{T}_\nu}(F_{\mathcal{T}_\nu}^{-1}(F_{\mathcal{T}_\nu}(y_n)))}\right),$$

$$= f_{\mathcal{T}_\nu, R}(y_1, \ldots, y_N) \left(\prod_{n=1}^{N} \frac{\cancel{f_{\mathcal{T}_\nu}(y_n)}}{\cancel{f_{\mathcal{T}_\nu}(y_n)}}\right),$$

$$= f_{\mathcal{T}_\nu, R}(y_1, \ldots, y_N),$$

$$= \frac{\Gamma\left(\frac{\nu+N}{2}\right)}{\Gamma\left(\frac{\nu}{2}\right)\sqrt{(\nu\pi)^N |R|}} \left(1 + \frac{y^T R^{-1} y}{\nu}\right)^{-\frac{(\nu+N)}{2}},$$

where y is now a vector of latent t-distributed state variables and $\Gamma(\cdot)$ denotes the gamma function. As suggested by Kotz and Nadarajah (2004), we will use the standard multivariate t-distribution—there are, it turns out, other possible choices. See Appendix A for more background, and useful references, on the t distribution.

In the final case, we follow a more complicated path. We impose a multivariate t joint distribution, but force the marginals to be Gaussian. The now familiar manipulation involves,

$$c(y_1, \ldots, y_N) = \breve{f}\left(\breve{F}_1^{-1}(F_1(y_1)), \ldots, \breve{F}_N^{-1}(F_N(y_N))\right) \quad (10.62)$$

$$\times \left(\prod_{n=1}^{N} \frac{f_n(y_n)}{\breve{f}_n\left(\breve{F}_n^{-1}(F_n(y_n))\right)}\right),$$

10.5 Transition Likelihood Approach

$$= f_{\mathcal{T}_\nu, R}(F_{\mathcal{T}_\nu}^{-1}(\Phi(y_1)), \ldots, F_{\mathcal{T}_\nu}^{-1}(\Phi(y_N))) \left(\prod_{n=1}^{N} \frac{\phi(y_n)}{f_{\mathcal{T}_\nu}(F_{\mathcal{T}_\nu}^{-1}(\Phi(y_n)))} \right),$$

$$= \frac{\Gamma\left(\frac{\nu+N}{2}\right)}{\Gamma\left(\frac{\nu}{2}\right)\sqrt{(\nu\pi)^N |R|}} \left(1 + \frac{y^T R^{-1} y}{\nu}\right)^{-\frac{(\nu+N)}{2}}$$

$$\times \left(\prod_{n=1}^{N} \frac{\frac{1}{\sqrt{2\pi}} e^{-\frac{y_n^2}{2}}}{\frac{\Gamma\left(\frac{\nu+1}{2}\right)}{\Gamma\left(\frac{\nu}{2}\right)\sqrt{\nu\pi}} \left(1 + \frac{\left(F_{\mathcal{T}_\nu}^{-1}(\Phi(y_n))\right)^2}{\nu}\right)^{-\frac{(\nu+1)}{2}}} \right),$$

$$= \frac{\Gamma\left(\frac{\nu+N}{2}\right)}{\Gamma\left(\frac{\nu}{2}\right)\sqrt{(\nu\pi)^N |R|}} \left(1 + \frac{y^T R^{-1} y}{\nu}\right)^{-\frac{(\nu+N)}{2}}$$

$$\times \left(\prod_{n=1}^{N} \left(\frac{\Gamma\left(\frac{\nu}{2}\right)\sqrt{\nu\pi}}{\sqrt{2\pi}\,\Gamma\left(\frac{\nu+1}{2}\right)} \right) e^{-\frac{y_n^2}{2}} \left(1 + \frac{\left(F_{\mathcal{T}_\nu}^{-1}(\Phi(y_n))\right)^2}{\nu}\right)^{\frac{(\nu+1)}{2}} \right),$$

$$= \frac{\Gamma\left(\frac{\nu+N}{2}\right)}{\Gamma\left(\frac{\nu}{2}\right)\sqrt{(\nu\pi)^N |R|}} \left(\frac{\Gamma\left(\frac{\nu}{2}\right)\sqrt{\frac{\nu}{2}}}{\Gamma\left(\frac{\nu+1}{2}\right)} \right)^N \exp\left(\frac{-\sum_{n=1}^{N} y_n^2}{2} \right)$$

$$\times \left(\prod_{n=1}^{N} \left(1 + \frac{y_n^2}{\nu}\right)^{\frac{(\nu+1)}{2}} \right) \left(1 + \frac{\tilde{y}^T R^{-1} \tilde{y}}{\nu}\right)^{-\frac{(\nu+N)}{2}},$$

$$= \delta |R|^{-\frac{1}{2}} \left(1 + \frac{\tilde{y}^T R^{-1} \tilde{y}}{\nu}\right)^{-\frac{(\nu+N)}{2}},$$

where y is unchanged, but

$$\tilde{y} = \begin{bmatrix} F_{\mathcal{T}_\nu}^{-1}(\Phi(y_1)) \\ \vdots \\ F_{\mathcal{T}_\nu}^{-1}(\Phi(y_N)) \end{bmatrix}, \qquad (10.63)$$

y_n denotes the nth element of y, and

$$\delta = \frac{\Gamma\left(\frac{\nu+N}{2}\right)\left(\Gamma\left(\frac{\nu}{2}\right)\right)^{N-1}}{\left(\Gamma\left(\frac{\nu+1}{2}\right)\right)^N \sqrt{(2\pi)^N}} \exp\left(\frac{-\sum_{n=1}^{N} y_n^2}{2}\right) \left(\prod_{n=1}^{N}\left(1 + \frac{y_n^2}{\nu}\right)^{\frac{(\nu+1)}{2}}\right). \quad (10.64)$$

The copula density, in the case of mixed joint and marginal distributions, is significantly more complex. The saving grace is that, in the log-likelihood framework, multiplicative constants can be ignored since they do not impact the optimum. This is the rationale for writing the copula density in equation 10.62 in terms of $R \equiv R_\theta$ and placing all other values—which do not vary with the choice of parameters—into the coefficient δ summarized in equation 10.64.

10.5.2 The Log-Likelihood Kernel

We will restrict our focus to the second example involving the joint t distribution with t-distributed marginals. This is a normal-variance mixture model exhibiting the positive characteristic of tail dependence, which makes it an interesting example. One of the shortcomings of this approach is that, in the fully t-distributed case, the joint and all marginal distributions need to share the same degrees-of-freedom parameter, ν.[28] Perhaps more disappointing is the need to treat this parameter as fixed. Technically, it is straightforward to include this parameter into the estimation algorithm. Practically, however, as will soon become clear, this approach is already quite fragile. Moreover, with relatively short time series, it is simply very difficult to identify this parameter choice.

To build the full likelihood function, it is necessary to add, as in the previous sections, the time dimension. We thus generalize our t-distributed, latent model-response variables as,

$$y_t = \begin{bmatrix} y_{1,t} \\ \vdots \\ y_{N,t} \end{bmatrix}, \quad (10.65)$$

for $t = 1, \ldots, T$. This is only one building block; it is useful to organize the full collection of inputs into a convenient form. The transition data is in panel format.

---

[28] This is a technical constraint associated with the form of multivariate t distribution employed in this analysis. See Kotz and Nadarajah (2004) for more details.

10.5 Transition Likelihood Approach

That implies that for all time increments $t = 1, \ldots, T$, there is a collection of credit-transition outcomes for each of the $n = 1, \ldots, N$ obligors in the dataset—that is, it is a combination of cross-sectional and time-series data. Let us denote the entire set of observations as,

$$Y \triangleq \{y_{n,t} \mid n = 1, \ldots, N; t = 1, \ldots, T\}, \tag{10.66}$$

implying $N \times T$ separate data-points. It is important to note that Y is not observable; it must be inferred from transition data. Its construction will require additional, fairly messy, computations.

The first order of business, however, is to determine the form of our likelihood function. Equation 10.65, along with our previous development in equation 10.61, allow us to restate the copula density at time t as,

$$c(y_{1,t}, \ldots, y_{N,t}) = \frac{\Gamma\left(\frac{\nu+N}{2}\right)}{\Gamma\left(\frac{\nu}{2}\right)\sqrt{(\nu\pi)^N |R|}} \left(1 + \frac{y_t^T R^{-1} y_t}{\nu}\right)^{-\frac{(\nu+N)}{2}}. \tag{10.67}$$

Equation 10.67 thus describes the copula density at a given point in time. To incorporate the desired time dimension, we use the serial independence of the observations over time to write the likelihood function in the following form,

$$\mathcal{L}(\theta|Y) = \prod_{t=1}^{T} c(y_{1,t}, \ldots, y_{N,t}). \tag{10.68}$$

Before proceeding to derive the precise form of the likelihood function, let us first apply the aforementioned logarithmic transformation to derive the log-likelihood function. This yields,

$$\ln(\mathcal{L}(\theta|Y)) = \ln\left(\underbrace{\prod_{t=1}^{T} c(y_{1,t}, \ldots, y_{N,t})}_{\text{Equation 10.68}}\right), \tag{10.69}$$

$$\ell(\theta|Y) = \sum_{t=1}^{T} \ln\left(\underbrace{c(y_{1,t}, \ldots, y_{N,t})}_{\text{Equation 10.67}}\right),$$

$$= \sum_{t=1}^{T} \ln\left(\frac{\Gamma\left(\frac{\nu+N}{2}\right)}{\Gamma\left(\frac{\nu}{2}\right)\sqrt{(\nu\pi)^N}} \cdot \frac{1}{\sqrt{|R|}} \cdot \left(1 + \frac{y_t^T R^{-1} y_t}{\nu}\right)^{-\frac{(\nu+N)}{2}}\right),$$

$$= \sum_{t=1}^{T} \ln\left(\frac{\Gamma\left(\frac{\nu+N}{2}\right)}{\Gamma\left(\frac{\nu}{2}\right)\sqrt{(\nu\pi)^N}}\right) + \ln\left(\frac{1}{\sqrt{|R|}}\right) + \ln\left(1 + \frac{y_t^T R^{-1} y_t}{\nu}\right)^{-\frac{(\nu+N)}{2}},$$

$$= T \ln\left(\frac{\Gamma\left(\frac{\nu+N}{2}\right)}{\Gamma\left(\frac{\nu}{2}\right)\sqrt{(\nu\pi)^N}}\right) - \frac{T}{2}\ln|R| - \frac{(\nu+N)}{2}\sum_{t=1}^{T}\ln\left(1 + \frac{y_t^T R^{-1} y_t}{\nu}\right).$$

This is a fairly involved expression with numerous terms. Observe, however, that not all terms are functions of the parameter vector, θ—indeed, only the correlation matrix, R, depends on the asset-correlation parameter. This implies that as we vary θ, these terms are constant. They show up in the likelihood function in a multiplicative form, but become additive in the log-likelihood representation. It is typical and advisable, for the purposes of maximization of the likelihood function, to ignore these terms. The log-likelihood function excluding additive constants is typically termed the likelihood kernel.[29] In our case, the likelihood kernel has the following, significantly streamlined, form

$$\ell(\theta|Y) = -\frac{T}{2}\ln|R| - \frac{(\nu+N)}{2}\sum_{t=1}^{T}\ln\left(1 + \frac{y_t^T R^{-1} y_t}{\nu}\right), \quad (10.70)$$

$$= -\frac{1}{2}\left(T\ln|R| + (\nu+N)\sum_{t=1}^{T}\ln\left(1 + \frac{y_t^T R^{-1} y_t}{\nu}\right)\right),$$

for $\theta \in \Theta$ where Θ denotes the parameter space.

Interestingly, the form of the log-likelihood kernel is equivalent for both t-distributed and Gaussian marginals. The difference arises from the specific form of the response variables. In our base case scenario, the y_t vector is defined in equation 10.65. If we move to Gaussian marginals, however, then one need only define y_t as,

$$\tilde{y}_t = \begin{bmatrix} F_{T_\nu}^{-1}(\Phi(y_{1,t})) \\ \vdots \\ F_{T_\nu}^{-1}(\Phi(y_{N,t})) \end{bmatrix}, \quad (10.71)$$

for $t = 1, \ldots, T$. The consequence, therefore, is a slight distortion of the latent-state variable vector, y_t, to accommodate the alternative marginal distribution.

---

[29] For more detail, see Held and Bové (2014, Chapter 2).

10.5 Transition Likelihood Approach

Algorithm 10.9 t-copula, transition, log-likelihood kernel

```
def logLCopula(x,Y,nu):
    M,T = Y.shape
    R = buildCorrelationMatrix1F(x,M)
    detR = anp.det(R)
    Rinv = anp.inv(R)
    V = 0
    for t in range(0,T):
        V += np.log(1+np.divide(np.dot(np.dot(Y[:,t],Rinv),Y[:,t]),nu))
    return -(-0.5*(T*np.log(detR)+(nu+M)*V))
```

Algorithm 10.10 Building a one-parameter, correlation matrix

```
def buildCorrelationMatrix1F(x,N):
    R = x*np.ones([N,N])+np.eye(N)
    R[R==1+x] = 1
    return R
```

The maximum likelihood estimator for the parameter vector is now merely,

$$\theta^* = \arg\max_{\theta \in \Theta} \underbrace{-\frac{1}{2}\left(T \ln |R_\theta| + (\nu + N)\sum_{t=1}^{T} \ln\left(1 + \frac{y_t^T R_\theta^{-1} y_t}{\nu}\right)\right)}_{\ell(\theta|X)}, \quad (10.72)$$

where R_θ explicitly illustrates the link to the parameter vector. Finding analytic representations of the first and second-order conditions of this optimization problem is not particularly practical. As a consequence, the optimal parameter vector, θ^*, is determined through the use of a non-linear optimization algorithm.[30]

It is not terribly difficult to transform equation 10.70 into a computer-based implementation. Algorithm 10.9 provides some sample Python code for the log-likelihood kernel. It requires a function call to construct the correlation matrix from the parameter values, but is otherwise reasonably straightforward. x represents the asset-correlation parameter, Y is the set of (unobservable) latent state variables, and nu, of course, is the degrees-of-freedom parameter. As usual, since we will employ a minimizer, we return the negative value of the objective function.

Algorithm 10.9 requires a correlation matrix, R, for its computation. Algorithm 10.10 highlights the, very simple, approach used to construct the $N \times N$ correlation matrix for this example. In a multi-factor setting, the dimensionality is unchanged, but its construction is a bit more involved since it requires multiple parameters and differentiation of one's obligors based on their region, credit rating, or industry.

---

[30]There are a variety of approaches—both with and without numerical derivatives—to solve this problem. Beck (2014) is an excellent reference in this area.

To this point, although a bit involved, this estimation approach is relatively manageable. The fundamental challenge, however, is inference of the required latent state variables. This is our next task.

10.5.3 Inferring the State Variables

Equation 10.72 provides a clear numerical recipe of how to determine the optimal dependence parameters. Less obvious, however, is the form of the input data. These is a sequence of response variables for each transition period and credit obligor in one's dataset. Ultimately, we need to make use of this data as an input into the log-likelihood kernel.

Return to our four credit-state setting with N credit counterparts over T periods. The key to the entire effort is the transition matrix. As in a previous section, we cumulate the transition probabilities from left to right to construct the matrix, Q, as described in equation 10.13. Each value in Q lies in the unit interval, $[0, 1]$. These values are then mapped back into the values of the marginal latent variable distribution using its inverse cumulative distribution function. In the case of t-distributed marginals, this leads to

$$\Delta = \begin{bmatrix} F_{\mathcal{T}_\nu}^{-1}(q_{11}) & F_{\mathcal{T}_\nu}^{-1}(q_{12}) & F_{\mathcal{T}_\nu}^{-1}(q_{13}) & F_{\mathcal{T}_\nu}^{-1}(q_{14}) \\ F_{\mathcal{T}_\nu}^{-1}(q_{23}) & F_{\mathcal{T}_\nu}^{-1}(q_{22}) & F_{\mathcal{T}_\nu}^{-1}(q_{23}) & F_{\mathcal{T}_\nu}^{-1}(q_{24}) \\ F_{\mathcal{T}_\nu}^{-1}(q_{31}) & F_{\mathcal{T}_\nu}^{-1}(q_{32}) & F_{\mathcal{T}_\nu}^{-1}(q_{33}) & F_{\mathcal{T}_\nu}^{-1}(q_{34}) \\ F_{\mathcal{T}_\nu}^{-1}(q_{41}) & F_{\mathcal{T}_\nu}^{-1}(q_{42}) & F_{\mathcal{T}_\nu}^{-1}(q_{43}) & F_{\mathcal{T}_\nu}^{-1}(q_{44}) \end{bmatrix}, \quad (10.73)$$

$$= \begin{bmatrix} \delta_{11} & \delta_{12} & \delta_{13} & \delta_{14} \\ \delta_{21} & \delta_{22} & \delta_{23} & \delta_{24} \\ \delta_{31} & \delta_{32} & \delta_{33} & \delta_{34} \\ \delta_{41} & \delta_{42} & \delta_{43} & \delta_{44} \end{bmatrix},$$

$$= \begin{bmatrix} 5.0000 & -1.8443 & -2.4834 & -3.5518 \\ 5.0000 & 1.3253 & -1.1848 & -2.8453 \\ 5.0000 & 1.2110 & 0.4186 & -2.5280 \\ 5.0000 & 5.0000 & 5.0000 & 5.0000 \end{bmatrix},$$

where the Markov-chain $\delta_{\text{from,to}}$ notation is used.

The matrix Δ delineates the boundaries between each credit state—recall Fig. 10.1 on page 583. Let $S_{n,t}$ denote the state of obligor n at time t. How, mechanically, does the obligor arrive in that position? It depends upon two things: the previous credit rating and the outcome of the global state variable. The rating at $t-1$—the *from* state—determines the appropriate *row* of Δ to be employed. The state-variable outcome of $y_{n,t}$ determines—along with partition of the space based on the appropriate row of Δ—the new credit state. Here arises the conundrum. This

10.5 Transition Likelihood Approach

works great if you observe $y_{n,t}$; unfortunately, since it is a latent variable, we do not. The best we can do is use rating outcomes, the $S_{n,t}$'s, to infer the value of the $y_{n,t}$ value within a given partition.

There is a direct relationship between $S_{n,t}$ and $y_{n,t}$, but functionally, it is not very convenient. The mapping from $y_{n,t}$ to $S_{n,t}$ is many-to-one. A broad range of possible y's yield a common S value; this is the entire point of the partition of the latent-variable space.[31] The inverse mapping, from $S_{n,t}$ to $y_{n,t}$, is, of course, one-to-many. We can, therefore, only identify a specific range, or set, of possible $y_{n,t}$'s associated with each $S_{n,t}$; we cannot pinpoint it with a higher degree of accuracy.

This is not a terribly obvious situation, but an example will perhaps help motivate the proposed solution. Imagine that obligor n finds itself in credit state B at time $t-1$. We then observe a movement to the first state, A, at time t. The rating transition, to be explicit, is $S_{n,t-1} = 2 \equiv B$ and $S_{n,t} = 1 \equiv A$. The $t-1$ value brings us to the second row of Δ, while the t value narrows it down to the interval between the first and second columns. The value for our dataset, $y_{n,t}$, falls in the interval,

$$y_{n,t} \in \left[y_{n,t}^-, y_{n,t}^+ \right] \equiv [\delta_{22}, \delta_{21}], \tag{10.74}$$

$$= [1.3253, 5.0000].$$

This makes the specification of the dataset somewhat ambiguous. One might try, for example, to estimate $y_{n,t}$ using the mid-point of the interval in equation 10.74

$$\hat{y}_{n,t} = y_{n,t}^- + \frac{\left(y_{n,t}^+ - y_{n,t}^- \right)}{2}, \tag{10.75}$$

but this would ignore the uncertainty in the true value of $y_{n,t}$.

To repeat, a precise value for $y_{n,t}$ cannot be assigned. One can only assign it to an interval. The typical approach to problems like this is to average, or integrate, over the many possible values associated with each $y_{n,t}$ given its observed credit-state transition.[32] We can unequivocally state that $\hat{y}_{n,t} \in \left[y_{n,t}^-, y_{n,t}^+ \right]$, where these boundaries are determined by the Δ matrix in equation 10.73. Practically, therefore, we perform the optimization many times with different randomly selected $\hat{y}_{n,t}$ values pulled from this interval. That is, we simulate our data as follows,

$$\hat{y}_{n,t}^{(k)} = y_{n,t}^- + u_k \left(y_{n,t}^+ - y_{n,t}^- \right), \tag{10.76}$$

for $n = 1, \ldots, N$, $t = 1, \ldots, T$ and where $u_k \in \mathcal{U}(0, 1)$. We generate K outcomes for this entire system leading to $\hat{Y}_k = \{\hat{y}_{n,t}^{(k)} | n = 1, \ldots, N; t = 1, \ldots, T\}$. The

---

[31] This discussion can be framed more formally in terms of injective and surjective functions, but this is perhaps too abstract for this setting. See Royden (1988, Chapter 1) or Bartle and Sherbert (1982, Chapter 1) for more background on these ideas.
[32] This approach is also described in detail in Berger et al. (1999).

boundaries of the $y_{n,t}$'s are often referred to as *nuisance* parameters. This approach is an attempt to numerically integrate out these parameters from the problem. See Held and Bové (2014), Berger et al. (1999), Berger (1985), and Pawitan (2001) for more detail on this concept.

A large number of different possible values of $\hat{Y}_k$ is produced using equation 10.76. This is performed for each rating transition in one's dataset and the optimization summarized in equation 10.72 is solved. This leads to a collection of optimal parameter vectors, $\{\theta_k^*; k = 1, \ldots, K\}$. The final estimate is thus simply,

$$\theta^* \approx \sum_{k=1}^{K} \frac{\theta_k^*}{K}. \qquad (10.77)$$

These imprecise estimates of $y_{n,t}$ permit the construction of a distribution of log-likelihood kernel maxima that may be used to identify the t-threshold asset-correlation parameter, $\theta \equiv \rho_T$.

Actually mapping the rating outcomes to the latent state variables, therefore, involves identifying the upper and lower bounds on the possible values of y. While not difficult, this book-keeping exercise requires a bit of care. Algorithm 10.11 includes two Python functions that manage this process. The first function, mapRating, takes as inputs the set of transitions, D, the *from* and *to* states, and the number of distinct credit states, K. It then handles the details, with particular care around the boundaries, by returning the upper and lower bound for a given rating transition. The second sub-routine, mapRatingData, simply loops over each obligor and time period to construct a matrix of partition boundaries.

These boundary values, d_low and d_up, form the basis for the simulation of the state variables in equation 10.76. Using these outcomes with scipy's minimize function and the logLCopula in Algorithm 10.9 yields a numerical estimate of the maximum-likelihood estimator. In a one-dimensional setting, this is relatively straightforward. As one adds more state variables and parameters, it becomes more complicated.

10.5.4 A Final Example

To examine the performance of this estimation approach, as in previous sections, we consider a practical example. Using a slight variation of Algorithm 10.2 on page 586, we simulate a new dataset. Indeed, the only difference is that the individual state variables follow a one-dimensional, t-threshold model with asset-correlation parameter, $\rho_T = 0.30$, and degrees-of-freedom $\nu = 20$. As usual, we follow $N = 100$ credit obligors over a 30-year period. The key distinction with the previous approaches—which focus exclusively on default outcomes—is an incorporation of all transition information into the estimation.

10.5 Transition Likelihood Approach

Algorithm 10.11 Inferring latent-state variables

```
def mapRating(D,from_value,to_value,K):
    if (to_value==K) & (from_value!=K):
        d_u = D[from_value-1,to_value-1]
        d_l = -5
    elif (to_value==K) & (from_value==K):
        d_u = -5
        d_l = -5
    else:
        d_u = D[from_value-1,to_value]
        d_l = D[from_value-1,to_value-1]
    return d_l, d_u
def mapRatingData(Y,D,K):
    N,T = Y.shape
    d_low = np.zeros([N,T-1])
    d_upp = np.zeros([N,T-1])
    for n in range(0,N):
        for m in range(1,T):
            d_low[n,m-1],d_upp[n,m-1] = mapRating(D,
                                        Y[n,m-1].astype(int),
                                        Y[n,m].astype(int),K)
    return d_low,d_upp
```

Our usual, and strongly advised, starting point is to examine the log-likelihood function to determine its regularity and, as a consequence, its suitability for treatment in the likelihood framework. In this case, however, we have an entire family of log-likelihood functions. Each member of the family is indexed to a simulated set of latent state variables given the boundaries determined by Algorithm 10.11. Figure 10.12 highlights 100 distinct raw and normalized log-likelihood functions for our simulated example.

While not particularly easy to interpret, Fig. 10.12 nonetheless provides important insight into this problem. The boundaries in each case are identical. The only difference is that we have randomly selected different constellations of state variables that are mathematically consistent with the observed ratings transitions. The maximum-likelihood estimates of ρ_T range from below 0.1 to above 0.5. This is a fairly sizable range.

Closer examination of Fig. 10.12 reveals that the absolute magnitude of the likelihood function, as seen in the raw values, varies fairly substantially. This would appear to indicate rather different probabilities, as a function of the latent state variables, of observing the parameter values given the data. It is important to mention that our example is particularly problematic with only four credit states; this implies that some of the partitions cover a huge range of possible latent state-

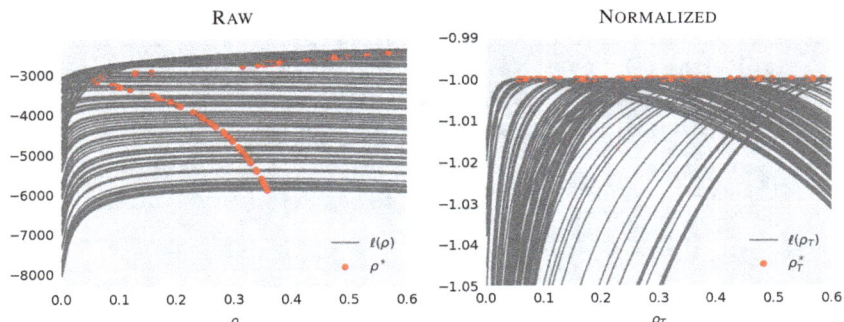

Fig. 10.12 *Multiple log-likelihoods*: Since we make K inference about the unobservable, latent, state variables, we generate K log-likelihood functions. It is still good practice to plot them and make a judgement on their granularity. The underlying graphics displays the consequences for $K = 100$.

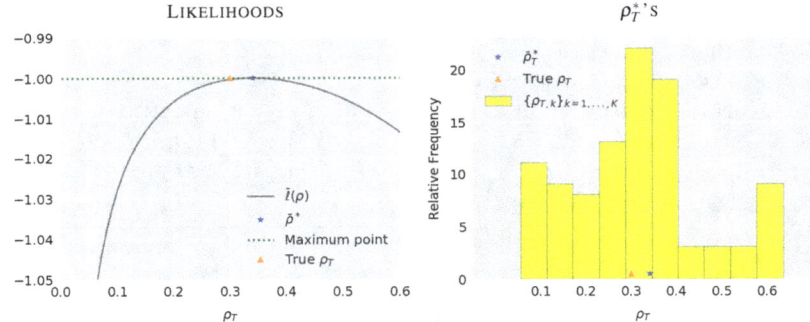

Fig. 10.13 *Averaging log-likelihoods*: Fig. 10.12 is interesting and informative, but it is also useful to consider the average result. The average, after all, forms our final estimate. The underlying graphics, therefore, illustrate the average log-likelihood function and the distribution of ρ_T estimates.

variable values. As we increase the number of credit states, we also raise the number of partitions and, in general, reduce the relative size of each interval. This possible shortcoming of our example should be kept in mind.

Despite its insight into the underlying problem, Fig. 10.12 is not particularly useful for assessing the regularity of the likelihood function. Figure 10.13 resolves this issue by examining the form of the *average*, normalized, log-likelihood across all of the 100 simulated functions in Fig. 10.12. It appears, as we would have hoped, to have a reasonably parabolic form. Although not definitive evidence of regularity, it does suggest that we have a reasonably well-behaved problem.

The right-hand side of Fig. 10.13 illustrates the distribution of asset-correlation estimates across the 100 individual sets of simulated latent-state variable outcomes. While a disproportionate number of estimates falls quite close to the true parameter, there is still a relatively large number of distant values. On average, as indicated

10.5 Transition Likelihood Approach

Table 10.9 $\mathcal{T}$-*threshold model results*: This table outlines the results of a one-factor, one-parameter $\mathcal{T}$-threshold model using the transition approach.

Quantity	Detail	Average
True parameter value	ρ_T	0.30000
Average grid search optimum	$\rho_{T,0}$	0.34202
Average hill-climber optimum	ρ_T^*	0.34174
Average score function	$s(\rho_T^*)$	−0.00002
Average observed Fisher information	$I(\rho_T^*)$	2136.9
Average standard error	$\sqrt{I^{-1}(\rho_T^*)}$	0.02163
95% confidence interval	$\rho_T^* \pm 1.96\sqrt{I^{-1}(\rho_T^*)}$	[0.30,0.38]
Total computational time	(in minutes)	0.1
Success ratio	(percent)	95.0

Table 10.10 *t-threshold model simulation study results*: This table illustrates the results of estimating the *three* alternative ρ_T parameters—0.15, 0.30, and 0.45—using the maximum-likelihood approach with 100 simulated datasets—and 250 nuisance-parameter integrations—for each parameter value.

Parameter						
ρ_T	$\hat{\rho}_T^*$	$\sigma(\hat{\rho}_T^*)$	Score function	Fisher information	Standard error	Success ratio
0.15	0.20	0.07	−0.03676	6241	0.014	94.3%
0.30	0.25	0.11	0.00015	4558	0.017	93.8%
0.45	0.31	0.16	0.00011	4053	0.017	93.1%

in Table 10.9, the maximum-likelihood estimator is about 0.34. Moreover, the true value lies within the surprisingly tight 95% confidence interval. Given the uncertainty across the distribution of simulated latent variables, the standard-error estimate of 0.02 looks artificially low; or, equivalently, the observed Fisher information seems to be overly large. This merits further investigation.

As we've learned in previous sections, a single simulated dataset does not provide a complete view of the efficacy of an estimator. This, along with questions about the veracity of the standard-error estimate, motivates our final simulation study. We will examine 100 simulated datasets for three different asset-correlation parameter levels: 0.15, 0.30, and 0.45. Each individual estimation will involve 250 integrations to eliminate the influence of the boundary nuisance parameters.

Table 10.10 summarizes the final results. The asset-correlation population parameters with values of 0.15 and 0.30 exhibit average estimates relatively close to their true settings. For $\rho_T = 0.45$, the accuracy clearly suffers. The standard error estimates, determined from the average Fisher information values, are unrealistic. This suggests that we are better to rely upon the standard-error of the parameter estimates from the simulation study. These values are increasing in the asset-correlation parameter value and, ultimately, are essentially equivalent to Efron (1979)'s bootstrap estimator.

Table 10.11 *t-threshold model full information estimators*: This table replicates the analysis from Table 10.10, but actually uses the true unobservable state variables to build and maximize the likelihood function. While unrealistic, and practically impossible, it helps us better understand the basic approach.

Parameter			Score function	Fisher information	Standard error	Success ratio
ρ_T	$\hat{\rho}_T^*$	$\sigma(\hat{\rho}_T^*)$				
0.15	0.15	0.03	0.00028	1145	0.030	98.0%
0.30	0.30	0.04	−0.00003	802	0.035	93.0%
0.45	0.45	0.03	−0.00005	1024	0.031	91.0%

To summarize, our simulation study shows evidence of bias in the parameter estimates and implausible observed Fisher information values. This is a bit disappointing, since by adding the transition data, we have significantly expanded the available information to assess asset and thus, indirectly, default correlation. What is the source of this problem? The likelihood function, derived directly from the model, is legitimate and the dataset stems from the true model. Neither seems to be the issue. The fundamental problem, it would appear, is the inference of the unobservable latent state variables. A simple experiment can be used to verify this point.

Table 10.11 summarizes the result of our experiment. It basically replicates the analysis from Table 10.10, but does so by breaking the rules. It actually looks into the simulated datasets and uses the true unobservable state variables to build and maximize the likelihood function. While unrealistic, practically impossible, and against the rules of mathematical statistics, it nevertheless helps us better understand the basic approach.

The first point from Table 10.11 is that the estimated parameters are virtually indistinguishable from the population parameters. Equally importantly, the bootstrap and likelihood standard-error estimates agree very strongly with one another. Finally, even in this virtually optimal case, the proportion of non-convergent optimizations is *not* zero. This highlights the practical challenges and difficulties of numerical non-linear optimization.

The larger implication stems from the fact that the only difference between the simulation studies outlined in Tables 10.10 and 10.11 is their treatment of the unobservable latent state variables. It thus seems reasonable to conclude that the challenges in the base algorithm stem from inferring these unobservable values. While this problem might seem insurmountable, there are a few mitigating factors. First, four distinct credit states are too few to be realistic and practically stable. As previously hinted, increasing the number of credit states to a more standard number—probably somewhere between 10 and 20—will reduce the size of the intervals in the partition thus tightening the range of simulation. A second point is that 100 obligors is relatively small. By focusing on transitions rather than defaults, this method should benefit even more from an expansion of the total number of obligors.

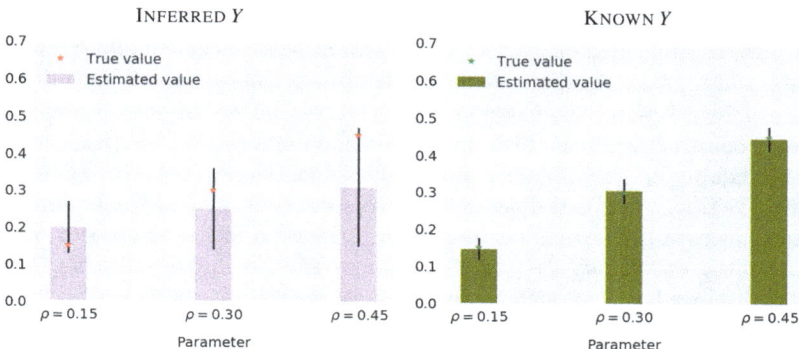

Fig. 10.14 *Lost in translation*: Although we do not observe the latent state variables, we can approximate the uncertainty with their inference. This is done by using the true, unobserved values; this is only possible, because we have simulated our underlying data.

Figure 10.14 closes out this section with a visual comparison of the inferred and (unrealistic) full information latent state-variable values. The differences are stark and they underscore the fragility of this method. Given the rareness of defaults and the challenges of estimating default-correlation parameters in our models, this is nonetheless a welcome addition to the quantitative analyst's toolkit. Despite its shortcomings, it does enjoy some benefits. The most important is that, because the parameters enter through the correlation matrix, R, only modest algorithmic complications are associated with incorporating additional state variables. One need only replace Algorithm 10.10 with something more involved. This does, of course, add new dimensions to one's numerical non-linear optimization making it more difficult to find global optima, but nothing in life is free.

10.6 Final Thoughts

In this final chapter, we have covered a significant amount of material. The investigation began with a simple and intuitive approach: the method of moments. This calibration-style technique can be applied, using a time series of default outcomes, in virtually any circumstance. Indeed, using a simulated dataset, we used this method to fit a large number of possible default-risk models. An alternative approach involves exploiting the common model property of conditional independence to construct a binomial likelihood function. With numerical integration and optimization, we find ourselves in the maximum-likelihood framework. This opens up a realm of statistical inference options and, in principle, permits treatment of more complex combinations of model parameters. A few simulation studies illustrate the general usefulness of this approach.

The default-only results are popular and generally quite useful, but suffer from the inherent rarity of default events. A final approach, which essentially applies only to threshold models, changes track and attempts to incorporate all transition data to inform the asset-correlation parameters. Through the underlying copula density, one may construct and maximize the log-likelihood function. There is, however, a catch. It requires inference of the unobservable latent state variables for each obligor and time period. While possible, this is a reasonably noisy and fragile procedure that introduces noise into one's estimates and creates problems for classical notions of parametric uncertainty. Nevertheless, the presence of a few mitigating factors and the dire need for any viable alternative estimation approach suggest that this technique merits consideration.

To conclude, like virtually all of the topics considered in the previous chapters, there exists neither a single dominant approach nor any straightforward easy solution. Parameter determination is not straightforward. On the contrary, computation and application of asset and default correlations is a complex and subtle undertaking and requires sound judgement. A judicious approach to this problem would involve simultaneous use of multiple estimation techniques, extensive sensitivity analysis, and expert judgement.

References

Bartle, R. G., & Sherbert, D. R. (1982). *Introduction to real analysis* (2nd edn.). New York, NY: John Wiley & Sons.
Beck, A. (2014). *Introduction to nonlinear optimization: Theory, algorithms, and applications with MATLAB* (1st edn.). Philadelphia, USA: Society for Industrial and Applied Mathematics.
Berger, J. O. (1985). *Statistical decision theory and Bayesian analysis*. Berlin: Springer-Verlag.
Berger, J. O., Liseo, B., & Wolpert, R. L. (1999). Integrated likelihood methods for eliminating nuisance parameters. *Statistical Science, 14*(1), 1–28.
Billingsley, P. (1995). *Probability and measure* (3rd edn.). Third Avenue, New York, NY: Wiley.
Bluhm, C., Overbeck, L., & Wagner, C. (2003). *An introduction to credit risk modelling* (1st edn.). Boca Raton: Chapman & Hall, CRC Press.
Bouyé, E., Durrleman, V., Nikeghbali, A., Riboulet, G., & Roncalli, T. (2000). Copulas for finance: A reading guide and some applications. Groupe de Recherche Opérationalle, Crédit Lyonnais.
Casella, G., & Berger, R. L. (1990). *Statistical inference*. Belmont, CA: Duxbury Press.
Chernick, M. R. (1999). *Bootstrap methods: A practitioner's guide*. New York: John Wiley & Sons.
Demey, P., Jouain, J.-F., Roget, C., & Roncalli, T. (2004). Maximum likelihood estimate of default correlations. *Risk*, 104–108.
Dhrymes, P. J. (1994). *Topics in advanced econometrics: Linear and nonlinear simultaneous equations* (Vol. II). Heidelberg, Germany: Springer-Verlag.
Efron, B. (1979). Bootstrap methods: Another look at the jackknife. *The Annals of Statistics, 7*(1), 1–26.
Embrechts, P., Lindskog, F., & McNeil, A. (2001). *Modelling dependence with copulas and applications to risk management*. ETH, Zürich: Department of Mathematics.
Fang, K., Kotz, S., & Ng, K. (1990). *Symmetric multivariate and related distributions. Monographs on statistics and applied probability*. Chapman and Hall.
Gordy, M. B. (1998). A comparative anatomy of credit risk models. Board of Governors of the Federal Reserve System.

References

Gordy, M. B., & Lütkebohmert, E. (2007). Granularity Adjustment for Basel II. Deutsche Bundesbank, Banking and Financial Studies, No 01–2007.

Gordy, M. B., & Lütkebohmert, E. (2013). *Granularity adjustment for regulatory capital assessment*. University of Freiburg.

Gordy, M., & Heitfield, E. (2002). *Estimating default correlations from short panels of credit rating performance data*. Federal Reserve Board.

Hamerle, A., Liebig, T., & Rösch, D. (1986). Benchmarking asset correlations. *Risk*, 77–81.

Hansen, L. P. (1982). Large sample properties of generalized method of moments estimators. *Econometrica, 50*(4), 1029–1054.

Hashimoto, T. (2009). *Asset correlation for credit risk analysis: Empirical study of default data for japanese companies*. Bank of Japan.

Held, L., & Bové, D. S. (2014). *Applied statistical inference*. Berlin, Germany: Springer-Verlag.

Judge, G. G., Griffiths, W., Hill, R. C., Lütkepohl, H., & Lee, T.-C. (1985). *The theory and practice of econometrics* (2nd edn.). New York, NY: John Wiley & Sons.

Kotz, S., & Nadarajah, S. (2004). *Multivariate t distributions and their applications*. New York: Cambridge University Press.

Koyluoglu, H. U., & Hickman, A. (1998). Reconcilable differences. *Risk, 11*(10), 56–62.

Magnello, M. E. (2009). Karl Pearson and the establishment of mathematical statistics. *International Statistical Review, 77*(1), 3–29.

Martin, L. (2013). *Analysis of the IRB asset correlation coefficient with an application to a credit portfolio*. University of Uppsala.

McNeil, A. J., Frey, R., & Embrechts, P. (2015). *Quantitative risk management: Concepts, tools and techniques*. Princeton, NJ: Princeton University Press.

Merton, R. (1974). On the pricing of corporate debt: The risk structure of interest rates. *Journal of Finance, 29*, 449–470.

Pawitan, Y. (2001). *In all likelihood: Statistical modelling and inference using likelihood*. Oxford: Clarendon Press.

Press, W. H., Teukolsky, S. A., Vetterling, W. T., & Flannery, B. P. (1992). *Numerical recipes in C: The art of scientific computing* (2nd edn.). Trumpington Street, Cambridge: Cambridge University Press.

Royden, H. L. (1988). *Real analysis*. Englewood Cliffs, NJ: Prentice-Hall Inc.

Ruiz, I.: 2015, *XVA Desks: A New Era for Risk Management*, Palgrave Macmillan, New York, NY.

Sklar, A. (1959). Fonctions de répartition à n dimensions et leurs marges. *Publications de l'Institut de Statistique de L'Université de Paris, 8*, 229–231.

Sklar, A. (1996). Random variables, distribution functions, and copulas: A personal look backward and forward. *Institute of Mathematical Statistics Lecture Notes–Monograph Series, 28*, 1–14.

Yan, J. (2007). Enjoy the joy of copulas: With a package copula. *Journal of Statistical Software, 21*(1), 1–21.

Appendix A
The *t*-Distribution

> A curve has been found representing the frequency distribution of standard deviations of samples drawn from a normal population.
>
> (William Sealy Gosset)

The *t*-distribution is used extensively in the *t*-threshold model. This appendix motivates, with a few brief derivations, the origin and form of this distribution. Let us begin our discussion with some basic facts from mathematical statistics. Given a random variable, $Z \sim \mathcal{N}(0, 1)$, then

$$Z^2 \sim \chi^2(1). \tag{A.1}$$

In words, a squared standard normal variable follows a χ^2-distribution with one degree of freedom. The support of the χ^2 distribution is $\mathbb{R}_+ \equiv [0, \infty)$, which makes sense given its definition as the squared value of a standard normal variate. Its density function has the following form for ν degrees of freedom,

$$f(x, \nu) = \frac{1}{\Gamma\left(\frac{\nu}{2}\right) 2^{\frac{\nu}{2}}} x^{\frac{\nu}{2}-1} e^{-\frac{x}{2}}, \tag{A.2}$$

where $\Gamma(\cdot)$ denote the gamma function,

$$\Gamma(u) = \int_0^\infty x^{u-1} e^{-x} dx. \tag{A.3}$$

The gamma function is, as discussed in the text of this book, an extension of the factorial function.[1] It has a number of useful and interesting properties, but, in this setting, one can think of it as the continuous analogue of the discrete binomial coefficient.

---
[1] Indeed, $\Gamma(n) = (n - 1)!$ for all $n \in \mathbb{Z}_+$.

A.1 The Chi-Squared Distribution

How is the density in equation A.2 derived? It requires a bit of effort and the underlying change of variables theorem; this result is also employed, without formal statement or proof, in Chap. 3.

Theorem A.1 (Changes of Variables) *Let Z be a random variable. If g is a monotonic function and g^{-1} exists, then for $Y = g(Z)$,*

$$f_Y(y) = f_Z(g^{-1}(y)) \left| \frac{dg^{-1}(y)}{dy} \right|. \tag{A.4}$$

For more background and insight into this theorem, please see Billingsley (1995, Chapter 16). We will not derive the result summarized in equation A.1 in full generality, but instead examine the one-dimensional case for illustrative purposes. In our case, $Z \sim \mathcal{N}(0, 1)$ and $g(Z) = Y = Z^2$. Moreover, $Z = g^{-1}(Y) = \sqrt{Y}$ and $\frac{dg^{-1}(y)}{dy} = \frac{1}{2\sqrt{Y}}$. Our choice of g is not monotonic, but it is symmetric with all negative values transformed to the positive domain. We can, therefore, multiple the right-hand-side of equation A.4 by 2 to account for this aspect of g. Applying the formula, and noting that $\Gamma\left(\frac{1}{2}\right) = \pi^{\frac{1}{2}}$, we have

$$f_Y(y) = 2 \cdot f_Z(\sqrt{y}) \left| -\frac{1}{2\sqrt{y}} \right|, \tag{A.5}$$

$$= 2 \cdot \frac{1}{\sqrt{2\pi}} e^{-\frac{(\sqrt{y})^2}{2}} \frac{1}{2\sqrt{y}},$$

$$= \frac{1}{\pi^{\frac{1}{2}} 2^{\frac{1}{2}}} y^{-\frac{1}{2}} e^{-\frac{y}{2}},$$

$$= \frac{1}{\Gamma\left(\frac{1}{2}\right) 2^{\frac{1}{2}}} y^{\frac{1}{2}-1} e^{-\frac{y}{2}},$$

$$= f(y, 1),$$

$$\sim \chi^2(1),$$

as desired. We have employed the fact that, rather surprisingly, $\Gamma\left(\frac{1}{2}\right) = \sqrt{\pi}$; this is a fascinating real-valued outcome of the Gamma function.

This result can be extended. In particular, given $\nu \in \mathbb{N}$ standard normal variates $Z_1, \ldots, Z_\nu$, their sum obeys the following distribution,

$$Y_\nu = Z_1^2 + Z_2^2 + \cdots + Z_\nu^2 \sim \chi^2(\nu). \tag{A.6}$$

A.2 Toward the t-Distribution

The generalized version of this χ^2-squared distribution with ν degrees of freedom is,

$$f_{Y_\nu}(y) = \frac{1}{\Gamma\left(\frac{\nu}{2}\right) 2^{\frac{\nu}{2}}} y^{\frac{\nu}{2}-1} e^{-\frac{y}{2}}, \tag{A.7}$$

$$\equiv \frac{1}{2\Gamma\left(\frac{\nu}{2}\right)} \left(\frac{y}{2}\right)^{\frac{\nu}{2}-1} e^{-\frac{y}{2}},$$

$$= f(y, \nu),$$

$$\sim \chi^2(\nu),$$

for $y \in [0, \infty)$ and $\nu \in \mathbb{N}_+$.

A.2 Toward the t-Distribution

This is one step in the right direction. If one generates an additional standard normal variate, Z_0, then we may construct the following quantity,

$$T = Z_0 \bigg/ \sqrt{\frac{1}{\nu} \left(Z_1^2 + Z_2^2 + \cdots + Z_\nu^2\right)}, \tag{A.8}$$

$$= \frac{Z_0}{\sqrt{\frac{U}{\nu}}},$$

$$= \sqrt{\frac{\nu}{U}} Z_0,$$

where $U \sim \chi^2(\nu)$ by equation A.6. The natural question is: what is the density of T? There are a number of good sources for this derivation, but the following illustration follows from Shaw (2006).[2] The trick is to make use of Bayes rule for densities:

$$f(x \mid y) = \frac{f(x, y)}{f(y)}, \tag{A.9}$$

$$f(x, y) = f(x \mid y) f(y).$$

---

[2] Shaw (2011) also provides a useful and interesting discussion of employing the assumption of t-distributed asset returns in a portfolio-optimization setting.

If one integrates the joint density with respect to y, then one will recover the marginal distribution as follows,

$$\int_a^b f(x, y) dy = \int_a^b f(x \mid y) f(y) dy, \qquad (A.10)$$
$$= f(x),$$

where $[a, b]$ denotes the support of $f(y)$.

Let's apply this trick to our situation. If we use equation A.8, we can determine the conditional density of T given a fixed value of $U \sim \chi^2(v)$,

$$f\left(t \mid U = u\right) = \frac{1}{\sqrt{2\pi \frac{v}{u}}} e^{-\frac{t^2}{2\frac{v}{u}}}, \qquad (A.11)$$

$$= \sqrt{\frac{u}{2\pi v}} e^{-\frac{t^2 u}{2v}},$$

which follows directly from the fact that if U is fixed at a value of u, then $T \sim \mathcal{N}\left(0, \frac{v}{u}\right)$. The marginal distribution for U is already provided in equation A.7, so applying the result in equation A.10, we have

$$\int_0^\infty f\left(t \mid U = u\right) f_U(u) du = \int_0^\infty \underbrace{\left(\sqrt{\frac{u}{2\pi v}} e^{-\frac{t^2 u}{2v}}\right)}_{f\left(t \mid U = u\right)} \underbrace{\left(\frac{1}{2\Gamma\left(\frac{v}{2}\right)} \left(\frac{u}{2}\right)^{\frac{v}{2}-1} e^{-\frac{u}{2}}\right)}_{f_U(u)} du, \qquad (A.12)$$

$$f_T(t; v) = \frac{1}{2\Gamma\left(\frac{v}{2}\right) \sqrt{\pi v}} \int_0^\infty \left(\frac{u}{2}\right)^{\frac{v}{2}-1} e^{-\frac{u}{2} - \frac{t^2 u}{2v}} \left(\frac{u}{2}\right)^{\frac{1}{2}} du,$$

$$= \frac{1}{2\Gamma\left(\frac{v}{2}\right) \sqrt{\pi v}} \int_0^\infty \left(\frac{u}{2}\right)^{\frac{v+1}{2}-1} e^{-\frac{u}{2}\left(1 + \frac{t^2}{v}\right)} du.$$

At this point, we will employ a change of variables. We set $x = \frac{u}{2}$ and note that $u = 2x$ and $du = 2dx$. This implies that,

$$f_T(t; v) = \frac{1}{2\Gamma\left(\frac{v}{2}\right) \sqrt{\pi v}} \int_0^\infty \left(\frac{2x}{2}\right)^{\frac{v+1}{2}-1} e^{-\frac{2x}{2}\left(1 + \frac{t^2}{v}\right)} 2dx, \qquad (A.13)$$

$$= \frac{1}{\Gamma\left(\frac{v}{2}\right) \sqrt{\pi v}} \int_0^\infty x^{\frac{v+1}{2}-1} e^{-x\left(1 + \frac{t^2}{v}\right)} dx.$$

A.2 Toward the t-Distribution

Now, if we set $b = \frac{v+1}{2} - 1$ and $a = 1 + \frac{t^2}{v}$, then we have

$$f_T(t; v) = \frac{1}{\Gamma\left(\frac{v}{2}\right)\sqrt{\pi v}} \int_0^\infty x^b e^{-ax} dx. \tag{A.14}$$

This feels like a dead-end, since it is not immediately obvious how one might analytically resolve the integral in equation A.14. Abramovitz and Stegun (1965, Section 6.1) come to the rescue with the following definition of the gamma function,

$$\Gamma(b+1) = \int_0^\infty x^{(b+1)-1} e^{-x} dx, \tag{A.15}$$

$$= a^{b+1} \int_0^\infty x^{(b+1)-1} e^{-ax} dx.$$

A quick manipulation immediately implies that,

$$\int_0^\infty x^{(b+1)-1} e^{-ax} dx = \frac{\Gamma(b+1)}{a^{b+1}}. \tag{A.16}$$

This allows us to simplify equation A.14 as follows,

$$f_T(t; v) = \frac{1}{\Gamma\left(\frac{v}{2}\right)\sqrt{\pi v}} \underbrace{\frac{\Gamma(b+1)}{a^{b+1}}}_{\text{Equation A.16}}, \tag{A.17}$$

$$= \frac{1}{\Gamma\left(\frac{v}{2}\right)\sqrt{\pi v}} \frac{\Gamma(\frac{v+1}{2} - \cancel{1} + \cancel{1})}{\left(1 + \frac{t^2}{v}\right)^{\frac{v+1}{2} - \cancel{1} + \cancel{1}}},$$

$$= \frac{\Gamma(\frac{v+1}{2})}{\Gamma\left(\frac{v}{2}\right)\sqrt{\pi v}} \left(1 + \frac{t^2}{v}\right)^{-\frac{(v+1)}{2}}.$$

This is the density function of the standard t-distribution, which has a support of $(-\infty, \infty)$. It is also a (pseudo) formal justification of the t-threshold model construction found in Chap. 4.[3] Its expectation is $\mathbb{E}(T) = 0$ for $v > 1$ and $\text{var}(T) = \frac{v}{v-2}$, which indicates that the variance is undefined when $v = 2$ and unreasonably negative when $v < 2$. Similarly, the skewness is undefined for $v = 3$ and the kurtosis is infinite when $v = 4$. For this reason, in financial applications, one typically restricts attention to the family of t-distributions for $v > 4$.

---

[3] A similar, albeit more complex, result is possible for other members of the normal-variance-mixture class of distributions.

If you transform a standard t-distributed variable, T, with ν degrees of freedom as follows,

$$\tilde{T} = \mu + \sigma T, \qquad (A.18)$$

then its density function has this form,

$$f_{\tilde{T}}(t; \nu, \mu, \sigma) = \frac{\Gamma(\frac{\nu+1}{2})}{\Gamma(\frac{\nu}{2})\sqrt{\pi\nu\sigma^2}} \left(1 + \frac{(t-\mu)^2}{\nu\sigma^2}\right)^{-\frac{(\nu+1)}{2}}. \qquad (A.19)$$

In this setting, as one might expect, the expected value is $\mathbb{E}(\tilde{T}) = \mu$ for $\nu > 1$, but the $\text{var}(\tilde{T}) = \left(\frac{\nu}{\nu-2}\right)\sigma^2$ for $\nu > 2$. This amounts to a change in both scale and location of T. As the degrees of freedom, ν, tends to infinity, the t-distribution converges to the Gaussian density. This is a particularly nice feature for risk applications.

The next step is to consider the multivariate version of the t-distribution. Unfortunately, the multivariate t-distribution is not uniquely specified. There are a number of possible choices. We will employ what Kotz and Nadarajah (2004) call the "most common and natural form." Let $X \in \mathbb{R}^d$ denote a random vector following a d-dimensional t-distribution with ν degrees of freedom. Its density has the following form,

$$f(x) = \frac{\Gamma\left(\frac{\nu+d}{2}\right)}{\Gamma\left(\frac{\nu}{2}\right)\nu^{\frac{d}{2}}\pi^{\frac{d}{2}}|\Sigma|^{\frac{1}{2}}} \left(1 + \frac{(x-\mu)^T \Sigma^{-1}(x-\mu)}{\nu}\right)^{-\frac{(\nu+d)}{2}}, \qquad (A.20)$$

where $|\cdot|$ denotes the determinant of a matrix, $\mu \in \mathbb{R}^d$ is the mean vector for X, and $\Sigma \in \mathbb{R}^{d \times d}$ is its covariance matrix. The expectation of X is $\mathbb{E}(X) = \mu$ and the variance of X, $\text{var}(X) = \left(\frac{\nu}{\nu-2}\right)\Sigma$. As a consequence, we can describe the distribution of $X \sim \mathcal{T}_\nu\left(\mu, \left(\frac{\nu}{\nu-2}\right)\Sigma\right)$.

The joint density specification in equation A.20 only makes sense if all of its marginals share the same number of degrees of freedom. In this case, the joint density belongs to the elliptical family of distributions, which may be described by a covariance matrix. Other choices of joint density for the multivariate t-distribution are possible, which allow for different degrees of freedom for each of the marginals, but they no longer belong to the elliptical family. While not necessarily problematic, it poses some practical problems in finance applications, where the covariance matrix is heavily employed to describe dependence. As a final note, the product of the marginal distributions does not, in general, yield the joint density. This implies that—unlike the Gaussian setting—uncorrelated t-distributed random variables are not necessarily independent. See Kotz and Nadarajah (2004) and Roth (2013) for a much more detailed description of the technical properties of this distribution.

A.3 Simulating Correlated t Variates

As a final contribution, this appendix will also briefly describe, and justify, an algorithm for generating a random t-distributed vector with mean μ and covariance matrix, Σ. This will form a useful link with the normal-variance mixture structure introduced in Chap. 4. Along the way, we will address the general question of generating correlated Gaussian random variables. First, we start by generating a random vector of standard normal variates, Z. This is easily accomplished in virtually any mathematical software.[4]

The next step is to transform the standard normal vector into a correlated random vector, $X \in \mathbb{R}^d$, with covariance matrix, $\Sigma \in \mathbb{R}^{d \times d}$. This requires manipulating the matrix, Σ. Since Σ is, by definition, symmetric, real-valued and positive-definite, there exists an assortment of useful decompositions that may be employed.[5] In particular, we may decompose our covariance matrix as follows,

$$\Sigma = UDU^{-1}, \tag{A.21}$$

where $D \in \mathbb{R}^{d \times d}$ is a diagonal matrix with its elements corresponding to the eigenvalues of Σ. Moreover, since Σ is a real-valued and symmetric matrix, the matrix $U \in \mathbb{R}^{d \times d}$ is orthogonal. Practically, this implies that $UU^T = I_d$. As a consequence, $U^T = U^{-1}$, which allows us to rewrite equation A.21 as,

$$\Sigma = UDU^T. \tag{A.22}$$

Since, D is diagonal, we may unequivocally write $\sqrt{D} \in \mathbb{R}^{d \times d}$ to denote an element-by-element square-root operation on each eigenvalue. Moreover, since D is positive-definite, again by definition, its eigenvalues are strictly positive implying that we need not enter the complex domain to evaluate $\sqrt{D}$. This permits us to re-write equation A.22 as,

$$\Sigma = U \underbrace{\sqrt{D}\sqrt{D}}_{D} U^T, \tag{A.23}$$

$$= U\sqrt{D}\left(U\sqrt{D}\right)^T.$$

To compute a vector of multivariate normally distributed variables with covariance matrix, Σ, we now need only perform the following multiplication,

$$X = U\sqrt{D}Z. \tag{A.24}$$

---

[4]In contrast, built-in software functions are often less straightforward for the t distribution. Hofert (2013), for example, provides a helpful discussion on the potential pitfalls of sampling from the multivariate t distribution in the R language.

[5]A matrix A is positive definite if $v \cdot A \cdot v > 0$ for all $v \in V$, where V is the appropriate vector space. In addition, a positive-definite matrix will have all postive eiqenvalues. See Golub and Loan (2012) for more details.

We can verify that this is correct by examining the variance of the vector, X, as follows

$$\text{var}(X) = \text{var}\left(U\sqrt{D}\mathbf{Z}\right), \tag{A.25}$$

$$= U\sqrt{D}\text{var}(Z)\left(U\sqrt{D}\right)^T,$$

$$= U\sqrt{D}I_d\left(U\sqrt{D}\right)^T,$$

$$= U\sqrt{D}\sqrt{D}U^T,$$

$$= \Sigma,$$

which is as desired.

This approach may also be performed through the use of the Cholesky decomposition on one's covariance matrix—see Press et al. (1992) and Golub and Loan (2012) for more information on this matrix operation. In theory, a covariance matrix must be positive definite, but in practice, it is certainly possible for a large empirically computed covariance matrix to be positive semi-definite leading to a failure of the Cholesky decomposition. This may also lead to practical problems with the eigenvalue decomposition. Practically, therefore, one should typically employ the singular-value-decomposition. In this case, the covariance matrix is decomposed as,

$$\Sigma = HSV^T, \tag{A.26}$$

where $S \in \mathbb{R}^{d\times d}$ is a diagonal matrix, whose elements are called the singular values. H and V both have dimensions $\mathbb{R}^{d\times d}$.[6] For a real-valued, symmetric matrix, the singular values and eigenvalues coincide. Moreover, in this setting $H = V$, which are both orthogonal and equal to U from equation A.22. Finally, should Σ, due to computational issues, be indefinite, then negative eigenvalues become positive in S. This leads to avoiding the problem associated with indefiniteness of Σ—the algorithm works precisely as before with H taking the role of U and S replacing D.

Armed with these correlated Gaussian random variables, we need only transform them into correlated t-distributed random variables. This requires the definition of the following random variable,

$$W = \sqrt{\frac{\nu}{U}}, \tag{A.27}$$

---

[6]The singular-value decomposition is very general and may be applied to non-square matrices. In these cases, H, S and V need not have the same dimensions. See Press et al. (1992) and Golub and Loan (2012) for much more detail.

A.3 Simulating Correlated t Variates

where $W \in \mathbb{R}_+$ and, as you will recall, ν is our choice of degrees of freedom. U denotes an independent randomly generated chi-squared random variate with ν degrees of freedom. We then proceed to construct the following random variate,

$$T = \mu + W \underbrace{U\sqrt{D}Z}_{X}, \tag{A.28}$$

where $T \sim \mathcal{T}_\nu(\mu, \Sigma)$. In other words, the expectation of T is $\mathbb{E}(T) = \mu$ and $\text{var}(T) = \left(\frac{\nu}{\nu-2}\right)\Sigma$. The expectation is easily verified,

$$\begin{aligned} \mathbb{E}(T) &= \mathbb{E}(\mu + WU\sqrt{D}Z), \\ &= \mu + \mathbb{E}(W)U\sqrt{D}\underbrace{\mathbb{E}(Z)}_{=0}, \\ &= \mu. \end{aligned} \tag{A.29}$$

Addition of μ amounts to adjusting the location of the t-distribution and works in precisely the same manner as in the Gaussian setting.

Demonstrating that the $\text{var}(T)$ has the desired value requires a bit more effort. Working directly with W is difficult—instead, we will use a trick. Let us first introduce a gamma-distributed variable, V, with parameters α and β—by definition, the $\mathbb{E}(V) = \alpha\beta$. More formally, we have that $V \sim \Gamma(\alpha, \beta)$. Then we re-define,

$$T = \mu + \sqrt{\frac{\nu}{V}} X, \tag{A.30}$$

where $\nu \in \mathbb{R}_+$ and, as before with the addition of a non-zero mean, $X \sim (\mu, \Sigma)$. Equation A.30 employs, in fact, the same trick used to construct a t-distributed default state variable in the threshold setting described in Chap. 4. Both X and V are independent. It is easy to see that the conditional distribution of T is relatively straightforward. In particular,

$$T|V = v \sim \mathcal{N}\left(\mu, \left(\frac{\nu}{v}\right)\Sigma\right). \tag{A.31}$$

Equation A.30 amounts, conditionally, to a scaling of the covariance matrix. It also follows, from independence, that $\mathbb{E}(T) = \mu$.

We now suggestively define, $\hat{W} = \frac{\nu}{V}$. $\hat{W}$ follows, by virtue of V, an *inverse-gamma* distribution—that is $\hat{W} \sim \Gamma^{-1}(\alpha, \nu\beta)$. This allows us to re-write equation A.30 as,

$$\begin{aligned} T &= \mu + \sqrt{\underbrace{\frac{\nu}{V}}_{\hat{W}} X}, \\ &= \mu + \sqrt{\hat{W}} X. \end{aligned} \tag{A.32}$$

By definition, $\mathbb{E}(\hat{W}) = \nu\left(\frac{\beta}{\alpha-1}\right)$. This may seem a bit odd, but in a moment, we will draw a link between the gamma, inverse-gamma, and chi-squared distributions. We are now ready to compute the variance of T. From first principles,

$$\text{var}(T) = \mathbb{E}\left(T^2 - \mathbb{E}(T)^2\right), \tag{A.33}$$

$$= \mathbb{E}\left(T^2 - \mu^T\mu\right),$$

$$= \mathbb{E}\left(T^2\right) - \mu^T\mu,$$

$$= \mathbb{E}\left(\left(\mu + \sqrt{\hat{W}}X\right)^T\left(\mu + \sqrt{\hat{W}}X\right)\right) - \mu^T\mu,$$

$$= \cancel{\mu^T\mu} + \mu^T\mathbb{E}\left(\sqrt{\hat{W}}X\right) + \mathbb{E}\left(\sqrt{\hat{W}}X^T\right)\mu + \mathbb{E}(\hat{W}X^TX) - \cancel{\mu^T\mu},$$

$$= \underbrace{\mathbb{E}(\sqrt{\hat{W}})\mu^T\overbrace{\mathbb{E}(X)}^{=0}}_{=0} + \underbrace{\mathbb{E}(\sqrt{\hat{W}})\overbrace{\mathbb{E}\left(X^T\right)}^{=0}\mu}_{=0} + \mathbb{E}(\hat{W}X^TX),$$

$$= \mathbb{E}(\hat{W})\mathbb{E}(X^TX),$$

$$= \nu\left(\frac{\beta}{\alpha-1}\right)\Sigma.$$

This might not seem like progress, but an inverse-gamma distributed variable with $\alpha = \frac{\nu}{2}$ and $\beta = \frac{1}{2}$ is, in fact, also an inverse χ^2-distributed variable with ν degrees of freedom. If we input these values into equation A.33, we have

$$\text{var}(T) = \nu\left(\frac{\beta}{\alpha-1}\right)\Sigma, \tag{A.34}$$

$$\text{var}(\mu + \sqrt{\hat{W}}X) = \nu\left(\frac{\frac{1}{2}}{\frac{\nu}{2}-1}\right)\Sigma,$$

$$\text{var}\left(\mu + \sqrt{\left(\frac{\nu}{V}\right)}X\right) = \nu\left(\frac{1}{\nu-2}\right)\Sigma,$$

$$\text{var}\left(\mu + \underbrace{\sqrt{\left(\frac{\nu}{U}\right)}X}_{W}\right) = \left(\frac{\nu}{\nu-2}\right)\Sigma,$$

where, the final step follows from the fact that although $\frac{1}{V}$ follows an inverse-χ^2 distribution with ν degrees of freedom, U follows a χ^2 distribution with ν

degrees of freedom. In other words, W and $\hat{W}$ are equivalent. Thus, equation A.34 demonstrates the correctness of the algorithm described in equation A.27 and A.28. This is a bit convoluted, but we have taken the scenic route to illustrate the close links between many of these fundamental statistical distributions.

This closes the loop between the construction of correlated t-distributed random variables and the inducement of a t-threshold model through multiplication of the Gaussian state-variable by the ratio of a chi-squared random variable and its degrees of freedom parameter. The link is the structure found in equation A.30. If one generalizes this approach, it leads to the class of normal-variance mixture models discussed in Chap. 4 and addressed in much more detail in Kotz et al. (2001).

A.4 A Quick Example

Practically, generation of correlated multivariate t-distributed variables can be performed through a small number of steps. It begins with the estimation, or specification, of the values of μ, Σ and the degrees-of-freedom parameter, ν. One then proceeds to generate a set of independent Gaussian distributed variables, $Z \sim \mathcal{N}(0, I)$. Given Σ, we compute its singular-value decomposition yielding $\Sigma = HSV^T$. This permits us to construct a collection of correlated Gaussian random variables as $X = \mu + H\sqrt{S}Z \sim \mathcal{N}(\mu, \Sigma)$. The final step is to simulate a chi-squared random variable $U \sim \chi^2(\nu)$ and use it to manufacture a set of t-distributed random variables as $T = \sqrt{\frac{\nu}{U}}X \sim \mathcal{T}_\nu(\mu, \Sigma)$.

To make this discussion more concrete, algorithm A.1 illustrates the Python subroutine used to simulate both t-distributed and Gaussian outcomes. It explicitly follows the previously outlined steps. N denotes the number of dimensions, M is the number of simulations, nu is the degrees of freedom, and myRho is degree of linear correlation between the individual random variables. To keep things simple, the location is set to zero and each random variable has unit variance.

Figure A.1 illustrates bivariate scatterplots for 10,000 draws from algorithm A.1 with myRho and nu set to 0.5 and 9, respectively. For comparison, the Gaussian

Algorithm A.1 Simulating correlated Gaussian and t-distributed random variates

```
def simulateT(N,M,myRho,nu):
    Z = np.random.normal(0,1,[N,M])
    Sigma = np.array([[1.0,myRho],[myRho,1.0]])
    H,S,V = anp.svd(Sigma)
    X = np.dot(np.dot(H,np.sqrt(np.diag(S))),Z)
    W = np.sqrt(nu/np.tile(np.random.chisquare(nu,M),(N,1)))
    T = W*X
    return X,T
```

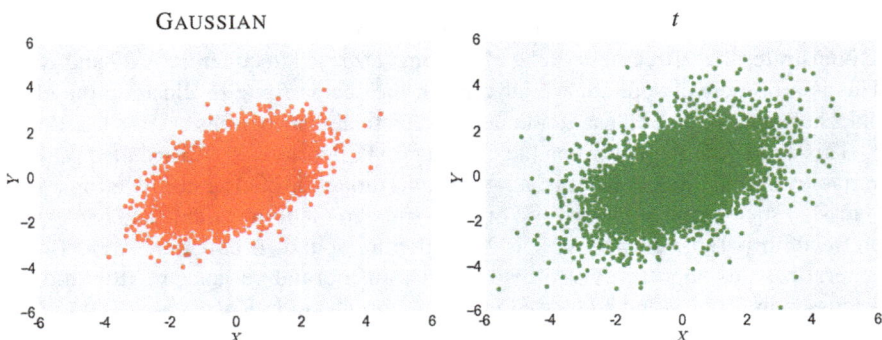

Fig. A.1 *Bivariate scatterplots*: The underlying figures use algorithm A.1 to illustrate the simulation of 10,000 bivariate Gaussian and t random-variate outcomes. The correlation coefficient, ρ, is set to 0.5 and the degrees of freedom parameter, $\nu = 9$.

outcomes are also provided. The elliptical form of both clouds of outcomes are obvious, but the number of extreme observations is clearly visible in the t setting. This is precisely the behaviour that makes it such a desirable and natural choice for financial applications.

To conclude, the t distribution is a normal-variance mixture distribution. In essence, the volatility is stochastic with an inverse-χ^2 form. The additional uncertainty associated with this specification of the volatility leads to a fascinating, and immensely useful, change in the structure of both the dependence and extreme outcomes in its joint distribution.

Appendix B
The Black-Scholes Formula

> *'Obvious' is the most dangerous word in mathematics.*
> (E.T. Bell)

In Chap. 5, the fundamentally important Merton (1974) model was introduced. Its development follows from the assumption of continuous-time asset and equity processes. This entails the use of Brownian motions and the associated stochastic calculus. This is not a trivial undertaking. Moreover, a critical result for the calibration of the model relies upon the celebrated Black and Scholes (1973) formula. For completeness, and to further one's insight into the inner workings of the structural Merton (1974) framework, this appendix works through the derivation of this result.

Harrison and Kreps (1979) and Harrison and Pliska (1981) are the foundational works underlying the general asset-pricing framework.[1] The principle idea is that the price of any contingent claim is its discounted expected value taken with respect to the equivalent martingale measure, $\mathbb{Q}$. Merton (1974) starts, however, with the base assumption of a firm's asset-value dynamics under $(\Omega, \mathcal{F}, \mathbb{P})$,

$$\frac{dA_t}{A_t} = \mu A_t dt + \sigma A_t d\hat{W}_t, \tag{B.1}$$

where μ and σ are scalar coefficients and $\{\hat{W}_t, \mathcal{F}_t\}_{t \geq 0}$ is a standard, scalar Wiener process on $(\Omega, \mathcal{F}, \mathbb{P})$. The first order of business, therefore, is to determine the asset dynamics under $\mathbb{Q}$.

B.1 Changing Probability Measures

Moving from the physical probability measure, $\mathbb{P}$, to the equivalent-martingale measure, $\mathbb{Q}$, requires the introduction of a numeraire asset. In principle, any positive-

---

[1] Duffie (1996) is also an excellent reference in this area.

valued asset will suffice, but not all choices are equally convenient. Our numeraire follows a (deterministic) process, termed the *money-market* account, and is defined as

$$dB_t = rB_t dt, \tag{B.2}$$

where r is the (constant) risk-free rate of interest.[2] Equation B.2 basically described a bank account earning, without uncertainty, the continuously compounded interest rate, r.

We can now define the following new process, which is a function of both the money-market account and our asset process,

$$Z_t = \frac{A_t}{B_t} = f(A_t, B_t), \tag{B.3}$$

for all $t \in [0, T]$. In essence, we have normalized the asset value by the money-market account.

We may employ Theorem 5.1—outlined in Chap. 5 and generally referred to as Itô's theorem, or occasionally, lemma—once again to describe the infinitesimal dynamics of our new process in equation B.3. For this, however, we require some partial derivatives. With respect to A, we have

$$\frac{\partial f}{\partial A} = \frac{1}{B_t}, \tag{B.4}$$

$$\frac{\partial^2 f}{\partial A^2} = 0.$$

The partial derivatives with reference to the money-market account, B, are

$$\frac{\partial f}{\partial B} = -\frac{A_t}{B_t^2}, \tag{B.5}$$

$$\frac{\partial^2 f}{\partial B^2} = \frac{2A_t}{B_t^3}.$$

This provides us with the raw ingredients required to implement Itô's famous result,

$$f(A_T, B_T) - f(A_t, B_t) = \int_t^T \frac{\partial f}{\partial A} dA_u + \int_t^T \frac{\partial f}{\partial B} dB_u \tag{B.6}$$

$$+ \frac{1}{2} \int_t^T \frac{\partial^2 f}{\partial A^2} \underbrace{d\langle A_u \rangle}_{=0} + \frac{1}{2} \int_t^T \frac{\partial^2 f}{\partial B^2} \underbrace{d\langle B_u \rangle}_{=0},$$

---

[2] r is assumed to be deterministic. This can be relaxed, but it complicates matters and, at this point, we are looking for the basic structure.

B.1 Changing Probability Measures

$$= \int_t^T \frac{1}{B_u} dA_u - \int_t^T \frac{A_u}{B_u^2} dB_u,$$

$$= \int_t^T \frac{1}{B_u}(\mu A_u du + \sigma A_u d\hat{W}(u)) - \int_t^T \frac{A_u}{B_u^2}(rB_u du),$$

$$Z_t - Z_t = \int_t^T (\mu - r)\frac{A_u}{B_u} du + \int_t^T \sigma \frac{A_u}{B_u} d\hat{W}_u,$$

$$dZ_t = (\mu - r)Z_t dt + \sigma Z_t d\hat{W}_t.$$

Conveniently, neither of the second-order terms contribute to the final results; in the first case, because the second derivative of f with respect to A vanishes and, in the second, because the quadratic-variation process, $\langle B \rangle$, is zero for deterministic B.

At this point, seeing that our new process also follows a geometric Brownian motion with a modified drift, we want to alter the underlying measure from $\mathbb{P}$ to the equivalent martingale measure, $\mathbb{Q}$, induced by the choice of the money-market account as the numeraire asset—this is also often termed the risk-neutral measure. It begins with the manipulation of the drift term,

$$dZ_t = (\mu - r)Z_t dt + \sigma Z_t d\hat{W}(t), \qquad (B.7)$$

$$= Z_t \left[(\mu - r)dt + \sigma d\hat{W}_t\right],$$

$$= Z_t \left[\sigma \left(\frac{\mu - r}{\sigma}\right) dt + \sigma d\hat{W}_t\right].$$

It is necessary to alter the original Wiener process, $\hat{W}(t)$, in order to alter the drift. In particular, we want to generate a new Wiener process, $W(t)$, which incorporates this change in drift, where $W(t)$ is also a standard, scalar Wiener process under the transformed probability measure, $\mathbb{Q}$. This involves adding the amount $\frac{\mu-r}{\sigma}$ to our original Wiener process, $\hat{W}(t)$. This quantity is typically termed the *market price of risk*. Our new Wiener process is defined as follows,[3]

$$dW_t = d\hat{W}_t + \frac{\mu - r}{\sigma} dt, \qquad (B.8)$$

$$d\hat{W}_t = dW_t - \frac{\mu - r}{\sigma} dt.$$

---

[3] There are several technical conditions for this operation to be permissible, which are satisfied easily given the nature of the integrand in the underlying expression. Fortunately, relatively little can go wrong in a model with constant coefficients.

Now let's substitute this expression into our process Z_t, defined in equation B.8, and simplify,

$$dZ_t = Z_t \left[\sigma \left(\frac{\mu - r}{\sigma} \right) dt + \sigma d\hat{W}_t \right], \tag{B.9}$$

$$= Z_t \left[\sigma \left(\frac{\mu - r}{\sigma} \right) dt + \sigma \overbrace{\left(dW_t - \frac{\mu - r}{\sigma} dt \right)}^{\text{Equation B.8}} \right],$$

$$= Z(u) \left[\sigma \cancel{\left(\frac{\mu - r}{\sigma} \right) dt} + \sigma dW_t - \sigma \cancel{\left(\frac{\mu - r}{\sigma} \right) dt} \right],$$

$$= \sigma Z_t dW(t).$$

Although it's not terribly obvious, this is the so-called risk-neutral process for the firm's asset-value process, A_t. To better see this, we may recall that $A_t = B_t Z_t = g(B_t, Z_t)$ and reapply Theorem 5.1 to aid us in writing down an expression for dA_t—we must also keep in mind, from equation B.3, the original definition of Z_t,

$$g(B_T, Z_T) - g(B_t, Z_t) = \int_t^T \frac{\partial g}{\partial B} dB_u + \int_t^T \frac{\partial g}{\partial Z} dZ_u \tag{B.10}$$

$$+ \frac{1}{2} \int_t^T \underbrace{\frac{\partial^2 g}{\partial Z^2}}_{=0} d\langle Z_u \rangle + \frac{1}{2} \int_t^T \underbrace{\frac{\partial^2 g}{\partial B^2}}_{=0} \underbrace{d\langle B_u \rangle}_{=0},$$

$$A_T - A_t = \int_t^T Z_u dB_u + \int_t^T B_u dZ_u,$$

$$= \int_t^T \frac{A_u}{\cancel{B_u}} (r\cancel{B_u} du) + \int_t^T \cancel{B_u} \underbrace{\left(\sigma \frac{A_u}{\cancel{B_u}} dW_u \right)}_{\text{Equation B.9}},$$

$$= r \int_t^T A_u du + \sigma \int_t^T A_u dW_u,$$

$$dA_t = rA_t dt + \sigma A_t dW_t.$$

Thus, the dynamics of the asset-price have a different drift—the risk-free interest rate, r—under $\mathbb{Q}$.[4] This is an important first step.

---

[4]This is the source of its common name, the risk-neutral probability measure.

B.2 Solving the Stochastic Differential Equation

Determination of the value of the firm's equity claim, in the Merton (1974) framework, reduces to evaluating the following expression,

$$E(t, A_t) = \mathbb{E}^{\mathbb{Q}} \left(e^{-\int_t^T r_u du} (A_T - K)^+ \right). \tag{B.11}$$

This amounts to a plain-vanilla European call option. Assuming a deterministic risk-free interest rate allows us to simplify this as,

$$E(t, A_t) = e^{-r(T-t)} \mathbb{E}^{\mathbb{Q}} \left((A_T - K)^+ \right). \tag{B.12}$$

Evaluating this expectation is not terribly simple. The task begins, however, by finding a solution to our risk-neutral stochastic differential equation, as summarized in equation B.10. We could use the usual approach from Theorem 5.1, but a more involved route is also worth examination. This thorny affair merits a new theorem.

Theorem B.1 (Girsanov) *Let X, Y be continuous, zero at the origin, semi-martingales on the probability space, $(\Omega, \mathcal{F}, \mathbb{P})$. Also let $\mathbb{H}_t$ be some $\mathcal{F}_t$-measurable random variable. Then the following stochastic differential equation,*

$$\mathbb{H}_T = \mathcal{E}(X_T) \left[\mathbb{H}_t + \int_t^T \mathcal{E}_s^{-1} dY_s - \int_t^T \mathcal{E}_s^{-1} d\langle X_s, Y_s \rangle \right], \tag{B.13}$$

has the solution,

$$\mathbb{H}_T = \mathbb{H}_t + Y_T + \int_t^T \mathbb{H}_s dX_s. \tag{B.14}$$

See Heunis (2011), Oksendal (1995), or Karatzas and Shreve (1998) for more background on this result.

To use this theorem, we need to choose convenient values for X and Y. We thus define the following two processes,

$$X_T = r \int_t^T du + \sigma \int_t^T W_u, \tag{B.15}$$

$$Y_T \equiv 0,$$

for all $t \in [0, T]$. Using these two processes, we construct A_t as follows

$$A_T = A_t + Y_T + \int_t^T A_u dX_u, \tag{B.16}$$

$$= A_t + \int_t^T A_u dX_u.$$

The final step involves evaluating the two integrals in equation B.13 to find the solution to our stochastic differential equation. It turns out, that because Y_T is identically zero, the resulting functions have very convenient forms. That is,

$$\int_t^T \mathcal{E}(X)_u^{-1} dY_u = 0, \tag{B.17}$$

and,

$$\int_t^T \mathcal{E}(X)_u^{-1} d\langle X_u, Y_u \rangle = 0, \tag{B.18}$$

and finally,

$$\mathcal{E}(X_T) = e^{X_T - \frac{1}{2} \langle X_T \rangle}, \tag{B.19}$$
$$= e^{r \int_t^T du + \sigma \int_t^T W_u - \frac{1}{2} \langle r \int_t^T du + \sigma \int_t^T W_u \rangle},$$
$$= e^{r(T-t) + \sigma(W_T - W_t) - \frac{\sigma^2}{2} \int_t^T du},$$
$$= e^{\left(r - \frac{\sigma^2}{2}\right)(T-t) + \sigma(W_T - W_t)}.$$

$\mathcal{E}(X_T)$—sometimes termed the Doléan or Itô exponential—is a strictly positive martingale for which a probability measure $\mathbb{Q}$ can be defined on $\{\Omega, \mathcal{F}\}$ whereby the Radon-Nikodym derivative is defined as,

$$\left. \frac{d\mathbb{Q}}{d\mathbb{P}} \right|_{\mathcal{F}_t} = \mathcal{E}(X_T). \tag{B.20}$$

This quantity permits us to change the probability measure on a measurable space. As a consequence, we can skip directly to our solution,

$$A_T = \mathcal{E}(X_T) \left[A_t + \underbrace{\int_t^T \mathcal{E}_s^{-1} dY_s}_{=0} - \underbrace{\int_t^T \mathcal{E}^{-1}(s) d\langle X, Y \rangle(s)}_{=0} \right], \tag{B.21}$$
$$= A_t e^{\left(r - \frac{\sigma^2}{2}\right)(T-t) + \sigma(W_T - W_t)},$$

which has basically the same form as the solution under the physical probability measure, $\mathbb{P}$, save the expected return of the asset has been replaced with the risk-free interest rate. Indeed, this is the point. In the risk-neutral world, all assets earn the risk-free rate.

B.3 Evaluating the Integral

We can also represent this, due to the normality of the independent increments in the Wiener process, in the following more convenient manner,

$$A_T = A_t e^Y, \tag{B.22}$$

where,

$$Y \sim \mathcal{N}\left(\left(r - \frac{\sigma^2}{2}\right)(T-t), \sigma^2(T-t)\right). \tag{B.23}$$

This last statement gives us the insight required to write a more workable expression for the price of our equity. This follows from equation B.12 and the formal definition of an expectation,

$$E(t, A_t) = e^{-r(T-t)} \int_{-\infty}^{\infty} (A_T - K)^+ \phi_Y(y) dy, \tag{B.24}$$

where $\phi_Y(y)$ is the probability density function for our normal random variate, Y, highlighted in equation B.23. The first term in the integrand, $(A_T - K)^+$, is simply a function of our random variable, A_T. With the use of Girsanov's theorem, we have solved the stochastic differential equation governing the risk-neutral dynamics of A_t. This solution is instrumental in reducing the equity value to a (reasonably) manageable integral.

B.3 Evaluating the Integral

Evaluating the integral on the right-hand-side of equation B.24 will yield the value of the firm's equity. In general, this must done using a numeric method, but in the case of a fairly large class of European derivative securities it may solved analytically. The steps involved are relatively tedious, but working through this problem in detail is a useful step in understanding the form of the final expression. In particular, we may use the form of the solution for this basic option and compare it to more complex solutions associated with exotic European derivatives—which may arise when one relaxes some of the less palatable assumptions associated with the base Merton model.

To attack equation B.24 in earnest, we introduce the following change of variables,

$$\bar{r} = \left(r - \frac{\sigma^2}{2}\right), \tag{B.25}$$

$$\tau = T - t,$$

and define $Z \sim \mathcal{N}(0, 1)$. We then introduce these transformed variables into our integral in equation B.24 and make a few substitutions and use the result from equation B.21,

$$\int_{-\infty}^{\infty} (A_T - K)^+ \phi_Y(y) dy = \int_{-\infty}^{\infty} \left(A_t e^{\left(r - \frac{\sigma^2}{2}\right)(T-t) + \sigma(\hat{W}_T - \hat{W}_t)} - K \right)^+ \phi_Y(y) dy, \quad \text{(B.26)}$$

$$= \int_{-\infty}^{\infty} \left(A_t e^{\bar{r}\tau + \sigma\sqrt{\tau}z} - K \right)^+ \phi(z) dz,$$

where we may now compute our expectation with respect to the standard normal density, $\phi(\cdot)$. We nevertheless need some appropriate way to deal with the positive function, $(\cdot)^+$, appearing in the integrand. We observe that if $S(T) < K$ then the entire integral vanishes. Working with this fact, we obtain the following lower bound

$$A_T < K, \quad \text{(B.27)}$$

$$A_t e^{\bar{r}\tau + \sigma\sqrt{\tau}z} < K,$$

$$\bar{r}\tau + \sigma\sqrt{\tau}z < \ln\left(\frac{K}{A_t}\right),$$

$$z < \frac{\ln\left(\frac{K}{A_t}\right) - \bar{r}\tau}{\sigma\sqrt{\tau}},$$

$$z < \bar{z},$$

where,

$$\bar{z} = \frac{\ln\left(\frac{K}{A_t}\right) - \bar{r}\tau}{\sigma\sqrt{\tau}}. \quad \text{(B.28)}$$

We have that our integral vanishes when our variable of integration, z, is less than the critical value $\bar{z}$. As a consequence, this will become our lower limit of integration and permits us to eliminate the $(\cdot)^+$ function. Making this change to equation B.26, we may make some immediate simplifications in the following manner,

$$\int_{-\infty}^{\infty} \left(A_t e^{\bar{r}\tau + \sigma\sqrt{\tau}z} - K \right)^+ \phi(z) dz = \int_{\bar{z}}^{\infty} \left(A_t e^{\bar{r}\tau + \sigma\sqrt{\tau}z} - K \right) \phi(z) dz, \quad \text{(B.29)}$$

$$= \int_{\bar{z}}^{\infty} A_t e^{\bar{r}\tau + \sigma\sqrt{\tau}z} \phi(z) dz - K \int_{\bar{z}}^{\infty} \phi(z) dz,$$

$$= \int_{\bar{z}}^{\infty} A_t e^{\bar{r}\tau + \sigma\sqrt{\tau}z} \frac{1}{\sqrt{2\pi}} e^{-\frac{z^2}{2}} dz - K\mathbb{P}(Z \geq \bar{z}),$$

$$= \frac{A_t e^{\bar{r}\tau}}{\sqrt{2\pi}} \int_{\bar{z}}^{\infty} e^{-\frac{z^2}{2} + \sigma\sqrt{\tau}z} dz - K\mathbb{P}(Z \leq -\bar{z}),$$

$$= \frac{A_t e^{\bar{r}\tau}}{\sqrt{2\pi}} \int_{\bar{z}}^{\infty} e^{-\frac{z^2}{2} + \sigma\sqrt{\tau}z} dz - K\Phi(-\bar{z}),$$

B.3 Evaluating the Integral

where, as per usual, $\Phi(\cdot)$ represents the cumulative distribution function of the standard normal distribution. The final few steps follow from the symmetry of the normal distribution.

A critical element in evaluating this integral involves dealing with the first expression. This moderately tedious task involves completing the square in the exponent of the integrand. To do this, observe the following,

$$\sigma\sqrt{\tau}z - \frac{z^2}{2} = -\frac{1}{2}\left(z^2 - 2\sigma\sqrt{\tau}z\right), \tag{B.30}$$

$$= -\frac{1}{2}\left(z - 2\sigma\sqrt{\tau} \underbrace{+\sigma^2\tau - \sigma^2\tau}_{=0}\right),$$

$$= -\frac{1}{2}\left((z - \sigma\sqrt{\tau})^2 - \sigma^2\tau\right),$$

$$= -\frac{(z - \sigma\sqrt{\tau})^2}{2} + \frac{\sigma^2\tau}{2}.$$

We may use this fact, in conjunction with equation B.29, to attain the underlying expression,

$$\frac{A_t e^{\bar{r}\tau}}{\sqrt{2\pi}} \int_{\bar{z}}^{\infty} e^{-\frac{z^2}{2} + \sigma\sqrt{\tau}z} dz = \frac{A_t e^{\bar{r}\tau}}{\sqrt{2\pi}} \int_{\bar{z}}^{\infty} e^{-\frac{(z-\sigma\sqrt{\tau})^2}{2} + \frac{\sigma^2\tau}{2}} dz, \tag{B.31}$$

$$= \frac{A_t e^{r\tau - \frac{\sigma^2\tau}{2} + \frac{\sigma^2\tau}{2}}}{\sqrt{2\pi}} \int_{\bar{z}}^{\infty} e^{-\frac{(z-\sigma\sqrt{\tau})^2}{2}} dz,$$

$$= A_t e^{r\tau} \int_{\bar{z}}^{\infty} \frac{1}{\sqrt{2\pi}} e^{-\frac{(z-\sigma\sqrt{\tau})^2}{2}} dz.$$

We have thus reduced things to a single random variable, which is now agreeably distributed as $\mathcal{N}(\sigma\sqrt{\tau}, 1)$. We would, of course, prefer to use the standard normal distribution to compute this value. This problem is easily solved by subtracting the mean from the random variable. We thus have a rather simple expression for our remaining integral in equation B.31,

$$A_t e^{r\tau} \int_{\bar{z}}^{\infty} \frac{1}{\sqrt{2\pi}} e^{-\frac{(z-\sigma\sqrt{\tau})^2}{2}} dz = A_t e^{r\tau} \mathbb{Q}(Z \geq \bar{z} - \sigma\sqrt{\tau}), \tag{B.32}$$

$$= A_t e^{r\tau} \mathbb{Q}(Z \leq -\bar{z} + \sigma\sqrt{\tau}),$$

$$= A_t e^{r\tau} \Phi(-\bar{z} + \sigma\sqrt{\tau}).$$

B.4 The Final Result

All that remains is to put all the pieces together to obtain the celebrated Black and Scholes (1973) formula for a European call option. Let's go back to our original equation B.26 and substitute to undo our variable transformations,

$$E(t, A_t) = e^{-r\tau} \left(\underbrace{A_t e^{r\tau} \Phi(-\bar{z} + \sigma\sqrt{\tau}) - K\Phi(-\bar{z})}_{\text{Equation B.32}} \right), \tag{B.33}$$

$$= A_t \Phi(-\bar{z} + \sigma\sqrt{\tau}) - e^{-r\tau} K\Phi(-\bar{z}),$$

$$= A_t \Phi \left(-\left(\frac{\ln\left(\frac{K}{A_t}\right) - \bar{r}\tau}{\sigma\sqrt{\tau}} \right) + \frac{\sigma^2 \tau}{\sigma\sqrt{\tau}} \right) - e^{-r\tau} K\Phi \left(-\left(\frac{\ln\left(\frac{K}{A_t}\right) - \bar{r}\tau}{\sigma\sqrt{\tau}} \right) \right),$$

$$= A_t \Phi \left(\frac{-\ln\left(\frac{K}{A_t}\right) + \left(r - \frac{\sigma^2}{2}\right)\tau + \sigma^2 \tau}{\sigma\sqrt{\tau}} \right) - e^{-r\tau} K\Phi \left(\frac{\ln\left(\frac{A_t}{K}\right) + \left(r - \frac{\sigma^2}{2}\right)\tau}{\sigma\sqrt{\tau}} \right),$$

$$= A_t \Phi \left(\frac{\ln\left(\frac{A_t}{K}\right) + \left(r + \frac{\sigma^2}{2}\right)(T - t)}{\sigma\sqrt{T - t}} \right) - e^{-r(T-t)} K\Phi \left(\frac{\ln\left(\frac{A_t}{K}\right) + \left(r - \frac{\sigma^2}{2}\right)(T - t)}{\sigma\sqrt{T - t}} \right).$$

This is a little unwieldy and thus, in the literature, it is generally quoted in the following form,

$$E(t, A_t) = A_t \Phi(d_1) - e^{-r(T-t)} K\Phi(d_2), \tag{B.34}$$

where,

$$d_1 = \frac{\ln\left(\frac{A_t}{K}\right) + \left(r + \frac{\sigma^2}{2}\right)(T - t)}{\sigma\sqrt{T - t}}, \tag{B.35}$$

$$d_2 = \frac{\ln\left(\frac{A_t}{K}\right) + \left(r - \frac{\sigma^2}{2}\right)(T - t)}{\sigma\sqrt{T - t}} = d1 - \sigma\sqrt{T - t}.$$

Thus for a plain vanilla European call option, we have, from the Merton model, a relatively straightforward formula to represent the value of its equity claim.

Appendix C
Markov Chains

> *War is the unfolding of miscalculations.*
>
> (Barbara W. Tuchman)

It is common practice to assume that the ratings transition behaviour for each obligor follows a Markov-chain process. This implies that this object plays a key role in credit-risk modelling. This appendix consequently provides some important Markov-chain properties and key results, which should prove useful for better understanding of this critical building block.

C.1 Some Background

A *Markov chain* is a discrete-state stochastic process that can be defined in either continuous or discrete time.[1] Our credit-risk application will focus on discrete-time Markov chains. Conceptually, this class of processes is similar to the more general (and subtle) theory relating to continuous-time, continuous-state Markov processes.[2]

To get started, let us consider an integer-valued, time-indexed sequence of random variables, $\{S_t, t \in \{1, \ldots, T\}\}$. Moreover, the probability that S_t takes on a value j, given that its current value is i, depends on that previous value alone. That is,

$$\mathbb{P}[S_t = j \mid S_{t-1} = i, S_{t-2} = k, \ldots] = \mathbb{P}[S_t = j \mid S_{t-1} = i], \quad (C.1)$$
$$= p_{ij}.$$

---

[1] This discussion follows, in more or less equal parts, from the excellent analysis in Hamilton (1994), Brémaud (1999), and Meyn and Tweedie (1993).

[2] The theory of Markov chains can also be presented in a measure-theoretic framework. Nevertheless, as our use of these processes will be very straightforward, this extra mathematical machinery does not seem necessary or even desirable.

This is the seminal property of a Markov chain. Loosely speaking, it is a process that has a very limited memory. At any given point in time, the distribution of outcomes in the subsequent state depends only on its current state. As stated in equation C.1, we let p_{ij} represent the probability that the process finds itself in state j given that it previously found itself in state i. These are called the transition probabilities of our Markov and we define them in the following matrix form,

$$\underbrace{\{p_{ij}\}_{i,j=1,2,\ldots,N}}_{\text{Set of transition probabilities}} \equiv \underbrace{\begin{bmatrix} p_{11} & p_{12} & \cdots & p_{1N} \\ p_{21} & p_{22} & \cdots & p_{2N} \\ \vdots & \vdots & \ddots & \vdots \\ p_{N1} & p_{N2} & \cdots & p_{NN} \end{bmatrix}}_{\text{Transition matrix: } P} \quad (C.2)$$

where,

$$p_{i1} + p_{i2} + \cdots + p_{iN} = \sum_{j=1}^{N} p_{ij} = 1 \quad (C.3)$$

for all $i = 1, \ldots, N$. Or, more simply, each row of the transition matrix must sum to unity.

A key feature of Markov chains is that the probability that an observation from state i will be followed, in m periods, by an observation from j is given by the ith row and jth column from the matrix P^m. Or, in mathematical terms,

$$\mathbb{P}[S_{t+m} = j \mid S_t = i] = \bar{p}_{ij}, \quad (C.4)$$

where,

$$P^m = \begin{bmatrix} \bar{p}_{11} & \cdots & \bar{p}_{1i} & \cdots & \bar{p}_{1N} \\ \bar{p}_{21} & \cdots & \bar{p}_{2i} & \cdots & \bar{p}_{2N} \\ \vdots & \vdots & \vdots & \ddots & \vdots \\ \bar{p}_{N1} & \cdots & \bar{p}_{Ni} & \cdots & \bar{p}_{NN} \end{bmatrix}. \quad (C.5)$$

This feature of Markov chains allows us to describe the future dynamics of the process using only its transition matrix. Additional details, on why this is actually true, are found in the next section.

A Markov chain is termed *irreducible* if all of its states can be reached for any given starting point. This basically means that every element in the transition matrix lies in the open interval, (0, 1). To see this more clearly, consider the following simple, first-order, two-state, Markov-chain transition matrix,

$$P = \begin{bmatrix} 0.75 & 0.25 \\ 0 & 1 \end{bmatrix}. \quad (C.6)$$

C.2 Some Useful Results

Once our stochastic process enters into the second state, of course, it will never exit. This is termed an *absorbing* state and it, may or may not, be a desirable property. It will depend on the dynamics underlying what one is trying to model. Default, for example, is a quite reasonable description of an absorbing state. A Markov chain that does not have any absorbing states is, in fact, irreducible. In a sense, these are well-behaved Markov chains. If we desire irreducibility in a first-order, two-state setting, we merely require the following inequalities to hold,

$$\mathbb{P}[S_t = 0 \mid S_{t-1} = 0] \text{ and } \mathbb{P}[S_t = 1 \mid S_{t-1} = 1] < 1. \tag{C.7}$$

One additional feature of a Markov chain worth reviewing relates to the first property we examined. Specifically, one might reasonably wonder whether, as $m \to \infty$, does the transition matrix, P^m, become stable? The answer to this rather vague question is yes. Slightly more formally, an irreducible Markov chain eventually converges to a limiting distribution.[3] For an N-state Markov chain the $1 \times N$ vector of ergodic (or steady-state) probabilities, denoted π, satisfies the following,

$$P\pi = \pi. \tag{C.8}$$

It is the vector, π, that describes the steady-state probabilities of a Markov chain. Moreover, we have the following result,

$$\lim_{m \to \infty} P^m = \mathbf{1}\pi, \tag{C.9}$$

where $\mathbf{1}$ is an $N \times 1$ column vector of ones. This result is discussed in the next section.

C.2 Some Useful Results

In the previous section, it was claimed that the transition probabilities for a Markov chain m periods in the future are described by the matrix, P^m. To demonstrate this concept, we introduce a $1 \times N$ random vector, ξ_t, which has the following form

$$\xi_t = \begin{cases} \begin{bmatrix} 1 & 0 & \cdots & 0 \end{bmatrix}, & \text{if } S_t = 1 \\ \begin{bmatrix} 0 & 1 & \cdots & 0 \end{bmatrix}, & \text{if } S_t = 2 \\ \vdots & \vdots \\ \begin{bmatrix} 0 & 0 & \cdots & 1 \end{bmatrix}, & \text{if } S_t = N \end{cases}. \tag{C.10}$$

---

[3] Technically, a Markov chain is termed *ergodic* if one of the eigenvalues of the transition matrix is unity and all other eigenvalues lie inside the unit circle.

This is a useful abstraction. It introduces the convenience of the indicator function into our analysis. More specifically, if $S_t = i$, then the jth element of ξ_t is equal to one with probability, p_{ij}, and zero otherwise. Moreover, if we write out the conditional expectation of ξ_{t+1} given that $S_t = i$, we have

$$\mathbb{E}[\xi_{t+1} \mid S_t = i] = \begin{bmatrix} p_{i1} & p_{i2} & \cdots & p_{iN} \end{bmatrix}. \tag{C.11}$$

This is merely the ith row of the transition matrix. Moreover, the actual realization of ξ_t is merely the ith row of an $N \times N$ identity matrix. To see this more clearly, consider a three-state example where

$$\mathbb{E}[\xi_{t+1} \mid S_t = 1] = \begin{bmatrix} p_{11} & p_{12} & p_{13} \end{bmatrix}, \tag{C.12}$$

and,

$$\xi_t = \begin{bmatrix} 1 & 0 & 0 \end{bmatrix}. \tag{C.13}$$

Exploiting the expression from equation C.11 and using the fact that conditioning on $S_t = 1$ or ξ_t is equivalent, we may conclude with a bit of simple matrix algebra that,

$$\mathbb{E}[\xi_{t+1} \mid S_t = 1] = \begin{bmatrix} p_{11} & p_{12} & p_{13} \end{bmatrix}, \tag{C.14}$$

$$\mathbb{E}[\xi_{t+1} \mid \xi_t] = \begin{bmatrix} 1 & 0 & 0 \end{bmatrix} \begin{bmatrix} p_{11} & p_{12} & p_{13} \\ p_{21} & p_{22} & p_{23} \\ p_{31} & p_{32} & p_{33} \end{bmatrix},$$

$$= \xi_t P.$$

Now, we may use the result in equation C.14, the Markov property, and a bit of manipulation, to obtain the following useful identity,

$$\mathbb{E}[\xi_{t+1} \mid \xi_t] = \xi_t P, \tag{C.15}$$

$$\mathbb{E}[\xi_{t+1} \mid \xi_t, \xi_{t-1}, \ldots] = \xi_t P,$$

$$\underbrace{\xi_{t+1} - \mathbb{E}[\xi_{t+1} \mid \xi_t, \xi_{t-1}, \ldots]}_{\text{Call this } \nu_{t+1}} = \xi_{t+1} - \xi_t P,$$

$$\xi_{t+1} = \xi_t P + \nu_{t+1}.$$

Thus, we have that ξ_t is an first-order autoregressive process and, given that ν_t is a sequence of martingale differences, it has a zero expectation (i.e., $\mathbb{E}[\nu_t] = 0$). Our

C.3 Ergodicity

original goal, however, was to see what happens m periods into the future. Consider, therefore, the simple case for $t+2$,

$$\xi_{t+2} = \xi_{t+1} P + v_{t+2}, \tag{C.16}$$

$$\xi_{t+2} = \underbrace{(\xi_t P + v_{t+1})}_{\text{Equation C.15}} P + v_{t+2},$$

$$\xi_{t+2} = \xi_t P^2 + v_{t+1} P + v_{t+2}.$$

If we generalize equation C.16 for an m-period forecast, we have

$$\xi_{t+m} = v_{t+m-0} P^0 + v_{t+m-1} P^1 + \cdots + v_{t+m-(m-1)} P^{m-1} + \xi_t P^m, \tag{C.17}$$

$$= \xi_t P^m + \sum_{i=0}^{m-1} v_{t+m-i} P^i.$$

This is a useful recursion relation, but it becomes even more handy when we add the expectation operator. Recall that the v_t terms have zero expectation. Therefore,

$$\mathbb{E}[\xi_{t+m} \mid \xi_t, \xi_{t-1}, \ldots] = \mathbb{E}\left[\xi_t P^m + \sum_{i=0}^{m-1} v_{t+m-i} P^i \,\middle|\, \xi_t, \xi_{t-1}, \ldots\right], \tag{C.18}$$

$$= \mathbb{E}[\xi_t P^m \mid \xi_t, \xi_{t-1}, \ldots] + \mathbb{E}\left[\sum_{i=0}^{m-1} v_{t+m-i} P^i \,\middle|\, \xi_t, \xi_{t-1}, \ldots\right],$$

$$= \underbrace{\mathbb{E}[\xi_t \mid \xi_t, \xi_{t-1}, \ldots]}_{\xi_t} P^m + \sum_{i=0}^{m-1} \underbrace{\mathbb{E}[v_{t+m-i} \mid \xi_t, \xi_{t-1}, \ldots]}_{=0} P^i,$$

$$\mathbb{E}[\xi_{t+m} \mid \xi_t] = \xi_t P^m.$$

This is a convenient result. It holds that—given initial conditions ξ_t—the transition probabilities for m periods in the future are provided by raising the single-period transition matrix to the mth power.

C.3 Ergodicity

In the early discussion, we also briefly addressed what happens to the transition probabilities of our Markov chain as m gets very large. Loosely speaking, as we consider the limit of m tending to infinity in equation C.9, the importance of the initial value—previously defined as ξ_t—decreases until we essentially lack any conditioning information. Thus, we can interpret these limiting values as the ergodic, or unconditional, probabilities of our Markov chain. Another way of thinking about these probabilities is to imagine a Markov chain achieving a steady set of transition probabilities after a large number of periods. To see how this works, let us consider a two-state example with the following transition matrix,

$$P = \begin{bmatrix} q & 1-q \\ 1-p & p \end{bmatrix}. \tag{C.19}$$

The first step is to find the eigenvalues of this matrix.[4] To accomplish this analytically, we need to find the roots of the characteristic polynomial,

$$\det(P - \lambda I) = \det\left(\begin{bmatrix} q - \lambda & 1-q \\ 1-p & p - \lambda \end{bmatrix}\right), \tag{C.20}$$

$$= (q - \lambda)(p - \lambda) - (1 - p)(1 - q),$$

$$= \lambda^2 - \lambda(q + p) + (q + p - 1),$$

$$= \lambda^2 - \lambda(q + p) + (q + p - 1) + \underbrace{\lambda - \lambda}_{=0},$$

$$= \lambda^2 - \lambda(q + p - 1) - \left(\lambda - (q + p - 1)\right),$$

$$= \lambda\left(\lambda - (q + p - 1)\right) - \left(\lambda - (q + p - 1)\right),$$

$$= (\lambda - 1)(\lambda - q - p + 1).$$

This implies directly that the eigenvalues are,

$$\lambda_1 = 1, \tag{C.21}$$

$$\lambda_2 = q + p - 1, \tag{C.22}$$

where I denotes the identity matrix. We can place conditions on q and p. In particular, we have one eigenvalue, λ_1, that is unity and thus the second eigenvalue must satisfy

$$|\lambda_2| < 1, \tag{C.23}$$

$$|q + p - 1| < 1,$$

$$-1 < q + p - 1 < 1,$$

$$0 < q + p < 2.$$

---

[4] See Golub and Loan (2012, Chapters 7–8) for much more detail and background on the eigenvalue problem.

C.3 Ergodicity

The next step is to compute the eigenvectors associated with our two eigenvectors. This is done in the usual way.[5] The results are,

$$x_{\lambda_1} = \left[\frac{1-p}{2-q-p} \quad \frac{1-q}{2-q-p}\right],\tag{C.25}$$

$$x_{\lambda_2} = \begin{bmatrix}-1 & 1\end{bmatrix}.\tag{C.26}$$

It is worth noting that the first eigenvector was normalized, not for the typical unit length, but rather such that the following is true,

$$x_{\lambda_1}\mathbf{1} = \left[\frac{1-p}{2-q-p} \quad \frac{1-q}{2-q-p}\right]\begin{bmatrix}1\\1\end{bmatrix} = 1.\tag{C.27}$$

That is, we force the sum of the eigenvector values associated with the unit eigenvalue to equal unity—this normalization essentially allows us to interpret them as probabilities. At this point, we make the claim that x_{λ_1} is, in fact, the vector of ergodic probabilities. That is, the unconditional probabilities are,

$$\mathbb{P}[S_t = 0] = \frac{1-p}{2-q-p},\tag{C.28}$$

$$\mathbb{P}[S_t = 1] = \frac{1-q}{2-q-p}.\tag{C.29}$$

While we will not prove this result, we will work through a simple exercise to provide some intuition.[6] First, we have shown that our matrix of transition probabilities, P, has two real-valued, distinct eigenvalues and thus we can safely perform the following spectral decomposition,

$$P = A\Lambda A^{-1},\tag{C.30}$$

$$= \begin{bmatrix} 1 & \frac{-(1-q)}{2-q-p} \\ 1 & \frac{1-p}{2-q-p} \end{bmatrix} \begin{bmatrix} 1 & 0 \\ 0 & q+p-1 \end{bmatrix} \begin{bmatrix} \frac{1-p}{2-q-p} & \frac{1-q}{2-q-p} \\ -1 & 1 \end{bmatrix}.$$

---

[5]That is, we must find a basis for the parameterized solution of the following two equations,

$$(P - \lambda_1 I)x_{\lambda_1} = 0,\tag{C.24}$$
$$(P - \lambda_2 I)x_{\lambda_2} = 0.$$

[6]This discussion is not a proof, because it lacks the necessary rigour and only considers the two-dimensional case where the eigenvalues in our spectral decomposition are real-valued and distinct. While it is possible to extend the following discussion to these cases, it is a level of generality that is not required in our analysis.

We have seen that if we want the m-period-ahead transition probabilities for an ergodic Markov chain, we need only to compute P^m. Clearly, the ergodic probabilities would be the limit of this quantity as m tends to infinity. Consider, therefore, the following,

$$\lim_{m \to \infty} P^m = \lim_{m \to \infty} \left(A \Lambda^m A^{-1} \right), \tag{C.31}$$

$$= A \lim_{m \to \infty} (\Lambda^m) A^{-1},$$

$$= A \lim_{m \to \infty} \left(\underbrace{\begin{bmatrix} 1 & 0 \\ 0 & q+p-1 \end{bmatrix}}_{\substack{\text{Recall } (q+p-1) < 1 \\ \text{by definition.}}} \right)^m A^{-1},$$

$$= A \begin{bmatrix} 1 & 0 \\ 0 & 0 \end{bmatrix} A^{-1},$$

$$= \begin{bmatrix} \frac{1-p}{2-q-p} & \frac{1-q}{2-q-p} \\ \frac{1-p}{2-q-p} & \frac{1-q}{2-q-p} \end{bmatrix},$$

$$= \begin{bmatrix} 1 \\ 1 \end{bmatrix} \begin{bmatrix} \frac{1-p}{2-q-p} & \frac{1-q}{2-q-p} \end{bmatrix},$$

$$= \mathbf{1}\pi,$$

where $\pi = x_{\lambda_1}$, as in equation C.25. Thus, we can see that, in a loose sense, the limiting probabilities are given by the eigenvector associated with the unit eigenvalue.[7]

---

[7]Furthermore, this gives some intuition as to why, to have an ergodic Markov chain, only one eigenvalue can be unity and the rest must lie inside the unit circle.

Appendix D
The Python Code Library

> *First, solve the problem. Then, write the code.*
>
> (John Johnson)

Practitioners, the main target audience for this work, ultimately need to transform theoretical relationships into concrete results. Throughout the entirety of this book, snippets of Python code, in the form of distinct generic computer algorithms, have been offered and discussed. The principal objective of this undertaking is to enhance the reader's understanding of the base material through explicit provision of implementation details. In short, the code examples are intended to enhance comprehension and be generally helpful.

Although the provided code snippets are largely self-explanatory, a few points require additional consideration and explanation. First, and perhaps most importantly, there are many possible ways to put any mathematical or statistical algorithm into action. We have elected for a highly procedural approach using the Python programming language.[1] Other computing paradigms and languages would also perform admirably and may, in many cases, actually be superior to what is presented. A clever class structure using object-oriented ideas would widely be considered much closer to the state of the art in computer programming. Moreover, a compiled professional programming language—such as C++, Java, or even C#— would most certainly provide significant speed advantages. There are good and defensible reasons for the choices taken in this book, but it is critical to be consciously aware of the many advantages offered by alternative implementations. In short, we provide a possible execution of the underlying algorithms, but there is no claim that it is either the optimal or most efficient route.

The second point is that, while a liberal sprinkling of Python code snippets adds significantly to the narrative in the previous chapters, it is neither practical nor desirable to provide an exhaustive description of all algorithms. On the one hand, it would dramatically lengthen many of the already quite long chapters and,

---
[1] There is a vast spectrum of opinion on appropriate programming languages. Dr. Yves Hilpisch, a tireless campaigner for Python and computational-finance practitioner, offers some compelling arguments supporting our choice in Hilpisch (2016).

on the other, it would prove extremely tedious for the reader. A complete code description, however, is necessary for the quantitative analyst. She will almost certainly wish to alter the structure and application of the presented algorithms, but she first requires a complete picture. The solution to this conundrum is to provide the entire code library—along with some simple illustrative examples—in a more natural form separate from the book. Decades ago, this would have taken the form of a CD-ROM—or even a floppy disk—affixed to back cover of the book. Today, a code repository offers a superior solution. Various platforms exist, but we ultimately selected GitHub. This appendix provides additional background, context, and direction associated with this code repository.

D.1 Explaining Some Choices

Before GitHub, there was Git. Git is a version-control, code-management software system—originally developed by Linus Torvalds, the creator of Linux—for use in the Linux operating system. Along the way, it was spun off from the Linux operating system and, following a complicated path, developed into a large-scale code repository. Given its size, scope, and general philosophy of its creation and ongoing development, it seemed like a natural code repository for this project.[2] There are, of course, other useful code repositories such as Sourceforge or GNU Savannah. We stress that our choice is, in no way, a recommendation nor a knock against such alternatives.

It is also useful to outline some of the specific software and hardware choices made for the production of the code and results found in the earlier chapters. All computations in the preceding pages were performed by a quite powerful laptop. Purchased in the autumn of 2016, it has a Intel©Core^{TM} i7-6700K CPU with 4.00 GHz clockspeed, four cores, and eight threads. This processor is, or was, sometimes referred to as part of the i7 Skylake series, but its origins and technical specifications are probably best left to computer-science experts. Along with a NVIDIA GM204M graphics card and 64 GB of cache memory, however, this system is certainly quite appealing to video-game aficionados. This is not a coincidence; the heavy computational demands of credit-risk modelling led to the purchase of one of the most powerful (affordable) laptops on the market. Almost invariably, those of us with high scientific-computing demands will find themselves looking enviously at gamer laptops. While this system is no longer, as of the time of this work's publication, at the cutting edge, it still is quite high performance. The reader

---

[2]There are myriad fascinating online references for GitHub, but since URL's tend to have short shelf-lives, it does not make sense to reference them. Fuller et al. (2017), however, is a interesting discussion of GitHub, its history, and its impact on the evolution of code repositories.

with an inferior system might, therefore, have difficulty in reproducing some of the computational times.³

Software choices are also consequential. Use of Python, for example, requires a number of decisions. This work was performed using the 3.x Anaconda distribution originally offered by Continuum Analytics. Other distributions are, of course, also excellent and one's ultimate choice is a question of personal taste and expediency. The reader is encouraged to experiment and identify the distribution that best satisfies their needs.⁴ The operating system used for this entire work—both the LaTeX word processing and Python development—was Linux Mint.⁵ If you opt for Linux, there are a multitude of choices; again, this is a matter of preference. Far from being the most hardcore release, Linux Mint was selected due to its stability, convenience of use, excellent tools, and strong similarities with both Debian and Ubuntu.⁶

D.2 The Library Structure

The organization of the Python libraries, in absence of a formal project or direct application, follows the individual chapters. There are nine libraries—each mapped to Chaps. 2 to 10—along with a utilities library used in most of the other libraries. The ten Python libraries are thus to be found in GitHub at the following location:

> https://github.com/djbolder/credit-risk-modelling

Figure D.1 provides a schematic description of these ten libraries along with their individual names and associated chapter affiliation. `cmUltilities.py`, as a collection of widely applicable functions, is not assigned to any specific chapter.

There are also, as one would certainly expect, interlinkages between the libraries. The model implementations from Chaps. 2 to 5 are recycled, where applicable and convenient, in the diagnostic-tool and parameter-estimation libraries. The conditional-default probability functions, for example, are referenced from the model libraries many times in the latter chapters. There are many other examples.

Each library is, to repeat, merely a collection of functions. There are no class definitions nor are any global variables employed. This may be restrictive, but

---

³With clever use of one's processor and graphic card—see, for example, Zaccone (2015)—one could also quite possibly outperform the presented computational expense.
⁴Gaining familiarity with the Python programming language involved a number of online resources, but VanderPlas (2017), McKinney (2013), and Hilpisch (2015) also offered much concrete and useful assistance.
⁵For those interested in such details, I am currently running Linux Mint 18 Cinnamon with the 4.4.0–50 generic kernel.
⁶Naturally, Microsoft or Mac operating systems are also perfectly acceptable choices. Sidestepping potentially charged issues and arguments for and against propriety and open-source operating systems, we offer an interesting anthropological perspective on Linux from Ratto (2005).

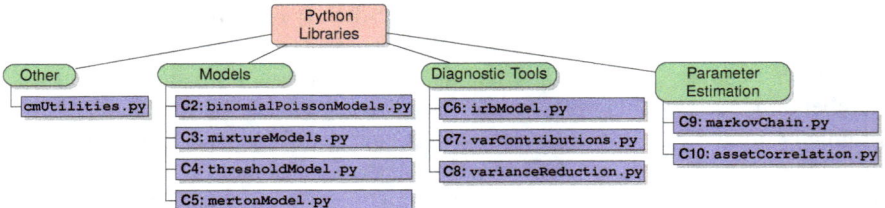

Fig. D.1 *The library structure*: The underlying figure organizes each of the 10 Python libraries by their thematic structure: models, diagnostic tools, and parameter estimation. There is also a utilities library, which finds use in numerous functions located in other libraries.

Table D.1 *Function count*: The underlying table highlights the number of distinct functions found in each of the ten code libraries summarized in Fig. D.1.

Chapter	Library	Functions
n/a	cmUtilities.py	24
2	binomialPoissonModels.py	6
3	mixtureModels.py	27
4	thresholdModels.py	44
5	mertonModels.py	19
6	irbModel.py	23
7	varContributions.py	39
8	varianceReduction.py	12
9	markovChain.py	25
10	assetCorrelation.py	45
Total		264

makes sense because this is a purely pedagogical exercise. The functions are not defined within the context of a system, a collection of tasks, or an organization's development framework. They are simply straightforward—and perhaps often rudimentary—implementations of the mathematical and statistical ideas presented in the preceding chapters. Enormous improvements in efficiency and cleverness are not only possible, but encouraged. Wisely using these ideas, enhancing them, and incorporating them into one's work is, as one so often reads in textbooks, left as an exercise for the reader.

Table D.1 outlines the total number of functions associated with each of the libraries introduced in Fig. D.1. While it varies by chapter and subject, there are, in total, more than 250 separate functions.[7] Thus, each library contains, on average, north of 25 Python functions. While certainly not on par with a large, or even, medium-scale systems, it is still non-trivial amount of computer code.

---

[7] It may also vary over time if changes are necessary to the original code.

D.3 An Example

With 250+ functions, it is not terribly meaningful to highlight and explain the role and usage of each individual element. This would be a long and monotonous exercise. It is nevertheless interesting to select one library and examine it in somewhat more detail. Figure D.2, therefore, illustrates the 24 functions found in the utilities library—otherwise known as cmUtilities.py—and breaks them into *five* distinct categories. These include statistical densities, printing, simulation of base underlying data, generic risk measures, and some mathematical functions. Logically, functions were placed in this category if they could be used in multiple other libraries and did not have any specific relationship to a given chapter.

This does not imply that only functions found in cmUlilities.py are employed outside of a given chapter. They are numerous examples of functions spanning multiple chapters. Numerous mixture and threshold functions are recycled

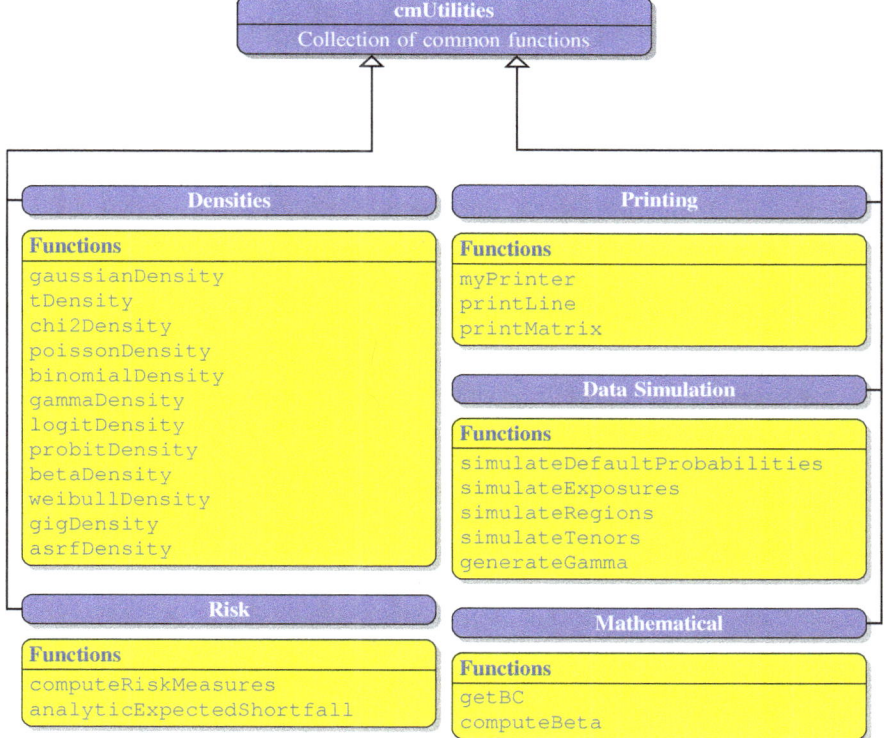

Fig. D.2 *A sample library*: Each library is essentially a collection of, typically related, Python functions used for a common purpose. The underlying schematic arranges the 24 functions found in the cmUtilities library into five categories describing their basic purposes. Each of the ten libraries summarized in Fig. D.1 can be analogously organized.

in the risk-attribution or parameter-estimation libraries. Moreover, given the link between the Esscher transform and Glasserman and Li (2005)'s importance-sampling technique, a number of functions from `varContributions.py` show up again in the `varianceReduction.py` library.

As a final note, the individual functions are sparsely documented. There is an occasional comment or clarification, but the majority of the code is intended to be self-documenting. This implies that, to the extent possible, functions and variables are assigned meaningful names and the individual algorithmic steps are (largely) procedurally related to the mathematical discussion in the text. I may not have succeeded in every instance, but in my experience, lengthy and detailed code documentation rarely adds much to one's understanding of the underlying algorithms.

D.4 Sample Exercises

Despite attempts to clearly organize and display the full set of algorithms, there is always the potential for misunderstanding and difficulty in its use. While this problem cannot be completely avoided, it can hopefully be mitigated. To this end, a collection of sample exercises is included in the model repository. The file `ex03.py`, for example, uses the functionality from `mixtureModels.py` to perform some illustrative computations and produce some tabular or graphical outputs. A similar file is provided, following from Table D.1, for each individual code library. These exercises are, it should be stressed, an illustrative application of a library's functionality, but not an exhaustive one.

To facilitate this process, we've also included a number of data files with exposure, default probability, regional-identity, and tenor information associated with the running example found in the preceding chapters.[8] This will not only provide a concrete portfolio, but allow one to replicate our computations.

As an author, of course, this is a potentially dangerous decision. I certainly open myself to readers unable to replicate my presented results due to—even potentially embarrassing—coding or logical errors. Not only do I fully accept this risk, but I completely expect there to be a few glitches and difficulties. Some of them may even prove unpleasant. It would nevertheless be the height of arrogance, not to mention folly, to assume that one can write 250 diverse and complex mathematical functions without committing some level of error or inconsistency. Putting this code and the data into the public domain is, quite probably, the most efficient and direct manner of improving it and resolving these inevitable issues.[9] As a practising model-risk

---

[8] Some of the simulated datasets used in Chaps. 9 and 10, for purposes of parameter estimation, are also included.

[9] This is, in fact, the true spirit behind the open software movement. In the words of Richard Stallman, free software implies *"the freedom to run it, to study and change it, and to redistribute*

manager and experienced model developer, I am keenly aware of the necessity—and positive externalities—associated with creating effective challenge for technical ideas and implementations.[10] I've preached this point throughout the entirety of this book; this gesture amounts to "walking the walk."

I plan, to the extent my other responsibilities permit, to eliminate errors and make necessary adjustments. Should the reader, however, be unhappy with the speed and nature of these changes, it is perfectly legitimate—and, indeed, encouraged, in the GitHub world—for the reader to copy the libraries and establish her own version. In conclusion, therefore, I hope these code libraries help expand the reader's understanding and mastery of this book's material and, most importantly, prove useful to fellow finance practitioners. Happy coding!

References

Abramovitz, M., & Stegun, I. A. (1965). *Handbook of mathematical functions*. New York: Dover Publications.
Billingsley, P. (1995). *Probability and measure* (3rd edn.). Third Avenue, New York, NY: Wiley.
Black, F., & Scholes, M. S. (1973). The pricing of options and corporate liabilities. *Journal of Political Economy, 81*, 637–654.
Brémaud, P. (1999). *Markov chains: Gibbs fields, Monte Carlo simulation and queues*. 175 Fifth Avenue, New York, NY: Springer-Verlag. *Texts in applied mathematics*.
Duffie, D. (1996). *Dynamic asset pricing theory* (2nd edn.). Princeton, NJ: Princeton University Press.
Fuller, M., Goffey, A., MacKenzie, A., Mills, R., & Sharples, S. (2017). Big diff, granularity, incoherence, and production in the github software repository. In I. Blom, T. Lundemo, & E. Røssaak (Eds.), *Memory in motion: Archives, technology, and the social*. John Wiley & Sons.
Glasserman, P., & Li, J. (2005). Importance sampling for portfolio credit risk. *Management Science, 51*(11), 1643–1656.
Golub, G. H., & Loan, C. F. V. (2012). *Matrix computations*. Baltimore, Maryland: The John Hopkins University Press.
Hamilton, J. D. (1994). *Time series analysis*. Princeton, NJ: Princeton University Press.
Harrison, J., & Kreps, D. (1979). Martingales and arbitrage in multiperiod security markets. *Journal of Economic Theory, 20*, 381–408.
Harrison, J., & Pliska, S. (1981). Martingales and stochastic integrals in the theory of continuous trading. *Stochastic Processes and Their Applications, 11*, 215–260.
Heunis, A. J. (2011). *Notes on stochastic calculus*. University of Waterloo.
Hilpisch, Y. (2015). *Derivative analytics with Python: Data analysis, models, simulation and hedging*. West Sussex, UK: Wiley.
Hilpisch, Y. (2016). Computational finance: Why Python is taking over. *Wilmott Magazine, 2016*, 43–44.
Hofert, M. (2013). On sampling from the multivariate *t* distribution. *The R Journal, 5*(2), 129–136.

---

copies with or without changes." This is my key objective in exposing these code libraries on GitHub.

[10]The interested reader is referred to OCC (2011) for more on this fundamental tenet of model-risk management.

Karatzas, I., & Shreve, S. E. (1998). *Methods of mathematical finance*. New York, NY: Springer-Verlag.

Kotz, S., Kozubowski, T., & Podgorski, K. (2001). *The Laplace distribution and generalizations: A revisit with applications to communications, economics, engineering, and finance*. Basel, Switzerland: Birkhäuser.

Kotz, S., & Nadarajah, S. (2004). *Multivariate t distributions and their applications*. New York: Cambridge University Press.

McKinney, W. (2013). *Python for data analysis*. Sebastopol, CA: O'Reilly Media, Inc.

Merton, R. (1974). On the pricing of corporate debt: The risk structure of interest rates. *Journal of Finance, 29*, 449–470.

Meyn, S. P., & Tweedie, R. L. (1993). *Markov chains and stochastic stability*. 175 Fifth Avenue, New York, NY: Springer-Verlag.

OCC. (2011). Supervisory guidance on model risk management. *Technical report*. Office of the Comptroller of the Currency.

Oksendal, B. K. (1995). *Stochastic differential equations* (4th edn.). Berlin: Springer-Verlag.

Press, W. H., Teukolsky, S. A., Vetterling, W. T., & Flannery, B. P. (1992). *Numerical recipes in C: The art of scientific computing* (2nd edn.). Trumpington Street, Cambridge: Cambridge University Press.

Ratto, M. (2005). "Don't Fear the Penguins": Negotiating the trans-local space of Linux development. *Current Anthropology, 46*(5), 827–834.

Roth, M. (2013). On the multivariate t distribution. Division of Automatic Control, Department of Electric Engineering, Linköpings Universitet, Sweden.

Shaw, W. T. (2006). New methods for simulating the student t-distribution–direct use of the inverse cumulative distribution. Nomura Centre for Quantitative Finance.

Shaw, W. T. (2011). Risk, var, cvar and their associated portfolio optimizations when asset returns have a multivariate student t distribution. Department of Mathematics and Computer Science, University College London.

VanderPlas, J. (2017). *Python data science handbook*. Sebastopol, CA: O'Reilly Media, Inc.

Zaccone, G. (2015). *Python parallel programming cookbook*. Birmingham, UK: Packt Publishing.

Index

Anaconda Python, 669
$A_{n,T}$
   correlation, 263
   covariance, 261
   dynamics, 234, 245
   expected value, 257
   variance, 259
   volatility, 260
Argonne National Laboratory, 110
Asset correlation, 158, 222, 275
Asymptotic Single Risk-Factor Model, *see* ASRF

Bank for International Settlements, 288
Basel Committee on Banking Supervision, 21, 288
Basel IRB approach, 22, 291–293, 304
   asset correlation, 296
   maturity adjustment, 298
   risk-capital coefficient, 292, 294, 300
   slope coefficient, 298
Bayes' rule, 177
Bayes' theorem, 551
BCBS, *see* Basel Committee on Banking Supervision
Bernoulli trial, 21, 43, 47, 52, 88, 166, 532
Beta distribution, 54, 92
Beta function, 93
Binomial coefficient, 47, 48, 65
Binomial distribution, 47, 80, 166, 532
Binomial theorem, 70
Bivariate Gaussian distribution, 158
Bivariate normal distribution, 264
Black-Scholes formula, 239
Bond pricing
   adding default risk, 547, 554
   risk-free, 544
Bootstrapping
   financial, 562
   statistical, 523, 531, 602
Borel-Cantelli theorem, 69, 72, 91

Calibration
   t, 188
   beta-binomial, 95
   Gaussian, 161
   generalized-hyperbolic, 203
   logit-binomial, 107
   matching one-factor Gaussian-threshold and CreditRisk+ models, 340
   Merton model, 265
   multivariate CreditRisk+, 143
   one-factor CreditRisk+, 133, 338
   Poisson Weibull, 127
   Poisson log-normal, 127
   Poisson-gamma, 117
   practical multivariate t, 222
   probit-binomial, 107
   variance-gamma, 201
Capital asset-pricing model, 11, 85, 133, 150, 309
CAPM, see Capital asset-pricing model, 11
Central-limit theorem, 368, 369, 372, 422, 442, 445, 446
Change-of-variables formula, 127, 638
Chebyshev's inequality, 68, 69, 71, 72, 91
Chi-squared distribution, *see* χ^2 distribution
Code repository, 668
Coin tossing, *see* Bernoulli trial
Completing the square, 258, 259, 261, 657

Complex integration, 377–379
Computation speed
    hardware choices, 668
    importance sampling, 465, 482
    raw Monte Carlo, 449
    saddlepoint approximation, 421
Concentration risk, 311
Conditional bivariate Gaussian distribution, 179
Conditional bivariate t distribution, 180
Conditional default probability, 86, 155
Conditional survival probability, 550
Conditional VaR, *see* Expected shortfall
Confidence interval, 22
    binomial distribution, 535
    default probabilities, 538
    Monte-Carlo method, 448
Contour integral, see Complex integration, 377
Convergence
    almost sure, 68, 72
    in distribution, 68, 91
    in probability, 68, 69, 72
Copula density
    Gaussian, 620
    t, 620
    t with Gaussian marginals, 620
Copula function, 160, 176, 177, 576, 616
    density, 616
    elliptical, 617
    Gaussian, 160
Credit-default swap, 24, 553, 555
    accrued premium-coupon payment, 559
    conditional protection payment, 560
    premium coupon payment, 558
Credit migration, 3, 582
Cumulant generating function, 377, 385, 386, 388, 424, 459
Curse of dimensionality, 439

Data visualization, 27, 418
Default correlation, 158, 186, 197, 577
    Merton model, 263, 277
Default dependence, 50, 152
Default event, *see* Default indicator
Default independence, 46, 50
Default indicator
    independent-default model, 43
    Merton model, 232
    threshold model, 154
Default probabilities
    empirical default rate, 587
    risk-neutral, 273, 565, 567
    theoretical moments, 590
    volatility, 590
Default probability risk premium, 568
Default time, 547
Distance-to-default, 237, 238, 246, 249, 277
Doléan exponential, 654

EAD, *see* Exposure-at-default
Economic capital, 22
Edgeworth expansion, 422
Elliptical copula, 617
Equity call option, 230, 231, 239, 653, 658
Equity correlation, 275
Esscher transform, 426
    calibration, 460, 469, 470
    χ^2-mixing variable, 476
    latent-variable mean, 468
    loss distribution, 458
Euler's theorem for homogeneous functions, 352, 370, 371, 396, 397, 403, 411
Expected positive exposure, 8
Expected shortfall, 32
    decomposition, 357, 359
    marginal, 355, 395
Exponential distribution, 553
Exposure-at-default, 53, 292

Feyman-Kac formula, 437
First-order homogeneous differential equation, 551
First-passage time, 232
Fisher, R.A., 523
Forward default probabilities, 549
Fundamental theorem of asset pricing, 6
Fundamental theorem of calculus, 241

Gamma distribution
    shape-rate parametrization, 115, 201, 475
    shape-scale parametrization, 329
Gamma function, 93, 115, 184, 333, 637
    properties, 116, 638
Gaussian distribution, 48, 243, 277, 422, 446, 468
Gaussian kernel, 374
Generalized inverse-Gaussian distribution, 200, 202
Geometric Brownian motion, 14, 16, 234, 649
    intuition, 240
    solution, 235
GIG distribution, 200. *See also* Generalized inverse-Gaussian distribution
Girsanov's theorem, 441, 653

Index 677

GitHub, 24, 668
Gradient vector, 526
Granularity adjustment, 311, 312, 604
  accuracy, 327, 344
  CreditRisk+ case, 328
  derivation, 312
  Gaussian-threshold case, 319
  general form, 317

Hazard function, 551, 552
  inference from market instruments, 557
  piecewise linear, 555
Hazard rate
  inference from market instruments, 561
  see Hazard function, 557
Heavyside step function, 409
Hessian matrix, 527
Homogeneity, 353

Ice hockey, xv
Importance sampling, see Variance-reduction techniques
Independent-default model
  binomial implementation, 50, 163, 304
  Poisson implementation, 73, 480
Inferring state variables, 626
Infinitely grained portfolio, 310
Internal ratings-based approach, see Basel IRB approach
Interval estimation, see Parameter uncertainty
Inverse-χ^2 distribution, 647
Inverse gamma distribution, 185, 647
Inverse Laplace transform, 379
IRB, see Basel IRB approach
Itô exponential, 654

Karl Pearson, 589
Kurtosis, 198

Laplace approximation, 374, 437
Law of iterated expectations, 92, 187, 197
Law of rare events, 82
Law of total variance, 469
Leptokurtosis, 97, 98, 101, 174, 182, 209, 543
LGD, see Loss-given-default
L'Hôpital's rule, 177
Likelihood function, 507
Likelihood ratio, see Radon-Nikodym derivative
Linux operating system, 669

Logistic function, 101
Logit-normal distribution, 104
Log-likelihood function, 507
  binomial distribution, 532
  binomial loss density, 598, 609
  copula density, 623
  Markov chain, 508
Log-likelihood kernel, 624, 625
Log-likelihood surface, 603, 604, 611
Log-normal distribution, 125, 243
Loss-given-default, 53, 293, 330
  Stochastic, 330

Market price of risk, 651
Market risk, 12, 16, 42
Markov, A.A, 501
Markov chain, 491, 499
  absorption, 505
  constraints, 501, 506
  cumulative transition matrix, 583, 626
  ergodicity, 504, 664
  forward probabilities, 503, 661
  properties, 505, 660
  transition matrix, 501, 506, 581
  transition probabilities, 500, 581, 659
Markov's inequality, 69, 72, 91
Matrix decomposition
  Cholesky, 253, 644
  eigenvalue, 253, 644, 664
  singular-value, 644
Matrix exponential, 511
Maximum-likelihood estimation, 508
  Fisher information, 527, 529, 531, 533, 599, 600, 612, 631
  flatness of likelihood function, 603
  regularity, 525, 528
  score function, 529, 600
Maximum-likelihood problem
  cohort approach, 508
  dependence parameters, 596, 607, 618
  hazard-rate approach, 511
Merton asset process, see $A_{n,T}$
Merton model, 247
  indirect approach, 247, 256
  other applications, 281
Method of Lagrange multipliers, 508
Method of moments, 589
  mixture models, 592
  threshold models, 593
MINPACK, 110, 564
Mixture models
  beta-binomial, 92, 163, 304, 416, 417
  general one-factor approach, 86, 578

logit-binomial, 101
multivariate CreditRisk+, 141
one-factor CreditRisk+, 132, 328, 416, 417, 480, 578
Poisson Weibull, 125, 416, 417
Poisson log-normal, 125
Poisson-gamma, 115
probit-binomial, 101
Model risk, 2, 29, 673
Modified Bessel function, 202
Moment-generating function, 69, 376, 385, 422
Monte-Carlo method, 56, 429
  beta-binomial model, 100
  binomial independent-default model, 59
  confidence intervals, 445, 448
  convergence, 442
  CreditRisk+ model, 139
  the eponymous casino, 439
  Gaussian threshold model, 162
  logit-binomial model, 111
  normal-variance-mixture threshold model, 207
  Poisson-gamma model, 124
  Poisson independent-default model, 76
  pricing contingent claims, 441
  probit-binomial model, 111
  shortcomings, 429
  t threshold model, 191
Multinomial distribution, 524
Multinomial trial, 514
Multivariate Gaussian distribution, 277

Negative binomial distribution, 116
Nelson-Siegel model, 562
Normal distribution, *see* Gaussian distribution
Nuisance parameters, 628
Numeraire asset, 441, 546, 649
Numerical integration, 106, 433
  Gaussian quadrature, 434
  Monte-Carlo integration, 437
  scipy's nquad algorithm, 189, 415
  scipy's quad algorithm, 126, 414, 434, 591, 598

Observed Fisher information, 529
ODE, *see* Ordinary differential equation
OIS discounting, 547
Option delta, 268
Option gamma, 268
Option *Greeks*, 351
Ordinary differential equation, 241

Panjer recursion, 139, 431
Parameter identification, 604
Parameter uncertainty
  default probabilities, 523
  dependence parameters, 601
Pascal's triangle, 48
Pascal, Blaise, 48
PFE, *see* Potential future exposure
Poisson distribution, 73, 113
  arrival intensity, 73
Poisson process, 553
Portfolio invariance, 291, 304
  benefits, 291
  correction, 309
  rationale, 291
  shortcomings, 302
Potential future exposure, 8, 53
Pricing credit risk, 3, 544
Probability measure
  equivalent martingale measure, 7, 239, 269, 441, 546, 649
  physical measure, 7, 23, 239
Probit function, 101
Probit-normal distribution, 105
Profile likelihood, 524, 611, 612
Programming languages, 25, 667
Python exercises, 672

Quadratic variation, 234, 242, 651

Radon-Nikodym derivative, 453, 457, 654
  mixing variable, 476
  shifted mean, 468
  twisted probabilities, 458
Random number generation
  beta-distributed outcomes, 98
  binomial outcomes, 56
  correlated Gaussian variates, 254, 647
  correlated t variates, 647
  dependent Markov chains, 580
  Gaussian state variables, 157
  generalized-hyperbolic state variables, 208
  importance sampling, 476
  independent Bernoulli trials, 56
  Markov chains, 514
  state variables, 191
  variance-gamma state variables, 208
Recovery rate, 53, 293, 554
Risk attribution, 34
  basel IRB approach, 306
  brute-force simulation, 361
  granularity adjustment, 347

Index   679

market risk, 354
normal approximation, 369
saddlepoint approximation, 372
saddlepoint expected-shortfall
    contributions, 403, 411
saddlepoint VaR contributions, 397, 410
VaR-matched expected shortfall, 367
Risk decomposition, *see* Risk attribution
Risk-neutral default probabilities, 24, 565
challenges, 570
comparison, 567
computation, 565
Risk-neutral measure, *see* Probability measure,
    Equivalent martingale measure
R programming language, 207

Saddlepoint approximation, 372, 430, 461, 462
alternative derivation, 421
density function, 378, 388, 407
expected-shortfall contributions, 403, 411
motivation, 373
shortfall integral, 381, 391, 409
tail probability, 379, 391, 408
VaR contributions, 397, 410
Sample portfolio files, 672
Sharpe ratio, 273, 274, 279
Simulation, *see* Random number generation
Simulation study, 514, 601, 614, 631
Skewness, 198
Spectral decomposition, *see* Matrix
    decomposition, Eigenvalue
Stallman, Richard, 672
State-variable correlation, 158
Stopping time, 232
Strong law of large numbers, 72
Student's t distribution, *see* t distribution
Survival function, 549

t distribution, 182, 446, 641
Tail dependence, 175
    t case, 180
    Gaussian case, 179

General normal-variance-mixture case, 198
Tail dependence coefficient, 176
Tail VaR, *see* Expected shortfall
Taylor-series expansion, 313, 314, 373, 377,
    378, 396, 401
Threshold models
    t, 183, 304, 480
    ASRF, 165, 290, 291, 304
    Canonical multivariate Gaussian, 211
    Canonical multivariate normal-variance-
    mixture, 215
    Equivalence with Merton approach, 250
    Gaussian, 151, 304, 319, 417, 419, 480
    Generalized-hyperbolic, 200
    Normal-variance-mixture, 193, 407, 408
    Variance-gamma, 200, 480
Twisting probabilities, *see* Esscher transform

Uniform distribution, 435

Value-at-Risk, *see* VaR
Decomposition, 356, 358
Marginal, 354–356, 395
Parametric market risk, 353
Under parameter uncertainty, 542
VaR, 32
VaR-matched expected shortfall, 367
Variance-reduction techniques, 426, 451
    Importance sampling, 451
Vasicek's limit-loss distribution, *see* Threshold
    models, ASRF

Weak law of large numbers, 69, 169
Weibull distribution, 26, 125
Wiener increment, 243, 246
Wiener process, 14, 269, 651
William Sealy Gosset, 446

χ^2 distribution, 26, 183, 475, 637

Author Index

Abramovitz, Milton, 93, 202, 641
Andersen, Per Kragh, 553
Ang, Andrew, 12
Antonov, A., 373, 382
Arnow, Barry J., 439
Artzner, Philippe, 353, 367
Aruoba, S. Boragan, 25
Asmussen, Søren, 476

Badger, Lee, 439
Balakrishnan, N., 202, 333, 475, 553
Barndorff-Nielsen, O., 198
Bartle, Robert G., 627
Beck, Amir, 473, 625
Bellman, Richard, 439, 485
Berger, James O., 627, 628
Berger, Roger L., 47, 54, 69, 78, 93, 125, 333, 376, 458, 508, 523, 553, 589
Bhatia, Mickey, 149
Bielecki, Tomasz R., 7, 544
Billingsley, Patrick, 14, 68, 72, 128, 316, 376, 377, 442, 617, 638
BIS, 7, 150, 291, 293, 302, 311, 429
Black, Fischer, 6, 232, 239, 240, 649, 658
Bluhm, Christian, 51, 367, 398, 549, 619
Bohn, Jeffrey R., 237
Bolder, David Jamieson, 18, 281, 308, 352, 501, 558, 562
Borgan, Ørnulf, 553
Bouyé, Eric, 617
Bové, D. Sabanés, 469, 523, 526, 624, 628
Box, George E. P., 285
Box, Joan Fisher, 446
Boyle, Phelim, 440
Brémaud, Pierre, 504, 659
Brereton, Tim J., 451

Brigo, Damiano, 544, 546
Broda, Simon A., 373

Casella, George, 47, 54, 69, 78, 93, 125, 333, 373, 376, 458, 508, 523, 553, 589
Chan, Joshua C., 451
Chernick, Michael R., 531, 602
Christoffersen, Peter F., 18
Coughlan, Guy, 499
Crosbie, Peter J., 237

Daniels, H. E., 373, 378, 379, 409
DasGupta, Anirban, 68, 69, 373, 378, 380, 409, 422, 528
Daul, Stéphane, 214
Davis, Philip J., 116
de Giorgi, Enrico, 214
de Kort, Jan, 176
de Servigny, Arnaud, 276
Delbaen, Freddy, 353, 367
Demey, Paul, 596, 600, 606
Derman, Emanual, 24
Devroye, Luc, 207
Dhrymes, Phoebus J., 604
Diebold, Francis X., 281
Duffie, Darrell, 7, 240, 441, 545, 548, 553, 649
Durrett, Rick, 68, 72, 242, 422, 442
Durrleman, Valdo, 617

Eber, Jean-Marc, 353, 367
Eberlein, Ernst, 195
Eckhardt, Roger, 439
Efron, B., 531, 537, 595, 602, 614, 631
Embrechts, Paul, 176, 178, 198, 451, 616, 617
Emmer, Susanne, 311

Fang, K., 617
Fernández-Villaverde, Jesús, 25
Filardo, Andrew J., 501
Finger, Christopher C., 149
Fishman, George S., 76, 93, 98, 253, 444, 445, 515, 516
Flannery, Brian P., 107, 109, 253, 274, 434, 600, 644
Fok, Pak-Wing, 150
Frey, Rüdiger, 51, 82, 92, 176, 178, 198, 451, 616
Fuh, C. D., 476, 531
Fuller, Matthew, 668

Geske, Robert, 256
Gil-Pelaez, J., 379
Gill, Richard D., 553
Glasserman, Paul, 52, 211, 361, 426, 430, 444, 445, 451, 453, 455–457, 460, 469, 470, 672
Glynn, Peter W., 476
Goffey, Andrew, 668
Golub, Gene H., 224, 253, 511, 643, 644, 664
Gordy, Michael B., 165, 172, 285, 291, 309–311, 313, 327–330, 334, 335, 337, 338, 340, 342, 345, 590, 596, 600, 604, 606, 608
Gourieroux, C., 314, 315, 355
Goutis, Constantinos, 373
Gregory, Jon, 55
Griffiths, W.E., 508, 523, 604
Gundlach, Matthias, 113, 138, 139, 141, 328, 431
Gupton, Greg M., 149, 499
Gusba, Scott, 558

Hallerbach, Winfried G., 373
Hamerle, Alfred, 596, 600
Hamilton, James D., 501, 659
Hansen, Lars Peter, 590
Harris, John W., 70
Harrison, J.M., 6, 239, 441, 545, 649
Hashimoto, Takashi, 615
Heath, David, 353, 367
Heidelberger, Philip, 451, 470
Heitfield, Erik, 596, 600, 606, 608
Held, L., 469, 523, 526, 624, 628
Heunis, Andrew J., 234, 243, 437, 453, 653
Hickman, Andrew, 590
Hill, R. Carter, 508, 523, 604
Hilpisch, Yves, 667, 669
Hofert, Marius, 643

Hörmann, Wolfgang, 207
Hu, Wenbo, 200
Huang, Xinzheng, 373, 396
Hull, John C., 266, 281, 567
Huzurbazar, S., 373, 409

Jafry, Yusuf, 506
Jarrow, Robert A., 499, 568, 570
Jeanblanc, Monique, 232, 547
Joe, Harry, 176
Johnson, Norman L., 202, 333, 475, 553
Jones, E. Philip, 266
Jorion, Philippe, 1
Jouain, Jean-Fredéric, 596, 600, 606
Judge, George G., 508, 523, 604

Kang, Wanmo, 451, 469, 470, 474, 476
Karatzas, Ioannis, 14, 234, 241, 242, 453, 653
Karlin, Samuel, 504, 550
Keiding, Niels, 553
Keller, Ulrich, 195
Kent, J., 198
Kercheval, Alex, 200
Klüppelberg, Claudia, 198
Kostadinov, Krassimir, 182
Kotz, Samuel, 198, 200, 202, 333, 475, 553, 617, 620, 622, 642, 647
Koyluoglu, H. Ugur, 590
Kozubowski, Tomasz, 198, 200, 647
Kreps, D.M., 6, 239, 441, 545, 649
Kroese, Dirk P., 451
Kuhn, Gabriel, 182
Kulperger, R. J., 531
Kupiec, Paul H., 18

Lando, David, 499, 506, 512
Langville, Amy N., 501
Laurent, J.P., 314, 315, 355
Lee, Tsoung-Chao, 508, 523, 604
Lehrbass, Frank, 113, 138, 139, 141, 328, 431
Leydold, Josef, 207
Li, Canlin, 281
Li, Jingyi, 426, 430, 451, 453, 455–457, 460, 469, 470, 672
Liebig, Thilo, 596, 600
Lindskog, Filip, 214, 617
Lintner, John, 11
Liseo, Brunero, 627, 628
Loan, Charles F. Van, 224, 253, 511, 643, 644, 664
Loomis, Lynn Harold, 353
Lugannani, Robert, 380, 409

Lütkebohmert, Eva, 311, 327–330, 334, 335, 337, 338, 340, 342, 345, 604
Lütkepohl, Helmet, 508, 523, 604

MacKenzie, Adrian, 668
MacKenzie, Donald, 182
Madan, Dilip B., 193, 200
Magnello, M. Eileen, 589
Martin, Lionel, 615
Martin, Richard J., 311, 373, 396, 421
Mason, Scott P., 266
McKinney, Wes, 669
McLeish, Don L., 451, 453
McNeil, Alexander J., 51, 82, 92, 176, 178, 198, 214, 451, 616, 617
Mechkov, S., 373, 382
Mercurio, Fabio, 544, 546
Merton, R.C., 16, 39, 150, 229–235, 238, 240, 243, 245, 247, 250, 252, 254–257, 264, 265, 270, 272, 275, 278–281, 296, 429, 548, 575, 649, 653
Mesters, Mâcé, 373, 396
Metropolis, Nicholas, 439
Meyn, S. P., 504, 659
Mikosch, Thomas, 198
Mills, Richard, 668
Mina, Jorge, 14
Misirpashaev, T., 373, 382
Morgan/Reuters, J.P., 14, 31
Muromachi, Yukio, 373
Murthy, Shashidar, 570
Musiela, Marek, 453

Nadarajah, Saralees, 620, 622, 642
Neftci, Salih N., 234
Nelken, Izzy, 266, 281
Nelsen, Roger B., 176
Nelson, C. R., 562
Ng, K.W., 617
Nikeghbali, Ashkan, 617

OCC, 673
Oksendal, Bernt K., 234, 242, 653
Oosterlee, Cornelis W., 373, 396
Ord, J. Keith, 54, 125, 523, 531, 550
Overbeck, Ludger, 51, 367, 398, 549, 619

Panjer, Harry H., 139, 431
Paolella, Marc S., 373
Pawitan, Yudi, 508, 523, 525, 528, 628

Pliska, S.R., 6, 239, 441, 545, 649
Podgorski, Krzysztof, 198, 200, 647
Predescu, Mirela, 567
Press, William H., 107, 109, 253, 274, 434, 600, 644

Rabinowitz, Philip, 107, 434
Ralston, Anthony, 107, 434
Rao, B. L. S. Prakasa, 531
Ratto, Matt, 669
Rau-Bredow, Hans, 314, 355, 359
Reesor, R. Mark, 453
Renault, Olivier, 276
Riboulet, Gaël, 617
Rice, Stephen, 380, 409
Roget, Céline, 596, 600, 606
Roncalli, Thierry, 596, 600, 606, 617
Rösch, Daniel, 596, 600
Rosenfeld, Eric, 266
Roth, Michael, 180, 642
Royden, H. L., 235, 627
Rudebusch, Glenn D., 281
Ruiz, Ignacio, 53
Rutkowski, Marek, 7, 453, 544

Savage, Leonard J., 523
Scaillet, O., 314, 315, 355
Schmidt, Rafael, 198
Schmitz, Volker, 178
Scholes, Myron S., 6, 232, 239, 240, 649, 658
Schönbucher, Philipp J., 51, 165
Schuermann, Til, 506
Scott, David, 207
Seneta, Eugene, 193, 200
Shahabuddin, Perwez, 451, 469, 470, 474, 476
Sharpe, William F., 11, 85, 150, 309
Sharples, Stuart, 668
Shaw, William T., 639
Sherbert, Donald R., 627
Shreve, Steven E., 14, 234, 241, 242, 453, 653
Siegel, A. F., 562
Singleton, Kenneth, 240, 548, 553
Sklar, A., 160, 616
Skødeberg, Torben M., 506, 512
Smith, Karl J., 48
Sørensen, M., 198
S&P, 231, 318, 329
Spears, Taylor, 182
Stegun, Irene A., 93, 202, 641
Sternberg, Shlomo, 353
Stocker, Horst, 70
Stréliski, David, 558

Stuart, Alan, 54, 125, 523, 531, 550
Sundaresan, Suresh, 230, 281

Tasche, Dirk, 311, 355, 359, 373
Taylor, Howard M., 504, 550
Teng, Huei-Wen, 476
Teukolsky, Saul A., 107, 109, 253, 274, 434, 600, 644
Thompson, Kevin, 373, 396
Tibshirani, R., 531, 537
Torell, Björn, 311, 316
Treynor, Jack L., 11
Tufte, Edward Rolfe, 2
Turnbull, Stuart M., 499
Tweedie, R. L., 504, 659

van der Weide, J.A.M., 373, 396
VanderPlas, Jake, 669
Vasicek, Olrich Alfons, 150, 165, 169, 171, 174, 211, 290, 310
Vetterling, William T., 107, 109, 253, 274, 434, 600, 644

von Hammerstein, Ernst August, 198
von Hilgers, Philipp, 501

Wagner, Christoph, 51, 367, 398, 549, 619
Wang, Reh-Her, 476
White, Allan D., 266, 281, 567
White, Lawrence J., 498
Wilde, Tom, 86, 113, 131, 132, 138, 201, 311, 328, 329, 373, 431
Wilson, Michael, 499
Wolpert, Robert L., 627, 628

Xiao, Jerry Yi, 14

Yago, Kazuhiko, 288
Yan, Jun, 617
Yan, Xiuling, 150
Yao, Guangmin, 150

Zaccone, Giancaro, 305, 485, 669

GPSR Compliance

The European Union's (EU) General Product Safety Regulation (GPSR) is a set of rules that requires consumer products to be safe and our obligations to ensure this.

If you have any concerns about our products, you can contact us on

ProductSafety@springernature.com

In case Publisher is established outside the EU, the EU authorized representative is:

Springer Nature Customer Service Center GmbH
Europaplatz 3
69115 Heidelberg, Germany

www.ingramcontent.com/pod-product-compliance
Lightning Source LLC
LaVergne TN
LVHW020338300426
837647LV00016B/1163